HANDBOOK OF GAME THEORY
with Economic Applications
VOLUME I

HANDBOOKS
IN
ECONOMICS

11

Series Editors

KENNETH J. ARROW
MICHAEL D. INTRILIGATOR

NORTH-HOLLAND
AMSTERDAM · LONDON · NEW YORK · TOKYO

HANDBOOK OF GAME THEORY
with Economic Applications

VOLUME I

Edited by

ROBERT J. AUMANN
The Hebrew University of Jerusalem

and

SERGIU HART
The Hebrew University of Jerusalem

1992

NORTH-HOLLAND
AMSTERDAM · LONDON · NEW YORK · TOKYO

ELSEVIER SCIENCE PUBLISHERS B.V.
Sara Burgerhartstraat 25
P.O. Box 211, 1000 AE Amsterdam, Netherlands

Library of Congress Cataloging-in-Publication Data

Handbook of game theory with economic applications/edited by Robert
  J. Aumann and Sergiu Hart.
      p.   cm. –– (Handbooks in economics : 11 -)
    Includes bibliographical references.
    ISBN 0-444-88098-4 (v. 1)
    1. Game theory, 2. Economics, Mathematical.   I. Aumann, Robert
  J.   II. Hart. Sergiu.   III. Series: Handbooks in economics ; bk.11,
  etc.
  HB144.H36   1992
  519.3––dc20   91-38429
                                                                        CIP

ISBN for this volume: 0 444 88098 4

INTRODUCTION TO THE SERIES

The aim of the *Handbooks in Economics* series is to produce Handbooks for various branches of economics, each of which is a definitive source, reference, and teaching supplement for use by professional researchers and advanced graduate students. Each Handbook provides self-contained surveys of the current state of a branch of economics in the form of chapters prepared by leading specialists on various aspects of this branch of economics. These surveys summarize not only received results but also newer developments, from recent journal articles and discussion papers. Some original material is also included, but the main goal is to provide comprehensive and accessible surveys. The Handbooks are intended to provide not only useful reference volumes for professional collections but also possible supplementary readings for advanced courses for graduate students in economics.

KENNETH J. ARROW and MICHAEL D. INTRILIGATOR

CONTENTS OF THE HANDBOOK*

VOLUME I

Preface

*Detailed contents of this volume (Volume I of the Handbook) may be found on p. xvii.

Chapter 18
The Bargaining Set, Kernel, and Nucleolus
MICHAEL MASCHLER

Chapter 19
Game and Decision Theoretic Models in Ethics
JOHN C. HARSANYI

CHAPTERS PLANNED FOR VOLUMES II–III

Games of incomplete information
Two-player games
Conceptual foundations of strategic equilibrium
Strategic equilibrium
Correlated and communication equilibria
Stochastic games
Non-cooperative games with many players
Differential games
Economic applications of differential games
Bargaining with incomplete information
Oligopoly
Implementation
Principal–agent models
Signalling
Search
Biological games
International conflict
Taxonomy of cooperative games
Cooperative models of bargaining
The Shapley value
Variations on the Shapley value
Values of large games
Values of non-transferable utility games
Values of perfectly competitive economies
Other economic applications of value theory
Power and stability in politics
Coalition structures
Cost allocation

PREFACE

Game Theory studies the behavior of decision-makers ("players") whose decisions affect each other. As in one-person decision theory, the analysis is from a rational, rather than a psychological or sociological viewpoint. The term "game" stems from the formal resemblance of these interactive decision problems to parlour games such as Chess, Bridge, Poker, Monopoly, Diplomacy, or Battleship. To date, the largest single area of application has been economics; other important connections are with political science (on both the national and international levels), evolutionary biology, computer science, the foundations of mathematics, statistics, accounting, social psychology, law, and branches of philosophy such as epistemology and ethics. The applications are supported by a sizeable body of pure theory that is significant and important in its own right. Needless to say, the relation is two-sided: the theory influences – and is influenced by – the applications, both in the questions asked and in the answers provided.

There is an important distinction between multi-person and one-person decision problems. In the one-person context, we are usually led to a well-defined optimization problem, like maximizing an objective function subject to some constraints. While this problem may be difficult to solve in practice, it involves no conceptual issue. The *meaning* of "optimal decision" is clear; we must only *find* one. But in the interactive multi-person context, the very meaning of "optimal decision" is unclear, since in general, no one player completely controls the final outcome. One must address the conceptual issue of *defining* the problem before one can start *solving* it. Game Theory is concerned with both matters: defining "solution concepts", and then investigating their properties, in general as well as in specific models coming from the various areas of application. This leads to mathematical theories that ultimately yield important and novel insights, quantitative as well as qualitative.

Game Theory may be viewed as a sort of umbrella or "unified field" theory for the rational side of social science, where "social" is interpreted broadly to include human individuals as well as other kinds of players (collectives such as corporations and nations, animals and plants, computers, etc.). Unlike other approaches to disciplines like economics or political science, Game Theory does not use different, ad-hoc constructs to deal with various specific issues, such as perfect competition, monopoly, oligopoly, international trade, taxation, voting, deterrence, animal behavior, and so on. Rather, it develops methodologies that apply in principle to *all* interactive situations, then sees where these methodologies lead in each specific application.

One may distinguish two approaches to Game Theory: the non-cooperative and the cooperative. A game is *cooperative* if commitments – agreements, promises, threats – are fully binding and enforceable.[1] It is *non-cooperative* if commitments are not enforceable. (Note that pre-play communication between the players does not imply that any agreements that may have been reached are enforceable.) Though this may not look like a basic distinction, it turns out that the two theories have quite different characters. The non-cooperative theory concentrates on the strategic choices of the individual – how each player plays the game, what strategies he chooses to achieve his goals. The cooperative theory, on the other hand, deals with the options available to the group – what coalitions form, how the available payoff is divided. It follows that the non-cooperative theory is intimately concerned with the details of the processes and rules defining a game; the cooperative theory usually abstracts away from such rules, and looks only at more general descriptions that specify only *what* each coalition can get, without saying *how*. A very rough analogy – not to be taken too literally – is the distinction between micro and macro, in economics as well as in biology and physics. Micro concerns minute details of process, whereas macro is concerned with how things look "on the whole". Needless to say, there is a close relation between the two approaches; they complement and strengthen one another.

This is the first volume of the *Handbook of Game Theory with Economic Applications*, to be followed by two additional volumes. Game Theory has burgeoned greatly in the last decade, and today it is an essential tool in much of economic theory. The vision laid out by the founding fathers, John von Neumann and Oskar Morgenstern, in their 1944 book *Theory of Games and Economic Behavior* has become a reality.

While it is no longer possible in three volumes even to survey Game Theory adequately, we have made an attempt to present the main features of the subject as they appear today. The three volumes will cover the fundamental theoretical aspects, a wide range of applications to economics, several chapters on applications to political science, and individual chapters on relations with other disciplines.

A list of the chapters planned for all the volumes is appended to this

---

[1] This definition is due to John C. Harsanyi ('A general theory of rational behavior in game situations', *Econometrica*, 34: 616 (1966)).

Preface.[2] We have organized this list roughly into "non-cooperative" and "cooperative"; there are also some "general" chapters. The boundary is often difficult to draw, as there are important connections between the categories; chapters may well contain aspects of both approaches. Within each category, some chapters are more theoretical, others more applicative; here again, the distinction is often hazy. It is to be noted that the division of the chapters of the Handbook into the three volumes was dictated only partly by considerations of substantive relationships; another, more mundane consideration was which chapters were available when the volume went to press.

We now provide a short overview of the organization of this volume. Chapters 1 through 11 may be viewed as "non-cooperative" and Chapters 12 through 18 as "cooperative". The final chapter, Chapter 19, is in the "general" category. Most of the chapters belong to conceptually well-defined groups, and require little further introduction. Others are not so clearly related to their neighbors, so a few more words are needed to put them in context. (Thus the space that this Preface devotes to a chapter is no indication of its importance.)

Historically, the first contribution to Game Theory was Zermelo's 1913 paper on chess, so it is fitting that the "overture" to the Handbook deals with this grand-daddy of all games. The chapter covers chess-playing computers. Though this is not mainstream game theory, the ability of modern computers to beat some of the best human chess players in the world constitutes a remarkable intellectual and technological achievement, which deserves to be recorded in this Handbook.

Chapter 2 provides an introduction to the non-cooperative theory. It describes the "tree" representation of extensive games, the fact that for many purposes one can limit oneself to consideration of strategies, and the related classical results. Unlike in most of the other chapters, there is no attempt here at adequate coverage (which is provided in later chapters); it only provides some basic tools.

Conceptually, the simplest games are those of perfect information: games like chess, in which all moves are open and "above board", in which there is no question of guessing what the other players have done or are doing. The fundamental fact in this area is the 1913 theorem of Zermelo (mentioned above), according to which each zero-sum game of perfect information has optimal pure strategies. In 1953 Gale and Stewart showed that this result does not always extend to infinite games of perfect information, and identified conditions under which it does. Chapter 3 deals with these results, and with the literature in the foundations of mathematics (set theory) that has grown from them.

---

[2]A fairly detailed historical survey of game theory, with cross-references to the chapters of the *Handbook*, is planned for a subsequent volume.

Repeated games model ongoing relationships; the theory "predicts" phe-nomena such as cooperation, communication, altruism, trust, threats, punish-ment, revenge, rewards, secrecy, signalling, transmission of information, and so on. Chapters 4, 5, and 6 are devoted to repeated games. Though this theory is basically "non-cooperative", it brings us to the interface with the cooperative theory; it may be viewed as a non-cooperative model that "justifies" the assumption of binding agreements that underlies cooperative theory.

Another such "bridge" between the non-cooperative and the cooperative is bargaining theory. Until the early eighties, most of bargaining theory belonged to the cooperative area. After the publication, in 1982, of Rubinstein's seminal paper on the subject, much of the emphasis shifted to the relation of non-cooperative models of bargaining to the older cooperative models. These and related developments are covered in Chapter 7.

Chapter 7 is also the first of five chapters in this volume dealing with economic applications of the non-cooperative theory. Chapters 8 through 11 are about auctions, location, entry deterrence, and patents. In each case, equilibrium analysis leads to important qualitative insights.

Starting with Chapter 12, we turn to the cooperative theory and its applica-tions. Chapters 12 through 16 offer a thorough coverage of what is perhaps the best known solution concept in cooperative game theory, the core. Chapters 12 and 13 provide theoretical foundations, while Chapters 14, 15, and 16 cover the best known economic applications.

Though the definition of the core is straightforward enough, it is perhaps somewhat simplistic; a careful consideration leads to some difficulties. Several solution concepts have been constructed to deal with these difficulties. One – historically the first cooperative solution concept – is the von Neumann–Morgenstern stable set; it is studied, together with some of its applications to economic and political models, in Chapter 17. Chapter 18 covers the extensive literature dealing with another class of "core-like" solutions: the bargaining set and the related concepts of kernel and nucleolus.

Though Game Theory makes no ethical recommendations – is ethically neutral – game-theoretic ideas nevertheless do play a role in ethics. A fitting conclusion to this first volume is Chapter 19, which deals with the relation between Game Theory and ethics.

ROBERT J. AUMANN and SERGIU HART

List of Chapters Planned for all the Volumes[3]

Non-Cooperative

The game of chess (I, 1)
Games in extensive and strategic forms (I, 2)
Games of perfect information (I, 3)
Games of incomplete information
Two-player games
Conceptual foundations of strategic equilibrium
Strategic equilibrium
Correlated and communication equilibria
Stochastic games
Repeated games of complete information (I, 4)
Repeated games of incomplete information: zero-sum (I, 5)
Repeated games of incomplete information: non-zero-sum (I, 6)
Non-cooperative games with many players
Differential games
Economic applications of differential games
Non-cooperative models of bargaining (I, 7)
Bargaining with incomplete information
Oligopoly
Implementation
Auctions (I, 8)
Principal-agent models
Signalling
Search
Location (I, 9)
Entry and exit (I, 10)
Patent licensing (I, 11)
Biological games
International conflict

Cooperative

Taxonomy of cooperative games
Cooperative models of bargaining

---

[3]"I, *n*" means that this is chapter *n* of volume I.

The core and balancedness (I, 12)
Axiomatizations of the core (I, 13)
The core in perfectly competitive economies (I, 14)
The core in imperfectly competitive economies (I, 15)
Two-sided matching (I, 16)
Von Neumann–Morgenstern stable sets (I, 17)
The bargaining set, kernel, and nucleolus (I, 18)
The Shapley value
Variations on the Shapley value
Values of large games
Values of non-transferable utility games
Values of perfectly competitive economies
Other economic applications of value theory
Power and stability in politics
Coalition structures
Cost allocation

General

History of game theory
Utility and subjective probability
Common knowledge
Computer science
Statistics
Social choice
Public economics
Voting methods
Experimentation
Psychology
Law
Ethics (I, 19)

CONTENTS OF VOLUME I

Chapter 8
Strategic Analysis of Auctions
ROBERT WILSON

Chapter 16
Two-Sided Matching
ALVIN E. ROTH and MARILDA SOTOMAYOR

Chapter 17
Von Neumann–Morgenstern Stable Sets
WILLIAM F. LUCAS

Chapter 18

The Bargaining Set, Kernel, and Nucleolus

MICHAEL MASCHLER

Chapter 19

Game and Decision Theoretic Models in Ethics

JOHN C. HARSANYI

Chapter 1

THE GAME OF CHESS

HERBERT A. SIMON

Carnegie-Mellon University

JONATHAN SCHAEFFER

University of Alberta

Contents

Handbook of Game Theory, Volume 1, Edited by R.J. Aumann and S. Hart

1. Introduction

The game of chess has sometimes been referred to as the Drosophila of artificial intelligence and cognitive science research – a standard task that serves as a test bed for ideas about the nature of intelligence and computational schemes for intelligent systems. Both machine intelligence – how to program a computer to play good chess (artificial intelligence) – and human intelligence – how to understand the processes that human masters use to play good chess (cognitive science) – are encompassed in the research, and we will comment on both in this chapter, but with emphasis on computers.

From the standpoint of von Neumann–Morgenstern game theory [von Neumann and Morgenstern (1944)] chess may be described as a trivial game. It is a two-person, zero-sum game of perfect information. Therefore the rational strategy for play is obvious: follow every branch in the game tree to a win, loss, or draw – the rules of the game guarantee that only a finite number of moves is required. Assign 1 to a win, 0 to a draw, and −1 to a loss, and minimax backwards to the present position.

For simple games, such as tic-tac-toe and cubic, the search space is small enough to be easily exhausted, and the games are readily solved by computer. Recently, the game of connect-four, with a search space of 10^{13} positions, was solved; with perfect play, the player moving first (white) will always win [Uiterwijk et al. (1989)]. This result was not obtained by a full search of the space, but by discovering properties of positions that, whenever present in a position, guarantee a win. In this way, large sub-trees of the game tree could be evaluated without search.

The only defect in trying to apply this optimal strategy to chess is that neither human beings nor the largest and fastest computers that exist (or that are in prospect) are capable of executing it. Estimates of the number of legally possible games of chess have ranged from 10^{43} to 10^{20}. Since even the smaller numbers in this range are comparable to the number of molecules in the universe, an exploration of this magnitude is not remotely within reach of achievable computing devices, human or machine, now or in the future.

In the literature of economics, the distinction has sometimes been made between *substantive rationally* and *procedural (a.k.a. computational) rationality*. Substantive rationality is concerned with the objectively correct or best action, given the goal, in the specified situation. Classical game theory has been preoccupied almost exclusively with substantive rationality.

Procedural rationality is concerned with procedures for finding good actions, taking into account not only the goal and objective situation, but also the knowledge and the computational capabilities and limits of the decision maker.

The only nontrivial theory of chess is a theory of procedural rationality in choosing moves. The study of procedural or computational rationality is relatively new, having been cultivated extensively only since the advent of the computer (but with precursors, e.g., numerical analysis). It is central to such disciplines as artificial intelligence and operations research.

Difficulty in chess, then, is computational difficulty. Playing a good game of chess consists in using the limited computational power (human or machine) that is available to do as well as possible. This might mean investing a great deal of computation in examining a few variations, or investing a little computation in each of a large number of variations. Neither strategy can come close to exhausting the whole game tree – to achieving substantive rationality.

2. Human chess play

To get an initial idea of how the task might be approached, we can look at what has been learned over the past half century about human chess play, which has been investigated in some depth by a number of psychologists. There is a considerable understanding today about how a grandmaster wins games, but not enough understanding, alas, to make it easy to become a grandmaster.

First, since the pioneering studies of the Dutch psychologist, Adriaan de Groot, we have known that a grandmaster, even in a difficult position, carries out a very modest amount of search of the game tree, probably seldom more than 100 branches [de Groot (1965)]. Even if this is an underestimate by an order of magnitude (it probably is not), 10^3 is a miniscule number compared with 10^{43}. In fact, de Groot found it very difficult to discriminate, from the statistics of search, between grandmasters and ordinary club players. The only reliable difference was that the grandmasters consistently searched more relevant variations and found better moves than the others – not a very informative result.

It should not surprise us that the skilled chess player makes such a limited search of the possibilities. The human brain is a very slow device (by modern electronic standards). It takes about a millisecond for a signal to cross a single synapse in the brain, hence ten to a hundred milliseconds for anything "interesting" to be accomplished. In a serious chess game, a player must make moves at the rate of twenty or thirty per hour, and only ten minutes or a quarter hour can be allotted to even a difficult move. If 100 branches in the game tree were examined in fifteen minutes, that would allow only nine seconds per branch, not a large amount of time for a system that operates at human speeds.

A second thing we know is that when grandmasters look at a chessboard in the course of a game, they see many familiar patterns or "chunks" (patterns they have often seen before in the course of play). Moreover, they not only immediately notice and recognize these chunks whenever they encounter them, but they have available in memory information that gives the chunks signifi- cance – tells what opportunities and dangers they signal, what moves they suggest. The capability for recognizing chunks is like a very powerful index to the grandmaster's encyclopedia of chess. When the chunk is recognized, the relevant information stored with it is evoked and recalled from memory.

The grandmaster's vast store of chunks seems to provide the main explana- tion for the ability to play many simultaneous games rapidly. Instead of searching the game tree, the grandmaster simply makes "positionally sound" moves until the opponent makes an inferior move, creating a (usually slight) weakness. The grandmaster at once notices this clue and draws on the associated memory to exploit the opponent's mistake.

Rough estimates place the grandmaster's store of chunks of chess knowledge at a minimum of 50,000, and there are perhaps even twice that many or more [Simon and Gilmartin (1973)]. This number is comparable in magnitude to the typical native language vocabularies of college-educated persons.

There is good evidence that it takes a minimum of ten years of intense application to reach a strong grandmaster level in chess. Even prodigies like Bobby Fischer have required that. (The same period of preparation is required for world class performance in all of the dozen other fields that have been examined.) Presumably, a large part of this decade of training is required to accumulate and index the 50,000 chunks.

It has sometimes been supposed that the grandmaster not only has special knowledge, but also special cognitive capabilities, especially the ability to grasp the patterns on chess boards in mental images. An interesting and simple experiment, which has been carried out by several investigators and is easily replicated, makes very clear the nature of chess perception [de Groot (1965)]. (The analogous experiment has been performed with other games, like bridge, with the same result.)

Allow a subject in the laboratory to view a chess position from a well-played game (with perhaps 25 pieces on the board) for five to ten seconds. Then remove the pieces and ask the subject to reconstruct the position. If the subject is a master or grandmaster, 90% or more of the pieces (23 out of 25, say) will be replaced correctly. If the subject is an ordinary player, only about six will be replaced correctly. This is an enormous and starting difference between excellent and indifferent chess players. Something about their eyes?

Next repeat the experiment, but place the same pieces on the board *at random* [Chase and Simon (1973)]. The ordinary player will again replace about six on the right squares. Now the master or grandmaster will also only replace about six correctly. Clearly the expert's pre-eminence in the first version of the experiment has nothing to do with visual memory. It has to do with familiarity. A chessboard from a well-played game contains 25 nearly unrelated pieces of information for the ordinary player, but only a half dozen familiar patterns for the master. The pieces group themselves for the expert into a small number of interrelated chunks.

Few familiar chunks will appear on the randomly arranged board; hence, in remembering that position, the expert will face the same 25 unrelated pieces of knowledge as the novice. Evidence from other domains shows that normal human adults can hold about six or seven unrelated items in short-term memory at the same time (equivalent to one unfamiliar telephone number). There is nothing special about the chess master's eyes, but a great deal that is special about his or her knowledge: the indexed encyclopedia stored in memory.

In summary, psychological research on chess thinking shows that it involves a modest amount of search in the game tree (a maximum of 100 branches, say) combined with a great deal of pattern recognition, drawing upon patterns stored in memory as a result of previous chess training and experience. The stored knowledge is used to guide search in the chess tree along the most profitable lines, and to evaluate the leaf positions that are reached in the search. These estimated values at the leaves of the miniature game trees are what are minimaxed (if anything is) in order to select the best move. For the grandmaster, vast amounts of chess knowledge compensate for the very limited ability of the slow human neurological "hardware" to conduct extensive searches in the time available for a move.

3. Computer chess: Origins

The idea of programming computers to play chess appeared almost simultaneously with the birth of the modern computer [see Newell and Simon (1972) for a detailed history to 1970 and Welsh and Baczynskyj (1985) for a more modern perspective]. Claude Shannon (1950), the creator of modern information theory, published a proposal for a computer chess program, and A.M. Turing (1953) published the score of a game played by a hand-simulated program. A substantial number of other designs were described and actually programmed and run in the succeeding decade.

There was a close family resemblance among most of the early programs. The task was viewed in a game-theoretic framework. Alternative moves were to be examined by search through the tree of legal moves, values were to be assigned to leaf nodes on the tree, and the values were to be minimaxed backwards through the tree to determine values for the initial moves. The move with the largest value was then selected and played. The same search and evaluation program represented both the player and the opponent.

Of course the programs represented only the crudest approximations to the optimal strategy of von Neumann–Morgenstern game theory. Because of the severe computational limits (and the early programs could examine at most a few thousand branches of the tree), a function for evaluating the (artificial) leaves had to be devised that could approximate the true game values of the positions at these nodes.

This evaluation function was, and remains, the most vulnerable Achilles heel in computer chess. We shall see that no one, up to the present time, has devised an evaluation function that does not make serious mistakes from time to time in choosing among alternative positions, and minor mistakes quite frequently. Although the best current chess programs seldom blunder outright, they often make moves that masters and grandmasters rightly regard as distinctly inferior to the best moves.

In addition to brute force search, two other ideas soon made their appearance in computer chess programs. The first [already proposed by Shannon (1950)] was to search selectively, rather than exhaustively, sacrificing completeness in the examination of alternatives in order to attain greater depth along the lines regarded as most important. Of course it could not be known with certainty which lines these were; the rules of selection were heuristic, rules of thumb that had no built-in guarantees of correctness. These rules could be expressed in terms of position evaluation functions, not necessarily identical with the function for evaluating leaf nodes.

Turing made the important distinction between "dead" positions, which could reasonably be evaluated in terms of their static features, and "live" positions having unresolved dynamic features (pieces en prise, and the like) that required further search before they could be evaluated. This distinction was incorporated in most subsequent chess programs.

The second departure from the basic game-theoretic analogue was to seek ways of reducing the magnitude of the computation by examining branches in the best order. Here there arose the idea of alpha–beta search, which dominated computational algorithms for chess during the next three decades.

The idea underlying alpha–beta search is roughly this. Suppose that, after a partial search of the tree from node A, a move, $M1$, has already been found that guarantees the player at A the value V. Suppose that after a partial search of one of the other moves at A, $M2$, a reply has been found for the opponent that guarantees him or her a value (for the player at A) of less than V. Then there is no need to carry on further search at the subnode, as $M2$ cannot be better than $M1$. Roughly speaking, use of this procedure reduces the amount of computation required to the square root of the amount without the algorithm, a substantial reduction. The alpha–beta algorithm, which has many variant forms, is a form of the branch-and-bound algorithm widely used in solving various combinatorial problems of management science.

While the alpha–beta algorithm provides a quite powerful tree-pruning principle, when it is combined (as it must be in chess) with an evaluation function that is only approximate, it is not without its subtleties and pitfalls. Nau (1983) first demonstrated that minimax trees with approximate evaluation may be *pathological*, in that deeper search may lead to poorer decisions. This apparent paradox is produced by the accumulation of errors in evaluation. Although this important result has been ignored in the design of chess programs, pathology probably does occur in tree search, but not to the extent of degrading performance seriously.

The dead position heuristic, alpha–beta search, and the other selective methods that served to modify brute-force game-theoretic minimaxing were all responses to the devastating computational load that limited depth of search, and the inability to devise, for these shallow searches, an evaluation function that yielded a sufficiently close approximation to the true value of the leaf positions. We must remember that during the first decade or two of this research, computers were so small and slow that full search could not be carried to depths of more than about two moves (four or five ply, or an average of about 1,000 to 30,000 branches).

We might mention two programs, one of 1958, the other of 1966, that departed rather widely from these general norms of design. Both were planned with the explicit view of approximating, more closely than did the game-theory approximation, the processes that human players used to choose moves.

The first of these programs had separate base-move generators and analysis-move generators for each of a series of goals: material balance, center control, and so on [Newell, Shaw and Simon (1958)]. The moves proposed by these generators were then evaluated by separate analysis procedures that determined their acceptability. The program, which was never developed beyond three goals for the opening, played a weak game of chess, but demonstrated that, in positions within its limited scope of knowledge, it could find reasonable moves with a very small amount of search (much less than 100 branches.)

Another important feature of the NSS (Newell, Shaw, and Simon) program was the idea of *satisficing*, that is, choosing the first move that was found to reach a specified value. In psychological terms, this value could be viewed as a level of aspiration, to be assigned on the basis of a preliminary evaluation of the current position, with perhaps a small optimistic upward bias to prevent too early termination of search. Satisficing is a powerful heuristic for reducing the amount of computation required, with a possible sacrifice of the quality of the moves chosen, but not necessarily a sacrifice when total time constraints are taken into account.

Another somewhat untypical program, MATER, was specialized to positions where the player has a tactical advantage that makes it profitable to look for a checkmating combination [see Newell and Simon (1972, pp. 762–775)]. By giving priority to forceful moves, and looking first at variations that maximally restricted the number of legal replies open to the opponent, the program was able to find very deep mating combinations (up to eight moves, 15 ply, deep) with search trees usually well under 100 branches.

An interesting successor to this program, which combines search and knowledge, with emphasis on the latter, is Wilkins' program PARADISE [Wilkins (1982)]. PARADISE built small "humanoid" trees, relying on extensive chess knowledge to guide its search. Within its domain it was quite powerful. However, like MATER, its capabilities were limited to tactical chess problems (checkmates and wins of material), and it could not perform in real time.

During this same period, stronger programs that adhered more closely to the "standard" design described earlier were designed by Kotok, Greenblatt, and others [described in Newell and Simon (1972, pp. 673–703)]. Gradual improvements in strength were being obtained, but this was due at least as much to advances in the sizes and speeds of computer hardware as to improvements in the chess knowledge provided to the programs or more efficient search algorithms.

4. Search versus knowledge

Perhaps the most fundamental and interesting issue in the design of chess programs is the trade-off between search and knowledge. As we have seen, human players stand at one extreme, using extensive knowledge to guide a search of perhaps a few hundred positions and then to evaluate the leaves of the search tree. Brute-force chess programs lie at the opposite extreme, using much less knowledge but considering millions of positions in their search. Why is there this enormous difference?

We can view the human brain as a machine with remarkable capabilities, and the computer, as usually programmed, as a machine with different, but also remarkable capabilities. In the current state of computer hardware and software technology, computers and people have quite different strengths and weaknesses. Most computer programs lack any capability for learning, while human brains are unable to sum a million numbers in a second. When we are solving a problem like chess on a computer, we cater to the strengths rather than the weaknesses of the machine. Consequently, chess programs have evolved in a vastly different direction from their human counterparts. They are not necessarily better or worse; just different.

4.1. Search

Progress in computer chess can be described subjectively as falling into three epochs, distinguished by the search methods the programs used. The pre-1975 period, described above, could be called the *pioneering era*. Chess programs were still a novelty and the people working on them were struggling to find a framework within which progress could be made (much like Go programmers today). By today's standards, many of the search and knowledge techniques used in the early programs were ad hoc with some emphasis on selective, knowledge-based search. Some (not all) programs sought to emulate the human approach to chess, but this strategy achieved only limited success.

The 1975–1985 period could be called the *technology era*. Independently, several researchers discovered the benefits of depth-first, alpha–beta search with iterative deepening. This was a reliable, predictable, and easily implemented procedure. At the same time, a strong correlation was observed between program speed (as measured by positions considered per second) and program performance. Initially, it was estimated that an additional ply of search (increasing the depth of search by a move by White or Black) could improve performance by as much as 250 rating points.[1] At this rate, overcoming the gap between the best programs (1800) of the middle 1970s and the best human players (2700) required only 4 ply of additional search. Since an extra ply costs roughly a factor of 4–8 in computing power, all one had to do was wait for technology to solve the problem by producing computers 4,000 times faster than those then available, something that would surely be achieved in a few years.

Unfortunately, the rating scale is not linear in amount of search, and later experience indicated that beyond the master level (2200 points), each ply was worth only an additional 100 points. And, of course, it is widely believed that the rate of gain will decrease even further as chess programs reach beyond the base of the grandmaster (2500) level.

All the top programs today reflect this fascination with brute-force alpha–beta search, having an insatiable appetite for speed. The major competitive chess programs run on super-computers (*Cray Blitz*), special purpose hardware (*Belle, Hitech, Deep Thought*), and multi-processors (*Phoenix*). The best chess programs reached the level of strong masters largely on the coat tails of technology rather than by means of major innovations in software or knowledge engineering.

Since 1985, computer chess has hit upon a whole host of new ideas that are producing an emerging *algorithm era*. The limit of efficiency in alpha–beta search had been reached; new approaches were called for [Schaeffer (1989)]. In quick succession, a number of innovative approaches to search appeared. Whereas traditional alpha–beta search methods examine the tree to a fixed

---

[1]Chess performance is usually measured on the so-called Elo scale, where a score of 2,000 represents Expert performance, 2,200 Master level performance, and 2,500 Grandmaster performance. There are both American and International chess ratings, which differ by perhaps 100 points, but the above approximation will be sufficient for our purposes.

depth (with some possible extensions), the new approaches attempt to expand the tree selectively, at places where additional effort is likely to produce a more accurate value at the root. Alpha–beta relies on depth as the stopping criterion; removal of this restriction is the most significant aspect of the new ideas in search.

Alpha–beta returns the maximum score achievable (relative to the fallible evaluation function) and a move that achieves this score. Little information is provided on the quality of other sibling moves. *Singular extensions* is an enhancement to alpha–beta to perform additional searches to determine when the best move in a position is *significantly* better than all the alternatives [Anantharaman et al. (1988)]. These *singular* or *forced* moves are then re-searched deeper than they normally would be. Consequently, a forcing sequence of moves will be searched to a greater depth than with conventional alpha–beta.

The *conspiracy numbers* algorithm maintains a count of the number of leaf nodes in the tree that must change value (or *conspire*) to cause a change in the root value [McAllester (1988)]. Once the number of conspirators required to cause a certain change in the root exceeds a prescribed *threshold*, that value is considered unlikely to occur and the corresponding move is removed from consideration. The search stops when one score for the root is *threshold* conspirators better than all other possible scores.

Min/max approximation uses mean value computations to replace the standard minimum and maximum operations of alpha–beta [Rivest (1988)]. The advantage of using mean values is that they have continuous derivatives. For each leaf node, a derivative is computed that measures the sensitivity of the root value to a change in value of that node. The leaf node that has the most influence on the root is selected for deeper searching. The search terminates when the influence on the root of all leaf nodes falls below a set minimum. When min/max approximation and conspiracy numbers are used, alpha–beta cut-offs are not possible.

Equi-potential search expands all leaf nodes with a marginal utility greater than a prescribed threshold [Anantharaman (1990)]. The utility of a node is a function of the search results known at all interior nodes along the path from the tree root to that node, together with any useful heuristic information provided by the knowledge of the program. The search terminates when all leaf nodes have a marginal utility below the prescribed threshold.

In addition to these procedures, a combination of selective search with brute-force search has re-emerged as a viable strategy.

Although many of these ideas have yet to make the transition from theory to practice, they are already strongly influencing the directions today's programs are taking. In some sense, it is unfortunate that computer chess received the gift of alpha–beta search so early in its infancy. Although it was a valuable and powerful tool, its very power may have inadvertently slowed progress in chess programs for a decade.

4.2. Knowledge

In view of the obvious importance of evaluation functions, it may seem surprising that more effort has not been devoted to integrating knowledge in a reliable manner into chess programs. The reason is simple: program performance is more easily enhanced through increases in speed (whether hardware or software) than through knowledge acquisition. With present techniques, capturing, encoding and tuning chess knowledge is a difficult, time-consuming and ad hoc process. Because performance has been the sole metric by which chess programs were judged, there has been little incentive for solving the knowledge problem.

Acquiring knowledge by having human grandmasters advise chess programmers on the weaknesses of their programs has not proved effective: the two sides talk different languages. Consequently, few chess programs have been developed in consultation with a strong human player.

Deeper search has had the unexpected effect of making chess programs appear more knowledgeable than they really are. As a trivial example, consider a chess program that has no knowledge of the concept of *fork*, a situation in which a single piece threatens two of the opponent's pieces simultaneously. Searching a move deeper reveals the captures without the need for representing the fork in the evaluation of the base position. In this manner deep searches can compensate for the absence of some important kinds of knowledge, by detecting and investigating the effects of the unnoticed features.

Arthur Samuel constructed, in the 1950s, a powerful program for playing checkers [Samuel (1967)]. The program's knowledge was embodied in a complex evaluation function (which was capable of self-improvement through learning processes), and the program's prowess rested squarely on the accuracy of its evaluations. The evaluation function was a weighted sum of terms, and the weight of each term could be altered on the basis of its influence on moves that proved (retrospectively) to have been good or bad.

The published descriptions of Samuel's program provide a complete description of the checkers knowledge incorporated in it. (Comparable detailed information is not available for most chess programs.) However, Samuel's program is now three decades old, and technology has improved machine speed by several orders of magnitude. Checkers programs today use less than half of the knowledge of Samuel, the programs depending on search to reveal the rest. The same holds true for chess programs.

That skilled human players process chess positions in terms of whole groups of pieces, or chunks, is recognized as important by chess programmers; but, with few exceptions, no one has shown how to acquire or use chunks in programs, particularly under the real-time constraints of tournament chess games. One of the exceptions is Campbell's CHUNKER program, which is able to group the pieces in king and pawn endgames and reason about the chunks in a meaningful way [Campbell and Berliner (1984)]. Although the search trees built by CHUNKER are larger than would be built by a skilled human player, CHUNKER, in the limited class of positions it could handle, achieved a level of performance comparable to that of a grandmaster.

Only recently has significant effort in the design of chess programs been re-directed towards the acquisition and use of chess knowledge. Knowledge is useful not only for position evaluation, but also to guide the direction of search effort. The *Hitech* chess program, one of the two or three strongest programs in existence today, employs high-speed, special-purpose parallel hardware, but also incorporates more complete and sophisticated chess knowledge than any other program built to date [Berliner and Ebeling (1989)]. Over a three-year period, and without any change in hardware to increase its speed, the program improved in strength from an expert to a strong master solely on the basis of software improvements – principally improvements of its chess knowledge.

4.3. A tale of two programs

It is interesting to compare the current two best chess programs. Both originate at Carnegie-Mellon University and both use special-purpose VLSI processors, but that is where the similarities end. *Deep Thought* concentrates on speed; a specially designed, single chip chess machine that can analyze 500,000 positions per second. Further, since the basic design is easily reproducible, one can run multiple copies in parallel, achieving even better performance (the system currently uses up to 6). As a consequence, under tournament conditions the program can look ahead almost 2 moves (or ply) deeper than any other program. However, considerable work remains to be done with its software to allow it to overcome the many glaring gaps in its chess knowledge.

Hitech also uses special-purpose chips; 64 of them, one for each square on the board. However, *Hitech*'s speed is less than that of a single *Deep Thought* processor. *Hitech*'s strength lies in its knowledge base. The hardware has been enhanced to include pattern recognizers that represent and manipulate knowledge at a level not possible in *Deep Thought*. Consequently, *Hitech* plays a more "human-like" game of chess, without many of the "machine" tendencies that usually characterize computer play. However, *Deep Thought*'s tremendous speed allows it to search the game tree to unprecedented depths, often finding in positions unexpected hidden resources that take both other computer programs and human players by surprise.

A game between *Hitech* and *Deep Thought* is analogous to a boxing match between fighters with quite different styles. *Hitech* has the greater finesse, and will win its share of rounds, whereas *Deep Thought* has the knock-out punch. Unfortunately for finesse, no one remembers that you out-boxed your opponent for ten rounds. All they remember is that you were knocked out in the eleventh.

5. Computer chess play

The current search-intensive approaches, using minimal chess knowledge, have not been able to eliminate glaring weaknesses from the machines' play. Now that machines can wage a strong battle against the best human players, their games are being studied and their soft spots uncovered. Chess masters, given the opportunity to examine the games of future opponents, are very good at identifying and exploiting weaknesses. Since existing computer programs do not improve their play through learning,[2] they are inflexible, unable to understand or prevent the repeated exploitation of a weakness perceived by an opponent.

Learning, in the current practice of computer chess, consists of the programmer observing the kinds of trouble the program gets into, identifying the causes, and then modifying the program to remove the problem. The programmer, not the program, does the learning!

The biggest shortcomings of current chess programs undoubtably lie in their knowledge bases. Programs are too quick to strive for short-term gains visible in their relatively shallow search trees, without taking into account the long-term considerations that lie beyond the horizons of the search. The absence of fundamental knowledge-intensive aspects of human chess play, such as strategic planning, constitute a major deficiency in chess programs. In addition, they are unable to learn from mistakes and to reason from analogies, so that they cannot improve dynamically from game to game.

On the other hand, chess programs are not susceptible to important human weaknesses. The ability to calculate deeply and without computational (as opposed to evaluational) errors are enormous advantages. They are also free from psychological problems that humans have to overcome. Humans, tuned to playing other humans, are frequently flustered by computer moves. Machines do not subscribe to the same preconceptions and aesthetics that humans do, and often discover moves that catch their opponents off guard. Many games have been lost by human opponents because of over-confidence induced by the incorrect perception that the machine's moves were weak.

---

[2] Other than to remember previous games played, so that they can avoid repeating the same mistake in the same opening position (but not in similar positions).

Further, a computer has no concept of fear and is quite content to walk along a precipice without worrying about the consequences of error. A human, always mindful of the final outcome of the game, can be frightened away from his best move because of fear of the unknown and the risk of losing.

Only recently have people questioned whether the human model of how to play chess well is the only good model or, indeed, even the best model. The knowledgeable chess player is indoctrinated with over 100 years of chess theory and practice, leaving little opportunity for breaking away from conventional thinking on how to play chess well. On the other hand, a computer program starts with no preconceptions. Increasingly, the evidence suggests that computer evaluation functions, combined with deep searches, may yield a different model of play that is *better* than the one humans currently use!

In fact, it is quite likely that computer performance is inhibited by the biases in the human knowledge provided to programs by the programmer. Since a machine selects moves according to different criteria than those used by people (these criteria may have little relation to anything a chess player "knows" about chess), it is not surprising that the machines' styles of play are "different" and that humans have trouble both in playing against them and improving their programs.

6. The future

In 1988, *Deep Thought* became the first chess program to defeat a grandmaster in a serious tournament game. Since then it has been shown that this was not a fluke, for two more grandmasters have fallen.[3] *Deep Thought* is acknowledged to be playing at the International Master level (2400+ rating) and *Hitech* is not far behind.

However, in 1989, World Champion Garry Kasparov convincingly demonstrated that computers are not yet a threat for the very best players. He studied all of *Deep Thought*'s published games and gained insight into the strengths and weaknesses of the program's play. In a two-game exhibition match, Kasparov decisively beat *Deep Thought* in both games.

It is difficult to predict when computers will defeat the best human player. This important event, so long sought since the initial optimism of Simon and Newell (1958) was disappointed, will be a landmark in the history of artificial intelligence. With the constantly improving technology, and the potential for massively parallel systems, it is not a question "if" this event will occur but

---

[3]The Grandmaster Bent Larsen was defeated in November 1988. Subsequently, Grandmasters Robert Byrne (twice), and Tony Miles have fallen. As of April 1990, *Deep Thought*'s record against Grandmasters under tournament conditions was 4 wins, 2 draws, 4 losses; against International Masters, 11 wins, 2 draws, 1 loss.

rather "when". As the brute-force programs search ever more deeply, the inadequacy of their knowledge is overcome by the discoveries made during search. It can only be a matter of a few years before technological advances end the human supremacy at chess.

Currently, computer chess research appears to be moving in two divergent directions. The first continues the brute-force approach, building faster and faster alpha–beta searchers. *Deep Thought*, with its amazing ability to consider millions of chess positions per second, best epitomizes this approach.

The second direction is a more knowledge-intensive approach. *Hitech* has made advances in the direction, combining extensive chess knowledge with fast, brute-force search. Commercial manufacturers of chess computers, such as *Mephisto*, limited to machines that consumers can afford, realized long ago that they could never employ hardware that would compete with the powerful research machines. To compensate for their lack of search depth (roughly 10,000 positions per second), they have devoted considerable attention to applying knowledge productively. The results have been impressive.

Which approach will win out? At the 1989 North American Computer Chess Championship, *Mephisto*, running on a commercially available micro-processor, defeated *Deep Thought*. (*Mephisto* is rated 2159 on the Swedish Rating List on the basis of 74 games against human opponents.) The verdict is not in.

7. Other games

Many of the lessons learned from writing computer chess programs have been substantiated by experience in constructing programs for other games: bridge bidding and play, Othello, backgammon, checkers, and Go, to mention a few. In both backgammon and Othello, the best programs seem to match or exceed the top human performances.

Backgammon is especially interesting since the interpolation of moves determined by the throw of dice makes tree search almost futile. As the basis for evaluating and choosing moves, programs rely on pattern recognition, as do their human counterparts, and exact probability calculations, for which the human usually approximates or uses intuition. Berliner's construction of a world-class backgammon program using pattern recognition [Berliner (1980)] paved the way for building a rich body of chess knowledge into *Hitech*.

8. Conclusion

We have seen that the theory of games that emerges from this research is quite remote in both its concerns and its findings from von Neumann–Morgenstern theory. To arrive at actual strategies for the playing of games as complex as chess, the game must be considered in extensive form, and its characteristic function is of no interest. The task is not to characterize optimality or substantive rationality, but to define strategies for finding good moves – procedural rationality.

Two major directions have been explored. On the one hand, one can replace the actual game by a simplified approximation, and seek the game-theoretical optimum for the approximation – which may or may not bear any close resemblance to the optimum for the real game. In a game like chess, usually it does not. On the other hand, one can depart more widely from exhaustive minimax search in the approximation and use a variety of pattern-recognition and selective search strategies to seek satisfactory moves.

Both of these directions produce at best satisfactory, not optimal, strategies for the actual game. There is no a priori basis for predicting which will perform better in a domain where exact optimization is computationally beyond reach. The experience thus far with chess suggests that a combination of the two may be best – with computers relying more (but not wholly) on speed of computation, humans relying much more on knowledge and selectivity.

What is emerging, therefore, from research on games like chess, is a computational theory of games: a theory of what it is reasonable to do when it is impossible to determine what is best – a theory of bounded rationality. The lessons taught by this research may be of considerable value for understanding and dealing with situations in real life that are even more complex than the situations we encounter in chess – in dealing, say, with large organizations, with the economy, or with relations among nations.

References

Anantharaman, T. (1990) 'A statistical study of selective min-max search', PhD thesis, Carnegie-Mellon University.

Anantharaman, T., M.S. Campbell and F.H. Hsu (1988) 'Singular extensions: Adding selectivity to brute-force searching', *Artificial Intelligence*, 4: 135–143.

Berliner, H.J. (1980) 'Computer backgammon', *Scientific American*, June: 64–72.

Berliner, H.J. and C. Ebeling (1989) 'Pattern knowledge and search: The SUPREM architecture', *Artificial Intelligence*, 38: 161–198.

Campbell, M.S. and H.J. Berliner (1984) 'Using chunking to play chess pawn endgames', *Artificial Intelligence*, 23: 97–120.

Chase, W.G. and H.A. Simon (1973) 'Perception in chess', *Cognitive Psychology*, 4: 55–81.

de Groot, A.D. (1965) *Thought and choice in chess*. The Hague: Mouton.

McAllester, D.A. (1988) 'Conspiracy numbers for min-max search', *Artificial Intelligence*, 35: 287–310.

Nau, D.S. (1983) 'Pathology in game trees revisited and an alternative to minimaxing', *Artificial Intelligence*, 21: 221–244.

Newell, A. and H.A. Simon (1972) *Human problem solving*. Englewood Cliffs, NJ: Prentice-Hall.

Newell, A., J.C. Shaw and H.A. Simon (1958) 'Chess-playing programs and the problem of complexity', *IBM Journal of Research and Development*, 2: 320–335.

Rivest, R.L. (1988) 'Game tree searching by min/max approximation', *Artificial Intelligence*, 34: 77–96.

Samuel, A.L. (1967) 'Some studies in machine learning using the game of checkers: II – Recent progress', *IBM Journal of Research and Development*, 11: 601–617.

Schaeffer, J. (1989) 'The history heuristic and alpha–beta search enhancements in practice', *IEEE Transactions on Pattern Analysis and Machine Intelligence*, 11: 1203–1212.

Shannon, C.E. (1950) 'Programming a digital computer for playing chess', *Philosophical Magazine*, 41: 356–375.

Simon, H.A. and K. Gilmartin (1973) 'A simulation memory for chess positions', *Cognitive Psychology*, 5: 29–46.

Simon, H.A. and A. Newell (1958) 'Heuristic problem solving: The next advance in operations research', *Operations Research*, 6: 1–10.

Turing, A.M. (1953) 'Digital computers applied to games', in: B.V. Bowden, ed., *Faster than thought*. London: Putnam.

Uiterwijk, J.W.J.M., H.J. van den Herik and L.V. Allis (1989) 'A knowledge-based approach to connect-four. The game is over: White to move wins!', *Heuristic programming in artificial intelligence*. New York: Wiley.

von Neumann, J. and O. Morgenstern (1944) *Theory of games and economic behavior*. Princeton, NJ: Princeton University Press.

Welsh, D. and B. Baczynskyj (1985) *Computer chess II*. Dubuque, IA: W.C. Brown Co.

Wilkins, D.E. (1982) 'Using knowledge to control tree searching', *Artificial Intelligence*, 18: 1–5.

Chapter 2

GAMES IN EXTENSIVE AND STRATEGIC FORMS

SERGIU HART*

The Hebrew University of Jerusalem

Contents

*Based on notes written by Ruth J. Williams following lectures given by the author at Stanford University in Spring 1979. The author thanks Robert J. Aumann and Salvatore Modica for some useful suggestions.

Handbook of Game Theory, Volume 1, Edited by R.J. Aumann and S. Hart

0. Introduction

This chapter serves as an introduction to some of the basic concepts that are used (mainly) in Part I ("Non-Cooperative") of this Handbook. It contains, first, formal definitions as well as a few illustrative examples, for the following notions: games in extensive form (Section 1), games in strategic form (Section 3), pure and mixed strategies (Sections 2 and 4, respectively), and equilibrium points (Section 5). Second, two classes of games that are of interest are presented: games of perfect information, which always possess equilibria in pure strategies (Section 6), and games with perfect recall, where mixed strategies may be replaced by behavior strategies (Section 7).

There is no attempt to cover the topics comprehensively. On the contrary, the purpose of this chapter is only to introduce the above basic concepts and results in as simple a form as possible. In particular, we deal throughout only with *finite* games. The reader is referred to the other chapters in this Handbook for applications, extensions, variations, and so on.

1. Games in extensive form

In this section we present a first basic way of describing a game, called the "extensive form". As the name suggests, this is a most detailed description of a game. It tells exactly which player should move, when, what are the choices, the outcomes, the information of the players at every stage, and so on.

We need to recall first the basic concept of a "tree" and a few related notions. The reader is referred to any book on Graph Theory for further details.

A (finite, undirected) *graph* consists of a finite set V together with a set A of unordered pairs of distinct members[1] of V. See Figure 1 for some examples of graphs. An element $v \in V$ is called a *vertex* or a *node*, and each $\{v_1, v_2\} \in A$ is an *arc*, a *branch* or an *edge* ("joining" or "connecting" the vertices v_1 and v_2). Note that A may be anything from the empty set (a graph with no arcs) to the set of all possible pairs (a "complete" graph[2]). An (open) *path* connecting the nodes v_1 and v_m is a sequence $v_1, v_2, \ldots, v_m$ of distinct vertices such that $\{v_1, v_2\}, \{v_2, v_3\}, \ldots, \{v_{m-1}, v_m\}$ are all arcs of the graph (i.e., belong to A). A *cycle* (or "closed path") is obtained when one allows $v_1 = v_m$ in the above definition.

---

[1] Note that neither multiple arcs (between the same two nodes) nor "loops" (arcs connecting a node with itself) are allowed.

[2] The complete graph with n nodes has $n(n-1)/2$ arcs.

Figure 1. (a) A graph: $V = \{a, b, c, d, e\}$; $A = \{\{a, b\}, \{a, c\}, \{c, d\}, \{c, e\}\}$. (b) A graph: $V = \{a, b, c, d\}$; $A = \{\{a, b\}, \{b, c\}\}$.

A *tree* is a graph where any two nodes are connected by exactly one path. See Figures 2 and 3 for some examples of trees and "non-trees", respectively. It is easy to see that a tree with n nodes has $n - 1$ arcs, that it is a connected graph, and that it has no cycles.

Let T be a tree, and let r be a given distinguished node of T, called the *root* of the tree. One may then uniquely "direct" all arcs so they will point away from the root. Indeed, given an "undirected" arc $\{v_1, v_2\}$, either the unique path from r to v_2 goes through v_1 – in which case the arc becomes the ordered pair (v_1, v_2) – or the unique path from r to v_1 goes through v_2 – and then the arc is directed as (v_2, v_1). The root has only "outgoing" branches. All nodes having only "incoming" arcs are called *leaves* or *terminal nodes*; we will denote by $L \equiv L(T)$ the set of leaves of the tree T. See Figure 4 for an example of a "rooted tree".

Figure 2. A tree.

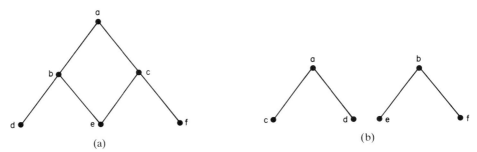

Figure 3. (a) Not a tree (two paths from a to e). (b) Not a tree (no path from a to b).

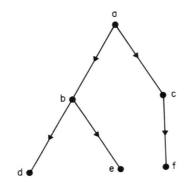

Figure 4. A rooted tree: root $= a$; leaves $= d, e, f$.

We can now formally define an *n-person game in extensive form*, Γ, as consisting of the following:[3]

(i) A set $N = \{1, 2, \ldots, n\}$ of *players*.

(ii) A rooted tree, T, called the *game tree*.

(iii) A partition of the set of non-terminal nodes[4] of T into $n + 1$ subsets denoted $P^0, P^1, P^2, \ldots, P^n$. The members of P^0 are called *chance* (or, *nature*) *nodes*; for each $i \in N$, the members of P^i are called the *nodes of player i*.

(iv) For each node in P^0, a probability distribution over its outgoing branches.

(v) For each $i \in N$, a partition of P^i into $k(i)$ *information sets*, $U^i_1, U^i_2, \ldots, U^i_{k(i)}$, such that, for each $j = 1, 2, \ldots, k(i)$:

---

[3]This definition is due to Kuhn (1953); it is more general than the earlier one of von Neumann (1928) [see Kuhn (1953, pp. 197–199) for a comparison between the two].

[4]A non-terminal node is sometimes called a "move".

    (a) all nodes in U^i_j have the same number of outgoing branches, and there is a given one-to-one correspondence between the sets of outgoing branches of different nodes in U^i_j;

    (b) every (directed) path in the tree from the root to a terminal node can cross each U^i_j at most once.

(vi) For each terminal node $t \in L(T)$, an n-dimensional vector $g(t) = (g^1(t), g^2(t), \ldots, g^n(t))$ of *payoffs*.

(vii) The complete description (i)–(vi) is common knowledge among the players.[5]

One can imagine this game Γ as being played in the following manner. Each player has a number of *agents*, one for each of his information sets [thus i has $k(i)$ agents]. The agents are isolated from one another, and the rules of the game [i.e., (i)–(vii)] are common knowledge among them too. A *play*[6] of Γ starts at the root of the tree T. Suppose by induction that the play has progressed to a non-terminal node, v. If v is a node of player i (i.e., $v \in P^i$), then the agent corresponding to the information set U^i_j that contains v chooses one of the branches going out of v, knowing only that he is choosing an outgoing branch at one of the nodes in U^i_j [recall (v) (a)]. If v is a chance node (i.e., $v \in P^0$), then a branch out of v is chosen according to the probability distribution specified for v [recall (iv); note that the choices at the various chance nodes are independent]. In this manner a unique path is constructed from the root to some terminal node t, where the game ends with each player i receiving a payoff $g^i(t)$.

Remark 1.1. The payoff vectors $g(t)$ are obtained as follows: to each terminal node $t \in L$ there corresponds a certain "outcome" of the game, call it $a(t)$. The payoff $g^i(t)$ is then defined as $u^i(a(t))$, where u^i is a von Neumann–Morgenstern utility function of player i. As will be seen below, the role of this assumption is to be able to evaluate a *random* outcome by its *expected* utility.

Example 1.2 ("Matching pennies"). See Figure 5: $N = \{1, 2\}$; root $= a$; $P^0 = \emptyset$; $P^1 = U^1_1 = \{a\}$; $P^2 = U^2_1 = \{b, c\}$; payoff vectors $(g^1(t), g^2(t))$ are written below each terminal node t. Note that player 2, when he has to make his choice, does not know the choice of player 1.[7] Thus both players are in a

---

[5]That is, all players know it, each one knows that everyone else knows it, and so on; see the chapter on 'common knowledge' in a forthcoming volume of this Handbook for a formal treatment of this notion.

[6]One distinguishes between a *game* and a *play*: the former is a complete description of the rules (i.e., the whole tree); the latter is a specific instance of the game being played (i.e., just one path in the tree).

[7]This shows the role of information sets; the game changes dramatically if player 2 knows at which node he is (b or c) when he has to make his choice – he can then always "win" (i.e., obtain a payoff of 1).

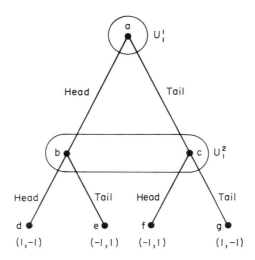

Figure 5. The game tree of Example 1.2.

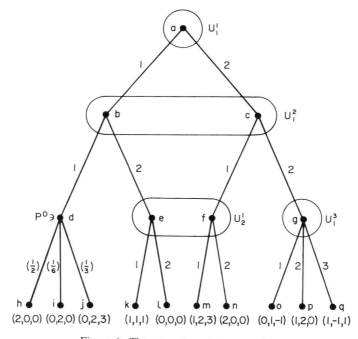

Figure 6. The game tree of Example 1.3.

similar situation: they do not know what is the choice of the other player; for instance, they may make their choices simultaneously. □

Example 1.3. See Figure 6: $N = \{1, 2, 3\}$; root $= a$; $P^0 = \{d\}$; $P^1 = \{a, e, f\}$; $U_1^1 = \{a\}$; $U_2^1 = \{e, f\}$; $P^2 = U_1^2 = \{b, c\}$; $P^3 = U_1^3 = \{g\}$; payoff vectors, the probability distribution at d, and the branches' correspondences [by (v) (a)] are all written on the tree. Note that at his second information set U_2^1, player 1 does not recall what his choice was at U_1^1; so player 1 consists of two agents (one for each information set), who do not communicate during the play. □

2. Pure strategies

Let $I^i := \{U_1^i, U_2^i, \ldots, U_{k(i)}^i\}$ be the set of information sets of player i; from now on we will simplify notation by using $U^i \in I^i$ to denote a generic element of I^i. For each information set U^i of player i, let $\nu \equiv \nu(U^i)$ be the number of branches going out of each node in U^i; number these branches from 1 through ν such that the one-to-one correspondence between the sets of outgoing branches of the different nodes of U^i is preserved. Thus, let $C(U^i) := \{1, 2, \ldots, \nu(U^i)\}$ be the set of choices available to player i at any node in U^i.

A *pure strategy* s^i *of player* i is a function

$$s^i : I^i \rightarrow \{1, 2, \ldots\},$$

such that

$$s^i(U^i) \in C(U^i) \quad \text{for all } U^i \in I^i.$$

That is, s^i specifies for every information set $U^i \in I^i$ of player i, a choice $s^i(U^i)$ there. Let S^i denote the set of pure strategies of player i, i.e., $S^i := \Pi_{U^i \in I^i} C(U^i)$. Let $S := S^1 \times S^2 \times \cdots \times S^n$ be the set of *n-tuples* (or *profiles*) *of pure strategies* of the players.

For an *n*-tuple $s = (s^1, s^2, \ldots, s^n) \in S$ of pure strategies, the (expected)[8] *payoff* $h^i(s)$ *to player* i is defined by

$$h^i(s) := \sum_{t \in L} p_s(t) g^i(t), \tag{2.1}$$

where, for each terminal node $t \in L(T)$, we denote by $p_s(t)$ the probability

---

[8]Recall Remark 1.1.

that the play of the game ends at t when the players use the strategies $s^1, s^2, \ldots, s^n$. This probability is computed as follows. Let $\pi \equiv \pi(t)$ be the (unique) path from the root to the terminal node t. If there exists a player $i \in N$ and a node of i on π at which s^i specifies a branch different from the one along π, then $p_s(t) := 0$. Otherwise, $p_s(t)$ equals the product of the probabilities, at all chance nodes on the path π, of choosing the branch which is along π. The function[9] $h^i : S \rightarrow \Re$ defined by (2.1) is called the *payoff function of player i.*

Example 2.2. Consider again the game of Example 1.3. Player 1 has four pure strategies: $\langle 1, 1 \rangle$, $\langle 1, 2 \rangle$, $\langle 2, 1 \rangle$ and $\langle 2, 2 \rangle$, where $\langle j_1, j_2 \rangle$ means that j_1 is chosen at U_1^1 and j_2 is chosen at U_2^1. Player 2 has two pure strategies: $\langle 1 \rangle$ and $\langle 2 \rangle$, and player 3 has three pure strategies: $\langle 1 \rangle$, $\langle 2 \rangle$ and $\langle 3 \rangle$. To see how payoffs are computed, let $s = (\langle 2, 1 \rangle, \langle 2 \rangle, \langle 3 \rangle)$; then the terminal node q is reached, thus $h^1(s) = 1$, $h^2(s) = -1$, and $h^3(s) = 1$. Next, let $s' = (\langle 1, 1 \rangle, \langle 1 \rangle, \langle 3 \rangle)$; then $h(s') = (1/2)(2, 0, 0) + (1/6)(0, 2, 0) + (1/3)(0, 2, 3) = (1, 1, 1)$. $\square$

3. Games in strategic form

A second basic way of describing a game is called the "strategic form" (also known as "normal form" or "matrix form").[10]

An *n-person game in strategic form* Γ consists of the following:
   (i) A set $N = \{1, 2, \ldots, n\}$ of *players.*
   (ii) For each player $i \in N$, a finite set S^i of (pure) *strategies.* Let $S := S^1 \times S^2 \times \cdots \times S^n$ denote the set of *n*-tuples of pure strategies.
   (iii) For each player $i \in N$, a function $h^i : S \rightarrow \Re$, called the *payoff function of player i.*

In the previous section we showed how the strategic form may be derived from the extensive form. Conversely, given a game in strategic form, one can always construct an extensive form as follows. Starting with the root as the single node of player 1, there are $|S^1|$ branches out of it, one for each strategy $s^1 \in S^1$ of player 1.[11] The $|S^1|$ end-nodes of these branches are the nodes of player 2, and they all form one information set. Each of these nodes has $|S^2|$ branches out of it, one for each strategy $s^2 \in S^2$ of player 2. All these $|S_1| \cdot |S_2|$ nodes form one information set of player 3. The construction of the tree is

---

[9]The real line is denoted $\Re$.
[10]We prefer "strategic form" since it is more suggestive.
[11]The number of elements of a finite set A is denoted by $|A|$.

Table 1

	s^2	
	Head	Tail
s^1 Head	1, −1	−1, 1
s^1 Tail	−1, 1	1, −1

continued in this manner: there are $|S^i|$ branches – one for each strategy $s^i \in S^i$ of player i – going out of every node of player i; the end-points of these branches are the nodes of player $i + 1$, and they all form one information set. The end-points of the branches out of the nodes of player n are the terminal nodes of the tree;[12] the payoff vector at such a terminal node t is defined as $(h^1(s), h^2(s), \ldots, h^n(s))$, where $s \equiv s(t)$ is the n-tuple of strategies of the players that correspond, by our construction, to the branches along the path from the root to t.

Example 3.1.   Let $N = \{1, 2\}$; $S^1 = S^2 = \{$Head, Tail$\}$; the payoff functions are given in Table 1, where each entry is $h^1(s^1, s^2), h^2(s^1, s^2)$. Plainly, the above construction yields precisely the extensive form of Example 1.2 ("matching pennies").   □

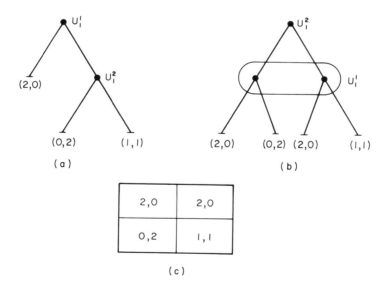

Figure 7. Two games in extensive form, (a) and (b), with the same strategic form, (c).

[12] There are $|S| = |S^1| \cdot |S^2| \cdots \cdots |S^n|$ terminal nodes.

It is clear that, in general, there may be many extensive forms with the same strategic form (up to "renaming" or "relabeling" of strategies). Such an example is presented in Figure 7. Thus, the extensive form contains more information about the game than the strategic form.

4. Mixed strategies

There are many situations in which a player's best behavior is to randomize when making his choice (recall, for instance, the game "matching pennies" of Examples 1.2 and 3.1). This leads to the concept of a "mixed strategy".

We need the following notation. Given a finite set A, the set of all probability distributions over A is denoted $\Delta(A)$. That is, $\Delta(A)$ is the $(|A| - 1)$-dimensional unit simplex

$$\Delta(A) := \left\{ x = (x(a))_{a \in A} : x(a) \geq 0 \text{ for all } a \in A \text{ and } \sum_{a \in A} x(a) = 1 \right\}.$$

The set of *mixed strategies* X^i *of player* i is defined as $X^i := \Delta(S^i)$, where S^i is the set of pure strategies of player i. Thus, a mixed strategy $x^i = (x(s^i))_{s^i \in S^i} \in X^i$ of player i means that i chooses each pure strategy s^i with probability $x^i(s^i)$. From now on we will identify a pure strategy $s^i \in S^i$ with the corresponding unit vector in X^i.

Let $X := X^1 \times X^2 \times \cdots \times X^n$ denote the set of *n-tuples of mixed strategies*. For every $x = (x^1, x^2, \ldots, x^n) \in X$, the (expected)[13] *payoff of player* i is

$$H^i(x) := \sum_{s \in S} x(s) h^i(s),$$

where $x(s) := \Pi_{j \in N} x^j(s^j)$ is the probability, under x, that the pure strategy n-tuple $s = (s^1, s^2, \ldots, s^n)$ is played. We have thus defined a *payoff function* $H^i : X \to \mathfrak{R}$ for player i. Note that $\Gamma^* := (N; (X^i)_{i \in N}; (H^i)_{i \in N})$ is an n-player (infinite)[14] game in strategic form, called the *mixed extension* of the original game $\Gamma = (N; (S^i)_{i \in N}; (h^i)_{i \in N})$.

If the game is given in extensive form, one obtains [from (2.1)] an equivalent expression for H^i:

$$H^i(x) = \sum_{t \in L} p_x(t) g^i(t), \tag{4.1}$$

where, for each terminal node $t \in L(T)$, we let $p_x(t)$ be the probability that the terminal node t is reached under x; i.e., $p_x(t) := \Sigma_{s \in S} x(s) p_s(t)$.

---

[13]Again, recall Remark 1.1.
[14]The strategy spaces are infinite.

5. Equilibrium points

We come now to the basic solution concept for non-cooperative games.

A (mixed) n-tuple of strategies $x = (x^1, x^2, \ldots, x^n) \in X$ is an *equilibrium point*[15] if

$$H^i(x) \geqslant H^i(x^{-i}, y^i)$$

for all players $i \in N$ and all strategies $y^i \in X^i$ of player i, where $x^{-i} := (x^1, \ldots, x^{i-1}, x^{i+1}, \ldots, x^n)$ denotes the $(n-1)$-tuple of strategies, in x, of all the players except i. Thus $x \in X$ is an equilibrium whenever no player i can gain by changing his own strategy (from x^i to y^i), assuming that all the other players do not change their strategies. Note that the notion of equilibrium point is based only on the strategic form of the game; various "refinements" of it may however depend on the additional data of the extensive form (see the chapters on 'strategic equilibrium' and 'conceptual foundations of strategic equilibrium' in a forthcoming volume of this Handbook for a comprehensive coverage of this issue).

The main result is

Theorem 5.1 [Nash (1950, 1951)]. *Every (finite) n-person game has an equilibrium point (in mixed strategies).*

The proof of this theorem relies on a fixed-point theorem (e.g., Brouwer's or Kakutani's).

6. Games of perfect information

This section deals with an important class of games for which equilibrium points in *pure* strategies always exist.

An n-person game Γ (in extensive form) is a *game of perfect information* if all information sets are singletons, i.e., $|U^i| = 1$ for each player $i \in N$ and each information set $U^i \in I^i$ of i. Thus, in a game of perfect information, every player, whenever called upon to make a choice, always knows exactly where he is in the game tree.

Examples of games of perfect information are Chess, Checkers, Backgammon (note that chance moves are allowed), Hex, Nim, and many others. In contrast, Poker, Bridge, Kriegsspiel (a variant of Chess where each player knows the position of his own pieces only) are games of *imperfect information*.

---

[15]Also referred to as "Nash equilibrium", "Cournot–Nash equilibrium", "non-cooperative equilibrium", and "strategic equilibrium".

(Another distinction, between *complete* and *incomplete information*, is presented and analyzed in the chapter on 'games of incomplete information' in a forthcoming volume of this Handbook.)

The historically first theorem of Game Theory deals with a game of perfect information.

Theorem 6.1 [Zermelo (1912)]. *In Chess, either*
  (i) *White can force a win, or*
 (ii) *Black can force a win, or*
(iii) *both players can force at least a draw.*

We say that a player can *force* an outcome if he has a strategy that makes the game terminate in that outcome, no matter what his opponent does. Zermelo's Theorem says that Chess is a so-called "determined" game: either there exists a pure strategy of one of the two players (White or Black) guaranteeing that he will always win, or each one of the two has a strategy guaranteeing at least a draw. Unfortunately, we do not know which of the three alternatives is the correct one (note that, in principle, this question is decidable in finite time, since the game tree of Chess is finite).[16]

The proof of Zermelo's Theorem is by induction, in a class of "Chess-like" games;[17,18] it is actually a special case of the following general result:

Theorem 6.2 [Kuhn (1953)]. *Every (finite) n-person game of perfect information has an equilibrium point in pure strategies.*

Proof. Assume by induction that the result is true for any game with less than m nodes. Consider a game Γ of perfect information with m nodes, and let r be the root of the game tree T. Let $v_1, v_2, \ldots, v_K$ denote the "sons" of r (i.e., those nodes that are connected to r by a branch), and let $T_1, T_2, \ldots, T_K$ (respectively), be the (disjoint) subtrees of T starting at these nodes. Each such T_k corresponds to a game Γ_k of perfect information (indeed, since Γ has perfect information, every information set is a singleton, thus completely included in one of the T_k's); Γ_k therefore possesses, by the induction hypothesis, an equilibrium point $s_k = (s_k^i)_{i \in N}$ in pure strategies. From these we construct a pure equilibrium point $s = (s^i)_{i \in N}$ for Γ, as follows. If r is a chance node, then s is just the "combination" (or "concatenation") of the s_k's i.e., $s^i(v) = s_k^i(v)$ for

---

[16]There are "Chess-like" games – for instance, Hex – where it can be proved that the first player can force a win, but nonetheless a winning strategy is not known. Other (simpler) games – e.g., Nim – have complete solutions (i.e., which player wins and by what strategy).

[17]For example, see Aumann (1989, pp. 1–4).

[18]Zermelo's Theorem 6.1 has been extended to two-person, zero-sum games by von Neumann and Morgenstern (1944, Section 15).

all nodes v of player i that belong to T_k. If r is a node of, say, player i, then, in addition to the above "combination", we put[19]

$$s^i(r) := \operatorname*{argmax}_{1 \leqslant k \leqslant K} h^i_k(s_k),$$

i.e., player i chooses at his first node r a branch k that leads to a subgame Γ_k where his equilibrium payoff is maximal. It is now straightforward to check that s is indeed a pure equilibrium point of Γ. $\square$

Remark 6.3. The above proof yields a construction of equilibrium points in pure strategies by "backwards induction", from the terminal nodes to the root:[20] at each node of a player, choose a branch which leads to a subtree with the highest equilibrium payoff for that player;[21] at each chance node, average the equilibrium payoffs of the subtrees. Note that the equilibrium points constructed in this manner, when restricted to any subgame of the original game, yield equilibria in the subgame as well; such equilibria are called "(subgame) perfect". The reader is referred to the chapters on 'strategic equilibrium' and 'conceptual foundations of strategic equilibrium' in a forth-

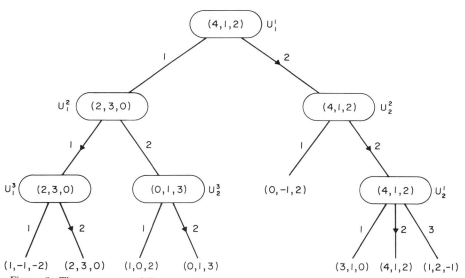

Figure 8. The game tree of Example 6.4 and the construction of the pure equilibrium point.

[19]We write h^i_k for the payoff function of player i in the subgame Γ_k, and "argmax" for a maximizer (if not unique, pick one arbitrarily).
[20]This is the standard procedure of "dynamic programming".
[21]Note that some of these choices need not be unique, in which case there is more than one such equilibrium.

coming volume of this Handbook for a discussion of these issues of backwards induction and perfection in relation to equilibrium points.

The following example illustrates the construction.

Example 6.4. See Figure 8: arrows indicate the choices forming the equilibrium strategies; the numbers in each node are the equilibrium payoffs for the subtree rooted at that node. The resulting equilibrium point is $s = (\langle 2, 2 \rangle, \langle 1, 2 \rangle, \langle 2, 2 \rangle)$, with payoffs $h(s) = (4, 1, 2)$. $\square$

The reader is referred to Chapter 3 in this volume for the development of the topic of games of perfect information, in particular in infinite games.

7. Behavior strategies and perfect recall

A pure strategy of a player is a complete plan for his choices in all possible contingencies in the game (i.e., at all his information sets). A mixed strategy means that the player chooses, *before* the beginning of the game, one such comprehensive plan at random (according to a certain probability distribution). An alternative method of randomization for the player is to make an independent random choice at each one of his information sets. That is, rather than selecting, for every information set, one definite choice – as in a pure strategy – he specifies instead a probability distribution over the set of choices there; moreover, the choices at different information sets are (stochastically) independent. These randomization procedures are called *behavior strategies*.

A useful way of viewing the difference between mixed and behavior strategies is as follows. One can think of each pure strategy as a book of instructions, where for each of the player's information sets there is one page which states what choice he should make at that information set. The player's set of pure strategies is a library of such books. A mixed strategy is a probability distribution on his library of books, so that, in playing according to a mixed strategy, the player chooses one book from his library by means of a chance device having the prescribed probability distribution. A behavior strategy is a single book of a different sort. Although each page still refers to a single information set of the player, it specifies a probability distribution over the choices at that set, not a specific choice.

We will see below that a behavior strategy is essentially a (special kind of) mixed strategy. Moreover, when a player has what is called "perfect recall", the converse also holds: every mixed strategy is fully "equivalent" to a behavior strategy.

We define a *behavior strategy* b^i *of player i* in the game Γ (in extensive form) as an element of

$$B^i := \prod_{U^i \in I^i} \Delta(C(U^i)), \tag{7.1}$$

that is, $b^i = (b^i(U^i))_{U^i \in I^i}$, where each $b^i(U^i)$ is a probability distribution over the set $C(U^i)$ of choices of player i at his information set U^i. We will write $b^i(U^i; c)$, rather than the more cumbersome $(b^i(U^i))(c)$, for the probability that the choice of player i at U^i is $c \in C(U^i)$; thus $\sum_{c \in C(U^i)} b^i(U^i; c) = 1$ and $b^i(U^i; c) \geqslant 0$.

Note that the linear dimension of the space of behavior strategies B^i of player i is $\Sigma_j (\nu^i_j - 1)$, whereas that of the space of mixed strategies X^i is $\prod_j \nu^i_j - 1$, where j ranges from 1 to $k(i) = |I^i|$ and $\nu^i_j = |C(U^i_j)|$. Therefore B^i is a much smaller set than X^i.

Actually, the set B^i of behavior strategies of player i can be identified with a subset of the set X^i of mixed strategies of i. Indeed, given a behavior strategy, one may perform all the randomizations (for all information sets) before the game starts, which yields a (random) pure strategy – i.e., a mixed strategy. Formally, the *mixed strategy* x^i *corresponding to the behavior strategy* $b^i \in B^i$ is defined by $x^i = (x^i(s^i))_{s^i \in S^i}$, where

$$x^i(s^i) := \prod_{U^i \in I^i} b^i(U^i; s^i(U^i)) \tag{7.2}$$

for each pure strategy $s^i \in S^i$. Since $b^i(U^i; s^i(U^i))$ is the probability, under b^i, that player i chooses $s^i(U^i)$ at the information set U^i, it follows that $x^i(s^i)$ is precisely the probability that all his (realized) choices are according to the pure strategy s^i; in short, $x^i(s^i)$ is the probability, under b^i, of using s^i. The following lemma is thus immediate.

Lemma 7.3. *For any behavior strategy* $b^i \in B^i$ *of player i, the corresponding* x^i *given by* (7.2) *is a mixed strategy of i that is equivalent to* b^i.

We call the two strategies y^i and z^i of player i *equivalent* if they yield the same payoffs[22] (to everyone) for any strategies of the other players, i.e., $H^j(y^i, x^{-i}) = H^j(z^i, x^{-i})$ for all x^{-i} and all $j \in N$. Note that the argument given above shows that a stronger statement is actually true: for each terminal node $t \in L$, the probabilities that t is reached under (b^i, x^{-i}) and under (x^i, x^{-i}) are the same, for any x^{-i}.

---

[22] We have defined the (expected) payoff functions H^i for n-tuples of mixed strategies (see Section 4). The definition may be trivially extended to behavior strategies as well: use (4.1) with the probabilities $p_x(t)$ computed accordingly.

The difference between behavior and mixed strategies can thus be viewed as *independent* vs. (possibly) *correlated* randomizations (at the various information sets). This may be also seen by comparing directly the two definitions: B^i is a product of probability spaces [see (7.1)], whereas X^i is the probability space on the product [i.e., $\Delta(\Pi_{U^i \in I^i} C(U^i))$]. The following example is most illuminating.

Example 7.4 [Kuhn (1953)]. Consider a two-player, zero-sum[23] game in which player 1 consists of two people,[24] Alice and her husband Bill, and player 2 is a single person, Zeno. Two cards, one marked "High" and the other "Low", are dealt at random to Alice and Zeno. The person with the High card receives \$1 from the person with the Low card, and then has the choice of stopping or continuing the play. If the play continues, Bill, *not knowing the outcome of the deal*, instructs Alice and Zeno either to exchange or to keep their cards. Again, the holder of the High card receives \$1 from the holder of the Low card, and the game ends. See Figure 9 for the game tree (A = Alice,

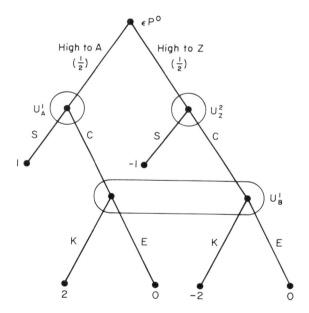

Figure 9. The game tree of Example 7.4.

---

[23]A two-player game is a *zero-sum* game if $h^1 + h^2 = 0$, i.e., what one player gains is what the other loses.
[24]With a joint bank account.

Table 2
Strategic form of Example 7.4

1 \ 2	$\langle S \rangle$	$\langle C \rangle$
$\langle S, K \rangle$	$\frac{1}{2} \cdot 1 + \frac{1}{2} \cdot (-1) = 0$	$\frac{1}{2} \cdot 1 + \frac{1}{2} \cdot (-2) = -\frac{1}{2}$
$\langle S, E \rangle$	$\frac{1}{2} \cdot 1 + \frac{1}{2} \cdot (-1) = 0$	$\frac{1}{2} \cdot 1 + \frac{1}{2} \cdot (0) = \frac{1}{2}$
$\langle C, K \rangle$	$\frac{1}{2} \cdot 2 + \frac{1}{2} \cdot (-1) = \frac{1}{2}$	$\frac{1}{2} \cdot 2 + \frac{1}{2} \cdot (-2) = 0$
$\langle C, E \rangle$	$\frac{1}{2} \cdot 0 + \frac{1}{2} \cdot (-1) = -\frac{1}{2}$	$\frac{1}{2} \cdot 0 + \frac{1}{2} \cdot (0) = 0$

B = Bill, Z = Zeno; S = Stop, C = Continue, K = Keep and E = Exchange; payoffs at the terminal nodes are those paid by player 2 to player 1).

The strategic form of this game is given in Table 2. Note that the strategies $\langle S, K \rangle$ and $\langle C, E \rangle$ of player 1 are strictly dominated (by $\langle C, K \rangle$ and $\langle S, E \rangle$, respectively). Eliminating them yields the reduced strategic form of Table 3.

It is now easy to see that the unique optimal (mixed) strategies of the players are $(0, 1/2, 1/2, 0)$ and $(1/2, 1/2)$, respectively;[25] the value of the game is $1/4$. Thus, in particular, player 1 can guarantee that his expected payoff will be at least $1/4$, regardless of what player 2 will do.

Suppose now that player 1 uses only behavior strategies. Let $b^1 = (b^1(U_A^1), b^1(U_B^1)) = ((\alpha, 1 - \alpha), (\beta, 1 - \beta)) \in B^1$, i.e., Alice chooses S with probability α and C with probability $1 - \alpha$, and Bill chooses K with probability β and E with probability $1 - \beta$. [Note that the mixed strategy corresponding to b^1 is $(\alpha\beta, \alpha(1 - \beta), (1 - \alpha)\beta, (1 - \alpha)(1 - \beta))$ – see (7.2).] Then player 1's expected payoff is[26] $(1 - \alpha)(\beta - 1/2)$ if player 2 plays S, and $\alpha(1/2 - \beta)$ if player 2 plays C. So the maximum payoff that player 1 can guarantee when restricted to behavior strategies is

Table 3
Reduced strategic form of Example 7.4

1 \ 2	$\langle S \rangle$	$\langle C \rangle$
$\langle S, E \rangle$	0	$\frac{1}{2}$
$\langle C, K \rangle$	$\frac{1}{2}$	0

[25]A mixed strategy of player 1 is written as the vector of probabilities for his pure strategies $\langle S, K \rangle, \langle S, E \rangle, \langle C, K \rangle, \langle C, E \rangle$, in that order; for player 2, the order is $\langle S \rangle, \langle C \rangle$.
[26]For example, the payoff if 2 plays S is computed as follows $(1/2) \cdot [\alpha \cdot 1 + (1 - \alpha) \cdot (\beta \cdot 2 + (1 - \beta) \cdot 0)] + (1/2) \cdot (-1)$.

$$\max_{0 \leqslant \alpha, \beta \leqslant 1} [\min\{(1-\alpha)(\beta-1/2), \alpha(1/2-\beta)\}],$$

which equals 0, since either $\beta - 1/2$ or $1/2 - \beta$ is always $\leqslant 0$. $\quad\square$

Thus, behavior strategies for player 1 do a poorer job in this example than mixed strategies: player 1 can guarantee 1/4 with the latter, but only 0 with the former. Indeed, there is no behavior strategy corresponding to the unique optimal mixed strategy $x^1 = (0, 1/2, 1/2, 0)$ of player 1, since x^1 requires the randomizations at his two information sets to be fully correlated (rather than independent).

The reason that behavior strategies are inadequate in Example 7.4 is that player 1 consists of two agents who are *not allowed to communicate* during the play. This implies that, in going from U_A^1 to U_B^1, player 1 "forgets" what he knew, namely the outcome of the initial draw. Therefore the player needs to correlate, before the game starts, the random choices of his agents at his two information sets. Conversely, if a player always remembers what he knew as well as what he chose at all his previous nodes – in which case we say that the player has "perfect recall" – then he has no need to correlate the choices at his different information sets: indeed, being at any information set uniquely determines what happened at all the previous ones.

Formally, given a game tree T and a node v of T, we will denote by $T(v)$ the subtree of T with root at v. For an information set U and a choice there $c \in C(U)$, we will write $T(U; c)$ for the union of the trees $T(w)$, where w is connected to some node $v \in U$ by a branch labeled c [i.e., $T(U; c)$ is the "remainder" of the game after the information set U has been reached, and the choice c has been made there by the corresponding player]. We will say that *player i has perfect recall in the game* Γ (in extensive form) if the following condition is satisfied.[27] Let $v_1, v_2 \in P^i$ be nodes of player i, let $U_1^i \ni v_1$ and $U_2^i \ni v_2$ be the corresponding information sets of i, and assume that $v_2 \in T(v_1)$ (i.e., there exists a play of the game – a path – where v_2 "comes after" v_1); then there exists a unique choice $c \in C(U_1^i)$ such that $U_2^i \subset T(U_1^i; c)$. A game Γ in which every player has perfect recall is called a *game of perfect recall*. Note that a player who is a single person has perfect recall;[28] isolated agents are not needed to play the game for him. This is the case for most parlor games (but not for Bridge, when viewed as a two-player game, with each player consisting of two partners).

The condition in the definition of perfect recall can be separated into two parts:

---

[27]The original definition of Kuhn (1953) is different but equivalent to the one presented here; the advantage of the latter is that it is stated (and may be checked) directly on the structure of the tree.

[28]Provided he is not too absent-minded.

(i) "Player i recalls what he knew": $U_2^i \subset T(U_1^i)$, i.e., each node in U_2^i has to be reachable from some node in U_1^i. Otherwise, let $v_2' \in U_2^i \backslash T(U_1^i)$;[29] then, when reaching the information set U_2^i, player i does not recall whether the play went through U_1^i (as in v_2) or not (as in v_2') – he forgot what he knew (when he was at U_1^i). If player i had perfect recall, then, at U_2^i, he would be able to distinguish between v_2 and v_2', according to whether or not he has already been called upon to make a choice at U_1^i; hence v_2 and v_2' would have lied in different information sets. In Example 7.4 above, player 1 does not have perfect recall, since, if the play is: 'player 1 gets "High" and decides "Continue" ', then at U_B^1 he no longer knows what he knew at U_A^1 (namely, the outcome of the draw).

(ii) "Player i recalls what he chose": $U_2^i \subset T(U_1^i; c)$ for a unique choice c at U_1^i. Otherwise, let $v_2' \in T(U_1^i; c') \cap U_2^i$ for some other choice $c' \neq c$; then, at U_2^i, player i does not recall whether his own choice at U_1^i was c (as is the case at v_2) or c' (as is the case at v_2') – he forgot what he chose (at U_1^i). If player i had perfect recall, then the nodes v_2 and v_2' would be distinguished by the choice he made at U_1^i, and would thus lie in different information sets. In Example 1.3 (in Section 1), for instance, condition (i) is satisfied but (ii) is not: U_2^1 contains two nodes (e and f) that follow different choices (1 and 2, respectively) of player 1 at U_1^1.

Example 7.5. In the game tree of Figure 10, assume that the root and the nodes at the third level do *not* belong to player 1 (it does not matter if they are chance nodes or personal nodes, and to which information sets they belong), and assume that all eight nodes at the fourth level are nodes of player 1. If player 1 has perfect recall, then condition (i) implies that $\{c, d, e, f\}$ is separated from $\{g, h, i, j\}$ (i.e., there can be no information set of player 1 containing nodes from both sets); condition (ii) separates $\{c, d\}$ from $\{e, f\}$, and $\{g, h\}$ from $\{i, j\}$. The dashed lines in Figure 10 show this partitioning. □

Example 7.6. Modify Example 7.5 by putting the two nodes a and b at the second level in one information set (of player 1). Then perfect recall for player 1 implies that $\{c, d, g, h\}$ must be separated from $\{e, f, i, j\}$ [condition (ii)]. □

We come now to the main result of this section.

Theorem 7.7 [Kuhn (1953)]. *Let Γ be a (finite) n-person game in which player i has perfect recall. Then for each mixed strategy $x^i \in X^i$ of player i there exists a corresponding behavior strategy $b^i \in B^i$ that is equivalent to x^i.*

---

[29]A "\" denotes set subtraction.

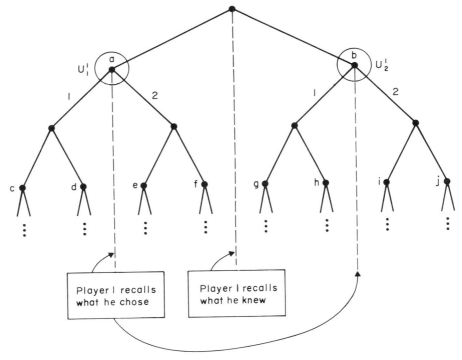

Figure 10. The game tree of Example 7.5.

Thus, having perfect recall is a sufficient[30] condition for restricting a player to behavior strategies instead of the (usually much) larger set of mixed strategies.

Proof (outline). Given a *pure* strategy $s^i \in S^i$ and an information set $U^i \in I^i$, we will say that U^i is *reachable under* s^i if there exists a play of the game that goes through U^i and is consistent with s^i, i.e., there exists a path π in the tree that intersects U^i and, at every node of the player i on π, the path π follows the choice dictated by s^i. Given $x^i \in X^i$ we define the *corresponding behavior strategy* $b^i \in B^i$ as follows. For every information set $U^i \in I^i$, let $\xi^i(U^i)$ be the probability that U^i is reachable under x^i, i.e., $\xi^i(U^i)$ is the sum of $x^i(s^i)$ over all $s^i \in S^i$ under which U^i is reachable. Similarly, for each choice $c \in C(U^i)$ at U^i, let $\xi^i(U^i; c)$ be the probability, under x^i, that U^i is reachable

---

[30]Kuhn (1953) shows that it is also *necessary*.

and the choice there is c [i.e., the sum of $x^i(s^i)$ over all s^i such that U^i is reachable and $s^i(U^i) = c$]. Finally, put

$$b^i(U^i; c) := \xi^i(U^i; c)/\xi^i(U^i) \tag{7.8}$$

if the denominator is positive, and define $b^i(U^i)$ arbitrarily when $\xi^i(U^i)$ vanishes (it will not matter, since then U^i is never reached when i plays x^i). One may interpret $b^i(U^i)$ as the "observed (random) behavior" of player i at U^i when he uses x^i.

Let x^{-i} be the strategies of the other players. We will show that

$$P_{(x^i, x^{-i})}(t) = P_{(b^i, x^{-i})}(t) \tag{7.9}$$

for each terminal node $t \in L$ (these are the probabilities that the play ends at t). Fix t, and denote by $\pi \equiv \pi(t)$ the path from the root to t. Let α be the probability that chance and all players except i always choose, at nodes on π, the branches along π; note that α depends on x^{-i} but not on the choices of player i. Similarly, let ξ (respectively, β) denote the probability under x^i (respectively, b^i) that all the choices of player i on π are along π. Then (7.9) becomes $\alpha \cdot \xi = \alpha \cdot \beta$, and it suffices to show that $\xi = \beta$ when $\alpha > 0$ (i.e., when t is reachable).

Let v_1 and v_2 be two consecutive nodes of i along π (i.e., there are no other nodes of i between them), and let $U_1^i \ni v_1$ and $U_2^i \ni v_2$ be the corresponding information sets; let c be the choice at v_1 along π. Player i has perfect recall, therefore $U_2^i \subset T(U_1^i; c)$, implying that U_2^i is reachable if and only if U_1^i is reachable and the choice there is c. Hence $\xi^i(U_2^i) = \xi^i(U_1^i; c)$; or, by (7.8), the denominator of $b^i(U_2^i; \cdot)$ equals the numerator of $b^i(U_1^i; c)$. To compute β we have to multiply the probabilities, under b^i, of all the choices of i along π. We thus obtain a telescoping product that simplifies to $\beta = \xi^i(U_m^i; c_m)$, where c_m is the choice of i along π at his last information set U_m^i on π. Now $\xi^i(U_m^i; c_m)$ is the probability, under x^i, that U_m^i is reachable and the choice there is c_m, or, equivalently (again by perfect recall, using induction), the probability that all the m choices of i on π are of the branches along π; but this probability is precisely ξ. Therefore $\beta = \xi$, and the proof is completed. $\square$

Theorem 7.7 has been generalized to infinite games by Aumann (1964). As is pointed out there (p. 630), the extension of Kuhn's proof to the case where the length of the game as well as the number of choices at all information sets are at most countable poses no problems; the difficulties arise when there are uncountably many choices at some information set(s). In addition, games in which "time" is continuous pose special problems of their own, and do not easily fit into the framework of this chapter (see Chapter 3 in this volume, and chapters on 'two-player games', 'differential games' and 'economic applications of differential games' in forthcoming volumes of this Handbook for some examples).

References

Aumann, R.J. (1964) 'Mixed and behavior strategies in infinite extensive games', in: M. Dresher, L.S. Shapley and A.W. Tucker, eds., *Advances in game theory*, Annals of Mathematics Studies 52. Princeton: Princeton University Press, pp. 627–650.

Aumann, R.J. (1989) *Lectures on game theory*. Boulder: Westview Press.

Kuhn, H.W. (1953) 'Extensive games and the problem of information', in: H.W. Kuhn and A.W. Tucker, eds., *Contributions to the theory of games*, Vol. II, Annals of Mathematics Studies 28. Princeton: Princeton University Press, pp. 193–216.

Nash, J.F. (1950) 'Equilibrium points in *n*-person games', *Proceedings of the National Academy of Sciences*, 36: 48–49.

Nash, J.F. (1951) 'Non-cooperative games', *Annals of Mathematics*, 54: 286–295

von Neumann, J. (1928) 'Zur Theorie der Gesellschaftsspiele', *Mathematische Annalen*, 100: 295–320. English translation: 'On the theory of games of strategy', in: A.W. Tucker and R.D. Luce, eds., *Contributions to the theory of games*, Vol. IV, Annals of Mathematics Studies 40. Princeton: Princeton University Press.

von Neumann, J. and O. Morgenstern (1944) *Theory of games and economic behavior*. Princeton: Princeton University Press.

Zermelo, E. (1912) 'Über eine Anwendung der Mengenlehre auf die Theorie des Schachspiels', *Proceedings of the Fifth International Congress of Mathematicians*, Vol. II, pp. 501–504.

Chapter 3

GAMES WITH PERFECT INFORMATION

JAN MYCIELSKI*

University of Colorado

Contents

*I am indebted to S. Swierczkowski, R. Telgarsky and the referees for many improvements in this chapter.

Handbook of Game Theory, Volume 1, Edited by R.J. Aumann and S. Hart

1. Introduction

The most seriously played games of perfect information (which we will call PI-games) are Chess and Go. But there are numerous other interesting PI-games: Checkers, Chinese Checkers, Halma, Nim, Hex, their misére variants, etc. Perfect information means that at each time only one of the players moves, that the game depends only on their choices, they remember the past, and in principle they know all possible futures of the game (a full definition is given in Section 2). For example War is not a PI-game since the generals move simultaneously, and Bridge and Backgammon are not PI-games because chance plays a role in them. However, as we shall see in Section 7, some cases of Pursuit and Evasion can be studied by means of PI-games, in spite of the simultaneity and continuity of the movements of the players. There exists a marvelous book, *Winning Ways* (Vols. 1 and 2), by Berlekamp, Conway and Guy (1982) which gives many old and new examples of PI-games and a deep (and light) development of their theories. Thus my first duty as a surveyor of this subject is to refer the reader to this book and to the literature quoted in its 24 sections of references. But, to the less assiduous reader, I will suggest Martin Gardner's several chapters on games [Gardner (1983, 1986)], and the *Boardgame Book* by Bell [Bell (1979) and the references therein], and other relevant chapters of this Handbook.

   This survey will not overlap much with the above literature since I will focus here on *infinite* PI-games. If one wanted to play such a game one would have to play infinitely many moves! So those games are not intended to be played in reality, and their theory has (as yet) no relevance for practical play of the finite games. The main chapters of game theory, which stem from von Neumann's Minimax Theorem, are much closer to real applications. But the theory of infinite PI-games is motivated by its beauty and manifold connections with other parts of mathematics. For example, it gave new insights or new points of view in descriptive set theory [see Kechris et al. (1977, 1979, 1981, 1985) and Moschovakis (1980)], general topology, some chapters of analysis developed by G. Choquet and others, and number theory [see the surveys of Piotrowski (1985) and Telgarsky (1987)]. As mentioned above they give also a natural mathematical theory for some games of pursuit and evasion (see Section 7 of this chapter). Finite and infinite PI-games are also used in model theory [see, for example, Ehrenfeucht (1961), Lynch (1985, 1992) and Hodges (1985), and references therein] and in recursion theory [see, for example, Yates (1974, 1976)].

   Before dropping the subject of finite PI-games (to which we return only briefly in Sections 5 and 6), let me emphasize the question: *What additions to the general theory would be needed to make it relevant for Chess or Go?* We can only say that in 1990 the state of the art is still confusing. On the one hand there exist machines playing Chess and Checkers well above the amateurish level, see Michie (1989) and references therein. Those machines rely essentially upon the high speed of special digital processors which allow them to examine a large number of possible future developments of the play and to choose a good move on account of this analysis. On the other hand we feel, and are told by masters, that the best human players do not think in this way! Moreover, those machines do not benefit from playing, they do not learn. So, in particular, we hesitate to call them intelligent since the ability to learn appears to us to be the single most important feature of intelligence. We think that a theory of long-range strategies or plans of attack and methods for the construction of a book or a classification of good moves will have to be developed for a more general and practical theory. We know only one general concept, called the temperature of a position, studied in Berlekamp et al. (1982), which appears to be practical; a similar concept is used in Chess-playing programs in order to decide if a given position should be analyzed further or not. In spite of the present shortcomings of the theory of finite PI-games its prospects are bright: a relevant general theory should yield a computer program such that anybody could code his favorite game (for me it would be Hex, see Section 5), and after playing a number of games with the computer, the machine should get better and better, and eventually display an overwhelming superiority. But this goal still appears to be far ahead in the future, especially for the game Go (see Chapter 1 of this volume).

This chapter on PI-games is self-contained in the sense that in principle the proofs given below do not require any specialized knowledge. The necessary background is given, for example, in the beautiful short text of Oxtoby (1971).

2. Basic concepts

A triple $\langle A, B, \varphi \rangle$, where A and B are abstract sets and $\varphi: A \times B \to \bar{\mathbb{R}}$, where $\bar{\mathbb{R}} = \mathbb{R} \cup \{-\infty, \infty\}$ and $\mathbb{R}$ is the set of real numbers, is called *a game*. A is called the set of *strategies* of player I and B the set of *strategies* of player II. This game is played as follows: player I chooses $a \in A$ and player II chooses $b \in B$. Both choices are made independently and without any knowledge about the choice of the other player. Then II pays to I the value $\varphi(a, b)$. [Of course $\varphi(a, b) < 0$ means that II gets from I the value $|\varphi(a, b)|$.] Occasionally it will be convenient to use also a dual definition in which I pays to II the value $\varphi(a, b)$.

As usual we denote the set $\{0, 1, 2, \ldots\}$ by ω and $\alpha = \{\xi: \xi < \alpha\}$ for all ordinal numbers α. In particular $\{0, 1\} = 2$. For any sets X and Y, Y^X denotes the set of all functions $f: X \to Y$.

The intuitive idea of an *infinite game of perfect information* is the following.

There is a set P called the set of *choices*. Player I chooses $p_0 \in P$, next player II chooses $p_1 \in P$, than I chooses $p_2 \in P$, etc. There is a function $f: P^\omega \to \bar{\mathbb{R}}$ such that "at the end" player II pays to I the value $f(p_0, p_1, \ldots)$. More precisely, and consistently with the previous general definition of a game, $\langle A, B, \varphi \rangle$ is said to be a *game of perfect information* (a PI-game) if there exists a set P such that A is the set of all functions

$$a: \bigcup_{n<\omega} P^n \to P, \quad \text{where } P^0 = \{\emptyset\},$$

B is the set of all functions

$$b: \bigcup_{0<n<\omega} P^n \to P$$

and there exists a function $f: P^\omega \to \bar{\mathbb{R}}$ such that

$$\varphi(a, b) = f(p_0, p_1, p_2, \ldots),$$

where $p_0 = a(\emptyset)$, $p_1 = b(p_0)$, $p_2 = a(p_1)$, $p_3 = b(p_0, p_2)$, $p_4 = a(p_1, p_3), \ldots$ (see Figure 1).

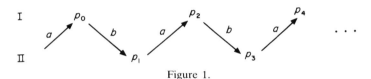

Figure 1.

From now on a game $\langle A, B, \varphi \rangle$ defined in this way will also be denoted by $\langle P^\omega, f \rangle$.

The sequence $p = (p_0, p_1, \ldots)$ defined as above will also be called *a game* and any finite sequence $(p_0, \ldots, p_{n-1}) \in P^n$ is called a *position*.

In the case when f is the *characteristic function of a set* $X \subseteq P^\omega$, i.e., $f(p) = 1$ if $p \in X$ and $f(p) = 0$ if $p \notin X$, the game $\langle P^\omega, f \rangle$ will also be denoted by $\langle P^\omega, X \rangle$. In this case we will say that player I *wins the game* p if $f(p) = 1$ and that II *wins the game* p if $f(p) = 0$.

A game $\langle A, B, \varphi \rangle$ is called *determined* if

$$\inf_{b \in B} \sup_{a \in A} \varphi(a, b) = \sup_{a \in A} \inf_{b \in B} \varphi(a, b), \tag{$*$}$$

and the common value of both sides of this equation is called *the value of the game* $\langle A, B, \varphi \rangle$. The assertion that a game is determined is also expressed by the phrase *the game has a value. Note: if the game is not determined, then the left-hand side of* $(*)$ *is larger than the right-hand side of* $(*)$.

If the game has the value V and there exists an a_0 such that $\varphi(a_0, b) \geq V$ for all b, then a_0 is called an *optimal strategy* for I. If $\varphi(a, b_0) \leq V$ for all a, then b_0 is called an *optimal strategy* for II.

A game can be determined but no optimal strategies need to exist. For example, this is so if $A = B =$ the open interval $(0, 1)$, and $\varphi(a, b) = a + b$. However, if the game is determined, then for every $\varepsilon > 0$ there exists a strategy a_0 which secures $\varphi(a_0, b) > V - \varepsilon$ for all $b \in B$ [or $\varphi(a_0, b) > \varepsilon$ if $V = +\infty$] and a strategy b_0 which secures $\varphi(a, b_0) < V + \varepsilon$ for all $a \in A$ [or $\varphi(a, b_0) < -\varepsilon$ if $V = -\infty$]. It follows that, *if the set of values of φ is finite and the game is determined, then both players have optimal strategies.* We will say that $\langle P^\omega, X \rangle$ *is a win for* I or *a win for* II if $\langle P^\omega, X \rangle$ has the value 1 or 0, respectively.

Why can games like Chess or Go be viewed as games of the form $\langle P^\omega, f \rangle$? The interpretation is the following: P is the set of all possible configurations of pieces on the board. Any infinite sequence of configurations is accepted as a game but the first player who violates the rules loses, unless the previous position is a win for one of the players or a draw. So f takes on three possible values: 1 (White wins), 0 (a draw) and -1 (Black wins), and $\langle P^\omega, f \rangle$ represents the desired game. We should add that this mathematical abstraction ignores some aspects of the reality. For example, the rules about timing are essential in most Chess tournaments but here they are ignored.

If $f: P^\omega \to \bar{\mathbb{R}}$ has the property that there exists an n such that $f(p_0, p_1, \ldots)$ does not depend on the choices p_i with $i > n$, then $\langle P^\omega, f \rangle$ is called a *finite game*. (Note that this does not imply that P is finite.)

Proposition 2.1. *Every finite game has a value.*

Proof. It is clear that the proposition is true for $n = 0$ and it is easy to see that, if it is true for $n = k$, then it is also true for $n = k + 1$. □ (For another proof see Proposition 3.2.)

Proposition 2.1 was first stated as a mathematical theorem by Zermelo (1913). As we shall see in the next section, it fails for some infinite PI-games. The main goal of this chapter, the theorems in Sections 3, 8 and 9, will be refinements of this proposition relaxing the condition of finiteness in various ways.

3. Open games are determined

The first published paper devoted to general infinite PI-games is due to Gale and Stewart (1953). The material of this section is contained in that paper.

For any set P we introduce the discrete topology in P and the corresponding *product topology* in P^ω. That is, the basic neighborhoods of $p = (p_0, p_1, \ldots) \in P^\omega$ are of the form

$$U(p_0, \ldots, p_{m-1}) = \{q \in P^\omega : q_i = p_i \text{ for } i < m\}.$$

In contrast to Proposition 2.1 we have:

Proposition 3.1. *There exist sets $X \subseteq \{0, 1\}^\omega$ such that the game $\langle \{0, 1\}^\omega, X \rangle$ is not determined.*

Proof. Note that if we fix one strategy for one of the players, then all games in $\{0, 1\}^\omega$ which remain possible constitute a perfect set, i.e., a set which is non-empty, closed and dense in itself. Now, it is an old and well known theorem of Bernstein [see Oxtoby (1971)] that there exists a partition of any Polish space S into two parts such that none of them includes a perfect set. This depends on the fact that every perfect set has no less elements than the set of all perfect subsets of S, and on the Axiom of Choice (a well ordering of the space and of the set of its perfect subsets). To conclude the proof it suffices to pick for X any of the parts of a Bernstein partition of $\{0, 1\}^\omega$. □

The existence of non-determined infinite PI-games follows also from each of the Theorems 4.1, 4.2, 4.4 and 4.5 of the next section.

However, we can rescue a part of Proposition 2.1 for the case of infinite games. The first step in this direction is the following.

Proposition 3.2. *If the set $X \subseteq P^\omega$ is closed or open, then the game $\langle P^\omega, X \rangle$ is determined.*

Proof. Assume that X is closed. If player II does not have a winning strategy (i.e., a strategy which secures $p \notin X$), then it is clear that player I can maintain that advantage, i.e., I has a strategy a_0 which secures that for every $n < \omega$ the position $(p_0, \ldots, p_{n-1})$ does not yield a winning strategy for II. In particular, a_0 guarantees that for all n we have

$$U(p_0, \ldots, p_{n-1}) \cap X \neq \emptyset.$$

Since X is closed, it follows that $(p_0, p_1, \ldots) \in X$, and so a_0 is a winning strategy for I.

If X is open the theorem follows by symmetry. □

Corollary 3.3. *If $f : P^\omega \to \bar{\mathbb{R}}$ has the property that for every $x \in \bar{\mathbb{R}}$ the set $\{p \in P^\omega : f(p) < x\}$ or the set $\{p \in P^\omega : f(p) \leq x\}$ is open or closed, then the game $\langle P^\omega, f \rangle$ is determined.*

Proof. By Proposition 3.2 there exists $v \in \bar{\mathbb{R}}$ such that for every $x < v$ the game $\langle P^{\omega}, A(x) \rangle$, where $A(x) = \{p \in P^{\omega}: f(p) \leq x\}$, is a win for II, and for every $x > v$ the game $\langle P^{\omega}, A(x) \rangle$ is a win for I. It follows that v is a value of $\langle P^{\omega}, f \rangle$. $\square$

Of course, if the game $\langle P^{\omega}, f \rangle$ is finite, then the function f is continuous and Corollary 3.3 yields Proposition 2.1. Much stronger results than Proposition 3.2 and Corollary 3.3 will be presented in Section 8.

4. Four classical infinite PI-games

We discuss here four games which are related to classical concepts of real analysis.

The first interesting infinite PI-game was invented by S. Mazur about 1935 [see Mauldin (1981, pp. 113–117)]. We define a slightly different (but now standard) version of that game which we call Γ_1. A set Q and a set $X \subseteq Q^{\omega}$ are given. The players choose alternately finite non-empty sequences of elements of Q. (As in Section 2, player I makes the first choice.) Those sequences are juxtaposed to form one sequence q in Q^{ω}. If $q \in X$, I wins. If $q \notin X$, II wins. We take Q with the discrete topology and Q^{ω} with the product topology.

Mazur pointed out that *if X is of the first category, then II has a winning strategy for Γ_1.* Then he asked if the converse is true and offered a bottle of wine for the solution. S. Banach won the prize proving the following theorem.

Theorem 4.1. *If II has a winning strategy for Γ_1 then X is of the first category.*

Proof. If $p_0, \ldots, p_n$ are finite sequences of elements of Q, let $p_0 p_1 \cdots p_n$ denote their juxtaposition. Let b_0 be a winning strategy for II. It is clear that in every neighborhood $U(p_0)$ there is a neighborhood of the form $U(p_0 p_1)$, where p_0 is the first choice of I and $p_1 = b_0(p_0)$. Hence, proceeding by transfinite recursion we can construct a family F_0 of disjoint neighborhoods of the form $U(p_0 p_1)$ such that their union is everywhere dense in Q^{ω}. Repeating the same construction within each neighborhood belonging to F_0 we obtain a family F_1 of disjoint neighborhoods $U(p_0 p_1 p_2 p_3)$ such that $p_1 = b_0(p_0)$, $p_3 = b_0(p_0, p_2)$, $U(p_0 p_1) \in F_0$ and $\bigcup(F_1)$ is everywhere dense in Q^{ω}. We continue in this way forming a sequence of families $F_0, F_1, \ldots$. It is clear from this construction that if $q \in \bigcap_{i < \omega} \bigcup(F_i)$, then q is the juxtaposition of a game consistent with b_0. Hence, since b_0 is a winning strategy we have $X \cap \bigcap_{i < \omega} \bigcup (F_i) = \emptyset$. And, since all $\bigcup(F_i)$ are dense and open in Q^{ω}, X is of the first category. $\square$

Our second example Γ_2 was invented by L. Dubins. The rules are similar to those of Γ_1 except that here it is assumed that $Q = \{0, 1\}$ and that the choices of II are sequences of length one, i.e., elements of Q, while the choices of I are still arbitrary finite sequences of elements of Q, but this time he can also choose the empty sequence. It is easy to see that *I has a winning strategy for Γ_2 iff X has a perfect subset* (perfect means non-empty, closed and dense in itself). [*Hint*: use again the fact that the set of all sequences which can occur when I plays any fixed strategy is perfect.] Furthermore, it is easy to see that *II has a winning strategy if X is at most countable*. Davis (1964) proved the converse:

Theorem 4.2. *If II has a winning strategy for Γ_2, then X is at most countable.*

Proof. Let b_0 be a winning strategy for II. We claim that for every $x \in X$ there exists a finite sequence $p_0, p_2, \ldots, p_{2k}$ (possibly empty) of choices of I, such that $x \in U(p_0 p_1 \cdots p_{2k+1})$, where $p_1 = b_0(p_0)$, $p_3 = b_0(p_0, p_2), \ldots$, $p_{2k+1} = b_0(p_0, \ldots, p_{2k})$, and such that for every choice p_{2k+2} of I we have $x \notin U(p_0 p_1 \cdots p_{2k+2} p_{2k+3})$, where $p_{2k+3} = b_0(p_0, p_2, \ldots, p_{2k+2})$. Indeed, if no such $p_0, p_2, \ldots, p_{2k}$ existed, then I could play forever in such a way that $x \in U(p_0 p_1 \cdots p_n)$, where $p_0, p_1, \ldots, p_n$ is determined by his choices p_{2k} and by b_0. Hence b_0 would not be a winning strategy.

Now, to prove that X is at most countable, it suffices to show that given $p_0, p_2, \ldots, p_{2k}$ there is at most one point $x \in X$ with the above property. So suppose to the contrary that there are two such points $x, x' \in U(p_0 p_1 \cdots p_{2k+1})$, where $p_1 = b_0(p_0), \ldots, p_{2k+1} = b_0(p_0, p_2, \ldots, p_{2k})$. Let q be the longest initial segment of x which equals an initial segment of x'. Then I can choose p_{2k+2} such that $p_0 p_1 \cdots p_{2k+1} p_{2k+2} = q$. Now, since $Q = \{0, 1\}$, either x or x' belongs to $U(p_0 p_1 \cdots p_{2k+3})$, where $p_{2k+3} = b_0(p_0, p_2, \ldots, p_{2k+2})$, contrary to our supposition about those points. This concludes the proof. $\square$

The third example Γ_3 is defined as follows. A set S is given. Player I splits S into two parts. Player II chooses one of them. Again I splits the chosen part into two disjoint parts and II choses one of them, etc. I wins iff the intersection of the chosen parts is not empty and II wins iff it is empty. It is easy to see that *I has a winning strategy iff* $|S| \geq 2^{\aleph_0}$, and that *II has a winning strategy if* $|S| \leq \aleph_0$. R.M. Solovay proved the converse.

Theorem 4.3. *If II has a winning strategy for Γ_3, then $|S| \leq \aleph_0$.*

Proof. The idea is similar to that of the former proof. It suffices to consider the case when $S \subseteq \mathbb{R}$, and to restrict player I to such partitions of S which are induced by the partition of $\mathbb{R}$ into the rays $\{x : x < r\}$ and $\{x : x \geq r\}$, where r is a rational number. Now we argue in the same way as in the proof for Γ_2 that if b_0 is a winning strategy for II, then for every $x \in S$ there exists a sequence of partitioning rationals $r_0, \ldots, r_n$ such that $x \in b_0(r_0, \ldots, r_n)$ but for every r_{n+1}, we have $x \notin b_0(r_0, \ldots, r_n, r_{n+1})$, and also that for every $r_0, \ldots, r_n$ there exists at most one such x. Of course this implies $|S| \leq \aleph_0$. $\square$

Our fourth example Γ_4 is due to L. Harrington and his analysis of Γ_4 given below simplifies former work of Mycielski and Swierczkowski (1964). We consider the Cantor set $\{0, 1\}^\omega$ with its natural Borel probability measure μ, given by $\mu(U(q)) = 1/2^n$ for any $q \in \{0, 1\}^n$. A set $X \subseteq \{0, 1\}^\omega$ is given and Γ_4 is played as follows. Player I chooses a rational number $\varepsilon > 0$ and a number $p_0 \in \{0, 1\}$. Then II chooses a clopen (i.e., closed and open) set $A_1 \subseteq \{0, 1\}^\omega$ with $\mu(A_1) < \varepsilon/4$. At stage n player I chooses $p_{n-1} \in \{0, 1\}$ and then II chooses a clopen set $A_n \subseteq \{0, 1\}^\omega$ with $\mu(A_n) < \varepsilon/4^n$. Player I wins iff $(p_0, p_1, \ldots) \in X \setminus \bigcup_{1 \leq i < \omega} A_i$. Player II wins otherwise.

Harrington has proved two facts about Γ_4.

Theorem 4.4. *If X has inner measure zero, then I has no winning strategy for Γ_4.*

Proof. Suppose to the contrary that I has a winning strategy a_0. Let A be the set of all sequences $(p_0, p_1, \ldots)$ which can occur when I uses a_0. Let $\tilde{A}$ be the set of all sequences of clopen sets $A_i \subseteq \{0, 1\}^\omega$ which can occur when I uses a_0. Since $\{0, 1\}^\omega$ has countably many clopen subsets we have a natural identification of $\tilde{A}$ with ω^ω. Providing ω^ω with the natural product topology we see that $a_0 : \tilde{A} \to A$ is a continuous surjection. Hence A is an analytic set (the definition is given at the beginning of Section 9). It follows that A is measurable [see Oxtoby (1971)] and, since $A \subseteq X$, $\mu(A) = 0$. One checks now that there is a sequence of clopen sets $A_1, A_2, \ldots$, with $\mu(A_n) < \varepsilon/4^n$ (ε being given to I by a_0) such that $A \subset A_1 \cup A_2 \cup \cdots$. This contradicts the assumption that a_0 was a winning strategy for I. $\square$

Theorem 4.5. *If II has a winning strategy for Γ_4, then $\mu(X) = 0$.*

Proof. Let b_0 be a winning strategy for II. Suppose to the contrary that X has outer measure $\alpha > 0$. Let I play $\varepsilon < \alpha$. Then for each n there are only 2^n plays $(p_0, \ldots, p_{n-1})$ of I, and so at most 2^n answers $A_n = b_0(\varepsilon, p_0, \ldots, p_{n-1})$. Hence $\mu(\bigcup_{(p_0, \ldots, p_{n-1})} A_n) \leq \varepsilon/2^n$. Let A be the union of all the sets A_n which II could play using b_0 given the above ε. So $\mu(A) \leq \varepsilon < \alpha$. Hence I could play $(p_0, p_1, \ldots) \in X \setminus A$, contradicting the assumption that b_0 was a winning strategy. $\square$

Three classical properties of sets are now seen to have a game theoretical role:

Corollary 4.6.   *Given a complete separable metric space M and a set $X \subseteq M$ which does not have the property of Baire, or, is uncountable but without any perfect subsets, or is not measurable relative to some Borel measure in M, one can define a game $\langle \{0,1\}^\omega, Y \rangle$ which is not determined.*

Proof.   By a well-known construction [see Oxtoby (1971)] we can assume without loss of generality that $M = \{0,1\}^\omega$, with its product topology and with the measure defined above. Let $X \subseteq M$ be a set without the property of Baire, U be the maximal open set in M such that $X \cap U$ is of the first category and V be the maximal open set such that $V \cap (M \backslash X)$ is of the first category. We see that the interior of $M \backslash (U \cup V)$ is not empty (otherwise X would have the property of Baire). Thus there is a basic neighborhood $W \subseteq M \backslash (U \cup V)$. We identify W with $\{0,1\}^\omega$ in the obvious way and define S to be the image of $X \cap W$ under this identification. So we see that S is not of the first category and, moreover, for each p_0 of I, $U(p_0) \cap (\{0,1\}^\omega \backslash S)$ is not of the first category. Thus, by the result of Banach (4.1), neither II nor I has a winning strategy for the game Γ_1.

Now Γ_1 is a game of the form $\langle P^\omega, Z \rangle$, where P is countable. We can turn such a game into one of the form $\langle \{0,1\}^\omega, Y \rangle$ using the fact that the number of consecutive 1's chosen by a player followed by his choice of 0 can code an element of P (the intermediate choices of the other player having no influence on the result of the game).

For the alternative assumptions about X considered in the corollary we apply the results of Davis and Harrington to obtain non-determined games Γ_2 and Γ_4. (For Γ_2 we use the fact that a perfect set in $\{0,1\}^\omega$ has cardinality $2^{\aleph_0}$. For Γ_4 we need a set S with inner measure 0 and outer measure 1. Its construction from X is similar to the above construction of S for Γ_1.) $\square$

All known constructions of M and X satisfying one of the conditions of Corollary 4.6 have used the Axiom of Choice, and, after the work of Paul Cohen, R.M. Solovay and others, it is known that indeed the Axiom of Choice is unavoidable in any such construction. Thus Corollary 4.6 suggested the stronger conjecture of Mycielski and Steinhaus (1962) that *the Axiom of Choice is essential in any proof of the existence of sets $X \subseteq \{0,1\}^\omega$ such that the game $\langle \{0,1\}^\omega, X \rangle$ is not determined.* This has been proved recently by Martin and Steel (1989) (see Section 8 below).

In the same order of ideas, Theorem 4.3 shows that *the Continuum Hypothesis is equivalent to the determinacy of a natural class of PI-games.*

For a study of some games similar to Γ_2 but with $|Q| > 2$, see Louveau (1980). Many other games related to Γ_2 and Γ_3 were studied by F. Galvin et al. (unpublished); see also the survey by Telgarsky (1987).

5. The game of Hex and its unsolved problem

Before plunging deeper into the theory of infinite games we discuss in this and the next section a few particularly interesting finite games. We begin with one of the simplest finite games of perfect information called Hex which has not been solved in a practical sense. Hex is played as follows. We use a board with a honeycomb pattern as in Figure 2. The players alternatively put white or black stones on the hexagons. White begins and he wins if the white stones connect the top of the board with the bottom. Black wins if the black stones connect the left edge with the right edge.

Theorem 5.1. (i) *When the board is filled with stones, then one of the players has won and the other has lost.*
    (ii) *White has a winning strategy.*

Proof (in outline).   (i) If White has not won and the board is full, then consider the black stones adjacent to the set of those white stones which are

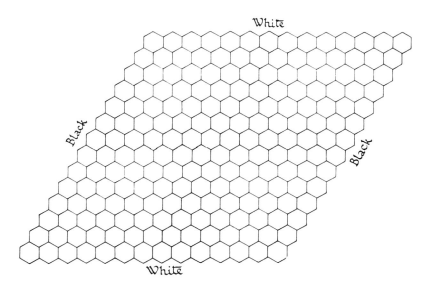

Figure 2. A 14×14 Hex board.

connected by some white path to the upper side of the board. Those stones are all black and together with the remaining black stones of the upper line of the board they contain a black path from the left edge to the right edge. Thus Black is the winner. [For more details, see Gale (1979).]

(ii) By (i) and Proposition 2.1 one of the players has a winning strategy. Suppose to the contrary that it is Black who has such a strategy b_0. Now it is easy to modify b_0 so that it becomes a winning strategy for White. (*Hint*: White forgets his first move and then he uses b_0.) Thus both players would have a winning strategy, which is a contradiction. □

Problem.   Find a useful description of a winning strategy for White!

This open problem is a good example of the general problem in the theory of finite PI-games which was discussed at the end of Section 1. In practice Hex on a board of size 14×14 is an interesting game and the advantage of White is hardly noticeable. It is surprising that such a very concrete finitistic existential theorem like Theorem 5.1(ii) can be meaningless from the point of view of applications. [Probably, strict constructivists would not accept our proof of Theorem 5.1(ii).]

Hex has a relative called Bridge-it for which a similar theorem is true. But for Bridge-it a useful description of a winning strategy for player I *has been found* [see Berlekamp et al. (1982, p. 680)]. However, this does not seem to help for the problem on Hex. Dual Hex, in which winning means losing in Hex, is also an interesting unsolved game. Here Black has a winning strategy.

Of course for Chess we do not know whether White or Black has a winning strategy or if (most probably) both have strategies that secure at least a draw.

It is proved in Even and Tarjan (1976) that some games of the same type as Hex are difficult in the sense that the problem of deciding if a position is a win for I or for II is complete in polynomial space (in the terminology of the theory of complexity of algorithms).

It is interesting that Theorem 5.1(i) implies easily the Brouwer fixed point theorem [see Gale (1979)].

6. An interplay between some finite and infinite games

Let G be a finite bipartite oriented graph. In other words G is a system $\langle P, Q, E \rangle$, where P and Q are finite disjoint sets and $E \subseteq (P \times Q) \cup (Q \times P)$ is called the set of arrows. We assume moreover that for each $(a, b) \in E$ there exists c such that $(b, c) \in E$. A function $\varphi : E \to \mathbb{R}$ is given and a point $p_{\text{first}} \in P$ is fixed. The players I and II pick alternately $p_0 = p_{\text{first}}$, $q_0 \in Q$, $p_1 \in P$, $q_1 \in$

$Q, \ldots$ such that $(p_i, q_i) \in E$ and $(q_i, p_{i+1}) \in E$, thereby defining a zig-zag path composed of arrows.

We define three PI-games.

G_1: player II pays to player I the value

$$\limsup_{n \to \infty} \frac{1}{2n} \sum_{i=0}^{n-1} (\varphi(p_i, q_i) + \varphi(q_i, p_{i+1})) .$$

G_2: player II pays to player I the value

$$\liminf_{n \to \infty} \frac{1}{2n} \sum_{i=0}^{n-1} (\varphi(p_i, q_i) + \varphi(q_i, p_{i+1})) .$$

G_3: the game ends as soon as a closed loop arises in the path defined by the players, i.e., as soon as I picks any $p_n \in \{p_0, \ldots, p_{n-1}\}$ or II picks $q_n \in \{q_0, \ldots, q_{n-1}\}$, whichever happens earlier. Then II pays to I the "loop average" v defined as follows. In the first case $p_n = p_m$ for some $m < n$, and then

$$v = \frac{1}{2(n-m)} \sum_{i=m}^{n-1} (\varphi(p_i, q_i) + \varphi(q_i, p_{i+1})) ;$$

in the second case $q_n = q_m$ with $m < n$ and then

$$v = \frac{1}{2(n-m)} \sum_{i=m}^{n-1} (\varphi(q_i, p_{i+1}) + \varphi(p_{i+1}, q_{i+1})) .$$

Thus in all three games the players are competing to minimize or maximize the means of some numbers which they encounter on the arrows of the graph.

Since the game G_3 is finite, by Proposition 2.1, it has a value V. Given a strategy σ of one of the players which secures V in G_3, each of the infinite games G_1 and G_2 can be played according to σ, by forgetting the loops (which necessarily arise). This also secures V [see Ehrenfeucht and Mycielski (1979) for details]. So it follows that *the games G_1 and G_2 are determined, and have the same value V as G_3.*

A strategy a for player I is called *positional* if $a(q_0, \ldots, q_n)$ depends only only on q_n. In a similar way a strategy b for II is *positional* if $b(p_0, \ldots, p_n)$ depends only on p_n.

Theorem 6.1. *Both players have positional strategies a_0 and b_0 which secure V for each of the games G_1, G_2 and G_3.*

This theorem was shown in Ehrenfeucht and Mycielski (1979). We shall not reproduce its proof here but only mention that it was helpful to use the infinite games G_1 and G_2 to prove the claim about the finite game G_3 and vice versa. In fact no direct proof is known. So, there is at least one example where infinite PI-games help us to analyze some finite PI-games.

An open problem related to the above games is the following: Is an appropriate version of Theorem 6.1, where P and Q are compact spaces and φ is continuous, still true?

7. Continuous PI-games

In this section we extend the theory of PI-games with countable sequences to a theory with functions over the interval $[0, \infty)$. R. Isaacs in the United States and H. Steinhaus and A. Zieba in Poland originated this development. Here are some examples of continuous games.

Two dogs try to catch a hare in an unbounded plane, or one dog tries to catch a hare in a half-plane. The purpose of the dogs is to minimize the time of the game and the purpose of the hare is to maximize it. We assume that each dog is faster than the hare and that only the velocities are bounded while the accelerations are not. There are neat solutions of those two special games: at each moment t the hare should run full speed toward any point a_t such that a_t is the most distant from him among all points which he can reach prior to any of the dogs (here "prior" is understood in the sense of $\leq$). And the dogs, at each instant t, should run full speed toward that same point a_t. (To achieve the best result the hare does not have to change the point a_t during the game.)

Now, how to turn the above statements into mathematical theorems? Notice that the sets of strategies have not been defined so we have not constructed any games in the sense of Section 2. The main point of this section is to build such definitions which may be useful for a wide variety of games. The literature of this subject is rich [see, for example, Behrand (1987), Hájek (1975), Kuhn and Szegö (1971), Mycielski (1988), and Rodin (1987)], but the games are rarely defined with full precision. The fundamentals of this theory presented in this section will not use the concepts of differentiation or integration.

Let P and Q be arbitrary sets and $F_{\mathrm{I}} \subseteq P^{[0, \infty)}$ and $F_{\mathrm{II}} \subseteq Q^{[0, \infty)}$ two sets of functions from $[0, \infty)$ to P or Q, respectively. We assume that F_X, for $X = \mathrm{I}, \mathrm{II}$, are *closed* in the following sense: if f is a function with domain $[0, \infty)$ such that for all $T > 0$ the restriction $f \upharpoonright [0, T)$ has an extension in F_X, then $f \in F_X$.

We will say that F_X is *saturated* if it is closed under the following operations. For every $\delta > 0$ if $f \in F_X$, then $f_\delta \in F_X$, where

$$f_\delta(t) = \begin{cases} f(0) & \text{for } 0 \le t < \delta, \text{ and} \\ f(t - \delta) & \text{for } t \ge \delta . \end{cases}$$

Let a function $\psi: F_I \times F_{II} \to \bar{\mathbb{R}}$ be given. We will define in terms of ψ the payoff functions of two PI-games G^+ and G^-. In order for those games to be convincing models of continuous games (like the games of the above examples with dogs and a hare) we will need that ψ satisfies at least one of the following two conditions of semicontinuity.

(S_1) The space F_I is saturated and for every $\varepsilon > 0$ there exists a $\Delta > 0$ such that for all $\delta \in [0, \Delta]$ and all $(p, q) \in F_I \times F_{II}$ we have

$$\psi(p_\delta, q) \le \psi(p, q) + \varepsilon .$$

We consider also a dual property for ψ:

(S_2) The space F_{II} is saturated and for every $\varepsilon > 0$ there exists a $\Delta > 0$ such that for all $\delta \in [0, \Delta]$ and all $(p, q) \in F_I \times F_{II}$ we have

$$\psi(p, q_\delta) \ge \psi(p, q) - \varepsilon .$$

The system $\langle F_I, F_{II}, \psi \rangle$ will be called *normal* iff F_I and F_{II} are closed and (S_1) or (S_2) holds.

Example 1. A metric space M with a distance function $d(x, y)$ and two points $p_0, q_0 \in M$ are given and $P = Q = M$. F_I is the set of all functions $p: [0, \infty) \to M$ such that $p(0) = p_0$, and

$$d(p(t_1), p(t_0)) \le |t_1 - t_0| \quad \text{for all } t_0, t_1 \ge 0 .$$

F_{II} is the set of all functions $q: [0, \infty) \to Q$ such that $q(0) = q_0$, and

$$d(q(t_1), q(t_0)) \le v|t_1 - t_0| \quad \text{for all } t_0, t_1 \ge 0 ,$$

where v is a constant in the interval $[0, 1]$.

Now ψ can be defined in many ways, e.g.

$$\psi(p, q) = d(p(1), q(1)) ,$$

or

$$\psi(p, q) = \limsup_{t \to \infty} d(p(t), q(t)) .$$

It is easy to prove that in these cases the system $\langle F_I, F_{II}, \psi \rangle$ is normal.

Example 2. The spaces F_I and F_{II} are defined as in the previous example but with further restrictions. For example, the total length of every p and/or every q is bounded, i.e., say for all $p \in F_I$,

$$\lim_{n \to \infty} \sum_{i=0}^{\infty} d\left(p\left(\frac{i}{n}\right), p\left(\frac{i+1}{n}\right) \right) \leq L ;$$

or P and/or Q is $\mathbb{R}^n$ and the acceleration of every p and/or every q is bounded, i.e., say for all $p \in F_I$,

$$\left\| p(t_0) - 2p\left(\frac{t_0 + t_1}{2}\right) + p(t_1) \right\| \leq A(t_1 - t_0)^2 \quad \text{for all } t_0, t_1 \geq 0 .$$

If the space F_X ($X = I, II$) represents the possible trajectories of a vehicle, the above conditions may correspond to limits of the available fuel or power. Conditions of this kind and functionals ψ as in the previous example are compatible with normality.

Example 3. F_I and F_{II} are the sets of all measurable functions $p: [0, \infty) \to B_I$ and $q: [0, \infty) \to B_{II}$, respectively, where B_I and B_{II} are some bounded sets in $\mathbb{R}^n$. And

$$\psi(p, q) = \left\| \int_0^1 (p(t) - q(t)) \, dt \right\| .$$

Such F_X are called spaces of *control functions*. Again it is easy to see that the system $\langle F_I, F_{II}, \psi \rangle$ is normal. Similar (and more complicated) normal systems are considered in the theory of differential games.

Given $\langle F_I, F_{II}, \psi \rangle$, with F_I and F_{II} closed in the sense defined above, we define two PI-games G^+ and G^-. In G^+ player I chooses some $\delta > 0$ and a path $p_0: [0, \delta) \to P$. Then II chooses $q_0: [0, \delta) \to Q$. Again I chooses $p_1: [\delta, 2\delta) \to P$ and II chooses $q_1: [\delta, 2\delta) \to Q$, etc. If $(\bigcup p_i, \bigcup q_i) \in F_I \times F_{II}$, then I pays to II the value $\psi(\bigcup p_i, \bigcup q_i)$. If $(\bigcup p_i, \bigcup q_i) \notin F_I \times F_{II}$, then there is at least an n such that $\bigcup_{i<n} p_i$ has no extension to a function in F_I or $\bigcup_{i<n} q_i$ has no extension to a function in F_{II}. If n satisfies the first alternative, I pays ∞ to II. Otherwise II pays ∞ to I.

The game G^- is defined in the same way except that now player II chooses $\delta > 0$ and $q_0: [0, \delta) \to Q$ and then I chooses $p_0: [0, \delta) \to P$, etc. Again if $(\bigcup p_i, \bigcup q_i) \in F_I \times F_{II}$, then I pays to II the value $\psi(\bigcup p_i, \bigcup q_i)$ and again, if $(\bigcup p_i, \bigcup q_i) \notin F_I \times F_{II}$, the player who made the first move causing this pays ∞ to the other.

Since both G^+ and G^- are PI-games, under very general conditions about ψ (see Corollary 3.3 and Theorem 8.1 in Section 8), both games G^+ and G^- have values. We denote those values by V^+ and V^-, respectively. By a proof similar to the proof of Theorem 5.1(ii), it follows from these definitions that

$$V^+ \geq V^- .$$

We claim that if $\langle F_\mathrm{I}, F_\mathrm{II}, \psi \rangle$ is normal, then G^+ and G^- represent essentially the same game. More precisely, we have the following theorem.

Theorem 7.1. *If V^+ and V^- exist and the system $\langle F_\mathrm{I}, F_\mathrm{II}, \psi \rangle$ is normal, then*

$$V^+ = V^- .$$

Proof. Suppose the condition (S_1) of normality holds. Choose $\varepsilon > 0$. Given a strategy σ^- for I in G^- which secures a payoff $\leq V^- + \varepsilon$ we will construct a strategy σ^+ for I in G^+ which secures a payoff $\leq V^- + 2\varepsilon$. Of course this implies $V^+ \leq V^-$ and so $V^+ = V^-$. Let σ^+ choose δ according to (S_1), and $p_0^+(t) = p_0$ for $t \in [0, \delta)$. When II answers with some $q_0: [0, \delta) \to Q$, then σ^+ chooses $p_1^+(t) = p_0^-(t - \delta)$ for $t \in [\delta, 2\delta)$, where $p_0^- = \sigma^-(q_0)$. Then II chooses $q_1: [\delta, 2\delta) \to Q$ and σ^+ chooses $p_2^+(t) = p_1^-(t - \delta)$ for $t \in [2\delta, 3\delta)$, where $p_1^- = \sigma^-(q_0, q_1)$, etc. Now the pair $(\bigcup p_i^-, \bigcup q_i)$ is consistent with a game in G^- where I uses σ^-. Also, we have $\bigcup p_i^+ = (\bigcup p_i^-)_\delta$. Hence, by (S_1),

$$\psi(\bigcup p_i^+, \bigcup q_i) \leq \psi(\bigcup p_i^-, \bigcup q_i) + \varepsilon \leq V^- + 2\varepsilon .$$

This concludes the proof in the case (S_1). In the case (S_2) the proof is symmetric. $\square$

The theorems about the existence of values presented in Section 8 plus the above Theorem 7.1 encompass the existential part of the theory of continuous PI-games over normal systems. However, we will consider an interesting case of continuous PI-games, called *pursuit and evasion*, which is not normal:

M, P, Q, F_I and F_II are defined as in Example 1, but now $\psi(p, q)$ is the least t such that $p(t) = q(t)$, if such a t exists, and $\psi(p, q) = \infty$ otherwise. It is easy to see that ψ violates (S_1) and (S_2). Still an interesting theory is possible. We will assume that the metric space M is complete, locally compact and connected by arcs of finite length. This is a natural assumption because under those conditions *for every two points of M there exists a shortest arc connecting them*. Then we can also assume without loss of generality that d is the geodesic metric, i.e., $d(x, y) = $ length of the shortest arc from x to y.

Now consider the game G^-. By Corollary 3.3, G^- has a value V^-. Of course I can be called the *pursuer* and II the *evader*, and G^- gives some tiny unfair advantage to the pursuer (because II has to declare first his trajectory over $[0, \delta)$, then $[\delta, 2\delta)$, etc.).

In this setting the dual game G^+ is uninteresting because, for trivial reasons, in most cases its value will be ∞. However, there exists a similar game G^{++} which gives a tiny unfair advantage to the evader. In G^{++} player II chooses first a number $\delta_0 > 0$, then I chooses δ and $\boldsymbol{p}_0 : [0, \delta) \rightarrow M$, then II chooses $\boldsymbol{q}_0 : [0, \delta) \rightarrow M$, and again I chooses $\boldsymbol{p}_1 : [\delta, 2\delta) \rightarrow M$, etc. Otherwise the rules are the same as in G^+, except that now I pays to II the least value t such that the distance from $\boldsymbol{p}(t)$ to $\boldsymbol{q}(t)$ is $\leq v\delta_0$, where $\boldsymbol{p} = \bigcup \boldsymbol{p}_i$ and $\boldsymbol{q} = \bigcup \boldsymbol{q}_i$. Again, by Corollary 3.3, it is clear that G^{++} has a value V^{++}. It is intuitively clear that $V^- \leq V^{++}$. Games very similar to G^- and G^{++} have been studied in Mycielski (1988) and the methods of that paper can be easily modified to prove the following theorems.

Theorem 7.2. *If $v < 1$, then $V^- = V^{++}$.*

(We do not know any example where $v = 1$ and $V^- < V^{++}$.)

By Theorem 7.2, for $v < 1$, it is legitimate to denote both V^- and V^{++} by V.

Now, given (M, d), it is interesting to study V as a function of p_0, q_0 and v (we will omit the argument v when its value is fixed). The function $V(p_0, q_0)$ is useful since the best strategy for I is to choose $\boldsymbol{p}_i : [i\delta, (i + 1)\delta) \rightarrow M$ such as to keep in F_{I} and to minimize $V(\boldsymbol{p}_i((i + 1)\delta), \boldsymbol{q}_i((i + 1)\delta))$, and the best strategy for II, after his choice of δ, is to choose $\boldsymbol{q}_i : [i\delta, (i + 1)\delta) \rightarrow M$ such as to keep in F_{II} and to maximize $\inf\{V(\boldsymbol{p}_i((i + 1)\delta), \boldsymbol{q}_i((i + 1\delta)) : \boldsymbol{p}_i$ is any choice of I$\}$.

The function $V(p, q, v)$ was studied in Mycielski (1988), where the following theorems are proved.

Theorem 7.3. *If $v < 1$, then*
  (i) $d(x, y) \leq V(x, y, v) \leq d(x, y)/(1 - v)$;
  (ii) $|V(x_1, y, v) - V(x_2, y, v)| \leq d(x_1, x_2)/(1 - v)$;
  (iii) $|V(x, y_1, v) - V(x, y_2, v)| \leq d(y_1, y_2)/(1 - v)$;
  (iv) *if $0 \leq v_1 < v_2 < 1$, then*

$$V(x, y, v_1) \leq V(x, y, v_2) \leq \frac{1 - v_1}{1 - v_2} V(x, y, v_1) .$$

We do not know if $V(x, y, v) \rightarrow V(x, y, 1)$ for $v \uparrow 1$.

Fixing $v < 1$, the next theorem gives a characterization of $V(x, y)$ which does not depend on any game theoretic concepts.

Theorem 7.4. *The function $V: M \times M \to \mathbb{R}$ satisfies, and is the only function satisfying, the following four conditions:*
   (i) $V(p, p) = 0$;
   (ii) $V(p, q) - d(p, x) \leqslant \sup\{V(x, y): d(y, q) \leqslant vd(x, p)\}$;
   (iii) $V(p, q) \geqslant d(p, q)$;
   (iv) $\max(0, V(p, q) - (1/v)d(q, y)) \geqslant \inf\{V(x, y): d(x, p) \leqslant (1/v)d(q, y)\}$.

The intuitive meaning of the inequality (ii) is the following: if I moves from p to x using the time $d(p, x)$, then II has an answer y using the same time such that after those moves the value $V(p, q)$ will not decrease by more than $d(p, x)$. The intuitive meaning of (iv) is the following: if II moves from q to y using the time $(1/v)d(q, y) \leqslant V(p, q)$, then I has an answer x using the same time such that the value $V(p, q)$ will decrease at least by $(1/v)d(q, y)$.

The above theorem implies the Isaacs equation [see Behrand (1987) and Mycielski (1988)].

Corollary 7.5. *If M is a Riemannian manifold with boundary, e.g., an n-dimensional polytope in $\mathbb{R}^n$, x_0 and y_0 are in the interior of M, and V is differentiable at (x_0, y_0), then V satisfies the Isaacs equation*

$$\|\nabla_x V(x_0, y_0)\| = 1 + v\|\nabla_y V(x_0, y_0)\| \, ,$$

where ∇ is the gradient operator.

In spite of all those facts and properties of $V(x, y)$, this function is still unknown, even for some simple spaces M such as a plane with the interior of a circle removed or if M is a circular disk. Those problems are discussed in Breakwell (1989) and Mycielski (1988).

The following function $W(x, y)$ could be useful:

$$W(x, y) = \sup\{d(x, z): d(x, z) > \frac{1}{v} d(y, z)\} \, .$$

Problem. Is it true that $V(x_1, y) < V(x_2, y)$ if $W(x_1, y) < W(x_2, y)$?

If the answer is yes, then the best strategy for I is to minimize W, which, as a rule, is much easier to compute than V. For the two games with dogs and a hare defined at the beginning of this section the answer is yes, and this is easy to prove by means of the games G^{++}.

8. The main results of the theory of infinite PI-games

The considerations of Sections 2 and 3 suggest the following general problem.

Problem. Let $g: A \times B \to C$ be a continuous function, where A, B and C are compact spaces. Suppose that for every continuous function $f: C \to \bar{\mathbb{R}}$ the game $\langle A, B, f \circ g \rangle$ is determined. Must it be also determined for every Borel measurable f?

This problem is open already for the case when C is the Cantor space $\{0, 1\}^{\omega}$, and instead of all Borel measurable functions we consider only characteristic functions of sets of class $\boldsymbol{F_\sigma}$ or $\boldsymbol{G_\sigma}$.

The only known results about this problem pertain to the case of PI-games and do not assume compactness of the spaces A, B and C. In this case $C = P^{\omega}$ with the product topology (see Section 3), A and B are defined as in Section 2 and $g(a, b) = (p_0, p_1, \ldots)$. Let us state immediately those results (which are the deepest theorems of the theory of PI-games), and explain later the concepts and terminology used in those statements. Part (ii) of Theorem 8.1 will be proved in Section 9.

Theorem 8.1. (i) *If $X \subseteq P^{\omega}$ is a Borel set, then the game $\langle P^{\omega}, X \rangle$ is determined (assuming the usual set theory ZFC).*

(ii) *If $X \subseteq P^{\omega}$ is an analytic set, then the game $\langle P^{\omega}, X \rangle$ is determined [assuming ZFC + there exists an Erdös cardinal $\kappa \to (\omega_1)_{\lambda}^{<\omega}$, where $\lambda = 2^{|P| + \aleph_0}$].*

(iii) *If $X \subseteq \omega^{\omega}$ and $X \in L(\mathbb{R})$, then the game $\langle \omega^{\omega}, X \rangle$ is determined (assuming ZFC + there exists a measurable cardinal with ω Woodin cardinals below it).*

A brief history and some outstanding qualities of these results are the following. Theorem 8.1(i) is due to Martin (1975, 1985). Thereby he solved a problem already stated by Gale and Stewart (1953). This theorem is remarkable not only because of its very ingeneous proof but also because it was the first theorem *in real analysis* the proof of which required the full power of the set theory ZFC. Indeed, Harvey Friedman proved that Theorem 8.1(i) depends on the axiom schema of replacement, while all the former theorems of real analysis could be proved from the weaker axiom schema of comprehension. We shall not include here any proof of Theorem 8.1(i) since it is not easier than that of 8.1(ii); the conclusion of 8.1(ii) is stronger, and we feel that the refinement of ZFC upon which 8.1(ii) depends is very natural.

Theorem 8.1(ii) is also due to Martin (1970). Again its proof is very remarkable since it is the simplest application of an axiom *beyond ZFC* to a theorem in real analysis. A set $X \subseteq P^{\omega}$ is called *analytic* if X is a projection of a closed subset of the product space $P^\omega \times \omega^\omega$ into P^ω, where ω^ω has also the product topology (ω^ω is homeomorphic to the set of irrational numbers of the real line). We will see in Section 9 that every Borel subset of P^ω is analytic but not vice versa. So, as mentioned above, the conclusion of Theorem 8.1(ii) is stronger than that of 8.1(i) (at the cost of a stronger set theoretic assumption). The Erdös cardinal numbers will be explained in Section 9. A measurable cardinal $\kappa > |P|$ would suffice since it satisfies the condition in 8.1(ii).

Theorem 8.1(iii) was proved by Martin and Steel (1989) using a former theorem of H. Woodin (the proof of the latter is still unpublished). $L(\mathbb{R})$ denotes the least class of sets which constitutes a model of *ZF* (i.e. *ZFC* without the Axiom of Choice), contains all the ordinal numbers and all the real numbers and is such that if $x \in L(\mathbb{R})$ and $y \in x$, then $y \in L(\mathbb{R})$. Theorem 8.1(iii) solves in the affirmative the problem raised in Mycielski and Steinhaus (1962) of showing that the Axiom of Choice is necessary to prove the existence of sets $X \subseteq \{0, 1\}^\omega$ such that the game $\langle \{0, 1\}^\omega, X \rangle$ is not determined. Also it yields a very large class F of sets $X \subseteq \omega^\omega$, namely $F = \mathscr{P}(\omega^\omega) \cap L(\mathbb{R})$, where $\mathscr{P}(S) = \{R: R \subseteq S\}$, such that all the games $\langle \omega^\omega, X \rangle$ with $X \in F$ are determined. This family F is closed under countable unions and complementation, under the Souslin operation (see Section 9) and many other set theoretic constructions. In particular, F includes all projective subsets of ω^ω. For the case $|P| = \omega$ the conclusion of Theorem 8.1(iii) is much stronger than that of 8.1(ii). But, as we shall see in Section 10, Theorem 8.1(iii) would fail if ω^ω was replaced by P^ω with an uncountable set P.

The concept of Woodin cardinals will not be explained here since it is rather technical. But there are several possible additions to *ZFC* which are simpler, intuitively well motivated and stronger than those of Theorem 8.1(iii). For example, the existence of 1-extendible cardinals [an axiom proposed by W. Reinhardt, see Solovay et al. (1978)] implies the existence of a measurable cardinal with ω Woodin cardinals below it. Again we cannot present here enough logic and set theory to explain the above axiom, but we can state an axiom proposed by P. Vopenka which is still stronger and hence also suffices for the conclusion of Theorem 8.1(iii).

(*V*) *If C is a proper class of graphs, then there are two graphs in C such that one is isomorphic to an induced subgraph of the other.*

The intuitive idea supporting (*V*) is the following: a proper class must be so large relative to the size of a set that a proper class of graphs must be repetitive in the sense expressed in (*V*). The proof of Theorem 8.1(iii) [even the part published in Martin and Steel (1989)] is much harder than the proof of 8.1(ii) given in the next section.

Let us add that H. Friedman, L. Harrington, D.A. Martin and H. Woodin have shown that the set theoretic axioms in Theorem 8.1(i), (ii) and (iii) are nearly as weak as possible for proving those theorems.

9. Proof of Theorem 8.1(ii)

For any topological space S, a set $X \subseteq S$ is called *analytic* if X is a projection of a closed subset of the product space $S \times \omega^\omega$ into S. For example, if $f: \omega^\omega \to S$ is a continuous function, then the image $f[\omega^\omega]$ is the projection of the graph of f into S, whence $f[\omega^\omega]$ is analytic. We list some elementary facts about analytic sets.

9.1. *A union of countably many analytic sets is analytic.*

   This follows immediately from the fact that ω^ω can be partitioned into ω clopen sets homeomorphic to ω^ω.

9.2. *An intersection of countably many analytic sets is analytic.*

Proof. Let $A_0, A_1, \ldots$ be analytic subsets of S. Let A_i be the projection of a closed set $C_i \subseteq S \times (\omega^\omega)_i$, where $(\omega^\omega)_i$ is a homeomorphic copy of ω^ω. Let C_i^* be the cylinder over C_i in the product space $S \times \Pi_{i<\omega} (\omega^\omega)_i$. Then C_i^* is closed and $\bigcap_{i<\omega} A_i =$ the projection of $\bigcap_{i<\omega} C_i^*$ into S. Since $\Pi_{i<\omega} (\omega^\omega)_i$ is homeomorphic to ω^ω and $\bigcap_{i<\omega} C_i^*$ is closed, it follows that $\bigcap_{i<\omega} A_i$ is analytic.

9.3. *Every closed set is analytic and every open set in P^ω is analytic.*

   For closed sets the assertion is obvious and for open sets it follows from the easy fact that in the space P^ω every open set is a countable union of clopen sets, and from 9.1.

Corollary 9.4. *All Borel subsets of P^ω are analytic.*

   This corollary is not true for all spaces. For example, the set ω_1 is open in the compact space $\omega_1 + 1$ with its interval topology, but ω_1 is not analytic.
   If $A \subseteq S$ is of the form

$$A = \bigcup_{q \in \omega^\omega} \bigcap_{n<\omega} F_{q \restriction n}, \tag{1}$$

where $q \restriction n = (q_0, \ldots, q_{n-1})$ and $F_{q \restriction n}$ are closed subsets of S, then A is analytic. In fact, A is the projection into S of the set

$$C = \bigcap_{n<\omega} \bigcup_{q \in \omega^\omega} F_{q \restriction n} \times U(q \restriction n).$$

It is easy to check that each union in this intersection is closed, and hence C itself is closed.

Also if $A \subseteq P^\omega$ is analytic, a projection of a closed set $C \subseteq P^\omega \times \omega^\omega$, then A is of the form (1), where

$$F_{q \upharpoonright n} = \{p \in P^\omega : (U(p \upharpoonright n) \times U(q \upharpoonright n)) \cap C \neq \emptyset\} \,.$$

However, there are spaces S where not every analytic set is of the form (1). [The form (1) is called the *Souslin operation* or the operation *(A)* upon the system $\langle F_{q \upharpoonright n} \rangle$.]

Finally, let us recall that *there exist analytic subsets of* $\{0, 1\}^\omega$ *which are not Borel* (for example, the set of all those sequences which code a subset of $\omega \times \omega$ which is not a well ordering of ω).

We now define the notion of an *Erdös cardinal* κ.

- $f \upharpoonright X$ denotes the restriction of a function f to a subset X of its domain.
- Every ordinal number is identified with the set of all smaller ordinals.
- Cardinals are identified with initial ordinals.
- For every cardinal α, α^+ denotes the cardinal successor of α.
- For any set X, $[X]^n$ denotes the set of all subsets of X of cardinality n.

For any cardinals κ, α and λ we write

$$\kappa \to (\alpha)_\lambda^{<\omega} \tag{2}$$

iff for every function $f: \bigcup_{n<\omega} [\kappa]^n \to \lambda$ there exists a set $H \subseteq \kappa$ of cardinality α such that, for every $n < \omega$, $f \upharpoonright [H]^n$ is constant. If (2) holds κ is called an *Erdös cardinal for* α *and* λ, and H is called a *homogeneous* set for f.

The relation (2) has many interesting properties, in particular:

Theorem 9.5. *If α is infinite and κ is the least cardinal such that $\kappa \to (\alpha)_2^{<\omega}$, then for every $\lambda < \kappa$ we have $\kappa \to (\alpha)_\lambda^{<\omega}$, and κ is strongly inaccessible.*

We refer the reader to Drake (1974, pp. 221 and 239) for the proof of the above theorem. For the proof of Theorem 8.1(ii) we need only a κ such that

$$\kappa \to (\omega_1)_\lambda^{<\omega}, \quad \text{where } \lambda = 2^{|P|+\aleph_0} \,. \tag{3}$$

By Theorem 9.5, if $|P|$ is less than the first strongly inaccessible cardinal, then (3) holds for the least κ such that $\kappa \to (\omega_1)_2^{<\omega}$. As mentioned above, every measurable cardinal $\kappa > |P|$ satisfies (3). The reader interested in those concepts should consult Drake (1974) and Solovay et al. (1978). Let us only recall that the condition (2) implies that κ is a very large cardinal number and its existence does not follow from the axioms of ZFC (not even for $\alpha = \aleph_0$ and $\lambda = 2$).

We still need a technical lemma.

Let T be the set of finite sequences of integers, i.e., $T = \bigcup_{n<\omega} \omega^n$. We define a linear ordering $<$ of T as follows: if a, $b \in T$ and a is a proper initial segment of b, then $b < a$, while, if there exists an i such that both a and b are of length $\geq i$ and $a_i \neq b_i$, then $a < b$ iff $a_i < b_i$ for the least such i. This is called the Brouwer–Kleene ordering of T.

Lemma 9.6. *If* $T_0 \subseteq T$ *and* T_0 *does not contain any infinite subset linearly ordered by the relation "a is an initial segment of b", then the Brouwer–Kleene ordering well orders* T_0.

Proof. We can assume without loss of generality that T_0 is saturated in the sense that it contains all the initial segments of its elements. Then T_0 partially ordered by the relation "a is an initial segment of b" is a tree without infinite branches, and has a root v_0. We define inductively a map $\rho: T_0 \to \omega_1$. We put $\rho(v) = 0$ if v is at the top of a branch. Let S_v be the set of immediate successors of v, i.e., if $v \in \omega^n$, then $S_v = \{w \in \omega^{n+1} \cap T_0: v \subseteq w\}$. Assuming $\rho \restriction S_v$ is already defined we put

$$\rho(v) = \sup\{\rho(s) + 1: s \in S_v\} .$$

It is easy to check that this defines a map ρ and that $\rho(v)$ increases as v runs towards the root along any branch. Now it is easy to prove Lemma 9.6 by induction on $\rho(v_0)$. It is clear that if the Brouwer–Kleene ordering restricted to any subtree T_s stemming from $s \in S_v$ is a well ordering, then it is also a well ordering of the subtree T_v stemming from v. $\square$

Proof of Theorem 8.1(ii). Let $A \subseteq P^\omega$ be analytic, a projection of the closed set $C \subseteq P^\omega \times \omega^\omega$, and let (3) hold. Suppose that II does not have a winning strategy for the game $\langle P^\omega, A \rangle$. We have to show that I has a winning strategy. First we define an auxiliary PI-game G defined by a closed set and show that I has a winning strategy for G. Then we deduce that I also has a winning strategy for $\langle P^\omega, A \rangle$.

Let T be (as above) the set of all finite sequences of integers and $t_1, t_2, \ldots$ be an ω-enumeration of T without repetitions. For $q \in P^{2n}$ and $t_n \in T$ we shall say that (q, t_n) is *insecure* (for II) if q and t_n are initial segments of some $p \in P^\omega$ and $r \in \omega^\omega$, respectively, such that $(p, r) \in C$. Otherwise we say that (q, t_n) is *secure* (for II).

We define G as follows. The choices of I are still elements $p_n \in P$ but the choices of II are pairs (q_n, α_n), where $q_n \in P$ and $\alpha_n \in \kappa$. Player II wins iff,

$$\alpha_n = 0 \text{ whenever } ((p_0, q_0, \ldots, p_{n-1}, q_{n-1}), t_n) \text{ is secure,} \tag{4}$$

and

$$\alpha_n < \alpha_m \text{ whenever both } ((p_0, q_0, \ldots, p_{n-1}, q_{n-1}), t_n) \text{ and}$$

$$((p_0, q_0, \ldots, p_{m-1}, q_{m-1}), t_m) \text{ are insecure and } t_m \text{ is a}$$

proper initial segment of t_n. $\tag{5}$

We claim that, if $p = (p_0, q_0, p_1, q_1, \ldots)$ and $p \in A$, then II must have lost in G. Indeed, if $p \in A$, there exists a $t \in \omega^\omega$ such that $(p, t) \in C$ and hence all the pairs $(p \upharpoonright 2m, t \upharpoonright n)$ are insecure. If $t \upharpoonright n = t_{k(n)}$ and if II had won G, then by (5), $\alpha_{k(0)} > \alpha_{k(1)} > \cdots$ and this is impossible since κ is well ordered. Hence, since we assumed that II has no winning strategy for $\langle P^\omega, A \rangle$, II has no winning strategy for G either. By (4) and (5) the set of player I in G is closed. Hence by Proposition 3.2, G is determined and I has a winning strategy, say s, for G.

Now we will modify s to get a winning strategy s^* for I for $\langle P^\omega, A \rangle$.

Let $L = P^D$ = the set of functions from D to P, where $D = \bigcup_{n < \omega} P^{2n}$. So we have $|L| = 2^{|P| + \aleph_0}$.

We define a map $f : \bigcup_{m < \omega} [\kappa]^m \to L$ as in the definition of (3). For $Q \in [\kappa]^m$, $f(Q) : D \to P$ is defined by

$$f(Q)(p_0, q_0, \ldots, p_{n-1}, q_{n-1}) = s((q_0, \alpha_0), \ldots, (q_{n-1}, \alpha_{n-1})),$$

where s is the winning strategy for I for G, and $\alpha_0, \ldots, \alpha_{n-1}$ are given by the following three conditions:

if $((p_0, q_0, \ldots, p_{i-1}, q_{i-1}), t_i)$ is secure, then $\alpha_i = 0$; $\tag{6}$

if the set $I = \{i < n : ((p_0, q_0, \ldots, p_{i-1}, q_{i-1}), t_i) \text{ is insecure}\}$ has exactly $|Q|$ elements, then the map $t_i \mapsto \alpha_i$ for $i \in I$ is the unique bijection into Q preserving the Brouwer–Kleene ordering; $\tag{7}$

if $|I| \neq |Q|$, then $\alpha_i = 0$ for all $i < n$. $\tag{8}$

Let $H \subseteq \kappa$ be a homogeneous set of order type ω_1 for f; its existence follows from the assumption (3). Then we can define a function $s^* : \bigcup_{n < \omega} P^n \to P$ inductively as follows:

$$s^*(q_0, \ldots, q_{n-1}) = f(Q)(p_0, q_0, \ldots, p_{n-1}, q_{n-1}),$$

$$p_i = s^*(q_0, \ldots, q_{i-1}) \quad \text{for } i < n,$$

where Q is any set such that $Q \subseteq H$ and $|Q| = |I|$ and I is as in (7). [Note that $p_0 = s^*(\emptyset)$, i.e., p_0 is the first choice of I by s.] Notice that, since H is homogeneous for f, this definition of s^* is correct because we can check by induction that the value of s^* written above does not depend on the choice of Q, as long as $Q \subseteq H$ and $|Q| = |I|$.

We claim that s^* is a winning strategy for I for $\langle P^\omega, A \rangle$. Suppose to the contrary that there exists a $p = (p_0, q_0, \ldots, p_n, q_n, \ldots) \in P^\omega \setminus A$ which is obtained when I plays by means of s^*. We shall derive from this a contradiction by showing that there exists a game in G in which I plays s and II wins. To define this game, let

$$T_0 = \{t_n : ((p_0, q_0, \ldots, p_{n-1}, q_{n-1}), t_n) \text{ is insecure for II}, n < \omega\} .$$

Since $p \notin A$, T_0 satisfies the assumption of Lemma 9.6, whence there exists a map $\alpha : T_0 \to H$ which preserves the Brouwer–Kleene ordering. Assign to p the game in G where, after each choice p_{n-1} of I, II chooses $(q_{n-1}, \alpha(t_n))$ if $t_n \in T_0$ and $(q_{n-1}, 0)$ if $t_n \notin T_0$. Clearly, conditions (4) and (5) are satisfied, which means that II wins, and this is the desired contradiction.

This concludes the proof of Theorem 8.1(ii).

10. The Axiom of Determinacy

The results of Section 4 suggest the study of an abstract theory $T = ZF + AD + DC$. Here ZF is the set theory ZFC without the *Axiom of Choice*. AD, called the *Axiom of Determinacy*, tells that for all $X \subseteq \omega^\omega$ the game $\langle \omega^\omega, X \rangle$ is determined; DC, called the *Axiom of Dependent Choices*, tells that for any binary relation $R \subseteq Y \times Y$, if $Y \neq \emptyset$ and $\forall a \in Y \, \exists b \in Y[(a, b) \in R]$, then there exists an ω-sequence $y_0, y_1, \ldots$ such that $\forall n < \omega[(y_n, y_{n+1}) \in R]$. Notice that, by the coding described in the proof of Corollary 4.6, AD is equivalent to the statement that all games of the form $\langle 2^\omega, X \rangle$ are determined. AD was proposed by Mycielski and Steinhaus (1962). By Theorem 8.1(iii) of Martin and Steel (1989), we know that T is consistent.

T is motivated by its deductive power, the coherence of its theorems and the interesting classes of sets which are known or conjectured to constitute models for T. For example, T proves (by the results of Section 4) that every uncountable set in a Polish space has a perfect subset and that every subset of a Polish space has the property of Baire and is measurable with respect to every Borel measure. T also yields many natural results about projective sets and projective well orderings of sets of real numbers [see Addison and Moschovakis (1968), Kechris et al. (1977, 1979, 1981, 1985), Martin (1968), and Moschovakis (1980)]. Those theorems solve problems which are not solvable in

ZFC, and admit only unnatural solutions in $ZF + (V = L)$. By Theorem 8.1(iii), T is true in $L(\mathbb{R})$ [the proof of *DC* is given in Kechris (1985)]. So all subsets of a Polish space in $L(\mathbb{R})$ which are in $L(\mathbb{R})$, in particular all projective subsets, have the above properties. The advantage of proving those theorems in T is that T may have other models. For example, the class $L(\mathcal{O}^\omega)$, where $\mathcal{O}$ is the class of all ordinal numbers. [$L(\mathcal{O}^\omega)$ is the least model of *ZF* which contains all members of its members, and all ω-sequences of ordinal numbers.] As much as we know $L(\mathcal{O}^\omega)$ may satisfy the following stronger version of *AD*. Let $X \subseteq P^\alpha$, where α is any ordinal number, and consider the game $\langle P^\alpha, X \rangle$ of length α defined in a way similar to $\langle P^\omega, X \rangle$, where player I makes all the even choices and player II all the odd choices. Perhaps, in the model $L(\mathcal{O}^\omega)$ for every $\alpha < \omega_1$ and every $X \subseteq \omega^\alpha$ the game $\langle \omega^\alpha, X \rangle$ is determined [see Mycielski (1964, p. 217)].

Let us still show *without using the Axiom of Choice* two facts which imply that the above refinement of *AD* and Theorem 8.1(iii) are the strongest possible in a certain sense.

10.1. *There exists an* $X \subseteq 2^{\omega_1}$ *such that* $\langle 2^{\omega_1}, X \rangle$ *is not determined.*

10.2. *There exists an* $X \subseteq \omega_1^\omega$ *such that* $\langle \omega_1^\omega, X \rangle$ *is not determined.*

Proof of 10.1. Assume to the contrary that all games of the form $\langle 2^{\omega_1}, X \rangle$ are determined. This implies *AD*. It follows that there is no injection $\omega_1 \to \mathbb{R}$. Indeed, if such an injection were to exist, Corollary 4.6 would imply that the image of ω_1 in $\mathbb{R}$ would have a perfect subset; hence there would be a well ordering of $\mathbb{R}$ and, again by Corollary 4.6, a non-determined game of the form $\langle \omega^\omega, X \rangle$.

Now for every ordinal number $\alpha < \omega_1$ there exists a well ordering of ω, i.e., a subset of $\omega \times \omega$, of type α. Consider the following game $\langle 2^{\omega_1}, X \rangle$. Player II wins iff I always chooses 0 or, if α being the first ordinal for which I chose 1, the ω choices of II following α constitute a sequence of 0's and 1's coding a subset of $\omega \times \omega$ which is a well ordering of ω of type α. It is clear that I has no winning strategy in that game. But the existence of a winning strategy for II implies the existence of an injection of ω_1 into $\mathbb{R}$. So we have a contradiction. $\square$

The proof of 10.2 is quite similar to that of 10.1.

Another conjecture is the following: if $\alpha < \omega_1$, $X \subseteq \omega^\alpha$ and X is definable from an ω-sequence of ordinal numbers, then the game $\langle \omega^\alpha, X \rangle$ is determined. [The class of sets which are definable from a sequence of ordinals is definable, so the above conjecture can be expressed in the language of *ZF*. For a related conjecture see Addison and Moschovakis (1968).]

Let us return to the theory T. In 1967 R.M. Solovay showed that this theory proves that *for every partition* $\omega_1 = A \cup B$ *either A or B has a subset which is closed in* ω_1 *and cofinal with* ω_1 [see Jech (1978); see also Martin (1968)]. This implies that T yields the consistency of the theory $ZFC+$ the existence of a measurable cardinal number, and hence that T is a very strong theory. In particular it implies that Theorem 8.1(iii) could not have been proved without some additional axiom.

We will not discuss here the consequences of T except to state the following important weak form of the Axiom of Choice or Selection Principle which is useful for the theory of capacities of Choquet [see Mycielski (1972), Srebny (1984, pp. 30–47), and Busch (1979)].

Theorem 10.3 (In the theory T).   *For every* $S \subseteq \mathbb{R} \times \mathbb{R}$, *there exists a function* $f \subseteq S$ *such that* $\mathrm{pr}_1(S) \backslash \mathrm{dom}(f)$ *is of Lebesgue measure zero and of the first category*.

This theorem was proved by R.M. Solovay around 1970; his proof is published in Busch (1979).

The literature about AD is large, see Kechris et al. (1977, 1979, 1981, 1985) and Moschovakis (1980), but let me add the following polemical remarks. First, Mycielski and Steinhaus (1962) overlooked that S. Ulam had already defined a game [Mauldin (1981, p. 113)] which is equivalent to their game $\langle 2^\omega, X \rangle$. Second, in the detailed monograph by Moschovakis (1980), the history of AD is skewed. Namely on pages 9 and 287 the role of the papers by Mycielski and Steinhaus (1962) and Mycielski (1964, 1966) is ignored, and on pages 378–379 their mathematical motivation is criticized. The reader may check that in fact the motivation expressed in Mycielski and Steinhaus (1962) and Mycielski (1964, 1966) is identical to that in Moschovakis (1980). The only difference is that Moschovakis adopts a philosophy which tells that in this area we study a pre-existing Platonic reality discovered by Cantor, while Mycielski and Steinhaus would have told that we study here some new human constructions. Be that as it may, the idea of infinite PI-games has proved to be very stimulating and still presents many challenging open problems.

References

Addison, J.W. and Y.N. Moschovakis (1968) 'Some consequences of definable determinateness', *Proceedings of the National Academy of Sciences, U.S.A.* 59: 708–712.

Behrand, P. (1987) 'Differential games: Isaacs' equation', in: M. Singh, ed., *Encyclopedia of systems and control*. Oxford: Pergamon Press.

Bell, R.C. (1979) *The boardgame book*. The Knapp Press, distributed by the Viking Press, London.

Berlekamp, E.R., J.H. Conway and R.K. Guy (1982) *Winning ways*, Vols. I and II. London: Academic Press; third printing with corrections, 1985.

Breakwell, J.W. (1989) 'Time optimal pursuit inside a circle', (manuscript to appear).

Busch, D.R. (1979) 'Capacitability and determinacy', *Fundamenta Mathematica*, 102: 195–202.

Davis, M. (1964) 'Infinite games of perfect information, advances in game theory', *Annals of Mathematical Studies*, 52: 85–101.

Drake, F.R. (1974) *Set theory*. Amsterdam: North-Holland.

Ehrenfeucht, A. (1961) 'An application of games to the completeness problems for formalized theories', *Fundamental Mathematics*, 49: 129–141.

Ehrenfeucht, A. and J. Mycielski (1979) 'Positional strategies for mean payoff games', *International Journal of Game Theory*, 8: 109–113.

Even, S. and R.E. Tarjan (1976) 'A combinatorial problem which is complete in polynomial space', *Journal of the Association of Computing Mathematics*, 23: 710–719.

Gale, D. (1979) 'The game of Hex and the Brouwer fixed-point theorem'. *American Mathematics Monthly*, 86: 818–827.

Gale, D. and F.M. Stewart (1953) 'Infinite games with perfect information, contributions to the theory of games', vol. II, *Annals of Mathematical Studies*, 28: 245–266.

Galvin F. et al. (unpublished) 'Notes on games of perfect information' (available from the author, University of Kansas, Dept. of Mathematics, Lawrence, KS 66045).

Gardner, M. (1983) *Wheels, life and other mathematical amusements*. New York: W.H. Freeman and Co.

Gardner, M. (1986) *Knotted doughnuts and other mathematical entertainments*. New York: W.H. Freeman and Co.

Hájek, O. (1975) *Pursuit games, an introduction to the theory and applications of differential games*. London: Academic Press.

Hodges, W. (1985) 'Building models by games', London Math. Soc. Student Text Series, No. 2. Cambridge: Cambridge University Press.

Jech, T. (1978) *Set theory*. London: Academic Press.

Kechris, A.S. 'Determinacy and the structure of $L(\mathbb{R})$', *Recursion Theory, Proceedings of Symposia on Pure Mathematics*, vol. 42. New York: American Mathematical Society, pp. 271–283.

Kechris, A.S., et al., eds. (1977, 1979, 1981, 1985) Cabal Seminar, *Springer Lecture Notes*, Nos. 689, 839, 1019, 1333, respectively.

Kuhn, H.W. and G.P. Szegö, eds. (1971) *Differential games and related topics*. Amsterdam: North-Holland.

Louveau, A. (1980) 'σ-idéaux engendrés par des ensembles fermés et théorèmes d'approximation', *Transactions of the American Mathematical Society*, 257: 143–169.

Lynch, J.F. (1985) '*Probabilities of first order sentences about unary functions*', *Transactions of the American Mathematical Society*, 287: 543–568.

Lynch, J.F. (1992) 'Probabilities of sentences about very sparse random graphs', (manuscript to appear).

Martin, D.A. (1968) 'The axiom of determinateness and the reduction principles of the analytic hierarchy', *Bulletin of the American Mathematical Society*, 74: 687–689.

Martin, D.A. (1970) 'Measurable cardinals and analytic games', *Fundamenta Mathematica*, 66: 287–291.

Martin, D.A. (1975) 'Borel determinacy', *Annals of Mathematics*, 102: 363–371.

Martin, D.A. (1985) 'A purely inductive proof of Borel determinacy', *Recursion Theory, Proceedings of Symposia on Pure Mathematics*, vol. 42. New York: American Mathematical Society, pp. 303–308.

Martin, D.A. and J.R. Steel (1989) 'A proof of projective determinacy', *Journal of the American Mathematical Society*, 2: 71–125.

Mauldin, R.D., ed. (1981) *The Scottish book*. Boston: Birkhauser.

Michie, D. (1989) 'Brute force in chess and science', *ICCA Journal of the International Computer Chess Association* 12: 127–143.

Moschovakis, Y.N. (1980) *Descriptive set theory*. Amsterdam: North-Holland.

Mycielski, J. (1964) 'On the axiom of determinateness', Part I, *Fundamenta Mathematica*, 53: 205–224.
Mycielski, J. (1966) 'On the axiom of determinateness', Part II, *Fundamenta Mathematica*, 59: 203–212.
Mycielski, J. (1972) 'Remarks on capacitability', *American Mathematical Society Notices* 19: A-765.
Mycielski, J. (1988) 'Theories of pursuit and evasion', *Journal of Optimization Theory and Applications*, 56: 271–284.
Mycielski, J. and H. Steinhaus (1962) 'A mathematical axiom contradicting the Axiom of Choice', *Bulletin de l' Academie Polonaise des Sciences*, 10: 1–3.
Mycielski J. and S. Swierczkowski (1964) 'On the Lebesgue measurability and the axiom of determinateness', *Fundamenta Mathematica*, 54: 67–71.
Oxtoby, J.C. (1971, second edition, 1980) *Measure and category*. Berlin: Springer-Verlag.
Piotrowski, Z. (1985) 'Separate and joint continuity'. *Real Analysis Exchange* 11, No. 2, 293–322.
Rodin, Y.E. (1987) 'A pursuit-evasion bibliography – version 1', *Comput. Math. Applic.* 13: 275–340.
Solovay, R.M., W. Reinhardt and A. Kanamori (1978) 'Strong axioms of infinity and elementary embeddings', *Annals of Mathematical Logic*, 13: 73–116.
Srebrny, M. (1984) 'Measurable selectors of PCA multifunctions with applications', *Memoirs of the American Mathematical Society*, 52: No. 311.
Telgarsky, R. (1987) 'Topological games: On the 50th anniversary of Banach–Mazur games, *Rocky Mountain Journal of Mathematics*, 17: 227–276.
Yates, C.E. (1974) 'Prioric games and minimal degrees below 0'', *Fundamenta Mathematica*, 82: 217–237.
Yates, C.E. (1976) 'Banach–Mazur games, comeager sets and degrees of unsolvability', *Mathematical Proceedings of the Cambridge Philosophical Society*, 79: 195–220.
Zermelo E. (1913) 'Über eine Anwendungen der Mengenlehre auf die Theorie der Schachspiels', *Proceedings of the International Fifth Congress of Mathematicians, Cambridge, 1912*, vol. II. Cambridge: Cambridge University Press, pp. 501–504.

Chapter 4

REPEATED GAMES WITH COMPLETE INFORMATION

SYLVAIN SORIN

Université Paris X and Ecole Normale Supérieure

Contents

Handbook of Game Theory, Volume 1, Edited by R.J. Aumann and S. Hart

0. Summary

The theory of repeated games is concerned with the analysis of behavior in long-term interactions as opposed to one-shot situations; in this framework new objects occur in the form of threats, cooperative plans, signals, etc. that are deeply related to "real life" phenomena like altruism, reputation or cooperation. More precisely, repeated games with complete information, also called supergames, describe situations where a play corresponds to a sequence of plays of the same stage game and where the payoffs are some long-run average of the stage payoffs. Note that unlike general repeated games [see, for example, Mertens, Sorin and Zamir (1992)] the stage game is the same (the state is constant; compare with stochastic games; see the chapter on 'stochastic games' in a forthcoming volume of this Handbook) and known to the players (the state is certain; compare with games of incomplete information, Chapters 5 and 6 in this Handbook).

1. Introduction and notation

A *repeated game* results when a given game is played a large number of times and, when deciding what to do at each stage, a player may take into account what happened at all previous stages (or more precisely what he knows about it). The payoff is an average of the stage payoffs.

More formally let $G = G_1$ be the following strategic form game: I is the finite set of players with generic element i (we also write I for its cardinality). Each player i has a finite non-empty set of moves (or actions) S^i and a payoff function g^i from $S = \Pi_{j \in I} S^j$ into $\mathbb{R}$. X^i will denote the set of randomized or mixed moves of i, i.e. probabilities on S^i. For x in $X = \Pi_i X^i$, $g(x)$ stands for the usual multilinear extension of g and is the expected vector payoff if each player i plays x^i.

To G is associated a *supergame* Γ, played in stages: at stage 1, all players choose a move simultaneously and independently, thus defining a move profile, that is an I-tuple $s_1 = \{s_1^i\}$ of moves in S. s_1 is then announced to all players and the game proceeds to stage 2. (Note that we are assuming full monitoring; all past behavior is observed by everyone. For a more general framework see Section 5.) Inductively at stage $n + 1$, knowing the previous sequence of move profiles $(s_1, s_2, \ldots, s_n)$, all players again choose their moves simultaneously and independently. This choice is then told to all and the game proceeds to the next stage.

A *history* (resp. a *play*) is a finite (resp. infinite) sequence of elements of S; and the set of such sequences will be denoted by H (resp. H_x). H_n is the subset of n-stage histories. Histories are the basic ingredients of repeated games; they allow the players to coordinate their behavior. Note that in the present framework histories are known by all players, but in more general models (see Section 5) they will lead to differentiated information.

A *pure strategy* for player i in Γ is, by the above description, a mapping from H to S^i, specifying after each history the action to select. A *mixed strategy* is a probability distribution on the set of pure strategies. Since Γ is a game with perfect recall, Kuhn's theorem implies that it is enough to work with behavioral strategies, a *behavioral strategy* σ^i of player i being a mapping from H to X^i. Alternatively, σ^i can be represented by a sequence $\{\sigma_n^i\}$, σ_n^i being a mapping from H_{n-1} to X^i that describes the "strategy of player i at stage n". Write Σ^i for the corresponding set and $\Sigma = \Pi \Sigma^i$.

Each pure strategy profile σ induces a play h_∞ in a natural way. Formally: $s_1 = \sigma(\emptyset)$, $s_{n+1} = \sigma(s_1, s_2, \ldots, s_n)$ and $h_\infty = (s_1, \ldots, s_n, \ldots)$. Accordingly, each σ in Σ (or in the set of mixed strategies) defines a probability, say P_σ, on $(H_\infty, \mathcal{H}_\infty)$, where $\mathcal{H}_\infty$ is the product σ-algebra on $H_\infty = S^\infty$ (and similarly $\mathcal{H}_n$ on H_n); we denote by E_σ the expectation operator corresponding to probability P_σ.

To complete the description of Γ it remains to define a payoff function φ from Σ to $\mathbb{R}^I$. The theory of repeated games deals with mappings that are some kind of average of the sequence of stage payoffs $(g_1 = g(s_1), \ldots, g_n = g(s_n), \ldots)$ associated with a play. This is (with the stationary structure of information) the main difference from multimove games where the payoff can be *any* function on plays. Three classes will be analyzed here.

(i) *The finite game G_n*. The payoff is the arithmetic average of the sum of the payoffs for the n first stages and is denoted by $\bar{\gamma}_n$; hence $\bar{\gamma}_n(\sigma) = E_\sigma(\bar{g}_n)$, where $\bar{g}_n = (1/n) \sum_{m=1}^n g(s_m)$, $n \in \mathbb{N}$. G_n is the usual n-stage game where we normalize the payoffs to allow for a comparative study as n varies.

(ii) *The discounted game G_λ*. Here φ is the geometric average of the infinite stream of payoffs; it is written $\bar{\gamma}_\lambda$ with $\bar{\gamma}_\lambda(\sigma) = E_\sigma(\sum_{m=1}^\infty \lambda(1-\lambda)^{m-1} \times g(s_m))$, $\lambda \in (0, 1]$. G_λ is thus the game with discount factor λ (where again the payoff is normalized).

In each of these two cases Γ is a well-defined game in strategic form, so that the usual concepts (like equilibrium) apply. The situation is a little more delicate in the final case.

(iii) *The infinite game G_∞*. The payoff is taken here as some limit of $\bar{g}_n$. Different definitions are possible, because the above limit may not exist and one may choose liminf or limsup or some Banach limit, and because one can take the expectation first or the limit first. Finally, especially if the infinite game is considered as an approximation of a long but finite game, some uniformity conditions may be required for equilibrium.

We will use mainly the following definitions: σ is a *lower* (resp. *upper*) equilibrium if $\bar{\gamma}_n(\sigma)$ converges to some $\bar{\gamma}(\sigma)$ as n goes to infinity, and for each

τ^i in Σ^i and each i one has: liminf (resp. limsup) $\bar{\gamma}_n^i(\tau^i, \sigma^{-i}) \leq \bar{\gamma}^i(\sigma)$, where as usual σ^{-i} stands for the $(I-1)$ tuple induced by σ on $I\backslash\{i\}$.

Similarly, σ is a *uniform* equilibrium if $\bar{\gamma}_n(\sigma)$ converges and, moreover, $\forall \varepsilon > 0, \exists N, n \geq N \Rightarrow \bar{\gamma}_n^i(\tau^i, \sigma^{-i}) \leq \bar{\gamma}_n^i(\sigma) + \varepsilon$, for each τ^i and each i. In words, for any positive ε, σ is an ε-equilibrium in any sufficiently long game G_n.

When the payoff function is unspecified, the result will be independent of its particular choice.

Remark. One can also work with the random variables $\bar{g}_n$ and say that a deviation is profitable if limsup $\bar{g}_n$ increases with probability one.

Recall, finally, that a subgame perfect equilibrium of Γ is a strategy profile σ such that for all h in H, $\sigma[h]$ is an equilibrium in Γ, where $\sigma[h]$ is defined on H by $\sigma[h](h') = \sigma(h, h')$ and (h, h') stands for the history h followed by h'.

The main aim of the theory is to study the behavior of long games. Hence, we will consider the asymptotic properties of G_n as n goes to infinity or G_λ as λ goes to 0, as well as the limit game G_∞.

(Note once and for all that the 0-sum case is trivial: each player can play his optimal strategy i.i.d. and the value is constant – compare with Chapter 5 and the chapter on 'stochastic games' in a forthcoming volume of this Handbook.)

Each of these approaches has its own advantages and drawbacks and to compare them is very instructive. G_n corresponds to the "real" finite game, but usually the actual length is unknown or not common knowledge (see Subsection 7.1.2). Here the existence of a last stage has a disturbing backwards effect. G_λ has some nice properties (compactness, stationary structure) but cannot be studied inductively and here the discount factor $\tilde{n}$ has to be known precisely. Note that G_λ can be viewed as some $G_{\tilde{n}}$, where $\tilde{n}$ is an integer-valued random variable, finite a.s., whose law (but not the actual value) is known by the players. On the other hand, the use of G_∞ is especially interesting if a uniform equilibrium exists.

A few more definitions are needed to state the results.

Given a normal form game $\Gamma = (\Sigma, \varphi)$, the set of achievable payoffs is $\Delta = \{d \in \mathbb{R}^I; \exists \sigma \in \Sigma, \varphi(\sigma) = d\} = \varphi(\Sigma)$; it is denoted by D_n, D_λ and D_∞ for G_n, G_λ and G_∞, respectively.

Similarly, the set of Nash equilibrium payoffs is $\mathscr{E} = \{d \in \mathbb{R}^I; \exists \sigma \in \Sigma$ that is an equilibrium in Γ with $\varphi(\sigma) = d\}$; it is denoted by E_n, E_λ or E_∞ in the respective cases. Finally, $\mathscr{E}'$ – and specifically, E_n', E_λ' and E_∞' – will denote the set of subgame perfect equilibrium payoffs.

D is the set of *feasible* payoffs with (public pure) correlated strategies in G_1, or equivalently, if Co denotes the convex hull: $D = Co \ D_1 = Co \ g(S)$. (This corresponds to the convex combination of payoffs in the original game G.) In fact we shall see that repetition will allow us to mimic this public correlation in a verifiable way (because of the pure support).

The *minimax* level is defined by $v^i = \min_{x^{-i}} \max_{x^i} g^i(x^i, x^{-i})$ (recall that $X^{-i} = \Pi_{j \neq i} X^j$ is the set of vectors of mixed actions of the opponents of i). If $x^{-i}(i)$ realizes the above minimum, it will be referred to as a punishing strategy of players in $I\setminus\{i\}$ against i, and $x^i(i)$ will be the best reply of player i to it. V with components v^i is the *threat point*.

Finally, E is the set of *individually rational* (i.r. for short) and feasible payoffs: $E = \{d \in D; \forall i \in I, d^i \geq v^i\}$.

We will be interested in studying the asymptotic behavior of the sets $D_n, D_\lambda, E_n, \ldots$ (all convergence of sets will be with respect to the Hausdorff topology) and in describing D_∞, E_∞ and E'_∞. We shall see that the sets D and E will play a crucial role.

Before letting the parameters vary, we note that the games (Σ, φ) for the first two classes (i) and (ii) have compact pure strategy spaces and jointly continuous payoffs; hence the following properties hold.

Proposition 1.1. D_n *and* D_λ *are non-empty, path-connected, compact sets.*

Proposition 1.2 (Nash). E_n, E'_n, E_λ *and* E'_λ *are non-empty, compact sets.*

Remarks. It is easy to see that neither D_n nor D_λ is necessarily convex, and neither E_n nor E_λ connected. On the other hand, both D and E are convex, compact, and non-empty, since E contains E_1.

The following easy result illustrates one aspect of repetition: the possibility of convexifying the joint payoffs.

Proposition 1.3. (i) D_n *converges to* D *as* n *goes to infinity.*
  (ii) *The same is true for* D_λ *as* λ *goes to* 0.
  (iii) $D_\infty = D$.

Proof. Note first that the random stage payoff takes its values in the closed convex set D and hence expectation, average and limits share the same property so that $\varphi(\sigma)$ belongs to D for all σ; thus $\Delta \subset D$ (but $Co(\Delta) = D$). Now for every $\varepsilon > 0$, there exists some integer p such that any point d in D can be ε-approximated by a barycentric rational combination of points in $g(S)$, say $d' = \Sigma_m (q_m/p)g(s_m)$. Thus the strategy profile σ defined as: play cycles of length p consisting of q_1 times s_1, q_2 times s_2, and so on, induces a payoff near d' in G_n for n large enough.

  (ii) follows from (i) since the above strategy satisfies $\bar{\gamma}_\lambda(\sigma) \to d'$ as $\lambda \to 0$.

(iii) is obtained by taking for σ a sequence of strategies σ_k, used during n_k stages, with $\|\bar{\gamma}^{\mu}_{nk}(\sigma_k) - d\| \leq 1/k$. $\square$

Note that D_n may differ from D for all n, but one can show that D_λ coincides with D as soon as $\lambda \leq 1/I$ [Sorin (1986a)].

It is worth noting that the previous construction associates a play with a payoff, and hence it is possible for the players to observe any deviation. This point will be crucial in the future analysis.

The next three sections are devoted to the study of various equilibrium concepts in the framework of repeated games, using both the asymptotic approach and that of limit games. Section 2 deals with strategic or Nash equilibria, Section 3 with subgame perfection, and Section 4 with correlated and communication equilibria.

2. Nash equilibria

To get a rough feeling for some of the ideas involved in the construction of equilibrium strategies, consider an example with two players having two strategies each, Friendly and Aggressive. In a repeated framework, an equilibrium will be composed of a plan, like playing (F, F) at each stage, and of a threat, like: "play A forever as soon as the other does so once". Note that in this way one can also sustain a plan like playing (F, F) on odd days and (A, A) otherwise, or even playing (F, F) at stage n, for n prime (which is very inefficient), as well as other convex combinations of payoffs. On the one hand new good equilibria (in the sense of being Pareto superior) will appear, but the set of all equilibrium payoffs will be much greater than in the one-shot game.

In a discounted game two new aspects arise. One is related to the relative weight of the present versus the future (some punishment may be too weak to prevent deviations), but this failure disappears when looking at asymptotic properties. The second one is due to the stationary structure of the game: the strategy induced by an equilibrium, given a history consistent with it, is again an equilibrium in the initial game. For example, if a "deviation" is ignored at one stage, then there is an equilibrium in which similar "deviations" at all stages are ignored. We shall nevertheless see that this constraint will generically not decrease the set of equilibrium payoffs.

In finite games, there cannot be any threat on the last day; hence by induction some constraints arise that may prevent some of the previous plan/threat combinations. Nevertheless in a large class of games, the asymptotic results are roughly similar to those above.

Let us now present the formal analysis.

A first result states that all equilibrium payoffs are in E; obviously they need to be achievable and i.r. Formally:

Proposition 2.0. $\mathscr{E} \subset E$.

Proof. Obviously $\mathscr{E} \subset D$. Now let d be in E and σ be an associated equilibrium strategy profile. Then player i can, after any history h, use a best reply to $\sigma^{-i}(h)$. This gives him a (stage, and hence total) payoff greater than v^i in G_n or G_λ. As for G_∞ (if the payoff is not defined through limits of expectations), let g_m^i denote the random payoff of player i at stage m, then the random variables $z_m = g_m^i - E(g_m^i | \mathscr{H}_{m-1})$ are bounded, uncorrelated and with zero mean and hence by an extension of the strong law of large numbers converge a.s. in Cesaro mean to 0. Since $E(g_m^i | \mathscr{H}_{m-1}) \geq v^i$, this implies that player i can guarantee v^i and hence $d^i \geq v^i$ as well. $\quad\square$

It follows that to prove the equality of the two sets, it will be sufficient to represent points in E as equilibrium payoffs.

We now consider the three models.

2.1. The infinitely repeated game G_∞

The following basic result is known as the Folk theorem and is the cornerstone of the theory of repeated games. It states that the set of Nash equilibrium payoffs in an infinitely repeated game coincides with the set of feasible and individually rational payoffs in the one-shot game so that the necessary condition for a payoff to be an equilibrium payoff obtained in Proposition 2.0 is also sufficient.

Most of the results in this field will correspond to similar statements but with other hypotheses regarding the kind of equilibria, the type of repeated game or the nature of the information for the players.

Theorem 2.1. $E_\infty = E$.

Proof. Let d be in E and h a play achieving it (Proposition 1.3). The equilibrium strategy is defined by two components: a *cooperative behavior* and *punishments* in the case of deviation. Explicitly, σ is: play according to h as long as h is followed; if the actual history differs from h for the first time at stage n, let player i be the first (in some order) among those whose move differs from the recommendation at that stage and switch to $x(i)$ i.i.d. from stage $n+1$ on. Note that it is crucial for defining σ that h is a play (not a probability distribution on plays). The corresponding payoff is obviously d. Assume now that player i does not follow h at some stage and denote by $N(s^i)$ the set of subsequent stages where he plays s^i. The law of large numbers implies that $(1/\#N(s^i)) \sum_{n \in N(s^i)} g_n^i$ converges a.s. to $g(s^i, x^{-i}(i)) \leq v^i$ as $\#N(s^i)$ goes to ∞ and hence $\limsup \bar{g}_n^i \leq v^i$, a.s. Moreover, it is easy to see that σ

defines a uniform equilibrium, since the total gain by deviation is uniformly bounded. This proves that $E \subset E_\infty$ and hence the result by the previous proposition. □

Note that since we are looking only for Nash equilibria, it may be better for one player not to punish. This point will be taken into account in the next section. For a nice interpretation of and comments on the Folk Theorem, see Kurz (1978). Conceptual problems arise when dealing with a continuum of players; see Kaneko (1982).

2.2. The discounted game G_λ

Note first that in this case the asymptotic set of equilibrium payoffs may differ from E, see Forges, Mertens and Neyman (1986). A simple example is the following three-person game, where player 3 is a dummy:

$$\begin{pmatrix} (1,0,0) & (0,1,0) \\ (0,1,0) & (1,0,1) \end{pmatrix}.$$

This being basically a constant-sum game between players 1 and 2, it is easy to see that for all values of the discount factor λ, the only equilibrium (optimal) strategies in G_λ are $(1/2, 1/2)$ i.i.d. for both, leading to the payoff $(1/2, 1/2, 1/4)$. Hence the point $(1/2, 1/2, 1/2)$ in E cannot be obtained. In particular this implies that Pareto payoffs cannot always be approached as equilibrium payoffs in repeated games even with low discount rates.

In fact this phenomenon does not occur in two-person games or when a generic condition is satisfied [Sorin (1986a)].

Theorem 2.2. *Assume $I = 2$ or that there exists a payoff vector d in E with $d^i > v^i$ for all i. Then E_λ converges to E.*

The idea, as in the Folk Theorem, is to define a play that the players should follow and to punish after a deviation. If $I \geq 3$, the play is cyclic and corresponds to a strictly i.r. payoff near the requested payoff. It follows that for λ small enough, the one-stage gain from deviating (coefficient λ) will be smaller than the loss (coefficient $1 - \lambda$) of getting at most the i.r. level in the future. If $I = 2$ and the additional condition is not satisfied, either $E = \{V\}$ or only one player can profitably deviate and the result follows. □

2.3. The n-stage game G_n

It is well known that E_n may not converge to E, the classical example being the

Prisoner's Dilemma described by the following two-person game:

$$\begin{pmatrix} (3,3) & (0,4) \\ (4,0) & (1,1) \end{pmatrix},$$

where $E_n = \{(1,1)\}$ for all n. This property is not related to the existence of dominant strategies; a similar one holds with a mixed equilibrium in

$$\begin{pmatrix} (2,0) & (0,1) \\ (0,1) & (1,0) \end{pmatrix}.$$

In fact, these games are representative of the following class [Sorin (1986a)]:

Proposition 2.3.1. *If $E_1 = \{V\}$, then $E_n = \{V\}$ for all n.*

Proof. Let σ be an equilibrium in G_n and denote by $H(\sigma)$ the set of histories having positive probability under σ. Note first that on all histories of length $(n-1)$ in $H(\sigma)$, σ induces V, by uniqueness of the equilibrium in G_1. Now let m be the smallest integer such that after each history in $H(\sigma)$ with length strictly greater than m, σ leads to V. Assume $m \geq 0$ and take a history, say h, of length m in $H(\sigma)$ with $\sigma(h)$ not inducing V. It follows that one player has a profitable deviation at that stage and cannot be punished in the future. $\square$

The following result is typical of the field and shows that a good equilibrium payoff can play a dissuasive role and prevent backwards induction effects:

Theorem 2.3.2 [Benoit and Krishna (1987)]. *Assume that for all i there exists $e(i)$ in E_1 with $e^i(i) > v^i$. Then E_n converges to E.*

Proof. The idea is to split the stages into a cooperative phase at the beginning and a reward/punishment phase of fixed length at the end. During the first part the players are requested to follow a cyclic history leading to a strictly i.r. payoff approximating the required point in E. The second phase corresponds to playing a sequence of R cycles of length I, leading to $(e(1), \ldots, e(I))$. Note that this part consists of equilibria and hence no deviation is profitable. On the other hand, a deviation during the first period is observable and the players are then requested to switch to $x(i)$ for the remaining stages if i deviates. It follows that, by choosing R large enough, the one-shot gain is less than $R \times (e^i(i) - v^i)$ and hence the above strategy is an equilibrium. Letting n grow sufficiently large gives the result. $\square$

Note that the above proof also shows the following: if E contains a strictly i.r. payoff, a necessary and sufficient condition for E_n to converge to E is that for all i there exists n_i and $e^i(i)$ in E_{n_i} with $e^i(i) > v^i$.

In conclusion, repetition allows for coordination (and hence new payoffs) and threats (new equilibria). Moreover, for a large class of games, the set of equilibria increases drastically with repetition and one has continuity at ∞: $\lim E_n = \lim E_\lambda = E_\infty = E$; every feasible i.r. payoff can be sustained by an equilibrium. On the other hand, this set seems too large (it includes the threat point V) and a first attempt to reduce it is to ask for subgame perfection.

3. Subgame perfect equilibria

The introduction of the requirement of perfection will basically not change the basic results concerning the limit game. Going back to the example at the beginning of Section 2, the length of the punishment (playing A) can be adapted to the deviation, but can remain finite and hence its impact on the payoff is zero.

On the other hand, the specific features of the discounted game (fixed point property) and of the finite game (backwards induction) will have a much larger impact, being applied on each history. For example, if A is a dominant move, playing A at each stage will be the only subgame perfect equilibrium strategy of the finite repeated game.

As in the previous section we will consider each type of game (and recall that $\mathscr{E}' \subset \mathscr{E}$).

3.1. G_∞

The first result is an analog of the Folk Theorem, showing that the equilibrium set is not reduced by requiring perfection. In fact, the possibly incredible threat of everlasting punishment can be adapted so that the same play will still be supported by a perfect equilibrium.

Theorem 3.1 [Aumann and Shapley (1976), Rubinstein (1976)]. $E'_\infty = E$.

Proof. The cooperative aspect of the equilibrium is like in the Folk Theorem. The main difference is in the punishment phase; if the payoff is defined through some limiting average it is enough to punish a deviator during a finite number of stages and then to come back to the original cooperative play. It is not advantageous to deviate; it does not harm to punish. Explicitly, if a deviation happens at stage n, punish until the deviator's average payoff is within $1/n$ of the required payoff. Deviations during the punishment phase are ignored. (To get more in the spirit of subgame perfection, one might require inductively the punisher to be punished if he is not punishing. For this to be done, since a deviation may not be directly observable during the punishment phase, some statistical test has to be used.) $\square$

The interpretation of the "Perfect Folk Theorem" is that punishments can be enforced either because they do not hurt the punisher or because higher levels of punishment are available against a player who would not punish. This second idea will be used below.

Remarks. (1) Note that a priori the previous construction will not work in G_n or G_λ since there a profitable deviation during a finite set of stages counts, and on the other hand the hierarchy of punishment phases may lead to longer and longer phases.

(2) For similar results with different payoffs or concepts, see Rubinstein (1979a, 1980).

3.2. G_λ

A simple and useful result in this framework, which is due to Friedman (1985), states that any payoff that strictly dominates a one-shot equilibrium payoff is in E'_λ for λ small enough. (The idea is, as usual, to follow a play that generates the payoff and to switch to the equilibrium if a deviation occurs.)

In order to get the analog of Theorem 2.2, not only is an interior condition needed (recall the example in Subsection 2.2), but also a dimensional condition, as shown by the following example due to Fudenberg and Maskin (1986a). Player 1 chooses the row, player 2 the column and player 3 the matrix in the game with payoffs:

$$\begin{pmatrix} (1,1,1) & (0,0,0) \\ (0,0,0) & (0,0,0) \end{pmatrix} \quad \text{and} \quad \begin{pmatrix} (0,0,0) & (0,0,0) \\ (0,0,0) & (1,1,1) \end{pmatrix}.$$

Let w be the worst subgame perfect equilibrium payoff in G_λ. Then one has $w \geq \lambda g_1 + (1 - \lambda)w$, where g_1 is any payoff achievable at stage 1 when two of the players are using their equilibrium strategies. It is easily seen that for any triple of randomized moves there exists one player's best reply that achieves at least $1/4$, i.e. $g_1 \geq 1/4$; hence $w \geq 1/4$ so that $(0,0,0)$ cannot be approached in E'_λ.

A generic result is due to Fudenberg and Maskin (1986a):

Theorem 3.2. *If E has a non-empty interior, then E'_λ converges to E.*

Proof. This involves some nice new ideas and can be presented as follows. First define a play leading to the payoff, then a family of plans, indexed by I, consisting of some punishment phase [play $x(i)$] and some reward phase [play $h(i)$ inducing an i.r. payoff $f(i)$]. Now if at some stage of the game player i is the first (in some order) deviator, the plan i is played from then on until a new possible deviation.

To get the equilibrium condition, the length R of the punishment phase has to be adapted and the the rewards must provide an incentive for punishing, i.e. for all i, j one needs $f^i(j) > f^i(i)$ (here the dimensional condition is used). Finally, if the discount factor is small enough, the loss in punishing is compensated by the future bonus.

The proof itself is much more intricate. Care has to be taken in the choice of the play leading to a given payoff; it has to be smooth in the following sense: given any initial finite history the remaining play has to induce a neighboring payoff. Moreover, during the punishment phase some profitable and non-observable deviation may occur [recall that $x(i)$ consists of mixed actions] so that the actual play following this phase will have to be a random variable $h'(i)$ with the following property: for all players j, $j \neq i$, the payoff corresponding to R times $x(i)$, then $h(i)$ is equal to the one actually obtained during the punishment phase followed by $h'(i)$. At this point we use a stronger version of Proposition 1.3 which asserts that for all λ small enough, any payoff in D can be exactly achieved by a smooth play in G_λ. [Note that $h'(i)$ has also to satisfy the previous conditions on $h(i)$.]   □

Remarks. (1) The original proof deals with public correlation and hence the plays can be assumed "stationary". Extensions can be found in Fudenberg and Maskin (1991), Neyman (1988) (for the more general class of irreducible stochastic games) or Sorin (1990).

(2) Note that for two players the result holds under weaker conditions; see Fudenberg and Maskin (1986a).

3.3. G_n

More conditions are needed in G_n than in G_λ to get a Folk Theorem-like result. In fact, to increase the set of subgame perfect equilibria by repeating the game finitely many times, it is necessary to start with a game having multiple equilibrium payoffs.

Lemma 3.3.1. *If $E'_1 = E_1$ has exactly one point, then $E'_n = E'_1$ for all n.*

Proof. By the perfection requirement, the equilibrium strategy at the last stage leads to the same payoff, whatever the history, and hence backwards induction gives the result.   □

Moreover, a dimension condition is also needed, as the following example due to Benoit and Krishna (1985) shows. Player 1 chooses the row, player 2 the column and player 3 the matrix, with payoffs as follows:

$$\begin{pmatrix} (3,3,3) & (0,0,0) \\ (0,0,0) & (0,0,0) \\ (0,1,1) & (0,0,0) \end{pmatrix} \quad \text{and} \quad \begin{pmatrix} (1,1,1) & (2,2,2) \\ (0,1,1) & (0,1,1) \\ (0,1,1) & (0,0,0) \end{pmatrix}.$$

One has $V = (0,0,0)$; $(2,2,2)$ and $(3,3,3)$ are in E_1 but players 2 and 3 have the same payoffs. Let w_n be the worst subgame perfect equilibrium payoff for them in G_n. Then by induction $w_n \geqslant 1/2$ since for every strategy profile one of the two can, by deviating, get at least $1/2$. (If player 1 plays middle with probability less than $1/2$, player 2 plays left; otherwise, player 3 chooses right.) Hence E'_n remains far from E.

A general result concerning pure equilibria (with compact action spaces) is the following:

Theorem 3.3.2 [Benoit and Krishna (1985)]. *Assume that for each i there exists $e(i)$ and $f(i)$ in E_1 (or in some E_n) with $e^i(i) > f^i(i)$, and that E has a non-empty interior. Then E'_n converges to E.*

Proof. One proof can be constructed by mixing the ideas of the proofs in Subsections 2.3 and 3.2. Basically the set of stages is split into three phases; during the last phase, as in Subsection 2.3, cycles of $(e(1), \dots, e(I))$ will be played. Hence no deviations will occur in phase 3 and one will be able to punish "late" deviations (i.e. in phase 2) of player i, say, by switching to $f(i)$ for the remaining stages. In order to take care of deviations that may occur before and to be able to decrease the payoff to V, a family of plans as in Subsection 3.2 is used. One first determines the length of the punishment phase, then the reward phase; this gives a bound on the duration of phase 2 and hence on the length of the last phase. Finally, one gets a lower bound on the number of stages to approximate the required payoff. □

As in Subsection 3.2 more precise results hold for $I = 2$; see Benoit and Krishna (1985) or Krishna (1988).

An extension of this result to mixed strategies seems possible if public correlation is allowed. Otherwise the ideas of Theorem 3.2 may not apply, because the set of achievable payoffs in the finite game is not convex and hence future equalizing payoffs cannot be found.

3.4. The recursive structure

When studying subgame perfect equilibria (SPE for short) in G_λ, one can use the fact that after any history, the equilibrium conditions are similar to the initial ones, in order to get further results on E_λ while keeping λ fixed.

The first property arising from dynamic programming tools and using only the continuity in the payoffs due to the discount factor (and hence true in any multistage game with continuous payoffs) can be written as follows:

Proposition 3.4.1.   *A strategy profile is a SPE in G_λ iff there is no one-stage profitable deviation.*

Proof.   The condition is obviously necessary. Assume now that player i has a profitable deviation against the given strategy σ, say τ^i. Then there exists some integer N, such that θ^i defined as "play τ^i on histories of length less than N and σ^i otherwise", is still better than σ^i. Consider now the last stage of a history of length less than N, where the deviation from σ^i to θ^i increase i's payoff. It is then clear that to always play σ^i, except at that stage of this history where τ^i is played, is still a profitable deviation; hence the claim.   □

This criterion is useful to characterize all SPE payoffs.

We first need some notation. Given a bounded set F of $\mathbb{R}^I$, let $\Phi_\lambda(F)$ be the set of Nash equilibrium payoffs of all one-shot games with payoff $\lambda g + (1 - \lambda)f$, where f is any mapping from S to F.

Proposition 3.4.2. E_λ' *is the largest (in terms of set inclusion) bounded fixed point of Φ_λ.*

Proof.   Assume first $F \subset \Phi_\lambda(F)$. Then, at each stage n, the future expected payoff given the history, say f_n in F, can be supported by an equilibrium leading to a present payoff according to g and some future payoff f_{n+1} in F. Let σ be the strategy defined by the above family of equilibria. It is clear that in G_λ σ yields the sequence f_n of payoffs, and hence by construction no one-stage deviation is profitable. Then, using the previous proposition, $\Phi_\lambda(F) \subset E_\lambda'$. On the other hand, the equilibrium condition for SPE implies $E_\lambda' \subset \Phi(E_\lambda')$ and hence the result.   □

Along the same lines one has $E_\lambda' = \bigcap_n \Phi_\lambda^n(D')$ for any bounded set D' that contains D. These ideas can be extended to a much more general setup; see the following sections.

Note that when working with Nash equilibria the recursive structure is available only on the equilibrium path and that when dealing with G_n one loses the stationarity.

Restricting the analysis to pure strategies and using the compactness of the equilibrium set (strategies and payoffs) allows for nice representations of all pure SPE; see Abreu (1988). Tools similar to the following, introduced by Abreu, were in fact used in the previous section.

Given $(I + 1)$ plays $[h; h(i), i \in I]$, a *simple strategy profile* is defined by requiring the players to follow h and inductively to switch to $h(i)$ from stage $n + 1$ on, if the last deviation occurred at stage n and was due to player i.

Lemma 3.4.3. $[h(0); h(i), i \in I]$ *induces a SPE in* G_λ *iff for all* $j = 0, \ldots, I$, $[h(j); h(i), i \in I]$ *defines an equilibrium in* G_λ.

Proof. The condition is obviously necessary and sufficiency comes from Proposition 3.4.1. □

Define $\sigma(i)$ as the pure SPE leading to the worst payoff for i in G_λ and denote by $h^*(i)$ the corresponding cooperative play.

Lemma 3.4.4. $[h^*(j); h^*(i), i \in I]$ *induces a SPE.*

Proof. Since $h^*(j)$ corresponds to a SPE, no deviation [leading, by $\sigma(j)$, to some other SPE] is profitable a fortiori if it is followed by the worst SPE payoff for the deviator. Hence the claim by the previous lemma. □

We then obtain:

Theorem 3.4.5 [Abreu (1988)]. *Let* σ *be a pure SPE in* G_λ *and* h *be the corresponding play. Then* $[h; h^*(i), i \in I]$ *is a pure SPE leading to the same play.*

These results show that extremely simple strategies are sufficient to represent all pure SPE; only $(I + 1)$ plays are relevant and the punishments depend only on the deviator, not on his action or on the stage.

3.5. *Final comments*

In a sense it appears that to get robust results that do not depend on the exact specification of the length of the game (assumed finite or with finite mean), the approach using the limit game is more useful. Note nevertheless that the counterpart of an "equilibrium" in G_∞ is an ε-equilibrium in the finite or discounted game (see also Subsection 7.1.1). The same phenomena of "discontinuity" occur in stochastic games (see the chapter on 'stochastic games' in a forthcoming volume of this Handbook) and even in the zero-sum case for games with incomplete information (Chapter 5 in this Handbook).

4. Correlated and communication equilibria

We now consider the more general situation where the players can observe signals. In the framework of repeated games (or multimove games) several such extensions are possible depending on whether the signals are given once or at each stage, and whether their law is controlled by the players or not. These mechanisms increase the set of equilibrium payoffs, but under the hypothesis of full monitoring and complete information lead to the same results. (Compare with Chapter 6 in this Handbook.)

Recall that given a normal form game $\Gamma = (\Sigma, \varphi)$ and a correlation device $C = (\Omega, \mathcal{A}, P; \mathcal{A}^i)$, $i \in I$, consisting of a probability space and sub σ-algebras of $\mathcal{A}$, a correlated equilibrium is an equilibrium of the extended game Γ_C having as strategies, say μ^i for i, $\mathcal{A}^i$-measurable mappings from Ω to Σ^i, and as payoff $\Phi(\mu) = \int \varphi(\mu(\omega)) \, P(d\omega)$. In words, ω is chosen according to P and $\mathcal{A}^i$ is i's information structure. Similarly, in a multimove game the notion of an extensive form correlated equilibrium can be defined with the help of private filtrations, say $\mathcal{A}^i_n$ for player i – i.e. there is new information on ω at each stage – and by requiring μ^i_n to be $\mathcal{A}^i_n \otimes \mathcal{H}_n$ measurable on $\Omega \times H_n$. Finally, for communication equilibria [see Forges (1986)], the probability induced by P on $\mathcal{A}_{n+1}$ is $\mathcal{A}^i_n \otimes \mathcal{H}_n$ measurable, i.e. the law of the signal at each stage depends on the past history, including the moves of the players.

Let us consider repeated games with a correlation device (resp. extensive correlation device; communication device). We first remark that the set of feasible payoffs is the same in any extended game and hence the analog of Proposition 1.3 holds.

For any of these classes we consider the union of the sets of equilibrium payoffs when the device varies and we shall denote it by cE_∞, CE_∞ and KE_∞, respectively. It is clear that the main difference from the previous analysis (without information scheme) comes from the threat point, since now any player can have his payoff reduced to $w^i = \min_{Y^{-i}} \max_{X^i} g^i(x^i, y^{-i})$, where Y^{-i} stands for the probabilities on S^{-i} (correlated moves of the opponent to i) and this set is strictly larger than X^{-i} for more than two players. Hence the new threat point W will usually differ from V and the set to consider will be $CE = \{d \in D : \forall i \in I, d^i \geq w^i\}$. Then one shows easily that $cE_\infty = CE_\infty = KE_\infty = CE$.

There is a deep relationship between these concepts and repeated games (or multimove games) in the sense that given a strategy profile σ, $C_n = (H_n, \mathcal{H}_n, P_\sigma)$ is a correlation device at stage n (where in the framework of Sections 1–3, the private σ-algebra is $\mathcal{H}_n$ for all players). This was first explicitly used in games with incomplete information when constructing a *jointly controlled lottery* [see Aumann, Maschler and Stearns (1968)]. For extensions of these tools under partial monitoring, see the next section.

5. Partial monitoring

Only partial results are available when one drops the assumption of full monitoring, namely that after each stage all players are told (or can observe) the previous moves of their opponents. In fact the first models in this direction are due to Radner and Rubinstein and also incorporate some randomness in the payoffs (moral hazard problems). We shall first cover results along these lines. Basically one looks for sufficient conditions to get results similar to the Folk Theorem or for Pareto payoffs to be achievable. In a second part we will present recent results of Lehrer, where the structure of the game is basically as in Section 1 except for the signalling function, and one looks for a characterization of E_∞ in terms of the one-stage game data.

5.1. Partial monitoring and random payoffs (see also the chapter on 'principal-agent models' in a forthcoming volume of this Handbook)

5.1.1. One-sided moral hazard

The basic model arises from *principal–agent* situations and can be represented as follows. Two players play sequentially; the first player (principal) chooses a *reward function* and then with that knowledge the second player (agent) chooses a move. The outcome is random but becomes *common knowledge* and depends only on the choice of player 2, which player 1 does not know. Formally, let Ω be the set of outcomes. The actions of player 1 are measurable mappings from Ω to some set S. Denote by T the actions set of player 2 and by Q_t the corresponding probabilities on Ω. The payoff functions are real continuous bounded measurable mappings, f on $\Omega \times S$ for player 1 and g on $\Omega \times S \times T$ for player 2. Assume, moreover, some *revelation condition* (RC), namely that there exists some positive constant K such that, for all positive ε, if $E_{s,t} g \geqslant E_{s,t'} g + \varepsilon$, then $|\int \omega \, dQ_t - \int \omega \, dQ_{t'}| \geqslant K\varepsilon$. In words, this means that profitable deviations of player 2 generate a different distribution of outcomes.

It is easy to see that generically one-shot Nash equilibria are not efficient in such games. The interest of repetition is then made clear by the following result:

Theorem 5.1.1 [Radner (1981)]. *Assume that a feasible payoff d strictly dominates a one-shot Nash equilibrium payoff e. Then $d \in E_\infty$.*

Proof. The idea of the proof is to require both players to use the strategy combination leading to d, as in the Folk Theorem. A deviation from player 1 is observable and one then requires that both players switch to the equilibrium payoff e. The main difficulty arises from the fact that the deviations of player 2 are typically non-observable (even if he is using a pure strategy the Q_t may have non-disjoint support). Both players have to use some statistical test, based for example on the law of large numbers, to check with probability one whether player 2 was playing a profitable deviation, using RC. In such a case they again both switch to e. □

By requiring the above punishment to last for a finite number of stages (adapted to the precision of the test), one may even obtain a form of "subgame perfection" (note that there are no subgames, but one may ask for an equilibrium condition given any common knowledge history); see again Radner (1981).

Similar results with alternative economic content have been obtained by Rubinstein (1979a, 1979b) and Rubinstein and Yaari (1983).

Going back to the previous model, it can also be shown [Radner (1985)] that the modified strategies described above lead to an equilibrium in G_λ if the discount factor is small enough, and that they approach the initial payoff d. A similar remark about perfection applies and hence formally the following holds:

Theorem 5.1.2. *Let d be feasible, $e \in E_1$, and assume $d \gg e$. Then for all $\varepsilon > 0$ there exists λ^* such that for all $\lambda \leq \lambda^*$, d is ε-close to E'_λ.*

Other classes of strategies with related properties have been introduced and studied by Radner (1986c).

5.1.2. Two sided moral hazard

A model where both players have private information on the history has been introduced and studied by Radner (1986a) under the name *partnership game*. Here the players are simultaneously choosing moves in some sets S and T. The outcome is again random with some law Q_{st}. At each stage the information of each player consists of his move and of the outcome; moreover, his own stage payoff depends only on this information and the revelation condition is still required. Then the analogy of the previous Theorem 5.1.1 holds [Radner (1986a)]. Here also the construction of the strategies is based on some statistical test and uses review and punishment phases.

Nevertheless, if one studies G_λ the previous arguments are no longer valid. More precisely, since none of the moves is observable it may be worthwhile for one player to deviate from the prescribed strategy when the sequence of records of outcomes starts to differ significantly from the mean and to try to "correct" it in order to avoid the punishment phase. (Note that when the payoff is not discounted, by the strong law of large numbers, there is no gain in doing so.)

In fact, an example of a partnership game due to Radner, Myerson and Maskin (1986) shows that E_λ may be uniformly (in λ) bounded away from the Pareto boundary. Schematically, the payoffs depend upon an outcome that may be good or bad and the game is symmetrical. If, at equilibrium, the future payoff is independent of the outcome one obtains only one-shot equilibrium payoffs. Thus, this future payoff has to be discriminating (higher for a good outcome than for a bad) and hence cannot be Pareto optimal in expectation. (See also the example in the next section.)

5.1.3. Public signals and recursive structure

We now turn to results that are not based on the use of statistical tests but rather on the recursive structure.

A first model due to Abreu, Pearce and Stachetti (1986, 1990) considers an oligopoly with compact pure strategy sets where the I firms are only told, after each stage, the price, which is a random function of the moves with a fixed support. One can see that in this case Nash and "subgame perfect" equilibria coincide and, moreover, the recursive properties still hold. This allows us to give a nice description of the set of equilibrium payoffs by using its extreme points.

Finally, in a recent work, Fudenberg, Levine and Maskin (1989) succeed in getting a theorem analogous to Theorem 3.2 in the following framework. Consider a game where after each stage each player gets some private information on a random signal depending on the moves of all players at that stage. We call public those strategies that depend only on events known to all players. Note first that an equilibrium in the discounted game restricted to public strategies is an equilibrium in the original game (given a best reply to public strategies, taking its conditional expectation on public events, is still a best reply) and that one can define "subgame perfect public equilibria" by introducing subgames related to public events. The tools of Subsection 3.4 are then applicable, and sufficient conditions are given, basically on the independence of the conditional laws of the signals as function of the moves of each player – the strategy of the others being fixed – to ensure that the corresponding set of payoffs converges to E as λ goes to 0. More precisely, it is shown that a smooth convex set F of payoffs included in E and at a small Hausdorff distance from it satisfies $F \subset \Phi_\lambda(F)$ for λ small enough. The main difficulty is to check the inclusion on extreme points. In fact, the above conditions allow us to compute explicitly the future payoffs by solving linear equations.

Note that here a dimension condition is needed, even in the two-player case. Let us consider the following game, due to Fudenberg and Maskin (1986b).

The payoff matrix is

$$\begin{pmatrix} (1,1) & (0,0) \\ (0,0) & (-1,-1) \end{pmatrix},$$

the moves of player 2 are announced and a public signal with values (α, β) has the following distribution:

$$\begin{pmatrix} (3/4,1/4) & (1/2,1/2) \\ (1/2,1/2) & (1/4,3/4) \end{pmatrix}.$$

Then $(1,1)$ is the only point in E'_λ. In fact, denote by w the worst SPE, by s and t the corresponding random moves of both players at stage 1, and by $w_{L\alpha}(\geq w)$ the expected payoff after Left and α, and so on. We note first that if $s = 1$ one has $w \geq \lambda + (1-\lambda)(3/4w_{L\alpha} + 1/4w_{L\beta})$, and hence $w \geq 1$. Otherwise one has:

$$w = t((1-\lambda)(1/2w_{L\alpha} + 1/2w_{L\beta}))$$
$$+ (1-t)(-\lambda + (1-\lambda)(1/4w_{R\alpha} + 3/4w_{R\beta}))$$

$$\geq t(\lambda + (1-\lambda)(3/4w_{L\alpha} + 1/4w_{L\beta}))$$
$$+ (1-t)((1-\lambda)(1/2w_{R\alpha} + 1/2w_{R\beta})),$$

so that $t(1-\lambda)w_{L\beta} + (1-t)(1-\lambda)w_{R\beta} \geq 4\lambda + (1-\lambda)w$. Substituting this into the first equality yields $w \geq t + 1$.

Note that 0 is a subgame perfect public equilibrium payoff in G_α (even if 2's moves are not announced) by asking the players to use their dominated move at each stage where the empirical past frequency of α is greater than $1/2$. [Compare with Sorin (1986b).]

On the other hand, 0 can be obtained as a perfect equilibrium in G_λ if 1's moves are observable by asking him to follow a history consisting of a sequence of 1 and -1 inducing a payoff increasing to 0 and playing again the same move in the case of a deviation from -1 to 0. In the previous framework player 1 could pretend to punish even if he did not and hence the punishment was not credible and player 2 would deviate.

It is important to remark that in these games the signals can be used as a correlation device or an extensive correlation device (recall Section 4 and see also Subsection 5.2). In particular, the set of equilibria can be larger than the set of public equilibria and can contain payoffs that are not i.r. (but in CE), if for example a subgroup of players get some common signal, unknown to the others. (But if there are two players and one is more informed than the other one can always assume public strategies.)

Finally, similar results are used in the framework of games with long-run and short-run players [Fudenberg and Levine (1989b)].

5.2. Signalling functions

The results of this section are due mainly to Lehrer. We consider the infinitely repeated game G_∞ of Section 1, but after each stage n, each player i is only told $q_n^i = Q^i(s_n)$, s_n being the I-action at that stage and Q^i being i's signalling (deterministic) function, defined on S with values in some set Q. Each player's strategy is then required to be measurable with respect to his private information. Hence a pure strategy σ_n^i is a mapping from sequences $(q_1^i, \ldots, q_{n-1}^i)$ to S^i and perfect recall is assumed.

Let us first consider the case of two players and a general signalling function (we shall assume in this section non-trivial information, namely that each player may, by playing some move, get some information about his opponent's move, so that the players can communicate through their actions – the other case is much simpler to analyze).

It is easy to see that, since the signals are not common knowledge, equilibrium strategies do not induce, after finitely many stages, an equilibrium in the remaining game but rather a correlated equilibrium (see Section 4).

One is thus led to consider extensive form correlated equilibria and in fact these are much easier to characterize.

We first define two relations on actions by

$$s^i \sim t^i \Leftrightarrow Q^{-i}(t^i, s^{-i}) = Q^{-i}(s^i, s^{-i}) \quad \text{for all } s^{-i}$$

(in words, in a one-shot game player $-i$ has no way to distinguish whether player i is playing s^i or t^i); and

$$s^i > t^i \Leftrightarrow s^i \sim t^i \text{ and } Q^i(t^i, s^{-i}) \neq Q^i(t^i, t^{-i}) \text{ implies}$$

$$Q^i(s^i, s^{-i}) \neq Q^i(s^i, t^{-i}) \quad \text{for all } s^{-i}, t^{-i}$$

(player i gets more information on $-i$'s move by playing s^i than t^i).

The crucial point is that player i can mimic a pure strategy, say τ^i, by any other σ^i with $\sigma^i(h) > \tau^i(h)$, for all h, without being detected. [Inductively, at each stage n he uses an action $s_n^i > \tau^i(h)$, h being the history that would have occurred had he used $\{t_m^i\}$, $m < n$, up to now.]

Let P be the set of probabilities on S (correlated moves). The set of equilibrium payoffs will be characterized through the following sets (note that, as in the Folk Theorem, they depend only on the one-shot game):

$$A^i = \{ p \in P: \sum_{s^{-i}} p(s^i, s^{-i}) g^i(s^i, s^{-i}) \geq \sum_{s^{-i}} p(s^i, s^{-i}) g^i(t^i, s^{-i}) \text{ for all } s^i$$

and all t^i with $t^i > s^i \}$.

$$B^i = A^i \cap X = \{ x \in X: g^i(s^i, x^{-i}) \geq g^i(t^i, x^{-i}) \text{ for all } s^i, t^i \text{ with } x^i(s^i) > 0$$

and $t^i > s^i \}$.

Write IR for the set of i.r. payoffs and $E_\propto$ (resp. $cE_\propto$, $CE_\propto$, $KE_\propto$) for the set of Nash (resp. correlated, extensive form correlated, communication) equilibrium payoffs in the sense of upper, $\mathscr{L}$ or uniform. $lE_\propto$ and $lCE_\propto$ will denote lower equilibrium payoffs [recall paragraph (iii) in Section 1].

Theorem 5.2.1 [Lehrer (1992a)].   (i) $cE_\propto = CE_\propto = KE_\propto = g(\bigcap_i A^i) \cap IR$.
(ii) $lCE_\propto = \bigcap_i g(A^i) \cap IR$.

Proof.   The proof of this result (and of the following) is quite involved and introduces new and promising ideas. Only a few hints will be presented here.

For (ii), the inclusion from left to right is due to the fact that given correlated strategies, each player can modify his behavior in a non-revealing way to force the correlated moves at each stage to belong to A^i.

Similarly, for the corresponding inclusion in (i) one obtains by convexity that if a payoff does not belong to the right-hand set, one player can profitably deviate on a set of stages with positive density. To prove the opposite inclusion in (i) consider p in $\bigcap_i A^i$. We define a probability on histories by a product $\otimes p_n$; each player is told his own sequence of moves and is requested to follow it. p_n is a perturbation of p, converging to p as $n \to \infty$, such that each I-move has a positive probability and independently each recommended move to one player is announced with a positive probability to his opponent. It follows that a profitable deviation, say from the recommended s^i to t^i, will eventually be detected if $t^i \not\sim s^i$. To control the other deviations ($t^i \sim s^i$ but $t^i \not> s^i$), note first that, since the players can communicate through their moves, one can define a code, i.e. a mapping from histories to messages. The correlated device can then be used to generate, at infinitely many fixed stages, say n_k, random times m_k in (n_{k-1}, n_k): at the stages following n_k the players use a finite code to report the signal they got at time m_k. In this case also a deviation, if used with a positive density, will eventually occur at some stage m_k where moreover the opponent is playing a revealing move and hence will be detected. Obviously from then on the deviator is punished to his minimax. To obtain the same result for correlated equilibria, let the players use their moves as signals to generate themselves the random times m_k [see Sorin (1990)].

Finally, the last inclusion in (ii) follows from the next result. $\square$

Theorem 5.2.2 [Lehrer (1989)]. $IE_x = \bigcap_i Co\ g(B^i) \cap IR(=ICE_x)$.

Proof. It is easy to see that $Co\ g(B^i) = g(A^i)$ and hence a first inclusion by part (ii) of the previous theorem. To obtain the other direction let us approximate the reference payoff by playing on larger and larger blocks M_k, cycles consisting of extreme points in B^i [if $k \equiv i$ (mod 2)]. On each block, alternatively, one of the players is then playing a sequence of pure moves; thus a procedure like in the previous proof can be used. $\square$

A simpler framework in which the results can be extended to more than two players is the following: each action set S^i is equipped with a partition $\mathbf{S}^i$ and each player is informed only about the elements of the partitions to which the other players' actions belong. Note that in this case the signal received by a player is independent of his identity and of his own move. The above sets B^i can now be written as

$$C^i = \{x \in X: g^i(x) \geqslant g(x^{-i}, y^i) \text{ for all } y^i \text{ with } y^i = x^i\}$$

where x^i is the probability induced by x^i on $\mathbf{S}^i$.

Theorem 5.2.3 [Lehrer (1990)]. (i) $E_x = Co\ g(\bigcap_i C^i) \cap IR$. (ii) $IE_x = \bigcap_i Co\ g(C^i) \cap IR$

Proof. It already follows in this case that the two sets may differ. On the other hand, they increase as the partitions get finer (the deviations are easier to detect) leading to the Folk Theorem for discrete partitions – full monitoring.

For (ii), given a strategy profile σ, note that at each stage n, conditional to $h_n = (x_1, \ldots, x_{n-1})$, the choices of the players are independent and hence each player i can force the payoff to be in $g(C^i)$; hence the inclusion of IE_x in the right-hand set. On the other hand, as in Theorem 5.2.2, by playing alternately in large blocks to reach extreme points in C^1, then $C^2, \ldots$, one can construct the required equilibrium.

As for E_x, by convexity if a payoff does not belong to the right-hand set, there is for some i a set of stages with positive density where, with positive probability, the expected move profiles, conditioned on h_n, are not in C^i. Since h_n is common knowledge, player i can profitably deviate.

To obtain an equilibrium one constructs a sequence of increasing blocks on each of which the players are requested to play alternately the right strategies in $\bigcap_i C^i$ to approach the convex hull of the payoffs. These strategies may induce random signals so that the players use some statistical test to punish during the following block if some deviation appears. $\square$

For the extension to correlated equilibria, see Naudé (1990).

Finally a complete characterization is available when the signals include the payoffs:

Theorem 5.2.4 [Lehrer (1992b)]. *If $g^i(s) \neq g^i(t)$ implies $Q^i(s) \neq Q^i(t)$ for all i, s, t, then $E_\infty = lE_\infty = Co\ g(\bigcap_i C^i) \cap IR$.*

Proof. To obtain this result we first prove that the signalling structure implies $\bigcap_i Co\ g(B^i) \cap IR = Co\ g(\bigcap_i B^i) \cap IR$. Then one uses the structure of the extreme points of this set to construct equilibrium strategies. Basically, one player is required to play a pure strategy and can be monitored as in the proof of Theorem 5.2.1(i); the other player's behavior is controlled through some statistical test. $\square$

While it is clear that the above ideas will be useful in getting a general formula for E_∞, this one is still not available. For results in this direction, see Lehrer (1991, 1992b).

When dealing with more than two players new difficulties arise since a deviation, even when detected by one player, has first to be attributed to the actual deviator and then this fact has to become common knowledge among the non-deviators to induce a punishment.

For non-atomic games results have been obtained by Kaneko (1982), Dubey and Kaneko (1984) and Masso and Rosenthal (1989).

6. Approachability and strong equilibria

In this section we review the basic works that deal with other equilibrium concepts.

6.1. Blackwell's theorem

The following results, due to Blackwell (1956), are of fundamental importance in many fields of game theory, including repeated games and games with incomplete information. [A simple version will be presented here; for extensions see Mertens, Sorin and Zamir (1992).]

Consider a two-person game G_1 with finite action sets S and T and a random payoff function g on $S \times T$ with values in $\mathbb{R}^k$, having a finite second-order moment (write f for its expectation). We are looking for an extension of the minimax theorem to this framework in G_∞ (assuming full monitoring) and hence for conditions for a player to be able to approach a (closed) set C in $\mathbb{R}^k$ – namely to have a strategy such that the average payoff will remain, in expectation and with probability one, close to C, after a finite number of stages. C is excludable if the complement of some neighborhood of it is approachable by the opponent.

To state the result we introduce, for each mixed action x of player 1, $P(x) = Co\{f(x,t): t \in T\}$ and similarly $Q(y) = Co\{f(s, y): s \in S\}$ for each mixed action y of player 2.

Theorem 6.1.1 *Assume that, for each point $d \notin C$ there exists x such that if c is a closest point to d in C, the hyperplane orthogonal to $[cd]$ through c separates d from $P(x)$. Then C is approachable by player 1.*

An optimal strategy is to use at each stage n a mixed action having the above property, with $d = \bar{g}_{n-1}$.

Proof. This is proved by showing by induction that, if d_n denotes the distance from $\bar{g}_n$, the average payoff at stage n, to C, then $E(d_n^2)$ is bounded by some K/n. Furthermore, one constructs a positive supermartingale converging to zero, which majorizes d_n^2. $\square$

If the set C is convex we get a minimax theorem, due to the following:

Theorem 6.1.2. *A convex set C is either approachable or excludable; in the second case there exists y with $Q(y) \cap C = \emptyset$.*

Proof. Note that the following sketch of the proof shows that the result is actually stronger: if $Q(y) \cap C = \emptyset$ for some y, C is clearly excludable (by playing y i.i.d.). Otherwise, by looking at the game with real payoff $\langle d - c, f \rangle$, the minimax theorem implies that the condition for approachability in the previous theorem holds. $\square$

Blackwell also showed that Theorem 6.1.2 is true for any set in $\mathbb{R}$, but that there exist sets in $\mathbb{R}^2$ that are neither approachable nor excludable, leading to the problem of "weak approachability", recently solved by Vieille (1989) which showed that every set is asymptotically approachable or excludable by a family of strategies that depend on the length of the game. This is related to the definitions of $\lim v_n$ and v_∞ in zero-sum games (see Chapter 5 and the chapter on "stochastic games" in a forthcoming volume of this Handbook).

6.2. Strong equilibria

As seen previously, the Folk Theorem relates non-cooperative behavior (Nash equilibria) in G_∞ to cooperative concepts (feasible and i.r. payoffs) in the one-shot game. One may try to obtain a smaller cooperative set in G_1, such as the Core, and to investigate what its counterpart in G_∞ would be. This problem has been proposed and solved in Aumann (1959) using his notion of strong equilibrium, i.e., a strategy profile such that no coalition can profitably deviate.

Theorem 6.2.1. *The strong equilibrium payoffs in G_∞ coincide with the β-Core of G_1.*

Proof. First, if d is a payoff in the β-Core, there exists some (correlated) action achieving it that the players are requested to play in G_∞. Now for each subset $I \backslash J$ of potential deviators, there exists a correlated action σ^J of their opponent that prevents them from obtaining more than $d^{I \backslash J}$, and this will be used as a punishment in the case of deviation.

On the other hand, if d does not belong to the β-Core there exists a coalition J that possesses, given each history and each corresponding correlated move $I \backslash J$ tuple of its complement, a reply giving a better payoff to its members. $\square$

Note the similarity with the Folk Theorem, with the β-characteristic function here playing the role of the minimax (as opposed to the α-one and the maximin).

If one works with games with perfect information, one has the counterpart of the classical result regarding the sufficiency of pure strategies:

Theorem 6.2.2 [Aumann (1961)]. *If G_1 has perfect information the strong equilibria of G_∞ can be obtained with pure strategies.*

Proof. The result, based on the convexity of the β-characteristic function and on Zermelo's theorem, emphasizes again the relationship between repetition and convexity. $\square$

Finally, Mertens (1980) uses Blackwell's theorem to obtain the convexity and superadditivity of the β-characteristic function of G_1 by proving that it coincides with the α-characteristic function (and also the β-characteristic function) of G_∞.

7. Bounded rationality and repetition

As we have already pointed out, repetition alone, when finite, may not be enough to give rise to cooperation (i.e., Nash equilibria and a fortiori subgame perfect equilibria of the repeated game may not achieve the Pareto boundary). On the other hand, empirical data as well as experiments have shown that some cooperation may occur in this context [for a comprehensive analysis, see Axelrod (1984)].

We will review here some models that are consistent with this phenomenon. Most of the discussion below will focus on the Prisoner's Dilemma but can be easily extended to any finite game.

7.1. Approximate rationality

7.1.1. ε-equilibria

The intuitive idea behind this concept is that deviations that induce a small gain can be ignored. More precisely, σ will be an ε-equilibrium in the repeated game if, given any history (or any history consistent with σ), no deviation will be more than ε-profitable in the remaining game [see Radner (1980, 1986b)]. Consider the Prisoner's Dilemma (cf. Subsection 2.3):

Theorem 7.1. $\forall \varepsilon > 0, \forall \delta > 0, \exists N$ *such that for all* $n \geq N$ *there exists an* ε-*equilibrium in* G_n *inducing a payoff within* δ *of the Pareto point* $(3, 3)$.

Proof. Define σ as playing cooperatively until the last N_0 stages (with $N_0 \geq 1/\varepsilon$), where both players defect. Moreover, each player defects forever as soon as the other does so once. It is easy to see that any defection will induce an (average) gain less than ε, and hence the result for N large enough. $\square$

The above view implicitly contains some approximate rationality in the behavior of the players (they neglect small mistakes).

7.1.2. Lack of common knowledge

This approach deals with games where there is lack of common knowledge on some specific data (strategy or payoff), but common knowledge of this uncertainty. Then even if all players know the true data, the outcome may differ from the usual framework by a contamination effect – each player considers the information that the others may have.

The following analysis of repeated games is due to Neyman (1989). Consider again the finitely repeated Prisoner's Dilemma and assume that the length of the game is a random variable whose law P is common knowledge among the players. (We consider here a closed model, including common knowledge of rationality.) If P is the point mass at n we obtain G_n and "$E_n = \{1, 1\}$" is common knowledge. On the other hand, for any λ there exists P_λ such that the corresponding game is G_λ if the players get no information on the actual length of the game. Consider now non-symmetric situations and hence a general information scheme, i.e. a correlation device with a mapping $\omega \mapsto n(\omega)$ corresponding to the length of the game at ω.

Recall that an event A is of mutual knowledge of order k [say $mk(k)$] at ω if $K^{i_0} \circ \cdots \circ K^{i_k}(\omega) \subset A$, for all sequences $i_0, \ldots, i_k$, where K^i is the knowledge operator of player i (for simplicity, assume Ω is countable and then $K^i(B) = \cap \{C: B \subset C, C$ is $\mathscr{A}^i$-measurable$\}$; hence K^i is independent of P). Thus $mk(0)$ is public knowledge and $mk(\infty)$ common knowledge.

It is easy to see that at any ω where "$n(\omega)$" is $mk(k)$, $(1, 1)$ will be played during the last $k + 1$ stages [and this fact is even $mk(0)$], but Neyman has constructed an example where even if $n(\omega) = n$ is $mk(k)$ at ω, cooperation can occur during $n - k - 1$ stages, so that even with large k, the payoff converges to Pareto as $n \mapsto \infty$.

The inductive hierarchy of K^i at ω will eventually reach games with length larger than $n(\omega)$, where the strategy of the opponent justifies the initial sequence of cooperative moves.

Thus, replacing a closed model with common knowledge by a local one with large mutual knowledge leads to a much richer and very promising framework.

7.2. Restricted strategies

Another approach, initiated by Aumann, Kurz and Cave [see Aumann (1981)], requests the players to use subclasses of "simple" strategies, as in the next two subsections.

7.2.1. Finite automata

In this model the players are required to use strategies that can be implemented by finite automata. The formal description is as follows: A finite automaton (say for player i) is defined by a finite set of states K^i and two mappings, α from $K^i \times S^{-i}$ to K^i and β from K^i to S^i. α models the way the internal memory or state is updated as a function of the old memory and of the previous moves of the opponents. β defines the move of the player as a function of his internal state. Note that given the state and β, the action of i is known, so it is not necessary to define α as a function of S^i.

To represent the play induced by an automaton, we need in addition to specify the initial state k_0^i. Then the actions are constructed inductively by $\alpha(k_0^i)$, $\alpha(\beta(k_0^i, s_1^{-i})) = \alpha(k_1^i)$, ...

Games where both players are using automata have been introduced by Neyman (1985) and Rubinstein (1986).

Define the size of an automaton as the cardinality of its set of states and denote by $G(\kappa)$ the game where each player i is using as pure strategies automata of size less than κ^i.

Consider again the n-stage Prisoner's Dilemma. It is straightforward to check that given Tit for Tat (start with the the cooperative move and then at each following stage use the move used by the opponent at the previous stage) for both players, the only profitable deviation is to defect at the last stage. Now if $\kappa^i < n$, none of the players can "count" until the last stage, so if the opponent plays stationary, any move actually played at the last stage has to be played before then. It follows that for $2 \leqslant \kappa^i < n$, Tit for Tat is an equilibrium in G_n. Actually a much stronger result is available:

Theorem 7.2.1 [Neyman (1985)]. *For each integer m, $\exists N$ such that $n \geqslant N$ and $n^{1/m} \leqslant \kappa^i \leqslant n^m$ implies the existence of a Nash equilibrium in $G_n(\kappa^1, \kappa^2)$ with payoff greater than $3 - 1/m$ for each player.*

Proof. Especially in large games, even if the memory of the players is much larger than the length of the game (namely polynomial), Pareto optimality is almost achievable.

The idea of the proof relies on the observation that the cardinality of the set of histories is an exponential function of the length of the game. It is now possible to "fill" all the memory states by requiring both players to remember "small" histories, i.e. by answering in a prespecified way after such histories (otherwise the opponent defects for ever) and then by playing cooperatively during the remaining stages. Note that no internal state will be available to count the stages and that cooperative play arises during most of the game. $\square$

It is easy to see that in this framework an analog of Theorem 2.3 is available.

Similar results using Turing machines have been obtained by Megiddo and Widgerson (1986); see also Zemel (1989).

The model introduced in Rubinstein (1986) is different and we shall discuss the related version of Abreu and Rubinstein (1988). Both players are required to use finite automata (and no mixture is allowed) but there is no fixed bound on the memory. The main change is in the preference function, which is strictly increasing in the payoff and strictly decreasing in the size [in Rubinstein (1986) some lexicographic order is used]. A complete structure of the corresponding set of equilibria is then obtained with the following striking aspect: $\kappa^1 = \kappa^2$; moreover, during the cycle induced by the automata each state is used only once; and finally both players change their moves simultaneously. In particular, this implies that in 2×2 two-person games the equilibrium payoffs have to lie on the "diagonals".

Considering now two-person, zero-sum games, an interesting question is to determine the worth of having a memory much larger than the memory of the other player: note that the payoff in $G_x(\kappa^1, \kappa^2)$ is well defined, hence also its value $V(\kappa', \kappa^2)$. Denote by V the value of the original G_1 and by $\bar{V}$ the minimax in pure strategies. This problem has been solved by Ben Porath (1986):

Theorem 7.2.2. *For any polynomial P, $\lim_{\kappa^2 \to \infty} V(P(\kappa^2), \kappa^2) = V$. There exists some exponential function Ψ such that $\lim_{\kappa^2 \to \infty} V(\Psi(\kappa^2), \kappa^2) = \bar{V}$.*

Proof. The second part is not difficult to prove, player 1 can identify player 2's automaton within $\Psi(\kappa^2)$ stages.

For the first part, player 2 uses an optimal strategy in the one-shot game to generate κ^2 random moves and then follows the corresponding distribution to choose an automaton generating these moves. The key point is, using large deviation tools, to show that the probability, with this procedure, of producing a sequence of κ^2 pairs of moves biased by more than ε is some exponential function, ψ, of $-\kappa^2 \cdot \varepsilon^2$. Since player 1 can have at most κ^1 different behaviors, the average payoff will be greater than $V + \varepsilon$ with a probability less than $P(\kappa^2)\psi(-\kappa^2 \cdot \varepsilon^2)$. □

7.2.2. Strategies with bounded recall

Another way to approach bounded rationality is to assume that players have bounded recall. Two classes of strategies can be introduced according to the following definitions: σ^i is of I- (resp. II)-bounded recall (BR) of size k if, for all histories h, $\sigma^i(h)$ depends only upon the last k components of h (resp. the last k moves of player $-i$).

It is easy to see that Tit for Tat can be implemented by a II-bounded recall strategy with $k = 1$; to punish forever after a deviation can be reached by a I-BR but not by a II-BR, and to punish forever after two deviations cannot be achieved with BR strategies (if the first deviation occurred a long time ago, the player will not remember it). Note nevertheless that with II-BR strategies the players can maintain the average frequency of deviations as low as required.

Using I-BR strategies Lehrer (1988) proves a result similar to Theorem 7.2.2

by using tools from information theory. (Note that in both cases player 1 does not need to know the moves of player 2.)

This area is currently very active and new results include the study of the complexity of a strategy and its relation with the size of an equivalent automaton [Kalai and Stanford (1988)], an analog of Theorems 3.1 and 3.2 in pure strategies for finite automata [Ben Porath and Peleg (1987)], and the works of Ben Porath, Gilboa, Kalai, Megiddo, Samet, Stearns and others on complexity. For a recent survey, see Kalai (1990).

To end these two subsections one should also mention the work of Smale (1980) on the Prisonner's Dilemma, in which the players are restricted to strategies where the actions at each stage depend continuously on some vector-valued parameter. The analysis is then performed in relation to dynamical systems.

7.3. Pareto optimality and perturbed games

The previous results, as well as sections 2 and 3, have shown that under quite general conditions a kind of Folk Theorem emerges; rationality and repetition enables cooperation. Note nevertheless that the previous procedures lead to a huge set of equilibrium payoffs (including all one-shot Nash equilibrium payoffs and even the threat point V). A natural and serious question was then to ask under which conditions would long-term interaction and utility maximizing behavior lead to cooperation; in other words, whether we would necessarily achieve Pareto points as equilibrium payoffs.

It is clear from the previous results that repetition is necessary and that complete rationality or bounded rationality alone would not be sufficient. In fact, one more ingredient – perturbation or uncertainty – is needed. Note that a similar approach was initiated by Selten (1975) in his work on perfect equilibria.

A first result in this direction was obtained in a very stimulating paper by Kreps, Milgrom, Roberts and Wilson (1982). Consider the finitely repeated Prisoner's Dilemma and assume that with some arbitrarily small but positive probability one of the players is a kind of automaton: he always uses Tit for Tat rather than maximizing. Then for sufficiently long games all the sequential equilibrium payoffs will be close to the cooperative outcome. The proof relies in particular on the following two facts: first, if the equilibrium strategies were non-cooperative, the perturbed player may play Tit for Tat thus pretending to be the automaton and thereby convincing his opponent that this is in fact the case; second, Tit for Tat induces payoffs that are close to the diagonal.

These suggestive and important ideas will be needed when trying to extend this result by dropping some of the conditions. The above result in fact depends crucially on Tit for Tat (inducing itself almost the cooperative outcome as the best reply) being the only perturbation. More precisely a result of Fudenberg and Maskin (1986a) indicates that by choosing the perturbation in an adequate way the set of sequential equilibrium payoffs of a sufficiently long but finitely repeated game would approach any prespecified payoff. Now if all perturbations are allowed, each of the players may pretend to be a different automaton, advantageous from his own point of view.

One is thus lead to consider two-person games with common interest: one payoff strongly Pareto dominates all the others. Assume then that each player's strategy is ε-perturbed by some probability distribution having as support the set of II-BR strategies of some size k. Then the associated repeated game possesses equilibria in pure strategies and all the corresponding payoffs are close to the cooperative (Pareto) outcome $P(G)$. Formally, if pE_n^ε(resp. pE_λ^ε) denotes the set of pure equilibria payoffs in the n-stage (resp. λ-discounted) perturbed game, one has:

Theorem 7.3 [Aumann and Sorin (1989)]. $\lim_{\varepsilon \to 0} \lim_{n \to \infty} pE_n^\varepsilon = \lim_{\varepsilon \to 0} \lim_{\lambda \to 0} pE_\lambda^\varepsilon = P(G)$.

Proof. To prove the existence of a pure equilibrium, one considers Pareto points in the payoff space generated by pure strategies in the perturbed game. One then shows that these are sustained by equilibrium strategies.

Now assuming the equilibrium to be not optimal, one player could deviate and mimic his best BR perturbation. Note that the corresponding history has positive probability under the initial strategies. Moreover, for n large enough (or λ small enough) a best reply on histories inconsistent with the "main" strategy is to identify the BR strategy used and then to maximize against it. For this to hold it is crucial to use II-BR perturbations: the moves used during this identification phase will eventually be forgiven and hence no punishment forever can arise. Finally, the game being with common interest a high payoff for one player implies the same for the other so that the above procedure would lead to a payoff close to the cooperative outcome; hence the contradiction. □

The crucial properties of the set of perturbations used in the proof are: (1) identifiability (each player has a strategy such that, after finitely many stages he can predict the behavior of his opponent, if this opponent is in the perturbed mode); (2) the asymptotic payoff corresponding to a best reply to a perturbation is history independent. [For example, irreducible automata could be used; see Gilboa and Samet (1989).]

The extension to more than two players requires new tools since, even with bounded recall, two players can build everlasting events (e.g. punish during two stages if the other did so at the previous stage).

To avoid non-Pareto mixed equilibria one has to ask for some kind of perfection (or equivalently more perturbation) to avoid events of common knowledge of rationality (i.e. histories in which the probability of facing an opponent who is in the perturbed mode is 0 and common knowledge).

More recently, similar results, when a long-run player can build a reputation leading to Pareto payoffs against a sequence of short-run opponents, have been obtained by Fudenberg and Levine (1989a).

8. Concluding remarks

Before ending let us mention a connected field, multimove games, where similar features (especially the recursive structure) can be observed (and in fact were sometimes analyzed previously in specific examples). In this class of games the strategy sets have the same structure as in repeated games but the payoff is defined only on the set of plays and does not necessarily come from a stage payoff. A nice sampling can be founded in *Contributions to the Theory of Games, Vol. III* [Dresher, Tucker and Wolfe (1957)], and deals mainly with two-person games.

A game with two-move information lag was extensively studied by Dubins (1957), Karlin (1957), Ferguson (1967) and others, introducing new ideas and tools. The case with three-move information lag is still open. A general formulation and basic properties of games with information lag can be found in Scarf and Shapley (1957). A deep analysis of games of survival (or ruin) in the general case can be found in Milnor and Shapley (1957), using some related works of Everett (1957) on "recursive games". [For some results in the non-zero-sum case and ideas of the difficulties there, see Rosenthal and Rubinstein (1984).]

The properties of multimove games with perfect information are studied in Chapter 3 of this Handbook. The extension of those to general games seems very difficult [see, for example the very elegant proof of Blackwell (1969) for $\mathcal{G}_\delta$ games] and many problems are still open.

To conclude, we make two observations.

The first is that it is quite difficult to draw a well-defined frontier for the field of repeated games. Games with random payoffs are related to stochastic games; games with partial monitoring, as well as perturbed games, are related to games with incomplete information; sequential bargaining problems and games with multiple opponents are very close, ... To get a full overview of the field the reader should also consult Chapters 5, 6 and 7, and the chapter on 'stochastic games' in a forthcoming volume of this Handbook.

The second comment is that not only has the domain been very active in the last twenty years but that it is still extremely attractive. The numerous recent ideas and results allow us to unify the field and a global approach seems conceivable [see the nice survey of Mertens (1987)]. Moreover, many concepts that are now of fundamental importance in other areas originate from repeated games problems (like selection of equilibria, plans, signals and threats, approachability, reputation, bounded complexity, and so on). In particular, the applications to economics (see, for example, Chapters 7, 8, 9, 10 and 11 in this Handbook) as well as to biology (see the chapter on 'biological games' in a forthcoming volume of this Handbook) have been very successful.

Bibliography

Abreu, D. (1986) 'Extremal equilibria of oligopolistic supergames', *Journal of Economic Theory*, 39: 191–225.

Abreu, D. (1988) 'On theory of infinitely repeated games with discounting', *Econometrica*, 56: 383–396.

Abreu, D. and A. Rubinstein (1988) 'The structure of Nash equilibria in repeated games with finite automata', *Econometrica*, 56: 1259–1282.

Abreu, D., D. Pearce and E. Stacchetti (1986) 'Optimal cartel equilibria with imperfect monitoring', *Journal of Economic Theory*, 39: 251–269.

Abreu, D., D. Pearce and E. Stacchetti (1990) 'Toward a theory of discounted repeated games with imperfect monitoring', *Econometrica*, 58: 1041–1063.

Aumann, R.J. (1959) 'Acceptable points in general cooperative *n*-person games', in: A.W. Tucker and R. Luce, eds., *Contributions to the theory of games*, Vol. IV, A.M.S. 40. Princeton: Princeton University Press, pp. 287–324.

Aumann, R.J. (1960) 'Acceptable points in games of perfect information', *Pacific Journal of Mathematics*, 10: 381–387.

Aumann, R.J. (1961) 'The core of a cooperative game without side payments', *Transactions of the American Mathematical Society* 98: 539–552.

Aumann, R.J. (1967) 'A survey of cooperative games without side payments', in: M. Shubik, ed., *Essays in mathematical economics in honor of Oskar Morgenstern*. Princeton: Princeton University Press, pp. 3–27.

Aumann, R.J. (1981) 'Survey of repeated games', *Essays in game theory and mathematical economics in honor of Oskar Morgenstern*. Mannheim: Bibliographisches Institüt, pp. 11–42.

Aumann, R.J. (1986) 'Repeated games', in: G.R. Feiwel, ed., *Issues in contemporary microeconomics and welfare*. London: Macmillan, pp. 209–242.

Aumann, R.J. and L.S. Shapley (1976) 'Long-term competition – A game theoretic analysis', preprint.

Aumann, R.J. and S. Sorin (1989) 'Cooperation and bounded recall', *Games and Economic Behavior*, 1: 5–39.

Aumann, R.J., M. Maschler and R. Stearns (1968) 'Repeated games of incomplete information: An approach to the non-zero sum case', *Report to the U.S. A.C.D.A. ST-143*, Chapter IV, pp. 117–216, prepared by Mathematica.

Axelrod, R. (1984) *The evolution of cooperation*. New York: Basic Books.

Benoit, J.P. and V. Krishna (1985) 'Finitely repeated games', *Econometrica*, 53: 905–922.

Benoit, J.P. and V. Krishna (1987) 'Nash equilibria of finitely repeated games', *International Journal of Game Theory*, 16: 197–204.

Ben-Porath, E. (1986) 'Repeated games with finite automata', preprint.

Ben-Porath, E. and B. Peleg (1987) 'On the folk theorem and finite automata', preprint.

Blackwell, D. (1956) 'An analog of the minimax theorem for vector payoff's', *Pacific Journal of Mathematics*, 6: 1–8.

Blackwell, D. (1969) 'Infinite G_δ games with imperfect information', *Applicationes Mathematicae* X: 99–101.

Cave, J. (1987) 'Equilibrium and perfection in discounted supergames', *International Journal of Game Theory*, 16: 15–41.

Dresher, M., A.W. Tucker and P. Wolfe, eds. (1957) *Contributions to the theory of games*, Vol. III, A.M.S. 39. Princeton: Princeton University Press.

Dubey, P. and M. Kaneko (1984) 'Information patterns and Nash equilibria in extensive games: I', *Mathematical Social Sciences*, 8: 11–139.

Dubins, L.E. (1957) 'A discrete evasion game', in: M. Dresher, A.W. Tucker and P. Wolfe, eds., *Contributions to the theory of games III*, A.M.S. 39. Princeton: Princeton University Press, pp. 231-255.

Everett, H. (1957) 'Recursive games', in: M. Dresher, A.W. Tucker and P. Wolfe, eds., *Contributions to the theory of games III*, A.M.S. 39. Princeton: Princeton University Press, pp. 47–78.

Ferguson, T.S. (1967) 'On discrete evasion games with a two-move information lag', *Proceedings of the Fifth Berkeley Symposium on Mathematical Statistics and Probability*, Vol. I: pp. 453–462. Berkeley U.P.

Forges, F., (1986) 'An approach to communication equilibria', *Econometrica*, 54: 1375-1385.

Forges, F., J.-F. Mertens and A. Neyman (1986) 'A counterexample to the folk theorem with discounting', *Economics Letters*, 20: 7.

Friedman, J. (1971) 'A noncooperative equilibrium for supergames', *Review of Economic Studies*, 38: 1–12.

Friedman, J. (1985) 'Cooperative equilibria in finite horizon noncooperative supergames', *Journal of Economic Theory*, 35: 390–398.

Fudenberg, D., D. Kreps and E. Maskin (1990) 'Repeated games with long-run and short-run players', *Review of Economic Studies*, 57: 555–573.

Fudenberg, D. and D. Levine (1989a) 'Reputation and equilibrium selection in games with a patient player', *Econometrica*, 57: 759–778.

Fudenberg, D. and D. Levine (1989b) 'Equilibrium payoffs with long-run and short-run players and imperfect public information', preprint.

Fudenberg, D. and D. Levine (1991) 'An approximate folk theorem with imperfect private information', *Journal of Economic Theory*, 54: 26–47.

Fudenberg, D. and E. Maskin (1986a) 'The folk theorem in repeated games with discounting and with incomplete information', *Econometrica*, 54: 533–554.

Fudenberg, D. and E. Maskin (1986b) 'Discounted repeated games with unobservable actions I: One-sided moral hazard', preprint.

Fudenberg, D. and E. Maskin (1991) 'On the dispensability of public randomizations in discounted repeated games', *Journal of Economic Theory*, 53: 428–438.

Fudenberg, D., D. Levine and E. Maskin (1989) 'The folk theorem with imperfect public information', preprint.

Gilboa, I. and D. Samet (1989) 'Bounded versus unbounded rationality: The tyranny of the weak', *Games and Economic Behavior*, 1: 213–221.

Hart, S. (1979) 'Lecture notes on special topics in game theory', IMSSS-Economics, Stanford University.

Kalai, E. (1990) 'Bounded rationality and strategic complexity in repeated games', in: T. Ichiishi, A. Neyman and Y. Tauman, eds., *Game theory and applications*. New York: Academic Press, pp. 131–157.

Kalai, E. and W. Stanford (1988) 'Finite rationality and interpersonal complexity in repeated games', *Econometrica* 56: 397–410.

Kaneko, M. (1982) 'Some remarks on the folk theorem in game theory', *Mathematical Social Sciences*, 3: 281–290; also Erratum in 5, 233 (1983).

Karlin, S. (1957) 'An infinite move game with a lag', in: M. Dresher, A.W. Tucker and P. Wolfe, eds., *Contributions to the theory of games III*, A.M.S. 39. Princeton: Princeton University Press, pp. 255–272.

Kreps, D., P. Milgrom, J. Roberts and R. Wilson (1982) 'Rational cooperation in the finitely repeated prisoner's dilemma', *Journal of Economic Theory*, 27: 245–252.

Krishna, V. (1988) 'The folk theorems for repeated games', Proceedings of the NATO-ASI conference: Models of incomplete information and bounded rationality, Anacapri, 1987, to appear.

Kurz, M. (1978) 'Altruism as an outcome of social interaction', *American Economic Review*, 68: 216–222.

Lehrer, E. (1988) 'Repeated games with stationary bounded recall strategies', *Journal of Economic Theory*, 46: 130–144.

Lehrer, E. (1989) 'Lower equilibrium payoffs in two-player repeated games with non-observable actions', *International Journal of Game Theory*, 18: 57–89.

Lehrer, E. (1990) 'Nash equilibria of *n*-player repeated games with semistandard information', *International Journal of Game Theory*, 19: 191–217.

Lehrer, E. (1991) 'Internal correlation in repeated games', *International Journal of Game Theory*, 19: 431–456.

Lehrer, E. (1992a) 'Correlated equilibria in two-player repeated games with non-observable actions', *Mathematics of Operations Research*, 17: 175–199.

Lehrer, E. (1992b) 'Two-player repeated games with non-observable actions and observable payoffs', *Mathematics of Operations Research*, 200–224.

Lehrer, E. (1992c) 'On the equilibrium payoffs set of two player repeated games with imperfect monitoring', *International Journal of Game Theory*, 20: 211–226.

Masso J. and R. Rosenthal (1989) 'More on the "Anti-Folk Theorem"', *Journal of Mathematical Economics*, 18: 281–290.

Megiddo, N. and A. Widgerson (1986) 'On plays by means of computing machines', in: J.Y. Halpern, ed., *Theoretical aspects of reasoning about knowledge*. Morgan Kaufman Publishers, pp. 259–274.

Mertens, J.-F. (1980) 'A note on the characteristic function of supergames', *International Journal of Game Theory*, 9: 189–190.

Mertens, J.-F. (1987) 'Repeated Games', *Proceedings of the International Congress of Mathematicians, Berkeley 1986*. New York: American Mathematical Society, pp. 1528–1577.

Mertens, J.-F., S. Sorin and S. Zamir (1992) *Repeated games*, book to appear.

Milnor, J. and L.S. Shapley (1957) 'On games of survival', in: M. Dresher, A.W. Tucker and P. Wolfe, eds., *Contributions to the theory of games III*, A.M.S. 39. Princeton: Princeton University Press, pp. 15–45.

Myerson, R. (1986) 'Multistage games with communication', *Econometrica*, 54: 323–358.

Naudé, D. (1990) 'Correlated equilibria with semi-standard information', preprint.

Neyman, A. (1985) 'Bounded complexity justifies cooperation in the finitely repeated prisoner's dilemma', *Economics Letters*, 19: 227–229.

Neyman, A. (1988) 'Stochastic games', preprint.

Neyman, A. (1989) 'Games without common knowledge', preprint.

Radner, R. (1980) 'Collusive behavior in non-cooperative epsilon-equilibria in oligopolies with long but finite lives', *Journal of Economic Theory*, 22: 136–154.

Radner, R. (1981) 'Monitoring cooperative agreements in a repeated principal-agent relationship', *Econometrica*, 49: 1127–1147.

Radner, R. (1985) 'Repeated principal–agent games with discounting', *Econometrica*, 53: 1173–1198.

Radner, R. (1986a) 'Repeated partnership games with imperfect monitoring and no discounting', *Review of Economic Studies*, 53: 43–57.

Radner, R. (1986b) 'Can bounded rationality resolve the prisoner's dilemma', in: A. Mas-Colell and W. Hildenbrand, eds., *Essays in honor of Gérard Debreu*. Amsterdam: North-Holland, pp. 387–399.

Radner, R. (1986c) 'Repeated moral hazard with low discount rates', in: W. Heller, R. Starr and D. Starrett, eds., *Essays in honor of Kenneth J. Arrow*. Cambridge: Cambridge University Press, pp. 25–63.

Radner, R., R.B. Myerson and E. Maskin (1986) 'An example of a repeated partnership game with discounting and with uniformly inefficient equilibria', *Review of Economic Studies*, 53: 59–69.

Rosenthal, R. and A. Rubinstein (1984) 'Repeated two-player games with ruin', *International Journal of Game Theory*, 13: 155–177.

Rubinstein, A. (1976) 'Equilibrium in supergames', preprint.

Rubinstein, A. (1979a) 'Equilibrium in supergames with the overtaking criterion', *Journal of Economic Theory*, 21: 1–9.

Rubinstein, A. (1979b) 'Offenses that may have been committed by accident – an optimal policy of redistribution', in: S.J. Brams, A. Schotter and G. Schwödiauer, eds., *Applied game theory*, Berlin: Physica-Verlag, pp. 406–413.

Rubinstein, A. (1980) 'Strong perfect equilibrium in supergames', *International Journal of Game Theory*, 9: 1–12.

Rubinstein, A. (1986) 'Finite automata play the repeated prisoner's dilemma', *Journal of Economic Theory*, 39: 83–96.

Rubinstein, A. and M. Yaari (1983) 'Repeated insurance contracts and moral hazard', *Journal of Economic Theory*, 30: 74–97.

Samuelson, L. (1987) 'A note on uncertainty and cooperation in a finitely repeated prisoner's dilemma', *International Journal of Game Theory*, 16: 187–195.

Scarf, H. and L.S. Shapley (1957) 'Games with partial information', in: M. Dresher, A.W. Tucker and P. Wolfe, eds., *Contributions to the theory of games III*, A.M.S. 39. Princeton: Princeton University Press, pp. 213–229.

Selten, R. (1975) 'Reexamination of the perfectness concept for equilibrium points in extensive games', *International Journal of Game Theory*, 4: 25–55.

Smale, S. (1980) 'The prisoner's dilemma and dynamical systems associated to non-cooperative games', *Econometrica*, 48: 1617–1634.

Sorin, S. (1986a) 'On repeated games with complete information', *Mathematics of Operations Research*, 11, 147–160.

Sorin, S. (1986b) 'Asymptotic properties of a non-zero stochastic game', *International Journal of Game Theory*, 15: 101–107.

Sorin, S. (1988) 'Repeated games with bounded rationality', Proceedings of the NATO-ASI Conference: Models of incomplete information and bounded rationality, Anacapri, 1987, to appear.

Sorin, S. (1990) 'Supergames', in: T. Ichiishi, A. Neyman and Y. Tauman, eds., *Game theory and applications*. New York: Academic Press, pp. 46–63.

Vieille, N. (1989) 'Weak approachability', to appear in *Mathematics of Operations Research*.

Zemel, E. (1989) 'Small talk and cooperation: a note on bounded rationality', *Journal of Economic Theory*, 49: 1–9.

Chapter 5

REPEATED GAMES OF INCOMPLETE INFORMATION: ZERO-SUM

SHMUEL ZAMIR

The Hebrew University of Jerusalem

Contents

Handbook of Game Theory, Volume 1, Edited by R.J. Aumann and S. Hart

1. Introduction

This chapter and the next apply the framework of repeated games, developed in the previous chapter, to games of incomplete information. The aim of this combination is to analyze the strategic aspects of information: When and at what rate to reveal information? When and how should information be concealed? What resources should be allocated to acquiring information? Repeated games provide the natural paradigm for dealing with these dynamic aspects of information. The repetitions of the game serve as a signaling mechanism which is the channel through which information is transmitted from one period to another.

It may be appropriate at this point to clarify the relation of repeated incomplete information games to stochastic games, treated in a forthcoming volume of this Handbook. Both are dynamic models in which payoffs at each stage are determined by the state of nature (and the player's moves). However, in stochastic games the state of nature changes in time but is common knowledge to all players, while *in repeated games of incomplete information the state of nature is fixed but not known to all players*. What changes in time is each player's knowledge about the other players' past actions, which affects his beliefs about the (fixed) state of nature. But it should be mentioned that it is possible to provide a general model which has both stochastic games and repeated games of incomplete information as special cases [see Mertens, Sorin and Zamir (1993, ch. IV), henceforth MSZ].

An important feature of the analysis is that when treating any specific game one has to consider a whole *family of games*, parameterized by the prior distribution on the states of nature. This is so because the state of information, which is basically part of the initial data of the game, changes during the play of the repeated game.

Most of the work on repeated games of incomplete information was done for two-person, zero-sum games, which is also the scope of this chapter. This is not only because it is the simplest and most natural case to start with, but also because it captures the main problems and aspects of strategic transmission of information, which can therefore be studied "isolated" from the phenomena of cooperation, punishments, incentives, etc. Furthermore, the theory of non-zero-sum repeated games of incomplete information makes extensive use of the notion of punishment, which is based on the minmax value borrowed from the zero-sum case.

1.1. Illustrative examples

Before starting our formal representation let us look at a few examples illustrating some of the main issues of strategic aspects of information. We start with an example, studied very extensively by Aumann and Maschler (1966, 1967):

Example 1.1. Imagine two players I (the maximizer) and II (the minimizer) playing repeatedly a zero-sum game given by a 2×2 payoff matrix. This matrix is chosen (once and for all) at the beginning to be either G^1 or G^2 where

$$G^1 = \begin{pmatrix} 1 & 0 \\ 0 & 0 \end{pmatrix}, \qquad G^2 = \begin{pmatrix} 0 & 0 \\ 0 & 1 \end{pmatrix}.$$

Player I is told which game was chosen but player II is not; he only knows that it is either G^1 or G^2 with equal probabilities and that player I knows which one is it. After the matrix is chosen the players repeatedly do the following: player I chooses a row, player II chooses a column (simultaneously). A referee announces these moves and records the resulting payoff (according to the matrix chosen at the beginning). He *does not announce the payoffs* (though player I of course knows them, knowing the moves and the true matrix). The game consists of n such stages and we assume that n is very large. At the end of the nth stage player I receives from player II the total payoff recorded by the referee divided by n (to get an average payoff per stage in order to be able to compare games of different length).

So player I has the advantage of knowing the real state (the real payoff matrix). How should he play in this game?

A first possibility is to choose the dominating move in each state: always play Top if the game is G^1 and always play Bottom if it is G^2. Assuming that I announces this strategy (which we may as well assume by the minmax theorem), it is a *completely revealing* strategy since player II will find out which matrix has been chosen by observing whether player I is playing Top or Bottom. Having found this, he will then choose the appropriate column (Right in G^1 and Left in G^2) to pay only 0 from then on. Thus, a completely revealing strategy yields the informed player an average payoff of almost 0 (that is a total payoff of at most 1, at the first stage after which the matrix is revealed, thus an average of at most $1/n$).

Another possible strategy for player I is to play *completely non-revealing*, that is, ignoring his private information and playing as if he, just like player II, does not know the matrix chosen. This situation is equivalent to repeatedly playing the average matrix game:

$$D = \begin{pmatrix} 1/2 & 0 \\ 0 & 1/2 \end{pmatrix} .$$

This game (which may also be called the non-revealing game) has a value $1/4$ and player I can guarantee this value (in the original game) by always playing Top and Bottom with equal probabilities ($1/2$ each) *independently of what the true matrix is*.

So, strangely enough, in this specific example the informed player is better off not using his information than using it. As we shall see below, the completely non-revealing strategy is in fact the (asymptotically) optimal strategy for player I; in very long games he cannot guarantee significantly more than $1/4$ per stage.

Example 1.2. The second example has the same description as the first one except that the two possible matrices are now

$$G^1 = \begin{pmatrix} -1 & 0 \\ 0 & 0 \end{pmatrix} , \qquad G^2 = \begin{pmatrix} 0 & 0 \\ 0 & -1 \end{pmatrix} .$$

Following the line of discussion of the previous example, if player I uses his dominating move at each state, Bottom in G^1 and Top in G^2, he will guarantee a payoff of 0 at each stage and again this will be a completely revealing strategy. Unlike in the previous example, here this strategy is an optimal strategy for the informed player. This is readily seen without even checking other strategies: 0 is the highest payoff in both matrices. Just for comparison, the completely non-revealing strategy would yield the value of the non-revealing game

$$D = \begin{pmatrix} -1/2 & 0 \\ 0 & -1/2 \end{pmatrix} ,$$

which is $-1/4$.

Example 1.3. Consider again a game with the same description as the previous two examples, this time with the two possible matrices given by

$$G^1 = \begin{pmatrix} 4 & 0 & 2 \\ 4 & 0 & -2 \end{pmatrix} , \qquad G^2 = \begin{pmatrix} 0 & 4 & -2 \\ 0 & 4 & 2 \end{pmatrix} .$$

Playing the dominant rows (Top in G_1 and Bottom in G^2) is again a completely revealing strategy which leads to a long-run average payoff of (almost) 0 (the value of each of the matrices is 0). Playing completely non-revealing leads to the non-revealing game

$$D = \begin{pmatrix} 2 & 2 & 0 \\ 2 & 2 & 0 \end{pmatrix},$$

which has a value 0. So, both completely revealing and completely non-revealing strategies yield the informed player an average payoff of 0. Is there still another, more clever, way of using the information to guarantee more than 0? If there is, it must be a strategy which *partially reveals the information*. In fact such a strategy exists. Here is how an average payoff of 1 per stage can be guaranteed by the informed player.

Player I prepares two non-symmetric coins, both with sides (T, B). In coin C^1 the corresponding probabilities are $(3/4, 1/4)$ while in coin C^2 they are $(1/4, 3/4)$. Then he plays the following strategy: if the true matrix is G^k, $k = 1, 2$, flip coin C^k. Whichever coin was used, if the outcome is T play Top in all stages, if it is B, play Bottom in all stages.

To see what this strategy does, let us assume that even player II knows it. He will then, right after the first stage, know whether the outcome of the coin was T or B (by observing whether player I played Top or Bottom). He will not know which coin was flipped (since he does not know the state k). However, he can update his beliefs about the probability of each matrix in view of a given outcome of the coin. Using Bayes' formula we find

$$P(G^1 \mid T) = 3/4 \quad \text{and} \quad P(G^1 \mid B) = 1/4.$$

Given that player I is playing Top, the payoffs will be either according to the line $(4, 0, 2)$ (this with probability $3/4$) or according to the line $(0, 4, -2)$ (and this with probability $1/4$). The expected payoffs given Top are therefore (depending on the move of player II)

$$(3/4)(4, 0, 2) + (1/4)(0, 4, -2) = (3, 1, 1).$$

Similarly, given that player I is playing Bottom, the expected payoffs are

$$(1/4)(4, 0, -2) + (3/4)(0, 4, 2) = (1, 3, 1).$$

We conclude that in any event, and no matter what player II does, the conditional expected payoff is at least 1 per stage. Therefore the expected average payoff for player I is at least 1. We shall see below that this is the most player I can guarantee in this game. So the optimal strategy of the informed player in this example is partially to reveal his information.

Let us have a closer look at this strategy. In what sense is it partially revealing? Player I, when being in G^1, *will more likely* (namely with probability $3/4$) play Top and when being in G^2, he will more likely play Bottom. Therefore when Top is played it becomes more likely that the matrix is G^1, while when Bottom is played it becomes more likely that it is G^2. Generally, player I is giving player II information in the right direction, but it is not definite; player II will adjust his *beliefs* about the true matrix from $(1/2, 1/2)$ to either $(3/4, 1/4)$ or $(1/4, 3/4)$ and with probability $3/4$ this adjustment will be *in the right direction*, increasing the subjective probability for the true game. This idea of changing a player's beliefs by giving him a signal which is partially correlated with the true state is undoubtedly the heart of the theory of games with incomplete information.

There is one point we wish to add about the notion of revealing. In all three examples we discussed the informed player was *revealing* information whenever he was *using* it. However, in principle, and in fact in the general model which will be presented below, these are two distinct concepts. Using information means to play differently in two different informational states; for instance, in our first example player I was using his information when he was playing Top in G^1 and Bottom in G^2. Revealing information is changing the beliefs of the uninformed player. Clearly, when the move of the informed player is observed by the uninformed player – which we shall later call *the full monitoring case* – the two concepts are two expressions of the same thing; the only way to play non-revealing is to play the same way, independently of one's information, i.e. not to use the information. This was the case in all our examples. More generally, the move of the informed player need not be observable. The uninformed player receives some *signal* which is correlated in an arbitrary way with the move of the informed player. It may then well be that in order to play non-revealing, a player *has to use his information*. Similarly, he may be revealing his information by not using it.[1]

2. A general model

A repeated game of incomplete information consists of the following elements:
- A finite set I, the set of players.
- A finite set K, the set of states of nature.
- A probability distribution p on K, the *prior* probability distribution on the states of nature.
- For each $i \in I$, a partition K^i of K, the initial information of player i.
- For each $i \in I$ and $k \in K$, a finite set S_k^i, which is the same for all k in the same partition element of K^i. This is the set of moves available to player I at state k. By taking the Cartesian product $\Pi_k S_k^i$ we may assume, without loss of generality, that the sets of moves S^i are state independent. Let $S = \Pi_i S^i$.

---

[1]For examples, see MSZ (1993, ch. V, section 3.b).

- For each $k \in K$, a payoff function $G^k: S \to R^I$. That is $G^k(s)$ is the vector of payoffs to the players when they play moves s and the state is k.
- For each $i \in I$, a finite alphabet A^i, the set of *signals* to player i. Let $A = \Pi_i A^i$ and $\mathscr{A} = \Delta(A)$ be the set of probability distributions on A.
- A transition probability Q from $K \times S$ to $\mathscr{A}$, the signaling probability distribution [we use the notation $Q^k(s)$ for the image of (k, s)].

On the basis of these elements the repeated game (or supergame) is played in stages as follows. At stage 0 a chance move chooses an element $k \in K$ according to the probability distribution p. Each player i is informed of the element of K^i containing the chosen k. Then at each stage m ($m = 1, 2, \ldots$), each player i chooses $s_m^i \in S^i$, a vector of signals $a \in A$ is chosen according to the probability distribution $Q^k(s)$ and a^i is communicated to player i. This is the signal to player i at stage m.

Notice that, as mentioned in the Introduction, the state k is chosen at stage 0 and remains fixed for the rest of the game. This is in contrast to stochastic games (to be discussed in a forthcoming volume of this Handbook) in which the state may change along the play. Also note that the payoff g_m at stage m is not explicitly announced to the players. In general, on the basis of the signals he receives, a player will be able to deduce only partial information about his payoffs.

2.1. Classification

Games of incomplete information are usually classified according to the nature of the three important elements of the model, namely players and payoffs, prior information, and signaling structure.

(1) *Players and payoffs.* Here we have the usual categories of two-person and n-person games. Within the two-person games one has the zero-sum games treated in this chapter and the non-zero-sum games treated in Chapter 6 of this Handbook.

(2) *Prior information.* Within two-person games the main classification is games with incomplete information on *one side*, versus incomplete information on *two sides*. In the first class are games in which one player knows the state chosen at stage 0 (i.e. his prior information partition consists of the singletons in K) while the other player gets no direct information at all about it (i.e. his prior information partition consists of one element $\{K\}$). More general prior information may sometimes be reduced to this case, for example if one of the player's partition is a refinement of the other's partition, and the signaling distribution Q, as a function of k, is measurable with respect to the coarser partition.

(3) *Signaling structure.* The simplest and most manageable signaling struc-

ture is that of *full monitoring*. This is the case in which the moves at each stage
are the only observed signals by all players, that is $A^i = S$ for all i, and for all k
and all s, $Q^k(s)$ is a probability with mass 1 at $(s, \ldots, s)$. The next level of
generality is that of *state independent signals*. This is the case in which $Q^k(s)$ is
constant in k, and consequently the signals do not reveal any direct information
about the state but only about the moves. Hence the only way for a player to
get information about k is by deducing it from other players' moves, about
which he learns something through the signals he receives. There is no
established classification beyond that, although two other special classes will be
treated separately: the case in which the signals are the same for all players and
include full monitoring (and possibly more information), and the case in which
the signal either fully reveals the state to all players or is totally non-
informative.

3. Incomplete information on one side

In this section we consider repeated two-person, zero-sum games in which only
one player knows the actual state of nature. These games were first studied by
Aumann, Maschler and Stearns, who proved the main results. Later contributions are due to Kohlberg, Mertens and Zamir.

Since in this chapter we consider only two-person, zero-sum games it is
convenient to slightly modify our notation for this case by referring to the two
players as player I (the maximizer) and player II (the minimizer). Their sets of
pure actions (or moves) are S and T, respectively, and their corresponding
mixed moves are $X = \Delta(S)$ and $Y = \Delta(T)$. The payoff matrix (to I) at state
$k \in K$ is denoted by G^k with elements G^k_{st}. The notation for general signaling
will be introduced later.

3.1. General properties

We shall consider the games $\Gamma_n(p)$, $\Gamma_\lambda(p)$ and $\Gamma_\infty(p)$ which are defined with
the appropriate valuation of the payoffs sequence $(g_m)^\infty_{m=1} = (G^k(s_m))^\infty_{m=1}$.
Before defining and analyzing these we shall first establish some general
properties common to a large family of games with incomplete information on
one side.

The games considered in this section will all be zero-sum, two person games
of the following form: chance chooses an element k from the set K of states
(games) according to some $p \in \Delta(K)$. Player I is informed which k was chosen
but player II is not. Players I and II then, simultaneously, choose $\sigma^k \in \Sigma$ and
$\tau \in \mathcal{T}$, respectively, and finally $G^k(\sigma^k, \tau)$ is paid to player I by player II. The

sets Σ and $\mathcal{T}$ are convex sets of strategies, and the payoff functions $G^k(\sigma^k, \tau)$ are bilinear and uniformly bounded on $\Sigma \times \mathcal{T}$. We may think of k as the *type* of player I which is some private information known only to him and could attain various values in K. This is thus a game of *incomplete information on one side*, on the side of player II.

Even though the strategies in Σ and $\mathcal{T}$ will usually be strategies in some repeated game (finite or infinite), it is useful at this point to consider the above described game as a one-shot game in strategic form in which the strategies are $\sigma \in \Sigma^K$ and $\tau \in \mathcal{T}$, respectively, and the payoff function is $G^p(\sigma, \tau) = \Sigma_k \, p^k G^k(\sigma^k, \tau)$. Denote this game by $\Gamma(p)$.

Theorem 3.1. *The functions* $\bar{w}(p) = \inf_\tau \sup_\sigma G^p(\sigma, \tau)$ *and* $\underline{w}(p) = \sup_\sigma \inf_\tau G^p(\sigma, \tau)$ *are concave.*

Proof. The argument is the same for both functions. We will show it for $\bar{w}(p)$. Let $p = (p_e)_{e \in E}$ be finitely many points in $\Delta(K)$, and let $\alpha = (\alpha_e)_{e \in E}$ be a point in $\Delta(E)$ such that $\Sigma_{e \in E} \, \alpha_e p_e = p$. We claim that $\bar{w}(p) \geq \Sigma_{e \in E} \, \alpha_e \bar{w}(p_e)$. To see this, consider the following two-stage game: chance chooses $e \in E$ according to the probability distribution α, then $k \in K$ is chosen according to p_e, the players then choose $\sigma^k \in \Sigma$ and $\tau \in \mathcal{T}$, respectively, and the payoff is $G^k(\sigma^k, \tau)$. We consider two versions, in both of which player I is informed of everything (both e and k) while player II may or may not be informed of the value of e (in any case he is not informed of the value of k).

Now, if player II is informed of the outcome e the situation following the first lottery is equivalent to $\Gamma(p_e)$. Thus, the $\inf_\tau \sup_\sigma$ for the game in which player II is informed of the outcome of the first stage is $\Sigma_{e \in E} \, \alpha_e \bar{w}(p_e)$. This game is more favorable to II than the game in which he is not informed of the value of e, which is equivalent to $\Gamma(\Sigma_e \, \alpha_e p_e) = \Gamma(p)$. Therefore we have:

$$\bar{w}(p) \geq \sum_{e \in E} \alpha_e \bar{w}(p_e). \quad \square$$

Remark. Although this theorem is formulated for games with incomplete information on one side it has an important consequence for games with incomplete information on both sides. This is because we did not assume anything about the strategy set of player II. In a situation of incomplete information on both sides, when a pair of types, one for each player, is chosen at random and each player is informed of his type only, we can still think of player II as being "uninformed" (of the type of I) but with strategies consisting of choosing an action after observing a chance move (the chance move choosing his type). When doing this, we can use Theorem 3.1 to obtain the concavity of $\bar{w}(p)$ and $\underline{w}(p)$ in games with incomplete information on both sides, when p (the joint probability distribution on the pairs of types) is restricted to the subset of the simplex where player I's conditional probability on the state k, given his own type, is fixed.

The concavity of $\underline{w}(p)$ can also be proved constructively by means of the following useful proposition, which we shall refer to as the *splitting procedure*.

Proposition 3.2. *Let p and $(p_e)_{e \in E}$ be finitely many points in $\Delta(K)$, and let $\alpha = (\alpha_e)_{e \in E}$ be a point in $\Delta(E)$ such that $\Sigma_{e \in E} \alpha_e p_e = p$. Then there are vectors $(\mu^k)_{k \in K}$ in $\Delta(E)$ such that the probability distribution P on $K \times E$ obtained by the composition of p and $(\mu^k)_{k \in K}$ (that is $k \in K$ is chosen according to p and then $e \in E$ is chosen according to μ^k) satisfies*

$$P(\cdot \mid e) = p_e \quad and \quad P(e) = \alpha_e , \quad for\ all\ e \in E .$$

Proof. In fact, if $p^k = 0$, μ^k can be chosen arbitrarily in $\Delta(E)$. If $p^k > 0$, μ_k is given by $\mu^k(e) = \alpha_e p_e^k / p^k$. □

Let player I use the above described lottery and then guarantee $\underline{w}(p_e)$ (up to ϵ). In this way he guarantees $\Sigma_e \alpha_e \underline{w}(p_e)$, even if player II were informed of the outcome of the lottery. So $\underline{w}(p)$ is certainly not smaller than that. Consequently the function $\underline{w}(p)$ is concave.

The idea of splitting is the following. Recall that the informed player, I, knows the state k while the uninformed player, II, knows only the probability distribution p according to which it was chosen. Player I can design a state dependent lottery so that if player II observes only the outcome e of the lottery, his conditional distribution (i.e. his new "beliefs") on the states will be p_e. Let us illustrate this using Example 1.3. At $p = (1/2, 1/2)$ player I wants to "split" the beliefs of II to become $p_1 = (3/4, 1/4)$ or $p_2 = (1/4, 3/4)$ (note that $p = 1/2 p_1 + 1/2 p_2$.) He does this by the state dependent lottery on $\{T, B\}$: $\mu^1 = (3/4, 1/4)$ and $\mu^2 = (1/4, 3/4)$.

Another general property worth mentioning is the Lipschitz property of all functions of interest (such as the value functions of the discounted game, the finitely repeated game, etc.), in particular $\bar{w}(p)$. This follows from the uniform boundedness of the payoffs, and hence is valid for all repeated games discussed in this chapter.

Theorem 3.3. *The function $\bar{w}(p)$ is Lipschitz with constant C (the bound on the absolute value of the payoffs).*

Proof. Indeed, the payoff functions of two games $\Gamma(p_1)$ and $\Gamma(p_2)$ differ by at most $C\|p_1 - p_2\|_1$. □

Let us turn now to the special structure of the repeated game. Given the basic data $(K, p, (G^k)_{k \in K}, A, B, Q)$ (here A and B are the signal sets of I and II, respectively), any play of the game yields a payoff sequence $(g_m)_{m=1}^{\infty} = (G^k(s_m t_m))_{m=1}^{\infty}$. On the basis of various valuations of the payoff sequence, we shall consider the following games (as usual, E denotes expectation with respect to the probability induced by p, Q, and the strategies).

The *n-stage game*, $\Gamma_n(p)$, is the game in which the payoff is $\bar{\gamma}_n = E(\bar{g}_n) = E((1/n) \sum_{m=1}^{n} g_m)$. Its value is denoted by $v_\lambda(p)$.

The *λ-discounted game*, $\Gamma_\lambda(p)$ (for $\lambda \in (0, 1]$), is the game in which the payoff is $E(\sum_{m=1}^{\infty} \lambda(1-\lambda)^{m-1} g_m)$. Its value is denoted by $v_\lambda(p)$.

The values $v_n(p)$ and $v_\lambda(p)$ clearly exist and are Lipschitz by Theorem 3.3.

As in the previous section, the infinite game $\Gamma_\infty(p)$ is the game in which the payoff is some limit of $\bar{g}_n$ such as lim sup, lim inf or, more generally, any Banach limit $\mathcal{L}$. It turns out that the results in this chapter are independent of the particular limit function chosen as a payoff. The definition of the value for $\Gamma_\infty(p)$ is based on a notion of guaranteeing.

Definition 3.4. (i) Player I can *guarantee* α if

$$\forall \epsilon > 0, \exists \sigma_\epsilon, \exists N_\epsilon, \text{ such that } \bar{\gamma}_n(\sigma_\epsilon, \tau) \geq \alpha - \epsilon, \forall \tau, \forall n \geq N_\epsilon .$$

(ii) Player II can *defend* α if

$$\forall \epsilon > 0, \forall \sigma, \exists \tau, \exists N, \text{ such that } \bar{\gamma}_n(\sigma, \tau) \leq \alpha + \epsilon, \forall n \geq N .$$

$\underline{v}(p)$ is the *maxmin* of $\Gamma_\infty(p)$ if it can be guaranteed by player I and can be defended by player II. In this case a strategy σ_ϵ associated with $\underline{v}(p)$ is called *ϵ-optimal*. The *minmax* $\bar{v}(p)$ and ϵ-optimal strategies for player II are defined in a dual way. A strategy is optimal if it is ϵ-optimal for all ϵ.

The game $\Gamma_\infty(p)$ has a *value* $v_\infty(p)$ iff $\underline{v}(p) = \bar{v}(p) = v_\infty(p)$. It follows readily from these definitions that:

Proposition 3.5. *If $\Gamma_\infty(p)$ has a value $v_\infty(p)$, then both $\lim_{n \to \infty} v_n(p)$ and $\lim_{\lambda \to 0} v_\lambda(p)$ exist and they are both equal to $v_\infty(p)$. An ϵ-optimal strategy in $\Gamma_\infty(p)$ is an ϵ-optimal strategy in all $\Gamma_n(p)$ with sufficiently large n and in all $\Gamma_\lambda(p)$ with sufficiently small λ.*

By the same argument used in Theorem 3.1 or by using the splitting procedure of Proposition 3.2 we have:

Proposition 3.6. *In any version of the repeated game ($\Gamma_n(p)$, $\Gamma_\lambda(p)$ or $\Gamma_\infty(p)$), if player I can guarantee $f(p)$ then he can also guarantee Cav $f(p)$.*

Here Cav f is the (pointwise) smallest concave function g on $\Delta(K)$ satisfying $g(p) \geq f(p)$, $\forall p \in \Delta(K)$. We now have:

Theorem 3.7. $v_n(p)$ and $v_\lambda(p)$ converge uniformly (as $n \to \infty$ and $\lambda \to 0$, respectively) to the same limit which can be defended by player II in $\Gamma_\infty(p)$.

Proof. Let τ_n be an ϵ-optimal strategy of player II in $\Gamma_n(p)$ with $\epsilon = 1/n$ and let $v_{n_i}(p)$ converge to $\lim \inf_{n \to \infty} v_n(p)$. Now let player II play τ_{n_i} for n_{i+1} times $(i = 1, 2. \ldots)$ – thus for $n_i \cdot n_{i+1}$ periods – before increasing i by 1. By this strategy player II guarantees $\lim \inf_{n \to \infty} v_n(p)$. Since player II certainly cannot guarantee less than $\lim \sup_{n \to \infty} v_n(p)$, it follows that $v_n(p)$ converges (uniformly by Theorem 3.3).

As for the convergence of v_λ, since clearly player II cannot guarantee less than $\lim \sup_{\lambda \to 0} v_\lambda(p)$, the above described strategy of player II proves that

$$\lim_{\lambda \to 0} \sup v_\lambda(p) \le \lim_{n \to \infty} v_n(p) .$$

To complete the proof we shall prove that $\lim_{n \to \infty} v_n(p) \le \lim \inf_{\lambda \to 0} v_\lambda(p)$ by showing that $\lim_{n \to \infty} v_n(p) \le v_\lambda(p)$ for any $\lambda > 0$. In fact, given $\lambda > 0$ let τ_λ be an optimal strategy of player II in the λ-discounted game and consider the following strategy (for player II): start playing τ_λ and at each stage restart τ_λ with probability λ and with probability $(1 - \lambda)$ continue playing the previously started τ_λ. With this strategy, for any $\epsilon > 0$, we have $E(\bar{g}_n) \le v_\lambda + \epsilon$ for all n sufficiently large (compared with $1/\lambda$). It follows that $\lim_{n \to \infty} v_n(p) \le v_\lambda(p)$. $\square$

Remark. If we interpret the discounted game as a repeated game with a probability λ of stopping after each stage, then the convergence of v_λ can be generalized as follows. Let $a = \{a_n\}_{n=1}^\infty$ be a probability distribution, with finite expectation, of T – the stopping time of the game – and let $v_a(p)$ be the value of this repeated game. If $\{a^l\}_{l=1}^\infty$ is a sequence of such distributions with mean going to infinity, then $\lim_{l \to \infty} v_{a^l}(p) = \text{Cav } u(p)$.

3.2. Full monitoring

The first model we consider is that of incomplete information on one side and with full monitoring. This is the case when the moves of the players at each stage are observed by both of them and hence they serve as the (only) device for transmitting information about the state of nature. The repeated game with the data $(K, p, S, T, (G^k)_{k \in K})$ is denoted by $\Gamma(p)$ and is played as follows.

At stage 0 a chance move chooses $k \in K$ with probability distribution $p \in \Delta(K)$, i.e. p^k is the probability of k. The result is told to player I, the row chooser, but not to the column chooser, player II who knows only the initial probability distribution p.

At stage $m = 1, 2, \ldots$, player I chooses $s_m \in S$ and II chooses, simultaneously and independently, $t_m \in T$ and then (s_m, t_m) is announced (i.e. it becomes common knowledge).

Actually $\Gamma(p)$ is not a completely defined game since the payoff is not yet specified. This will be done later; according to the specific form of the payoff, we will be speaking of $\Gamma_n(p)$ (the n-stage game), $\Gamma_\lambda(p)$ (the discounted game) or $\Gamma_\infty(p)$ (the infinitely repeated game).

The main feature of these games is that the informed player's moves will typically depend on (among other things) his information (i.e. on the value of k). Since these moves are observed by the uninformed player, they serve as a channel which can transfer information about the state k. This must be taken into account by player I when choosing his strategy. In Example 1.1., for instance, playing the move $s = 1$ if $k = 1$ and $s = 2$ if $k = 2$ is a dominant strategy as far as the single-stage payoff is concerned. However, such behavior will reveal the value of k to player II and by that enable him to reduce the payoffs to 0 in all subsequent stages. This is of course very disadvantageous in the long run and player I would be better off even by simply ignoring his information. In fact, playing the mixed move $(1/2, 1/2)$ at each stage independently of the value of k guarantees an expected payoff of at least $1/4$ per stage. We shall see that this is indeed the best he can do in the long run.

3.2.1. Posterior probabilities and nonrevealing strategies

For $n = 1, 2, \ldots$, let $H_n^{II} = [S \times T]^{n-1}$ be the set of possible histories for player II at stage n (that is, an element $h_n \in H_n^{II}$ is a sequence $(s_1, t_1; s_2, t_2; \ldots; s_{n-1}, t_{n-1})$ of the moves of the two players in the first $n-1$ stages of the game). Similarly, H_∞^{II} denotes the set of all infinite histories (i.e. plays) in the game. The set of all histories is $H^{II} = \bigcup_{n>1} H_n^{II}$. Let $\mathcal{H}_n^{II}$ be the σ-algebra on H_∞^{II} generated by the cylinders above H_n^{II} and let $\mathcal{H}_\infty^{II} = \bigvee_{n>1} \mathcal{H}_n^{II}$.

A pure strategy for player I in the supergame $\Gamma(p)$ is $\sigma = (\sigma_1, \sigma_2, \ldots)$, where for each n, σ_n is a mapping from $K \times \mathcal{H}_n^{II}$ to S. Mixed strategies are, as usual, probability distributions over pure strategies. However, since $\Gamma(p)$ is a game of perfect recall, we may (by Aumann's generalization of Kuhn's Theorem; see Aumann (1964)) equivalently consider only behavior strategies that are sequences of mappings from $K \times \mathcal{H}_n^{II}$ to X or equivalently from $\mathcal{H}_n^{II}$ to X^K. Similarly, a behavior strategy of player II is a sequence of mappings from $\mathcal{H}_n^{II}$ (since he does not know the value of k) to Y. Unless otherwise specified the word "strategy" will stand for behavior strategy. A strategy of player I is denoted by σ and one of player II by τ.

Any strategies σ and τ of players I and II, respectively, and $p \in \Delta(K)$ induce a joint probability distribution on states and histories – formally, a probability distribution on the measurable space $(K \times H_\infty^{II}, 2^K \otimes \mathcal{H}_\infty^{II})$. This will be our basic probability space and we will simply write P or E for probability or expectation when no confusion can arise.

Let $p_1 \equiv p$ and for $n \geq 2$ define

$$p_n^k = P(k \mid \mathcal{H}_n^{II}) \quad \forall k \in K .$$

These random variables on $\mathcal{H}_n^{II}$ have a clear interpretation: p_n is player II's *posterior probability distribution* on K at stage n given the history of moves up to that stage. These posterior probabilities turn out to be the natural *state variable* of the game and therefore play a central role in our analysis.

Observe first that the sequence $(p_n)_{n=1}^{\infty}$ is a $(\mathcal{H}_n^{II})_{n=1}^{\infty}$ martingale, being a sequence of conditional probabilities with respect to an increasing sequence of σ-algebras, i.e.

$$E(p_{n+1} \mid \mathcal{H}_n^{II}) = p_n \quad \forall n = 1, 2, \dots .$$

In particular, $E(p_n) = p$ $\forall n$. Furthermore, since this martingale is uniformly bounded, we have the following bound on its L_1 variation (derived directly from the martingale property and the Cauchy–Schwartz inequality):

Proposition 3.8.

$$\frac{1}{n} \sum_{m=1}^{n} E\| p_{m+1} - p_m \| \leq \sum_k \sqrt{\frac{p^k(1 - p^k)}{n}} .$$

Note that $\sum_k \sqrt{p^k(1 - p^k)} \leq \sqrt{\#K - 1}$ since the left-hand side is maximized for $p^k = 1/(\#K)$ for all k. Intuitively, Proposition 3.8 means that in "most of the stages" p_{m+1} cannot be very different from p_m.

The explicit expression of p_m is obtained inductively by Bayes' formula: given a strategy σ of player I, for any stage n and any history $h_n \in H_n^{II}$, let $\sigma(h_n) = (x_n^k)_{k \in K}$ denote the vector of mixed moves of player I at that stage. That is, he uses the mixed move $x_n^k = (x_n^k(s))_{s \in S} \in X = \Delta(S)$ in the game G^k. Given $p_n(h_n) = p_n$, let $\bar{x}_n = \sum_{k \in K} p_n^k x_n^k$ be the (conditional) average mixed move of player I at stage n. The (conditional) probability distribution of p_{n+1} can now be written as follows: $\forall s \in S$ such that $\bar{x}_n(s) > 0$ and $\forall k \in K$,

$$p_{n+1}^k(s) = P(k \mid \mathcal{H}_n^{II}, s_n = s) = \frac{p_n^k x_n^k(s)}{\bar{x}_n(s)} . \tag{5.1}$$

It follows that if $x_n^k = \bar{x}_n$ whenever $p_n^k > 0$, then $p_{n+1} = p_n$, that is:

Proposition 3.9. *Given any player II's history h_n, the posterior probabilities do not change at stage n if player I's mixed move at that stage is independent of k over all values of k for which $p_n^k > 0$.*

In such a case we shall say that player I plays *non-revealing* at stage n and, motivated by that, we define the corresponding set

$$NR = \{x \in X^K \mid x^k = x^{k'} \quad \forall k, k' \in K\} \,.$$

We see here, because of the full monitoring assumption, that not revealing the information is equivalent to not using the information. But then the outcome of the initial chance move (choosing k) is not needed during the game. This lottery can also be made at the end, just to compute the payoff.

Definition 3.10. For $p \in \Delta(K)$ the non-revealing game at p, denoted by $D(p)$, is the (one-stage) two-person, zero-sum game with payoff matrix

$$D(p) = \sum_{k \in K} p^k G^k \,.$$

Let $u(p)$ denote the value of $D(p)$. Clearly, u is a continuous function on $\Delta(K)$ (it is, furthermore, Lipschitz with constant $C = \max_{k,s,t} |G_{st}^k|$).

So if player I uses NR moves at all stages, the posterior probabilities remain constant. Hence the (conditional) payoff at each stage can be computed from $D(p)$. In particular, by playing an optimal strategy in $D(p)$ player I can guarantee an expected payoff of $u(p)$ at each stage. Thus we have:

Proposition 3.11. *Player I can guarantee $u(p)$ in $\Gamma_n(p)$, in $\Gamma_\lambda(p)$, and in $\Gamma_\infty(p)$ by playing i.i.d. an optimal strategy in $D(p)$.*

Combined with Proposition 3.6 this yields:

Corollary 3.12. *The previous proposition holds if we replace $u(p)$ by Cav $u(p)$.*

Given a strategy σ of player I, let $\sigma_n = (x_n^k)_{k \in K}$ be "the strategy at state n" [see MSZ (1993, ch. IV, section 1.6)]. Its average (over K) is the random variable $\bar{\sigma}_n = \mathrm{E}(\sigma_n \mid \mathcal{H}_n^{\mathrm{II}}) = \Sigma_k \, p_n^k \sigma_n^k$. Note that $\bar{\sigma}_n \in NR$.

A crucial element in the theory is the following intuitive property. If the σ_n^k are close (i.e. all near $\bar{\sigma}_n$), p_{n+1} will be close to p_n. In fact a much more precise relation is valid; namely, if the distance between two points in a simplex $[\Delta(S)$

or $\Delta(K)$] is defined as the L_1 norm of their difference, then the expectations of these two distances are equal. Formally,

Proposition 3.13. *For any strategies σ and τ of the two players*

$$\mathrm{E}(\|\sigma_n - \bar{\sigma}_n\| \mid \mathcal{H}_n^{\mathrm{II}}) = \mathrm{E}(\|p_{n+1} - p_n\| \mid \mathcal{H}_n^{\mathrm{II}}) \,.$$

This is directly verified using expression (5.1) for p_{n+1} in terms of σ_n.

Next we observe that the distance between payoffs is bounded by the distance between the corresponding strategies. In fact, given σ and τ let $\rho_n(\sigma, \tau) = \mathrm{E}(g_n \mid \mathcal{H}_n^{\mathrm{II}})$, and define $\tilde{\sigma}(n)$ to be the same as the strategy σ except for stage n where $\tilde{\sigma}_n(n) = \bar{\sigma}_n$, we then have:

Proposition 3.14. *For any σ and τ,*

$$|\rho_n(\sigma, \tau) - \rho_n(\tilde{\sigma}(n), \tau)| \leq CE(\|\sigma_n - \bar{\sigma}_n\| \mid \mathcal{H}_n^{\mathrm{II}}) \,.$$

Proof. In fact, since p_n is the same under σ and under $\tilde{\sigma}(n)$, we have (for any ω in H_∞^{II}):

$$|\rho_n(\sigma, \tau) - \rho_n(\tilde{\sigma}(n), \tau)| (\omega) \leq C \sum_k p_n^k(\omega) \|\sigma_n^k - \bar{\sigma}_n\|$$

$$= CE(\|\sigma_n - \bar{\sigma}_n\| \mid \mathcal{H}_n^{\mathrm{II}})(\omega) \,. \quad \square$$

3.2.2. The limit values $\lim v_n(p)$ *and* $v_\infty(p)$

The main consequence of the bound derived so far is:

Proposition 3.15. *For all $p \in \Delta(K)$ and all n,*

$$v_n(p) \leq \mathrm{Cav}\, u(p) + \frac{C}{\sqrt{n}} \sum_k \sqrt{p^k(1 - p^k)} \,.$$

Proof. Making use of the minmax theorem, it is sufficient to prove that for any strategy σ of player I in $\Gamma_n(p)$, there exists a strategy τ of player II such that

$$E_{\sigma,\tau}(\bar{g}_n) \leq \mathrm{Cav}\, u(p) + \frac{C}{\sqrt{n}} \sum_k \sqrt{p^k(1 - p^k)} \,.$$

Given σ let τ be the following strategy of player II: at stage m; $m = 1, \ldots, n$, compute p_m and play a mixed move τ_m which is optimal in $D(p_m)$.

By Proposition 3.14 and Proposition 3.13, for $m = 1, \ldots, n$:

$$\rho_m(\sigma, \tau) \le \rho_m(\tilde{\sigma}(m), \tau) + CE(\| p_{m+1} - p_m \| \mid \mathcal{H}_m^{\mathrm{II}}) \ .$$

Now

$$\rho_m(\tilde{\sigma}_m, \tau) = \sum_k p_m^k \, \bar{\sigma}_m G^k \tau_m \ ,$$

with $\bar{\sigma}_m \in NR$ and τ_m optimal in $D(p_m)$. Hence

$$\rho_m(\tilde{\sigma}(m), \tau) \le u(p_m) \le \mathrm{Cav}\, u(p_m) \ ,$$

which yields

$$\rho_m(\sigma, \tau) \le \mathrm{Cav}\, u(p_m) + CE(\| p_{m+1} - p_m \| \mid \mathcal{H}_m^{\mathrm{II}}) \ .$$

Averaging on $m = 1, \ldots, n$ and over all possible histories $\omega \in H_n^{\mathrm{II}}$ we obtain [using $E(\mathrm{Cav}\, u(p_m(\omega))) \le \mathrm{Cav}\, u(p)$ by Jensen's inequality]:

$$E_{\sigma,\tau}(\bar{g}_n) \le \mathrm{Cav}\, u(p) + \frac{C}{n} \sum_{m=1}^{n} E\| p_{m+1} - p_m \| \ .$$

The claimed inequality now follows from Proposition 3.8. $\quad\square$

Combining Proposition 3.15 with Corollary 3.12 we obtain [Aumann and Maschler (1967)]:

Theorem 3.16. *For all $p \in \Delta(K)$, $\lim_{n \to \infty} v_n(p)$ exists and equals* $\mathrm{Cav}\, u(p)$. *Furthermore, the speed of convergence is bounded by*

$$0 \le v_n(p) - \mathrm{Cav}\, u(p) \le \frac{C}{\sqrt{n}} \sum_n \sqrt{p^k(1 - p^k)} \ .$$

The strategy in Proposition 3.15 yields also:

Corollary 3.17. $\lim_{\lambda \to 0} v_\lambda(p)$ *exists and equals* $\mathrm{Cav}\, u(p)$ *and the speed of convergence satisfies*

$$0 \le v_\lambda(p) - \mathrm{Cav}\, u(p) \le C \sqrt{\frac{\lambda}{2 - \lambda}} \sum_k \sqrt{p^k(1 - p^k)} \ .$$

This follows using

$$\sum_{m=1}^{\infty} \lambda(1 - \lambda)^{m-1} E\| p_{m+1}^k - p_m^k \| \le \sqrt{\frac{\lambda}{2 - \lambda}} \sqrt{p^k(1 - p^k)} \ ,$$

which is a consequence of the Cauchy–Schwartz inequality and Proposition 3.8.

Combining now Corollary 3.12, Theorem 3.16 and Theorem 3.7 establishes the existence of the value of the infinite game $\Gamma_\infty(p)$ [Aumann, Maschler and Stearns (1968)]:

Theorem 3.18. *For all* $p \in \Delta(K)$ *the value* $v_\infty(p)$ *of* $\Gamma_\infty(p)$ *exists and equals* Cav $u(p)$.

3.2.3. Recursive formula for $v_n(p)$

The convergence of $v_n(p)$ is actually a monotone convergence. This follows from the following recursive formula for $v_n(p)$. Recall that $x = (x^k)_{k \in K} \in [\Delta(S)]^K$ is a one-stage strategy of player I (i.e. he plays the mixed move x^k in game G^k), then we have

$$v_{n+1}(p) = \frac{1}{n+1} \max_x \left\{ \min_t \sum_k p^k x^k G^k_{\cdot, t} + n \sum_{s \in S} \bar{x}_s v_n(p_s) \right\},$$

where $\bar{x} = \Sigma_k \, p^k x^k$ and for each s in S for which $\bar{x}_s > 0$, p_s is the probability on K given by $p_s^k = p^k x_s^k / \bar{x}_s$, and $G^k_{\cdot, t}$ denotes the t-th column of the matrix G^k.

By this recursive formula it can be proved inductively, using the concavity of $v_n(p)$, that:

Proposition 3.19. *For all* $p \in P$, *the sequence* $v_n(p)$ *is monotonically decreasing.*

The above recursive formula and the monotonicity are valid much more generally than in the full monitoring case under consideration. They hold (with the appropriate notation) in any signalling structure in which the signal received by player I includes the signal received by player II. However, when this condition is not satisfied, $v_n(p)$ may not be monotone. In fact, Lehrer (1987) has exhibited an example of a game with incomplete information on one side in which $v_1 \geqslant v_2 < v_3$.

3.2.4. Approachability strategy

Corollary 3.12 provides an explicit simple optimal strategy for player I in $\Gamma_\infty(p)$ which is played as follows. Express p as a convex combination $p = \Sigma_{e \in E} \alpha_e p_e$ of points $(p_e)_{e \in E}$ in $\Delta(E)$ such that Cav $u(p) = \Sigma_{e \in E} \alpha_e u(p_e)$. Perform the appropriate lottery described in Proposition 3.2 to choose $e \in E$, and then play i.i.d. at each stage an optimal strategy of the non-revealing game $D(p_e)$ (with the chosen e.)

On the other hand, the optimal strategy for player II provided by Theorem 3.7 is far from easy to compute. We now describe a simple optimal strategy for player II, making use of Blackwell's approachability theory for vector payoff games.

At any stage $(n + 1)$, given the history $h_{n+1} = (s_1, t_1, \ldots, s_n, t_n)$, player II can compute $\bar{g}_n^k = \frac{1}{n} \sum_{m=1}^{n} G_{s_m t_m}^k$, which is what his average payoff would be up to that stage if the state was k. Since the prior distribution on the states is p, his expected average payoff is $\langle p, \bar{g}_n \rangle = \sum_{k \in K} p^k \bar{g}_n^k$. We shall show that player II can play in such a way that, with probability one, the quantity $\langle p, \bar{g}_n \rangle$ will be arbitrarily close to Cav $u(p)$, for n sufficiently large.

Having focused our attention on the vector of averages $\bar{g}_n = (\bar{g}_n^k)_{k \in K}$, it is natural to consider the game with vector payoffs in the Euclidean space $\mathbb{R}^K$. So when moves (s, t) are played, the resulting vector payoff is $G_{st} \in \mathbb{R}^K$.

Consider the game $\Gamma_\infty(p)$. Let $u(p)$ be its NR-value function, and let $l = (l^k)_{k \in K}$ be the vector of intercepts of a supporting hyperplane to Cav u at p (see Figure 1); that is,

$$\text{Cav } u(p) = \langle l, p \rangle = \sum_k l^k p^k \quad \text{and} \quad u(q) \leq \langle l, q \rangle \quad \text{for all } q \text{ in } \Delta(K) .$$

If player II can play so that the average vector payoff $\bar{g}_n$ will *approach*[2] the "corner set" $C = \{x \in \mathbb{R}^K \mid x \leq l\}$, it will mean that $\forall \epsilon \geq 0$, $\langle p, \bar{g}_n \rangle \leq \langle l, p \rangle + \epsilon = \text{Cav } u(p) + \epsilon$, both in expectation and with probability one, for n sufficiently large. This is precisely the optimal strategy we are looking for.

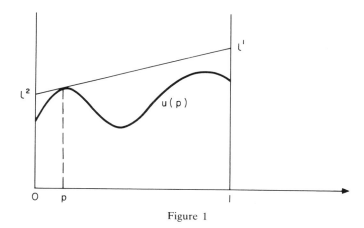

Figure 1

[2]That is, for any strategy of player I, the distance $d(\bar{g}_n, C)$ tends to 0 with probability 1. See Blackwell (1956).

For any mixed move $y \in Y$ of player II let $Q(y) = \text{Co}\{\Sigma_t\, G_{st}y_t \mid s \in S\}$, where Co A denotes the convex hull of A. (This is the set in which lies the expected vector payoff when y is played.) A sufficient condition for the approachability of C by player II [Blackwell (1956)] is that for any $g \not\in C$, he has a mixed move y such that if c is the closest point to g in C, then the hyperplane H orthogonal to $[cg]$ through c separates g from $Q(y)$ (see Figure 2).

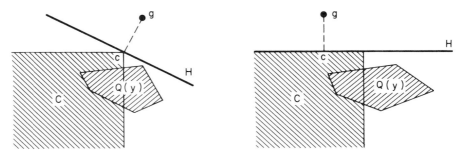

Figure 2

To verify this condition in our case let $q \in \Delta(K)$ be the unit vector in the direction $(g - c)$ and let $H = \{x \in \mathbb{R}^K \mid \langle q, x \rangle = \langle q, c \rangle\}$. Note that since C is a corner set, $q \geq 0$ and therefore $\langle q, c \rangle \geq \langle q, c' \rangle$ for all points c' in C, in particular $\langle q, c \rangle \geq \langle q, l \rangle$. Since $c \in C$, we also have $c \leq l$, which implies $\langle q, c \rangle \leq \langle q, l \rangle$; hence $l \in H$ and C lay in the half space defined by H, which can be written as $H_1 = \{x \in \mathbb{R}^K \mid \langle q, x \rangle \leq \langle q, l \rangle\}$. Now, by playing optimally in the non-revealing game $D(q)$, player II guarantees

$$\sum_{k \in K} q^k \sigma G^k y \leq u(q) \quad \forall \sigma \in \Delta(S).$$

This means that for any mixed strategy σ of player I, the vector payoff $x = (\sigma G^k y)_{k \in K}$ satisfies $\langle q, x \rangle \leq u(q) \leq \langle q, l \rangle$, i.e. $x \in H_1$, establishing the approachability of C.

The optimal strategy of player II in $\Gamma_\infty(p)$ can now be summarized as follows:

(1) Choose $l \in \mathbb{R}^K$ such that $\langle p, x \rangle = \langle p, l \rangle$ is the supporting hyperplane to the graph of Cav u at p.

(2) Define the corner set $C = \{x \in \mathbb{R}^K \mid x \leq l\}$, and at each stage n compute the average vector payoff $\bar{g}_n$ up to that stage.

(3) At stage $(n + 1)$, $n = 1, 2, \ldots$, if $\bar{g}_n \in C$, play arbitrarily. If $\bar{g}_n \not\in C$, let $c \in C$ be the closest point to $\bar{g}_n$ in C, compute $q = (\bar{g}_n - c)/\|\bar{g}_n - c\| \in \Delta(K)$ and play an optimal mixed move in $D(q)$.

3.2.5. *The examples revisited*

Let us look again at the examples we discussed in the Introduction in view of the general results.

In Example 1.1, $D(p)$ is the matrix game:

$$p\begin{pmatrix} 1 & 0 \\ 0 & 0 \end{pmatrix} + (1-p)\begin{pmatrix} 0 & 0 \\ 0 & 1 \end{pmatrix} = \begin{pmatrix} p & 0 \\ 0 & 1-p \end{pmatrix},$$

and its value is $u(p) = p(1-p)$. Since this is a concave function, Cav $u(p) = u(p) = p(1-p)$ and we have (see Figure 3)

$$\lim_{n\to\infty} v_n(p) = \lim_{\lambda\to 0} v_\lambda(p) = v_\infty(p) = p(1-p).$$

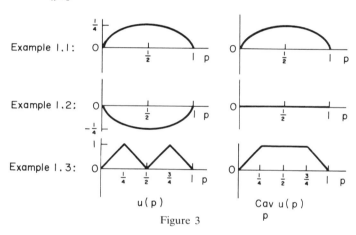

u(p) Cav u(p)

p

Figure 3

In particular, for $p = 1/2$ this limit is $1/4$. So asymptotically the value is that of the game in which no player is informed about the value of k. In other words, the informed player has an advantage only in games of finite length. This advantage may be measured by $v_n(p) - v_\infty(p)$. By Theorem 3.16 this is bounded by

$$v_n(p) - p(1-p) \le \frac{2\sqrt{p(1-p)}}{\sqrt{n}} \le \frac{1}{\sqrt{n}}.$$

It turns out that for this specific game this bound can be improved and the actual speed of convergence is [see Zamir (1971–72)]

$$v_n(p) - p(1-p) = O\left(\frac{\ln n}{n}\right).$$

In Example 1.2,

$$D(p) = p\begin{pmatrix} -1 & 0 \\ 0 & 0 \end{pmatrix} + (1-p)\begin{pmatrix} 0 & 0 \\ 0 & -1 \end{pmatrix} = \begin{pmatrix} -p & 0 \\ 0 & -(1-p) \end{pmatrix},$$

and its value is $u(p) = -p(1-p)$. Since this is a convex function and its concavification is the constant function 0, Cav $u(p) = 0$ $\forall p \in [0, 1]$ and we have (see Figure 3)

$$\lim_{n \to \infty} v_n(p) = \lim_{\lambda \to 0} v_\lambda(p) = v_\infty(p) = 0.$$

For $p = 1/2$ (as we had in our example), the value is 0.

In Example 1.3,

$$D(p) = p\begin{pmatrix} 4 & 0 & 2 \\ 4 & 0 & -2 \end{pmatrix} + (1-p)\begin{pmatrix} 0 & 4 & -2 \\ 0 & 4 & 2 \end{pmatrix} = \begin{pmatrix} 4p & 4-4p & 4p-2 \\ 4p & 4-4p & 2-4p \end{pmatrix},$$

and its value is (see Figure 3)

$$u(p) = \begin{cases} 4p, & 0 \leq p \leq 1/4, \\ 2-4p, & 1/4 \leq p \leq 1/2, \\ 4p-2, & 1/2 \leq p \leq 3/4, \\ 4-4p, & 3/4 \leq p \leq 1. \end{cases}$$

Therefore

$$\lim_{n \to \infty} v_n(p) = \lim_{\lambda \to 0} v_\lambda(p) = v_\infty(p) = \text{Cav } u(p),$$

where Cav $u(p)$ is (see Figure 3)

$$\text{Cav } u(p) = \begin{cases} 4p, & 0 \leq p \leq 1/4, \\ 1, & 1/4 \leq p \leq 3/4, \\ 4-4p, & 3/4 \leq p \leq 1. \end{cases}$$

For $p = 1/2$ (as we had in our example), the value is 1.

Remark. To all results so far there are of course corresponding dual results for the case in which the informed player is player II (while player I is uninformed). In particular the dual to Theorem 3.18 is:

Theorem 3.20. *In the game in which player II is informed and player I is not, for all $p \in \Delta(k)$ the value $v_\infty(p)$ of $\Gamma_\infty(p)$ exists and equals* Vex $u(p)$.

Here Vex $u(p)$ is the maximal convex function pointwise majorized by $u(p)$.

3.3. The general case

The main results so far, specifically the existence of lim v_n and of v_∞, extend to the general case without full monitoring, so we no longer assume that the moves are announced after each stage but rather that some individual message is transmitted to each of the players. This model was also treated by Aumann, Maschler and Stearns (1968), who proved the main result about the existence and the formula of $v_\infty(p)$. The generalization of the strategy for the un-informed player, using Blackwell approachability, is due to Kohlberg (1975a). Although the analysis follows the lines developed for the case of full monitoring, the mathematical details require several new ideas. These will be only outlined in this section [for the detailed proofs see, for example, MSZ (1993, Ch. V).

Recalling the general model, we add a signaling structure – two finite sets of signals A and B and transition probability Q from $K \times S \times T$ to $\Delta(A) \times \Delta(B)$. We denote by Q^k_{st} the probability distribution at (k, s, t). The repeated game is played as in the previous model with the following modification. At each stage n, $n \geq 1$, instead of announcing the moves (s_n, t_n), the signal a_n is announced to player I and b_n is announced to player II, where (a_n, b_n) is chosen according to the distribution $Q^k_{s_n t_n}$. It turns out that the value of $\Phi_\infty(p)$ does not depend on the signaling structure to the informed player, so by abuse of notation we denote the marginal of Q on B also by $Q^k_{s_n t_n}$.

The generalization of the notion of non-revealing utilizes the property that when a non-revealing strategy is played by player I at a certain stage, the conditional probability on K does not change at that stage. This is equivalent to:

Definition 3.21. $x \in X^K$ is said to be *non-revealing* at $p \in \Delta(K)$ if, for each move $t \in T$ of player II, the distribution of b (induced by t and x^k) in the kth state is the same for all k for which $p^k > 0$.

We denote by $NR(p)$ the set of non-revealing strategies in $\Gamma_1(p)$. For $p \in \Delta(K)$ let $K(p) \subseteq K$ denote the support of p. Then

$$NR(p) = \{x \in X^K \mid x^k Q^k = x^{k'} Q^{k'} \quad \forall (k, k') \in K(p) \times K(p)\}$$

Note that $NR(p) \subseteq NR(\tilde{p})$ whenever $K(p) \supseteq K(\tilde{p})$. Therefore $NR(p)$ is a "step set-valued function" on $\Delta(K)$ with possible discontinuities at the intersections of two (or more) facets. $NR(p)$ may be empty for some $p \in \Delta(K)$; however, if p is an extreme point of $\Delta(K)$, then $NR(p) = X^K$. This is intuitive – an extreme point of $\Delta(K)$ corresponds to a situation of *complete information*, where k is known to both players and hence every strategy of **I** is non-revealing since there is nothing to reveal.

The non-revealing game (*NR*-game), denoted by $D(p)$, is the (one-stage) two-person, zero-sum game in which player I's strategy set is $NR(p)$, player II's strategy set is $\Delta(T)$ and for $x = (x^k)_{k \in K} \in NR(p)$ and $y \in \Delta(T)$ the payoff is $\Sigma_{k \in K} p^k x^k G^k y$.

We denote by $u(p)$ the value of $D(p)$ and refer to it as the *NR*-value. If $NR(p) = \emptyset$ [hence $D(p)$ is undefined] we define $u(p) = -\infty$. Since $u(p)$ is finite at least on the extreme points, it follows that Cav $u(p)$ is well defined [and Lipschitz on $\Delta(K)$ with constant C].

Theorem 3.18 can now be proved for the general signaling case with this Cav $u(p)$. The proof that player I can guarantee $v(p)$ in $\Gamma_\infty(p)$ is the same as in the full monitoring case, that is, by applying an appropriate "splitting" followed by a non-revealing strategy. The major difficulty is in generalizing the optimal strategy of the uninformed player. The problem is that the above described optimal strategy for player II is based on the *statistics* $\bar{g}_n = (1/n) \Sigma_{m=1}^n g_n$. This is the vector whose kth coordinate is $(1/n) \Sigma_{m=1}^n G^k_{s_m t_m}$ which is *observable* by player II in the full monitoring case since he observes the moves (s_m, t_m). In the general case $\bar{g}_n$ is not observable by player II. Another optimal strategy is to be provided which is based only on the history $h_n = (b_1, \ldots, b_n)$ available to player II at each stage.

For any signal $b \in B$, and any move $t \in T$ at any stage n, let ρ_n^{tb} be the proportion of stages, up to stage n, in which b was received by player II following a move t, out of all stages in which move t was played, i.e.

$$\rho_n^{tb} = \frac{\#\{m \mid m \leq n, b_m = b, t_m = t\}}{\#\{m \mid m \leq n, t_m = t\}}.$$

The vector $\rho_n = (\rho_n^{tb})_{t \in T, b \in B}$, which is observable by player II after each stage n, is the basis for his strategy. There is also a vector payoff ξ_n which plays the role of the non-observable g_n. We do not define it formally here; it is, roughly speaking, the worst vector payoff which is compatible (up to a small deviation δ) with the observed vectors $\rho_1, \ldots, \rho_n$. To this vector payoff one applies Blackwell's approachability theory. The definition of ξ_n and the strategy of player II are such that [for the details see Kohlberg (1975a, 1975b) or MSZ (1993, ch. V)]:

- The ξ-payoff, i.e. $\langle p, \xi \rangle$, will be as close as we wish to Cav $u(p)$.
- The actual unobserved payoff will not exceed the observed ξ-payoff by more than an arbitrarily small ϵ.
- Player II plays each mixed move in a large block of stages so that, using (an appropriate version of) the strong law of large numbers, both the signals distribution and the (unobserved) payoffs are close to their means.

4. Incomplete information on two sides

The case of incomplete information on two sides is that in which each of the two players initially has only partial information about the state of nature, represented by a general partition of K. We denote these partitions by K^{I} and K^{II}. (The case in which one of the partitions is $\{\{1\}, \{2\}, \ldots, \{\#K\}\}$ is the case of incomplete information on one side treated in the previous section.) By common terminology, the elements of K^{I} and K^{II} are called the *types* of players I and II, respectively. The initial probability p can then be thought of as a joint prior probability distribution on the pairs of types.

A special case is that in which the types of the two players are *independent*, i.e. there exist two probability vectors q^{I} and q^{II} on the elements of K^{I} and K^{II}, respectively, such that

$$p(\kappa_j^{\mathrm{I}} \cap \kappa_l^{\mathrm{II}}) = q_j^{\mathrm{I}} q_l^{\mathrm{II}} \quad \forall \kappa_j^{\mathrm{I}} \in K^{\mathrm{I}} \quad \text{and} \quad \kappa_l^{\mathrm{II}} \in K^{\mathrm{II}}.$$

No general results are available for the whole class of these games. Most of this section is devoted to the special case in which Q^k is independent of k. This will be called the case of *state independent signaling*. That is, the information gained at each stage does not depend on the state of nature and it is determined completely by the players' moves at that stage. We omit the index k and denote the signaling mechanism by one transition probability from $S \times T$ to $A \times B$.

4.1. Minmax and maxmin

Let $\mathcal{H}^{\mathrm{I}}$ and $\mathcal{H}^{\mathrm{II}}$ be the σ-fields generated by K^{I} and K^{II}, respectively. A one-stage strategy $x = (x^k)_{k \in K}$ of player I in X^K is *non-revealing* if it is $\mathcal{H}^{\mathrm{I}}$ measurable and $\Sigma_{s \in S} x^k(s) Q_{s,t}(b)$ is independent of k for all t in T and b in B. In words, for each column of Q, the marginal probability distribution on B induced on the letters of that column is independent of the state of nature k. The set of non-revealing one-stage strategies of player I is denoted by NR^{I}. The set of non-revealing one-stage strategies of player II is defined in a dual way and is denoted by NR^{II}. These sets are obviously non-empty; they contain, for instance, the strategies constant on K. Denote by $D(p)$ the one-stage game in which players I and II are restricted to strategies in NR^{I} and NR^{II}, respectively. Let $u(p)$ be the value of $D(p)$.

Remarks. (i) The above definition of non-revealing strategy differs formally from Definition 3.21 in that there we required the induced distribution on B

(resp. on A) to be constant in k only on $K(p)$, while here we require it over all of K. However, it is easily seen that in this case the two definitions lead to the same $u(p)$. Since all results are formulated in terms of $u(p)$, we prefer to use here the above introduced definitions which have the advantage of making NR^{I} and NR^{II} independent of p.

(ii) Note that $u(p)$ is continuous in p on the simplex $\Delta(K)$ of prior probabilities.

We need now to generalize the notion of concavity and convexity:

A function on $\Delta(K)$ is said to be *concave with respect to* I (abbreviated w.r.t. I) if for every $p = (p^k)_{k \in K}$ it has a concave restriction on the subset $\Pi^{\mathrm{I}}(p)$ defined by

$$\Pi^{\mathrm{I}}(p) = \left\{ (\alpha^k p^k)_{k \in K} \mid \alpha^k \geq 0 \quad \forall k, \sum_k \alpha^k p^k = 1 \text{ and } (\alpha^k)_{k \in K} \right.$$

$$\left. \text{is } \mathcal{K}^{\mathrm{I}}\text{-measurable} \right\}.$$

Interpretation: Given the prior probability distribution p on K and given any one stage strategy of player I (which is hence $\mathcal{K}^{\mathrm{I}}$-measurable), the conditional probability distribution on K given the move of player I is an element of $\Pi^{\mathrm{I}}(p)$. In other words, when updating the distribution on K in view of observations on player I's moves only (knowing his strategy), the range of the posterior distribution is $\Pi^{\mathrm{I}}(p)$.

A function on $\Delta(K)$ is said to be *convex with respect to* II (abbreviated w.r.t. II) if for every $p = (p^k)_{k \in K}$ it has a convex restriction on the subset $\Pi^{\mathrm{II}}(p)$ defined by

$$\Pi^{\mathrm{II}}(p) = \left\{ (\beta^k p^k)_{k \in K} \mid \beta^k \geq 0 \quad \forall k, \sum_k \beta^k p^k = 1 \text{ and } (\beta^k)_{k \in K} \right.$$

$$\left. \text{is } \mathcal{K}^{\mathrm{II}}\text{-measurable} \right\}.$$

Note that for any p in $\Delta(K)$ both $\Pi^{\mathrm{I}}(p)$ and $\Pi^{\mathrm{II}}(p)$ are convex and compact subsets of $\Delta(K)$ containing p, which justify the above definitions of concavity w.r.t. I and convexity w.r.t. II.

In the independent case it is more convenient to work not with p in $\Pi = \Delta(K)$ but rather with the product probability $(q^{\mathrm{I}}, q^{\mathrm{II}}) \in \Delta^{\mathrm{I}} \times \Delta^{\mathrm{II}}$, where Δ^{I} and Δ^{II} are the simplices of probability distributions on the types of player I (i.e. the elements of K^{I}) and of player II, respectively. In this case,

$$\Pi^{\mathrm{I}}(q^{\mathrm{I}}, q^{\mathrm{II}}) = \{q^{\mathrm{I}}\} \times \Delta^{\mathrm{II}} \quad \text{and} \quad \Pi^{\mathrm{II}}(q^{\mathrm{I}}, q^{\mathrm{II}}) = \Delta^{\mathrm{I}} \times \{q^{\mathrm{II}}\}.$$

Thus, concavity w.r.t. I means simply concavity in the first variable q^I (for any value of q^{II}), and similarly for convexity w.r.t. II.

Given any function g on $\Delta(K)$, the *concavification* of g w.r.t. I (denoted by $\mathrm{Cav}_I \, g$) is the (pointwise) minimal function which is concave w.r.t. I and is greater than or equal to g on $\Delta(K)$. Similarly, the *convexification* of g w.r.t. II (denoted by $\mathrm{Vex}_{II} \, g$) is the (pointwise) minimal function which is convex w.r.t. II and is less than or equal to g on $\Delta(K)$.

Remark. Note that in the special case of incomplete information on one side $(K^I = \{\{1\}, \{2\}, \ldots, \{\#K\}\}$ and $K^{II} = \{K\})$, $\mathrm{Cav}_I \, g$ is the usual $\mathrm{Cav} \, g$ and $\mathrm{Vex}_{II} \, g$ is g.

Theorem 4.1. *The minmax of $\Gamma_\infty(p)$ exists and is given by*

$$\bar{v}(p) = \mathrm{Vex}_{II} \, \mathrm{Cav}_I \, u(p) \, .$$

Similarly, $\mathrm{Cav}_I \, \mathrm{Vex}_{II} \, u(p)$ is the maxmin of $\Gamma_\infty(p)$.

Proof. The heuristic arguments of the proof are as follows.[3] Proving that the minmax of $\Gamma_\infty(p)$ is $\mathrm{Vex}_{II} \, \mathrm{Cav}_I \, u(p)$ consists of two parts.

Part (i): Player II can guarantee $\mathrm{Vex}_{II} \, \mathrm{Cav}_I \, u(p)$. If player II ignores his private information (i.e. K^{II}) and if for each $\kappa^I \in K^I$ let $q^{\kappa^I} = \Sigma_{k \in \kappa^I} \, p^k$ and take as payoffs $A^{\kappa^I} = (1/q^{\kappa^I}) \Sigma_{k \in \kappa^I} \, p^k G^k$ (keeping the same distribution on signals), we obtain a game $\Gamma(q)$ with incomplete information on one side, with K^I as the set of states of nature, with initial probability distribution q on it, and player I informed. In this game, denoting the value of the non-revealing game by $w(q)$, player II can guarantee $\mathrm{Cav} \, w(q)$. Now by our construction $w(q) = u(p)$ and $\mathrm{Cav} \, w(q) = \mathrm{Cav}_I \, u(p)$. Finally, by the dual of Proposition 3.6, player II can also guarantee the Vex_{II} of this function, namely $\mathrm{Vex}_{II} \, \mathrm{Cav}_I \, u(p)$ (by applying the appropriate splitting procedure established in Proposition 3.2).

Part (ii): Player I can defend $\mathrm{Vex}_{II} \, \mathrm{Cav}_I \, u(p)$. Any pair of strategies σ and τ of the two players induces a martingale of posterior probability distributions $\{p_n\}_{n=1}^\infty$, converging with probability one and hence having a bounded total variation: $\mathrm{E}_{\sigma,\tau}[\Sigma_{k \in K} \, \Sigma_{n=1}^\infty \, (p_n^k - p_{n-1}^k)^2]$. Given any strategy τ of player II, define σ_0 as a non-revealing strategy of player I inducing a martingale with N-stage variation ϵ-close to the supremum, over all his non-revealing

---

[3]The first proof of this result, for a less general model (namely "the independent case"), is due to Aumann, Maschler and Stearns (1968), who also gave the first example of such a game in which $\mathrm{Cav} \, \mathrm{Vex} \, u(p) \neq \mathrm{Vex} \, \mathrm{Cav} \, u(p)$ and hence has no value.

strategies, of the total variation. Denoting by NR_∞^1 the set of all non-revealing strategies of player I, this means that σ_0 and N are defined by

$$E_{\sigma_0,\tau}\left[\sum_{k\in K}\sum_{n=1}^N (p_n^k - p_{n-1}^k)^2\right] > \sup_{\sigma\in NR_\infty^1} E_{\sigma,\tau}\left[\sum_{k\in K}\sum_{n=1}^\infty (p_n^k - p_{n-1}^k)^2\right] - \epsilon .$$

By playing this σ_0 against τ up to stage N, player I "exhausts" almost all the variation of the martingale, i.e. player II will be playing "practically non-revealing" from that stage on. Thus, the situation is almost that of incomplete information on one side in which player I is informed and he can then guarantee $\text{Cav}_1 u(p_N)$ (where p_N is the posterior probability at stage N). Finally, since up to stage N player I is playing non-revealing, we have $p_N \in \Pi^{11}(p)$ and $E(p_N) = p$ implying that the expected average payoff to player I is at least $E(\text{Cav}_1 u(p_N)) \geqslant \text{Vex}_{11} \text{Cav}_1 u(p)$. $\square$

It should be noted that the formal proof of the above outlined arguments is quite intricate and non-trivial mainly because in a general signaling structure "exhausting" the information from the other player's strategy usually involves revealing the player's own information. Another general difficulty in all proofs involving the posterior probabilities p_m of a certain player is that they have to be assumed computable by the other player as well, which is usually not the case when there is general signaling. The way to overcome these difficulties is the following. Assume that we want to prove that player I can guarantee a certain payoff level. We perturb the game to make it slightly more disadvantageous to him. This perturbation consists of not giving player I his signal according to Q unless he buys it for an amount C. Furthermore, he is restricted to use this option of buying information exactly with probability $\delta > 0$ while with probability $(1 - \delta)$ he gets no information whatsoever. Whenever he does receive non-trivial information, his signal is completely known to player II. This implies that $\mathcal{H}_m^{11} \supset \mathcal{H}_m^1$ and hence p_m, the posterior distribution of player I, is also computable by player II. If, despite the disadvantageous modifications, player I can guarantee a certain amount (for sufficiently small δ) then he can also certainly guarantee it in the original game.

A corollary of Theorem 4.1 is that the infinite game $\Gamma_\infty(p)$ has a value if and only if

$$\text{Cav}_1 \text{Vex}_{11} u(p) = \text{Vex}_{11} \text{Cav}_1 u(p) .$$

An example of a game without a value is the following game with independent types and full monitoring in which there are two types of each player with the payoff matrices given by

$$G^{11} = \begin{pmatrix} 0 & 0 & 0 & 0 \\ -1 & 1 & 1 & -1 \end{pmatrix}, \quad G^{12} = \begin{pmatrix} 1 & -1 & 1 & -1 \\ 0 & 0 & 0 & 0 \end{pmatrix},$$

$$G^{21} = \begin{pmatrix} -1 & 1 & -1 & 1 \\ 0 & 0 & 0 & 0 \end{pmatrix}, \quad G^{22} = \begin{pmatrix} 0 & 0 & 0 & 0 \\ 1 & -1 & -1 & 1 \end{pmatrix}.$$

That is, the set of states is $K = \{11, 12, 21, 22\}$ and the partitions of initial information are $K^I = \{11, 12\}\{21, 22\}$ and $K^{II} = \{11, 21\}\{12, 22\}$. If, for instance, the initial probability distributions on types are $q^I = q^{II} = (1/2, 1/2)$, then[4] $\underline{v}_\infty(1/2, 1/2) = \text{Cav}_I \text{Vex}_{II} u(1/2, 1/2) = -1/4$ and $\bar{v}_\infty(1/2, 1/2) = \text{Vex}_{II} \text{Cav}_I u(1/2, 1/2) = 0$.

4.2. *The asymptotic value* $\lim_{n\to\infty} v_n(p)$

The non-existence of a value for infinite games with incomplete information on both sides is a very important feature of these games which, among other things, exemplifies the difference between repeated games with incomplete information and stochastic games, in which the value always exists. Given this result, the next natural question is that of the existence of the asymptotic value $\lim_{n\to\infty} v(p)$. Here the result is positive [Mertens and Zamir (1971–72)]:

Theorem 4.2. $v(p) = \lim_{n\to\infty} v_n(p)$ *exists for all $p \in \Delta(K)$ and is the unique solution to the following set of functional equations*:
   (1) $f(p) = \text{Vex}_{II} \max\{u(p), f(p)\}$;
   (2) $f(p) = \text{Cav}_I \min\{u(p), f(p)\}$.

Proof.   To outline the main arguments let $\underline{v}(p)$ and $\bar{v}(p)$ be, respectively, the lim inf and lim sup of $\{v_n\}_{n=1}^\infty$. Both functions are Lipschitz, $\underline{v}$ is concave w.r.t. I and $\bar{v}$ is convex w.r.t. II. For any strategy of player II consider the following response strategy of player I (actually a sequence of strategies one for each finite game): play optimally in $D(p_m)$ at stage m as long as $u(p_m) \geqslant \underline{v}(p_m)$. As soon as $u(p_m) < \underline{v}(p_m)$ play optimally in the remaining subgame (p_m is the posterior probability distribution on K at stage m). This strategy guarantees player I an expected average payoff arbitrarily close to the maximum of $u(p)$ and $\underline{v}(p)$, for n large enough, proving that:

   (1') $\underline{v}(p) \geqslant \text{Vex}_{II} \max\{u(p), \underline{v}(p)\}$,

and similarly:

   (2') $\bar{v}(p) \leqslant \text{Cav}_I \min\{u(p), \bar{v}(p)\}$.

[4]For the detailed computations see Mertens and Zamir (1971–72).

Actually, this argument shows that if player I can guarantee $f(p)$ in $\Gamma_n(p)$ for large enough n, he can also guarantee $\text{Vex}_{II} \max\{u(p), f(p)\}$.

Now, since $\bar{v}(p)$ is convex w.r.t. II, it follows from (2') that

(2'') $\bar{v}(p) \le \text{Vex}_{II} \text{Cav}_I \min\{u(p), \bar{v}(p)\}$.

Next, when a player plays an optimal strategy in $D(p_m)$ at stage m, his expected payoff at that stage differs from $u(p_m)$ by at most a constant times $|p_{m+1} - p_m|$. Combining this with Proposition 3.8 one shows that any function $f(p)$ satisfying $f(p) \le \text{Vex}_{II} \text{Cav}_I \min\{u(p), f(p)\}$ must satisfy

$$(f(p) - v_n(p))^+ \le R \frac{\Sigma_k \sqrt{p^k(1-p^k)}}{n}, \tag{5.2}$$

for some constant R. In particular, letting $n \to \infty$, this implies $f(p) \le \underline{v}$. It follows now from (2'') that $\bar{v}(p) \le \underline{v}(p)$ and hence $v_n(p)$ converge to, say, $v(p)$ with the speed of convergence of $1/\sqrt{n}$. The limit is the smallest solution to

$$f(p) \ge \text{Cav}_I \text{Vex}_{II} \max\{u(p), f(p)\},$$

and the largest solution to

$$f(p) \le \text{Vex}_{II} \text{Cav}_I \min\{u(p), f(p)\}.$$

It is then the only simultaneous solution to both. Finally, since $v(p)$ is both concave w.r.t. I and convex w.r.t. II, it must also satisfy (1) and (2), and is the only solution to this system. $\square$

The above outline can be made a precise proof for the case of full monitoring. For the general signaling case, one has to use a sequence of δ-perturbations of the game. This provides the same results as far as the functional equations are concerned but with different bound on the speed of convergence for $v_n(p)$, namely [see MSZ (1993)]

$$|v_n(p) - v(p)| \le \tilde{C} \frac{[\Sigma_{k \in K} \sqrt{p^k(1-p^k)}]^{2/3}}{\sqrt[3]{n}}, \tag{5.3}$$

for some constant $\tilde{C}$ which depends only on the game.

4.3. Existence and uniqueness of the solution of the functional equations

The pair of dual equations (1) and (2) that determine $v(p)$ are of interest and can be analyzed without reference to game theoretic context and techniques. This was in fact done [see Mertens and Zamir (1977b), Sorin (1984b)] and the results can be summarized as follows:

Denote by $\mathscr{C}(\Delta)$ the space of all continuous functions on the simplex Δ, and by U the subset of $\mathscr{C}(\Delta)$ consisting of those functions that are "*u*-functions": values of $D(p)$, for some two-person, zero-sum game with incomplete information $\Gamma(p)$ with full monitoring. Denote by φ the mapping from U to $\mathscr{C}(\Delta)$ defined by $\varphi(u) = v = \lim v_n$ [using Theorem 4.2, this mapping is well defined since $\lim v_n$ is the same for all games $\Gamma(p)$ having the same *u*-function]. Let $\mathscr{C}(\Delta)$ be endowed with the topology of uniform convergence.

Proposition 4.3. (a) *U is a vector lattice[5] and a vector algebra[6] which contains all the affine functions.*

(b) *U is dense in $\mathscr{C}(\Delta)$.*

Proposition 4.4. *The mapping $\varphi: U \to \mathscr{C}(\Delta)$ has a unique continuous extension $\varphi: \mathscr{C}(\Delta) \to \mathscr{C}(\Delta)$. This extension is monotone and Lipschitz with constant 1 [or non-expansive, i.e. $\|\varphi(f) - \varphi(g)\| \leq \|f - g\|$].*

Theorem 4.5. *Consider the following functional inequalities and equations in which u, f and g denote arbitrary functions on the simplex Δ:*

(α) $f \geq \mathrm{Cav}_{\mathrm{I}} \, \mathrm{Vex}_{\mathrm{II}} \, \max\{u, f\}$;

(β) $f \leq \mathrm{Vex}_{\mathrm{II}} \, \mathrm{Cav}_{\mathrm{I}} \, \min\{u, f\}$;

(α') $g = \mathrm{Vex}_{\mathrm{II}} \, \max\{u, g\}$;

(β') $g = \mathrm{Cav}_{\mathrm{I}} \, \min\{u, g\}$.

There exists a monotone non-expansive mapping $\varphi: C(\Delta) \to C(\Delta)$ such that, for any $u \in \mathscr{C}(\Delta)$:

(i) *$\varphi(u)$ is the smallest f satisfying (α) and the largest f satisfying (β), and thus in particular it is the only solution f of the system (α)–(β).*

(ii) *$\varphi(u)$ is also the only solution g of the system (α')–(β').*

Theorem 4.6 [An approximation procedure for $\varphi(u)$]. *Define $\underline{v}_0 = -\infty$, $\bar{v}_0 = +\infty$, and for $n = 1, 2, \ldots$ let $\underline{v}_{n+1} = \mathrm{Cav}_{\mathrm{I}} \, \mathrm{Vex}_{\mathrm{II}} \, \max\{u, \underline{v}_n\}$ and $\bar{v}_{n+1} = \mathrm{Vex}_{\mathrm{II}} \, \mathrm{Cav}_{\mathrm{I}} \, \min\{u, \bar{v}_n\}$. Then $\{\underline{v}_n\}_{n=1}^{\infty}$ is monotonically increasing, $\{\bar{v}_n\}_{n=1}^{\infty}$ is monotonically decreasing and both sequences converge uniformly to $\varphi(u)$.*

Note that $\underline{v}_1$ (resp. $\bar{v}_1$) is the maxmin (resp. minmax) of $\Gamma_{\infty}(p)$ if $u(p)$ is the value of $D(p)$.

---

[5] That is, an ordered vector space V such that the maximum and the minimum of two elements of V exist (in V).

[6] That is, the product of two elements of U is in U.

4.4. The speed of convergence of $v_n(p)$

As mentioned in previous sections, the proofs for the convergence of $v_n(p)$ yield as a byproduct a bound for the speed of convergence: $1/\sqrt{n}$ for the full monitoring case [inequality (5.2)] and $1/\sqrt[3]{n}$ for the general signaling case [inequality (5.3)]. It turns out that these bounds are the best possible. In fact, games with these orders of speed of convergence can be found in the special case of incomplete information on one side.

Example 4.7. Consider the following game in which $k = \{1, 2\}$, player I is informed of the value of k, with full monitoring and payoff matrices:

$$G^1 = \begin{pmatrix} 3 & -1 \\ -3 & 1 \end{pmatrix}, \qquad G^2 = \begin{pmatrix} 2 & -2 \\ -2 & 2 \end{pmatrix},$$

and the prior probability distribution on K is $(p, 1-p)$.

For this game it is easily verified that $v_\infty(p) \equiv 0$. More precisely, we have [see Zamir (1971–72)]

$$\frac{p(1-p)}{\sqrt{n}} \leqslant v_n(p) \leqslant \frac{\sqrt{p(1-p)}}{\sqrt{n}}, \tag{5.4}$$

for all n and for all $p \in [0, 1]$.

Remaining in the framework of the previous example we change the payoffs and signals to obtain:

Example 4.8. Let $K = \{1, 2\}$. The payoff matrices G^1 and G^2 and the signaling matrices Q^1 and Q^2 (to player II) are given by

$$G^1 = \begin{pmatrix} 8 & 3 & -1 \\ 8 & -3 & 1 \end{pmatrix}, \qquad G^2 = \begin{pmatrix} 8 & 2 & -2 \\ 8 & -2 & 2 \end{pmatrix}$$

and

$$Q^1 = Q^2 = \begin{pmatrix} a & c & d \\ b & c & d \end{pmatrix}.$$

Here the signals (to player II) are deterministic (e.g. if top and middle are played the signal is c, etc.). Observe that deleting the left strategy of player II and changing the signaling matrices to provide full monitoring we obtain the game in Example 4.7. It is easily verified that the differences between the examples do not affect the value and in the game of this example we still have $v_\infty(p) \equiv 0$. However, in the present example player II is not informed of the last move of his opponent unless he chooses his left strategy which is strictly dominated in terms of payoffs. In other words, player II has to pay 8 units whenever he wants to observe his opponent's move. Since observing the moves of the informed player is his only way to collect information about the state k, it is not surprising that his learning process will be slower and more costly than in Example 4.7. This yields a slower rate of convergence of v_n to v_∞, [see Zamir (1973a)]

$$\frac{p(1-p)}{\sqrt[3]{n}} \le v_n(p) \le \frac{\alpha\sqrt{p(1-p)}}{\sqrt[3]{n}}$$

for some positive constant α, for all n and for all $p \in [0, 1]$.

The speed of convergence of $v_n(p)$ can also be of lower order, such as $(\ln n)/n$, $1/n$. There are some partial results for classification of games according to those speeds [see Zamir (1971–72, 1973a)].

The special role of the normal distribution

One of the interesting, and still quite puzzling results in the study of the speed of convergence of $v_n(p)$ is the appearance of the normal distribution. Consider again the game in Example 4.7. It follows from inequality (5.4) that for any $0 < p < 1$, $\sqrt{n}v_n(p)$ is bounded between $p(1-p)$ and $\sqrt{p(1-p)}$. A natural question is then: Does this sequence converge? If it does, the limit is the coefficient of the leading term (i.e. $1/\sqrt{n}$) in the expansion of $v_n(p) - v_\infty(p)$ in fractional powers of n [recall that $v_\infty(p) \equiv 0$]. The sequence does turn out to converge and the limit is the well-known standard normal distribution function:

Theorem 4.9. *For all* $p \in [0, 1]$,

$$\lim_{n\to\infty} \sqrt{n}\, v_n(p) = \phi(p)\,,$$

where

$$\phi(p) = \frac{1}{\sqrt{2\pi}}\, e^{-(1/2)x_p^2} \quad \text{and} \quad \frac{1}{\sqrt{2\pi}} \int_{-\infty}^{x_p} e^{-(1/2)x^2}\, dx = p\,.$$

In words: the limit of $\sqrt{n}\, v_n(p)$ *is the standard normal density function evaluated at its p-quantile.*

The proof is rather technical [see Mertens and Zamir (1976b)] and does not give the intuition behind the result. It is based on a general result about the variation of martingales in $[0, 1]$ [Mertens and Zamir (1977a)]. Let $\mathscr{X}^n_p = \{X_m\}^n_{m=1}$ denote an n-martingale bounded in $[0, 1]$ with $E(X_1) = p$, and let $V(\mathscr{X}^n_p)$ denote its L_1 variation, i.e.

$$V(\mathscr{X}^n_p) = \sum_{m=1}^{n-1} E(|X_{m+1} - X_m|) .$$

Then we have

Theorem 4.10. (The L_1 variation of a bounded martingale).

$$\lim_{n \to \infty} \sup_{\mathscr{X}^n_p} \left[\frac{1}{\sqrt{n}} V(\mathscr{X}^n_p) \right] = \phi(p) .$$

It turns out that this is not an isolated incident for this specific example but rather part of a general phenomenon. Consider a game with incomplete information on one side and two states of nature, each with a 2×2 payoff matrix. If the error term is of the order of $1/\sqrt{n}$, then $\sqrt{n}[v_n(p) - v_\infty(p)]$ tends (as $n \to \infty$) to an appropriately scaled normal density function [Mertens and Zamir (1990)]. This result was recently further generalized by De Meyer (1989) to any (finite) number of states of nature and any (finite) number of strategies for each player.

5. Incomplete information on two sides: The symmetric case

In the general situation of incomplete information on two sides, the case of state independent signaling treated in the previous section is the case with the most complete analysis. In this section we consider a special case in which signaling may be state dependent but it is *symmetric* in the sense that at each stage both players get the same signal.

Formally, we are given a finite collection of $S \times T$ payoff matrices $\{G^k\}_{k \in K}$ with initial probability p in $\Pi = \Delta(K)$, and the players have no initial information about the true state k except the prior distribution p. We denote by A the finite set of signals and by A^k the signaling matrix for state k. Given k and a pair of moves (s, t), a signal a is announced to both players according to a given probability distribution A^k_{st} on A. Assuming perfect recall means in this framework that for all k and k' in K, $s \neq s'$ or $t \neq t'$ implies that A^k_{st} and $A^{k'}_{s't'}$ have disjoint support.

Denoting the above described infinite game by $\Gamma_\infty(p)$ the result is:

Theorem 5.1. $\Gamma_\infty(p)$ *has a value.*

Proof. To see the idea of the proof consider first the special case in which the signals are deterministic – the support of $A_{s,t}^k$ consists of a single element of A (which will also be denoted by A_{st}^k). Define the set of non-revealing moves:

$$NR = \{(s, t) \in S \times T \mid A_{st}^k = A_{st}^{k'} \quad \forall k, k' \in K\}.$$

That is, a non-revealing move is one which gives no additional information about the state k and hence after a non-revealing move, the players face the same (infinite) game as the one they faced before that move. Whenever a move $(s, t) \notin NR$ is played and a certain signal a is announced, a non-empty subset of K is eliminated from the set of possible states, namely all states k for which $A_{st}^k \neq a$. The resulting situation is a game having the same data as the original one but with K replaced by a proper subset of itself, and the prior probability distribution on this smaller set is the normalization of its marginal according to p. Now if we prove our theorem by induction on $\#K$, then by the induction hypothesis the game resulting from a move not in NR has a value which can be guaranteed by both players from that stage on. In other words, using stochastic games terminology, the result of such a move is an *absorbing* state with payoff equal to that value.

Writing this formally, for each move (s, t) and for each signal a let $K_{st}(a) = \{k \in K \mid A_{st}^k = a\}$ and $p_{st}(a) = \Sigma_{k' \in K_{st}(a)} p^{k'}$. Let p_a be the probability distribution on $K_{st}(a)$ given by $p_a^k = p^k / p_{st}(a)$. Finally, denote by $v_{st}(a)$ the value of the game obtained from $\Gamma_\infty(p)$ when replacing K by $K_{st}(a)$ and p by p_a. The game $\Gamma_\infty(p)$ is equivalent to an $S \times T$ game with absorbing states in which the payoffs are given by (x^* indicates an absorbing state with payoff x):

$$\tilde{G}_{st} = \begin{cases} \Sigma_{k \in K} \, p^k G_{st}^k & \text{if } (s, t) \in NR, \\ (\Sigma_{a \in A} \, p_{st}(a) v_{st}(a))^* & \text{otherwise}. \end{cases}$$

Since this game (like any finite stochastic game) has a value, the original game $\Gamma_\infty(p)$ also has a value, completing the inductive step of the proof. $\square$

Remark. It is worth noting that historically the reduction of symmetric games of incomplete information to games with absorbing states was done before the latter were known to have a value [see Kohlberg and Zamir (1974)]. In fact, this focused attention on games with absorbing states and on the particular example of *The big match* treated by Blackwell and Ferguson (1968). The general solution of these games by Kohlberg (1974) then led to the solution of general stochastic games [Bewley and Kohlberg (1976a, 1976b, 1978), Mertens and Neyman (1981)].

In the general signaling case a "revealing" signal need not eliminate elements of K as impossible but rather it leads to a new (posterior) probability distribution $p_1 \neq p$ on K. The value function is then a continuous function on $\pi = \Delta(K)$ and its existence is proved by induction on the dimension of this simplex [see Forges (1982) and MSZ (1993)].

6. Games with no signals

We consider here a class of games which was introduced by Mertens and Zamir (1976a) under the name "repeated games without a recursive structure". These games consist again of a finite collection of $S \times T$ payoff matrices G^k, $k \in K$, with an initial probability distribution p on K. No player is informed of the initial state. The signals are defined by a family of matrices A^k with deterministic entries (the extension to random signals is simple). Moreover, we assume that in each matrix A^k there are only two possible signals; either both players receive a "white" (totally uninformative) signal (0) or the game is completely revealed to both players. We can thus assume in the second case that the payoff is absorbing and equal to the value of the revealed game from this time on. It is then enough to define the strategies on the "white" histories; hence the name "game with no signals". Note that unlike the games considered in the previous section, the signal 0 does not include the moves of the players. By Dalkey's theorem [Dalkey (1953)], each player may be assumed to remember his own move, and hence the "white" signal is actually asymmetric information.

For a typical simple example of such games consider a game with two states, $\#S = \#T = 2$ and signaling matrices given by

$$A^1 = \begin{pmatrix} a & 0 \\ 0 & 0 \end{pmatrix}, \qquad A^2 = \begin{pmatrix} 0 & 0 \\ 0 & b \end{pmatrix}.$$

To see the special feature of these games assume that in our example the prior probability distribution is $(1/2, 1/2)$, both players play at the first stage the mixed move $(1/2, 1/2)$ which results in (*top*, *right*) and a white signal (an event of probability $3/4$). Consequently, the posterior probability distribution of player I is $(1/3, 2/3)$ while that of player II is $(2/3, 1/3)$. The "state variable" of the problem can no longer be just a probability distribution on K. A larger and actually unbounded dimensional space is needed, and hence the name "games without a recursive structure". The above mentioned example was introduced and solved in Mertens and Zamir (1976a). The general case (i.e. general K and general size A^k) was solved by Waternaux (1983a, 1983b).

The analysis of these games brought about a new tool. The minmax and the maxmin of the game Γ_∞ are equal, respectively, to the values of two auxiliary one-shot games $\overline{G}$ and $\underline{G}$ in strategic form. The pure strategies in each of these

games mimic strategies in Γ_∞, and the payoffs are defined to be the corresponding asymptotic payoffs in Γ_∞ to the strategies which are mimicked. Fix p and
write Γ for $\Gamma_\infty(p)$, and $\overline{v}(\Gamma)$ and $\underline{v}(\Gamma)$ for its minmax and maxmin, respectively.

Formally, define the (pure) strategy sets $\overline{X}$ and $\overline{Y}$, in the one-shot game $\overline{G}$,
by

$$\overline{X} = \bigcup_{S' \subset S} \Delta(S') \times \mathbb{N}^{S \backslash S'} \times S' \,,$$

$$\overline{Y} = \bigcup_{T' \subset T} \Delta(T') \times \mathbb{N}^{T \backslash T'} \,,$$

where $\mathbb{N}$ is the set of positive integers. Given x in $\overline{X}$ (resp. y in $\overline{Y}$) we denote
the corresponding subset S' by S^x, the first component by α^x, the second by c^x
and the third by s^x.

The heuristic representation of the strategies in $\overline{G}$ is given by the following
strategies in Γ_∞. For x in $\overline{X}$, player I plays i.i.d. the mixed move $\alpha^x \in \Delta(S^x)$
except for $c(x) = \Sigma_s c_s^x$ exceptional moves; each move s which is not part of the
mixed move α^x (i.e. $s \notin S^x$) is played c_s^x times uniformly distributed before
some large state N_0. From stage N_0 on, player I uses the (pure) move s^x.

A strategy $y \in \overline{Y}$ of player II has a similar meaning with the difference that,
after stage N_0, he continues playing i.i.d. his mixed move (with no exceptional
moves).

Note that these (behavioral) strategies are specified only for uninformative
histories of the type $0 \ldots 0$. As soon as a signal other than 0 appears, both
players know the true payoff matrix G^k and the payoff stream in the supergame is assumed to be "absorbed" at $v(G^k)$ from that stage on.

The payoffs in $\overline{G}$ when x and y are used is defined as the asymptotic payoff
corresponding to these strategies in the finite game [for formal definitions see
Waternaux (1983a, 1983b) or MSZ (1993)].

Note that the players are not symmetric in $\overline{G}$ since this game is designed to
provide the upper value $\overline{v}(\Gamma)$. In a dual way we define the strategy sets $\underline{X}$ and
$\underline{Y}$ for the game $\underline{G}$ which provides the lower value $\underline{v}(\Gamma)$.

We first have:

Proposition 6.1. *The game* $\overline{G}$ *has a value* $v(\overline{G})$ *and both players have*
ϵ-*optimal strategies.*

Theorem 6.2. (i) $\overline{v}(\Gamma)$ *exists and equals* $v(\overline{G})$.

(ii) *Player II has an* ϵ-*optimal strategy which is a finite mixture of i.i.d.
sequences, each of which is associated with a finite number of exceptional
moves, uniformly distributed before some stage* N_0.

(iii) *Dual results hold for* $\underline{v}(\Gamma)$.

It follows that the game under consideration has a value iff $v(\overline{G}) = v(\underline{G})$, which is generally not the case [examples of games where $v(\overline{G}) \neq v(\underline{G})$, were exhibited by Mertens and Zamir (1976a))]. In view of this, one is led to study $\lim v_n$ and $\lim v_\lambda$ [Sorin (1989)]. Sorin's approach is similar to that adopted in the study of Γ_∞; that is, using more manageable auxiliary "approximating games" as follows. For each L in $\mathbb{N}$ we construct a game G_L. The heuristic interpretation of G_L is Γ_n played in L large blocks, during each of which both players use stationary strategies, except for some singular moves. The strategy sets in G_L are $\overline{X}^L$ and $\overline{Y}^L$, where

$$\overline{X} = \bigcup_{S' \subset S} \Delta(S') \times \mathbb{N}^{S \setminus S'},$$

$$\overline{Y} = \bigcup_{T' \subset T} \Delta(T') \times \mathbb{N}^{T \setminus T'}.$$

Again, the payoff to a pair of strategies is defined as the corresponding asymptotic average payoff (as the block size tends to ∞). Then we have first:

Proposition 6.3. G_L *has a value* w_L *and both players have* ϵ-*optimal strategies.*

Theorem 6.4. $\lim_{n \to \infty} v_n$ *and* $\lim_{L \to \infty} w_L$ *exist and coincide.*

Then, a similar construction gives:

Theorem 6.5. $\lim v_\lambda$ *exists and* $\lim_{\lambda \to 0} v_\lambda = \lim_{L \to \infty} w_L$.

7. A game with state dependent signaling

For games with incomplete information on two sides, the general results so far are mainly those described in Section 4. In that section we considered the special case in which the signals provided to the players after each stage do not depend on the state k (but only on the player's moves). When the signals depend also on the states, we have results only for two special cases: the symmetric case (Section 5), and "games with no signals" (Section 6).

In this section we briefly introduce another game with state dependent signals which was studied by Sorin (1985b). This work illustrates an example of a game at the forefront of the research in games with incomplete information. It is not only strongly related to stochastic games (as were the games studied in the previous two sections), but it involves what may be called *stochastic games with incomplete information*.

Consider the class of games with lack of information on both sides (and state dependent signaling) given by the following data: $K = \{0, 1\}^2 = L \times M$ [we write $k = (l, m)$], and the probability on K is the product $p \otimes q$ of its marginals.

At stage 0, player I is informed about l and player II about m. The payoffs are defined by 2×2 payoffs matrices G^{lm}, and the signaling matrices are given by

$$A^{11} = \begin{pmatrix} T & L \\ c & d \end{pmatrix}, \qquad A^{10} = \begin{pmatrix} T & R \\ c & d \end{pmatrix},$$

$$A^{01} = \begin{pmatrix} B & L \\ c & d \end{pmatrix}, \qquad A^{00} = \begin{pmatrix} B & R \\ c & d \end{pmatrix}.$$

The special features of this information structure to be noted are:

(a) The signals include the moves.

(b) As soon as player I plays Top, the "type" of one of the players is revealed: l if player II played Left at that stage, m if he played Right.

Denoting this game by $\Gamma(p, q)$, we note first that as soon as $p^1 p^0 q^1 q^0 = 0$, it is reduced to a game with incomplete information on one side (treated in Section 2). In particular it has a value $v(p, q)$.

Sorin has given explicit expressions for the minmax and maxmin of these games, which will not be given here. We just mention that these also rely on a family of auxiliary games which are of the form:

$$G^1 = \begin{pmatrix} a_{11}^* & a_{12}^* \\ a_{21} & a_{22} \end{pmatrix}, \qquad G^0 = \begin{pmatrix} b_{11}^* & b_{12}^* \\ b_{21} & b_{22} \end{pmatrix}, \qquad p = (p^1, p^0).$$

That is, the auxiliary games are repeated games of incomplete information on one side in which the games G^1 and G^0 are stochastic games with absorbing states (more specifically, "Big match" type games). In fact, when studying the game under consideration Sorin found the minmax and the maxmin of this family of games and of the dual family in which the absorbing states are in the columns, i.e. in the control of the uninformed player:

$$G^1 = \begin{pmatrix} a_{11}^* & a_{12} \\ a_{21}^* & a_{22} \end{pmatrix}, \qquad G^0 = \begin{pmatrix} b_{11}^* & b_{12} \\ b_{21}^* & b_{22} \end{pmatrix}, \qquad p = (p^1, p^0).$$

The analysis of these games, which is beyond the scope of this review, is rather deep and involves new ideas and tools developed specifically for this purpose.

8. Miscellaneous results

In this section we mention some interesting results which somehow remained isolated and were not followed by further research.

8.1. Discounted repeated games with incomplete information

Mayberry (1967) studied a game with incomplete information on one side, full monitoring, and λ-discounted payoff. Specifically, he considered the game in Example 1.1. Denoting this game by $\Gamma_\lambda(p)$ and its value by $v_\lambda(p)$, he first derived the following formula:

$$v_\lambda(p) = \max_{s,t}\{\lambda \min(ps, p't') + (1-\lambda)(\bar{s}v_\lambda(ps/\bar{s}) + \bar{s}'v_\lambda(ps'/\bar{s}'))\}. \quad (5.5)$$

Here, for $x \in [0, 1]$, x' stands for $(1-x)$, and $(s, t) \in X^2$ is the pair of mixed moves used by player I in one stage [that is, play $(s, 1-s)$ in game G^1 and $(t, 1-t)$ in game G^2], and $\bar{s} = ps + p't$.

Using this formula and the concavity of v_λ, it can be proved that the value of v_λ at any rational $p = n/m \le 1/2$ is given in terms of v_λ at some other rational numbers q with smaller denominator.

By differentiating (5.5) we obtain (letting $v'_\lambda = \mathrm{d}v_\lambda/\mathrm{d}p$):

$$v'_\lambda(p) = (1-\lambda)(1-p/p')v'_\lambda(p/p') - (1-\lambda)v_\lambda(p/p'). \quad (5.6)$$

From this it follows (using the symmetry of v_λ) that for $2/3 < \lambda < 1$, the function has a left derivative and a right derivative at $p = 1/2$, *but they are not equal*.

By induction on the denominator, one can then prove that for any rational p, the sequence of derivatives obtained by repeated use of equation (5.6) leads to an expression for $v'_\lambda(p)$ in terms of $v'_\lambda(0)$, $v'_\lambda(1)$ and $v'_\lambda(1/2)$.

Combining the last two results we conclude, for $2/3 < \lambda < 1$, that although v_λ is concave, it has discontinuous derivatives at every rational point.

8.2. Sequential games

Sequential games with incomplete information were first studied by Ponssard in a series of papers [Ponssard (1975a, 1975b, 1976)] and also by Ponssard and Zamir (1973), Ponssard and Sorin (1980a, 1980b) and Sorin (1979). The basic model is the following. The players' type sets are $K = \{1, \ldots, L\}$ for player I and $R = \{1, \ldots, M\}$ for player II. For each pair of types (k, r) the corresponding payoff matrix is $G^{kr} = (G^{kr}_{st})_{s \in S, t \in T}$. For each $p \in P = \Delta(K)$ and $q \in Q = \Delta(R)$, the n-stage sequential game $\Gamma_n(p, q)$ is played as follows:

- At stage 0, a chance move chooses independently k according to p and r according to q. Player I is informed of (his type) k and player II is informed of r.

- At stage m ($m = 1, \ldots, n$), knowing $h_m = (s_1, t_1, \ldots, s_{m-1}, t_{m-1})$, the history of the moves up to that stage, player I chooses s_m in S. This is told to player II who then, knowing $h'_m = (s_1, t_1, \ldots, s_{m-1}, t_{m-1}, s_m)$, chooses t_m.
- At the end of n stages player II pays player I the amount $(1/n) \sum_{m=1}^{n} G^{kr}_{s_m t_m}$.

Let $v_n(p, q)$ denote the value of $\Gamma_n(p, q)$ and $v(p, q) = \lim_{n \to \infty} v_n(p, q)$. Clearly, by normalizing the strategies of player II at each stage, this is shown to be a special case of the simultaneous repeated games discussed in Section 3, in which the payoff matrices are of size $|S| \times |T|^{|S|}$ (a move of player II is an element of T depending on the choice of player I at that stage). However, it turns out that stronger results hold for this case because of its special structure.

The (behavior) strategies σ_n and τ_n of $\Gamma_n(p, q)$ are defined in the natural way as sequences of mappings from the player's type and available history to the set of his mixed moves [$\Delta(S)$ and $\Delta(T)$, respectively]. The non-revealing game is again the one-shot sequential game with payoff matrix $G(p, q) = \sum_{k,r} p^k q^r G^{k,r}$, and its value is therefore

$$u(p, q) = \max_{s} \min_{t} \sum_{k,r} p^k q^r G^{kr}.$$

8.2.1. Incomplete information on one side

For incomplete information on the side of player II (the minimizer and the second to move), it was proved by Ponssard and Zamir (1973) that:

Proposition 8.1. *For all $p \in P$, $v_1(p) = \text{Cav}_p u(p)$.*

Using the monotonicity of $v_n(p)$ (Proposition 3.19), one has:

Corollary 8.2. $v_n(p) = \text{Cav}_p u(p)$, *for all n and all $p \in P$. Consequently* $\lim_{n \to \infty} v_1(p) = \text{Cav}_p u(p)$.

8.2.2. Incomplete information on two sides

In this case one can prove a recursive formula for $v_n(p, q)$ which is much simpler than the corresponding formula for the general simultaneous move game:

$$v_{n+1}(p, q) = \frac{1}{n+1} \text{Cav}_p \max_{s} \text{Vex}_q \min_{t} \left\{ \sum_{k,r} p^k q^r G^{kr}_{st} + n v_n(p, q) \right\}.$$

Using this, it was proved by Sorin (1979) that for all p and all q the sequence $v_n(p, q)$ is increasing (and therefore it converges), the speed of convergence is bounded by

$$0 \leqslant v(p, q) - v_n(p, q) \leqslant \frac{C}{n}$$

for some positive constant C, and that this is the best bound.

8.3. A game with incomplete information played by "non-Bayesian players"

Megiddo (1980) considered a game with incomplete information on one side in which there is no given prior on the states of nature. More specifically, the uninformed player II knows only the set of his moves (columns) and is told his payoff at each stage. Megiddo provided an algorithm to construct an optimal strategy for the uninformed "non-Bayesian" player. Basically the algorithm considered a dense grid of games with a given number of columns, and tested statistically the performance of each strategy which is optimal in one of these games.

Looking carefully at the problem, it turns out that this result can be derived as a consequence of the general results in Section 2 along the following lines [Mertens (1987)].

(a) Assume first that the unknown payoff matrix is an element of a finite set $(G^k)_{k \in K}$ of matrices having the same set J of columns and any (finite or infinite) number of rows.

(b) Since player II is told his payoff at each stage, any non-revealing strategy $\sigma \in NR(p)$ yields the same distribution of payoffs, and a fortiori the same expected payoff, in all games G^k in the support of p, for all columns $j \in J$. It follows that $u(p)$ is constant in the interior of each facet of the simplex $\Delta(K)$. Since u is upper-semicontinuous, this implies that Cav u *is linear in p* on $\Delta(K)$.

(c) Since Cav $u(p)$ is linear, player II has a strategy (in Γ_∞) which guarantees $v(G^k)$ if the true state is k, for all k. In fact, any optimal strategy $\tau(p)$ of player II at some interior point p has this property [otherwise player I could obtain against $\tau(p)$ strictly more than $v(G^{k_0})$ at some state k_0 and, by playing optimally at each other state, he could get strictly more than $\Sigma_k p^k v(G^k) =$ Cav $u(p)$, contradicting the optimality of $\tau(p)$].

(d) These results are valid not only for a finite state set K but also for a countable K. In particular, if we consider the countable set $\mathcal{G}$ of all finite matrices with J columns and rational entries, it follows that if the true game is in $\mathcal{G}$, then player II has a strategy τ' which guarantees its value. To extend this to any real entries, we perform the following approximation procedure.

(e) For any $\epsilon > 0$ let τ'_ϵ be the strategy of player II which consists of playing τ' while "rationalizing" the histories as follows: if the announced payoff (at some stage) is α, replace it by a rational number $r(\alpha) \geq \alpha$ such that $r(\alpha) - \alpha < \epsilon$. Clearly, for any play of the game induced by τ_ϵ^r there is a $G^\epsilon \in \mathcal{G}$ with $\|G^\epsilon - G\| < \epsilon$ such that if it was the true game instead of G, it would have induced the same play when player II is using τ^r, and hence the expected payoff would be at most $v(G^\epsilon)$, which is at most $v(G) + \epsilon$.

We conclude that for any $\epsilon > 0$ player II has a strategy τ_ϵ^r which guarantees $v(G) + \epsilon$.

(f) Finally, choose a sequence $\{\epsilon_n\}_{n=1}^\infty$ decreasing to 0 and play successively $\tau_{\epsilon_n}^r$ in large blocks with appropriately increasing sizes so that the resulting strategy guarantees $v(G) + \epsilon_n$ for all n and hence it guarantees $v(G)$. $\square$

The main idea of this argument is that the announcement of the payoffs induces the linearity of Cav u, which in turn implies the existence of a strategy for the uninformed player which is uniformly optimal for all prior distributions on K. This is the sense in which the player is non-Bayesian, since he does not need any prior in order to play his optimal strategy.

8.4. A stochastic game with signals

Ferguson, Shapley and Weber (1970) considered the following game which was the first treated example of a stochastic game with incomplete information.

We are given two states of nature with the following payoff matrices.:

$$G^1 = \begin{pmatrix} 1 & 0 \\ 1 & 0 \end{pmatrix}, \qquad G^2 = \begin{pmatrix} 0 & 1 \\ 0 & 0 \end{pmatrix}.$$

The transition probability from state 1 to state 2 is a constant $(1 - \pi) \in (0, 1)$, independent of the moves. The reverse transition, from state 2 to state 1, takes place if and only if player I plays Bottom. Player I knows everything while player II is told only the times of the transition from 2 to 1.

Let us consider Γ_λ, the discounted game starting from $k = 1$, and write v_λ for its value. It can be shown that

$$v_\lambda = \frac{[1 - (1 - \lambda)^k][1 - \pi(1 - \lambda)] - \lambda\{[1 - 2[\pi(1 - \lambda)]^k\}}{[1 - (1 - \lambda)^{k+1}][1 - \pi(1 - \lambda)] + 2(1 - \lambda)^{k+1}\pi^k\lambda}.$$

Letting

$$v_0 = \lim_{\lambda \to 0} v_\lambda = \frac{r(1 - \pi) - (1 - 2\pi^r)}{(r + 1)(1 - \pi) + 2\pi^r}$$

where r is the positive integer satisfying $\pi^{r-1} > 1/2$, and $\pi^r \leq 1/2$, one can then find optimal strategies σ^* and τ^* for the two players such that for each ϵ, each player (with his optimal strategy), can ϵ-guarantee v_0 in all Γ_λ with sufficiently small λ [for details see MSZ (1993)].

Bibliography

Aumann, R.J. (1964) 'Mixed and behavior strategies in infinite extensive games', in: M. Dresher et al., eds., *Advances in game theory*, Annals of Mathematical Studies, Vol. 52: Princeton: Princeton University Press, pp. 627–650.

Aumann, R.J. (1985) 'Repeated games', in: G.R. Feiwel, ed., *Issues in contemporary micro-economics and welfare*. New York: Macmillan, pp. 109–242.

Aumann, R.J. and M. Maschler (1966) *Game-theoretic aspects of gradual disarmament*, Reports to the U.S. Arms Control and Disarmament Agency, ST-80, Chapter V, pp. 1–55.

Aumann, R.J. and M. Maschler (1967) *Repeated games with incomplete information*: *A survey of recent results*, Reports to the U.S. Arms Control and Disarmament Agency, ST-116, Chapter III, pp. 287–403.

Aumann, R.J. and M. Maschler (1968) *Repeated games of incomplete information*: *The zero-sum extensive case*, Reports to the U.S. Arms Control and Disarmament Agency ST-143, Chapter II, pp. 37–116.

Aumann, R.J., M. Maschler and R. Stearns (1968) *Repeated games of incomplete information*: *An approach to the nonzero sum case*, Reports to the U.S. Arms Control and Disarmament Agency, ST-143, Chapter IV, pp. 117–216.

Bewley, T. and E. Kohlberg (1976a) 'The asymptotic theory of stochastic games', *Mathematics of Operations Research*, 1: 197–208.

Bewley, T. and E. Kohlberg (1976b) 'The asymptotic solution of a recursion equation occurring in stochastic games', *Mathematics of Operations Research*, 1: 321–336.

Bewley, T. and E. Kohlberg (1978) 'On stochastic games with stationary optimal strategies', *Mathematics of Operations Research*, 3: 104–125.

Blackwell, D. (1956) 'An analog of the minmax theorem for vector payoffs', *Pacific Journal of Mathematics*, 65: 1–8.

Blackwell, D. and T.S. Ferguson (1968) 'The big match', *Annals of Mathematical Statistics*, 39: 159–163.

Dalkey, N. (1953) 'Equivalence of information patterns and essentially determinate games', in: H.W. Kuhn and A.W. Tucker, eds., *Contributions to the theory of games*, Vol. II, Annals of Mathematics Study, 28: Princeton: Princeton University Press, pp. 217–243.

Ferguson, T.S., L.S. Shapley and R. Weber (1970) 'A stochastic game with incomplete information'. Mimeograph.

Forges, F. (1982) 'Infinitely repeated games of incomplete information: symmetric case with random signals', *International Journal of Game Theory*, 11: 203–213.

Harsanyi, J.C. (1967–68) 'Games of incomplete information played by Bayesian players. Parts I, II, III', *Management Science*, 14: 159–182, 320–334, 486–502.

Kohlberg, E. (1974) 'Repeated games with absorbing states', *Annals of Statistics*, 2: 724–738.

Kohlberg, E. (1975a) 'Optimal strategies in repeated games with incomplete information', *International Journal of Game Theory*, 4: 7–24.

Kohlberg, E. (1975b) 'The information revealed in infinitely-repeated games of incomplete information', *International Journal of Game Theory*, 4: 57–59.

Kohlberg, E. and S. Zamir (1974) 'Repeated games of incomplete information: the symmetric case', *Annals of Statistics*, 2: 1040–1041.

Lehrer, E. (1987) 'A note on the monotonicity of v_n', *Economic Letters*, 23: 341–342.

Mayberry, J.-P. (1967) 'Discounted repeated games with incomplete information', *Mathematica*, ST-116, Ch. V, pp. 435–461.

Megiddo, N. (1980) 'On repeated games with incomplete information played by non-Bayesian players', *International Journal of Game Theory*, 9: 157–167.

Mertens, J.-F. (1972) 'The value of two-person zero-sum repeated games: The extensive case', *International Journal of Game Theory*, 1: 217–225.

Mertens, J.-F. (1973) 'A note on "the value of two-person zero-sum repeated games: The extensive case"', *International Journal of Game Theory*, 9: 189–190.

Mertens, J.-F. (1982) 'Repeated games: An overview of the zero-sum case', in: W. Hildenbrand, ed., *Advances in economic theory*, Cambridge: Cambridge University Press.

Mertens, J.-F. (1986) 'The minmax theorem for u.s.c.-l.s.c. payoff functions', *International Journal of Game Theory*, 10(2): 53–56.

Mertens, J.-F. (1987) 'Repeated games', *Proceedings of the International Congress of Mathematicians* (Berkeley), 1986, American Mathematical Society. pp. 1528–1577.

Mertens, J.-F. and A. Neyman (1981) 'Stochastic games', *International Journal of Game Theory*, 10(2): 53–56.

Mertens, J.-F. and A. Neyman (1982) 'Stochastic games have a value', *Proceedings of the National Academy of Sciences of the U.S.A.*, 79: 2145–2146.

Mertens, J.-F. and S. Zamir (1971–72) 'The value of two-person zero-sum repeated games with lack of information on both sides', *International Journal of Game Theory*, 1: 39–64.

Mertens, J.-F. and S. Zamir (1976a) 'On a repeated game without a recursive structure', *International Journal of Game Theory*, 5: 173–182.

Mertens, J.-F. and S. Zamir (1976b) 'The normal distribution and repeated games', *International Journal of Game Theory*, 5: 187–197.

Mertens, J.-F. and S. Zamir (1977a) 'The maximal variation of a bounded martingale', *Israel Journal of Mathematics*, 27: 252–276.

Mertens, J.-F. and S. Zamir (1977b) 'A duality theorem on a pair of simultaneous functional equations', *Journal of Mathematical Analysis and Application*, 60: 550–558.

Mertens, J.-F. and S. Zamir (1980) 'Minmax and Maxmin of repeated games with incomplete information', *International Journal of Game Theory*, 9: 201–215.

Mertens, J.-F. and S. Zamir (1981) 'Incomplete information games with transcendental values', *Mathematics of Operations Research*, 6: 313–318.

Mertens, J.-F. and S. Zamir (1985) 'Formulation of Bayesian analysis for games with incomplete information', *International Journal of Game Theory*, 14: 1–29.

Mertens, J.-F. and S. Zamir (1990) 'Incomplete information games and the normal distribution', mimeograph.

Mertens, J.-F., S. Sorin and S. Zamir (1993) *Repeated games*, forthcoming.

De Meyer, B. (1989) 'Repeated games and multidimensional normal distribution', CORE Discussion Paper 89, Université Catholique de Louvain, Louvain-la-Neuve, Belgium.

Ponssard, J.-P. (1975a) 'Zero-sum games with "almost" perfect information', *Management Science*, 21: 794–805.

Ponssard, J.-P. (1975b) 'A note on the L-P formulation of zero-sum sequential games with incomplete information', *International Journal of Game Theory*, 4: 1–5

Ponssard, J.-P. (1976) 'On the subject of nonoptimal play in zero-sum extensive games: The trap phenomenon', *International Journal of Game Theory*, 5: 107–115.

Ponssard, J.-P. and S. Sorin (1980a) 'The LP formulation of finite zero-sum games with incomplete information', *International Journal of Game Theory*, 9: 99–105.

Ponssard, J.-P. and S. Sorin (1980b) 'Some results on zero-sum games with incomplete information: The dependent case', *International Journal of Game Theory*, 9: 233–245.

Ponssard, J.-P. and S. Sorin (1982) 'Optimal behavioral strategies in zero-sum games with almost perfect information', *Mathematics of Operations Research*, 7: 14–31.

Ponssard, J.-P. and S. Zamir (1973) 'Zero-sum sequential games with incomplete information', *International Journal of Game Theory*, 2: 99–107.

Sorin, S. (1979) 'A note on the value of zero-sum sequential repeated games with incomplete information', *International Journal of Game Theory*, 8: 217–223.

Sorin, S. (1980) 'An introduction to two-person zero-sum repeated games with incomplete information', IMSS-Economics, Stanford University, TR 312. French version in *Cahiers du Groupe de Mathématiques Economiques*, 1: Paris (1979).

Sorin, S. (1984a) '"Big match" with lack of information on one side (part I)', *International Journal of Game Theory*, 13: 201–255.

Sorin, S. (1984b) 'On a pair of simultaneous functional equations', *Journal of Mathematical Analysis and Applications*, 98(1): 296–303.

Sorin, S. (1985a) '"Big match" with lack of information on one side (part II)', *International Journal of Game Theory*, 14: 173–204.

Sorin, S. (1985b) 'On a repeated game with state dependent signaling matrices', *International Journal of Game Theory*, 14: 249–272.

Sorin, S. (1986) 'On repeated games with complete information', *Mathematics of Operations Research*, 11: 147–160.

Sorin, S. (1989) 'On repeated games without a recursive structure: Existence of lim v_n', *International Journal of Game Theory*, 18: 45–55.

Sorin, S. and S. Zamir (1985) 'A 2-person game with lack of information on $1\frac{1}{2}$ sides', *Mathematics of Operations Research*, 10: 17–23.

Sorin, S. and S. Zamir (1991) '"Big match" with lack of information on one side (III)', in: T.E.S. Raghavan, T.S. Ferguson, T. Parthasarthy and O.J. Vrieze, eds., *Stochastic games and related topics in honor of Professor L.S. Shapley*. Kluwer Academic Publishers, pp. 101–112.

Stearns, R.E. (1967) 'A formal information concept for games with incomplete information', *Mathematica*, ST-116, Ch. IV, pp. 405–433.

Waternaux, C. (1983a) 'Solution for a class of games without recursive structure', *International Journal of Game Theory*, 12: 129–160.

Waternaux, C. (1983b) 'Minmax and maxmin of repeated games without a recursive structure', Core Discussion Paper 8313, Université Catholique de Louvain, Louvain-la-Neuve, Belgium, to appear in *International Journal of Game Theory*.

Zamir, S. (1971–72) 'On the relation between finitely and infinitely repeated games with incomplete information', *International Journal of Game Theory*, 1: 179–198.

Zamir, S. (1973a) 'On repeated games with general information function', *International Journal of Game Theory*, 2: 215–229.

Zamir, S. (1973b) 'On the notion of value for games with infinitely many stages', *Annals of Statistics*, 1: 791–796.

Chapter 6

REPEATED GAMES OF INCOMPLETE INFORMATION: NON-ZERO-SUM

FRANÇOISE FORGES

C.O.R.E., Université Catholique de Louvain

Contents

Handbook of Game Theory, Volume 1, Edited by R.J. Aumann and S. Hart

1. Introduction

Non-zero-sum infinitely repeated games with incomplete information are an appropriate model to analyze durable relationships among individuals whose information is not symmetric. Two particular cases have been the subject of the two previous chapters. The study of infinitely repeated games of *complete information* (Chapter 4) shows that cooperation may result from the threat of punishment in the future. Repetition appears as an *enforcement mechanism*. In *zero-sum* infinitely repeated games of incomplete information (Chapter 5), the problems of strategic information transmission can be investigated on their own, independently of any cooperation effect. Repetition appears as a *signalling mechanism*. The results in the zero-sum case will be used explicitly in the non-zero-sum case. Indeed, in the tradition of the Folk theorem, the characterization of equilibria will make use of individual rationality conditions.

In this chapter the repetition of the game will have the effects of an enforcement mechanism and of a signalling mechanism at the same time. As in the zero-sum case, the pioneering work has been done in the Mathematica papers [Aumann, Maschler and Stearns (1968)]. Most results available at the moment concern a particular model: there are two players, exactly one of whom is *completely informed* of the situation (this is called "*lack of information on one side*"), with each player observing the actions of the other after every stage ("*full monitoring*"). Nash equilibria have been studied, as have extensions of this solution concept, like the correlated equilibria introduced by Aumann (1974) (see a forthcoming volume).

The model and its basic properties are presented in Section 2. The main result of Section 3 (presented in Subsection 3.1) is a characterization of Nash equilibria in infinitely repeated games with lack of information on one side [Hart (1985), Aumann and Hart (1986)]; the results of the zero-sum case are used to evaluate the individually rational levels of each player. Subsection 3.2 deals with a special class of games, referred to as games "with known own payoffs". In these, equilibria are characterized in a very transparent way. The model provides insights for the study of repeated games of *complete* information [Shalev (1988), Koren (1988), Israeli (1989)]. The main open question in the present context is the *existence* of equilibrium. It is treated in Subsection 3.3. The only "general" statement applies to repeated games with lack of information on one side and two states of nature [Sorin (1983)].

In Section 4, communication equilibria (which include correlated equilibria as a particular case) are studied. The main theme there is that an infinitely repeated game contains enough communication possibilities to obtain the equivalence of different solution concepts. At best, one could hope that the observation of private correlated signals before the beginning of the repeated game (as in Aumann's correlated equilibrium), together with the structure of the repeated game itself, would be sufficient to obtain the effect of any coordinating device, acting at every stage of the game (preserving the non-cooperative character). Results in this direction have been obtained in Forges (1985, 1988). In relating this chapter to the previous one, it is worth mentioning that to obtain the appropriate individual rationality conditions in the context of correlated equilibria of infinitely repeated games with lack of information *on one side*, one is led to use the theory of infinitely repeated zero-sum games with lack of information *on both sides*. This shows again that a deep knowledge of the zero-sum case may be necessary before starting the study of the non-zero-sum case.

2. Basic definitions

Let us introduce the following terminology:

K = finite set of states of nature (or of types of player 1):
p = probability distribution on K; $p \in \Delta^K$, the unit simplex of $\mathbb{R}^K$; it is assumed that $p^k > 0$, $\forall k \in K$;
I, J = finite sets of actions of player 1 and player 2, respectively (containing at least two elements, i.e. $|I|, |J| \geqslant 2$);
A^k, B^k = payoff matrices (of dimensions $|I| \times |J|$) for player 1 and player 2, respectively, in state $k \in K$.

The two-person infinitely repeated game $\Gamma(p)$ is described as follows. Once k is chosen according to p, it is told to player 1 only and kept fixed throughout the game; then at every stage t ($t = 1, 2, \ldots$), player 1 and player 2 simultaneously make a move i_t in I and j_t in J, respectively. The pair of moves (i_t, j_t) [but not the stage payoffs $A^k(i_t, j_t)$, $B^k(i_t, j_t)$] is announced to both players.

Strategies are defined in the natural way [as in the zero-sum case, see Chapter 5; see also Aumann (1964)]. In the present context, it is convenient to define a workable payoff function, for instance using the limit of means criterion and a Banach limit L (in that respect, the approach is similar in repeated games of complete information, see Chapter 4). Given a sequence of moves $(i_t, j_t)_{t \geqslant 1}$, the average payoffs for the n first stages are

$$a_n^k = \frac{1}{n} \sum_{t=1}^{n} A^k(i_t, j_t), \quad a_n = (a_n^k)_{k \in K},$$

$$b_n^k = \frac{1}{n} \sum_{t=1}^{n} B^k(i_t, j_t), \quad b_n = (b_n^k)_{k \in K}.$$

The prior probability distribution p on K and a pair of strategies (σ, τ) in $\Gamma(p)$ induce a probability distribution on these average payoffs, with expectation denoted $E_{p,\sigma,\tau}$. Using a transparent notation, write $\sigma = (\sigma^k)_{k \in K}$. The payoff associated with (σ, τ) is $(a(\sigma, \tau), \beta(\sigma, \tau))$, where $a(\sigma, \tau)$ is *player 1's* limit expected *vector payoff* and $\beta(\sigma, \tau)$ is *player 2's* limit expected *payoff*:

$$a(\sigma, \tau) = (a^k(\sigma, \tau))_{k \in K} , \quad a^k(\sigma, \tau) = L[E_{p,\sigma,\tau}(a_n^k \mid k)] ,$$

$$\beta(\sigma, \tau) = L[E_{p,\sigma,\tau}(b_n^\kappa)] = \sum_{k \in K} p^k L[E_{p,\sigma,\tau}(b_n^k \mid k)] ,$$

where κ stands for the state of nature as a random variable. Observe that the conditional expectation given k corresponds to the probability distribution induced by σ^k and τ. It is necessary to refer to the conditional expected payoff of player 1, given type k, to express individual rationality or incentive compatibility conditions.

Let γ be a constant bounding the payoffs ($\gamma = \max_{k,i,j}\{|A^k(i, j)|, |B^k(i, j)|\}$) and let $\mathbb{R}_\gamma = [-\gamma, \gamma]$. Throughout the chapter, (a, β) is used for a payoff in $\mathbb{R}_\gamma^K \times \mathbb{R}_\gamma$; state variables of the form $(p, a, \beta) \in \Delta^K \times \mathbb{R}_\gamma^K \times \mathbb{R}_\gamma$ have also to be considered, with (a, β) as the payoff in $\Gamma(p)$.

Let $F \subseteq \mathbb{R}_\gamma^K \times \mathbb{R}_\gamma^K$ be the set of *feasible* vector payoffs in the one-shot game, using a correlated strategy (i.e. a joint distribution over $I \times J$):

$$F = \mathrm{co}\{((A^k(i, j))_{k \in K}, (B^k(i, j))_{k \in K}) : (i, j) \in I \times J\} ,$$

where co denotes the convex hull. Let $\pi \in \Delta^{I \times J}$. A typical element of F is defined by

$$A^k(\pi) = \sum_{i,j} \pi_{ij} A^k(i, j) , \qquad B^k(\pi) = \sum_{i,j} \pi_{ij} B^k(i, j) . \tag{1}$$

Let $a(p)$ [resp. $b(p)$] be the value for player 1 (resp. player 2) of the one-shot game with payoff matrix $p \cdot A = \Sigma_k p^k A^k$ (resp. $p \cdot B$).

A vector payoff $x = (x^k)_{k \in K} \in \mathbb{R}_\gamma^K$ is *individually rational* for player 1 in $\Gamma(p)$ if

$$q \cdot x \geq a(q) , \quad \forall q \in \Delta^K . \tag{2}$$

This definition is justified by Blackwell's approachability theorem [Blackwell (1956)]. Consider an infinitely repeated zero-sum game with vector payoffs, described by matrices C^k, $k \in K$. A set $S \subseteq \mathbb{R}^K$ is said to be "approachable" by the minimizing player if he can force the other player's payoff to belong to S.

By Blackwell's characterization, a closed convex set S is approachable if and only if

$$\max_{s \in S}(q \cdot s) \geq c(q), \quad \forall q \in \Delta^K,$$

where $c(q)$ is the value of the expected one-shot game with payoff matrix $q \cdot C = \Sigma_k q^k C^k$. (2) is thus a necessary and sufficient condition for player 2 to have a strategy τ in $\Gamma(p)$ such that for every $k \in K$, the payoff of player 1 of type k does not exceed x^k, whatever his strategy (see also Chapter 5, Section 2.)

Let X be the set of all vector payoffs $x \in \mathbb{R}_\gamma^K$ such that (2) is satisfied. X can be interpreted as a set of punishments of player 2 against player 1.

In an analogous way, a payoff $\beta \in \mathbb{R}_\gamma$ is *individually rational* for player 2 in $\Gamma(p)$ if

$$\beta \geq \text{vex } b(p) \tag{3}$$

where vex b is the greatest convex function on Δ^K below b.

Like (2), this is justified by the results of Aumann and Maschler (1966) (see Chapter 5); the value of the zero-sum infinitely repeated game with payoff matrices $-B^k$ to player 1 (the informed player) is $\text{cav}(-b(p)) = -\text{vex } b(p)$. Hence (3) is a necessary and sufficient condition for player 1 to have a strategy σ in $\Gamma(p)$ such that player 2's expected payoff does not exceed β, no matter what his strategy is.

The strategy σ uses player 1's information and depends on p. In the development of the chapter, it will be useful to know punishments that player 1 can apply against player 2, whatever the probability distribution of the latter over K. An analog of the set X above can be introduced.

Let Φ be the set of all mappings $\varphi : \Delta^k \to \mathbb{R}_\gamma$ which are convex, Lipschitz of constant γ and such that $\varphi \geq b$.

From Blackwell's theorem mentioned earlier $\varphi \in \Phi$ is a necessary and sufficient condition for the set

$$\{y \in \mathbb{R}_\gamma^K : q \cdot y \leq \varphi(q) \quad \forall q \in \Delta^K\}$$

of vector payoffs of player 2 to be approachable by player 1 in the game with vector payoff matrix $(B^k)_{k \in K}$. If $\varphi \in \Phi$, player 1 has a non-revealing strategy in $\Gamma(p)$ such that for every $q \in \Delta^K$, the expected payoff of player 2 does not exceed $\varphi(q)$, no matter what he does. Φ can thus be interpreted as a set of non-revealing punishments by player 1 against player 2. Obviously, player 1 can use his information by choosing φ as a function of the state of nature. The important property is that punishments in Φ do not require that player 1 knows player 2's beliefs over K.

3. Nash equilibria

3.1. The "standard one-sided information case"

Combining the ideas of the Folk theorem for repeated games of complete information (see Chapter 4) with the definitions of individual rationality derived in the previous section, a description of the *non-revealing* Nash equilibrium payoffs of $\Gamma(p)$ is easily obtained. In these, as long as player 2 does not deviate, player 1's strategy is independent of the state of nature. This set will play a crucial role for the characterization of *all* Nash equilibrium payoffs.

Let G be the set of all triples $(p, a, \beta) \in \Delta^K \times \mathbb{R}_\gamma^K \times \mathbb{R}_\gamma$ such that
- (a, β) is *feasible*: $\exists (c, d) \in F$ such that $a \geq c$ and $p \cdot a = p \cdot c$; $\beta = p \cdot d$.
- a (resp. β) is *individually rational* for player 1 (resp. player 2) in $\Gamma(p)$ [in the sense of (2) and (3)].

This definition of feasibility is adopted (instead of the more natural $a = c$) for later use in the characterization of all Nash equilibrium payoffs. Towards this aim, all values of p are considered because later, p will vary with the revelation of information.

Payoffs (a, β) satisfying the above two properties will be referred to as "non-revealing Nash equilibrium payoffs" of $\Gamma(p)$.

Let us consider a few examples of games with two states of nature; $p \in [0, 1]$ denotes the probability of state 1. Some strategies have been duplicated so that $|I| \geq 2$ and $|J| \geq 2$. In particular, the last examples are *games of information transmission*, where the informed player has no direct influence on the payoffs [these games are akin to the sender–receiver games (see the chapter on 'correlated and communication equilibria' in a forthcoming volume); here, the moves at each stage are used as signals].

Example 1.

$$[A^1, B^1] = \begin{bmatrix} 1,0 & 1,0 \\ 0,0 & 0,0 \end{bmatrix}, \qquad [A^2, B^2] = \begin{bmatrix} 0,0 & 0,0 \\ 1,0 & 1,0 \end{bmatrix}.$$

This game has no non-revealing equilibrium for $0 < p < 1$. Indeed, player 2's moves cannot affect the payoff of player 1 and the latter's (strictly) best move is $i_t = k$ at every stage t, which immediately reveals his type k.

Example 2.

$$[A^1, B^1] = \begin{bmatrix} 1,1 & 0,0 \\ 1,1 & 0,0 \end{bmatrix}, \qquad [A^2, B^2] = \begin{bmatrix} 0,0 & 1,1 \\ 0,0 & 1,1 \end{bmatrix}.$$

For every p, $\Gamma(p)$ has a non-revealing equilibrium payoff (described by player 1's vector payoff and player 2's payoff): $((1,0), p)$ for $p < \frac{1}{2}$, $((0,1), 1-p)$ for $p > \frac{1}{2}$ and all $(a, \frac{1}{2})$ with a on the segment $[(1,0),(0,1)]$ for $p = \frac{1}{2}$. $\Gamma(p)$ also has a completely revealing equilibrium for every p: player 1 chooses $i_1 = k$ at the first stage and player 2 chooses $j_t = i_1$ at every subsequent stage; this yields the payoff $((1,1),1)$. Player 1 has no reason to lie about the state of nature. This is no longer true in

Example 3.

$$[A^1, B^1] = \begin{bmatrix} 1,1 & 0,0 \\ 1,1 & 0,0 \end{bmatrix}, \qquad [A^2, B^2] = \begin{bmatrix} 1,0 & 0,1 \\ 1,0 & 0,1 \end{bmatrix}.$$

Here, independently of the true state, player 1 would pretend that the first state of nature obtained, in order to make player 2 choose $j = 1$.

The next example shows a partially revealing equilibrium, where player 1 uses a type-dependent lottery over his moves (as in the "splitting procedure" described in Chapter 5).

Example 4.

$$[A^1, B^1] = \begin{bmatrix} 1,-3 & 0,0 & 0,1 & 1,2 \\ 1,-3 & 0,0 & 0,1 & 1,2 \end{bmatrix},$$

$$[A^2, B^2] = \begin{bmatrix} 0,2 & 1,1 & 0,0 & 1,-3 \\ 0,2 & 1,1 & 0,0 & 1,-3 \end{bmatrix}.$$

This game has no completely revealing equilibrium. However, $(\frac{1}{4}, (\frac{1}{2}, \frac{1}{2}), \frac{3}{4})$ and $(\frac{3}{4}, (\frac{1}{2}, \frac{1}{2}), \frac{3}{4})$ belong to G (according to the definition introduced at the beginning of Subsection 3.1, $(\frac{1}{2}, \frac{1}{2})$ denotes the vector payoff of player 1 and $\frac{3}{4}$ the payoff of player 2). Take $p = \frac{1}{2}$. If player 1 plays $i_1 = 1$ with probability $\frac{1}{4}$ (resp. $\frac{3}{4}$) if $k = 1$ (resp. $k = 2$) at stage 1, player 2's posterior probability that $k = 1$ is $\frac{1}{4}$ when $i_1 = 1$, $\frac{3}{4}$ when $i_1 = 2$. If no more information is sent to him, player 2 can, at all subsequent stages, choose $j_t = 1$, $j_t = 2$ (resp. $j_t = 3$, $j_t = 4$) with equal probability $\frac{1}{2}$ if $i_1 = 1$ (resp. $i_1 = 2$). This describes an equilibrium because player 1's expected payoff is $(\frac{1}{2}, \frac{1}{2})$, independently of the signal i_1 that he sends.

In Examples 2 and 4 we described equilibria with one single phase of signalling, followed by payoff accumulation. Such scenarios were introduced by Aumann, Maschler, and Stearns (1968) under the name *joint plan*. Formally, a joint plan consists of a set of signals S (a subset of I^t for some t), a signalling strategy (conditional probability distributions $q(\cdot \mid k)$ on S given k, for every $k \in K$), and a correlated strategy $\pi(s) \in \Delta^{I \times J}$ for each $s \in S$; $\pi_{ij}(s)$ is interpreted as the frequency of the pair of moves (i, j) to be achieved after signal s

has been sent. Recalling (1), let $a(s)$ [resp. $b(s)$] be the vector payoff with components $a^k(s) = A^k(\pi(s))$ [resp. $b^k(s) = B^k(\pi(s))$]; $(a(s), b(s)) \in F$. Set also

$$\beta(s) = \sum_{k \in K} p^k b^k(s).$$

To be in equilibrium, a joint plan must be *incentive compatible* for player 1. This means that any signal s with $q(s \mid k) > 0$ must give the *same* expected payoff $a^k(s)$ to player 1 of type k [otherwise, player 1 would send the signal s yielding the highest payoff in state k with probability one instead of $q(s \mid k)$]. We may therefore set $a^k(s) = a^k$ for every s such that $q(s \mid k) > 0$. Obviously, signals of zero probability must only yield an inferior payoff: $a^k(s') \leq a^k$ for every s' such that $q(s' \mid k) = 0$. Let $p(s)$ be the posterior probability distribution over K given s [we implicitly assumed that each signal s has a positive (total) probability]. We have $p(s) \cdot a(s) = p(s) \cdot a$ and $p(s) \cdot b(s) = \beta(s)$ for every $s \in S$.

In order that the players do not deviate from the joint plan after the signalling phase, we must still require that the payoffs $(a, \beta(s))$, to be reached if s has been sent, are *individually rational* [in the sense of (2) and (3), with $p(s)$]. Indeed, when communication is over, one can proceed as for games with complete information (traditional Folk theorem): the deviations of one player are detected by the other, who can punish him at his individually rational level. Of course, player 1 may use his information (and thus reveal it further) in punishing player 2 and the latter must take account of the different possible types of his opponent.

The above reasoning shows that (a, β) is a joint plan equilibrium payoff of $\Gamma(p)$ if (p, a, β) is a convex combination of elements $(p(s), a, \beta(s))$, $s \in S$, of G, all with the same payoff, a, for player 1. New equilibrium payoffs can thus be obtained from G by convexifying it in (p, β) for every fixed a.

Observe that G is convex in (a, β) for every fixed p. More generally, the set of *all* equilibrium payoffs of $\Gamma(p)$ is *convex*. To establish this property, Aumann, Maschler and Stearns (1968) observed that signalling was not the only form of communication available to the players of a repeated game with incomplete information.

In a *jointly controlled lottery*, no information is revealed by player 1 but the players decide together how to continue the play in such a way that unilateral deviations are not profitable. Suppose the players want to select one of M possible values for the subsequent expected payoff, according to a probability distribution $\rho = (\rho_1, \ldots, \rho_r, \ldots, \rho_M) \in \Delta^M$. This can be decided in finitely many stages. At each stage each player chooses one of his two first actions at random with the same probability $\frac{1}{2}$ (recall that I and J contain at least two elements); identical (resp. different) choices of the players are interpreted as a 1 (resp. 0), so that the sequence of moves can be seen as the binary expansion of a uniform random variable u on $[0, 1]$. Set $\rho_0 = 0$; if $u \in [\Sigma_{r=0}^m \rho_r, \Sigma_{r=0}^{m+1} \rho_r)$, $m = 0, \ldots, M - 1$, the players then decide on the $(m + 1)$st outcome, which happens after a finite number of stages. A unilateral deviation of one of the players cannot modify the probability distribution of u. Observe that jointly controlled lotteries enable us to convexify the set of Nash equilibrium payoffs, at any fixed p. Observe also that the roles of the two players are symmetric and that this procedure was not needed in the zero-sum case (see Chapter 5).

At this point we can easily conceive equilibrium payoffs achieved as convex combinations of non-revealing equilibrium payoffs and/or joint plan equilibrium payoffs. They all involve at most a single stage of signalling.

Aumann, Maschler and Stearns (1968) showed in an example that more equilibrium payoffs become available if an additional stage of signalling is permitted [simpler examples illustrating this can also be found in Forges (1984)].

The characterization of the whole set of Nash equilibrium payoffs makes use of sequences of communications where signalling phases alternate with jointly controlled lotteries. To state the result precisely, we need the following concept: a process $(g_t)_{t \geq 1} = (p_t, a_t, \beta_t)_{t \geq 1}$ of $(\Delta^K \times \mathbb{R}_\gamma^K \times \mathbb{R}_\gamma)$-valued random variables (on some probability space) is called a *bi-martingale* if it is a martingale [i.e. $E(g_{t+1} \mid H_t) = g_t$ a.s., $t = 1, 2, \ldots$, for a sequence $(H_t)_{t \geq 1}$ of finite sub σ-fields] such that for each $t = 1, 2, \ldots$, either $p_{t+1} = p_t$ or $a_{t+1} = a_t$ a.s. Let G^* be the set of all $g = (p, a, \beta) \in \Delta^K \times \mathbb{R}_\gamma^K \times \mathbb{R}_\gamma$ for which there exists a bi-martingale $g_t = (p_t, a_t, \beta_t)_{t \geq 1}$ as above, starting at g (i.e. $g_1 = g$ a.s.) and converging a.s. to $g_\infty \in G$.

Theorem 1 [Hart (1985)]. *Let $(a, \beta) \in \mathbb{R}_\gamma^K \times \mathbb{R}_\gamma$; (a, β) is a Nash equilibrium payoff of $\Gamma(p)$ if and only if $(p, a, \beta) \in G^*$.*

The theorem first states that a bi-martingale converging a.s. to an element of G can be associated with any equilibrium payoff in $\Gamma(p)$. Let (σ, τ) be an equilibrium achieving the payoff (a, β) in $\Gamma(p)$. Define $p_t(h_t)$, $a_t(h_t)$, and $\beta_t(h_t)$, respectively, as the conditional probability distribution over K, the expected vector payoff of player 1, and the expected payoff of player 2, before stage t, given the past history (i.e. the sequence of moves) h_t up to stage t [expectations are with respect to the probability distribution induced by (σ, τ)]. Stage t can be split into two half-stages, with the interpretation that player 1 (resp. player 2) makes his move i_t (resp. j_t) at the first (resp. second) half-stage; we thus have $h_{t+1} = (h_t, i_t, j_t)$; let $p_t(h_t, i_t)$, $a_t(h_t, i_t)$, and $\beta_t(h_t, i_t)$ be defined in the same way as above. The process indexed by the half-stages forms a martingale. The *bi* property follows from the *incentive compatibility conditions* for player 1. Assume that at stage t the posterior probability distribution moves from $p_t(h_t)$ to $p_t(h_t, i_t)$. This means that player 1 chooses his move i_t at stage t according to a probability distribution depending on his type k (though it may also depend on the past history). As observed above, the equilibrium condition implies $a_t^k(h_t, i_t) = a^k(h_t)$ for every move i_t of positive probability (given state k). No change in p_t can occur when player 2 makes a move [hence, $p_{t+1}(h_{t+1}) = p_t(h_t, i_t)$]; in this case, a_t can vary. As a bounded martingale, (p_t, a_t, β_t) converges a.s. as $t \to \infty$, say to $(p_\infty, a_\infty, \beta_\infty)$. To see that this must belong to G, observe first that at every stage t, a_t (resp. β_t) must satisfy the individual rationality condition since otherwise player 1 (resp. player 2) would deviate from his equilibrium strategy to obtain his minmax level. This property goes to the limit. Finally, the limit payoff must be feasible in non-revealing strategies. Imagine that the martingale reaches its limit after a finite number of stages, T: $(p_\infty, a_\infty, \beta_\infty) = (p_T, a_T, \beta_T)$; then the game played from stage T on is $\Gamma(p_T)$ and (a_T, β_T) must be a non-revealing payoff in this game. In general, the convergence of p_t shows that less and less information is revealed.

The converse part of the theorem states that the players can achieve any Nash equilibrium payoff (a, β) by applying strategies of a simple form, which generalizes the joint plans. To see this, let us first construct an equilibrium yielding a payoff (a, β) associated with a bi-martingale converging in a finite number of stages, T, i.e. $(p_\infty, a_\infty, \beta_\infty) = (p_T, a_T, \beta_T) \in G$. The first $T - 1$ stages are used for *communication*; from stage T on, the players play for *payoff accumulation*, i.e. they play a non-revealing equilibrium of $\Gamma(p_T)$, with payoff (a_T, β_T).

To decide on which non-revealing equilibrium to settle, the players use the two procedures of communication described above: *signalling* and *jointly controlled lotteries*. At the stages t where $p_{t+1} \neq p_t$, player 1 sends signals according to the appropriate type-dependent lottery, so as to reach the probability distribution p_{t+1}; the incentive compatibility conditions are satisfied since at these stages $a_{t+1} = a_t$. At the other stages t, $p_{t+1} = p_t$; the players perform a jointly controlled lottery in order that the conditional expected payoffs correspond to the bi-martingale. More precisely,

$$(a_t(h_t), \beta_t(h_t)) = \mathrm{E}((a_{t+1}, \beta_{t+1}) \mid h_t)$$

and the jointly controlled lottery is described by the probability distribution appearing on the right-hand side.

To construct a Nash equilibrium given an arbitrary bi-martingale, one uses the same ideas as above but phases of communication must alternate with phases of payoff accumulation in order to achieve the suitable expected payoff.

Theorem 1 is one step in the characterization of Nash equilibrium payoffs; it is completed by a geometric characterization in terms of separation properties. For this, a few definitions are needed.

Let B be a subset of $\Delta^K \times \mathbb{R}_\gamma^K \times \mathbb{R}_\gamma$; an element of B is denoted as (p, a, β). For any fixed $p \in \Delta^K$, the section B_p of B is defined as the set of all points (a, β) of $\mathbb{R}_\gamma^K \times \mathbb{R}_\gamma$ such that $(p, a, \beta) \in B$. The sections B_a are defined similarly for every $a \in \mathbb{R}^K$. B is *bi-convex* if for every p and a, the sets B_p, and B_a are convex. A real *function f* on such a bi-convex set B is *bi-convex* if for every p and a, the function $f(p, \cdot, \cdot)$ is convex on B_p and $f(\cdot, a, \cdot)$ is convex on B_a. Let B be a bi-convex set containing G; let $\mathrm{nsc}(B)$ be the set of all $z \in B$ that cannot be separated from G by any bounded bi-convex function f on B which is continuous at each point of G [namely $f(z) \leq \sup f(G) = \sup\{f(g) \mid g \in G\}$ for every $z \in B$ and f with the properties just listed].

Theorem 2 [Aumann and Hart (1986)]. G^* *is the largest set B such that* $\mathrm{nsc}(B) = B$.

Aumann and Hart (1986) also showed that without the condition of continuity on the separating bi-convex functions, one obtains the subset $G^\#$ of G^* of all triples (p, a, β) at the starting point of a bi-martingale reaching G in *finite* (random) time. If one adds the requirement that this reaching time be bounded, one obtains an even smaller subset of G^*: $\mathrm{bi\text{-}co}(G)$, the smallest bi-convex set containing G. This corresponds to all Nash equilibrium payoffs for which the number of steps of the communication phase can be determined in advance. Even in games of information transmission (see Examples 2, 3, and 4) with two states of nature, $\mathrm{bi\text{-}co}(G)$ may be *strictly* included in $G^\#$ [see Forges (1984, 1990)]; there exist Nash equilibrium payoffs which require an unbounded (though a.s. finite) number of signalling stages. In principle, one may have $G^\# \subsetneq G^*$, but the available examples [Aumann and Hart (1986)] are not in the game-theoretic context.

3.2. Games with known own payoffs

In the previous model, player 2 generally does not observe his stage payoffs, which depend on player 1's type. Although this is not unrealistic [see, for example, Mertens (1986, p. 1531)] the particular case where player 2's payoffs are independent of the state of nature ($B_k = B$ for every $k \in K$) deserves a special attention. In such a game $\Gamma(p)$ *with known own payoffs*, the characterization of Nash equilibrium payoffs is dramatically simplified: *all* can be achieved through a *completely revealing joint plan*. Denote by $\mathrm{val}_2 B$ the value for player 2 of the one-shot game with payoff matrix B. Observe that in the present model, (3) reduces to $\beta \geq \mathrm{val}_2 B$. Recall also definition (1).

Proposition 1 [Shalev (1988)].   *Let $\Gamma(p)$ be such that $B^k = B$ for every $k \in K$; let $(a, \beta) \in \mathbb{R}_\gamma^K \times \mathbb{R}_\gamma$. Then (a, β) is a Nash equilibrium payoff of $\Gamma(p)$ if and only if there exist $\pi^k \in \Delta^{I \times J}$, $k \in K$, such that*
   (i) $A^k(\pi^k) = a^k$, $\forall k \in K$ and $\Sigma_k \, p^k B^k(\pi^k) = \beta$;
   (ii) *a is individually rational for player 1 [i.e. (2)] and $\beta^k = B^k(\pi^k)$ is individually rational for player 2, $\forall k \in K$ (i.e. $\beta^k \geqslant \mathrm{val}_2 \, B$, $\forall k \in K$); and*
   (iii) $A^k(\pi^k) \geqslant A^k(\pi^{k'})$, $\forall k, k' \in K$.

This statement is obtained by particularizing the conditions for a joint plan to be in equilibrium in the case of complete revelation of the state of nature. π^k contains the frequencies of moves to be achieved if state k is revealed; equalities (i) express that (a, β) is the expected payoff; (ii) contains the individual rationality conditions; and the incentive compatibility conditions for player 1 take the simple form (iii) because of the particular signalling strategy.

Proposition 1 extends to repeated games $\Gamma(p, q)$ with lack of information on *both* sides and known own payoffs. Notice that without such a specific assumption, Nash equilibria are not yet characterized in this model. Assume therefore that besides K, another set L of states of nature is given; let $p \in \Delta^K$, $q \in \Delta^L$, and let A^k, $k \in K$, and B^l, $l \in L$, be the payoff matrices (of dimensions $|I| \times |J|$, as above) for player 1 and player 2, respectively. Let $\Gamma(p, q)$ be the infinitely repeated game with lack of information on both sides where k and l are chosen independently (according to p and q, respectively), k is only told to player 1, and l to player 2. Assume further that $|I| \geqslant |K|$ and $|J| \geqslant |L|$. Then we have

Proposition 2 [Koren (1988)].   *Every Nash equilibrium of $\Gamma(p, q)$ is payoff-equivalent to a completely revealing equilibrium.*

This result may be strengthened by deriving the explicit equilibrium conditions as in Proposition 1, in terms of $\pi^{kl} \in \Delta^{I \times J}$, $(k, l) \in K \times L$ [see Koren (1988)].

Infinitely repeated games with known own payoffs may be a useful tool to study infinitely repeated games with *complete* information. The approach is similar to the models of "reputation" (see Chapter 10).

Let Γ_0 consist of the infinite repetition of the game with payoff matrices A for player 1 and B for player 2. Let F_0 be the set of feasible payoffs; by the Folk theorem, the equilibrium payoffs of Γ_0 are described by

$$(\alpha, \beta) \in F_0 \colon \alpha \geqslant \mathrm{val}_1 \, A, \; \beta \geqslant \mathrm{val}_2 \, B \; .$$

For instance, in the "battle of the sexes",

$$[A, B] = \begin{bmatrix} 2,1 & 0,0 \\ 0,0 & 1,2 \end{bmatrix},$$

$\text{val}_1 A = \text{val}_2 B = \frac{2}{3}$, so that the projection of the set of equilibrium payoffs of Γ_0 on player 1's payoffs is $[\frac{2}{3}, 2]$.

The example is due to Aumann (1981). Suppose that player 2 is unsure of player 1's preferences and that player 1 realizes it. A game $\Gamma(p)$ with observable payoffs, with $A^1 = A$ and

$$A^2 = \begin{bmatrix} 2 & 2 \\ 1 & 1 \end{bmatrix},$$

may represent the situation (player 2's payoff is described by B in either case). By Proposition 1, all the equilibria of this game are payoff-equivalent to completely revealing ones. It is easy to see that for any interior p, the payoff a^1 of player 1 of type 1 is greater than $\frac{4}{3}$. Indeed, if $k = 2$ is revealed, the individual rationality conditions imply that the correlated strategy π^2 satisfies $\pi_{11}^2 \geqslant \frac{2}{3}$. Hence, if $a^1 < \frac{4}{3}$, player 1 of type 1 would gain by pretending to be of type 2. Thus, the introduction of even the slightest uncertainty reduces considerably the set of equilibrium payoffs.

To state a general result, let us call "a repeated game with lack of information on one side and known own payoffs derived from Γ_0," any such game $\Gamma(p)$ with $|K|$ states of nature, $p^k > 0$, $\forall k \in K$, and payoff matrices A^k, $k \in K$, $A^1 = A$, for player 1, and B for player 2. The set $G^*(p)$ of all equilibrium payoffs (a, β) of $\Gamma(p)$ is characterized by Proposition 1. Let us denote by $G^*(p)|_{k=1}$ its projection on the (a^1, β)-coordinates.

Proposition 3 [Shalev (1988), Israeli (1989)]. *For every repeated game with lack of information on one side and known own payoffs $\Gamma(p)$ derived from Γ_0, there exists a number v (depending only on the payoff matrices A^k and B but not on p) such that*

$$G^*(p)|_{k=1} = \{(a^1, \beta) \in F_0 : a^1 \geqslant v, \beta \geqslant \text{val}_2 B\}.$$

The maximal value v^* of v is achieved for $|K| = 2$ and $A^2 = -B$; then

$$v^* = \max_{x \in \Delta^I} \min_{y \in \Delta^J(x)} \left(\sum_{i,j} x_i y_j A(i, j) \right).$$

where

$$\Delta^J(x) = \left\{ y \in \Delta^J : \sum_{i,j} x_i y_j B(i, j) \geqslant \text{val}_2 B \right\}.$$

One can check that in the battle of the sexes, $v^* = \frac{4}{3}$. Proposition 3 suggests that player 1 should sow the doubt that he is not maximizing his original payoff (described by A) but is actually trying to minimize player 2's payoff, as if his own payoffs were described by $-B$. This increases the individually rational level of player 1 optimally, to v^*.

3.3. Existence

The existence of a Nash equilibrium in repeated games with incomplete information is still a central issue. In the zero-sum case, under the information structure mainly treated in this chapter ("standard one-sided information") a value does exist (see Chapter 5); this is no longer true in the case of lack of information on both sides (see Chapter 5). For the general model $\Gamma(p)$ of Section 2, a partial answer is given by the next theorem.

Theorem 3 [Sorin (1983)]. *If the number of states of nature is two ($|K| = 2$), then $\Gamma(p)$ has a Nash equilibrium for every p.*

Observe that the existence of a Nash equilibrium in $\Gamma(p)$ for every p amounts to the non-emptiness of the sections $G^*(p)$ of G^* for every p. The proof of Theorem 3 does not, however, use the characterization of Theorem 1. It is constructive and exhibits equilibria of the form introduced by Aumann, Maschler and Stearns (1968). Since $|K| = 2$, let $p \in [0, 1]$ be the probability of state 1. If $\Gamma(p)$ has no non-revealing equilibrium, then p belongs to an interval $(p(1), p(2))$ such that $\Gamma(p(s))$, $s = 1, 2$, has a non-revealing equilibrium [at "worst", $p(1) = 0$ and $p(2) = 1$; recall Example 1]. A joint plan equilibrium is proved to exist, reaching the posterior probabilities $p(1)$ and $p(2)$. It has the further property that after the signalling phase, the players play mixed strategies in Δ^I and Δ^J, respectively, independently of each other. The proof uses the fact that connected and convex subsets of $\Delta^K = [0, 1]$ coincide (with subintervals), which obviously does not hold in higher dimensions.

For an arbitrary number of states of nature, no general result is available. It is observed in Sorin (1983) that if the value of player 1's one-shot game (a, defined in Section 2) is concave, then $\Gamma(p)$ has a non-revealing equilibrium at every p. This arises in particular in games of information transmission (see Examples 2, 3, and 4). One also has the following result for the model of Subsection 3.2.

Proposition 4 [Shalev (1988)]. *Let $\Gamma(p)$ be a game with lack of information on one side and known own payoffs. Then $\Gamma(p)$ has a Nash equilibrium for every p.*

A counterexample of Koren (1988) shows that similar games $\Gamma(p, q)$ with lack of information on both sides (as in Proposition 2) may fail to have an equilibrium for some $(p, q) \in \Delta^K \times \Delta^L$.

Remark. Throughout Section 3, the *limit of means* criterion for the payoffs made it possible to use infinitely many stages for communication, without having to worry about the payoffs at these moments. Such an approach is no longer possible when payoffs are *discounted*. This criterion is used in Bergin (1989) to evaluate the payoffs corresponding to *sequential* equilibria [see Kreps and Wilson (1982)] in two-person, non-zero-sum games with lack of information *on both sides* (with independent states but without any restriction on the payoff matrices). With any sequential equilibrium, one can associate a Markov chain satisfying certain incentive compatibility conditions. The state variables consist of the distribution over players' types and the vector of payoffs. Conversely, any Markov chain on that state space which satisfies the incentive compatibility conditions defines a sequential equilibrium. Thus, without loss of generality (as far as payoffs are concerned), we can restrict ourselves to sequential equilibrium strategies where each player chooses his move at stage t according to a probability distribution depending on his type and the current state.

4. Communication equilibria

The underlying model in this section is the game described in Section 2. Here, we can write Γ for $\Gamma(p)$.

Definition [Aumann (1974, 1987); see also the chapter on 'correlated and communication equilibria' in a forthcoming volume]. A (strategic form) correlated equilibrium for Γ is a Nash equilibrium of an extension of Γ of the following form: before the beginning of Γ (in particular, before the choice of the state of nature), the players can observe correlated private signals (which can be thought of as outputs selected by a *correlation device*).

Let C be the set of correlated equilibrium payoffs of Γ. Observe that if one applies the original definition of Aumann (1974) to Γ, conceived as a game in strategic form, one readily obtains the definition above. In particular, player 1 receives an extraneous signal from the correlation device before observing the state of nature and he does not make any report to the device. The extended game is played as Γ, except that each player can choose his moves (i_t or j_t) as a function of his extraneous signal. Now, Γ is a game with incomplete information and it is tempting to extend the Nash equilibrium concept by allowing player 1 to report information to the device, as in Myerson (1982). Γ being a repeated game, even more general devices can be used.

Definition. A *communication equilibrium* for Γ is a Nash equilibrium of an extension of Γ where, at every stage, the players send inputs to a *communication device* which selects a vector of outputs, one for each player, according to a probability distribution depending on all past inputs and outputs. An *r-device* ($r = 0, 1, \ldots$) is a communication device which cannot receive inputs after stage r; the *r-communication equilibrium* can be associated with the *r-device*.

With $r = \infty$, one obviously obtains the notion of a communication equilibrium. 0-devices only send outputs to the players at *every stage* and hence can be called *autonomous*; observe that they are more general than correlation devices, which act only at stage 0. Recalling the concept of a (strategic form) correlated equilibrium, it is appropriate to refer to the 0-communication equilibrium as an *extensive form correlated equilibrium*. Let D_r ($r = 0, 1, \ldots, \infty$) be the set of all payoffs to *r*-communication equilibria in Γ. The sets are ordered by inclusion as follows:

$$C \subseteq D_0 \subseteq \cdots \subseteq D_r \subseteq D_{r+1} \subseteq \cdots D_\infty .$$

These equilibrium concepts can be defined in any multistage game [Forges (1986b)]. In general, the sequence is strictly increasing: every extension of the Nash equilibrium requires a wider interpretation of the rules of the game. However, infinitely repeated games will appear to involve enough communication possibilities to obtain the (payoff) equivalence of several equilibrium concepts.

Remark. Appropriate versions of the "revelation principle" [see e.g. Myerson (1982)] apply here: the sets C and D_r ($r = 0, \ldots, \infty$) have a "canonical representation" [see Forges (1986b, 1988)].

We begin with a characterization of D_∞, stating in particular that any communication equilibrium payoff can be achieved by means of a communication device of a simple form. Player 1 is asked to reveal his type, as a first input to the device; if he announces k (which need not be his "true type"), the device selects $(c, d) \in F$ and $x \in X$ (recall the definitions of Section 2) according to a probability distribution P^k. The pair (c, d) is transmitted to both players immediately, as a first output. The device is also equipped with an input "alarm", which can be sent by any player at any stage (to obtain a characterization of D_∞, it is natural to exploit its specific properties: the device that we are describing is not an *r*-device for any finite r). If the alarm is given, the output x is sent to both players (formally, two inputs are available to the players at every stage, one is interpreted as "everything is all right" and the other as an alert).

Let D'_x be the set of payoffs associated with equilibria "adapted" to the special devices above, consisting of specific strategies described as follows: first, player 1 truthfully reveals his type; then, the players play a sequence of moves yielding the suggested (c, d) [in a similar way as in the Folk theorem or in Section 3, an infinite sequence of moves yielding the payoff (c, d) can be associated with every $(c, d) \in F$]. If one of the players deviates, the other gives the alarm; player 2 can punish player 1 by approaching x using a Blackwell strategy (see Section 2); player 1 can punish player 2 at his minimax level in $\Gamma(p(\cdot | c, d, x))$, namely vex $b(p(\cdot | c, d, x))$, where $p(\cdot | c, d, x)$ denotes the posterior probability distribution over K given (c, d, x). Indeed, at every stage the game remains an infinitely repeated game with incomplete information on one side; it simply moves from $\Gamma(p)$ to $\Gamma(p(\cdot | c, d))$ and possibly to $\Gamma(p(\cdot | c, d, x))$ if the alarm is given. Although this constant structure should not be surprising, we will see below that communication devices may very well modify the basic information pattern.

Observe that by definition D'_x is a subset of D_x. D'_x is easily characterized: $(a, \beta) \in D'_x \Leftrightarrow$ there exist probability distributions P^k, $k \in K$, on $F \times X$ such that

$$E^k(c^k) = a^k , \quad \forall k \in K , \tag{4}$$

$$E(p(\cdot | c, d) \cdot d) = \beta , \tag{5}$$

$$E^k(\max\{c^l, E^k(x^l | c, d)\}) \leqslant a^l , \quad \forall (k, l) \in K \times K , \tag{6}$$

$$p(\cdot | c, d) \cdot d \geqslant E(\text{vex } b(p(\cdot | c, d, x)) | c, d) \quad \text{a.s.} , \tag{7}$$

where E^k is the expectation with respect to P^k, $k \in K$; E is the expectation with respect to the probability distribution P induced on $K \times F \times X$ by p and the P^k's; and $p(\cdot | c, d)$ [resp. $p(\cdot | c, d, x)$] is the conditional probability distribution (under P) on K given (c, d) [resp. (c, d, x)].

Expressions (4)–(7) follow from the specific device and equilibrium strategies introduced to define D'_x. (4) and (5) express that (a, β) is the corresponding equilibrium payoff. (6) is the equilibrium condition for player 1; if his type is $l \in K$ and his input is $k \in K$ (which may or may not coincide with l), (c, d, x) is selected according to P^k by the device and (c, d) is the output to player 1. He can either follow, which yields c^l, or deviate, which is detected by player 2 who gives the alarm and approaches x; hence an expected payoff of $E^k(x^l | c, d)$ to player 1. Observe that by (4) we have for $k = l$:

$$E^k(x^k \mid c, d) \leq c^k, \quad P^k - \text{a.s.}$$

Having told the truth, player 1 cannot gain by deviating from (c, d). However to take account of all possible deviations of player 1, combining lying and non-obedience, a condition like (6) must hold. (7) is, similarly, the equilibrium condition for player 2; following (c, d), he obtains the left-hand side. If he deviates, player 1 gives the alarm which moves the posterior to $p(\cdot \mid c, d, x)$; player 1 punishes player 2 at his minimax level $\text{vex } b(p(\cdot \mid c, d, x))$, which gives the right-hand side of (7) as expected payoff given (c, d) (i.e. at the time player 2 considers the possibility of deviating).

As in the case of Nash equilibria, the main idea is to settle on a payoff in F satisfying appropriate individual rationality conditions, so that punishments can be applied in the case of deviation. The information of player 1 is used to choose the non-revealing equilibrium to be played. When a communication device is used, the incentive compatibility conditions of player 1 take a simpler form, essentially because his choices are limited to $|K|$ possible inputs (instead of all probability distributions over the signals, in the case of Nash equilibria). With a communication device, jointly controlled lotteries can be dispensed with. The device considered here is quite powerful. It combines different functions of information transmission, coordination of strategies, and threats, so that in particular the punishment against player 1 is chosen as a function of the type that he reports as an input. Although the game remains a game with lack of information on one side at every stage, the conditions (4)–(7) *do not imply that the vector payoff c satisfies the individual rationality condition* (2); x does. Without the threat of being punished at level x, player 1 could want to deviate from (c, d). The use of communication devices deeply modifies the structure of the game; not only the incentive compatibility conditions but also the individual rationality levels differ from those obtained in Section 3.

The next theorem states that *all* communication equilibrium payoffs (a, β) of Γ can be obtained in the specific way described above. Some properties of D_∞ can be deduced. Let D^f_∞ be the subset of D'_∞ where the probability distributions P^k can be chosen to have finite support.

Theorem 4 [Forges (1988)].

$$D_\infty = D'_\infty = D^f_\infty,$$

D_∞ *is compact and convex* .

Caratheodory's theorem is used to prove that $D'_\infty \subseteq D^f_\infty$; in particular, the finite support depends on the underlying equilibrium payoff [this contrasts with games of information transmission (see below)].

The difficult part of the proof is $D_\infty \subseteq D'_\infty$; we will only indicate the sort of arguments that are used. Let $(a, \beta) \in D_\infty$; without loss of generality, by an extended "revelation principle", one can assume that (a, β) corresponds to a canonical communication equilibrium where player 1 first reveals his type to the device and then, at every stage, inputs consist of reported past moves and outputs consist of suggested moves. Appropriate processes (c_t, d_t), x_t are constructed in the same spirit as the bi-martingale of Section 3, the limits of which satisfy (4)–(7) (observe that here *only the limits*, not the specific form of the trajectory, intervene in the characterization). One of the problems is that x_∞ must satisfy (6) and (7) at the same time, i.e. be a threat for player 1, which player 2 can learn by playing an appropriate strategy (in the scenario of an arbitrary communication equilibrium, there is no alarm system; one has to exhibit one).

x_t^l is defined as the best expected payoff player 1 can guarantee from stage t on, in state $l \in K$, without taking his first input (reported type $k \in K$) into account. Having forgotten his input k, player 1 can use a strategy σ' exhausting a maximum of information about k (on which the outputs of the device, and thus also player 2's moves, depend). Such strategies were introduced in Stearns (1967) and are used in Mertens and Zamir (1980). To become aware of x_∞, player 2 can mimic σ' when he reports the moves of player 1 to the communication device (which cannot distinguish between deviation in the play by one player and deviation in the report by the other).

One of the underlying features in the previous characterization is that player 1 can always stay as informed as player 2. This property is a by-product of the equality between D_∞ and D'_∞ and depends on the ∞-devices used to define D'_∞. To see that the information pattern may change, consider a correlation device selecting a pair of correlated signals, ζ^1 and ζ^2, that it sends to player 1 and player 2, respectively, before the beginning of the game. "Signalling" from player 1 to player 2 has the following form: player 1 selects his moves according to a probability distribution depending on ζ^1 and his type k and player 2 computes his posterior probability distribution on K, given player 1's moves and ζ^2. Since it depends on ζ^2, this distribution may in particular be *unknown* to player 1. The addition of the correlation device has transformed the game Γ, with lack of information *on one side*, into a game with lack of information *on $1\frac{1}{2}$ sides* [using the terminology of Sorin and Zamir (1985)], where player 1 knows the state of nature but is ignorant of the beliefs of player 2 on it. As illustrated by Sorin and Zamir (1985), these games really belong to the class of infinitely repeated games with lack of information *on both sides*, especially studied (in the zero-sum case) by Mertens and Zamir (1971–72, 1980). In particular, they may fail to have a value (recall Subsection 3.3). This will not be cumbersome here: we only need an expression for the *minimax* levels. However, (2) and (3) will have to be modified using the results of Mertens and Zamir (1980).

The next result is a characterization of the sets D_r ($r = 1, 2, \ldots$), *all* of which coincide with a set D_1' of particular 1-communication equilibrium payoffs. The corresponding 1-devices lead to games with incomplete information on both sides. As above, for D_∞', the set of inputs of player 1 at the first stage is K. If k is sent, a probability distribution P^k is used by the device to select $((c, d), \varphi, x) \in F \times \Phi \times X$ (see Section 2); the output to player 1 (resp. player 2) at the first stage is $((c, d), \varphi)$ [resp. $((c, d), x)$]. Payoffs in D_1' are achieved by adapted strategies. Player 1 truthfully reveals his type; both players follow (c, d) unless a deviation occurs (this is similar to D_∞'); then, a Blackwell strategy corresponding to φ or x is applied as a punishment against the deviator (see again Section 2). Unlike the device constructed for D_∞', this one sends all outputs at once, at the beginning of the game; however, this one transmits *private* outputs to the players so that both of them have incomplete information.

In the same way as for D_∞' (and with similar notations), we have: $(a, \beta) \in D_1' \Leftrightarrow$ there exist probability distributions P^k, $k \in K$, on $F \times \Phi \times X$, such that

$$E^k(c^k) = a^k , \quad \forall k \in K , \tag{8}$$

$$E(p(\cdot \,|\, c, d, x) \cdot d) = \beta , \tag{9}$$

$$E^k(\max\{c^l, E^k(x^l \,|\, c, d, \varphi)\}) \le a^l , \quad \forall (k, l) \in K \times K , \tag{10}$$

$$p(\cdot \,|\, c, d, x) \cdot d \ge E(\varphi(p(\cdot \, c, d, \varphi, x)) \,|\, c, d, x) . \tag{11}$$

These conditions can be justified in the same way as (4)–(7). Player 1 now evaluates his expected punishment, given his information (c, d, φ). If player 2 deviates from (c, d), player 1 punishes him using φ, which can reveal further information to player 2 and changes his posterior probability distribution into $p(\cdot \,|\, c, d, \varphi, x)$. Thus, if φ is selected, player 2 cannot obtain more than $\varphi(p(\cdot \,|\, c, d, \varphi, x))$, which yields the expected level on the right-hand side of (11).

Theorem 5 [Forges (1988)].

$$D_r = D_1' \qquad\qquad (r = 1, 2, \ldots) ,$$

$$D_r \text{ is compact and convex} \quad (r = 1, 2, \ldots) .$$

Let D_1^f denote, like D_∞^f, the subset of D_1' where the P^k's have finite support. One cannot proceed as above to obtain $D_r = D_1^f$. This holds nevertheless in particular cases (see below).

Let us briefly illustrate how the results of zero-sum infinitely repeated games with incomplete information on both sides intervene in the proof of Theorem 5. Since an r-device does not receive any input after stage r, it can select at once all future outputs at stage $r+1$. To construct a random variable x satisfying (8)–(11), one is concerned with the best player 1 can guarantee from stage t on. For $t \geq r + 1$, the game starting at stage t can be related with an auxiliary game Γ_t with lack of information on $1\frac{1}{2}$ sides where player 1 has the same information as in the original game, but player 2 has already received the whole sequence of all his outputs. Player 1 can certainly do as well in Γ from stage t than in Γ_t and adequate expressions of punishment levels in Γ_t can be determined using Mertens and Zamir's (1980) results.

For instance, let L and M be finite sets and $q \in \Delta^{L \times M}$; let $G(q)$ be the zero-sum infinitely repeated game, where player 1's type belongs to L, player 2's type belongs to M, and the payoff matrix for player 1 in state (l, m) is C^{lm}. By definition, $y \in \mathbb{R}^L_\gamma$ (γ is the maximum payoff) is *approachable* by player 2 in $G(q)$ if player 2 has a strategy guaranteeing that, for every type l, the expected payoff of player 1 will not exceed y^l, whatever his strategy. Mertens and Zamir's (1980) results imply that y is approachable by player 2 in $G(q)$ if and only if it is approachable using a "standard" strategy τ, consisting of a type-dependent lottery followed by the usual non-revealing, Blackwell (1956) strategy. More precisely, let S be a finite set ($|S| \leq |L| \, |M|$), let $\pi(\cdot \mid m)$ be a probability distribution on S for every $m \in M$, and let $x_s \in \mathbb{R}^L_\gamma$, $s \in S$, be approachable [in the sense of Blackwell (1956)] in the game with payoff matrices $C^l_s = \Sigma_m (q * \pi)(m \mid l, s) C^{lm}$, where $q * \pi$ denotes the probability distribution on $L \times M \times S$ generated by q and π. Let $\tau = (\tau^m)_{m \in M}$, where τ^m consists of choosing $s \in S$ according to $\pi(\cdot \mid m)$ and of applying a Blackwell strategy to approach x_s if s is realized. τ guarantees that player 1's expected payoff in state l cannot exceed

$$y^l = \sum_{s \in S} (q * \pi)(s \mid l) x^l_s .$$

In the case of lack of information on $1\frac{1}{2}$ sides ($C^{lm} = A^l$, $\forall (l, m) \in L \times M$), standard strategies take an even simpler form: player 2 first reveals his true type $m \in M$ and then approaches a vector $x_m \in \mathbb{R}^L$ in the game with payoff matrices $A^l, l \in L$; in other words a vector $x_m \in X$. The same kind of approach is used to describe the vectors of $\mathbb{R}^m$ that are approachable by player 1, and enables us to exhibit punishments in Φ. To apply this in auxiliary games like Γ_t above, one has to extend the results to games where the "approaching player" has infinitely many types.

Theorem 5 shows the equivalence of all r-devices ($r = 1, 2, \ldots$) in the repeated game Γ. Can we go one step further and show that the subsets D_0 and C also coincide with D_1? A partial answer is provided by the next statement.

Theorem 6 [Forges (1988)].

   (I) *Let* $(a, \beta) \in D_1^f$ *be such that a is strictly individually rational for player* 1 *(i.e.* $q \cdot a > a(q)$, $\forall q \in \Delta^K$). *Then* $(a, \beta) \in D_0$.

   (II) *Suppose that b is convex or concave* (vex *b linear*). *Then* $D_1 = D_1^f = C$.

In a slightly different context, an example indicates that a condition of strict individual rationality as in (I) may be necessary to obtain the result [see Forges (1986a)].

In (II), observe that all sets but D_∞ are equal. The assumptions on b guarantee that player 1 can punish player 2 at the same level, knowing or not player 2's actual beliefs on K. If b is convex, then b is the best punishment in Φ, and player 1 can hold player 2 at $b(q)$ for every $q \in \Delta^K$ by playing a non-revealing strategy. If vex b is linear, player 1 can punish player 2 by revealing his type $k \in K$ and playing a minmax strategy in the corresponding game k, with payoff matrix B^k.

This happens in particular in games with known own payoffs (see Subsection 3.2) and games of information transmission (see Section 2). In these cases, *all* sets, from C to D_∞, coincide. More precisely, Propositions 1 and 2 apply to communication equilibria [Koren (1988)], while in games of information transmission, $C = D_\infty$ coincides with the set of equilibrium payoffs associated with simple 1-devices, represented by conditional probabilities on J given $k \in K$ [Forges (1985)]. Two properties can be deduced: $C(= D_\infty)$ is a convex polyhedron and every payoff in C can be achieved by a correlated equilibrium requiring one single phase of signalling from player 1 to player 2 (notice that this is not true for Nash equilibria).

References

Aumann, R.J. (1964) 'Mixed and behaviour strategies in infinite extensive games', in: M. Dresher et al., eds., *Advances in game theory*, Annals of Mathematics Studies 52. Princeton: Princeton University Press, pp. 627–650.

Aumann, R.J. (1974) 'Subjectivity and correlation in randomized strategies', *Journal of Mathematical Economics*, 1: 67–95.

Aumann, R.J. (1981) 'Repetition as a paradigm for cooperation in games of incomplete information', mimeo, The Hebrew University of Jerusalem.

Aumann, R.J. (1987), 'Correlated equilibria as an expression of Bayesian rationality', *Econometrica*, 55: 1–18.

Aumann, R.J. and S. Hart (1986) 'Bi-convexity and bi-martingales', *Israel Journal of Mathematics*, 54: 159–180.

Aumann, R.J. and M. Maschler (1966) *Game-theoretic aspects of gradual disarmament*. Princeton: Mathematica ST-80, Ch. V, pp. 1–55.

Aumann, R.J., M. Maschler and R.E. Stearns (1968) *Repeated games of incomplete information: An approach to the non-zero-sum case*. Princeton: Mathematica ST-143, Ch. IV, pp. 117–216.

Bergin, J. (1989) 'A characterization of sequential equilibrium strategies in infinitely repeated incomplete information games', *Journal of Economic Theory*, 47: 51–65.

Blackwell, D. (1956) 'An analog of the minmax theorem for vector payoffs', *Pacific Journal of Mathematics*, 6: 1–8.

Forges, F. (1984) 'A note on Nash equilibria in repeated games with incomplete information', *International Journal of Game Theory*, 13: 179–187.

Forges, F. (1985) 'Correlated equilibria in a class of repeated games with incomplete information', *International Journal of Game Theory*, 14: 129–150.

Forges, F. (1986a) 'Correlated equilibria in repeated games with lack of information on one side: A model with verifiable types', *International Journal of Game Theory*, 15: pp. 65–82.

Forges, F. (1986b) 'An approach to communication equilibria', *Econometrica*, 54: 1375–1385.

Forges, F. (1988) 'Communication equilibria in repeated games with incomplete information', *Mathematics of Operations Research*, 13: 191–231.

Forges, F. (1990) 'Equilibria with communication in a job market example', *Quarterly Journal of Economics*, CV, pp. 375–398.

Hart, S. (1985) 'Nonzero-sum two-person repeated games with incomplete information', *Mathematics of Operations Research*, 10: 117–153.

Israeli, E. (1989) Sowing doubt optimally in two-person repeated games, M.Sc. thesis, Tel-Aviv University [in Hebrew].

Koren, G. (1988) Two-person repeated games with incomplete information and observable payoffs, M.Sc. thesis, Tel-Aviv University.

Kreps, D. and Wilson, R. (1982) 'Sequential equilibria', *Econometrica*, 50: 443–459.

Mathematica (1966, 1967, 1968), *Reports to the U.S. Arms Control and Disarmament Agency*, prepared by Mathematica, Inc., Princeton: ST-80 (1966), ST-116 (1967), ST-140 (1968).

Mertens, J.-F. (1986) *Proceedings of the International Congress of Mathematicians*. Berkeley, California, 1986, pp. 1528–1577.

Mertens, J.-F. and S. Zamir (1971–72) 'The value of two-person zero-sum repeated games with lack of information on both sides', *International Journal of Game Theory*, 1: 39–64.

Mertens, J.-F. and S. Zamir (1980) 'Minmax and maxmin of repeated games with incomplete information', *International Journal of Game Theory*, 9: 201–215.

Myerson, R.B. (1982) 'Optimal coordination mechanisms in generalized principal agent problems', *Journal of Mathematical Economics*, 10: 67–81.

Shalev, J. (1988) 'Nonzero-sum two-person repeated games with incomplete information and observable payoffs', The Israel Institute of Business Research, Working paper 964/88, Tel-Aviv University.

Sorin, S. (1983) 'Some results on the existence of Nash equilibria for non-zero-sum games with incomplete information, *International Journal of Game Theory*, 12: 193–205.

Sorin, S. and S. Zamir (1985) 'A 2-person game with lack of information on $1\frac{1}{2}$ sides', *Mathematics of Operations Research*, 10: 17–23.

Stearns, R.E. (1967) *A formal information concept for games with incomplete information*. Princeton: Mathematica ST-116, Ch. IV, pp. 405–433.

Chapter 7

NONCOOPERATIVE MODELS OF BARGAINING

KEN BINMORE

University of Michigan and University College London

MARTIN J. OSBORNE

McMaster University

ARIEL RUBINSTEIN*

Tel Aviv University and Princeton University

Contents

*This chapter was written in Fall 1988. Parts of it use material from Rubinstein (1987), a survey of sequential bargaining models. Parts of Sections 5, 7, 8, and 9 are based on a draft of parts of Osborne and Rubinstein (1990). The first author wishes to thank Avner Shaked and John Sutton and the third author wishes to thank Asher Wolinsky for long and fruitful collaborations. The second author gratefully acknowledges financial support from the Natural Sciences and Engineering Research Council of Canada and from the Social Sciences and Humanities Research Council of Canada.

Handbook of Game Theory, Volume 1, Edited by R.J. Aumann and S. Hart

1. Introduction

John Nash's (1950) path-breaking paper introduces the bargaining problem as follows:

> A two-person bargaining situation involves two individuals who have the opportunity to collaborate for mutual benefit in more than one way (p. 155).

Under such a definition, nearly all human interaction can be seen as bargaining of one form or another. To say anything meaningful on the subject, it is necessary to narrow the scope of the inquiry. We follow Nash in assuming that

> the two individuals are highly rational, ... each can accurately compare his desires for various things, ... they are equal in bargaining skill ....

In addition we assume that the procedure by means of which agreement is reached is both clear-cut and unambiguous. This allows the bargaining problem to be modeled and analyzed as a noncooperative game.

The target of such a noncooperative theory of bargaining is to find theoretical predictions of what agreement, if any, will be reached by the bargainers. One hopes thereby to explain the manner in which the bargaining outcome depends on the parameters of the bargaining problem and to shed light on the meaning of some of the verbal concepts that are used when bargaining is discussed in ordinary language. However, the theory has only peripheral relevance to such questions as: What is a just agreement? How would a reasonable arbiter settle a dispute? What is the socially optimal deal? Nor is the theory likely to be of more than background interest to those who write manuals on practical bargaining techniques. Such questions as "How can I improve my bargaining skills'? and "How do bargainers determine what is jointly feasible?" are psychological issues that the narrowing of the scope of the inquiry is designed to exclude.

Cooperative bargaining theory (see the chapter on 'cooperative models of bargaining' in a forthcoming volume of this Handbook) differs mainly in that the bargaining *procedure* is left unmodeled. Cooperative theory therefore has to operate from a poorer informational base and hence its fundamental assumptions are necessarily abstract in character. As a consequence, cooperative solution concepts are often difficult to evaluate. Sometimes they may have more than one viable interpretation, and this can lead to confusion if distinct interpretations are not clearly separated. In this chapter we follow Nash in adopting an interpretation of cooperative solution concepts that attributes the same basic aims to cooperative as to noncooperative theory. That is to say, we focus on interpretations in which, to quote Nash (1953), "the two approaches to the [bargaining] problem . . . are complementary; each helps to justify and clarify the other" (p. 129). This means in particular that what we have to say on cooperative solution concepts is not relevant to interpretations that seek to address questions like those given above which are specifically excluded from our study.

Notice that we do not see cooperative and noncooperative theory as rivals. It is true that there is a sense in which cooperative theory is "too general"; but equally there is a sense in which noncooperative theory is "too special". Only rarely will the very concrete procedures studied in noncooperative theory be observed in practice. As Nash (1953) observes,

> Of course, one cannot represent all possible bargaining devices as moves in the non-cooperative game. The negotiation process must be formalized and restricted, but in such a way that each participant is still able to utilize all the essential strengths of his position (p. 129).

Even if one makes good judgments in modeling the essentials of the bargaining process, the result may be too cumbersome to serve as a tool in applications, where what is required is a reasonably simple mapping from the parameters of the problem to a solution outcome. This is what cooperative theory supplies. But which of the many cooperative solution concepts is appropriate in a given context, and how should it be applied? For answers to such questions, one may look to noncooperative theory for guidance. It is in this sense that we see cooperative and noncooperative theory as complementary.

2. A sequential bargaining model

The archetypal bargaining problem is that of "dividing the dollar" between two players. However, the discussion can be easily interpreted broadly to fit a large class of bargaining situations. The set of feasible agreements is identified with $A = [0, 1]$. The two bargainers, players 1 and 2, have opposing preferences over A. When $a > b$, 1 prefers a to b and 2 prefers b to a. Who gets how much?

The idea that the information so far specified is not sufficient to determine the bargaining outcome is very old. For years, economists tended to agree that further specification of a bargaining solution would need to depend on the vague notion of "bargaining ability". Even von Neumann and Morgenstern (1944) suggested that the bargaining outcome would necessarily be determined by unmodeled psychological properties of the players.

Nash (1950, 1953) broke away from this tradition. His agents are fully rational. Once their preferences are given, other psychological issues are irrelevant. The bargaining outcome in Nash's model is determined by the players' attitudes towards risk – i.e. their preferences over lotteries in which the prizes are taken from the set of possible agreements together with a predetermined "disagreement point".

A sequential bargaining theory attempts to resolve the indeterminacy by explicitly modeling the bargaining procedure as a sequence of offers and counteroffers. In the context of such models, Cross (1965, p. 72) remarks, "If it did not matter when people agreed, it would not matter whether or not they agreed at all." This suggests that the players' time preferences may be highly relevant to the outcome. In what follows, who gets what depends exclusively on how patient each player is.

The following procedure is familiar from street markets and bazaars all over the world. The bargaining consists simply of a repeated exchange of offers. Formally, we study a model in which all events take place at one of the times t in a prespecified set $T = (0, t_1, t_2, . . .)$, where (t_n) is strictly increasing. The players alternate in making offers, starting with player 1. An offer x, made at time t_n, may be accepted or rejected by the other player. If it is accepted, the game ends with the agreed deal being implemented at time t_n. This outcome is denoted by (x, t_n). If the offer is rejected, the rejecting player makes a counteroffer at time t_{n+1}. And so on. Nothing binds the players to offers they have made in the past, and no predetermined limit is placed on the time that may be expended in bargaining. In principle, a possible outcome of the game is therefore perpetual disagreement or *impasse*. We denoted this outcome by D.

Suppose that, in this model, player 1 could make a commitment to hold out for a or more. Player 2 could then do no better than to make a commitment to hold out for $1 - a$ or better. The result would be a Nash equilibrium sustaining an agreement on a. The indeterminacy problem would therefore remain. However, we follow Schelling (1960) in being skeptical about the extent to which such commitments can genuinely be made. A player may make threats about his last offer being final, but the opponent can dismiss such threats as mere bombast unless it would actually be in the interests of the threatening player to carry out his threat if his implicit ultimatum were disregarded. In such situations, where threats need to be credible to be effective, we replace Nash equilibrium by Selten's notion of subgame-perfect equilibrium (see the chapters on 'strategic equilibrium' and 'conceptual foundations of strategic equilibrium' in a forthcoming volume of this Handbook).

The first to investigate the alternating offer procedure was Ståhl (1967, 1972, 1988). He studied the subgame-perfect equilibria of such time-structured models by using backwards induction in finite horizon models. Where the horizons in his models are infinite, he postulates nonstationary time preferences that lead to the existence of a "critical period" at which one player prefers to yield rather than to continue, independently of what might happen next. This creates a "last interesting period" from which one can start the backwards induction. [For further comment, see Ståhl (1988).] In the infinite horizon models studied below, which were first investigated by Rubinstein

(1982), different techniques are required to establish the existence of a unique subgame-perfect equilibrium.

Much has been written on procedures in which all the offers are made by only one of the two bargainers. These models assign all the bargaining power to the party who makes the offers. Such an asymmetric set-up does not fit very comfortably within the bargaining paradigm as usually understood and so we do not consider models of this type here.

2.1. Impatience

Players are assumed to be impatient with the unproductive passage of time. The times in the set T at which offers are made are restricted to $t_n = \eta\tau$ $(n = 0, 1, 2, \ldots)$, where $\tau > 0$ is the length of one period of negotiation. Except where specifically noted, we take $\tau = 1$ to simplify algebraic expressions. Rubinstein (1982) imposes the following conditions on the players' (complete, transitive) time preferences. For a and b in A, s and t in T, and $i = 1, 2$:

(TP1) $a > b$ implies $(a, t) >_1 (b, t)$ and $(b, t) >_2 (a, t)$.
(TP2) $0 < a < 1$ and $s < t$ imply that $(a, s) >_i (a, t) >_i D$.
(TP3) $(a, s) \gtrsim_i (b, s + \tau)$ if and only if $(a, t) \gtrsim_i (b, t + \tau)$.
(TP4) the graphs of the relations $\gtrsim_i$ are closed.

These conditions are sufficient to imply that for any $0 < \delta_1 < 1$ and any $0 < \delta_2 < 1$ the preferences can be represented by utility functions Φ_1 and Φ_2 for which $\Phi_1(D) = \Phi_2(D) = 0$, $\Phi_1(a, t) = \phi_1(a)\delta_1^t$,   and $\Phi_2(a, t) = \phi_2(1 - a)\delta_2^t$, where the functions $\phi_i \colon [0, 1] \to [0, 1]$ are strictly increasing and continuous [see Fishburn and Rubinstein (1982)]. Sometimes we may take as primitives the "discount factors" δ_i. However, note that if we start, as above, with the preferences as primitives, then the numbers δ_i may be chosen arbitrarily in the range $(0, 1)$. The associated discount rates ρ_i are given by $\delta_i = \mathrm{e}^{-\rho_i}$.

To these conditions, we add the requirement:

(TP0) for each $a \in A$ there exists $b \in A$ such that $(b, 0) \sim_i (a, \tau)$.

By (TP0) we have $\phi_i(0) = 0$; without loss of generality, we take $\phi_i(1) = 1$.

The function $f \colon [0, 1] \to [0, 1]$ defined by $f(u_1) = \phi_2(1 - \phi_1^{-1}(u_1))$ is useful. A deal reached at time 0 that assigns utility u_1 to player 1 assigns $u_2 = f(u_1)$ to player 2. More generally, the set U^t of utility pairs available at time t is

$$U^t = \{(u_1\delta_1^t, \ f(u_1)\delta_2^t) \colon 0 \leq u_1 \leq 1\} \,. \tag{1}$$

Note that a feature of this model is that all subgames in which a given player makes the first offer have the *same* strategic structure.

Our goal is to characterize the subgame-perfect equilibria of this game. We begin by examining a pair of stationary strategies, in which both players always plan to do the same in strategically equivalent subgames, regardless of the history of events that must have taken place for the subgame to have been reached. Consider two possible agreements a^* and b^*, and let u^* and v^* be the utility pairs that result from the implementation of these agreements at time 0. Let s_1 be the strategy of player 1 that requires him always to offer a^* and to accept an offer of b if and only if $b \geqslant b^*$. Similarly, let s_2 be the strategy of player 2 that requires him always to offer b^* and to accept an offer of a if and only if $a \leqslant a^*$. The pair (s_1, s_2) is a subgame-perfect equilibrium if and only if

$$v_1^* = \delta_1 u_1^* \quad \text{and} \quad u_2^* = \delta_2 v_2^*. \tag{2}$$

In checking that (s_1, s_2) is a subgame-perfect equilibrium, observe that each player is always offered precisely the utility that he will get if he refuses the offer and s_1 and s_2 continue to be used in the subgame that ensues.

Notice that (2) admits a solution if and only if the equation

$$f(x) = \delta_2 f(x\delta_1) \tag{3}$$

has a solution. This is assured under our assumptions because f is continuous, $f(0) = 1$ and $f(1) = 0$.

Each solution to (2) generates a different subgame-perfect equilibrium. Thus, the uniqueness of a solution to (2) is a necessary condition for the uniqueness of a subgame-perfect equilibrium in the game.

In the following we will assume that

(TP5) (2) has a unique solution.

A condition that ensures this is

(TP5*) $(a + \alpha, \tau) \sim_i (a, 0)$, $(b + \beta, \tau) \sim_i (b, 0)$, and $a < b$ imply that $\alpha < \beta$.

This has the interpretation that the more you get, the more your have to be compensated for delay in getting it. A weak sufficient condition for the uniqueness of the solution for (2) is that ϕ_1 and ϕ_2 be concave. (This condition is far from necessary. It is enough, for example, that $\log \phi_1$ and $\log \phi_2$ be concave.)

Result 1 [Rubinstein (1982)]. Under assumptions (TP0)–(TP5) the bargaining game has a unique subgame-perfect equilibrium. In this equilibrium, agreement is reached immediately, and the players' utilities satisfy (2).

Alternative versions of Rubinstein's proof appear in Binmore (1987b) and Shaked and Sutton (1984). The following proof of Shaked and Sutton is especially useful for extensions and modifications of the theorem.

Proof.   Without loss of generality, we take $\tau = 1$. Let the supremum of all subgame-perfect equilibrium payoffs to player 1 be M_1 and the infimum be m_1. Let the corresponding quantities for player 2 in the companion game, in which the roles of 1 and 2 are reversed, be M_2 and m_2. We will show that $m_1 = u_1^*$ and $M_2 = v_2^*$, where u_1^* and v_2^* are uniquely defined by (2). An analogous argument shows that $M_1 = u_1^*$ and $m_2 = v_2^*$. It follows that the equilibrium *payoffs* are uniquely determined. To see that this implies that the equilibrium *strategies* are unique, notice that, after every history, the proposer's offer must be accepted in equilibrium. If, for example, player 1's demand of u_1^* were rejected, he would get at most $\delta_1 v_1^* < u_1^*$.

As explained earlier, u^* is a subgame-perfect equilibrium pair of payoffs. Thus $m_1 \leqslant u_1^*$ and $M_2 \geqslant v_2^*$. We now show that (i) $\delta_2 M_2 \geqslant f(m_1)$ and (ii) $M_2 \leqslant f(\delta_1 m_1)$.

(i) Observe that if player 2 rejects the opening offer, then the companion game is played from time 1. If equilibrium strategies are played in this game, player 2 gets no more than $\delta_2 M_2$. Therefore in any equilibrium player 2 must accept at time $t = 0$ any offer that assigns him a payoff strictly greater than $\delta_2 M_2$. Thus player 1 can guarantee himself any payoff less than $f^{-1}(\delta_2 M_2)$. Hence $m_1 \geqslant f^{-1}(\delta_2 M_2)$.

(ii) In the companion game, player 1 can guarantee himself any payoff less

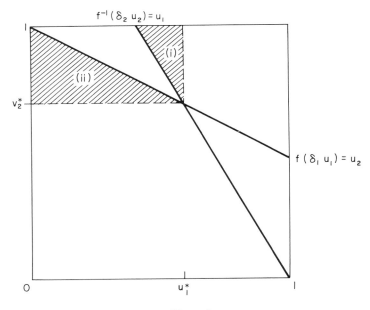

Figure 1.

than $\delta_1 m_1$ by rejecting player 2's opening offer (provided equilibrium strategies are used thereafter). Thus $M_2 \leqslant f(\delta_1 m_1)$.

The uniqueness of (u_1^*, v_2^*) satisfying (2) is expressed in Figure 1 by the fact that the curves $f(\delta_1 u_1) = u_2$ and $f^{-1}(\delta_2 u_2) = u_1$ intersect only at (u_1^*, v_2^*). From (i) and $m_1 \leqslant u_1^*$, (m_1, M_2) lies in region (i). From (ii) and $M_2 \geqslant v_2^*$, (m_1, M_2) lies in region (ii). Hence $(m_1, M_2) = (u_1^*, v_2^*)$. Similarly $(M_1, m_2) = (u_1^*, v_2^*)$. □

2.2. Shrinking cakes

Binmore's (1987b) geometric characterization (see Figure 2) applies to preferences that do not necessarily satisfy the stationarity assumption (TP3). The "cake" available at time t is identified with a set U^t of utility pairs that is assumed to be closed, bounded above, and to have a connected Pareto frontier. It is also assumed to shrink over time. This means that if $s \leqslant t$, then, for each $y \in U^t$, there exists $x \in U^s$ satisfying $x \geqslant y$. The construction begins by truncating the game to a finite number n of stages. Figure 2 shows how a set E_{t_n}

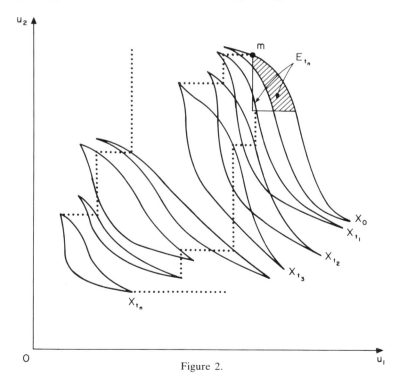

Figure 2.

of payoff vectors is constructed from the truncated game in the case when n is odd. The construction when n is even is similar. The set of all subgame-perfect equilibrium payoff vectors is shown to be the intersection of all such E_{t_n}. Since the sets E_{t_n} are nested, their intersection is also their limit as $n \to \infty$. The methodology reveals that, when there is a unique equilibrium outcome, this must be the limit of the equilibrium outcomes in the finite horizon models obtained by calling a halt to the bargaining process at some predetermined time t_n. In fact, the finite horizon equilibrium outcome in Figure 2 is the point m.

2.3. Discounting

A very special case of the time preferences covered by Result 1 occurs when $\phi_1(a) = \phi_2(a) = a$ $(0 \le a \le 1)$. Reverting to the case of an arbitrary $\tau > 0$, we have $u_1^* = (1 - \delta_2^\tau)/(1 - \delta_1^\tau \delta_2^\tau) \to \rho_2/(\rho_1 + \rho_2)$ as $\tau \to 0+$. When $\delta_1 = \delta_2$, it follows that the players share the available surplus of 1 equally in the limiting case when the interval between successive proposals is negligible. If δ_1 decreases, so does player 1's share. This is a general result in the model: it always pays to be more patient. More precisely, define the preference relation $\succsim_1$ to be at least as patient as $\succsim_1'$ if $(y, 0) \succsim_1 (x, 1)$ implies that $(y, 0) \succsim_1' (x, 1)$. Then player 1 always gets at least as much in equilibrium when his preference relation is $\succsim_1$ as when it is $\succsim_1'$ [Rubinstein (1987)].

2.4. Fixed costs

Rubinstein (1982) characterizes the subgame-perfect equilibrium outcomes in the alternating offers model under the hypotheses (TP1)–(TP4) and a version of (TP5*) in which the last inequality is weak. These conditions cover the interesting case in which each player i incurs a fixed cost $c_i > 0$ for each unit of time that elapses without an agreement being reached. Suppose, in particular, that their respective utilities for the outcome (a, t) are $a - c_1 t$ and $1 - a - c_2 t$. It follows from Rubinstein (1982) that, if $c_1 < c_2$, the only subgame-perfect equilibrium assigns the whole surplus to player 1. If $c_1 > c_2$, then 1 obtains only c_2 in equilibrium. If $c_1 = c_2 = c < 1$, then *many* subgame-perfect equilibria exist. If c is small $(c < 1/3)$, some of these equilibria involve *delay* in agreement being reached. That is, equilibria exist in which one or more offers get *rejected*. It should be noted that, even when the interval τ between successive proposals becomes negligible $(\tau \to 0+)$, the equilibrium delays do not necessarily become negligible.

2.5. *Stationarity, efficiency, and uniqueness*

We have seen that, when (2) has a unique solution, the game has a *unique* subgame-perfect equilibrium which is *stationary* and that its use results in the game ending *immediately*.

The efficiency of the equilibrium is not a consequence of the requirement of perfection by itself. As we have just seen, when multiple equilibria exist [that is, when (2) has more than one solution], some of these may call for some offers to be rejected before agreement is reached, so that the final outcome need not be Pareto efficient. It is sometimes suggested that rational players with complete information must necessarily reach a Pareto-efficient outcome when bargaining costs are negligible. This example shows that the suggestion is questionable.

Some authors consider it adequate to restrict attention to stationary equilibria on the grounds of simplicity. We do *not* make any such restriction, since we believe that, for the current model, such a restriction is hard to justify. A strategy in a sequential game is more than a plan of how to play the game. A strategy of player i includes a description of player i's beliefs about what the other players think player i would do were he to deviate from his plan of action. (We are not talking here about beliefs as formalized in the notion of sequential equilibrium, but of the beliefs built into the definition of a strategy in an extensive form game.) Therefore, a stationarity assumption does more than attribute simplicity of behavior to the players: it also makes players' beliefs insensitive to past events. For example, stationarity requires that, if player 1 is supposed to offer a 50:50 split in equilibrium, but has always demanded an out-of-equilibrium 60:40 split in the past, then player 2 still continues to hold the belief that player 1 will offer the 50:50 split in the future. For a more detailed discussion of this point, see Rubinstein (1991).

Finally, it should be noted that the uniqueness condition of Result 1 can fail if the set A from which players choose their offers is sufficiently restricted. Suppose, for example, that the dollar to be divided can be split only into whole numbers of cents, so that $A = \{0, 0.01, \ldots, 0.99, 1\}$. If $\phi_1(a) = \phi_2(a) = a$ and $\delta_1 = \delta_2 = \delta > 0.99$, then *any* division of the dollar can be supported as the outcome of a subgame-perfect equilibrium [see, for example, Muthoo (1991) and van Damme, Selten and Winter (1990)]. Does this conclusion obviate the usefulness of Result 1? This depends on the circumstances in which it is proposed to apply the result. If the grid on which the players locate values of δ is finer than that on which they locate values of a, then the bargaining problem remains indeterminate. Our judgment, however, is that the reverse is usually the case.

2.6. Outside options

When bargaining takes place it is usually open to either player to abandon the negotiation table if things are not going well, to take up the best option available elsewhere. This feature can easily be incorporated into the model analyzed in Result 1 by allowing each player to opt out whenever he has just rejected an offer. If a player opts out at time t, then the players obtain the payoffs $\delta_1^t e_1$ and $\delta_2^t e_2$, respectively. The important point is that, under the conditions of Result 1, the introduction of such exit opportunities is *irrelevant* to the equilibrium bargaining outcome when $e_1 < \delta_1 u_1^*$ and $e_2 < \delta_2 v_2^*$. In this case the players always prefer to continue bargaining rather than to opt out. The next result exemplifies this point.

Result 2 [Binmore, Shaked and Sutton (1988)].   Take $\phi_1(a) = \phi_2(a) = a$ $(0 \le a \le 1)$ and $\delta_1 = \delta_2 = \delta$. If $e_i \ge 0$ for $i = 1, 2$ and $e_1 + e_2 < 1$, then there exists a unique subgame-perfect equilibrium outcome, in which neither player exercises his outside option. The equilibrium payoffs are

$$
\begin{cases}
\left(\dfrac{1}{1+\delta}, \dfrac{\delta}{1+\delta} \right) & \text{if } e_i \le \dfrac{\delta}{1+\delta} \quad \text{for } i = 1, 2, \\[4mm]
(1 - \delta(1 - e_1), \delta(1 - e_1)) & \text{if } e_1 > \dfrac{\delta}{1+\delta} \quad \text{and } e_2 \le \delta(1 - e_1), \\[2mm]
(1 - e_2, e_2) & \text{otherwise.}
\end{cases}
$$

As modeled above, a player cannot leave the table without first listening to an offer from his opponent, who therefore always has a last chance to save the situation. This seems to capture the essence of traditional face-to-face bargaining. Shaked (1987) finds multiple equilibria if a player's opportunity for exit occurs not after a rejection by himself, but after a rejection by his opponent. He has in mind "high tech" markets in which binding deals are made quickly over the telephone. Intuitively, a player then has the opportunity to accompany the offer with a threat that the offer is final. Shaked shows that equilibria exist in which the threat is treated as credible and others in which it is not. When outside options are mentioned later, it is the face-to-face model that is intended. But it is important to bear in mind how sensitive the model can be to apparently minor changes in the structure of the game. For further discussion of the "outside option" issue in the alternating-offers model, see Sutton (1986) and Bester (1988).

Harsanyi and Selten (1988, ch. 6) study a model of simultaneous demands in which one player has an outside option. Player 1 either claims a fraction of the pie or opts out, and simultaneously player 2 claims a fraction of the pie. If player 1 opts out, then he receives a fraction α of the pie and player 2 receives nothing. If the sum of the players' claims is one, then each receives his claim. Otherwise each receives nothing. The game has a multitude of Nash equilibria. That selected by the Harsanyi and Selten theory results in the division $(1/2, 1/2)$ if $\alpha < 1/4$ and the division $(\sqrt{\alpha}, 1 - \sqrt{\alpha})$ if $\alpha \geqslant 1/4$. Thus the model leads to a conclusion about the effect of outside options on the outcome of bargaining that is strikingly different from that of the alternating-offers model. Clearly further research on the many possible bargaining models that can be constructed in this context is much needed.

2.7. Risk

Binmore, Rubinstein and Wolinsky (1986) consider a variation on the alternating-offers model in which the players are indifferent to the passage of time but face a probability p that any rejected offer will be the last that can be made. The fear of getting trapped in a bargaining *impasse* is then replaced by the possibility that intransigence will lead to a *breakdown* of the negotiating process owing to the intervention of some external factor. The extensive form in the new situation is somewhat different from the one described above: at the end of each period the game ends with the breakdown outcome with probability p. Moreover, the functions ϕ_1 and ϕ_2 need to be reinterpreted as von Neumann and Morgenstern utility functions. That is to say, they are derived from the players' attitudes to risk rather than from their attitudes to time. The conclusion is essentially the same as in the time-based model. We denote the breakdown payoff vector by b and replace the discount factors by $1 - p$. The fact that b may be nonzero means that (2) must be replaced by

$$v_1^* = pb_1 + (1 - p)u_1^* \quad \text{and} \quad u_2^* = pb_2 + (1 - p)v_2^*, \tag{4}$$

where, as before, u^* is the agreement payoff vector when player 1 makes the first offer and v^* is its analog for the case in which it is 2 who makes the first offer.

2.8. More than two players

Result 1 does *not* extend to the case when there are more than two players, as the following three-player example of Shaked demonstrates.

Three players rotate in making proposals $a = (a_1, a_2, a_3)$ on how to split a cake of size one. We require that $a_1 + a_2 + a_3 = 1$ and $a_i \geqslant 0$ for $i = 1, 2, 3$. A proposal a accepted at time t is evaluated as worth $a_i \delta^t$ by player i. A proposal *a* made by player *j* at time *t* is first considered by player $j + 1$ (mod 3), who may accept or reject it. If he accepts it, then player $j + 2$ (mod 3) may accept or reject it. If both accept it then the game ends and the proposal *a* is implemented. Otherwise player $j + 1$ (mod 3) makes the next proposal at time $t + 1$.

Let $1/2 \leqslant \delta < 1$. Then, for *every* proposal *a*, there exists a subgame-perfect equilibrium in which *a* is accepted immediately. We describe the equilibrium in terms of the four commonly held "states (of mind)" *a*, e^1, e^2, and e^3, where e^i is the *i*th unit vector. In state *y*, each player *i* makes the proposal *y* and accepts the proposal *z* if and only if $z_i \geqslant \delta y_i$. The initial state is *a*. Transitions occur only after a proposal has been made, *before* the response. If, in state *y*, player *i* proposes *z* with $z_i > y_i$, then the state becomes e^j, where $j \neq i$ is the player with the lowest index for whom $z_j < 1/2$. Such a player *j* exists, and the requirement that $\delta \geqslant 1/2$ guarantees that it is optimal for him to reject player *i*'s proposal.

Efforts have been made to reduce the indeterminacy in the *n*-player case by changing the game or the solution concept. One obvious result is that, if attention is confined to stationary (one-state) strategies, then the unique subgame-perfect equilibrium assigns the cake in the proportions $1 : \delta : \ldots : \delta^{n-1}$. The same result follows from restricting the players to have continuous expectations about the future [Binmore (1987d)].

2.9. Related work

Perry and Reny (1992) study a model in which time runs continuously and players choose when to make offers. Muthoo (1990) studies a model in which each player can withdraw from an offer if his opponent accepts it; he shows that all partitions can be supported by subgame perfect equilibria in this case. Haller (1991), Fernandez and Glazer (1991) and Haller and Holden (1990) [see also Jones and McKenna (1990)] study a model of wage bargaining in which after any offer is rejected the union has to decide whether or not to strike or continue working at the current wage. [See also the general model of Okada (1991a, 1991b).] The model of Admati and Perry (1991) can be interpreted as a variant of the alternating offers model in which neither player can retreat from concessions he made in the past.

Models in which offers are made simultaneously are discussed, and compared with the model of alternating offers, by Stahl (1990), and Chatterjee and Samuelson (1990). Chikte and Deshmukh (1987), Wolinsky (1987) and Muthoo (1989b) study models in which players may search for outside options while bargaining.

3. The Nash program

The ultimate aim of what is now called the "Nash program" [see Nash (1953)] is to classify the various institutional frameworks within which negotiation takes place and to provide a suitable "bargaining solution" for each class. As a test of the suitability of a particular solution concept for a given type of institutional framework, Nash proposed that attempts be made to reduce the available negotiation ploys within that framework to moves within a formal bargaining game. If the rules of the bargaining game adequately capture the salient features of the relevant bargaining institutions, then a "bargaining solution" proposed for use in the presence of these institutions should appear as an equilibrium outcome of the bargaining game.

The leading solution concept for bargaining situations in the Nash bargaining solution [see Nash (1950)]. The idea belongs in cooperative game theory. A "bargaining problem" is a pair (U, q) in which U is a set of pairs of von Neumann and Morgenstern utilities representing the possible deals available to the bargainers, and q is a point in U interpreted by Nash as the status quo. The Nash bargaining solution of (U, q) is a point at which the Nash product

$$(u_1 - q_1)(u_2 - q_2) \tag{5}$$

is maximized subject to the constraints $u \in U$ and $u \geq q$. Usually it is assumed that u is convex, closed, and bounded above to ensure that the Nash bargaining solution is uniquely defined, but convexity is not strictly essential in what follows.

When is such a Nash bargaining solution appropriate for a two-player bargaining environment involving alternating offers? Consider the model we studied in Section 2.7, in which there is a probability p of breakdown after any rejection. We have the following result. [See also Moulin (1982), Binmore, Rubinstein and Wolinsky (1986) and McLennan (1988).]

Result 3 [Binmore (1987a)]. When a unique subgame-perfect equilibrium exists for each p sufficiently close to one, the bargaining problem (U, q), in which U is the set of available utility pairs at time 0 and $q = b$ is the breakdown utility pair, has a unique Nash bargaining solution. This is the limiting value of the subgame-perfect equilibrium payoff pair as $p \to 0+$.

Proof. To prove the concluding sentence, it is necessary only to observe from (4) that $u^* \in U$ and $v^* \in U$ lie on the same contour of $(u_1 - b_1)(u_2 - b_2)$ and that $u^* - v^* \to (0, 0)$ as $p \to 0+$. $\square$

We can obtain a similar result in the time-based alternating-offers model when the length τ of a bargaining period approaches 0. One is led to this case by considering two objections to the alternating-offers model. The first is based on the fact that the equilibrium outcome favors player 1 in that $u_1^* > v_1^*$ and $u_2^* < v_2^*$. This reflects players 1's first-mover advantage. The objection evaporates when τ is small, so that "bargaining frictions" are negligible. It then becomes irrelevant who goes first. The second objection concerns also the reasons why players abide by the rules. Why should a player who has just rejected an offer patiently wait for a period of length $\tau > 0$ before making a counteroffer? If he were able to abbreviate the waiting time, he would respond immediately. Considering the limit as $\tau \to 0+$ removes some of the bite of the second objection in that the players need no longer be envisaged as being constrained by a rigid, exogenously determined timetable.

Figure 3 illustrates the solution u^* and v^* of equations (2) in the case when δ_1 and δ_2 are replaced by δ_1^τ and δ_2^τ and $\rho_i = -\log \delta_i$. It is clear from the figure that, when τ approaches zero, both u^* and v^* approach the point in U at which $u_1^{1/\rho_1} u_2^{1/\rho_2}$ is maximized. Although we are not dealing with von Neumann and Morgenstern utilities, it is convenient to describe this point as being located at an *asymmetric* Nash bargaining solution of U relative to a status quo q located

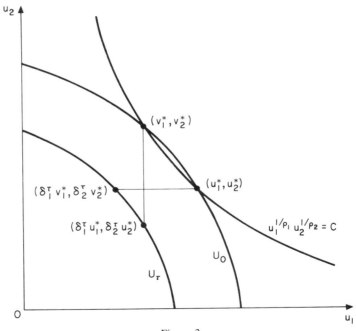

Figure 3.

at the impasse payoff pair $(0, 0)$. [See the chapter on 'cooperative models of bargaining' in a forthcoming volume of this Handbook and Roth (1977).]

Such an interpretation should not be pushed beyond its limitations. In particular, with our assumptions on time preferences, it has already been pointed out that, for *any* δ in $(0, 1)$, there exist functions w_1 and w_2 such that $w_1(a)\delta^t$ and $w_2(1 - a)\delta^t$ are utility representations of the players' time preferences. Thus if the utility representation is tailored to the bargaining problem, then the equilibrium outcome in the limiting case as $\tau \to 0+$ is the *symmetric* Nash bargaining solution for the utility functions w_1 and w_2.

This discussion of how the Nash bargaining solution may be implemented by considering limiting cases of sequential noncooperative bargaining models makes it natural to ask whether other bargaining solutions from cooperative game theory can be implemented using parallel techniques. We mention only Moulin's (1984) work on implementing the Kalai–Smorodinsky solution. [See the chapter on 'cooperative models of bargaining' in a forthcoming volume of this Handbook and Kalai and Smorodinsky (1975).] Moulin's model begins with an auction to determine who makes the first proposal. The players simultaneously announce probabilities p_1 and p_2. If $p_1 \geqslant p_2$, then player 1 begins by proposing an outcome a. If player 2 rejects a, that he makes a counterproposal, b. If player 1 rejects b, then the status quo q results. If player 1 accepts b, then the outcome is a lottery that yields b with probability p_1 and q with probability $1 - p_1$. (If $p_2 > p_1$ then it is player 2 who proposes an outcome, and player 1 who responds.) The natural criticism is that it is not clear to what extent such an "auctioning of fractions of a dictatorship" qualifies as bargaining in the sense that this is normally understood.

3.1. Economic modeling

The preceding section provides some support for the use of the Nash bargaining solution in economic modeling. One advantage of a noncooperative approach is that it offers some insight into *how* the various economic parameters that may be relevant should be assimilated into the bargaining model when the environment within which bargaining takes place is complicated [Binmore, Rubinstein and Wolinsky (1986)]. In what follows we draw together some of the relevant considerations.

Assume that the players have von Neumann and Morgenstern utilities of the form $\delta_i^t u_i(a)$. (Note that this is a very restrictive assumption.) Consider the placing of the status quo. In cooperative bargaining theory this is interpreted as the utility pair that results from a failure to agree. But such a failure to agree may arise in more than one way. We shall, in fact, distinguish three possible ways:

(a) A player may choose to abandon the negotiations at time t. Both players are then assumed to seek out their best outside opportunities, thereby deriving utilities $e_i \delta_i^t$. Notice that it is commonplace in modeling wage negotiations to ignore timing considerations and to use the Nash bargaining solution with the status quo placed at the "exit point" e.

(b) The negotiations may be interrupted by the intervention of an exogenous random event that occurs in each period of length τ with probability $\lambda \tau$. If the negotiations get broken off in this way at time t, each player i obtains utility $b_i \delta_i^t$.

(c) The negotiations may continue for ever without interruption or agreement, which is the outcome denoted by D in Section 2. As in Section 2, utilities are normalized so that each player then gets $d_i = 0$.

Assume that the three utility pairs e, b, and d satisfy $0 = d < b < e$.

When contemplating the use of an asymmetric Nash bargaining solution in the context of an alternating offers model for the "frictionless" limiting case when $\tau \rightarrow 0+$, the principle is that the status quo is placed at the utility pair q that results from the use of impasse strategies. Thus, if we ignore the exit point e then the relevant disagreement point is q with

$$q_i = \lim_{\tau \rightarrow 0+} \sum_{j=0}^{\infty} b_i \delta_i^{\tau j} \lambda \tau (1 - \lambda \tau)^j = \lambda b_i / (\lambda + \rho_i) \quad \text{for } i = 1, 2,$$

where $\rho_i = -\log \delta_i$. The (symmetric) Nash bargaining solution of the problem in which q is the status quo point is the maximizer of $u_1^{\alpha_1} u_2^{\alpha_2}$, where $\alpha_i = 1/(\lambda + \rho_i)$ (i.e. it is the asymmetric Nash bargaining solution in which the "bargaining power" of player i is α_i). This reflects the fact that both time and risk are instrumental in forcing an agreement.

It is instructive to look at two extreme cases. The first occurs when λ is small compared with the discount rates ρ_1 and ρ_2 so that it is the *time* costs of disagreement that dominate. The status quo goes to d (=0) and the bargaining powers become $1/\rho_i$. The second case occurs when ρ_1 and ρ_2 are both small compared with λ so that *risk* costs dominate. This leads to a situation closer to that originally envisaged by Nash (1950). The status quo goes to the breakdown point b and the bargaining powers approach equality so that the Nash bargaining solution becomes symmetric.

As for the exit point, the principle is that its value is *irrelevant* unless at least one player's outside option e_i exceeds the appropriate Nash bargaining payoff. There will be no agreement if this is true for both players. When it is true for just one of the players, he gets his outside option and the other gets what remains of the surplus. (See Result 2 in the case that $\delta \rightarrow 1$.)

Note finally that the above considerations concerning bargaining over *stocks* translate immediately to the case of bargaining over *flows*. In bargaining over the wage *rate* during a strike, for example, the status quo payoffs should be the impasse flows to the two parties during the strike (when the parties' primary motivation to reach agreement is their impatience with delay).

4. Commitment and concession

A *commitment* is understood to be an action available to an agent that constrains his choice set at future times in a manner beyond his power to revise. Schelling (1960) has emphasized, with many convincing examples, how difficult it is to make genuine commitments in the real world to take-it-or-leave-it bargaining positions. It is for such reasons that subgame-perfect equilibrium and other refinements now supplement Nash equilibrium as the basic tool in noncooperative game theory. However, when it is realistic to consider take-it-or-leave-it offers or threats, these will clearly be overwhelmingly important. Nash's (1953) demand game epitomizes the essence of what is involved when both sides can make commitments.

In this model, the set U of feasible utility pairs is assumed to be convex, closed, and bounded above, and to have a nonempty interior. A point $q \in U$ is designated as the status quo. The two players simultaneously make take-it-or-leave-it demands u_1 and u_2. If $u \in U$, each receives his demand. Otherwise each gets his status quo payoff.

Any point of $V = \{u \geq q: u \text{ is Pareto efficient in } U\}$ is a Nash equilibrium. Other equilibria result in disagreement. Nash (1953) dealt with this indeterminacy by introducing a precursor of modern refinement ideas. He assumed some shared uncertainty about the location of the frontier of U embodied in a quasi-concave, differentiable function $p: \mathbf{R}^2 \to [0, 1]$ such that $p(u) > 0$ if u is in the interior of U and $p(u) = 0$ if $u \notin U$. One interprets $p(u)$ as the probability that the players commonly assign to the event $u \in U$. The modified model is called the *smoothed* Nash demand game. Interest centers on the case in which the amount of uncertainty in the smoothed game is small. For all small enough $\epsilon > 0$, choose a function $p = p^\epsilon$ such that $p^\epsilon(u) = 1$ for all $u \in U$ whose distance from V exceeds ϵ. The existence of a Nash equilibrium that leads to agreement with positive probability for the smoothed Nash demand game for $p = p^\epsilon$ follows from the observation that the maximizer of $u_1 u_2 p^\epsilon(u_1, u_2)$ is such a Nash equilibrium.

Result 4 [Nash (1953)]. Let u^ϵ be a Nash equilibrium of the smoothed Nash demand game associated with the function p^ϵ that leads to agreement with positive probability. When $\epsilon \to 0$, u^ϵ converges to the Nash bargaining solution for the problem (U, q).

Proof. The following sketch follows Binmore (1987c). Player i seeks to maximize $u_i p^\epsilon(u) + q_i(1 - p^\epsilon(u))$. The first-order conditions for $u^\epsilon > q$ to be a Nash equilibrium are therefore

$$(u_i^\epsilon - q_i)p_i^\epsilon(u^\epsilon) + p^\epsilon(u^\epsilon) = 0 \quad \text{for } i = 1, 2 , \tag{6}$$

where p_i^ϵ is the partial derivative of p^ϵ with respect to u_i. Suppose that $p(u^\epsilon) = c > 0$. From condition (6) it follows that the vector u^ϵ must be a maximizer of $H(u_1, u_2) = (u_1 - q_1)(u_2 - q_2)$ subject to the constraint that $p(u) = c$. Let w^ϵ be the maximizer of $H(u_1, u_2)$ subject to the constraint that $p(u) = 1$. Then $H(u^\epsilon) \geqslant H(w^\epsilon)$. By the choice of p^ϵ the sequence w^ϵ converges to the Nash bargaining solution and therefore the sequence u^ϵ converges to the Nash bargaining solution as well. $\square$

There has been much recent interest in the Nash demand game with incomplete information, in which context it is referred to as a "sealed-bid double-auction" [see, for example, Leininger, Linhart and Radner (1989), Matthews and Postlewaite (1989), Williams (1987) and Wilson (1987a)]. It is therefore worth noting that the smoothing technique carries over to the case of incomplete information and provides a noncooperative defense of the Harsanyi and Selten (1972) axiomatic characterization of the $(M + N)$-player asymmetric Nash bargaining solution in which the bargaining powers β_i $(i = 1, \dots, M)$ are the (commonly known) probabilities that player 2 attributes to player 1's being of type i and β_j $(j = M + 1, \dots, M + N)$ are the probabilities attributed by player 1 to player 2's being of type j. If attention is confined to pooling equilibria in the smoothed demand game, the predicted deal $a \in A$ is the maximizer of $\Pi_{i=1,\dots,M}(\phi_i(a))^{\beta_i}\Pi_{j=M+1,\dots,M+N}(\phi_j(a))^{\beta_j}$, where $\phi_i: A \to R$ is the von Neumann and Morgenstern utility function of the player of type i [Binmore (1987c)].

4.1. Nash's threat game

In the Nash demand game, the status quo q is given. Nash (1953) extended his model in an attempt to endogenize the choice of q. In this later model, the underlying reality is seen as a finite two-person game, G. The bargaining activity begins with each player making a *binding* threat as to the (possibly mixed) strategy for G that he will use *if* agreement is not reached in the negotiations that follow. The ensuing negotiations consist simply of the Nash demand game being played. If the latter is appropriately smoothed, the choice of threats t_1 and t_2 at the first stage serves to determine a status quo $q(t_1, t_2)$ for the use of the Nash bargaining solution at the second stage. The players can write contracts specifying the use of lotteries, and hence we identify the set U of feasible deals with the convex hull of the set of payoff pairs available in G when this is played noncooperatively. This analysis generates a reduced game in which the payoff pair $n(t)$ that results from the choice of the strategy pair t is the Nash bargaining solution for U relative to the status quo $q(t)$.

Result 5 [Nash (1953)]. The Nash threat game has an equilibrium, and all equilibria yield the same agreement payoffs in U.

The threat game is strictly competitive in that the players' preferences over the possible outcomes are diametrically opposed. The result is therefore related to von Neumann's maximin theorem for two-person, zero-sum games. In particular, the equilibrium strategies are the players' security strategies and the equilibrium outcome gives each player his security level. For a further discussion of the Nash threat game, see Owen (1982).

The model described above, together with Nash's (1953) axiomatic defense of the same result, is often called his *variable threats* theory. The earlier model, in which q is given, is then called the *fixed threat* theory and q itself is called the *threat point*. It needs to be remembered, in appealing to either theory, that the threats need to have the character of conditional *commitments* for the conclusions to be meaningful.

4.2. The Harsanyi–Zeuthen model

In what Harsanyi (1977) calls the "compressed Zeuthen model", the first stage consists of Nash's simple demand game (with no smoothing). If the opening demands are incompatible, a second stage is introduced in which the players simultaneously decide whether to concede or to hold out. If both concede, they each get only what their opponent offered them. If both hold out, they get their status quo payoffs, which we normalize to be zero.

The concession subgame has three Nash equilibria. Harsanyi (1977) ingeniously marshals a collection of "semi-cooperative" rationality principles in defense of the use of Zeuthen's (1930) principle in making a selection from these three equilibria. Denoting by r_i the ratio between i's utility gain if j concedes and i's utility loss if there is disagreement, Zeuthen's principle is that, if $r_i > r_j$, then player i concedes. When translated into familiar terms, this calls for the selection of the equilibrium at which the Nash product of the payoffs is biggest. When this selection is made in the concession subgames, the equilibrium pair of opening demands is then simply the Nash bargaining solution.

The full Harsanyi–Zeuthen model envisages not one sudden-death encounter but a sequence of concessions over small amounts. However, the strategic situation is very similar and the final conclusion concerning the implementation of the Nash bargaining solution is identical.

4.3. Making commitments stick

Crawford (1982) offers what can be seen as an elaboration of the compressed Harsanyi–Zeuthen model with a more complicated second stage in which making a concession (backing down from the "commitment") is costly to an extent that is uncertain at the time the original demands are made. He finds not only that impasse can occur with positive probability, but that this probability need not decrease as commitment is made more costly.

   More recent work has concentrated on incomplete information about preferences as an explanation of disagreement between rational bargainers (see Section 8). In consequence, Schelling's (1960) view of bargaining as a "struggle to establish commitments to favorable bargaining positions" remains largely unexplored as regards formal modeling.

5. Pairwise bargaining with few agents

In many economic environments the parameters of one bargaining problem are determined by the forecast outcomes of other bargaining problems. In such situations the result of the bargaining is highly sensitive to the detailed structure of the institutional framework that governs how and when agents can communicate with each other. The literature on this topic remains exploratory at this stage, concentrating on a few examples with a view to isolating the crucial institutional features. We examine subgame-perfect equilibria of some elaborations of the model of Section 2.

5.1. One seller and two buyers

An indivisible good is owned by a seller S whose reservation value is $v_S = 0$. It may be sold to one and only one of two buyers, H and L, with reservation values $v = v_H \geq v_L = 1$. In the language of cooperative game theory, we have a three-player game with value function V satisfying $V(\{S, H\}) = V(\{S, H, L\}) = v$, $V(\{S, L\}) = 1$, and $V(C) = 0$ otherwise. The game has a nonempty core in which the object is sold to H for a price $p \geq 1$ when $v > 1$. (When $v = 1$, it may be sold to either of the buyers at the price $p = 1$.) The Shapley value is $(1/6 + v/2, v/2 - 1/3, 1/6)$ (where the payoffs are given in the order S, H, L).

How instructive are such conclusions from cooperative theory? The following noncooperative models are intended to provide some insight. In these models, if the object changes hands at price p at time t, then the seller gets $p\delta^t$ and the successful buyer gets $(v_B - p)\delta^t$, where v_B is his valuation and $0 < \delta < 1$. An agent who does not participate in a transaction gets zero. Information is always perfect.

5.1.1. *Auctioning* [Wilson (1984), Binmore (1985)]

The seller begins at time 0 by announcing a price, which both buyers hear. Buyer H either accepts the offer, in which case he trades with the seller and the game ends, or rejects it. In the latter case, buyer L then either accepts or rejects the seller's offer. If both buyers reject the offer, then there is a delay of length τ, after which both buyers simultaneously announce counteroffers; the seller may either accept one of these offers or reject both. If both are rejected, then there is a delay of length τ, after which the seller makes a new offer; and so on.

5.1.2. *Telephoning* [Wilson (1984, Section 4), Binmore (1985)]

The seller begins by *choosing* a buyer to call. During their conversation, the seller and buyer alternate in making offers, a delay of length τ elapsing after each rejection. Whenever it is the seller's turn to make an offer, he can hang up, call the other buyer and make an offer to him instead. An excluded buyer is not allowed to interrupt the seller's conversation with the other buyer.

5.1.3. *Random matching* [Rubinstein and Wolinsky (1990)]

At the beginning of each period, the seller is randomly matched with one of the two buyers with equal probability. Each member of a matched pair then has an equal chance of getting to make a proposal which the other can then accept or reject. If the proposal is rejected, the whole process is repeated after a period of length τ has elapsed.

5.1.4. *Acquiring property rights* [Gul (1989)]

The players may acquire property rights originally vested with other players. An individual who has acquired the property rights of all members of the coalition C enjoys an income of $V(C)$ while he remains in possession. Property rights may change hands as a consequence of pairwise bargaining. In each period, any pair of agents retaining property rights has an equal chance of being chosen to bargain. Each member of the matched pair then has an equal chance of getting to make a proposal to the other about the rate at which he is willing to rent the property rights of the other. If the responder agrees, he leaves the game and the remaining player enjoys the income derived from coalescing the property rights of both. If the responder refuses, both are returned to the pool of available bargainers. In this model a strategy is to be understood as stationary if the behavior for which it calls depends only on the current distribution of property rights and not explicitly on time or other variables.

Result 6

(a) [Binmore (1985)]. If, in the *auctioning* model, $\delta^\tau v/(1+\delta^\tau)<1$, then there is a subgame-perfect equilibrium, and in all such equilibria the good is sold immediately (to H if $v>1$) at the price $\delta^\tau+(1-\delta^\tau)v$. If $\delta^\tau v/(1+\delta^\tau)>1$, then the only subgame-perfect equilibrium outcome is that the good is sold to H at the bilateral bargaining price (of approximately $v/2$ if τ is sufficiently small) that would obtain if L were absent altogether.

(b) [Binmore (1985)]. In any subgame-perfect equilibrium of the *telephoning* model immediate agreement is reached on the bilateral bargaining price (approximately $v/2$ when τ is small) that would obtain if L were absent altogether. If $v>1$ then the good is sold to H.

(c) [Rubinstein and Wolinsky (1990)]. If, in the *random matching* model, $v=1$, then there is a unique subgame-perfect equilibrium in which the good is sold to the first matched buyer at a price of approximately 1 when τ is small.

(d) [Gul (1989)]. For the *acquiring property rights* model, among the class of stationary subgame-perfect equilibria there is a unique equilibrium in which all matched pairs reach immediate agreement. When τ is small, this equilibrium assigns each player an expected income approximately equal to his Shapley value allocation.

5.2. Related work

Shaked and Sutton (1984) and Bester (1989a) study variations of the "telephoning" model, in which the delay before the seller can make an offer to a *new* buyer may differ from the delay between any two successive periods of bargaining. [See also Bester (1988) and Muthoo (1989a).] The case $v>1$ in the "random matching" model is analyzed by Hendon and Tranæs (1991). An implementation of the Shapley value that is distinct but related to that in the "acquiring property rights" model is given by Dow (1989). Gale (1988) and Peters (1988, 1989, 1991, 1992) study the relation between the equilibria of models in which sellers announce prices ("auctioning", or "ex ante pricing"), and the equilibria of models in which prices are determined by bargaining after a match is made ("ex post pricing"). Horn and Wolinsky (1988a, 1988b) analyze a three-player cooperative game in which $V(1, 2, 3) > V(1, 2) = V(1, 3) > 0$ and for $V(C) = 0$ all other coalitions C. [See also Davidson (1988), Jun (1989), and Fernandez and Glazer (1990).] In this case, the game does not end as soon as one agreement is reached and the question of whether the first agreement is implemented immediately becomes an important factor. [Related papers are Jun (1987) and Chae and Yang (1988).]

6. Noncooperative bargaining models for coalitional games

In Section 8.2 we showed that Result 1 does not directly extend to situations in which more than two players have to split a pie. The difficulties are compounded if we wish to provide a noncooperative model for an arbitrary coalitional game.

Selten (1981) studies a model that generalizes the alternating-offers model. He restricts attention to coalitional games (N, v) with the "one-stage property": $v(C) > 0$ implies $v(N \setminus C) = 0$. In such a game, let d be an n-vector, and let $F_i(d)$ be the set of coalitions that contain i and satisfy $\Sigma_{i \in C} d_i = v(C)$. Then d is a "stable demand vector" if $\Sigma_{i \in C} d_i \geq v(C)$ for all coalitions C, and no $F_i(d)$ is a proper subset of $F_j(d)$ for any j.

Selten's game is the following. Before play begins, one of the players, say i, is assigned the initiative. In any period, the initiator can either pass the initiative to some other player, or make a proposal of the form (C, d, j), where $C \ni i$ is a coalition, d is a division of $v(C)$ among the members of C, and $j \in C$ is the member of C designated to be the responder. Player j either accepts the proposal and selects one of the remaining members of C to become the next responder, or rejects the proposal. In the latter case, play passes to the next period, with j holding the initiative. If all the members of C accept the proposal, then it is executed, and the game ends. The players are indifferent to the passage of time.

The game has many stationary subgame-perfect equilibria. However, Selten restricts attention to equilibria in which (i) players do not needlessly delay agreement, (ii) the initiator assigns positive probability to all optimal choices that lead to agreement with probability 1, and (iii) whenever some player i has a deviation $x = (C, d, j)$ with the properties that d_i exceeds player i's equilibrium payoff, d_j is less than player j's equilibrium payoff, and player i is included in all the coalitions that may eventually form and obtains his equilibrium payoff, then player j has a deviation (C', d', i) that satisfies the same conditions with the roles of i and j reversed. Selten shows that such equilibria generate stable demand vectors, in the sense that a stable demand vector d is obtained by taking d_i to be player i's expected payoff in such an equilibrium conditional on his having the initiative.

Chatterjee, Dutta, Ray and Sengupta (1992) study a variant of Selten's game in which the players are impatient, and the underlying coalitional game does not necessarily satisfy the one-stage property. They show that for convex games, stationary subgame-perfect equilibria in which agreement is reached immediately on an allocation for the grand coalition converge, as the degree of impatience diminishes, to the egalitarian allocation [in the sense of Dutta and Ray (1989)].

Harsanyi (1981) studies two noncooperative models of bargaining that implement the Shapley value in certain games. We briefly discuss one of these. In every period each player proposes a vector of "dividends" for each coalition of which he is a member. If all members of a coalition propose the same dividend vector, then they receive these dividends if this is feasible. At the end of each period there is a small probability that the negotiations break down. If it is not ended by chance, then the game ends when the players unanimously agree on their dividend proposals. Harsanyi shows that in decomposable games – games that are the sums of unanimity games – the outcome of the bargaining game selected by the Harsanyi and Selten (1988) equilibrium selection procedure is precisely the Shapley value. [For a related, "semi-cooperative" interpretation of the Shapley value, see Harsanyi (1977).]

Various implementations of the solution sets of von Neumann and Morgenstern are also known, notably that of Harsanyi (1974). In each period t some feasible payoff vector x^t is "on the floor". A referee chooses some coalition S to make a counterproposal. The members of S simultaneously propose alternative payoff vectors. If they all propose the same vector y, and y dominates x^t through S, then y is on the floor in period $t + 1$; otherwise the game ends, and x^t is the payoff vector the players receive. The solution concept Harsanyi applies is a variant of the set of stationary subgame perfect equilibrium.

As things stand, these models demonstrate only that various cooperative solution concepts can emerge as equilibrium outcomes from suitably designed noncooperative or semi-cooperative bargaining models. However, these pioneering papers provide little guidance as to which of the available cooperative solution concepts, if any, it is appropriate to employ in an applied model. For this purpose, bargaining models need to be studied that are not hand-picked to generate the solution concept they implement. But it is difficult to see how to proceed while the simple alternating-offers model with three players remains open. Presumably, as in the case of incomplete information considered in Section 8, progress must await progress in noncooperative equilibrium theory.

7. Bargaining in markets

Bargaining theory provides a natural framework within which to study price formation in markets where transactions are made in a decentralized manner via interaction between pairs of agents rather than being organized centrally through the use of a formal trading institution like an auctioneer. One might describe the aim of investigations in this area as that of providing "mini-micro" foundations for the microeconomic analysis of markets and, in particular, of determining the range of validity of the Walrasian paradigm. Such a program represents something of a challenge for game theorists in that its success will presumably generate new solution concepts for market situations intermediate between those developed for bilateral bargaining and the notion of a Walrasian equilibrium.

Early studies of matching and bargaining models are Diamond and Maskin (1979), Diamond (1981) and Mortensen (1982a, 1982b) in which bargaining is modeled using cooperative game theory. This approach is to be contrasted with the noncooperative approach of the models that follow. A pioneering paper in this direction is Butters (1977).

The models that exist differ in their treatment of several key issues. First, there is the information structure. What does a player know about the events in other bargaining sessions? Second, there is the question of the detailed structure of the pairwise bargaining games. In particular, when can a player opt out? Third, there is the modeling of the search technology through which the bargainers get matched. Finally, there is the nature of the data given about agents in the market. Sometimes, for example, it relates to stocks of agents in the market, and sometimes to flows of entrants or potential entrants.

7.1. Markets in steady state [Rubinstein and Wolinsky (1985)]

Most of the literature has concentrated on a market for an individual good in which agents are divided into two groups, sellers and buyers. All the sellers have reservation value 0 for the good and all the buyers have reservation value 1. A matched seller and buyer can agree on any price p, with $0 \leqslant p \leqslant 1$. If agreement is reached at time t, then the seller leaves the market with a von Neumann and Morgenstern utility of $p\delta^t$ and the buyer leaves with $(1-p)\delta^t$.

The first event in each period is a matching session in which all agents in the market participate, including those who may be matched already. Any seller has a probability σ of being matched with a buyer and any buyer has a probability β of being matched with a seller. The numbers σ and β are assumed to be constant so that the economic environment remains in a steady state.

Bargaining can take place only between individuals in a matched pair. After the matching session, each member of a matched pair is equally likely to be chosen to make the first offer. This may be accepted or rejected by the proposer's partner. If it is accepted, both leave the market. In either case, the next period commences after time τ has elapsed.

Pairs who are matched at time t but do not reach agreement remain matched at time $t + \tau$, unless one or both partners gets matched elsewhere. An agent *must* abandon his old partner when matched with a new one. Thus, for example, a seller with a partner at time t who does not reach agreement at time t has probability $(1 - \sigma)\beta$ of being without a partner at time $t + \tau$. (A story can be told about the circumstances under which it would always be optimal to abandon the current partner if this decision were the subject of strategic choice, but this issue is neglected here.)

The model is not a game in the strict sense. For example, the set of players is not specified. Nevertheless, a game-theoretic analysis makes sense using a solution concept that is referred to as a "market equilibrium". This is a pair of strategies, one for buyers and one for sellers, that satisfies:

(1) *Semi-stationarity*. The strategies prescribe the same bargaining tactics for all buyers (or sellers) independently of their personal histories.

(2) *Sequential rationality*. The strategies are optimal after all possible histories.

Result 7 [Rubinstein and Wolinsky (1985)].   There is a unique market equilibrium. As $\tau \rightarrow 0+$, the price at which the good changes hands converges to $\sigma / (\sigma + \beta)$.

The probabilities σ and β depend on the matching technology, which depends in turn on how search is modeled. Let S and B be the steady-state measures of sellers and buyers, respectively, and consider the most naïve of search models in which $\sigma = c\tau B / (B + S)$ and $\beta = c\tau S / (B + S)$, where the constant c represents a "search friction". In the limit as $\tau \rightarrow 0+$, the market equilibrium price approaches $B / (B + S)$. Thus, for example, if there are few sellers and many buyers, the price is high.

Notice that the short side of the market does not appropriate the entire surplus even in the case when several frictions become negligible. Gale (1987) points out that, if this conclusion seems paradoxical, it is as a consequence of thinking of supply and demand in terms of the *stocks* S and B of agents in the market at any time. To keep the market in a steady state, the *flows* of buyers and sellers into the market at any time have to be equal. If supply and demand are measured in terms of these flows, then *any* selling price is Walrasian. For further discussion, see Rubinstein (1987, 1989).

7.2. Unsteady states [Binmore and Herrero (1988a, 1988b)]

Binmore and Herrero (1988a, 1988b) generalize the preceding model in two directions. The informational difficulties finessed by Rubinstein and Wolinsky's "semi-stationarity" condition are tackled by observing that subgame-perfect equilibria in alternating-offers models can be replaced by "security equilibria" without losing the uniqueness conclusion. A security equilibrium is related to the notion of "rationalizability" introduced by Bernheim (1984) and Pearce (1984). Their requirement about its being common knowledge that strictly dominated strategies are never played is replaced by a similar requirement concerning security levels. It is assumed to be common knowledge that no player takes an action under any contingency that yields less than he calculates his security level to be, given the occurrence of the contingency. Any equilibrium notion normally considered is also a security equilibrium. A proof of uniqueness for security equilibria therefore entails uniqueness for more conventional equilibria also. However, in markets with a continuum of traders, security equilibria are insensitive to the players' personal histories. The immediate point is that stationarity restrictions on the equilibrium concept used in the Rubinstein and Wolinsky model and its relatives are not crucial in obtaining a uniqueness result (provided $\delta < 1$).

The second generalization of the Rubinstein and Wolinsky model results from applying the technique to markets that are not necessarily in a steady state in that the equilibrium measures of traders may vary with time as a consequence of satisfied traders leaving the market without there being an exactly counterbalancing inflow of new traders. Closed-form conclusions are obtained for the continuous time case obtained by considering the limit as $\tau \to 0+$. In particular, the equilibrium deal can be expressed as an integral involving the equilibrium probabilities that a buyer or a seller is matched at all future times.

Aside from the steady-state model, the simplest special case occurs when no new traders enter the market after time 0. There is then no replacement of those traders present at time 0 when they finally conclude a successful deal and leave the market. With the naïve search technology considered in the Rubinstein and Wolinsky model, the following Walrasian conclusion is obtained:

Result 8 [Binmore and Herrero (1988a, 1988b)]. There is a unique security equilibrium. As search frictions become negligible, the equilibrium deal approximates that in which the entire surplus is assigned to agents on the short side of the market.

Among many other results, Gale (1987) has extended versions of both Results 7 and 8 to the case in which there is a spectrum of reservation prices on both sides of the market.

7.3. Divisible goods with multiple trading [Gale (1986c)]

Gale (1986a, 1986b, 1986c) studies traditional barter markets in which many divisible goods are traded and agents can transact many times before leaving the market. We now describe one of the models from Gale (1986c) [which is a simplification of the earlier paper Gale (1986a)]. [The existence of market equilibrium is established in Gale (1986b), and the relation between Gale's work and general equilibrium theory is explored in McLennan and Sonnenschein (1991).]

All agents, of which there are K types, enter the market at time zero. Initially, there is a measure n_k of agents of each type $k = 1, 2, \ldots, K$. An agent of type k is characterized by his initial commodity bundle w_k and his utility function $u_k: \boldsymbol{R}_+^m \cup \{D\} \to \boldsymbol{R} \cup \{-\infty\}$, where $\boldsymbol{R}_+^m$ is the space of commodity bundles with which he might leave the market and D is the event of his remaining in the market for ever. Agents are not impatient ($\delta = 1$) and bundles may be stored costlessly.

Each period begins with a matching session which operates independently of past events. In particular, no matches survive from previous periods, The probability of a given agent getting matched with an agent with specified characteristics is proportional to the current measure of such agents in the population. Once a match is established, each of the paired agents learns the type of his partner and his partner's commodity bundle. Bargaining then begins. Each member of a matched pair is equally likely to be chosen to make a proposal. This must consist of a vector representing a feasible transfer of goods from himself to his bargaining partner. This proposal may be accepted or rejected. If it is rejected, the responding agent then decides whether or not to leave the market. An important assumption is that agents do not leave the market except after such a rejection.

As trade occurs, the bundle held by each agent changes. Given the restrictions on strategies imposed below, the number of different bundles held is always finite. Thus, in any period the *state of the market* can be characterized by a finite list $(c_i, k_i, \nu_i)_{i=1,\ldots,I}$, where c_i is a feasible holding and ν_i is the measure of agents of type k_i holding c_i.

A *market equilibrium* is defined to be a K-tuple σ^* of strategies, one for each type, that satisfies the following conditions:

(1) *Semi-stationarity*. The bargaining tactics prescribed by the strategy depend only on time, the agent's current bundle and the opponent's type and current bundle.

(2) *Sequential rationality*. Whenever an agent makes a decision, his strategy calls for an optimal decision, given the strategies of the other types and given that the agent believes that the state of the market is that which occurs when all agents use σ^*.

A K-tuple of bundles $(x_1, \ldots, x_K)$ is an allocation if $\Sigma_{k=1}^K n_k x_k = \Sigma_{k=1}^K n_k w_k$. If there exists a price vector p such that, for all k, the bundle x_k maximizes u_k subject to the budget constraint $px \leq pw_k$, then the allocation is Walrasian. Gale's concern is with the circumstances under which the equilibrium outcome is Walrasian.

For technical reasons, Gale restricts the utility functions to be considered. Here [as in the presentation in Osborne and Rubinstein (1990)] we require the existence of an increasing and continuous function $\phi_k: R_+^m \to R$ that is zero on the boundary of R_+^m and strictly concave in its interior. For a given $\phi > 0$, it is then required that

$$u_k(x) = \begin{cases} \phi_k(x) & \text{if } \phi_k(x) \geq \phi, \\ -\infty & \text{otherwise.} \end{cases}$$

In addition, a regularity condition has to be imposed on the indifference curves: their curvature has to be uniformly bounded.

Result 9 [Gale 1986a, 1986b]. For every market equilibrium, there is a Walrasian allocation $(x_1, \ldots, x_K)$ such that each agent of type k leaves the market holding bundle x_k with probability one.

The constraint that the strategies be semi-stationary may reflect an assumption about the information available to the agents. The role of the informational structure in such models is explored in Rubinstein and Wolinsky (1990). In particular, it is shown that, in a model with $\delta = 1$ and a finite number of traders, *any* price can be supported as a sequential equilibrium, provided that agents are permitted perfect knowledge of the events in the market, or even if the agents are able to recall only their personal histories.

7.4. Related work

Wolinsky (1987) studies a model in which each agent chooses the intensity with which to search for an alternative partner. Wolinsky (1988) analyzes the case in which transactions are made by auction, rather than by matching and bargaining. A model in which some agents are middlemen who buy from sellers and resell to buyers (and do not themselves consume the good) is studied by Rubinstein and Wolinsky (1987).

Wolinsky (1990) initiates an investigation of the extension of the models to include asymmetric information. In Wolinsky's model the equilibrium outcome of a decentralized trading process may not approximate the rational expectations equilibrium of the corresponding trading process, even when the market is "approximately frictionless". For related models see Rosenthal and Landau (1981), Green (1991) and Samuelson (1992).

Models of decentralized trade that explicitly specify the trading procedure provide a vehicle by which to analyze the role and value of money in a market.

Gale (1986d) and Kiyotaki and Wright (1989) initiate an investigation of the issues that arise.

8. Bargaining with incomplete information

This section presents some attempts to build theories of bargaining when the information available to the bargainers about their opponents is incomplete. The proposals and responses in an alternating-offers model then do more than register a player's willingness to settle on a particular deal: they also serve as signals by means of which the players may communicate information to each other about their private characteristics. Such signals need not be "truthful". A player in a weak bargaining position may find it worthwhile to imitate the bargaining behavior that he would use if he were strong with a view to getting the same deal as a strong player would get. A strong player must therefore consider whether or not to choose a bargaining strategy that it would be too costly for a weak player to imitate lest the opponent fail to recognize that he is strong. Such issues are studied in the literature on signaling games (see the chapter on 'signalling' in a forthcoming volume of this Handbook) which is therefore central to what follows.

A central goal in studying bargaining with incomplete information is to explain the delays in reaching agreement that we observe in real-life bargaining. (Recall that the alternating-offers model of Result 1, in which information is complete, predicts no delay at all.) Much has been learned in pursuing this goal, but its attainment remains elusive. In this section we propose to do no more than indicate the scope of the difficulties as currently seen.

The literature uses the Kreps and Wilson (1982) notion of a sequential equilibrium after reducing the bargaining situation with incomplete information to a game with imperfect information in accordance with Harsanyi's (1967/68) theory, within which each player is seen as being chosen by a chance move from a set of "types" of player that he might have been. Although subgame-perfection is a satisfactory concept for some complete information bargaining games, the set of sequential equilibria for bargaining games with incomplete information is typically enormously large. It is therefore necessary, if informative results are to emerge, to refine the notion of sequential equilibrium. Progress in the study of bargaining games of incomplete information, as with signaling games in general, is therefore closely tied to developments in the literature on refinements of sequential equilibrium. It should be noted, however, that advances in refinement theory have only a tentative character. Although one idea or another may seem intuitively plausible in a particular context, the theory lacks any firmly grounded guiding principles. Until these problems in the foundations of game theory are better understood, it therefore seems premature to advocate any of the proposed resolutions of the problem of bargaining under incomplete information for general use in economic theory.

8.1. An alternating-offers model with incomplete information [Rubinstein (1985a, 1985b)]

We return to the problem of "dividing the dollar" in which the set of feasible agreements is identified with $A = [0, 1]$. For simplicity, we confine attention to the case of fixed costs per unit time of delay. Recall that the players' preferences over the possible deals (a, t), in which 1 gets a and 2 gets $1 - a$ at time t, may then be represented by $a - c_1 t$ and $1 - a - c_2 t$, where $c_i > 0$ $(i = 1, 2)$. Player 1's cost $c_1 = c$ per unit time of delay is taken to be common knowledge, but 2's cost c_2 is known for certain only by 2. It is common knowledge only that player 1 initially believes that c_2 must take one of the two values c_W or c_S and that the probability of the former is π_W. It is assumed that $c_S < c < c_W$ and the costs are small enough that $c + c_W + c_S < 1$. The interval between successive proposals is fixed at $\tau = 1$ except where otherwise noted.

Having a high cost rate is a source of weakness in one's bargaining position. For example, if $\pi_W = 1$, so that it is certain that 2 has a higher cost rate than 1, then we have seen that 1 gets the entire surplus in equilibrium. On the other hand, if $\pi_W = 0$, so that it is certain that 2 has a lower cost rate than 1, then 1 gets only c_S. For this reason, a high cost type of 2 is said to be *weak* and a low cost type to be *strong*.

In the context of this model, a sequential equilibrium is a strategy triple, one for player 1 and one each for the two possible types of player 2, combined with a belief function that assigns, to every possible history after which player 1 has to move, the probability that player 1 attaches to the event that player 2 is weak. The beliefs have to be updated using Bayes' Rule whenever this is possible, and the initial belief has to be π_W. The strategy of each player must be optimal after *every* history (sequential rationality). We impose two auxiliary requirements. First, if the probability that player 1 attaches to the event that player 2 is weak is zero (one) for some history, it remains zero (one) subsequently. Thus, once player 1 is convinced of the identity of his opponent, he is never dissuaded of this view. Second, when he makes an offer, player 1's belief is the same as it was when he rejected the previous offer of player 2.

As is shown in Rubinstein (1985a, 1985b), many sequential equilibria may exist:

(1) If $\pi_W > 2c/(c + c_W)$, then in all sequential equilibria player 1's expected payoff is at least $\pi_W + (1 - \pi_W)(1 - c_W - c)$.

(2) If $\pi_W \leq 2c/(c + c_W)$, then, for any a^* between c and $1 - c + c_S$, there exists a ("pooling") sequential equilibrium in which player 1's opening demand is a^*, which both a weak and a strong player 2 accept.

(3) If $(c + c_S)/(c + c_W) \leq \pi_W \leq 2c/(c + c_W)$, then for any $a^* \geq c_W$ there exists a ("separating") sequential equilibrium in which player 1's opening demand is a^*. A weak player 2 accepts this demand, while a strong player 2 rejects it and makes the counteroffer $a^* - c_W$, which player 1 accepts.

The multiplicity of equilibria arises because of the freedom permitted by the concept of sequential equilibrium in attributing beliefs to players after they have observed a deviation from equilibrium. Such deviations are zero probability events and so cannot be dealt with by Bayesian updating.

We illustrate the ideas underlying these results by considering case (2). Let $c \leq a^* \leq 1 - c + c_S$. We construct a sequential equilibrium in terms of three commonly held states-of-mind labeled I (for initial), O (for optimistic), and S (for strong). In state I it is common knowledge that 1 believes that 2 is weak with probability π_W. In state W it is common knowledge that 1 believes that 2 is weak for sure, while in state S it is common knowledge that 1 believes that 2 is strong for sure. In state W player 1 and the weak type of player 2 behave precisely as in the complete information case when it is certain that 2 has the high cost rate c_W; the strong type of player 2 uses a best response against player 1's strategy. In state S player 1 and the strong type of player 2 behave precisely as in the complete information case when it is certain that 2 has the low cost rate c_S; the weak type of player 2 uses a best response against player 1's strategy. In state I:

(1) Player 1 demands a^* and accepts an offer of a if and only if $a \geq a^* - c$.
(2) A strong player 2 offers $a^* - c$ and accepts only a demand of $a \leq a^*$.
(3) A weak player 2 offers $a^* - c$ and accepts only a demand of $a \leq a^* - c + c_W$.

The players continue in state I until either (i) player 2 rejects a demand a with $a^* < a < a^* + c_W - c$ and counteroffers $a^* - c$, in which case there is a transition to state S, or (ii) player 2 takes an action inconsistent with the strategies of both the weak and the strong player 2, in which case they switch to state W. The second transition occurs immediately after the inconsistent action. Once in state W or state S they remain there no matter what. (The conjectures that lead the players to move to state W after a deviation are called "optimistic". They are useful in rendering deviations unattractive and hence in constructing multiple equilibria.)

Some comments on why the parameters need to be restricted in order to sustain the equilibrium may be helpful. Notice that if 1 demands more than a^* and less than $a^* - c + c_W$ at time 0, then a weak 2 accepts this demand, while a strong 2 rejects it and proposes $a^* - c$, the state changes to S, and 1 accepts this counteroffer. Thus by deviating in this way 1 obtains at most $\pi_W(a^* - c + c_W) + (1 - \pi_W)(a^* - 2c)$. The condition that this quantity not exceed a^* is that $\pi_W \leq 2c/(c + c_W)$. The requirement that $a^* \geq c$ is simply to ensure that the offer $a^* - c$ be feasible. Finally, observe that if a strong 2 rejects an opening demand of a^*, then the state changes to W, in which a strong 2 obtains $c - c_S$. The condition $a^* \leq 1 - c + c_S$ ensures that this payoff is no more than $1 - a^*$.

8.2. Prolonged disagreement

We now use case (2) from the preceding subsection to construct a sequential equilibrium in which the bargaining may be prolonged for many periods before agreement is achieved.

Choose three numbers, x^*, y^*, and z^*, that satisfy $c \leq x^* < y^* < z^* \leq 1 - c + c_S$. The time that elapses in equilibrium before agreement is reached is denoted by N, where N is chosen to be the largest even integer smaller than

$$\min\{(y^* - x^*)/c, (z^* - y^* + c_W - c)/c_W, (z^* - y^* + c_S - c)/c_S\}.$$

Until period N, player 1 and both types of player 2 hold out for the entire surplus, and player 1 retains his initial belief that player 2 is weak with probability π_W, so long as no deviation occurs. If period N is reached without a deviation then the players switch to a sequential equilibrium with $a^* = y^*$. If there is a deviation in period $t \leq N - 1$ then *immediately* after the deviation (i.e. before a response if the deviation is in the offer made) the players switch to a sequential equilibrium as described in case 2 of the previous subsection as follows: $a^* = x^*$ if player 1 deviates and $a^* = z^*$ if player 2 deviates.

The bound on N ensures that 1 does not deviate at time 0. The prescribed play yields him a payoff of $y^* - Nc$ as opposed to his best alternative, which is to demand x^*. The bound also ensures that neither type of player 2 deviates in the second period: the prescribed play yields type I a payoff of $1 - y^* - Nc_I$ as opposed to his best alternative, which is to offer $z^* - c$, whose acceptance yields a payoff of $1 - z^* + c - c_I$, $I = W, S$.

When the length of a period is τ, the parameters c, c_S, and c_W in the above must be multiplied by τ and the delay time to agreement becomes $\tau N(\tau)$. The limit of the latter as $\tau \to 0+$ is positive. Thus, there may be significant delay in reaching agreement, even when τ is small, although no information is revealed along the equilibrium path after a deviation occurs. Any deviation is interpreted as signaling weakness and leads to an equilibrium that favors the nondeviant.

Gul and Sonnenschein (1988) do not accept that such nonstationary equilibria are reasonable. In this context, stationarity refers to the assumption that players do not change their behavior so long as 1 does not change his belief about 2's type. A version of their result for the fixed costs model that we have been using as an example is that any sequential equilibrium in which 2's strategies are stationary must lead to an agreement no later than the second period.

In their paper, Gul and Sonnenschein analyze a more complex bargaining model between a seller and a buyer in which the seller's reservation value is 0 and the buyer's reservation value has a continuous distribution F with support $[l, h]$. They impose two properties in addition to stationarity on sequential equilibrium. The *monotonicity* property requires that, for histories after which the seller's posterior distribution for the buyer's reservation value is the conditional distribution of F given $[l, x]$, the seller's offer must be increasing in x. The *no free screening* property requires that the buyer's offer can influence the seller's beliefs only after histories in which at least one of the buyer's equilibrium offers is supposed to be accepted by the seller.

Result 10 [Gul and Sonnenschein (1988)]. For all $\epsilon > 0$ there is $\tau^* > 0$ such that for all positive $\tau < \tau^*$, in every sequential equilibrium that satisfies stationarity, monotonicity and no free screening, the probability that bargaining continues after time ϵ is at most ϵ.

Gul and Sonnenschein conclude from Result 10 that bargaining with one-sided uncertainty leads to vanishingly small delays when the interval between successive proposals becomes sufficiently small. We are not convinced that such a sweeping conclusion is legitimate, although we do not deny that actual delays in real-life bargaining must often be caused by factors that are more complex than the uncertainties about the tastes or beliefs of a player as we have modeled them. Uncertainties about how rational or irrational an opponent is are probably at least as important. The reason for our skepticism lies in the fact that, as is shown by Ausubel and Deneckere (1989) and others, the result relies heavily on the stationarity assumption. As explained in Section 2, stationarity assumptions do more than attribute simplicity of behavior to the players: they also make players' beliefs insensitive to past events.

Note that Result 10 and that of Gul, Sonnenschein and Wilson (1986) have an importance beyond bargaining theory because of their significance for the "Coase conjecture". Note also that Vincent (1989) demonstrates that, if the seller and the buyer have *correlated* valuations for the traded item, then delay is possible when the time between offers goes to zero even under stationarity assumptions.

8.3. Refinements of sequential equilibrium in bargaining models [Rubinstein (1985a, 1985b)]

Our study of the fixed costs model shows that the concept of sequential equilibrium needs to be refined if unique equilibrium outcomes are to be obtained. To motivate the refinement that we propose, consider the following situation. Player 1 makes a demand of a, which is rejected by 2 who makes a counteroffer of b, where $a - c_W < b < a - c_S$. If the rejection and the counteroffer are out of equilibrium, then the sequential equilibrium concept does not preclude 1 from assigning probability one to the event that 2 is weak. Is this reasonable? Observe that 2 rejects the demand a in favor of an offer of b which, if accepted, leads to a payoff of $1 - b - c_S > 1 - a$ for the strong 2, but only $1 - b - c_W < 1 - a$ for the weak 2. One can therefore "rationalize" the offer of b on the part of the strong player but not on the part of the weak player. Should not this offer therefore convince 1 that his opponent is strong?

The next result, which is a version of that of Rubinstein (1985a, 1985b), explores the hypothesis that players' beliefs incorporate such "rationalizations" about their opponents. The precise requirements for *rationalizing conjectures* are that, in any history after which player 1 is not certain that he faces the weak type of player 2:

(1) If 2 rejects the offer a and makes a counteroffer b satisfying $a - c_W < b < a - c_S$, then 1 assigns probability one to the event that his opponent is strong.

(2) If 2 rejects the offer a and makes a counteroffer b satisfying $a - c_W > b$, then 1 does not increase the probability he attaches to 2's being strong.

Result 11 [Rubinstein (1985a, 1985b)].   For any sequential equilibrium with rationalizing conjectures:

(1) If $2c/(c + c_W) < \pi_W < 1$, then if 2 is weak there is an immediate agreement in which 1 gets the entire surplus, while if 2 is strong the agreement is delayed by one period, at which time 1 gets $1 - c_W$.

(2) If $(c + c_S)/(c + c_W) < \pi_W < 2c/(c + c_W)$, then if 2 is weak there is an immediate agreement in which 1 gets c_W, while if 2 is strong the agreement is delayed by one period, at which time 1 gets nothing at all.

(3) If $0 < \pi_W < (c + c_S)/(c + c_W)$, then there is an immediate agreement in which 1 gets c_S whatever 2's type.

Rubinstein (1985a) provides a more general result applied to the family of time preferences explored in Section 2. Various refinements of a similar nature have been proposed by numerous authors. In particular, Grossman and Perry (1986) propose a refinement they call "perfect sequential equilibrium", which seems to lead to plausible outcomes in bargaining models for which it exists.

8.4. Strategic delay [Admati and Perry (1987)]

One may modify the previous model by allowing a responding player to *choose* how much time may pass before he makes his counteroffer. He may either immediately accept the proposal with which he is currently faced or he may reject the demand and choose a pair (a, Δ), where $a \in A$ is his counterproposal and $\Delta \geqslant \tau$ is the length of the delay during which no player may make a new offer. (Without incomplete information, this modification has no bite. In equilibrium, each player minimizes the delay and chooses $\Delta = \tau$.)

The refinement of sequential equilibrium described here [which is somewhat stronger than that offered by Admati and Perry (1987)] is similar to that of the preceding section:

(1) After any history that does not convince 1 that 2 is weak for sure, suppose that 1 demands a and that this demand is rejected by 2 who then counters with an offer of b after a delay of $\Delta \geqslant \tau$. If $1 - b - c_w\Delta < 1 - a \leqslant 1 - b - c_s\Delta$, then 1 concludes that 2 is strong for sure.

(2) Suppose that 1 is planning to accept an offer a if this is delayed by Δ but that 2 delays a further $d > 0$ before making an offer b satisfying $1 - b - c_w d < 1 - a \leqslant 1 - b - c_s d$. Then, whatever the previous history, 1 concludes that 2 is strong for sure.

For $2c/(c + c_w) < \pi_w$ Admati and Perry (1987) show that any sequential equilibrium satisfying these additional assumptions has player 1 demanding the entire surplus at time 0. A weak player 2 accepts, but a strong player 2 rejects and makes a counteroffer of 0 after a delay of $1/c_w$, which player 1 accepts.

The result is to be compared with case (1) of Result 11 in which agreement is delayed by a vanishingly small amount when $\tau \rightarrow 0+$. Here, the delay in reaching agreement when 2 is strong does not depend on τ, and hence the delay persists in the limiting case as $\tau \rightarrow 0+$. [The constraint on π_w is necessary. See Admati and Perry (1987) for details.]

8.5. Related work

Strategic sequential bargaining models with incomplete information are surveyed by Wilson (1987b) and by several contributors to Roth (1985). We have dealt only with one-sided uncertainty. Cramton (1992) constructs a sequential equilibrium for the alternating offers model with two-sided uncertainty; see also Ausubel and Deneckere (1992), Chatterjee and Samuelson (1988), and Cho (1989).

Bikhchandani (1992) points out that the sensitivity of the results on prolonged disagreement to certain changes in the bargaining procedure and in the solution concept employed. Grossman and Perry (1986) propose a refinement of sequential equilibrium in the case that there are many types (not just two) of player 2. Perry (1986) seeks to endogenize the choice of the initial proposer.

The complexity of the analysis is reduced substantially if only two possible agreements are available; sharp results can then be obtained. See Chatterjee and Samuelson (1987). Notice that this case is strongly related to games of attrition as studied in other game-theoretic contexts.

Many of the issues in bargaining with incomplete information that we have studied arise also in models in which only the uninformed party is allowed to make offers. Fudenberg and Tirole (1983) and Sobel and Takahashi (1983) study such models; see also, for example, Fudenberg, Levine and Tirole (1985).

9. Bargaining and mechanism design

The mechanism design literature regards a theory of bargaining as providing a mapping from the space of problem parameters to a solution to the bargaining problem. Attention is focused on the mappings or *mechanisms* that satisfy certain interesting properties, the aim being to study simultaneously the Nash equilibria for a large class of bargaining games of incomplete information without the need to specify each of the bargaining games in detail.

The rest of the section follows ideas appearing in the path-breaking paper by Myerson and Satterthwaite (1983). The idea is explained in the context of a particularly simple case analyzed by Matsuo (1989).

A seller and a buyer of a single indivisible good have to negotiate a price. Both buyer and seller may be strong or weak, it being common knowledge that the prior probability of each possible pairing of types is the same. A player's strength or weakness depends on his reservation value, which may be $s_1, s_2, b_1,$ or b_2, where $0 = s_1 < b_1 < s_2 < b_2$. We let $s_2 = b_1 + \eta$ and assume that $b_1 = s_1 + \alpha$, and $b_2 = s_2 + \alpha$.

A *mechanism* M in this context is a mapping that assigns an outcome to each realization of (s, b). An outcome is a pair consisting of a price and a probability. Thus a mechanism is a pair of functions (p, π). The interpretation is that when the realization is (s, b), then with probability $\pi(s, b)$ agreement is reached on the price $p(s, b)$, and with probability $1 - \pi(s, b)$ there is disagreement. The expected utility gain to a seller with reservation value s from the use of the mechanism M is $U(s) = E_b \pi(s, b)(p(s, b) - s)$. The expected utility gain to a buyer with reservation utility b is $V(b) = E_s \pi(s, b)(b - p(s, b))$.

Suppose that the buyer and the seller negotiate by choosing strategies in a noncooperative bargaining game. A mechanism M can then be constructed by making a selection from the Nash equilibria of this game. It should be noted that the restriction of the set of outcomes to consist of a price and a probability significantly limits the scope of bargaining games to which the current investigation is applicable.

If the bargaining game has the properties that each player's security level is at least as large as his reservation value and that the action spaces are independent of the type of a buyer or a seller, then the mechanism must satisfy the following constraints:

Individual rationality.   For all s and b we have $U(s) \geq 0$ and $V(b) \geq 0$.

Incentive compatibility.   For all s, s', b, and b' we have $U(s) \geq E_b \pi(s', b)(p(s', b) - s)$ and $V(b) \geq E_s \pi(s, b')(b - p(s, b'))$.

If the mechanism represents the outcome of a game, the second condition asserts that no player prefers to use the strategy employed by another player. [Note that we are not necessarily discussing a *direct* mechanism and so the strategies need not consist simply of an announcement of a player's type. however, one could, of course, apply the "revelation principle" (see the chapter on 'correlated and communication equilibria' in a forthcoming volume of this Handbook) and thereby study only direct mechanisms without loss of generality.]

An *efficient mechanism* is a mechanism that induces an agreement whenever a surplus exists (i.e. $b > s$). In our example, a surplus exists except when a low reservation value buyer confronts a high reservation value seller (i.e. $s = s_2$ and $b = b_1$).

We now explain why an efficient mechanism satisfying individual rationality and incentive compatibility exists if and only if $2\alpha \geq \eta$.

Assume first that $2\alpha < \eta$. Let $\sigma(s)$ denote the probability with which a seller with reservation value s reaches agreement. Let $\beta(b)$ be similarly defined for buyers. The incentive compatibility constraints can then be rewritten as $(s_2 - s_1)\sigma(s_2) \leq U(s_1) - U(s_2) \leq (s_2 - s_1)\sigma(s_1)$ and $(b_2 - b_1)\beta(b_1) \leq V(b_2) - V(b_1) \leq (b_2 - b_1)\beta(b_2)$.

If an efficient mechanism exists, then $\sigma(s_2) = \beta(b_1) = 1/2$. It follows that $U(s_1) \geq U(s_1) - U(s_2) \geq (s_2 - s_1)/2$ and $V(b_2) \geq V(b_2) - V(b_1) \geq (b_2 - b_1)/2$.

The sum of the expected gains to a strong buyer and a strong seller is then $U(s_1)/2 + V(b_2)/2 \geq (s_2 - s_1 + b_2 - b_1)/4 = (\alpha + \eta)/2$, but the total expected surplus is only $[(b_2 - s_1) + (b_2 - s_2) + (b_1 - s_1) + 0]/4 = \alpha + \eta/4 < (\alpha + \eta)/2$. No efficient mechanism can therefore exist.

Next, assume that $\eta \leq 2\alpha$. We now construct a game in which there is a Nash equilibrium that induces an efficient mechanism. In the game, the seller announces either s_1 or s_2 and the buyer announces either b_1 or b_2. Table 1 indicates the prices (*not* payoffs) that are then enforced (D means disagreement).

This game has a Nash equilibrium in which all types tell the truth and in which an efficient outcome is achieved. Notice in particular that if a weak seller

Table 1

	b_1	b_2
s_1	b_1	$(s_1 + b_2)/2$
s_2	D	s_2

is honest and reports s_1, he obtains a price of $(s_1 + b_2 + 2b_1)/4$, while if he is dishonest and reports s_2, he gets $(s_1 + s_2)/2$. But $(s_1 + b_2 + 2b_1)/4 - (s_1 + s_2)/2 = (2\alpha - \eta)/4 \geqslant 0$.

The above example illustrates some of the ideas of Myerson and Satterthwaite (1983). They offer some elegant characterization results for incentive-compatible mechanisms from which they are able to deduce a number of interesting conclusions. In particular:

Result 12 [Myerson and Satterthwaite (1983)]. Let $\bar{s} \leqslant \underline{b} < \bar{s} \leqslant \bar{b}$. If s is distributed with positive density over the interval $[\underline{s}, \bar{s}]$ and b is independently distributed with positive density over the interval $[\underline{b}, \bar{b}]$, then *no* incentive-compatible, individually rational mechanism is efficient.

Given this result, it is natural to ask what can be said about the mechanisms that maximize expected total gains from trade. The conclusion of Myerson and Satterthwaite in the case when both s and b are uniformly distributed on $[0, 1]$ is a neat one: the expected gains from trade are maximized by a mechanism that transfers the object if and only if $b \geqslant s + 1/4$. Chatterjee and Samuelson (1983) had previously shown that the sealed-bid double auction, in which the object is sold to the buyer at the average of the two bid prices whenever the buyer's bid exceeds the seller's, admits an equilibrium in which this maximal gain from trade is achieved. (The seller proposes the price $2s/3 + 1/4$ and the buyer proposes $2b/3 + 1/12$.)

The mechanism design approach is more general than that of noncooperative bargaining theory with which this chapter has been mostly concerned. However, the above mechanism design results, although wide in the scope of the situations to which they apply, do no more than to classify scenarios in which efficient outcomes are or are not achievable in equilibrium. Even when an efficient outcome is achievable, it need not be the realized outcome in the class of noncooperative games that is actually relevant in a particular applied context. This trade-off between generality and immediate applicability is one that we noted before in comparing cooperative and noncooperative game theory. As in that case, the two approaches should be seen as complementary, each providing insights where the other is silent.

10. Final comments

In the past decade Nash's (1950, 1953) pioneering work on noncooperative bargaining theory has been taken up again and developed by numerous authors. We see three directions in which progress has been particularly fruitful:

(1) sequential models have been introduced in studying specific bargaining procedures;

(2) refinements of Nash equilibrium have been applied; and

(3) bargaining models have been embedded in market situations to provide insights into markets with decentralized trading.

In spite of this progress, important challenges are still ahead. The most pressing is that of establishing a properly founded theory of bargaining under incomplete information. A resolution of this difficulty must presumably await a major breakthrough in the general theory of games of incomplete information. From the perspective of economic theory in general, the main challenge remains the modeling of trading institutions (with the nature of "money" the most obvious target).

Because many of the results of noncooperative bargaining theory are relatively recent, there are few sources of a general nature that can be recommended for further reading. Harsanyi (1977) provides an interesting early analysis of some of the topics covered in the chapter. Roth (1985) and Binmore and Dasgupta (1987) are collections of papers the scope of which coincides with that of this chapter. Sutton (1986), Rubinstein (1987), and Bester (1989b) are survey papers. Osborne and Rubinstein (1990) contains a more detailed presentation of much of the material in this chapter.

References

Admati, A.R. and M. Perry (1987) 'Strategic delay in bargaining', *Review of Economic Studies*, 54: 345–364.

Admati, A.R. and M. Perry (1991) 'Joint projects without commitment', *Review of Economic Studies*, 58: 259–276.

Ausubel, L.M. and R.J. Deneckere (1989) 'Reputation in bargaining and durable goods monopoly', *Econometrica*, 57: 511–531.

Ausubel, L.M. and R.J. Deneckere (1992) 'Durable goods monopoly with incomplete information', unpublished paper, Center for Mathematical Studies in Economics and Management Science, Northwestern University.

Bernheim, B.D. (1984) 'Rationalizable stategic behavior', *Econometrica*, 52: 1007–1028.

Bester, H. (1988) 'Bargaining, search costs and equilibrium price distributions', *Review of Economic Studies*, 55: 201–214.

Bester, H. (1989a) 'Noncooperative bargaining and spatial competition', *Econometrica*, 57: 97–113.

Bester, H. (1989b) 'Non-cooperative bargaining and imperfect competition: A survey', *Zeitschrift für Wirtschafts- und Sozialwissenschaften*, 109: 265–286.

Bikhchandani, S. (1992) 'A bargaining model with incomplete information', *Review of Economic Studies*, 59: 187–203.

Binmore, K.G. (1985) 'Bargaining and coalitions', in: A.E. Roth, ed., *Game-theoretic models of bargaining*. Cambridge: Cambridge University Press, pp. 269–304.

Binmore, K.G. (1987a) 'Nash bargaining theory I, II, III', in: K.G. Binmore and P. Dasgupta, eds., *The economics of bargaining*. Oxford: Blackwell, pp. 27–46, 61–76, 239–256.

Binmore, K.G. (1987b) 'Perfect equilibria in bargaining models', in: K.G. Binmore and P. Dasgupta, eds., *The economics of bargaining*. Oxford: Blackwell, pp. 77–105.

Binmore, K.G. (1987c) 'Nash bargaining and incomplete information', in: K.G. Binmore and P. Dasgupta, (1987). eds, *The economics of bargaining*. Oxford: Blackwell, pp. 155–192.

Binmore, K.G. (1987d) 'Modeling rational players', Parts I and II, *Economics and Philosopy*, 3: 179–214 and 4: 9–55.

Binmore, K.G. and P. Dasgupta (1987) *The economics of bargaining*. Oxford: Blackwell.

Binmore, K.G. and M.J. Herrero (1988a) 'Matching and bargaining in dynamic markets', *Review of Economic Studies*, 55: 17–31.

Binmore, K.G. and M.J. Herrero (1988b) 'Security equilibrium', *Review of Economic Studies*, 55: 33–48.

Binmore, K.G., A. Rubinstein and A. Wolinsky (1986) 'The Nash bargaining solution in economic modelling', *Rand Journal of Economics*, 17: 176–188.

Binmore, K.G., A. Shaked and J. Sutton (1988) 'An outside option experiment', *Quarterly Journal of Economics*, 104: 753–770.

Butters, G.R. (1977) 'Equilibrium price distributions in a random meetings market', unpublished paper, Princeton University.

Chae, S. and J.-A. Yang (1988) 'The unique perfect equilibrium of an N-person bargaining game', *Economics Letters*, 28: 221–223.

Chatterjee, K. and L. Samuelson (1987) 'Bargaining with two-sided incomplete information: an infinite horizon model with alternating offers', *Review of Economic Studies*, 54: 175–192.

Chatterjee, K. and L. Samuelson (1988) 'Bargaining under two-sided incomplete information: The unrestricted offers case', *Operations Research*, 36: 605–618.

Chatterjee, K. and L. Samuelson (1990) 'Perfect equilibria in simultaneous-offers bargaining', *International Journal of Game Theory*, 19: 237–267.

Chatterjee, K. and W. Samuelson (1983) 'Bargaining under incomplete information', *Operations Research*, 31: 835–851.

Chatterjee, K., B. Dutta, D. Ray and K. Sengupta (1992) 'A non-cooperative theory of coalitional bargaining', *Review of Economic Studies*, to appear.

Chikte, S.D. and S.D. Deshmukh (1987) 'The role of external search in bilateral bargaining', *Operations Research*, 35: 198–205.

Cho, I.-K. (1989) 'Characterization of stationary equilibria in bargaining models with incomplete information', unpublished paper, Department of Economics, University of Chicago.

Cramton, P.C. (1992) 'Strategic delay in bargaining with two-sided uncertainty', *Review of Economic Studies*, 59: 205–225.

Crawford, V.P. (1982) 'A theory of disagreement in bargaining', *Econometrica*, 50: 607–637.

Cross, J.G. (1965) 'A theory of the bargaining process', *American Economic Review*, 55: 67–94.

Davidson, C. (1988) 'Multiunit bargaining in oligopolistic industries', *Journal of Labor Economics*, 6: 397–422.

Diamond, P.A. (1981) 'Mobility costs, frictional unemployment and efficiency', *Journal of Political Economy*, 89: 798–812.

Diamond, P.A. and E. Maskin (1979) 'An equilibrium analysis of search and breach of contract, I: steady states', *Bell Journal of Economics*, 10: 282–316.

Dow, G.K. (1989) 'Knowledge is power: informational precommitment in the capitalist firm', *European Journal of Political Economy*, 5: 161–176.

Dutta, B. and D. Ray (1989) 'A concept of egalitarianism under participation constraints', *Econometrica*, 57: 615–635.

Fernandez, R. and J. Glazer (1990) 'The scope for collusive behavior among debtor countries', *Journal of Development Economics*, 32: 297–313.

Fernandez, R. and J. Glazer (1991) 'Striking for a bargain between two completely informed agents', *American Economic Review*, 81: 240–252.

Fishburn, P.C. and A. Rubinstein (1982) 'Time preference', *International Economic Review*, 23: 677–694.

Fudenberg, D. and J. Tirole (1983) 'Sequential bargaining with incomplete information', *Review of Economic Studies*, 50: 221–247.

Fudenberg, D., D. Levine and J. Tirole (1985) 'Infinite-horizon models of bargaining with one-sided incomplete information', in: A.E. Roth, ed., *Game-theoretic models of bargaining.* Cambridge: Cambridge University Press, pp. 73–98.

Gale, D. (1986a) 'Bargaining and competition. Part I: Characterization', *Econometrica*, 54: 785–806.

Gale, D. (1986b) 'Bargaining and competition. Part 2: Existence', *Econometrica*, 54: 807–818.

Gale, D. (1986c) 'A simple characterization of bargaining equilibrium in a large market without the assumption of dispersed characteristics', Working Paper 86-05, Center for Analytical Research in Economics and the Social Sciences, University of Pennsylvania.

Gale, D. (1986d) 'A strategic model of trade with money as a medium of exchange', Working Paper 86-04, Center for Analytic Research in Economics and the Social Sciences, University of Pennsylvania.

Gale, D. (1987) 'Limit theorems for markets with sequential bargaining', *Journal of Economic Theory*, 43: 20–54.

Gale, D. (1988) 'Price setting and competition in a simple duopoly model', *Quarterly Journal of Economics*, 103: 729–739.

Green, E.J. (1991) 'Eliciting traders' knowledge in 'frictionless' asset market', Staff Report 144, Federal Reserve Bank of Minneapolis.

Grossman, S.J. and M. Perry (1986) 'Sequential bargaining under asymmetric information', *Journal of Economic Theory*, 39: 120–154.

Gul, F. (1989) 'Bargaining foundations of Shapley value', *Econometrica*, 57: 81–95.

Gul, F. and H. Sonnenschein (1988) 'On delay in bargaining with one-sided uncertainty', *Econometrica*, 56: 601–611.

Gul, F., H. Sonnenschein and R. Wilson (1986) 'Foundations of dynamic monopoly and the Coase conjecture', *Journal of Economic Theory*, 39: 155–190.

Haller, H. (1991) 'Wage bargaining as a strategic game', in: R. Selten, ed., *Game equilibrium models III: Strategic bargaining.* Berlin: Springer-Verlag, pp. 230–241.

Haller, H. and S. Holden (1990), 'A letter to the editor on wage bargaining', *Journal of Economic Theory*, 52: 232–236.

Harsanyi, J.C. (1967/68) 'Games with incomplete information played by 'Bayesian' players', Parts I, II, III, *Management Science*, 14: 159–182, 320–334, 486–502.

Harsanyi, J.C. (1974) 'An equilibrium-point interpretation of stable sets and a proposed alternative definition', *Management Science* (Theory Series), 20: 1472–1495.

Harsanyi, J.C. (1977) *Rational behavior and bargaining equilibrium in games and social situations.* Cambridge: Cambridge University Press.

Harsanyi, J.C. (1981) 'The Shapley value and the risk-dominance solutions of two bargaining models for characteristic-function games', in: R.J. Aumann, J.C. Harsanyi, W. Hildenbrand, M. Maschler, M.A. Perles, J. Rosenmüller, R. Selten, M. Shubik and G.L. Thompson, *Essays in game theory and mathematical economics.* Mannheim: Bibliographisches Institut, pp. 43–68.

Harsanyi, J.C. and R. Selten (1972) 'A generalized Nash solution for two-person bargaining games with incomplete information', *Management Science*, 18: P-80–P-106.

Harsanyi, J.C. and R. Selten (1988) *A general theory of equilibrium selection in games.* Cambridge, Mass.: MIT Press.

Hendon, E. and T. Tranæs (1991) 'Sequential bargaining in a market with one seller and two different buyers', *Games and Economic Behavior*, 3: 453–466.

Horn, H. and A. Wolinsky (1988a) 'Bilateral monopolies and incentives for merger', *Rand Journal of Economics*, 19: 408–419.

Horn, H. and A. Wolinsky (1988b) 'Worker substitutability and patterns of unionisation', *Economic Journal*, 98: 484–497.

Jones, S.R.G. and C.J. McKenna (1990), 'Bargaining with fallback reserves', unpublished paper, McMaster University.

Jun, B.H. (1987) 'A strategic model of 3-person bargaining', unpublished paper, State University of New York at Stony Brook.

Jun, B.H. (1989) 'Noncooperative bargaining and union formation', *Review of Economic Studies*, 56: 59–76.

Kalai, E. and M. Smorodinsky (1975) 'Other solutions to Nash's bargaining problem', *Econometrica*, 43: 513–518.

Kiyotaki, N. and R. Wright (1989) 'On money as a medium of exchange', *Journal of Political Economy*, 97: 927–954.

Kreps, D.M. and R. Wilson (1982) 'Sequential equilibria', *Econometrica*, 50: 863–894.

Leininger, W., P.B. Linhart and R. Radner (1989) 'Equilibria of the sealed-bid mechanism for bargaining with incomplete information', *Journal of Economic Theory*, 48: 63–106.

Matsuo, T. (1989) 'On incentive compatible, individually rational, and ex post efficient mechanisms for bilateral trading', *Journal of Economic Theory*, 49: 189–194.

Matthews, S.A. and A. Postlewaite (1989) 'Pre-play communication in two-person sealed-bid double auctions', *Journal of Economic Theory*, 48: 238–263.

McLennan, A. (1988) 'Bargaining between two symmetically informed agents', unpublished paper, University of Minnesota.

McLennan, A. and H. Sonnenschein (1991) 'Sequential bargaining as a noncooperative foundation for Walrasian equilibrium', *Econometrica*, 59: 1395–1424.

Mortensen, D.T. (1982a) 'Property rights and efficiency in mating, racing, and related games', *American Economic Review*, 72: 968–979.

Mortensen, D.T. (1982b) 'The matching process as a noncooperative bargaining game', in: J.J. McCall, ed., *The economics of information and uncertainty*. Chicago: University of Chicago Press, pp. 233–254.

Moulin, H. (1982) 'Bargaining and noncooperative implementation', Discussion Paper A239 0282, Laboratorie d'Econométrie, Paris.

Moulin, H. (1984) 'Implementing the Kalai–Smorodinsky bargaining solution', *Journal of Economic Theory*, 33: 32–45.

Muthoo, A. (1989a) 'Sequential bargaining and competition', unpublished paper, Department of Economics, London School of Economics.

Muthoo, A. (1989b) 'A note on the strategic role of outside options in bilateral bargaining', unpublished paper, Department of Economics, London School of Economics.

Muthoo, A. (1990) 'Bargaining without commitment', *Games and Economic Behavior*, 2: 291–297.

Muthoo, A. (1991) 'A note on bargaining over a finite number of feasible agreements', *Economic Theory*, 1: 290–292.

Myerson, R.B. and M.A. Satterthwaite (1983) 'Efficient mechanisms for bilateral trading', *Journal of Economic Theory*, 29: 265–281.

Nash, J.F. (1950) 'The bargaining problem', *Econometrica*, 18: 155–162.

Nash, J.F. (1953) 'Two-person cooperative games', *Econometrica*, 21: 128–140.

Okada, A. (1991a) 'A two-person repeated bargaining game with long-term contracts', in: R. Selten, ed., *Game equilibrium models III: Strategic bargaining*. Berlin: Springer-Verlag, pp. 34–47.

Okada, A. (1991b) 'A noncooperative approach to the Nash bargaining problem', in: R. Selten, ed., *Game equilibrium models III: Strategic bargaining*. Berlin: Springer-Verlag, pp. 7–33.

Osborne, M.J. and A. Rubinstein (1990) *Bargaining and markets*. San Diego: Academic Press.

Owen, G. (1982) *Game theory*, 2nd edn. New York: Academic Press.

Pearce, D.G. (1984) 'Rationalizable strategic behavior and the problem of perfection', *Econometrica*, 52: 1029–1050.

Perry, M. (1986) 'An example of price formation in bilateral situations: a bargaining model with incomplete information', *Econometrica*, 54: 313–321.

Perry, M. and P.J. Reny (1992) 'A non-cooperative bargaining model with strategically timed offers', *Journal of Economic Theory*, to appear.

Peters, M. (1988) 'Ex ante pricing and bargaining', unpublished paper, University of Toronto.

Peters, M. (1989) 'Stable pricing institutions are Walrasian', unpublished paper, University of Toronto.

Peters, M. (1991) 'Ex ante price offers in matching games: Non-steady states', *Econometrica*, 59: 1425–1454.

Peters, M. (1992) 'On the efficiency of ex ante and ex post pricing institutions', *Economic Theory*, 2: 85–101.

Rosenthal, R.W. and H.J. Landau (1981) 'Repeated bargaining with opportunities for learning', *Journal of Mathematical Sociology*, 8: 61–74.

Roth, A.E. (1977) 'The Nash solution and the utility of bargaining', *Econometrica*, 45: 657–664.

Roth, A.E. (ed.) (1985) *Game-theoretic models of bargaining*. Cambridge: Cambridge University Press.

Rubinstein, A. (1982) 'Perfect equilibrium in a bargaining model,' *Econometrica*, 50: 97–109.

Rubinstein, A. (1985a) 'A bargaining model with incomplete information about time preferences', *Econometrica*, 53: 1151–1172.

Rubinstein, A. (1985b) 'Choice of conjectures in a bargaining game with incomplete information', in: A.E. Roth ed., *Game-theoretic models of bargaining*. Cambridge: Cambridge University Press, pp. 99–114.

Rubinstein, A. (1987) 'A sequential strategic theory of bargaining', in: T.F. Bewley, ed., *Advances in economic theory*. Cambridge: Cambridge University Press, pp. 197–224.

Rubinstein, A. (1989) 'Competitive equilibrium in a market with decentralized trade and strategic behavior: An introduction', in: G.R. Feiwel, ed., *The economics of imperfect competition and employment*. Basingstoke: Macmillan, pp. 243–259.

Rubinstein, A. (1991) 'Comments on the interpretation of game theory', *Econometrica*, 59: 909–924.

Rubinstein, A. and A. Wolinsky (1985) 'Equilibrium in a market with sequential bargaining', *Econometrica*, 53: 1133–1150.

Rubinstein, A. and A. Wolinsky (1987) 'Middlemen', *Quarterly Journal of Economics*, 102: 581–593.

Rubinstein, A. and A. Wolinsky (1990) 'Decentralized trading, stategic behaviour and the Walrasian outcome', *Review of Economic Studies*, 57: 63–78.

Samuelson, L. (1992) 'Disagreement in markets with matching and bargaining', *Review of Economic Studies*, 59: 177–185.

Schelling, T.C. (1960) *The strategy of conflict*. Cambridge, Mass.: Harvard University Press.

Selten, R. (1981) 'A noncooperative model of characteristic-function bargaining', in: R.J. Aumann, J.C. Harsanyi, W. Hildenbrand, M. Maschler, M.A. Perles, J. Rosenmüller, R. Selten, M. Shubik and G.L. Thompson, *Essays in game theory and mathematical economics*. Mannheim: Bibliographisches Institut, pp. 131–151.

Shaked, A. (1987) 'Opting out: Bazaars versus "hi tech" markets', Discussion Paper 87/159, Suntory Toyota International Centre for Economics and Related Disciplines, London School of Economics.

Shaked, A. and J. Sutton (1984) 'Involuntary unemployment as a perfect equilibrium in a bargaining model', *Econometrica*, 52: 1351–1364.

Sobel, J. and I. Takahashi (1983) 'A multistage model of bargaining', *Review of Economic Studies*, 50: 411–426.

Stahl, D.O., II (1990) 'Bargaining with durable offers and endogenous timing', *Games and Economic Behavior*, 2: 173–187.

Ståhl, I. (1967) 'Studier i bilaterala monopolets teori', Licentiat thesis, HHS, Stockholm.

Ståhl, I. (1972) *Bargaining theory*. Stockholm: Economics Research Institute, Stockholm School of Economics.

Ståhl, I. (1988) 'A comparison between the Rubinstein and Ståhl bargaining models', Research Report 6347, Stockholm School of Economics.

Sutton, J. (1986) 'Non-cooperative bargaining theory: An introduction', *Review of Economic Studies*, 53: 709–724.

van Damme, E., R. Selten and E. Winter (1990) 'Alternating bid bargaining with a smallest money unit', *Games and Economic Behavior*, 2: 188–201.

Vincent, D.R. (1989) 'Bargaining with common values', *Journal of Economic Theory*, 48: 47–62.

von Neumann, J. and O. Morgenstern (1944) *Theory of games and economic behavior*. Princeton: Princeton University Press.

Williams, S.R. (1987) 'Efficient performance in two agent bargaining', *Journal of Economic Theory*, 41: 154–172.

Wilson, R. (1984) 'Notes on market games with complete information', unpublished paper, Graduate School of Business, Stanford University.

Wilson, R. (1987a) 'Game-theoretic analysis of trading processes', in: T.F. Bewley, ed., *Advances in economic theory*. Cambridge: Cambridge University Press, pp. 33–70.

Wilson, R. (1987b) 'Bilateral bargaining', unpublished paper, Graduate School of Business, Stanford University.

Wolinsky, A. (1987) 'Matching, search, and bargaining', *Journal of Economic Theory*, 42: 311–333.

Wolinsky, A. (1988) 'Dynamic markets with competitive bidding', *Review of Economic Studies*, 55: 71–84.

Wolinsky, A. (1990) 'Information revelation in a market with pairwise meetings', *Econometrica*, 58: 1–23.

Zeuthen, F. (1930) *Problems of monopoly and economic warfare*. London: George Routledge and Sons.

Chapter 8

STRATEGIC ANALYSIS OF AUCTIONS

ROBERT WILSON*

Stanford Business School

Contents

*Assistance received from NSF grant SES8908269.

Handbook of Game Theory, Volume 1, Edited by R.J. Aumann and S. Hart

1. Introduction

In many markets, transaction prices are determined in auctions. In the most common form, prospective buyers compete by submitting bids to a seller. Each bid is an offer to buy that states a quantity and a maximum price. The seller then allocates the available supply among those offering the highest prices exceeding the seller's asking price. The actual price paid by a successful bidder depends on a pricing rule, usually selected by the seller: two common pricing rules are that each successful bidder pays the price bid; or they all pay the same price, usually the highest rejected bid or the lowest accepted bid.

Auctions have been used for millennia, and remain the simplest and most familiar means of price determination for multilateral trading without intermediary "market makers" such as brokers and specialists. Their trading procedures, which simply process bids and offers, are direct extensions of the usual forms of bilateral bargaining. Auctions also implement directly the demand submission procedures used in Walrasian models of markets. They therefore have prominent roles in the theory of exchange and in studies of the effects of economic institutions on the volume and terms of trade. Their allocative efficiency in many contexts ensures their continued prominence in economic theory. They are also favored in experimental designs investigating the predictive power of economic theories.

Auctions are apt subjects for applications of game theory because they present explicit trading rules that largely fix the "rules of the game". Moreover, they present substantive problems of strategic behavior of practical importance. They are particularly valuable as illustrations of games of incomplete information because bidders' private information is the main factor affecting strategic behavior. The simpler forms of auctions induce normal-form games that are essentially "solved" by applying directly the basic equilibrium concepts of noncooperative game theory, such as the Nash equilibrium, without recourse to criteria for selecting among multiple equilibria. The common-knowledge assumption on which game theory relies is often tenable or innocuous applied to an auction.

In this chapter we describe several forms of auctions, present the formulations used in the main models, review some of the general results and empirical findings, and indicate a few applications. The aim is to acquaint readers with the contributions to a subject in which game theory has had notable success in addressing significant practical problems – and many challenging problems remain. Several other surveys of auction theory are also available, including Engelbrecht-Wiggans (1980), Milgrom (1985, 1987, 1989), Wilson (1985b, 1987a, 1987b), McAfee and McMillan (1987a), Smith (1987), Rothkopf (1990) and Kagel (1991), as well as the collection of articles in Engelbrecht-Wiggans,

Shubik and Stark (1983), and the bibliography by Stark and Rothkopf (1979); in addition, Cassady (1967) surveys the history and practice of auctions. The origin of the subject is the seminal work by Vickrey (1961, 1962), and later the important contributions by Griesmer, Levitan and Shubik (1967) and Ortega-Reichert (1968), who initiated formulations in terms of games with incomplete information. There is also a literature on games with complete information emphasizing multi-market and general equilibrium formulations that is not reviewed here; cf. Shapley and Shubik (1977), Wilson (1978), Schmeidler (1980), Dubey (1982), and Milgrom (1987).

2. Varieties of auctions

The diverse trading rules used in auctions share a common feature. Each player's feasible actions specify offered net trades. The trades accepted are selected by an explicit procedure, and the transaction prices are calculated from the offered trades by an explicit formula. In effect, each trader reports a demand or supply function and then prices are chosen to clear the market. Two main categories of auctions differ according to whether the process is static or dynamic. In static versions, traders submit sealed bids: each acts in ignorance of others' bids. In dynamic versions, traders observe others' bids and they can revise their bids sequentially. Repeated auctions can be further complicated by linkages such as reputational effects.

Static versions allow a useful distinction between *discriminating* pricing in which trades are accepted at differing prices, usually the prices offered for the trades accepted, and *nondiscriminating* pricing, in which for identical items a single price applies to all transactions, such as the highest rejected bid.

An important example of a static auction proceeds as follows. Each seller announces a supply function; for example, if a single seller offers q identical indivisible items at an ask price a, then the supply function is $s(p) = q\mathbf{1}_{\{p \geq a\}}$. Then each buyer $i = 1, \ldots, n$ submits a demand function $d_i(p)$: this might be piecewise constant, indicating the maximum number of items desired at each price; or in the inverse form $b_i(x)$ it indicates the maximum price offered for the xth item. The pricing rule then selects one price p^0 from among the interval of "clearing prices" that equate aggregate demand and supply; for example, the maximum (the "first price" rule), the minimum (the "second price" or "highest rejected bid" rule), or the midpoint. Nondiscriminating pricing assigns this price to all transactions, namely if $d_i(p^0) = x_i$, then i obtains x_i items at the uniform price p^0 for each item, whereas purely discriminating pricing imposes the price $b_i(x)$ for the xth item, for each $x \leq x_i$. The process is similar if a buyer solicits offers from sellers. If there are multiple sellers and multiple buyers, then nondiscriminating pricing is the usual rule in static auctions, although there are important exceptions. A variant, proposed by Vickrey

(1961), assigns to each bidder a clearing price that would have resulted if he had not participated.

In the simplest case, a single seller offers a single item and each buyer submits a single bid. The item is sold to the bidder submitting the highest bid at either the price he bid (discriminating) or the greater of the ask price and the second-highest bid (nondiscriminating). Many variations occur in practice; cf. Cassady (1967). The seller may impose an entry fee and need not announce the ask price in advance, the number and characteristics of the participating bidders may be uncertain, etc. If nonidentical items are offered, the seller may solicit bids for each item as well as bids for lots. The process might have a trivial dynamic element, as in the case of a "Dutch" auction in which the seller lowers the price until some buyer accepts. This is evidently a version of discriminating pricing if players are not impatient and the seller's minimal ask price is fixed in advance, because the induced game is strategically equivalent to a static discriminating auction. Auctions in which bidders have continual opportunities to raise their bids have significant dynamic elements because the bids signal information. In an "English" auction, a seller offers a single item and she accepts the highest bid offered above her ask price as in a static auction: the dynamic feature is that buyers can repeatedly raise their bids. The Dutch variant has the seller raising her ask price until a single buyer remains willing to buy. Possibly remaining bidders do not observe the prices at which others drop out – and indeed tracking bids can be difficult if anonymity is feasible. If such observations are precluded, then an English auction resembles a sealed-bid second-price auction, and indeed is strategically equivalent if the players are not impatient. An especially important example of a dynamic auction is a bid–ask market in which traders continually make public offers of bids (to buy) and asks (to sell) that can be accepted or withdrawn at any time.

Auctions are used mostly to exchange one or several identical items for money, but in principle auctions could be used to obtain core allocations or Walrasian equilibria of barter economies involving many goods. The familiar auctions employ invariant trading rules that are unrelated to the participants' information and preferences, and even the numbers of buyers and sellers, but we shall see later that the theory of efficient auctions finds it advantageous to adapt the trading rule to the characteristics of the participants. Conversely, implementations of demand-revelation mechanisms designed to achieve efficient allocations often resemble auctions.

3. Auctions as games

To formulate an auction as an extensive game, in principle one first takes the trading rule as specifying an extensive form that applies to the special case that the numbers of sellers and buyers are common knowledge, as well as their characteristics (preferences, endowments, etc.) and any choices by nature. That is, the procedural steps described by the trading rule generate a list of the possible complete histories of the process, and this list matches the possible plays generated by a tree in which the order of players' moves and their feasible actions at each move are specified. The procedure also specifies the information sets of each player, consisting of minimal sets of moves that cannot be distinguished and for which the feasible actions are the same; perfect recall is assumed. In practice, lacunae in the trading rule leave gaps that are filled with specifications chosen to meet the requirements of behavioral accuracy or modeling tractability.

For example, a static auction might be modeled by a tree representing simultaneous moves by all participants: each seller selects a supply function from a feasible set, and each buyer selects a demand function. Or, if sellers first announce their supplies and ask prices before the buyers move, then two stages are required. An English auction, on the other hand, requires specification of the mechanism (such as rotation or random selection) that determines the order in which buyers obtain opportunities to raise the previous high bid, and their opportunities to observe others' bids. For tractability a reduced form of the tree may be used, as in the approximation that has the seller raising her asking price.

If all information is common knowledge, this extensive form becomes an extensive game by adding specifications of the probabilities of nature's moves and the players' payoffs for each play. The players' payoffs are determined by applying their preferences to the allocation determined by the trading rule; that is, the trading rule, including its pricing rule, assigns to each play an allocation that indicates for each player the (possibly random) transfers of goods and money obtained. For example, in a single-item discriminating auction, a buyer who assigns a value v to the item may obtain the net payoff $v - p$ if he receives the item after bidding p, and zero otherwise.

In practice, however, participants' preferences are rarely common knowledge, and indeed a major motivation for using auctions is to elicit revelation of preferences so that maximal gains from trade can be realized. Thus, the actual extensive game to be studied derives from a larger extensive form in which initially nature chooses an assignment of "types" (e.g., preferences or other private information) for the players and possibly also the set of participating players. In this larger form, each players' information sets are the unions of the corresponding information sets in the various versions of the smaller extensive form that he cannot distinguish based on his private information about nature's initial choice. This larger extensive form becomes an extensive game by specifying a probability distribution of nature's choices of assignments. Thus, the induced extensive game has incomplete information.

For example, for a static single-item auction a bidder might learn those features that are common knowledge, such as the number of bidders, as well as private information represented by the valuation v he assigns to the item. In this case, his possible pure strategies are functions $b(v)$ that assign a bid to each contingency v. If the pricing is discriminating (and ignoring ties) then his expected payoff (absent risk aversion) from such a strategy is $\mathscr{E}\{[v - b(v)]\mathbf{1}_{\{b(v)>B\}}\}$, where B is the maximum of the others' bids and the ask price. Note that B is a random variable even if the other bidders' stategies are specified, since their valuations are not known. The second-price rule, on the other hand, yields the payoff $\mathscr{E}\{[v - B]\mathbf{1}_{\{b(v)>B\}}\}$ because the winning bidder pays the highest rejected price, which is B. Note that the pricing rule affects the extensive game and its normal form only via the expected payoffs assigned to strategy combinations of the players.

Static auctions with simultaneous moves conform exactly to their normal-form representation so the usual equilibrium concept is the Nash equilibrium. It is desirable, however, to enforce perfection to obtain the constraint $b(v) \le v$ on bids with no chance of succeeding, and one focuses naturally on equilibria that preserve symmetries among the players. Dynamic elements require sequential equilibria. For example, if the seller sets an ask price first, then a Nash equilibrium might allow the seller to be deterred from setting a high ask price by expectations of lower bids or fewer bidders; or in a Dutch auction a Nash equilibrium allows bidders to expect the seller to withdraw the item before the price drops to her valuation. The role of equilibrium selection criteria in truly dynamic auctions, such as bid–ask markets, has not been studied.

4. Static single-item symmetric auctions

In this section we review a portion of the basic theory of static auctions in which a seller offers a single item and the bidders are symmetric. Our aim is to indicate the formulation and methods used, because they are indicative of the approach taken in more elaborate problems. We present Milgrom and Weber's (1982a) characterization of the symmetric Nash equilibrium. The number n of bidders, the seller's ask price a, and the probability distribution of bidders' private information are assumed to be common knowledge. All parties are assumed to be risk-neutral.

The model supposes that nature assigns each bidder i a pair (x_i, v_i) of numbers. These have the following roles: bidder i observes the real-valued "signal" x_i before submitting his bid and if he wins the item at price p, then his payoff is $v_i - p$. In practice, x_i is interpreted as bidder i's sample observation, from which he constructs an initial estimate $\mathscr{E}\{v_i \mid x_i\}$ of his subsequent valuation v_i of the item, which may be observed only after the auction. Let $z = (x_i, v_i)_{i=1,\ldots,n}$ indicate nature's choice and use $F(z)$ to denote its joint distribution function, which we assume has an associated density $f(z)$ on a support that is a rectangular cell $Z = \{z \mid \underline{z}\mathbf{1} \le z \le \bar{z}\mathbf{1}\}$. Let $z \vee z'$ and $z \wedge z'$ be the elements of Z that are the component-wise maximum and minimum of z and z'. And, let $x^1, \ldots, x^n$ be the components of x arranged in nonincreasing order. The conditional distribution function of x^2 given $x^1 = s$ is denoted by $\hat{F}(\cdot \mid s)$, and its density by $\hat{f}(\cdot \mid s)$.

By symmetry of the bidders we mean that F is invariant under permutations of the bidders. In particular, the conditional distribution of v_i given $x_1, \ldots, x_n$ is invariant under permutations of those bidders $j \ne i$. A further technical assumption is called *affiliation*:

$$(\forall z, z' \in Z) \quad f(z \vee z')f(z \wedge z') \ge f(z)f(z'). \tag{1}$$

This assumption states essentially that on every subcell of Z the components of z are non-negatively correlated random variables. Its useful consequence is that the conditional expectation of a nondecreasing function of z given that z lies in a subcell, is a nondecreasing function of the boundaries of that subcell. Affiliation implies also that (x^1, x^2) has an affiliated density and that $\hat{F}$ has the monotone likelihood ratio property (MLRP); furthermore, $\hat{f}(t \mid s)/\hat{F}(t \mid s)$ is a nondecreasing function of s.

Symmetry implies that the function

$$v(s, t) = \mathscr{E}\{v_i \mid x_i = x^1 = s \ \& \ x^2 = t\} \tag{2}$$

is well defined (i.e., it does not depend on i nor on the $j \ne i$ for which $x_j = x^2$). Moreover, affiliation implies that it is a nondecreasing function. Let $v(\underline{x}, \underline{x}) = \underline{v}$, where $\underline{z} = (\underline{x}, \underline{v})$. The central role of this function in the analysis is easily anticipated. If the symmetric equilibrium strategy makes each bidder's bid an increasing function of his signal, then the one, say i, observing the highest signal $x^1 = s$ will win and obtain the conditional expected payoff

$$V_n(s) \equiv \mathscr{E}\{v_i \mid x_i = x^1 = s \ \& \ x^2 < s\} = \mathscr{E}\{v(s, x^2) \mid x^2 < s\}, \tag{3}$$

given that he wins after observing the signal s, gross of the price he pays. Thus, $V_n(x_i)$ as well as $v(x_i, x_i)$ are upper bounds on the profitable bids that bidder i can submit after observing the signal x_i.

Theorem 1 (Milgrom and Weber). *Assume symmetry and affiliation, and suppose that $a \le \underline{v}$. Then the symmetric equilibrium strategy σ in a discriminating (first-price) auction prescribes the bid*

$$\sigma(x) = \left[\underline{v}\theta(\underline{x}) + \int_{\underline{x}}^{x} v(t, t)\, d\theta(t) \right] \bigg/ \theta(x)$$

$$= v(x, x) - \int_{\underline{x}}^{x} \theta(t)\, dv(t, t)/\theta(x) , \tag{4}$$

if the bidder's observed signal is x, where

$$\theta(x) = \exp\left\{ \int_{\underline{v}}^{x} \frac{\hat{f}(t \mid t)}{\hat{F}(t \mid t)}\, dt \right\} . \tag{5}$$

In a nondiscriminating (second-price) auction, $\sigma(x) = v(x, x)$.

Sketch of proof.   In a first-price auction, if the strategy σ is increasing and differentiable with an inverse function X, then the optimal bid b must maximize the expected profit

$$\int_{\underline{x}}^{X(b)} [v(x, t) - b]\, d\hat{F}(t \mid x) . \tag{6}$$

It must therefore satisfy the necessary condition

$$0 = -\hat{F}(X(b) \mid x) + [v(x, X(b)) - b]\hat{f}(X(b) \mid x)X'(b) ,$$

so

$$0 = \{-\sigma'(x)\hat{F}(x \mid x)/\hat{f}(x \mid x) + [v(x, x) - \sigma(x)]\} \frac{\hat{f}(x \mid x)}{\sigma'(x)} , \tag{7}$$

where the second equality uses the equilibrium condition that the optimal bid must be $b = \sigma(x)$, and $X'(b) = 1/\sigma'(x)$. It is also necessary that $\sigma(x)$ does not exceed $v(x, x)$ (otherwise winning is unprofitable), and that $\sigma(\underline{x}) \geq \underline{v}$ (otherwise a larger bid would be profitable when $x = \underline{x}$), and therefore this differential equation is subject to the boundary condition that $\sigma(\underline{x}) = \underline{v}$. The formula in the theorem simply states the solution of the differential equation subject to the boundary condition. Verification that this solution is indeed increasing is obtained by recalling that $v(t, t)$ is an increasing function of t and noting that as x increases the weighting function $\theta(t)/\theta(x)$ puts greater weight on higher values of t. The second version of the necessary condition implies that the expression in curly brackets is zero at the bid $b = \sigma(x)$, and since $\hat{f}(\hat{x} \mid x)/$

$\hat{F}(\hat{x} \mid x)$ is nondecreasing in x as noted earlier, for a bid $\hat{b} = \sigma(\hat{x})$ it would be non-negative if $\hat{x} < x$ and nonpositive if $\hat{x} > x$; thus the expected profit is a unimodal function of the bid and it follows that the necessary condition is also sufficient. The assumed differentiability of the strategy is innocuous if it is continuous since affiliation is preserved under monotone transformations of the bidders' observations x_i. Consequently, the remainder of the proof consists of showing that in general the strategy must be continuous on each of several disjoint intervals, in each of which it is common knowledge among the bidders that all observations lie in that interval. This last step also invokes affiliation to show that the domains of continuity are intervals.[1] The argument is analogous for a second-price auction except that the preferred bid is the maximum profitable one, $\sigma(x) = v(x, x)$, since his bid does not affect the price a bidder pays.

The assumption that the distribution F has a density is crucial to the proof because it assures that the probability of tied bids is zero.[2] If the distribution F is not symmetric, then generally one obtains a system of interrelated differential equations that characterize the bidders' strategies.

The theorem allows various extensions. For example, if the seller's ask price is $a > \underline{v}$ and $x(a)$ solves $a = v(x(a), x(a))$, then

$$\sigma(x) = \min\left\{ v(x, x), \frac{a\theta(x(a)) + \int_{x(a)}^{x} v(t, t) \, d\theta(t)}{\theta(x)} \right\}. \tag{8}$$

In this form the theorem allows a random number of *active* bidders submitting bids exceeding the ask price. That is, the effect of an ask price (or bid preparation costs) is to attract a number of bidders that is affiliated with the bidders' signals; thus high participation is associated with high valuations for participants.

4.1. The independent private-values model

If each bidder observes directly his valuation, namely the support of F is restricted to the domain where ($\forall i$) $x_i \equiv v_i$, then $v(s, t) = s$. One possible source of correlation among the bidders' valuations is that, even though the

---

[1]Milgrom and Weber (1982a, fn. 21) mention an example with two domains; at their common boundary the strategy is discontinuous.

[2]Milgrom (1979a, p. 56) and Milgrom and Weber (1985a, fn. 9) mention asymmetric examples of auctions with no equilibria. In one, each bidder knows which of two possible valuations he has and these are independently but *not* identically distributed. An equilibrium must entail a positive probability of tied bids, and yet if ties are resolved by a coin flip, then each bidder's best response must avoid ties.

bidders' valuations are independently and identically distributed, the bidders are unsure about the parameters of the distribution. If their valuations are actually independent, say $F(z) = \Pi_i G(x_i)$, then the bidders are said to have *independent private values*. For this model, $\theta(x) = \hat{F}(x) = G(x)^{n-1}$ is just the distribution of the maximum of the others' valuations; hence

$$\sigma(x) = \min\{x, \mathscr{E}\{\max[a, x^2]|x^1 = x\}\} \tag{9}$$

in a first-price auction. In a second-price auction, $\sigma(x) = x$, which is actually a dominant strategy. The seller's expected revenue is therefore the expectation of $\max\{a, x^2\}\boldsymbol{1}_{\{x^1 \geq a\}}$ for either pricing rule – a result often called the *revenue equivalence* theorem. This result applies also if the number of bidders is independently distributed; for example, if each participating bidder assigns the Poisson distribution $q_m = e^{-\lambda}\lambda^m/m!$ to the number $m = n - 1$ of other participating bidders, then $\theta(x) = e^{\lambda F(x)}$.[3] It also illustrates the general feature that the seller's ask price a can be regarded as another bid; for example, if the seller has an independent private valuation $v_0 \geq 0$, then this plays the role of an extra bid, although presumably it has a different distribution. If the seller can commit beforehand to an announced ask price, however, then she prefers to set $a > v_0$. For example, if the bidders' valuations are uniformly distributed on the unit interval, namely $G(x) = x$, then their symmetric strategy and the seller's expected revenue are

$$\sigma(x) = \min\left\{x, \frac{n-1}{n}x + \frac{1}{n}a^n/x^{n-1}\right\}$$

and

$$R_n(a) = \frac{n-1}{n+1} + a^n\left[1 - \frac{2n}{n+1}a\right], \tag{10}$$

from which it follows that for every n the seller's optimal ask price is $a = \frac{1}{2}[1 + v_0]$.

---

[3]More generally, if k identical items are offered and each bidder demands at most one, then the unique symmetric equilibrium strategy (ignoring the ask price a) for discriminating pricing is $\sigma(x) = \mathscr{E}\{x^{k+1}|x^{k+1} < x\}$, and the seller's expected revenue is $k\mathscr{E}\{x^{k+1}\}$ for either pricing rule. Milgrom and Weber (1982a) and Weber (1983) demonstrate revenue equivalence whenever the k bidders with the highest valuations receive the items and the bidder with the lowest valuation gets a payoff of zero. This is true even if the items are auctioned sequentially, in which case the successive sale prices have the Martingale property with the unconditional mean $\mathscr{E}\{x^{k+1}\}$. Harstad, Kagel and Levin (1990) demonstrate revenue equivalence among five auctions: The two pricing rules combined with known or unknown numbers (with a symmetric distribution of numbers) of bidders, plus the English auction. For example, a bidder's strategy in the symmetric equilibrium of a first-price auction is the expectation of what he would bid knowing there are n bidders, conditional on winning; that is, each bid $\sigma_n(x)$ with n bidders is weighted by $w_n(x)$, which is the posterior probability of n given that x is the largest among the bidders' valuations.

4.2. The common-value model

In a situation of practical importance the bidders' valuations are identical but not observed directly before the auction. In the associated *common-value* model, $(\forall i)$ $v_i = v$, and conditional on this common value v their samples x_i are independently and identically distributed. In this case, an optimal bid must be less than the conditional expectation of the value given that the bidder's sample is the maximum of all the bidders' samples, since this is the circumstance in which the bid is anticipated to win.

For example, suppose that the marginal distribution of the common value has the Pareto distribution $F(v) = 1 - v^{-\alpha}$ for $v \geq 1$ and $\alpha > 2$, so that $\mathscr{E}\{v\} = \alpha/[\alpha - 1]$. If the conditional distribution of each sample is $G(x_i \mid v) = [x_i/v]^{\beta}$ for $0 \leq x_i \leq v$, then the conditional distribution of the value given that an observed sample x is the maximum of n samples is

$$V_n(x) = \frac{\alpha + n\beta}{\alpha + n\beta - 1} \max\{1, x\} . \tag{11}$$

The symmetric equilibrium strategy in this case, assuming that the seller's ask price is not binding and using $B = [n - 1]\beta$, is

$$\sigma(x) = \frac{B + [\max\{1, x\}]^{-B-1}}{B + 1} V_n(x) . \tag{12}$$

Ex ante, each bidder's and the seller's expected profit are

$$\frac{\beta}{[\alpha + B][\alpha + B + \beta - 1]} \mathscr{E}\{v\}$$

and

$$\left[1 - \frac{B + \beta}{[\alpha + B][\alpha + B + \beta - 1]}\right] \mathscr{E}\{v\} . \tag{13}$$

To take another example of the common-value model, suppose that the marginal distribution of the value is a gamma distribution with mean m/k and variance m/k^2, and the conditional distribution of a sample is the Weibull distribution $G(x_i \mid v) = e^{vy(x_i)}$, where $y(x_i) = -x_i^{-\beta}$. Then the symmetric equilibrium bidding strategy is

$$\sigma(x) = \frac{m + 2}{m + 1 + \dfrac{n}{n - 1}} V_n(x) , \quad \text{where } V_n(x) = \frac{m + 1}{k - ny(x)} . \tag{14}$$

In practice, bidding strategies are often constructed on the assumption that the marginal distribution of the common value has a large variance. For example, suppose that each estimate x_i has a normal conditional distribution with mean v and variation s_1^2, and that the marginal distribution of v has a normal distribution with variance s_0^2. If $a = -\infty$, then the limit of the symmetric equilibrium bidding strategy as $s_0 \to \infty$ is $\sigma(x) = x - \alpha_n s_1$, where

$$\alpha_n = \frac{\int_{-\infty}^{\infty} \xi^2 \, \mathrm{d}N(\xi)^n}{\int_{-\infty}^{\infty} \xi \, \mathrm{d}N(\xi)^n} , \tag{15}$$

using the standard normal distribution function N with mean 0 and variance 1; that is, α_n is the ratio of the second to the first moment of the distribution of the maximum of n standard normal variables.[4]

For the lognormal distribution, suppose the conditional distribution of $\ln(x_i)$ has a normal distribution with mean $\ln(v)$ and variance s_1^2 and the marginal distribution of $\ln(v)$ has a normal distribution with variance s_0^2. If $a \leq 0$, then the limit of the bidding strategy as $s_0 \to \infty$ is $\sigma(x) = \beta_n(s_1)x$, where

$$\beta_n(s) = \frac{\int_{-\infty}^{\infty} [s + \xi] \, e^{-s\xi} \, \mathrm{d}N(\xi)^n}{\int_{-\infty}^{\infty} [s + \xi] \, \mathrm{d}N(\xi)^n} . \tag{16}$$

In this case, $\mathscr{E}\{v \mid x_i\} = x_i B(s_1)$, where $B(s) = e^{0.5s^2}$, so it is useful to correct for bias by taking the estimate to be $x_i' = x_i B(s_1)$ and the strategy to be $\sigma(x') = [\beta_n(s_1)/B(s_1)]x'$. Table 1 tabulates a few values of α_n and $\beta_n(s)/B(s)$. Notice that as n increases, α_n first decreases due to increasing competition, and then increases due to the decline in the expected value of the item conditional on winning. That is, the supposition that x is the maximum of n unbiased

Table 1
Normal and lognormal bid factors

n:	2	3	4	6	8	10	15	20
α_n:	1.772	1.507	1.507	1.595	1.686	1.763	1.909	2.014
s				$\beta_n(s)/B(s)$				
0.01	0.983	0.985	0.985	0.984	0.983	0.983	0.981	0.980
0.10	0.847	0.863	0.861	0.852	0.844	0.837	0.824	0.815
0.25	0.682	0.698	0.689	0.667	0.650	0.636	0.611	0.594
0.50	0.498	0.495	0.474	0.437	0.411	0.392	0.360	0.339
1.00	0.281	0.244	0.212	0.173	0.150	0.134	0.111	0.098

[4]An erroneous statement of this result in Thiel (1988) is corrected by Levin and Smith (1991), who show also that if $s_0 \equiv \infty$, then additional equilibria exist with strategies having an additional nonlinear term.

estimates of v implies that x is biased by an amount that increases with n. A similar effect can be seen in the behavior of $\beta_n(s)/B(s)$.

A model with wide applicability assumes that a bidder's valuation of the item has the form $p_i v$, where p_i is a private factor specific to bidder i and v is a common factor. Before bidding, each bidder i observes (p_i, x_i), where x_i is interpreted as an estimate of the common value v. Conditional on unobserved parameters (p, v), the bidders' observations are independent, and the private and common factor components are independent. Wilson (1981) studies such a model adapted to bidding for oil leases. Assume that $\ln(p_i)$ and $\ln(x_i)$ have conditional distributions that are normal with means $\ln(p)$ and $\ln(v)$, and variances t^2 and s^2, respectively. If the marginal variances of the unobserved parameters are infinite, then the symmetric equilibrium strategy is linear of the form $\sigma(p_i, x_i) = \gamma_n p_i \mathscr{E}\{v \mid x_i\}$. In this case the variance of the natural logarithm of the bids is $t^2 + s^2$, which for auctions of leases on the U.S. Outer Continental Shelf is usually about 1.0. This model allows the further interpretation that there is variance also in the bid factors used by the bidders (included in t^2) due perhaps to differences in the models and methods used by the bidders to prepare their bids. Table 2 displays the bid factor γ_n and the percentage expected profit of the winning bidder for several cases. These conform roughly to the one-third and one-quarter maxims used in the oil industry [Levinson (1987)]: bid a third, profits average a quarter. These two fractions add to less than one because winning indicates that the estimate is biased too high.

A key feature of the equilibrium strategies identified by Theorem 1 is that each bidder takes account of the information that would be revealed by the event that his bid wins. That is, in the symmetric case, winning reveals that the bidder's sample is the maximum among those observed by all bidders. Even if each sample is an unbiased estimate of the item's value, it is a biased estimate conditional on the event that the bid wins. This feature is inconsequential if bidders directly observe their valuations, but it crucially affects the expected profitability of winning in a common-value model. Failure to take account of estimating bias conditional on winning has been called the "winner's curse"; cf. Thaler (1988). In Section 9 we report some of the experimental evidence on its prevalence.

Table 2
Private and common factors model: bid factor γ_n and winner's expected profit percent

t^2	s^2	Bid factor				Profit percentage			
		n:2	4	8	16	2	4	8	16
0.75	0.25	0.307	0.314	0.271	0.228	37	17	9	6
0.50	0.50	0.312	0.366	0.355	0.329	47	27	17	12
0.25	0.75	0.307	0.410	0.444	0.451	59	40	30	23

The following subsections provide a sampling of further results regarding static single-item auctions. The first examines the effect of increasing competition in a symmetric common-value auction.

4.3. Auctions with many bidders

The main results are due to Wilson (1977), Milgrom (1979a, 1979b), and Wang (1990), who study the case of an unobserved common value v and signals that are conditionally independent and identically distributed given v. Wang assumes a discrete distribution for the signals, whereas the others assume the signals' conditional distribution has a positive density $f(\cdot \mid v)$. Consider a sequence of symmetric auctions with discriminating pricing, all with the same common value, in which the nth auction has bidders $i = 1, \ldots, n$ who observe the signals $x_1, \ldots, x_n$. Say that the sequence of signals is an extremal-consistent estimator of v if there exists a sequence of functions g_n such that $g_n(\max_{i \leqslant n} x_i)$ converges in probability to v as $n \to \infty$. Milgrom shows that the signal sequence is an extremal-consistent estimator of the common value if and only if $v < v'$ implies $\inf_x \{ f(x \mid v) / f(x \mid v') \} = 0$.

Theorem 2 (Milgrom).   *For a sequence of discriminating (first-price) symmetric auctions with increasing numbers of bidders, the winning bid converges in probability to the common value if and only if the signal sequence is an extremal-consistent estimator of the common value.*

Thus, whenever a consistent estimator of the common value can be based on the maximum signal, the winning bid is one such estimator, using $g_n \equiv \sigma_n$, where σ_n is the symmetric equilibrium strategy when there are n bidders.

Allowing risk aversion as well, Milgrom (1981) demonstrates comparable results for auctions with nondiscriminating pricing. The dependence on equilibrium strategies is relaxed by Levin and Harstad (1990) using the more restrictive model in Wilson (1977); e.g., the support of the bidders' signals moves monotonely as the value changes. They show that convergence obtains if bidders' strategies are restricted only by single-iteration elimination of dominated strategies: each bidder uses a stategy that is undominated if other bidders use undominated strategies.

An extremal-consistent estimator exists for most of the familiar distributions, such as the normal or lognormal. But there are important exceptions, such as the exponential distribution with mean v, for which the conditional distribution of the maximum bid is nondegenerate in the limit, although the limit distribution does have v as its mean. There seem to be no general results on the rate of convergence, but convergence is of order $1/n$ in examples – although these all have the property that this is the rate of convergence of extremal-consistent estimators.

Matthews (1984a) studies the special case of the common-value model in which the conditional distribution of a bidder's signal is $F_i(x_i \mid v) = [x_i/v]^{m_i}$; that is, the signal x_i is the maximum of m_i samples uniformly distributed on the interval $[0, v]$. However, the formulation is enriched to allow that each bidder chooses both the number $m_i \in \Re_+$ of samples, at a cost $c(m_i)$, and his bid depending on the signal x_i observed. In a symmetric pure-strategy equilibrium, of course, all bidders choose the same number m of samples. Assume that c is convex and increasing, with $c(0) = 0$. Matthews establishes that as the number n of bidders increases, $m \to 0$ but the total number nm of samples purchased is bounded and bounded away from zero. Moreover, ex ante the expectation of the difference between the common value and the sum of the maximum bid and the sampling costs $nc(m)$ of *all* the bidders converges to zero.[5] Thus, pure-strategy equilibria necessarily entail limits on aggregate expenditures for information, and the seller expects to reimburse these expenditures. Indeed, if $c(m) = \hat{c}m$, then the maximum bid converges in probability to the common value only if $\hat{c} = 0$.

4.4. Superior information

The familiar auction rules treat the bidders symmetrically. Consequently, the principal asymmetries among bidders are due to differences in payoffs and differences in information. Here we describe briefly some of the features that occur when some bidders have superior information.

The basic result about the effect of superior information is due to Milgrom (1979a) and Milgrom and Weber (1982b). Consider a static auction of a single item with risk-neutral bidders and, as in Theorem 1, let x_i and v_i be the signal and valuation of bidder i.

Theorem 3 (Milgrom). *In an equilibrium of a first-price (discriminating) auction, a bidder i's expected payoff is zero if there is another bidder j whose information is superior (x_j reveals x_i) and whose valuation is never less ($v_j \geq v_i$). In a second-price (nondiscriminating) auction, a bidder i's equilibrium expected payoff is zero if there is another bidder j such that i's signal is a garbling of j's, j's valuation is never less, and j's bids have positive probabilities of winning in equilibrium.*

---

[5]Harstad (1990) shows that with equilibrium strategies this expectation is exactly zero.

Sketch of proof. In a first-price auction, given x_j and therefore x_i, if i's strategy allows a positive expected payoff for a bid b, then j profits by bidding slightly more than b whenever he would have bid the same or less. In a second-price auction, j's (serious) bid is a conditional expectation of v_j that is independent of i's strategy (because i's bid reveals no additional information) and that implies a nonpositive expected payoff for j, and hence also i because i must pay at least j's bid to win.

To examine the implications of this result, consider a common-value model in which pricing is discriminating, the seller's ask price is a, and there are $m + n$ bidders. Suppose that m bidders know the common value v and n know only its probability distribution F, which has a positive density on an interval support. If $m > 1$ and $v > a$, then in any equilibrium all of the informed players bid v; consequently, an uninformed player expects to win the item only by paying more than v and incurring a loss. An equilibrium therefore requires that the uninformed players have no chance to win and obtain zero payoffs. If $m = 1$, then the informed player's unique equilibrium strategy is to bid

$$\sigma(x) = \max[a, \mathscr{E}\{v \mid v \leqslant x\}] \tag{17}$$

when he observes $v = x > a$. The uninformed players use mixed strategies such that the maximum b of their n bids has the distribution function

$$H(b) = \max\{F(v(a)), F(\sigma^{-1}(b))\} \tag{18}$$

if $b \geqslant a$, where $a = \mathscr{E}\{v \mid v \leqslant v(a)\}$ defines $v(a)$. That is, the uninformed players' mixed strategies replicate the distribution of the informed player's bids on the support where $\sigma(x) > a$; in addition, there may be a probability $F(v(a))$ of submitting bids sure to lose (or not bidding).[6] If the equilibrium is symmetric, then each uninformed player uses the distribution function $H(b)^{1/n}$. The informed player's strategy ensures that each uninformed player's expected payoff is zero; the expected payoff of the informed player is

$$\int\limits_{a}^{\infty} [v - \sigma(v)]H(\sigma(v))\,\mathrm{d}F(v) = \left\{[1 - F(v(a))][v(a) - a] + \int\limits_{-\infty}^{a} F(v)\,\mathrm{d}v\right\}F(v(a))$$

$$+ \int\limits_{v(a)}^{\infty} [1 - F(v)]F(v)\,\mathrm{d}v . \tag{19}$$

---

[6]This feature can be proved directly using the methods of distributional strategies in Section 5; cf. Englebrecht-Wiggans, Milgrom and Weber (1983) and Milgrom and Weber (1985).

Examples. (1) If $F(v) = v$, then $\sigma(x) = \max[a, x/2]$ if $x > a$, $v(a) = 2a$, and $H(b) = \max[2a, 2b]$; if $a = 0$ then the informed player's expected payoff is $1/6$. (2) If $a = -\infty$ and v has the normal distribution $F(v) = N([v - m]/s)$, then $\sigma(x) = m - sN'(\xi)/N(\xi)$, where $\xi = [x - m]/s$. (3) If $a = 0$ and $F(v) = N([\ln(v) - m]/s)$, then $\sigma(x) = \mu N(\xi - s)/N(\xi)$, where $\mu = \exp(m + 0.5s^2)$ and $\xi = [\ln(x) - m]/s$.

These results extend to the case that the uninformed bidders value the item less. Suppose they all assign value $u(v) \leq v$, and for simplicity assume that $a \leq \underline{v} = u(\underline{v})$, where $F(\underline{v}) = 0$. Then

$$\sigma(x) = \int_{\underline{v}}^{x} u(v)\, dF(v)/F(x) \quad \text{and} \quad H(\sigma(x)) = \exp\left\{ -\int_{x}^{\infty} \left[\frac{u(v) - \sigma(v)}{v - \sigma(v)} \right] \frac{dv}{v} \right\}.$$

(20)

Milgrom and Weber (1982b) show that these results imply several conclusions about the participants' incentives to acquire or reveal information. Assume that the informed bidder's valuation is actually the conditional expectation $v = \mathscr{E}\{V \mid X\}$ of the ultimate common value V, given an observation X. Assume also that $a = -\infty$ and that the value is the same for all bidders.

- The bidder with superior information gains by acquiring additional information, and more so if this is done overtly; i.e., the uninformed bidders know he acquires this information.
- An uninformed bidder gains by acquiring some of the informed bidder's information, provided this is done covertly.
- The seller gains (in expectation) by publicizing any part of the informed bidder's information, or any information that is jointly affiliated with *both* the informed bidder's observation X and the common value V.[7]

Alternatively, suppose one bidder, the "insider", knows precisely the common value v and the n other bidders obtain informative signals x_i about this common value. In this case, if the insider's strategy is ρ, then the appropriate extension of Theorem 1 to characterize the symmetric equilibrium strategies of those bidders other than the insider uses the revised function

$$v(s, t) = \mathscr{E}\{v \mid x_i = x^1 = s \ \& \ x^2 = t \ \& \ \rho(v) < \sigma(s)\} .$$

(21)

We illustrate with an example in Wilson (1975). Suppose the common value has a uniform distribution on the unit interval and the estimates x_i are uniformly distributed between zero and $2v$. The symmetric equilibrium in this

---

[7]Milgrom and Weber note that joint affiliation of the triplet is necessary; otherwise the seller's information can be "complementary" to the informed bidder's information and increase his expected profit.

Table 3
Bidding strategies – one perfectly informed bidder

	n	
	1	2
α	0.5000	0.6667
β	0.6796	0.5349
Expected profit (informed)	0.0920	0.0647
Expected profit (estimator)	0.0244	0.0178
Expected revenue (seller)	0.3836	0.3998

case has $\rho(v) = \alpha v$ and $\sigma(x) = \min\{\alpha, \beta x\}$. Table 3 tabulates α and β when there are $n = 1$ or 2 imperfectly informed bidders. Generally, $\alpha = n/[n + 1]$.

Observe that if $n = 0$ [or 1] then an additional uninformed bidder, who would otherwise submit a bid of zero, is willing to pay at most 0.0244 [or 0.0178] to acquire a signal (thus making $n = 1$ [or 2]) and submit a positive bid.

If there are additional bidders with no private information, then their best strategy is to bid zero, or not to bid, since otherwise their expected profit is negative. If there were two perfectly informed bidders then they would each bid the value v (i.e., $\alpha = 1$) and then all imperfectly informed bidders prefer not to bid any positive amount.

4.5. Asymmetric payoffs

Strongly asymmetric outcomes can also occur if the bidders have identical information, but one values the item less. Milgrom (1979a) gives an example in which two bidders assign valuations $v_1 > v_2 > a$ that are common knowledge: all equilibria have the form that 1 uses a pure strategy $b_1 \in [v_2, v_1]$ and 2 uses a mixed strategy H such that b_1 is optimal against H and 1 surely wins. Bikhchandani (1988) provides an example of repeated nondiscriminating common-value auctions in which there is a chance that one of the two bidders, say 1, consistently values the items more. Because this feature accentuates the adverse selection encountered by the winning bidder, in the equilibrium bidder 1 surely wins every auction.

Maskin and Riley (1983) provide an example indicating that the seller's choice of the auction rules can be affected by asymmetries. They consider an auction with two bidders having independent private valuations that are uniformly distributed on two different intervals. Represent these intervals as the interval from zero to $1/(1 - \alpha)$ for bidder 1 and from zero to $1/(1 + \alpha)$ for bidder 2. The equilibrium strategies for a first-price auction in this case are

$$X_1(b) = \frac{2b}{1 - \alpha(2b)^2} \quad \text{and} \quad X_2(b) = \frac{2b}{1 + \alpha(2b)^2} \tag{22}$$

Table 4
Seller's expected revenue

	α					
	0.00	0.25	0.33	0.50	0.90	1.00
First-price auction	0.3333	0.3392	0.3443	0.3590	0.4411	0.5000
Second-price auction	0.3333	0.3290	0.3125	0.2962	0.2585	0.2500

for $b \leqslant 0.5$, indicating the valuations at which the bidders would submit the same bid b. Note that bidder 1, whose valuation is perceived by bidder 2 to be drawn from a more favorable distribution, requires a higher valuation than does bidder 2 to bid the same. For a second-price auction it suffices that each bidder submits his valuation. The seller's expected revenue in these two cases is shown in Table 4 for various values of the parameter α. Observe that the seller can realize an advantage from a first-price auction if there is substantial asymmetry.

4.6. Attrition games

Closely related to auctions are contests to acquire an item in which the winner is the player expending the greatest resources. Riley (1988a) describes several examples, including the "war of attrition" that is an important model of competition in biological [Riley (1980), Nalebuff and Riley (1985)] and political [Wilson (1989)] as well as economic [Holt and Sherman (1982)] contexts. In attrition games, a player's expenditures accumulate over time as long as he is engaged in the contest; when all other players have dropped out, the remaining player wins the prize. Unlike an ordinary auction in which only the winner pays, a player incurs costs whether he wins or not; moreover, there need not be a seller to benefit from the expenditure. Some models of bargaining and arms races have this form, and price wars between firms competing for survival in a natural monopoly are similar.

Huang and Li (1990) establish a general theorem regarding the existence of equilibria for such games. We follow Milgrom and Weber (1985) to illustrate the construction of equilibria. In the simplest symmetric model, the n players' privately known valuations of the prize, measured in terms of the maximum stopping time that makes the contest worth the prize, are independent and identically distributed, each according to the distribution function F. A player with the valuation v obtains the payoff $v - t$ if the last of the other players drops out at time t and he does not, and otherwise it is $-t$ if he drops out at time t. Adopting a formulation in terms of distributional strategies (see Section 5), let $V(x) = F^{-1}(x) \equiv \sup\{v \mid F(v) < x\}$ and interpret $x = F(v)$ as the type of a player with the valuation v. Thus, a strategy σ_n that assigns a stopping time to each type implies a distribution $\rho_n = \sigma_n^{-1}$ of his stopping times. If each other player has this distribution of stopping times, then a player with the type x prefers to stop at time t if his cost per unit time equals the corresponding conditional expectation of winning the prize. When k players remain, this yields the condition:

$$1 = V(x) \frac{[k-1]\rho_k(t)^{k-2}\rho_k'(t)}{1 - \rho_k(t)^{k-1}} , \tag{23}$$

since ρ_k^{k-1} is the distribution of the maximum of the other players' stopping times. The relevant case, however, is when only two players remain, which we now assume. Adding the equilibrium condition that $x = \rho_2(t)$ to the above condition yields a differential equation for the distribution ρ_2 that is subject to the boundary condition $\rho_2(0) = 0$. The solution in terms of the strategy is

$$\sigma_2(x) = \int_0^x \frac{V(z)}{1 - z} \, dz . \tag{24}$$

Milgrom and Weber note further the properties that the hazard rate of the duration of the game is a decreasing function of time; the distribution of stopping times increases stochastically with the distribution of valuations; and the equilibrium is in pure strategies if and only if the distribution of valuations is atomless, in which case the stopping time as a function of the player's valuation is

$$\beta(v) = \int_{\underline{v}}^v \frac{t}{1 - F(t)} \, dF(t) . \tag{25}$$

Extensions to asymmetric equilibria and to formulations with benefits or costs that vary nonlinearly with time are developed by Fudenberg and Tirole (1986) and Ghemawat and Nalebuff (1985).

5. Uniqueness and existence of equilibria

In this section we mention a few results regarding uniqueness of the symmetric equilibrium in the symmetric case, and regarding existence of equilibrium in general formulations. Affiliation is assumed unless mentioned.

5.1. Uniqueness

Addressing a significantly more general formulation (e.g., allowing risk aversion), but still requiring symmetry, Maskin and Riley (1986) derive a characterization similar to Theorem 1 of the symmetric equilibrium, and for the case of two bidders they establish that the symmetric equilibrium is the unique equilibrium. Thus, for symmetric first-price auctions there is some presumption that the symmetric equilibrium is the unique equilibrium. For symmetric second-price auctions, Matthews (1987) finds a symmetric equilibrium that with risk aversion generalizes Theorem 1 and for the common-value model this is shown to be unique by Levin and Harstad (1986). Milgrom (1981) shows, however, that symmetric second-price auctions can have many asymmetric equilibria, and Riley (1980) provides an example of a symmetric attrition game with a continuum of asymmetric equilibria. Bikhchandani and Riley (1991) provide sufficient conditions for uniqueness of the equilibrium in a second-price auction for the common-value model of preferences. For an "irrevocable exit" version (an English auction with publicly observed irreversible exits of losing bidders) they establish that, within the class of equilibria with non-decreasing strategies, the unique symmetric equilibrium is accompanied (when there are more than two bidders) by a continuum of asymmetric equilibria with increasing and continuous strategies.

For auctions with asymmetrically distributed valuations, Plum (1989) proves existence as well as uniqueness within the entire class of measurable strategies for the case that there are two bidders, their valuations are independent and uniformly distributed (the support of bidder i's valuation is an interval $[0, \beta_i]$), and the sale price is a convex combination of the higher and lower bids. Smoothness of the equilibrium strategies is established for general independent distributions, and then for the case of uniform distributions and positive weight assigned to the higher bid, the equilibrium strategies are characterized by differential equations having a unique solution. The first-price auction maximizes the seller's expected revenue, which depends on the pricing rule if $\beta_1 \neq \beta_2$.

There appear to be no lower semi-continuity results indicating that, say, the symmetric equilibrium is the limit of equilibria of nearby asymmetric games. In particular, Bikhchandani (1988) studies a slightly asymmetric second-price auction in which there is a small chance that one bidder values the item more, but otherwise the auction is symmetric with common values; and for this game he finds that the equilibrium is strongly asymmetric. In a related vein, Bikhchandani and Riley (1991) provide sufficient conditions for the seller's revenue to be higher at the symmetric equilibrium of a second-price common-value auction than at any other.

5.2. Existence of equilibria in distributional strategies

As usually formulated, games representing auctions pose special technical problems in establishing existence of equilibria. Primary among these is that such games are not finite, in that each bidder has an infinity of pure strategies. The source of this difficulty can be that infinitely many bids are feasible, or that an infinite variety of private information can condition the selection of a bid. It suffices in practice to suppose that only finitely many pure strategies are feasible, but this approach typically yields equilibria with mixed strategies, whereas often the corresponding game with a continuum of strategies has an equilibrium in pure strategies. A further characteristic feature of auctions is that payoffs are discontinuous in strategies. In Section 4, fn. 2, we mentioned an example of an auction with discriminating pricing that has no equilibrium, due essentially to the discontinuity of payoffs at tied bids. In simple formulations these features do not present difficulties, and as seen in Section 4, equilibria in pure strategies are characterized by differential equations.

Here we describe an alternative formulation that avoids some of these difficulties by generalizing the characterization in terms of differential equations for both auctions and attrition games. We follow Milgrom and Weber (1985) who introduced the formulation in terms of distributional strategies, but refer also to Balder (1988) who uses the standard formulation in terms of behavioral strategies.[8]

Standard formulations introduce pure strategies specifying actions at each information set, mixtures of pure strategies, and in extensive games, behavioral strategies that specify mixtures of actions at each information set. Static auctions have the special feature that for each play of the game each bidder takes an action (selection of a bid) at a single information set that represents his private information. Thus a pure strategy consists of a specification of a bid conditional on the observed private information. Alternatively, in terms of Harsanyi's (1967–68) description of games with incomplete information, given his *type* as represented by his private information, each bidder selects a bid (or a stopping time). Given a distribution of his private information and a strategy (pure, mixed, or behavioral), therefore, each bidder's behavior is summarized by a joint distribution on the pair consisting of his information and his bid. Indeed, from the viewpoint of other bidders, this is the relevant summary. In general, we shall say that a *distributional strategy* is such a joint distribution for which the marginal distribution on the bidder's type is the one specified by the information structure of the game.

---

[8]An alternative approach, applied especially to attrition games, focuses on a detailed analysis of the role of discontinuous payoffs and establishes sufficient conditions for the limit of equilibria of a sequence of finite approximating games to be an equilibrium of the limit game; cf. DasGupta and Maskin (1986) and Simon (1987).

The formulation is specified precisely as follows. The game has a finite set N of players indexed by $i = 1, \ldots, n$, each of whom observes a type in a complete and separable metric space T_i and then takes an action in a compact metric space A_i of feasible actions.[9] Allowing another complete and separable metric space T_0 for unobservable states, define $T = T_0 \times \cdots \times T_n$ and $A = A_1 \times \cdots \times A_n$. The game then specifies each player's payoff function U_i as a real-valued bounded measurable function on $T \times A$, and the information structure as a probability measure η on the (Borel) subsets of T having a specified marginal distribution η_i on each T_i (including T_0). A distributional strategy for player i is then a probability measure, say μ_i, on the (Borel) subsets of $T_i \times A_i$ having the marginal distribution η_i and T_i. Each distributional strategy induces a behavioral strategy that is just a regular conditional distribution of actions given the player's type. Specified distributional strategies for all players imply an expected payoff for each player; consequently, a Nash equilibrium in distributional strategies is defined as usual. Milgrom and Weber impose the following regularity conditions.

R1 *Equicontinuous payoffs.* For each player i and each $\epsilon > 0$ there exists a subset $E \subset T$ such that $\eta(E) > 1 - \epsilon$ and the family of functions $\{ U_i(t, \cdot) \mid t \in E \}$ is equicontinuous.

R2 *Absolutely continuous information.* The measure η is absolutely continuous with respect to the measure $\hat{\eta} \equiv \eta_0 \times \cdots \times \eta_n$.

R1 implies that each player's payoffs are continuous in his actions, and therefore excludes known examples of auctions with finite type spaces that have no Nash equilibria in mixed strategies [cf. Milgrom (1979a) and Milgrom and Weber (1985)]. However, it is sufficient for R1 that the action spaces are finite or that the payoff functions are uniformly continuous. R2 implies that η has a density with respect to $\hat{\eta}$. It is sufficient for R2 that the type spaces are finite or countable, or the players' types are independent, or that η is absolutely continuous with respect to some product measure on T.

Milgrom and Weber establish that with these assumptions there exists an equilibrium in distributional strategies, obtained as a fixed point of the best response mapping.[10] Moreover, with appropriate specifications of closeness for

[9]Milgrom and Weber (1985, section 6) define two natural metrics on the type spaces. Balder (1988) uses a formulation in terms of behavioral strategies and is able to dispense with topological restrictions on the type spaces.

[10]Balder (1988) extends this result by considering behavioral strategies. Without imposing topologies on the type spaces, and replacing R1 with the requirement that each player's payoff function conditional on each $t \in T$ is continuous on the space A of joint actions, Balder proves the existence of an equilibrium in behavioral strategies, and obtains an extension for a class of noncompact action spaces.

strategies, information structures, and payoffs, the graph of the equilibrium correspondence is closed (upper hemicontinuity).

The relevance of these results for existence of equilibria in pure strategies is established in further results for the case that the marginal measure η_i of each player's private information is atomless.[11] First, if the action spaces are compact, then for every $\epsilon > 0$ there exists an ϵ-equilibrium in pure strategies. For the second, we follow Radner and Rosenthal (1982) in saying that a pure strategy σ_i purifies the distributional strategy μ_i if (a) for almost all of i's types the action selected by ϕ_i lies in the support of the behavioral strategy induced by μ_i (thus, the action is an optimal response); and (b) player i's expected payoffs are unchanged if i uses σ_i rather than μ_i's behavioral strategy. Part (b) is interpreted strictly: it must hold for every combination of the other players' distributional strategies.

Theorem 4 (Milgrom and Weber). *If R1 is satisfied and* (i) *the players' types are conditionally independent given each state* $t_0 \in T_0$, *and* T_0 *is finite, and* (ii) *each player's payoff function is independent of the other players' types, then each distributional strategy of each player has a purification. Moreover, the game has an equilibrium in pure strategies.*

Except for the requirement that T_0 is finite, an application of this theorem is to the model discussed in Section 4, where the symmetric equilibrium in pure strategies was characterized exactly. Note that condition (ii) admits both the independent private-values model and the common-value model.

The intuitive motivation for Theorem 4 is simple. If a player's actions can depend on private information that is sufficiently fine (i.e., η_i is atomless), then from the viewpoint of other players his pure strategies are capable of generating all of the "unpredictability" that mixed or distributional strategies might entail. Harsanyi (1973) follows a similar program in interpreting mixed strategies in complete information games as equivalent to pure-strategy equilibria in the corresponding incomplete-information game with privately known payoff perturbations.

6. Share auctions

The theory of static auctions of several identical items provides direct generalizations of Theorem 1 that are reviewed by Milgrom (1981) and Weber (1983), some of which are summarized in Section 4, fn. 3, for the case that each bidder

---

[11]Without invoking R1 or R2, this case already implies that each player's set of pure strategies is dense in his set of distributional strategies; cf. Milgrom and Weber (1985, theorem 3).

values only a single item. On the other hand, versions in which bidders' demands are variable pose rather different problems. Maskin and Riley (1987) show that for the seller an optimal procedure employs discriminating pricing of the form used in nonlinear pricing schemes. Here we address an alternative formulation studied by Wilson (1979) that preserves the auction format using nondiscriminating pricing.

If the supply offered by the seller is divisible, then the rules of the auction can allow that each bidder submits a schedule indicating the quantity demanded at each price. For instance, if the seller's supply is 1 unit and each bidder i submits a (nonincreasing) demand schedule $D_i(p)$, then the clearing price p^0 is the (maximum) solution to the equation $D(p^0) = 1$, where $D(p) = \Sigma_i D_i(p)$.[12] Each bidder i receives the share $D_i(p^0)$ and pays $p^0 D_i(p^0)$ if the pricing is nondiscriminating. [He pays an additional amount $\int_{p_i}^{p^0} p \, dD_i(p)$ if the pricing is discriminating, or in a Vickrey auctions he receives a rebate $p^0 - p_i - \int_{p_i}^{p^0} [D(p) - D_i(p)] \, dp$, where $D(p_i) - D_i(p_i) = 1$.]

To illustrate, consider a symmetric common-value model in which each bidder i observes privately an estimate x_i and then submits a schedule $D_i(p; x_i)$. Allowing risk aversion described by a concave utility function u, his payoff is $u([v - p^0]D_i(p^0; x_i))$ if the realized value is v. Assuming the bidders' estimates are conditionally independent and identically distributed given v, he can predict that if each other bidder uses a strategy D that is a decreasing function of the price, then the conditional distribution of the clearing price given the value v and his share y is

$$H(p; v, y) = \Pr\{p^0 \leqslant p \,|\, v, y\} = \Pr\left\{\sum_{j \neq i} D(p; x_j) \leqslant 1 - y \,|\, v\right\}. \qquad (26)$$

Consequently, a symmetric equilibrium requires that for each of his estimates x_i the choice of the function $y(p)$ that maximizes his expected payoff

$$\mathscr{E}\left\{\int_{-\infty}^{\infty} u([v - p]y(p)) \, dH(p; v, y(p)) \,|\, x_i\right\} \qquad (27)$$

is $y(p) = D(p; x_i)$. The Euler condition for this maximization is

$$\mathscr{E}\{u' \cdot [(v - p)H_p + yH_y] \,|\, x_i\} = 0, \qquad (28)$$

omitting the arguments of functions. Often, however, this condition allows a

---

[12]This assumes no ask price a is imposed by the seller. Other allocation rules are possible; for example, the seller can choose the clearing price to maximize $[p - a]D(p)$ if pricing is nondiscriminating.

continuum of equilibrium strategies if the seller does not impose a minimum ask price.

An example is provided by omitting risk aversion ($u' = 1$) and assuming that (1) the marginal distribution of the common value v is Gamma with mean m/k and variance m/k^2, and (2) each observation x_i has the conditional distribution function e^{vx_i} on $(-\infty, 0)$. Then one equilibrium is

$$D(p; x) = \frac{1}{n-1}\left[1 - 2p\,\frac{k-nx}{n(n+m)}\right],$$

$$p^0 = \frac{1}{2}\,\frac{m+n}{k - \Sigma_i x_i} = \tfrac{1}{2}\mathscr{E}\{v \mid x_1, \ldots, x_n\}\,. \tag{29}$$

The clearing price is positive, as is each bidder's resulting share. Note that in this example the clearing price is half the conditional expectation of the common value, regardless of the number of bidders. Anomalies appear in many examples of share auctions; presumably better modeling of the seller's behavior is necessary to eliminate these peculiarities.

One motive for studying share auctions is to develop realistic formulations of "rational expectations" features in markets affected by agents' private information. The Walrasian assumption of price-taking behavior can be paradoxical in such markets: if demands reflect private information, then prices can be fully informative, but if agents take account of the information in prices, then their demands at each price are uninformative. By taking account of agents' effects on the clearing price, models of share auctions avoid this conundrum. Jackson (1988) develops this argument and shows further the incentives that agents have to obtain costly information.

A share auction in the case of a finite number of identical items offered for sale is just a multi-item auction with nondiscriminating pricing: bidders whose offers are accepted pay the amount of the highest rejected bid. This formulation is developed by Milgrom (1981), who uses a symmetric model and the symmetric equilibrium identified in Theorem 1 to establish the information-revealing properties of the transaction price. He shows that bidders nevertheless utilize their private information in selecting a bid, and have incentives initially to acquire information. Thus, this formulation provides a sensible alternative to the price-taking behavior assumed in Walrasian models of rational-expectations equilibria.

7. Double auctions

In a static double auction, both the sellers and the buyers submit supply and demand schedules. A clearing price is then selected that equates supply and demand at that price. If the pricing is nondiscriminating, then all trades are consummated at the selected clearing price. This procedure is sometimes called a demand-submission game.

Working with a complete information model of Walrasian general equilibrium, Roberts and Postlewaite (1976) anticipate the subsequent game-theoretic analyses of double auctions. They consider a sequence of finite exchange economies (each described by a simple measure μ_n on the set of agents' characteristics) converging to an infinite economy (required to be a measure) at which the Walrasian price correspondence is continuous. They establish the following property for each agent persisting in the sequence whose inverse utility function is continuous in a neighborhood of the Walrasian prices for the limit economy: for each $\epsilon > 0$ there exists N such that if $n > N$, then the agent cannot gain more than ϵ from submitting demands other than his Walrasian demands. The gist of this result is that in a large economy an agent's incentive to distort his demand to affect the clearing prices is small. Subsequent work has examined whether a comparable result might hold for Nash equilibria, especially if agents' characteristics are privately known; however, comparable generality in the formulation has not been attempted.

More detailed characterizations of static double auctions have been obtained only for the case that a single commodity is traded for money, each seller offers one indivisible unit and each buyer demands one unit, their valuations are independent and (among sellers and buyers separately) identically distributed on the same interval, the traders are risk neutral, and the numbers of sellers and buyers are common knowledge. If the clearing price is p, then a trader's payoff is $p - v$ or $v - p$ for a seller or buyer with the valuation v who trades, and zero otherwise. If the asks and bids submitted allow k units to be traded, then the maximum feasible clearing price is the minimum of the kth highest bid and the $k + 1$th lowest ask, and symmetrically for the minimal clearing price. A symmetric equilibrium comprises a strategy σ for each seller and a strategy ρ for each buyer, where each strategy specifies an offered ask or bid price depending on the trader's privately known valuation. Following Myerson (1981), to avoid mixed strategies it is useful to assume the "regular" case that the distribution, say F, of a trader's valuation has a positive density f and that $v + F(v)/f(v)$ is increasing for a seller or $v - [1 - F(v)]/f(v)$ is increasing for a buyer. The symmetric equilibrium pure strategies are characterized by differential equations in Wilson (1985a), Williams (1987), and Satterthwaite and Williams (1989a, 1989b, 1989c). The most general characterization and proof of existence, in terms of vector fields for generic data and symmetric equilibria, is by Williams (1988).

The basic characterization of the effect of many traders is due to Williams (1988) and Satterthwaite and Williams (1989a, 1989b, 1989c), who address the case that the clearing price used is the maximal (or symmetrically, the minimal) one. In this case the sellers' dominant strategy is the identify $\sigma(v) = v$, because a seller's ask cannot affect the price at which she trades. Their main result demonstrates for each buyer's valuation v that $v - \rho(v) = O(1/M)$, where M is the minimum of the number of sellers and buyers. Thus, in double auctions of this kind with many traders of both types, each trader asks or bids nearly his valuation; and, the resulting allocation is nearly efficient, since missed gains from trade are both small and unlikely.[13] For example, if all valuations are uniformly distributed on the unit interval and there are m buyers and n sellers, then the unique smooth symmetric equilibrium is linear, $\rho(v) = [m/(m + 1)]v$, independently of the number of sellers.

Williams (1988) shows further that if there is a single buyer, then his strategy is independent of the number of sellers, whereas if there is a single seller and a regularity condition is imposed then the buyers' strategy again implies bids that differ from their valuations by $O(1/m)$. These results indicate that asymmetry in the auction rule leads to competition among the buyers, which is the main explanation for the tendency towards ex post efficiency as the number of buyers increases.

Chatterjee and Samuelson (1983) construct a symmetric equilibrium for the case of one seller and one buyer with uniformly distributed valuations, taking the transaction price to be the midpoint of the interval of clearing prices: $\sigma(v) = \frac{1}{4} + \frac{2}{3}v$ and $\rho(v) = \frac{1}{12} + \frac{2}{3}v$. However, Leininger, Linhart and Radner (1989) show that this linear equilibrium is one among many nonlinear ones; indeed, the characterization by Satterthwaite and Williams (1989a, 1989b, 1989c) shows that this feature is entirely general. Myerson and Satterthwaite (1983) establish, nevertheless, that with this linear equilibrium the double auction is ex ante efficient; that is, no other (individually rational – each trader's conditional expected payoff given his valuation is non-negative) trading mechanism has an equilibrium yielding a greater sum of the two traders' expected payoffs. This conclusion does not extend to greater numbers of sellers and buyers, however; cf. Gresik (1991a).

Wilson (1985a, 1985b) examines the weaker criterion of *interim* efficiency defined by Holmström and Myerson (1983); namely there is no other trading mechanism having an equilibrium for which, conditional on each trader's valuation, it is common knowledge that every trader's expected payoff is greater. Using the model of Satterthwaite and Williams, except that the supports need not agree and the clearing price can be an arbitrary convex combination of the endpoints, and assuming that the derivatives of the strategies are uniformly bounded, he demonstrates that a double auction is *interim* efficient if M is sufficiently large.

---

[13] This shows also that this property must hold for any optimal trading mechanism, which strengthens a result in Gresik and Satterthwaite (1989).

Significantly stronger results are obtained by McAfee (1989) for double auction rules that allow a surplus of money to accumulate. In the simplest of the three versions he examines, the rules are as follows. Suppose the bids and offers submitted allow at most a quantity q to be traded; i.e., q is the maximum k such that the kth highest bid b_k exceeds the kth lowest offer s_k. Then $q - 1$ units are traded with the $q - 1$ highest bidders buying items at the price b_q, and the $q - 1$ lowest offerers selling items at the price s_q. Note that a monetary surplus of $(q - 1)(b_q - s_q)$ remains. If these rules are used, then the traders have dominant strategies, namely bid or offer one's valuation. Only the least valuable efficient trade is lost. In fact, for a slightly more complicated scheme, with n traders the realized prices differ from an efficient price by $O(1/n)$ and the loss in expected potential surplus is approximately $O(1/n^2)$ – in the sense that it is $O(1/n^\alpha)$ for all $\alpha < 2$.

7.1. Bid–ask markets

Dynamic double auctions in which buyers and sellers have repeated oppor-tunities to submit or accept bids and offers are commonly used in commodity markets and some financial markets. They have been intensively studied experimentally as we describe in Section 9, but few theoretical analyses have been published. Their remarkable efficiency in realizing gains from trade even when subjects have little information has motivated two studies arguing that simple heuristic behaviors [Easley and Ledyard (1982)] or arbitrage processes [Friedman (1984)] could explain the data.

Friedman's (1984) analysis supposes that the traders' strategies imply "no congestion" at the conclusion of trading; that is, if an extra static double auction were appended to the end of the bid–ask market, and for this auction each trader were to assign positive probability that the final maximum bid and minimal ask would be acceptable to some other traders, then no trader would actually want to accept one of these offered trades nor would any trader want to alter his bid to ask – thus, no trade would occur in the appended auction. Using this auxiliary assumption instead of the usual requirement of Nash equilibrium, he shows that if the commodity traded is divisible (and prefer-ences are regular) then the final bid and ask must agree and be a market clearing price; in particular, the attained allocation is efficient. If the items traded are indivisible, then this conclusion is slightly weakened: the allocation is within a single trade of being efficient. Although it is not explained precisely how traders' strategies achieve the no-congestion property, Friedman's results demonstrate that fairly weak properties of traders' strategies suffice to explain the remarkable efficiency observed experimentally, and that perhaps it is not necessary to appeal to complete analyses of Nash equilibria.

McAfee (1989) offers a different view based on a dynamic auction design derived from his modified double auction rules described previously [a similar design, but without monetary surpluses, is studied by McCabe et al. (1989, 1990a)]. Because this auction is plausibly similar to a bid–ask market, and its unrealized gains from trade are approximately of order $O(1/n^2)$, the experimental efficiency of bid–ask markets is perhaps unsurprising.

Wilson (1986) proposes a conjectured equilibrium and verifies that it satisfies various necessary conditions (as well as the no-congestion property). His "equilibrium" is a multilateral generalization of the equilibrium for bilateral bargaining constructed by Cramton (1984, 1990) in which at each time the buyer with the highest valuation and the seller with the lowest valuation are endogenously matched into a bargaining process that is affected by subsequent opportunities to trade and by the prospect that immediately profitable trades might be usurped by competing traders. In particular, the risk of usurpation plays the role of the interest rate commonly used in studies of bargaining to reflect impatience to trade early. During the bargaining process, each party delays making a serious bid or offer (i.e., one with positive chances of acceptance) sufficiently to signal credibly his valuation, and after a serious offer the other party also delays sufficiently to signal credibly before accepting or making an offer that is surely accepted. However, this model fares poorly in explaining the data that subjects often do not trade in order of their valuations, and often trades are completed by extra-marginal traders (e.g., buyers with valuations less than the clearing price). A variety of other bargaining models and associated equilibria are available [Kennan and Wilson (1989)], nevertheless, and perhaps one of these could fit the data better.

Markets conducted by intermediaries such as specialists have also been studied. One strand of research focuses on how a specialist can cope with traders having superior ("inside") information. Glosten and Milgrom (1985) characterize the specialist's bid and ask prices that account for the effects of adverse selection. For a double auction in which the specialist trades for his own account to set the clearing price, Kyle (1985) derives the equilibrium between the specialist's pricing strategy and the insider's strategy of modulating his trades to avoid revealing too much information too early. The finance literature includes many subsequent studies.

8. Applications

Auctions in which a single seller offers one or several items for sale are common; Cassady (1967) describes many examples. He also notes that different types of auctions tend to be associated with particular kinds of commodities. For example, oral auctions, either English (ascending bids) or Dutch (descending offers), are favored for animal stock and perishable commodities, perhaps to ensure rapid consideration of many lots with variable quality attributes. Most auctions of art and antiques use the oral format, as do sales of property, used machinery, and other producers' durables. On the other hand, in the United States new issues of corporate bonds and stock are usually sold to investment bankers via sealed bids, as are rights for timber and minerals, including coal and oil. Land and buildings are often sold via sealed bids also. In many countries, large firms and government agencies use sealed-bid auctions to select vendors and to procure services, especially construction. Rozek (1989) describes the increasing use of auctions to select providers of electric power supplies. In practice most auctions use discriminating pricing rules, but exceptions include the Exxon Corporation's auctions of its bonds [Levinson (1987)] and the auctions of privately placed preferred stock conducted by Goldman Sachs & Company (1987) and others. Holt (1979, 1980), Brown (1984, 1986, 1987, 1989), McAfee and McMillan (1985, 1987c), Engelbrecht-Wiggans (1987b), Laffont and Tirole (1987), Nti (1987), and Lang and Rosenthal (1990), as well as the seven contributed chapters in Part V of Engelbrecht-Wiggans, Shubik and Stark (1983), are indicative of studies of procurement contracting that take account of incentive effects in contract design as well as strategic behavior in the auction process. Kahn et al. (1990) report on a simulation study of auctions for procurement of power supplies in the electricity industry. The design of optimal auction rules for efficient procurement contracting is characterized by Riordan and Sappington (1987).

The sealed-bid auctions studied most thoroughly are those conducted periodically since 1954 in the United States to sell exploration and development leases for tracts on public lands and offshore on the outer continental shelf (OCS). These leases are unusual for their value (often sold for tens, and occasionally hundreds, of millions of dollars) and the auctions are notable for the evident intensity of strategic behavior; indeed, the larger oil companies maintain large permanent staffs for the preparation of bids. The leases have an evident common-value component because the amount and value of oil and gas is essentially the same for all bidders [Capen, Clapp and Campbell (1971)]; consequently, firms have increasingly realized that bidding strategies must be carefully designed to avoid the effects of adverse selection. That is, each bidder must take account of the so-called "winner's curse": the one who most overestimates the value of a lease is the one most likely to win. Experimental and empirical evidence on the incidence of overbidding in common-value auctions is summarized in Section 9. Strategic models of bidding have long been used by the Department of Interior to examine public policies regarding leasing [Reece (1978, 1979), Wilson (1981), DeBrock and Smith (1983)] and by the Department of Defense regarding procurement [Engelbrecht-Wiggans, Shubik and Stark (1983, Part V)]. Accounts of the oil companies' use of strategic analysis are unusual because of the extreme secrecy they maintain.

Share auctions are rare in practice. In most countries, public issues of firms' stocks are sold as single blocks (often via auctions) to investment banking firms who then resell the stock to investors, perhaps because marketing is an important ingredient. This has also been the practice for bonds [Christenson (1961)], which for public utilities are invariably sold via sealed-bid, first-price auctions. A recurring share auction is conducted by the Paris Bourse for stocks of firms newly listed on the exchange; however, it rations shares to bidders and rejects very high bids, which apparently is necessary because the transactions price is chosen to be less than the clearing price to attract bidders [Jacquillat and McDonald (1974)]. In the 1970s the International Monetary Fund sold its excess gold supplies via share auctions; and in this period there were recurrent proposals to sell shares of large "unitized" offshore oil leases. In the 1980s, shares of several privatized national corporations were sold via procedures resembling share auctions.

The U.S. Treasury conducts weekly sales of bonds via a multi-unit auction that closely resembles a share auction. Cammack (1991) describes the institutional aspects of this market and the secondary resale market; she also provides evidence that the market is affected by dispersed private information among the bidders. Except for a few trials, the Treasury Bill auction has used discriminating pricing. Plott (1982) reports experimental evidence that discriminating and nondiscriminating pricing in share auctions yield about the same revenue to the seller. Bikhchandani and Huang (1989) provide a novel analysis of the Treasury auction, for both the discriminating and nondiscriminating pricing rules. Two important institutional features are that (1) the Treasury accepts "noncompetitive bids" filed at the average price of the accepted competitive bids, and (2) the competitive bidders (who are dealers purchasing in order to resell in a secondary market) have incentives to *signal* their private information in order to influence the subsequent price in the secondary market. Whereas the first feature adds a noise component to the primary auction and the secondary market, the second feature increases the symmetric bidding strategy by an extra term for which Bikhchandani and Huang obtain a closed-form experession. For discriminating pricing, a stronger property than affiliation, interpreted as complementarity of information, is required to establish existence. Moreover, revelation of information by the seller need not increase expected revenues. They also establish sufficient conditions for the symmetric equilibrium of the nondiscriminating auction to be preferred by the seller.

Static double auctions, relying on sealed bids and offers, are employed frequently in security markets. They are used to determine the opening price in stock exchanges and markets for precious metals. They are also used in periodic markets for trading privately placed preferred stock [Goldman Sachs & Co. (1987)]. Proposals to automate trading on the major exchanges for financial instruments have focused on using periodic (e.g., hourly) double auctions to accomplish market clearing. The Stockholm Exchange's automated procedures resemble an English auction, however.

In terms of trading volume, however, dynamic versions of double auctions are more widely used. In most organized exchanges, markets for storable commodities, industrial metals, and crude oil are conducted via open outcry of bids and asks from floor traders and brokers acting for clients. Summaries of experimental results on the extraordinary efficiency of these markets are reported by Plott (1982) and Smith (1982). Financial markets mostly rely on intermediaries such as specialists who maintain inventories and order books of bids and offers in order to sustain continual trading opportunities and price stability, but some markets for options and futures contracts use oral bid-ask auctions; cf. Brady (1988) and Miller (1988) for details of the organization of security markets in the United States.

Labor markets involve interesting variants of double auctions to match workers with positions in firms, students with openings at schools, etc. Called the "marriage problem", this version differs in that each buyer or seller offers an item with unique quality attributes valued differently by each party on the other side of the market. Much of the literature focuses on a procedure that provides one side of the market a dominant strategy. One dynamic procedure has buyers apply to sellers, who reject or tentatively accept each application, and this continues until each buyer is accepted or would prefer not to trade with any remaining seller. Excluding strategic behavior by the sellers (i.e., they accept their preferred applicants), this procedure provides a dominant strategy for the buyers, and yields for them their best allocation in the core of the associated cooperative game. Roth (1984a, 1984b, 1984c) and Roth and Sotomayor (1990) describe the history of the market in the United States that matches medical interns (sellers) with positions in hospitals (buyers), which has developed a similar procedure to cope with hospitals' incentives to act strategically. For a version in which prices as well as the matching are determined by the process, see Demange and Gale (1985). Game-theoretical analyses of Nash equilibria have not been fully developed.

9. Experimental and empirical evidence

Game-theoretic models of auctions make strong assumptions about the information and behavior of participants. Correspondingly, the predictions obtained from the models imply severe restrictions on the outcomes of auctions in practice and in experimental settings. Students of economic behavior relying on empirical data or experimental observations have therefore found auctions to be rich sources of evidence. In this section we review briefly several principal studies. The discussion divides between experimental studies in which the objective is to examine behavior conditional on controlled environments, and empirical studies in which much of the relevant data about the environment are inaccessible.

9.1. Experimental studies

The objective of experimental studies is to examine subjects' bidding behavior in settings in which the experimentor has complete information about the economic data (but not subjective aspects of subjects' preferences, such as risk aversion), and the procedural rules and informational conditions are controlled. Results from experiments allow comparisons of procedural rules and other features, and most importantly, they allow tests of the hypothesis that *all* participants simultaneously use equilibrium strategies. A large portion of the experimental evidence is summarized in three surveys by Smith (1982, 1987) and Plott (1982) that we distill even further. Roth's (1988) survey is an incisive methodological critique.

Smith's (1987) summary of over 1500 single-item auction experiments in which subjects have independent private valuations concludes that English (oral discriminating ascending) and second-price (static nondiscriminating) auctions achieve high efficiency and have about the same mean observed prices, whereas first-price (static discriminating) and Dutch (oral discriminating descending) auctions have appreciably lower efficiency and appreciably higher prices (first-price auctions have higher efficiency measures and higher prices than Dutch auctions). These differences are attributed to risk aversion, and with this proviso, judged to be consistent with Nash equilibrium; further qualified support is found in experiments with multi-item discriminating auctions (but not for nondiscriminating versions).[14] Smith's (1982) summary emphasizes further that heterogeneity of risk aversion among subjects is necessary to explain the data, which exhibit considerable dispersion; a critique and alternative view is suggested by Harrison (1989). Plott (1982, p. 1505) describes experiments with multi-unit static discriminating auctions and concludes that the results "provide support for Nash equilibrium bidding models when there are several (three or four) bidders", and that "after convergence [of subjects' behaviors] take place, [discriminating and nondiscriminating auctions] generate about the same revenue" as implied by theoretical results.

The extensive literature on double auctions, especially bid–ask markets (oral discriminating multi-item), is partially reviewed by Smith (1982, section 3B) and Plott (1982, section II); for more recent results see Friedman and Ostroy

---

[14]Kagel and Levin (1988) study the role of risk aversion in experiments using a different experimental design based on a "third-price" auction that provides a stronger test.

(1989). Although portions of this work involve a variety of particular institutional features, the main finding emphasized in all studies is that transaction prices usually converge rapidly (say, four repetitions), even with few participants (eight), to Walrasian clearing prices.[15] In addition, the efficiency of trading is very high.[16] This striking finding is quite robust, but its conformity to the predictions of game theory is moot owing to the dearth of theoretical results.

Single-item static discriminating auctions with common values have been studied experimentally by Kagel and Levin (1986). Their main conclusions are that "in auctions involving a limited number of bidders (3–4 bidders), average profits are consistently positive and closer to the Nash equilibrium bidding outcome than to the winner's curse hypothesis"; in particular, profits average about two-thirds of the amount predicted by the equilibrium strategies for risk-neutral bidders. However, "bids are found to be an increasing function of the number of rivals faced, in clear violation of risk-neutral Nash equilibrium bidding theory", contributing to a "reemergence of the winner's curse, with bankruptcies and negative profits, in auctions with large numbers (6–7) of bidders".[17] They further observe that providing public information about the common value increased the seller's revenue in the former case (few bidders) as predicted, but decreased it in the latter. All of these conclusions refer to auctions with experienced bidders, and in particular they emphasize that learning is evident and partially successful in repeated auctions with few bidders, but not in auctions with many bidders; moreover, the learning is specific, in that it is not entirely carried over to new situations. They conclude, therefore, that the probable explanation of the results is persistent errors in judgment, manifested in an inability to fully comprehend the adverse selection that afflicts bidders in common value auctions.

9.2. Empirical studies

Empirical studies must contend with less complete data and few controls on the auction environment are possible. On the other hand, they have the advantage that the data pertain to practical situations in which the stakes are often large

[15]This is also true when rational expectations aspects are involved, provided a sufficiently rich set of securities are traded, as shown by Plott and Sunder (1988).

[16]Asymmetric versions, say with a single seller, often do not achieve monopoly outcomes as might be predicted, but rather approximate the more nearly Walrasian outcome predicted by the Coase property; cf. Gul, Sonnenschein and Wilson (1986).

[17]These results are replicated for subjects who were professional managers of construction firms in Dyer, Kagel and Levin (1989a), and for second-price auctions, by Kagel, Levin and Harstad (1988).

and the participants are skilled and experienced. In the case of auctions of offshore oil leases, millions of dollars are at stake and the firms bidding for leases have large staffs, data bases, and computer facilities devoted to the task of preparing bids. The procedural rules are simple, since the lease is awarded to the high bidder at the price offered by sealed tender, but the information structure is not, and in important cases is very asymmetric. In view of the extensive review of empirical studies in McAfee and McMillan (1987a) we review only studies omitted from their survey.

We mentioned above the experimental finding that even experienced bidders tend to overbid in common-value auctions with six or more bidders. Hendricks, Porter and Boudreau (1987) examine a similar situation in auctions of leases for wildcat tracts (i.e., in unexplored areas) for the years 1954–69. One of their main conclusions is that winning bidders' average realized net profits were negative for auctions with more than six bidders. The authors point out that the data can be explained by nonoptimal bidding strategies that account inadequately for adverse selection in valuation estimation, or equally, by adverse selection in estimating the number of bidders. That is, most tracts receive fewer than six bids (the average was 3.5 in the sample) and supposing that firms expect this, profits will be less on those tracts receiving more bids. Overall, winning bidders captured about a quarter of the value of the tracts, which is consistent with a supposition that active bidders expected three or four bids to be submitted. Overall, the authors conclude that "the data are consistent with both the assumptions and predictions of the [common value] model", allowing for bidders' uncertainty about the number of active bidders in each auction.[18]

These data are from the period before the publication of Capen, Clapp and Campbell's (1971) influential article suggesting that adverse selection (the winner's curse) might account for the low returns realized by winning bidders in such auctions. Helfat (1987), using ex ante data on expected returns and allowing risk aversion in a portfolio model of firm's decisions, finds that returns remained low until the OPEC oil embargo of 1974, but rose substantially thereafter due to lower average bids. Whether this effect is associated with better bidding strategies or is an incidental consequence of the altered structure of the oil market is unclear.

Auctions of "drainage" tracts adjacent to explored tracts typically involve substantial asymmetries of information, since firms who have explored neighboring tracts have superior information. Hendricks and Porter (1988) derive seven implications of the equilibrium bidding strategies for the case that one bidder has superior information, as in the common-value model presented in Section 4. They test these predictions using data on 114 drainage tracts leased

---

[18]This conclusion is stronger than in previous studies, where mixed results were often reported; for example, see Gilley, Karels and Leone (1986) and the references therein.

in the period 1959–69. Their conclusion, subject to one proviso, is that essentially all seven of these predictions are confirmed by the data. The proviso is that multiple neighbors act as a single cartel to submit a single serious bid; indeed, at the time cartels were not prohibited and firms routinely cooperated in other activities, and Hendricks and Porter provide substantial evidence of coordinated bidding by neighboring firms. The sharpest test is the prediction that non-neighbor firms' profits should be zero on average, compounded from positive profits when a neighbor bid and lost and negative profits when neighbors chose not to bid: this prediction is confirmed, and in particular average profits differed from zero by only one-quarter standard deviation of the sample mean. Overall, the authors "find that the data strongly support the hypotheses that . . . firms bid strategically in accordance with the Bayesian–Nash equilibrium model". Hendricks, Porter and Spady (1988) obtain similar results for data from the period 1970–79, using however a somewhat richer model that assumes (realistically) that the government uses a random reservation price. The conclusion is again that "the hypothesis that neighbor and non-neighbor firms bid strategically in accordance with the theory of auctions with asymmetric information is strongly supported by the data". This analysis is extended further by Hendricks, Porter and Wilson (1992) to take account of the informational content of the government's reservation price: assuming affiliation, they show that the distribution of the informed (neighbor) bidder's bid stochastically dominates the distribution of the highest bid among those submitted by uninformed bidders, but conditional on the bid being high enough these two distributions are identical. The data from drainage tracts favor rejection of a null hypothesis that the two marginal bid distributions are identical, and acceptance of the null hypothesis that the two conditional distributions are identical. These positive conclusions are reversed for less risky "development" tracts, providing further support for the supposition that informational differences account for the results.

Thiel (1988) uses data from 130 auctions of highway construction contracts by 28 state governments in the United States to test the common-value model with symmetric information, assuming normal probability distributions. He supposes that the state engineer's estimate of the cost of fulfilling the contract, revealed after the auction, is an unbiased estimator of the true (but unobserved) cost. However, his main conclusion that "the model fits the data reasonably well" is moot since the equilibrium bidding strategies are misspecified, as shown by Levin and Smith (1991).

10. Comparisons of auction rules

Several studies have compared the distributional effects of various types of auctions. Here we provide a sample of results that indicate major themes.

In Section 4 we mentioned single-item auctions in which bidders are risk-neutral and have independently and identically distributed private valuations, and that the seller and the bidders are indifferent among the standard procedural rules, such as first-price or second-price and oral or sealed bids. Matthews (1987) shows that this feature persists for bidders in the case that all bidders have identical exponential utility functions, so that each has the same constant Arrow–Pratt measure of risk aversion, even if the number of bidders is uncertain – although the expected price is higher in a first-price auction if the number is not revealed, as shown by McAfee and McMillan (1987b) and studied experimentally by Dyer, Kagel and Levin (1989b). However, if the risk-aversion measure is decreasing, then the bidders prefer a second-price (SP) auction to a first-price auction with the number of bidders revealed (FPR) to a first price auction with the number not revealed (FPU), in that order; moreover, the seller's preference is the reverse ordering if the seller is risk-neutral.[19] However, if their valuations are affiliated, then the bidders' preferences are biased away from the second-price auction; e.g., if utilities are exponential then they prefer FPR to FPU to SP, whereas the seller prefers FPU to FPR to SP.

Milgrom and Weber (1982a), assuming that the number of bidders is known and that bidders are symmetric and their valuations are affiliated, establish that an English oral ascending auction has a higher expected price than a second-price auction, and in turn if bidders are risk-neutral, the latter is higher than the expected price in a first-price auction. In all three auctions, moreover, a risk-neutral seller prefers to (acquire and) disclose any private information it can if doing so is costless.[20] The comparisons are applications of the "linkage principle": since bidders' profits are returns to their private information, procedures that reveal more of their private information during the auction (such as an English auction), or that dilute their information (such as revelation of information known to the seller), tend to increase the expected price

---

[19]These results assume that the number of active bidders is uninformative about their valuations and the seller's ask price is not binding. Holt (1980), Harris and Raviv (1981b), Maskin and Riley (1984a), and Riley (1989) show that the seller prefers a first-price sealed-bid auction to an English auction if bidders are risk averse, a preference that is strengthened if the seller is also risk averse. Maskin and Riley also investigate a plethora of other schemes the seller can use to exploit the bidders' aversion to risk.

[20]These results are valid also for an ask price that is the same in all three auctions, and the seller's preference for an English auction over a second-price auction carries over to the case that bidders have exponential utilities. Riley (1989) shows further for the comparison between first- and second-price auctions, that of two rules using a price that is a weighted combination of the first and second bids the one giving greater weight to the second has the higher expected price; and indeed, the seller prefers rules that weight all bids as versus only the high bid. However, Maskin and Riley (1984b) show for the case of independent private valuations that wealth effects can make an English auction an inferior choice for the seller. Riley (1989) uses elementary methods to examine some of these features.

obtained. Milgrom and Weber also show for each of these auctions that between two such auctions that differ only in the ask price and an entry fee but attract the same set of bidders, the one with the lower ask price and higher entry fee obtains the higher expected price for the seller.

Assuming two bidders, Hausch (1987) shows that these results extend to common-value auctions in which the bidders' sampling distributions differ but their conditional distributions (of one bidder's sample given the other's observation) agree; moreover, an equilibrium with symmetric strategies exist. However, if the conditional distributions disagree, then the strategies may be asymmetric and the first-price auction can have a higher expected sale price.

For common-value auctions, Harstad (1990) extends Milgrom and Weber's results to contexts in which the number of bidders is determined endogenously by the bidders' participation cost (e.g., the cost of acquiring sample information) and the seller's selection of the auction type (including the information revealed, the entry fee, and the ask price), provided the induced probability of selling the item is sufficiently large. Again with this proviso, an auction format for which participation costs are recovered with fewer participants is preferable for the seller. Similarly, a reduction in participation cost is preferable if the number of participants responds inelastically. Both of these observations follow from the general principle that with endogenous participation, and conditional on a sale, the seller's expected revenue is the expected common value less the aggregate of participation costs. Thus, in the context the role of the linkage principle is played by the induced reduction in participation.

Applications to bidding for oil leases have also noted that royalties payable on the actual amount extracted can alleviate bidders' risk aversion and therefore increase the seller's expected revenue. Riley (1985) shows further, subject to regularity conditions, that even without risk aversion, if bidders' valuations are affiliated, then contingent payment schemes conditioned on ex post observations are optimal for the seller. Nevertheless, high royalties can diminish the winning bidder's incentive to pursue an efficient plan of oil recovery, and therefore the net effects are mixed.

The seller's ask price is affected by the power to commit in advance. To illustrate, consider only the case that the bidders have independent private valuations. In a static auction with a fixed number of bidders, the seller usually prefers to commit to an ask price above cost. But in the Dutch auction dynamic version, having failed to receive a bid the seller prefers to continue lowering the price so long as it remains above cost, which increases the efficiency of trade. Engelbrecht-Wiggans (1988) studies an example modeled after an historical incident: assume that the number of bidders increases until the expected gain ex ante from attending an auction is reduced to the expense incurred. In this case, the seller prefers ex ante to commit to an ask price that is lower than the expectation of the ask prices conditional on the number of attendees, chosen to take account of the option of reoffering the item in a fresh auction with a new sample of bidders. The reason is essentially that a low ask price ex ante attracts more bidders and raises the average sale price. Engelbrecht-Wiggans cites an incident in which taxation imposed on goods offered for sale, rather than those actually sold, allegedly increased tax revenues as well as benefited sellers – the reason being that taxation of offered goods provides a disincentive to reoffer an item, and therefore lowers conditional ask prices to levels closer to the ex ante optimum.

Auctions in which the seller has superior information about an item of common value to the bidders have novel features. Vincent (1990) studies a repeated auction in which only the seller knows the benefit each of two identical uniformed bidders would obtain from acquiring the item, and the seller's valuation is a fixed amount less than the bidders' valuation; also, all parties discount delayed payoffs. The sequential equilibrium in this case involves "screening" by the bidders: they offer an increasing sequence of bids until the seller accepts. The bidders' expected profits are zero of course, and the seller captures the difference, but the outcome is inefficient because of the delay. In Vincent's example, the seller prefers to exclude repetition of the auction, or to deal with only one bidder rather than two; in both cases the seller's motive is to reduce screening and the resulting delay.

11. Optimal auctions

In parallel with the analyses of specific auction forms, a large literature has addressed the design of auction rules that are efficient or optimal for the seller. In this section we provide a brief synopsis of the main results. If not mentioned, the default assumption is that bidders have independently and identically distributed valuations, according to a distribution function having an increasing hazard rate.

The basic results are due to Myerson (1981) and Myerson and Satterthwaite (1983), although here we follow the exposition in Wilson (1985a, 1985b).[21] An alternative approach by Bulow and Roberts (1989) shows that the methodology

---

[21] Our exposition excludes the most general case in Myerson by assuming, in effect, a monotone hazard rate. Other authors developing this methodology are Harris and Raviv (1981a), who derive an optimal priority pricing scheme similar to a Dutch auction; Harris and Raviv (1981b), who obtain specializations of Myerson's results; Harris and Townsend (1981), who study the general properties of revelation mechanisms; and Moore (1984). Extensions to the case that a seller offers a quantity of a (divisible) good and each buyer can select any amount to purchase are developed by Maskin and Riley (1986); in this case the seller's optimal mechanism entails a nonlinear price schedule. Applications to the design of auctions of procurement or franchise contracts are developed by Riordan and Sappington (1987).

is isomorphic to the theory of a monopolist offering prices that discriminate among multiple markets.

As in the method of distributional strategies in Section 5, assume that each participant i's valuation (for trade of a single indivisible item) depends on his privately known type t_i via a decreasing function $u_i(t_i)$ if i is a buyer and an increasing function $v_i(t_i)$ if i is a seller, where the types are independently and uniformly distributed on the unit interval. Assume further that $\bar{u}_i(t) = tu_i(t)$ and $\bar{v}_i(t) = tv_i(t)$ are concave and convex functions, respectively, as implied by the increasing hazard rate property. Party i's expected payoff from an equilibrium of a particular mechanism is denoted $V_i(t_i)$, and we consider a welfare measure

$$W = \mathcal{E}\left\{\sum_i \alpha_i(t_i)V_i(t_i)\right\}, \tag{30}$$

depending on non-negative welfare weights $\alpha_i(t_i)$ that may depend on the party's type, as in the case of *interim* incentive efficiency [Holmström and Myerson (1983)]. The "revelation principle" takes advantage of the property of an equilibrium that each party must prefer to act according to his true type to conclude, say for a buyer, that

$$V_i(t_i) = V_i(1) + \int_{t_i}^{1} P_i(\hat{t})\, d[u_i(0) - u_i(\hat{t})], \tag{31}$$

where $P_i(\hat{t})$ is the probability that i trades if the type is $\hat{t}$. Using this property, along with the feasibility condition that net trades of money and goods among the parties must sum to zero, enables one to rewrite the welfare measure as

$$W = \mathcal{E}\left\{\sum_{i \in B} \phi_i(t_i) - \sum_{i \in S} \psi_i(t_i)\right\} - \sum_i [1 - \bar{\alpha}_i(1)]V_i(1), \tag{32}$$

where

$$\bar{\alpha}_i(t) = \mathcal{E}\{\alpha_i(t_i) \mid t_i \leqslant t\}, \tag{33}$$

$$\phi_i(t) = u_i(t) + [1 - \bar{\alpha}_i(t)]tu_i'(t) \quad \text{and} \quad \psi_i(t) = v_i(t) + [1 - \bar{\alpha}_i(t)]tv_i'(t). \tag{34}$$

The functions ϕ_i and ψ_i for the buyers and sellers are called "virtual valuations" by Myerson. In the expression for the welfare measure, it is important to note that the expectation is taken with respect to both the traders' types and the sets B and S of buyers and sellers who trade conditional on the entire vector (t_i) of types (feasibility requires that these two sets have equal cardinality). This representation shows that the design problem is summarized by choosing welfare weights, and for specified weights, choosing the rules of the mechanism to maximize the welfare measure via the induced sets B and S of successful traders. The requirement of "individual rationality", in the form that each $V_i(t) \geq 0$, is satisfied by setting each $V_i(1) = 0$. Furthermore, the mechanism maximizes the welfare measure if there exists an increasing function f such that the equilibrium induces trading sets B and S that maximize

$$\sum_{i \in B} f(\phi_i(t_i)) - \sum_{i \in S} f(\psi_i(t_i)) \tag{35}$$

subject to $|B| = |S|$ for each realization (t_i). Gresik and Satterthwaite (1989) show that rules for monetary payments exist that actually realize the equilibrium with these rules for trades of goods.

To take a special case, suppose there is a single seller with a commonly known valuation for a single item. The seller's optimal mechanism is obtained by setting the buyers' welfare weights to zero; consequently, it should sell the item to the buyer with the largest among the virtual valuations

$$\phi_i(t_i) = u_i(t_i) + t_i u_i'(t_i) = \bar{u}_i'(t_i) , \tag{36}$$

provided it exceeds the seller's valuation. In the symmetric case $u_i \equiv u$, for example, this rule merely specifies that the buyer with the highest valuation should obtain the item if it is sold (since the actual and virtual valuations have the same ordering), and that the seller should use an optimal ask price. Thus, this mechanism conforms exactly to the usual auction formats. Bulow and Roberts (1989) observe that generally the virtual valuations are marginal revenues in the seller's calculations.

If the welfare weights are independent of the types, then the auction design is ex ante efficient, and further, if they are all the same, then the design maximizes the expected total surplus. An elaborate example of relevance to regulatory policy is worked out in detail by Riordan and Sappington (1987).

For the case of risk-averse buyers, Matthews (1983, 1984b) and Maskin and Riley (1984b) characterize the seller's design problem as an optimal control problem. They find that it is optimal for the seller to charge entry fees that decline with the magnitude of the bid submitted (negative for large bids), and to reject the high bid with positive probability (though small if the bid is large). The essential idea is to impose risk on the buyers to motivate higher bids, but a buyer with a very high valuation is nearly perfectly insured (marginal utility differs little between winning and losing). With extremely risk-averse buyers, the seller can attain nearly perfect price discrimination. Analogous results obtain in the case of risk-neutral bidders who have correlated private information, for example about an item of common value. Cremér and McLean (1985), and McAffee, McMillan and Reny (1989), McAfee and Reny (1992) provide conditions under which the seller can in principle extract nearly all of the potential profit. In all cases, however, full exploitation of these features requires that the rules of the auction depend crucially on the probability distribution of buyers' information.

Border (1991) studies the "reduced form" of an auction, interpreted as a function that assigns a probability of winning to each possible type of each bidder. He characterizes the set of all such functions that are implementable as auctions, and shows that this set can be represented as a convex polyhedron with extreme points that are associated with assignments that simply order the types. That is, the bidder whose type is ranked highest wins. This geometric characterization enables the implementation as an auction to be constructed from the solution to a linear programming problem.

Myerson and Satterthwaite (1983) examine the case of a single seller and a single buyer in which the welfare weights are identical constants, corresponding to ex ante incentive efficiency of the mechanism, as in Holmström and Myerson (1983). They note in the special case of valuations distributed uniformly on the same interval that an efficient mechanism is the static double auction in which the price is the average of the bid and offer submitted, assuming that the parties follow the linear equilibrium strategies identified by Chatterjee and Samuelson (1983) – although there are many nonlinear equilibria, only the linear one is efficient. Gresik and Satterthwaite (1989) construct ex ante efficient mechanisms for the general case of several sellers and several buyers with differing independent probability distributions of their valuations; cf. Gresik (1991c) for the case with correlated distributions. Their main result is that for ex ante efficient trading mechanisms the ex post inefficiency, as measured by the maximal difference between the valuations of a buyer and a seller who do not trade, is of order $\sqrt{\ln(M)}/M$ in terms of the minimum M of the numbers of buyers and sellers [as noted in Section 7 on double auctions, this bound is improved to $1/M$ by Satterthwaite and Williams (1989a, 1989b, 1989c)]. Generally, however, the rules of these efficient mechanisms depend on the distributions and therefore they do not conform to the usual forms of auctions; e.g., the payment rules they use (although they are not the only possible ones) often mandate payments by buyers who do not trade. Indeed, even for two sellers and two buyers with uniform distributions, the ordinary double auction that uses the price at the midpoint of the interval of clearing prices is inefficient. This difficulty motivates much of the work reported in Section 7 on double auctions, particularly those that demonstrate the efficiency of double auctions with many participants.

An alternative construction by Gresik (1991b) obtains stronger positive results. He strengthens the *interim* individual rationality constraint $V_i(t_i) \geq 0$ used above, to the ex post individual rationality constraint that each trader must obtain a non-negative net profit in every contingency. In particular, participants who do not trade do not make or receive payments, and those who do trade make or receive payments bounded by their valuations. His main result establishes that there exists an open set of trading problems (in the space of probability distributions) for which the ex ante efficient mechanism can be implemented with payment rules that satisfy these stronger individual rationality constraints. This set is characterized by problems for which certain functions have unique roots, which he interprets as a "single crossing property" of the sort assumed in many studies of incentive problems. The net result is the demonstration that mechanisms that enforce ex post rationality, and therefore conform more closely to standard auctions, but allow contingent selections of trading prices from the interval of clearing prices, are ex ante efficient in a nontrivial class of problems.

Similar methods can be applied to other contexts akin to auctions. We mention one among several examples in Kennan and Wilson (1992). Suppose that in a legal dispute a trial will cost each party c and yield a judgment $v = p - d$ paid to the plaintiff by the defendant, where initially the plaintiff knows p and the defendant knows d and they both know these have independent distribution functions F and G with densities f and g. Thus the gain from a pretrial settlement is $2c$. The incentive-compatible mechanism that maximizes the sum of the parties' ex ante expected payoffs can be derived using the methods above. One finds that they settle if

$$2c \geq \alpha \left[\frac{F(p)}{f(p)} + \frac{G(d)}{g(d)} \right], \tag{37}$$

provided the right-hand side is increasing, where α is a number chosen to ensure feasibility. In the case of uniform distributions, for example, if $c \leq \frac{1}{3}$ then $\alpha = \frac{2}{3}$ and they settle if $p + d \leq 3c$. Analogous to Myerson and Satterthwaite's example above, this optimal mechanism is implemented by a procedure in which the plaintiff asks P, the defendent offers D, and if $P < D$ then they settle on a payment $\frac{1}{2}[P + D]$, and otherwise go to trial. For this game the linear equilibrium has strategies $P = -2c + \frac{4}{3}p$ and $D = 2c + \frac{4}{3}d$. This procedure can also be used in a common-value model, although its optimality properties are unknown. Suppose (v, p, d) has a normal distribution such that, conditional on v, p and d are independent and identically distributed with mean v and variance $\frac{1}{2}s^2$, and consider the limiting case as the variance of the marginal distribution of v increases; at the limit, the linear equilibrium strategies are $P = \frac{1}{2}\Delta + p$ and $D = -\frac{1}{2}\Delta + d$, where $[2c/s]h(\Delta/s) = 1$ and $h = f/[1 - F]$ is the hazard function for the standard normal distribution function. As in Section 4, similar results are obtained with a lognormal distribution and trial costs that are proportional to the judgment.

12. Research frontiers

Strategic analyses of auctions have developed rapidly, but significant gaps remain. The theory relies mainly on static formulations that invoke strong assumptions, such as symmetries among bidders, common knowledge of probability distributions, absence of risk aversion, etc. The fundamental assumption that an equilibrium predicts behaviors has rarely been relaxed. These assumptions facilitate theoretical work but they hamper empirical and experimental studies, since they are never precisely true in practice and little has been done to establish the robustness of the predictions. Moreover, they thwart applications of the theory to practical affairs. Indeed, the paucity of reported applications and the occasional rejections of the theory [e.g., Levinson (1987)] by skilled practitioners indicate that more can be done to make it a useful tool.

Dynamic procedures, such as bid–ask markets, have received little attention although they have paramount importance in practice. The theory of efficient mechanisms remains a weak explanation for the prevalence of auction rules that are invariant to the characteristics of participants. Scant progress has been made in building theories with generality comparable to the Walrasian model of general equilibrium, even though the enigma of price formation in the Walrasian model is a prime motivation for studies of auctions.

Nevertheless, the methods of game theory have contributed substantially to the strategic analysis of auctions, and the main empirical studies [e.g., Hendricks and Porter (1988)] provide some support. This accomplishment stems partly from precise formulations and exact criteria for a solution, but most importantly it derives from explicit recognition of the effects of private information on strategic behavior. The emphasis on private information has brought game theory closer to practical affairs, and the resulting development of new techniques has enriched the methodology. One can hope that the emerging power of game theory to characterize market behavior will enable a general reformulation of economic models to include strategic behavior affected by private information.

Bibliography

Anton, James J. and Dennis A. Yao (1988) 'Coordination in split-award auctions', State University of New York at Stony Brook, D.P. #307; *Quarterly Journal of Economics*, to appear 1992.

Balder, Erik J. (1988) 'Generalized equilibrium results for games with incomplete information', *Mathematics of Operations Research*, 13: 265–276.

Bernheim, B. Douglas and Michael D. Whinston (1986) 'Menu auctions, resource allocation, and economic influence', *Quarterly Journal of Economics*, 101: 1–32.

Bikhchandani, Sushil (1988) 'Reputation in repeated second-price auctions', *Journal of Economic Theory*, 46: 97–119. Also appears in 'Market games with few traders', Ph.D. dissertation, Stanford University, 1986.

Bikhchandani, Sushil and Chi-Fu Huang (1989) 'Auctions with resale markets: An exploratory model of Treasury Bill auctions', *Review of Financial Sudies*, 2: 311–339.

Bikhchandani, Sushil and John G. Riley (1991) 'Equilibria in open common value auctions', *Journal of Economic Theory*, 53: 101–130.

Border, Kim (1991) 'Implementation of reduced form auctions: A geometric approach', *Econometrica*, 59: 1175–1187.

Brady, Nicolas (Chairman) (1988) *Report of the Presidential Task Force on market mechanisms.* Washington: U.S. Government Printing Office.

Brannman, L., J.D. Klein and L.W. Weiss (1987) 'The price effects of increased competition in auction markets', *Review of Economics and Statistics*, 69: 24–32.

Brown, Pamela C. (1984) 'Design of optimal procurement contracts that allow risk sharing', Ph.D. dissertation. Stanford CA: Stanford University.

Brown, Pamela C. (1986) 'Competitive procurement contracting when risk-preferences are uncertain', *Mathematical Modelling*, 7: 285–299.

Brown, Pamela C. (1987) 'Optimal risk-sharing when risk preferences are uncertain', *Journal of the Operational Research Society*, 38: 17–29.

Brown, Pamela C. (1989) 'A risk-averse buyer's contract design', *European Economic Review*, 33: 1527–1544.

Bulow, Jeremy and D. John Roberts (1989) 'The simple economics of optimal auctions', *Journal of Political Economy*, 97: 1060–1090.

Bulow, Jeremy and Paul Klemper (1991) 'Rational frenzies and crashes', Research Paper 1150, Stanford Business School.

Cammack, Elizabeth B. (1991) 'Evidence on bidding strategies and the information in Treasury bill auctions', *Journal of Political Economy*, 99: 100–130.

Capen, E.C., R.V. Clapp and W.M. Campbell (1971) 'Competitive bidding in high-risk situations', *Journal of Petroleum Technology*, 23: 641–653.

Cassady, Ralph, Jr. (1967) *Auctions and auctioneering.* Berkeley: University of California Press.

Chatterjee, Kalyan and Williams Samuelson (1983) 'Bargaining under incomplete information', *Operations Research*, 31: 835–851.

Christenson, Charles J. (1961) *Competitive bidding for corporate securities.* Boston: Harvard Business School Division of Research.

Cramton, Peter C. (1984) 'The role of time and information in bargaining', Ph.D. dissertation. Stanford CA: Stanford University.

Cramton, Peter C. (1990) 'Strategic delay in bargaining with two-sided uncertainty', *Review of Economic Studies*, 59: to appear.

Cremér, Jacques and Richard McLean (1985) 'Optimal selling strategies under uncertainty for a discriminating monopolist when demands are interdependent', *Econometrica*, 53: 345–361.

Cremér, Jacques and Richard McLean (1988), 'Full extraction of the surplus in Bayesian and dominant strategy auctions', *Econometrica*, 56: 1247–1257.

DasGupta, Partha and Eric Maskin (1986) 'The existence of equilibria in discontinuous economic games', *Review of Economic Studies*, 53: 1–42.

DeBrock, L. and J. Smith (1983) 'Joint bidding, information pooling, and the performance of petroleum lease auctions', *Bell Journal of Economics*, 14: 395–404.

Demange, Gabriel and David Gale (1983) 'The strategy structure of two-sided matching markets', *Econometrica*, 53: 873–888.

Dubey, Pradeep (1982) 'Price-quantity strategic market games', *Econometrica*, 50: 111–126.

Dyer, Douglas, John H. Kagel and Dan Levin (1989a) 'A comparison of naive and experienced bidders in common value offer auctions: A laboratory analysis', *Economic Journal*, 99: 108–115.

Dyer, Douglas, John H. Kagel and Dan Levin (1989b) 'Resolving uncertainty about the number of bidders in independent private-value auctions: An experimental analysis', *Rand Journal of Economics*, 20: 268–279.

Easley, David and John Ledyard (1982) 'A theory of price formation and exchange in oral auctions', Northwestern University, D.P. #461. To appear in: D. Friedman et al., eds., *The double action market*. Reading, MA: Addison-Wesley.

Engelbrecht-Wiggans, Richard (1980) 'Auctions and bidding models: A survey', *Management Science*, 26: 119–142.

Engelbrecht-Wiggans, Richard (1987a) 'Optimal reservation prices in auctions', *Management Science*, 33: 763–770.

Engelbrecht-Wiggans, Richard (1987b) 'On optimal competitive contracting', *Management Science*, 33: 1481–1488.

Engelbrecht-Wiggans, Richard (1988) 'An example of auction design: A theoretical basis for 19th century modifications to the port of New York imported goods market', University of Illinois, D.P. #1486.

Engelbrecht-Wiggans, Richard (1990) 'Optimal auctions revisited', University of Illinois, mimeo.

Engelbrecht-Wiggans, Richard and Charles Kahn (1991) 'Protecting the winner: second-price vs oral auctions', *Economics Letters*, 35: 243–248.

Engelbrecht-Wiggans, Richard and Robert J. Weber (1979) 'An example of a multi-object auction game', *Management Science*, 25: 1272–1277.

Engelbrecht-Wiggans, Richard, Paul Milgrom and Robert Weber (1983) 'Competitive bidding with proprietary information', *Journal of Mathematical Economics*, 11: 161–169.

Engelbrecht-Wiggans, Richard, Martin Shubik and Robert Stark, eds. (1983) *Auctions, bidding, and contracting: Uses and theory*. New York: New York University Press.

Forsythe, Robert, R. Mark Isaac and Thomas R. Palfrey (1989) 'Theories and tests of blind bidding in sealed bid auctions', *Rand Journal of Economics*, 20: 214–238.

Friedman, Daniel (1984) 'On the efficiency of double auction markets', *American Economic Review*, 74: 60–72.

Friedman, Daniel and Joseph Ostroy (1989) 'Competition in auction markets: An experimental and theoretical investigation', D.P. #202, University of California, Santa Cruz, CA.

Fudenberg, Drew and Jean Tirole (1986) *Dynamic models of oligopoly*. London: Harwood Academic Publishers.

Ghemawat, Pankaj and Barry Nalebuff (1985) 'Exit', *RAND Journal of Economics*, 16: 184–194.

Gilley, Otis, Gordon Karels and Robert Leone (1986) 'Uncertainty, experience, and the "Winner's Curse" in OCS lease bidding', *Management Science*, 32: 673–682.

Glosten, Lawrence and Paul R. Milgrom (1985) 'Bid, ask, and transaction prices in a specialist market with heterogeneously informed traders', *Journal of Financial Economics*, 14: 71–100.

Goldman, Sachs & Co. (1987) 'Auction preferred stock', New York.

Graham, Daniel and Robert Marshall (1987), 'Collusive bidder behavior at single-object second-price and English auctions', *Journal of Political Economy*, 95: 1217–1239.

Gresik, Thomas A. (1991a) 'The efficiency of linear equilibria of sealed-bid double auctions', *Journal of Economic Theory*, 56: to appear.

Gresik, Thomas A. (1991b) 'Ex Ante efficient, Ex Post individually rational trade', *Journal of Economic Theory*, 56: to appear.

Gresik, Thomas A. (1991c) 'Efficienct bilateral trade with statistically dependent beliefs', *Journal of Economic Theory*, 56: to appear.

Gresik, Thomas A. (1991d) 'Ex Ante incentive efficient trading mechanisms without the private valuation restriction', *Journal of Economic Theory*, 55: 41–63.

Gresik, Thomas A. and Mark A. Satterthwaite (1989) 'The rate at which a simple market becomes efficient as the number of traders increases: An asymptotic result for optimal trading mechanisms', *Journal of Economic Theory*, 48: 304–332.

Griesmer, J., R. Levitan and Martin Shubik (1967) 'Toward a study of bidding processes. Part IV: Games with unknown costs', *Naval Research Logistics Quarterly*, 14: 415–433.

Gul, Faruk, Hugo Sonnenschein and Robert Wilson (1986) 'Foundations of dynamic monopoly and the Coase conjecture', *Journal of Economic Theory*, 39: 155–190.

Harris, Milton and Artur Raviv (1981a) 'A theory of monopoly pricing schemes with demand uncertainty', *American Economic Review*, 71: 347–365.

Harris, Milton and Artur Raviv (1981b) 'Allocation mechanisms and the design of auctions', *Econometrica*, 49: 1477–1499.

Harris, Milton and Robert M. Townsend (1981) 'Resource allocation under asymmetric information', *Econometrica*, 49: 33–64.

Harrison, Glenn W. (1989) 'Theory and misbehavior in first-price auctions', *American Economic Review*, 79: 749–762.

Harsanyi, John C. (1967–68) 'Games with incomplete information played by Bayesian players', *Management Science*, 14: 159–182, 320–234, 486–502.

Harsanyi, John C. (1973) 'Games with randomly disturbed payoffs: A new rational for mixed strategy equilibrium points', *International Journal of Game Theory*, 2: 1–23.

Harstad, Ronald (1990) 'Alternative common-value auction procedures: revenue comparisons with free entry', *Journal of Political Economy*, 98: 421–429.

Harstad, Ronald (1991) 'Auctions with endogenous bidder participation', Virginia Commonwealth University, mimeo.

Harstad, Ronald and Dan Levin (1985) 'A class of dominance solvable common-value auctions', *Review of Economic Studies*, 52: 525–528.

Harstad, Ronald, John H. Kagel and Dan Levin (1990) 'Equilibrium bid functions for auctions with an uncertain number of bidders', *Economics Letters*, 33: 35–40.

Hausch, Donald B. (1987) 'An asymmetric common value auction model', *Rand Journal of Economics*, 18: 611–621.

Hausch, Donald B. and Lode Li (1990) 'A common value auction model with endogenous entry and information acquisition', University of Wisconsin, mimeo.

Helfat, Constance (1987) 'U.S. Offshore Oil Leasing', Northwestern University, mimeo.

Hendricks, Kenneth and Robert H. Porter (1988) 'An empirical study of an auction with asymmetric information', *American Economic Review*, 78: 865–883.

Hendricks, Kenneth and Robert H. Porter (1989) 'Collusion in auctions', *Annales d'Economie et de Statistique*, to appear.

Hendricks, Kenneth, Robert H. Porter and Bryan Boudreau (1987) 'Information, returns, and bidding behavior in OCS auctions: 1954–1969', *Journal of Industrial Economics*, 35: 517–542.

Hendricks, Kenneth, Robert H. Porter and Richard H. Spady (1988) 'Random reservation prices and bidding behavior in OCS drainage auctions', Northwestern University, D.P. #807.

Hendricks, Kenneth, Robert H. Porter and Charles A. Wilson (1992) 'First price auctions with an informed bidder and a random reservation price', *American Economic Review*, 82: to appear.

Holmström, Bengt R. and Roger B. Myerson (1983) 'Efficient and durable decision rules with incomplete information', *Econometrica*, 51: 1799–1819.

Holt, Charles A. (1979) 'Uncertainty and the bidding for incentive contracts', *American Economic Review*, 69: 697–705.

Holt, Charles A. (1980) 'Competitive bidding for contracts under alternative auction procedures', *Journal of Political Economy*, 88: 433–445.

Holt, Charles A. and Roger Sherman (1982) 'Waiting-line auctions', *Journal of Political Economy*, 90: 280–294.

Huang, Chi-fu and Lode Li (1990) 'Continuous time stopping games with monotone reward structures', *Mathematics of Operations Research*, 15: 496–507.

Jacquillat, Bertrand and John McDonald (1974) 'Pricing of initial equity issues: The French sealed-bid auction', *Journal of Business*, 47: 000–000.

Jackson, Matthew (1988) 'Private information and exchange: The implications of strategic behavior, Ph.D. dissertation, Stanford University.

Kagel, John H. (1991) 'Auctions: a survey of experimental research', in: *Handbook of experimental economics*. Amsterdam: Elsevier Science Publishers, to appear.

Kagel, John H. and Dan Levin (1986) 'The winner's curse and public information in common value auctions', *American Economic Review*, 76: 894–920.

Kagel, John H. and Dan Kevin (1988) 'Independent private value auctions: Bidder behavior in first, second, and third-price auctions with varying numbers of bidders', University of Houston, mimeo.

Kagel, John H., Ronald Harstad and Dan Levin (1987) 'Information impact and allocation rules in auctions with affiliated private values: A laboratory study', *Econometrica*, 55: 1275–1304.

Kagel, John, Dan Levin and Ronald Harstad (1988) 'Judgment, evaluation, and information processing in second-price common value auctions', University of Houston, mimeo.

Kahn, Edward P., Michael H. Rothkopf, Joseph H. Eto and Jean-Michel Nataf (1990) 'Auctions for PURPA purchases: A simulation study', *Journal of Regulatory Economics*, 2: 129–149.

Kennan, John and Robert Wilson (1989) 'Strategic analysis of bargaining and interpretation of strike data', *Journal of Applied Econometrics*, 4 (Supplement): S87–S130.

Kennan, John and Robert Wilson (1992) 'Bargaining with private information', *Journal of Economic Literature*, 30: to appear.

Krishna, Kala (1990) 'The case of the vanishing revenues: Auction quotas with monopoly', *American Economic Review*, 80: 828–836.

Kyle, Albert S. (1985) 'Continuous auctions and insider trading', *Econometrica*, 53: 1315–1336.

Laffont, Jean-Jacques and Eric Maskin (1980) 'Optimal reservation price in the Vickrey auction', *Economic Letters*, 6: 309–313.

Laffont, Jean-Jacques and Jean Tirole (1987) 'Auctioning incentive contracts', *Journal of Political Economy*, 95: 921–937.

Lang, Kevin and Robert W. Rosenthal (1990) 'The contractors' game', Boston University, mimeo.

Lebrun, Bernard (1991a) 'Repeat trade and asymmetric information: A principal agent analysis in the auction framework', Working Paper E91-05-01, Virginia Polytechnic Institute.

Lebrun, Bernard (1991b) 'Asymmetry in auctions and competition between groups: A discrete model', Working Paper E91-05-02, Virginia Polytechnic Institute.

Lebrun, Bernard (1991c) 'First price auction with two bidders: The asymmetric case', Working Paper E91-05-03, Virginia Polytechnic Institute.

Leininger, Wolfgang, Peter B. Linhart and Roy Radner (1989) 'Equilibria of the sealed bid mechanism for bargaining with incomplete information', *Journal of Economic Theory*, 48: 63–106.

Levin, Dan and Ronald Harstad (1986) 'Symmetric bidding in second-price, common-value auctions', *Economics Letters*, 20: 315–319.

Levin, Dan and Ronald Harstad (1990) 'Arriving at competitive prices through dominance arguments for large auctions', *Journal of Economic Theory*, to appear.

Levin, Dan and James L. Smith (1990) 'Comment on "some evidence on the winner's curse"', *American Economic Review*, 81: 370–375.

Levinson, Marc (1987) 'Using science to bid for business', *Business Month*, 129(4): 50–51.

Maskin, Eric S. and John G. Riley (1983) 'Auctions with asymmetric beliefs', University of California at Los Angeles, D.P. #254.

Maskin, Eric S. and John G. Riley (1984a) 'Optimal auctions with risk averse buyers', *Econometrica*, 52: 1473–1518.

Maskin, Eric S. and John G. Riley (1984b) 'Monopoly with incomplete information', *Rand Journal of Economics*, 15: 171–196.

Maskin, Eric S. and John G. Riley (1986) 'Existence and uniqueness of equilibrium in sealed high bid auctions', University of California, Los Angeles, D.P. #407.

Maskin, Eric S. and John G. Riley (1987) 'Optimal multi-unit auctions', in: F. Hahn, ed., *The economics of missing markets, information and games*. Oxford: Oxford University Press, 1989.

Matthews, Steven A. (1983) 'Selling to risk averse buyers with unobservable tastes', *Journal of Economic Theory*, 30: 370–400.

Matthews, Steven A. (1984a) 'Information acquisition in discriminating auctions', in: M. Boyer and R. Kihlstrom, eds., *Bayesian models in economic theory*. Amsterdam: North-Holland.

Matthews, Steven A. (1984b) 'On the implementability of reduced form auctions', *Econometrica*, 52: 1519–1522.

Matthews, Steven A. (1987) 'Comparing auctions for risk averse buyers: A buyer's point of view', *Econometrica*, 55: 633–646.

McAfee, R. Preston (1989) 'A dominant strategy double auction', SSWP #734, California Institute of Technology, revised 1990; *Journal of Economic Theory*, to appear.

McAfee, R. Preston (1991) 'Efficient allocation and continuous quantities', *Journal of Economic Theory*, 53: 51–73.

McAfee, R. Preston and John McMillan (1985) *Incentives in government contracting*. London, Ontario: University of Western Ontario and the Ontario Economic Council.

McAfee, R. Preston and John McMillan (1986) 'Bidding for contracts: A principal–agent analysis', *Rand Journal of Economics*, 17: 415–440.

McAfee, R. Preston and John McMillan (1987a) 'Auctions and bidding', *Journal of Economic Literature*, 25: 699–738.

McAfee, R. Preston and John McMillan (1987b) 'Auctions with a stochastic number of bidders', *Journal of Economic Theory*, 43: 1–19.

McAfee, R. Preston and John McMillan (1987c) 'Competition for agency contracts', *Rand Journal of Economics*, 18: 296–307.

McAfee, R. Preston and John McMillan (1987d) 'Auctions with entry', *Economic Letters*, 23: 343–347.

McAfee, R. Preston and Philip Reny (1992) 'Correlated information and mechanism design', *Econometrica*, 60: to appear.

McAfee, R. Preston, John McMillan and Philip Reny (1989) 'Extracting the surplus in the common value auction', *Econometrica*, 57: 1451–1460.

McCabe, Kevin A., Stephen J. Rassenti and Vernon L. Smith (1989) 'Designing call auction institutions: Is double Dutch the best?', University of Arizona, mimeo.

McCabe, Kevin A., Stephen J. Rassenti and Vernon L. Smith (1990a) 'Auction institutional design: Theory and behavior of simultaneous multiple-unit generalizations of Dutch and English auctions', *American Economic Review*, 80: 1276–1283.

McCabe, Kevin A., Stephen J. Rassenti and Vernon L. Smith (1990b) 'Auction design for composite goods: The natural gas industry', *Journal of Economic Behavior and Organization*, 14: 127–149.

Mead, W.J., A. Moseidjord and P.E. Sorenson (1984) 'Competitive bidding under asymmetrical information: behavior and performance in Gulf of Mexico drainage lease sales, 1959–1969', *Review of Economics and Statistics*, 66: 505–508.

Milgrom, Paul R. (1979a) *The structure of information in competitive bidding*. New York: Garland.

Milgrom, Paul R. (1979b) 'A convergence theorem for competitive bidding with differential information', *Econometrica*, 47: 679–688.

Milgrom, Paul R. (1981) 'Rational expectations, information acquisition, and competitive bidding', *Econometrica*, 49: 921–943.

Milgrom Paul R. (1985) 'The economics of competitive bidding: A selective survey', in: L. Hurwicz, D. Schmeidler and H. Sonnenschein, eds., *Social goals and social organization*. Cambridge: Cambridge University Press.

Milgrom, Paul R. (1987) 'Auction theory', in: T. Bewley, ed., *Advances in economic theory: Fifth World Congress*. Cambridge: Cambridge University Press.

Milgrom, Paul R. (1989) 'Auctions and bidding: A primer', *Journal of Economic Perspectives*, 3: 3–22.

Milgrom, Paul R. and Nancy Stokey (1982) 'Information, trade, and common knowledge', *Journal of Economic Theory*, 26: 17–27.

Milgrom, Paul R. and Robert J. Weber (1981) 'Topologies on information and strategies in games with incomplete information', in: O. Moeschlin and D. Pallaschke, eds., *Game theory and mathematical economics*. Amsterdam: North-Holland.

Milgrom, Paul R. and Robert J. Weber (1982a) 'A theory of auctions and competitive bidding', *Econometrica*, 50: 1089–1122.

Milgrom, Paul R. and Robert J. Weber (1982b) 'The value of information in a sealed bid auction', *Journal of Mathematical Economics*, 10: 105–114.

Milgrom, Paul R. and Robert J. Weber (1985) 'Distributional strategies for games with incomplete information', *Mathematics of Operations Research*, 10: 619–632.

Miller, Merton H. (Chairman) (1988) *Report of the committee of inquiry to examine the events surrounding 19 October 1987*. Chicago Mercantile Exchange.

Moore, John (1984) 'Global incentive constraints in auction design', *Econometrica*, 52: 1523–1535.

Myerson, Roger B. (1981) 'Optimal auction design', *Mathematics of Operations Research*, 6: 58–63.

Myerson, Roger B. and Mark A. Satterthwaite (1983) 'Efficient mechanisms for bilateral trade', *Journal of Economic Theory*, 29: 265–281.

Nalebuff, Barry and John G. Riley (1985) 'Asymmetric equilibria in the war of attrition', *Journal of Theoretical Biology*, 113: 517–527.

Nti, Kofi O. (1987) 'Competitive procurement under uncertainty', *Management Science*, 11: 1489–1500.

Ortega-Reichert, Armando (1968) 'Models for competitive bidding under uncertainty', Ph.D. dissertation, Stanford University; Technical Report #8, Operations Research Department.

Palfrey, Thomas R. (1985) 'Uncertainty resolution, private information aggregation and the Cournot competitive limit', *Review of Economic Studies*, 52: 69–83.

Pitchnik, Carolyn (1989), 'Budget-constrained sequential auctions with incomplete information', London School of Economics and Political Science, ICERD #89 201.

Plott, Charles R. (1982) 'Industrial organization theory and experimental economics', *Journal of Economic Literature*, 20: 1485–1527.

Plott, Charles R. and Shyam Sunder (1982) 'Efficiency of experimental security markets with insider information: An application of rational expectations models', *Journal of Political Economy*, 90: 663–698.

Plott, Charles R. and Shyam Sunder (1988), 'Rational expectations and the aggregation of diverse information in laboratory security markets', *Econometrica*, 56: 1085–1118.

Plum, Michael (1989) 'Continuous sealed-bid auctions with asymmetric incomplete information', Köln, FRG: Mathematisches Institut der Universität zu Köln.

Porter, Robert H. and J. Douglas Zona (1991) 'On the detection of bid rigging in procurement auction data', Northwestern University, mimeo.

Radner, Roy and Robert W. Rosenthal (1982) 'Private information and pure-strategy equilibria', *Mathematics of Operations Research*, 7: 401–409.

Reece, Douglas K. (1979) 'Competitive bidding for offshore petroleum leases', *Bell Journal of Economics*, 9: 369–384.

Reece, Douglas K. (1979) 'Alternative bidding mechanisms for offshore petroleum leases', *Bell Journal of Economics*, 10: 659–669.

Riley, John G. (1979), 'Evolutionary equilibrium strategies', *Journal of Theoretical Biology*, 72.

Riley, John G. (1980) 'Strong evolutionary equilibria in the war of attrition', *Journal of Theoretical Biology*, 82: 383–400.

Riley, John G. (1998a) 'An introduction to the theory of contests', University of California, Los Angeles, D.P. #469. To appear in: H.W. Kuhn and G. Szego, eds., *Incomplete information and bounded rationality decision models*. Berlin: Springer-Verlag, circa 1989.

Riley, John G. (1988b) 'Ex Post information in auctions', *Review of Economic Studies*, 55: 409–430.

Riley, John G. (1989), 'Expected revenue from open and sealed bid auctions', *Journal of Economic Perspectives*, 3: 41–50.

Riley, John G. and William F. Samuelson (1981) 'Optimal auctions', *American Economic Review*, 71: 381–392.

Riordan, Michael H. and David E.M. Sappington (1987) 'Awarding monopoly franchises', *American Economic Review*, 77: 375–387.

Roberts, D. John and Andrew Postlewaite (1976) 'The incentive for price taking behavior in large exchange economies', *Econometrica*, 44: 115–128.

Robinson, Douglas R. (1991) 'A framework for determining optimal petroleum leasing', PhD dissertation, Engineering-Economic Systems Department, Stanford University.

Rosenthal, Robert W. (1980) 'A model in which an increase in the number of sellers leads to a higher price', *Econometrica*, 48: 1575–1579.

Roth, Alvin (1984a) 'The evolution of the labor market for medical interns and residents: A case study in game theory', *Journal of Political Economy*, 92: 991–1016.

Roth, Alvin (1984b) 'Misrepresentation and stability in the marriage problem', *Journal of Economic Theory*, 34: 383–387.

Roth, Alvin (1984c) 'Stability and polarization of interests in job matching', *Econometrica*, 52: 47–58.

Roth, Alvin (1988) 'Laboratory experimentation in economics: A methodological overview', *Economic Journal*, 98: 974–1031.

Roth, Alvin (1991) 'Game theory as a part of empirical economics', *The Economic Journal*, 101: 107–114.

Roth, Alvin and Marilda Oliveira Sotomayor (1990) *Two-sided matching: A study in game-theoretic modelling and analysis*. Cambridge: Cambridge University Press.

Rothkopf, Michael H. (1969) 'A model of rational competitive bidding', *Management Science*, 15: 774–777.

Rothkopf, Michael H. (1990) 'Models of auctions and competitive bidding', in: *Handbook of operations research*. Amsterdam: Elsevier Science Publishers, to appear.

Rothkopf, Michael H. (1991) 'On auctions with withdrawable winning bids', *Marketing Science*, 10: 40–57.

Rothkopf, Michael H., Thomas J. Teisberg and Edward P. Kahn (1990) 'Why are Vickrey auctions rare?', *Journal of Political Economy*, 98: 94–109.

Rozek, Richard P. (1989) 'Competitive bidding in electricity markets: A survey', *Energy Journal*, 10: 117–138.

Rustichini, Aldo, Mark A. Satterthwaite and Steven R. Williams (1990) 'Convergence to price taking behavior in a simple market', Discussion Paper #914, Northwestern University, mimeo.

Satterthwaite, Mark A. and Steven R. Williams (1989a) 'Bilateral trade with the sealed bid k-double auction: Existence and efficiency', *Journal of Economic Theory*, 48: 107–133.

Satterthwaite, Mark A. and Steven R. Williams (1989b) The rate of convergence to efficiency in the buyer's bid double auction as the market becomes large', *Review of Economic Studies*, 56: 477–498.

Satterthwaite, Mark A. and Steven R. Williams (1989c) 'The rate at which a simple market converges to efficiency as the number of traders increases: An asymptotic result for optimal trading mechanisms', *Journal of Economic Theory*, 48: 304–332.

Schmeidler, David (1980) 'Walrasian analysis via strategic outcome functions', *Econometrica*, 48: 1585–1593.

Shapley, Lloyd and Martin Shubik (1977) 'Trade using one commodity as a means of payment', *Journal of Political Economy*, 85: 937–968.

Simon, Leo K. (1987) 'Games with discontinuous payoffs', *Review of Economic Studies*, 54: 569–598.

Smiley, A.K. (1979) *Competitive bidding under uncertainty: The case of offshore oil.* Cambridge, Mass.: Balinger.

Smith, Vernon (1982) 'Microeconomic systems as experimental science', *American Economic Review*, 72: 923–955.

Smith, Vernon (1987) 'Auctions', entry in: J. Eatwell, M. Milgate and P. Newman, eds., *The New Palgrave: A dictionary of economics.* London: The Macmillan Press Ltd.

Stark, Robert M. and Michael H. Rothkopf (1979) 'Competitive bidding: A comprehensive bibliography', *Operations Research*, 27: 364–390.

Thaler, Richard H. (1988) 'Anomalies: The winner's curse', *Journal of Economic Perspectives*, 2: 191–202.

Thiel, Stuart E. (1988) 'Some evidence on the winner's curse', *American Economic Review*, 78: 884–895.

Vickrey, William (1961) 'Counterspeculation, auctions, and sealed tenders', *Journal of Finance*, 16: 8–37.

Vickrey, William (1962) 'Auctions and bidding games', in: O. Morgenstern and A. Tucker, eds., *Recent advances in game theory.* Princeton: Princeton University Press.

Vincent, Daniel R. (1990) 'Dynamic Auctions', *Review of Economic Studies*, 57: 49–62.

Wang, Ruqu (1991a) 'Common-value auctions with discrete private information', *Journal of Economic Theory*, 54: 429–447.

Wang, Ruqu (1991b) 'Auctions versus posted-price selling', Queen's University, Kingston, Ontario, mimeo.

Weber, Robert J. (1983) 'Multiple-object auctions', in: R. Engelbrecht-Wiggans, M. Shubik and R. Stark, eds., *Auctions, bidding, and contracting.* New York: New York University Press, ch. 3.

Williams, Steven R. (1987) 'Efficient performance in two-agent bargaining', *Journal of Economic Theory*, 41: 154–172.

Williams, Steven R. (1991) 'Existence and convergence of equilibria in the buyer's bid double auction', *Review of Economic Studies*, 58: 351–374.

Wilson, Robert (1969) 'Competitive bidding with disparate information', *Management Science*, 15: 446–448.

Wilson, Robert (1975) 'On the incentive for information acquisition in competitive bidding with asymmetrical information', Stanford University, mimeo.

Wilson, Robert (1977) 'A bidding model of perfect competition', *Review of Economic Studies*, 44: 511–518.

Wilson, Robert (1978) 'Competitive exchange', *Econometrica*, 46: 557–85.

Wilson, Robert (1979) 'Auctions of shares', *Quarterly Journal of Economics*, 93: 675–689.

Wilson, Robert (1981) 'The basic model of competitive bidding', Office of Policy Analysis, U.S. Department of the Interior; text and computer programs.

Wilson, Robert (1985a) 'Incentive efficiency of double auctions', *Econometrica*, 53: 1101–1115.

Wilson, Robert (1985b) 'Efficient trading', in: G. Feiwel, ed., *Issues in contemporary microeconomics and welfare*. London: The Macmillan Press Ltd.

Wilson, Robert (1986) 'Equilibria of bid–ask markets', in: G. Feiwel, ed., *Arrow and the ascent of economic theory: Essays in honor of Kenneth J. Arrow*. London: The Macmillan Press Ltd.

Wilson, Robert (1987a) 'Game-theoretic analyses of trading processes', in: T. Bewley, ed., *Advances in economic theory: Fifth World Congress*. Cambridge: Cambridge University Press.

Wilson, Robert (1987b) 'Bidding', entry in: J. Eatwell, M. Milgate and P. Newman, eds., *The New Palgrave: A dictionary of economics*. London: The Macmillan Press Ltd.

Wilson, Robert (1989) 'Deterrence in oligopolistic competition', in: R. Axelrod, R. Radner and P. Stern, eds., *Perspectives on deterrence*. Oxford: Oxford University Press.

Zona, J.D. (1986) 'Bid-rigging and the competitive bidding process: Theory and evidence', PhD thesis, State University of New York, Stony Brook, New York.

Chapter 9

LOCATION

JEAN J. GABSZEWICZ[a] and JACQUES-FRANÇOIS THISSE[ab]*

[a]*C.O.R.E., Université Catholique de Louvain and* [b]*Université de Paris I–Sorbonne*

Contents

*The first version of this chapter was written when the second author was Visiting Research Professor at INSEAD. Financial support from CIM (Belgium) is gratefully acknowledged. We thank S. Anderson, A. de Palma, A. Kats, D. Neven, M. Osborne, J.-Ch. Rochet, X. Vives, R. Wilson, and the editors for helpful discussions and comments.

Handbook of Game Theory, Volume 1, Edited by R.J. Aumann and S. Hart

1. Introduction

Space, by its very nature, is a source of *market power*. Indeed, most markets operate over intricate networks of scattered buyers and sellers. Because the market activities are performed at dispersed points in space, each firm finds only a few rivals in its immediate neighborhood; further away there might be more competitors, but their influence is weakened by the existence of transportation costs. Similarly, not all consumers are alike to the firm; those who are far away will not buy from the firm because they have to pay too high a transportation cost. Accordingly, competition in space occurs "among the few", thus leading to an analysis of the problem as a game of strategy.

The model designed to describe that situation has come to be known as the model of *spatial competition*. In this model, a population of consumers is spread out over a geographical area, while firms selling a homogeneous product are (to be) located in the same space. Consumers have specific preferences regarding the commodity made available by the sellers either at the firms' or consumers' place (depending on who controls the transport). Since the product is homogeneous, a basic feature of consumers' behavior is that they buy from the firm charging the lowest full price, i.e. the price gross of the transportation costs. As a result, the number of customers patronizing a particular firm depends on its location and price policy, as well as on locations and price policies of competing firms established in the relevant area. This situation typically involves the basic ingredients of a *noncooperative game* in which the players are firms, strategies prices and/or locations, and the payoffs are profit functions.

The economic relevance of location games does not exclusively stem from their initial geographical set-up. Indeed, location problems are fundamentally related to many aspects of business competition in modern economies. Firstly, the spatially dispersed nature of markets has a direct analog in industrial economies under the form of an industry with *differentiated products*. In that set-up, product substitutes are dispersed in a space of characteristics à la Lancaster, and the seller of a particular variant enjoys a quasi-monopolistic position relative to the consumers who most prefer it. Moreover, the counterpart of the transportation costs is the utility loss incurred by a consumer who does not find his "ideal product" on the market. (In the geographical setting, this means that transport is under the control of the consumer.) Thus, the interest in modelling spatial competition extends immediately to the process of competition amongst firms producing differentiated commodities. In this domain, it was found useful to distinguish between market competition under *horizontal* versus *vertical* product differentiation. Two variants of a product are said to be horizontally differentiated whenever, sold at the same price, some consumers choose one variant while the others buy the alternative variant. Two variants are vertically differentiated whenever, sold at the same price, all consumers purchase the same variant (like in the case of a "standard" and a "luxury" product). Along several dimensions, the nature of competition turns out to be different under the two types of differentiation. Interestingly enough, these two forms of competition have precise counterparts in spatial competition. To horizontal product differentiation corresponds a process of spatial competition with firms locating within the sub-space where the consumers themselves are located. The typical case is provided by shops installed inside the residential area: the "Main Street" model of Hotelling (1929). In what follows, we call such games *inside location games*. The analog of vertical product differentiation in spatial competition corresponds to a situation where the sellers locate outside the residential area, like shopping centers set up along a road at the outskirts of a city. At the same price, all consumers prefer to buy from the shopping center which is the closest to the city. These games are called *outside location games*.

Secondly, another important issue in industrial economics is related to the practice of price discrimination. Since some sort of market segmentation is inherent to price discrimination, the spatial competition model offers a natural framework for the study of oligopolistic markets with price discriminating firms. Of course, price discrimination is possible only when firms can discern among customers. To this effect, we suppose that transport is under the firm's control, thus enabling discrimination with respect to location. If the difference between delivered prices at two different locations is not larger than the transportation costs between these points, arbitrage is never profitable and firms may exercise price discrimination. When two or more firms price discriminate in a spatial economy, the resulting game typically involves, as strategic variables, *price schedules* specifying the delivered prices at which each firm is willing to supply the customers located at each point. This gives rise to a new class of location games in which firms' decision variables are price functions instead of price scalars, i.e. discriminatory versus mill pricing.

Finally, the location model is also well suited for analyzing *nonprice competition*. In other words, firms are assumed to compete on other variables than prices; in particular products specification appears as a basic decision variable in such a competitive environment. Marketers view the product sold by a firm as a mix of goods in conjunction with an array of services. The spatial analog of a firm choosing the attributes of a product defined as such, given some competitive brands, is the choice by a shop-keeper of a location for his store, given some competing facilities. It is worth noting that this model may also be useful for dealing with collective decision-making processes, like voting or competition between political parties.

The remainder of this chapter is organized as follows. In Section 2 we study inside and outside location games assuming mill price competition; we first assume that firm's locations are given and subsequently allow for variable locations. Section 3 deals with the inside location game under discriminatory pricing. Location under non-price competition is taken up in Section 4. In each section we concentrate on the basic models and results, mentioning in footnotes several recent extensions and reinterpretations. Regarding historical details, we refer the reader to Ponsard (1983). Finally, we draw some conclusions in Section 5.

2. Location under mill price competition

2.1. Variable prices and parametric locations

2.1.1. The inside location game

The prototype model of spatial competition for the inside location game has been introduced by Hotelling (1929). On a line whose length is normalized to one by an adequate choice of the unit of length, two sellers A and B of a homogeneous product with zero production cost are installed at respective distances a and b from the endpoints of the line ($a + b \leqslant 1; a \geqslant 0, b \geqslant 0$). Customers are distributed along the unit interval according to a positive and continuously differentiable density f, and each customer consumes exactly one unit of the commodity. Since the product is homogeneous, a consumer will buy from the seller who quotes the lower full price, namely the mill price plus transportation cost. It is supposed that the transport is under the customer's control. We denote by $c(x)$ the transportation cost function, i.e. the cost in terms of a given numéraire of shipping one unit of the product over a distance of length x. The transportation cost function is assumed continuous, increasing and convex in x, with $c(0) = 0$. Let p_1 and p_2 denote, respectively, the *mill price* of A and B and denote by $m(p_1, p_2)$ the "marginal consumer" $y \in [0, 1]$ satisfying

$$p_1 + c(|y - a|) = p_2 + c(|1 - b - y|) ;$$

whenever it exists, it is unique.[1] If $m(p_1, p_2)$ does not exist, then either,

$$p_1 + c(|y - a|) < p_2 + c(|1 - b - y|) , \quad \text{for all } y \in [0, 1] ,$$

or,

$$p_1 + c(|y - a|) > p_2 + c(|1 - b - y|) , \quad \text{for all } y \in [0, 1] .$$

---

[1]When c is linear, price ties may occur over a positive measure subset of $[0, 1]$. We assume that they are broken in favor of the nearer firm so that $m(p_1, p_2)$ equals a or $1 - b$.

In the first case, the market is segmented at $m(p_1, p_2)$: customers located in $[0, m(p_1, p_2)]$ buy from seller A, those in $]m(p_1, p_2), 1]$ from seller B. In the second case, the whole market is served by seller A at prices (p_1, p_2) while the converse holds in the third case.

The situation described above gives rise to a two-person game with players A and B, strategies $p_1 \in [0, \infty[$ and $p_2 \in [0, \infty[$; the payoff function of seller A is given by

$$\pi_1(p_1, p_2; a, b) = p_1 \int_0^{m(p_1, p_2)} f(z)\, dz \,, \quad \text{if } m(p_1, p_2) \text{ exists} \,,$$

$$= p_1 \,, \qquad \qquad \text{if, for all } y \in [0, 1] \,,$$
$$\qquad\qquad p_1 + c(|y - a|) < p_2 + c(|1 - b - y|) \,,$$

$$= 0 \,, \qquad\qquad \text{if, for all } y \in [0, 1] \,,$$
$$\qquad\qquad p_1 + c(|y - a|) > p_2 + c(|1 - b - y|) \,.$$

The payoff function of seller B is defined similarly and is, therefore, omitted throughout the chapter.

Now we consider the problem of existence of a *noncooperative price equilibrium* in pure strategies for the class of inside location games described above, i.e. a pair of prices (p_1^*, p_2^*) such that $\pi_i(p_i^*, p_j^*; a, b) \geq \pi_i(p_i, p_j^*; a, b)$, $\forall p_i \geq 0$, $i = 1, 2$ and $i \neq j$. The difficulties raised by this problem are best illustrated by the specific model initially considered by Hotelling. This author assumes a uniform customer density and linear transportation costs:

$$c(x) = tx \,,$$

where the scalar $t > 0$ denotes the transportation rate.[2] In this case, $m(p_1, p_2)$ exists when $a \leq m(p_1, p_2) \leq 1 - b$ so that $m(p_1, p_2)$ must be the solution of the equation

---

[2] In measure-theoretic terms, the Hotelling model can be interpreted as follows: the distribution of consumers over space is continuous, whereas the distribution of transportation rates is atomic (there is a single atom since t is the same across consumers). Garella and Martinez-Giralt (1989) study what we may consider as the "dual" model: the distribution of consumers is atomic (there are two atoms called cities) and the transportation rates are distributed continuously over a compact interval. Demands are always continuous, but profits are not quasiconcave. A pure strategy price equilibrium exists when cities differ enough in size and when transportation rates range over a sufficiently wide interval.

$$p_1 + t(y - a) = p_2 + t(1 - b - y),$$

that is,

$$m(p_1, p_2) = \frac{p_2 - p_1}{2t} + \frac{1 - b + a}{2}.$$

It is easily seen that $m(p_1, p_2) \in [a, 1 - b]$ if and only if $|p_1 - p_2| \leqslant t(1 - a - b)$. Furthermore, since $p_1 < p_2 - t(1 - a - b)$ implies $p_1 + t|y - a| < p_2 + t|1 - b - y|$ for all $y \in [0, 1]$, and $p_2 < p_1 - t(1 - a - b)$ implies $p_2 + t|1 - b - y| < p_1 + t|y - a|$ for all y, in the linear case seller A's payoff function becomes

$$\pi_1(p_1, p_2; a, b)$$

$$= \left(\frac{1 - b + a}{2}\right)p_1 + \frac{1}{2t}(p_1 p_2 - p_1^2), \quad \text{if } |p_1 - p_2| \leqslant t(1 - a - b),$$

$$= p_1, \qquad\qquad\qquad\qquad\qquad\qquad\quad \text{if } p_1 < p_2 - t(1 - a - b),$$

$$= 0, \qquad\qquad\qquad\qquad\qquad\qquad\qquad \text{if } p_1 > p_2 + t(1 - a - b).$$

Thus, in the first case the market is split between the two firms; in the second, firm 1 captures firm 2's hinterland and serves the whole market; finally, in the third case, firm 1 loses its hinterland and has no demand. The profit function has, therefore, two discontinuities at the prices where the group of buyers located in either hinterland is indifferent between the two sellers (see

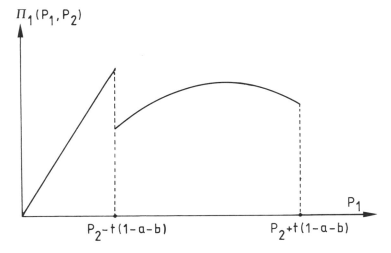

Figure 1. Firm 1's profit function.

Figure 1 for an illustration).[3] Notice also that this function is never quasiconcave (except when the two firms are located at the endpoints of the interval).

The following proposition, proven in d'Aspremont et al. (1979), provides the
necessary and sufficient conditions on the location parameters a and b guaranteeing the existence of a price equilibrium (p_1^*, p_2^*) in pure strategies for the
above game.

Proposition 1. *For $a + b = 1$, the unique price equilibrium is given by $p_1^* = p_2^* = 0$. For $a + b < 1$, there is a price equilibrium if and only if*

$$\left(1 + \frac{a - b}{3}\right)^2 \geq \tfrac{4}{3}(a + 2b),$$

$$\left(1 + \frac{b - a}{3}\right)^2 \geq \tfrac{4}{3}(b + 2a).$$

Whenever it exists, the price equilibrium is unique.

In words, there exists a price equilibrium when firms are located sufficiently
far apart (in the symmetric case, $a = b$, the above two inequalities impose that
the two firms are established outside the first and third quartiles). This is so
because otherwise at least one firm has an incentive to undercut its rival's price
in order to capture its hinterland: in Figure 1, the supremum of the linear piece
of the profit function lies above the maximum of the quadratic piece. Hence
the Hotelling example reveals that insufficient product differentiation may lead
to price instability.[4]

The above discussion may suggest that the discontinuities in the payoff
functions, observed under linear transportation costs, are responsible for the
absence of equilibrium. A reasonable conjecture, then, would be that the
assumption of *strictly* convex transportation cost functions – which guarantees
the continuity of the payoff functions – would restore the existence property in
the whole domain of (a, b) locations. This point of view is reinforced when the
quadratic transportation cost case is examined, i.e. when $c(x)$ is defined by

---

[3] These discontinuities vanish when the characteristics space is n-dimensional, with $n \geq 2$, and
when a l_p-metric is used, with $p \in]1, \infty[$ [see Economides (1986)]. However, even in the simple
case of the Euclidean metric ($p = 2$), there is still a lack of quasiconcavity in the payoff functions.
Economides (1986) has shown that a price equilibrium exists in the special case of two firms
located symmetrically on an axis passing through the center of a disk over which consumers are
evenly distributed.

[4] Economides (1984) shows that the introduction of a reservation price, i.e. the maximum full
price that a customer is willing to pay to obtain the product, reduces the (a, b) segment of
nonexistence but does not suppress it (except in the limit case of two separated monopolies).
Shilony (1981) reaches similar conclusions by considering symmetric, single-peaked distributions of
customers.

$$c(x) = sx^2, \quad s > 0.$$

It is readily verified that the payoff functions are then not only continuous but also quasiconcave. Accordingly, under quadratic transportation costs, there exists a price equilibrium in pure strategies wherever the locations a and b are. Furthermore, the pair of prices (p_1^*, p_2^*), defined by

$$p_1^* = s(1 - a - b)\left(1 + \frac{a - b}{3}\right), \qquad p_2^* = s(1 - a - b)\left(1 + \frac{b - a}{2}\right), \qquad (1)$$

is the unique equilibrium for fixed a and b.

Unfortunately, as shown by the following example, even if strictly convex transportation cost functions imply the continuity of the payoff functions, they are not sufficient to imply the existence of an equilibrium for every location pair (a, b). Assume, indeed, that the transportation cost function is of the "linear-quadratic" type, i.e.

$$c(x) = sx^2 + tx, \quad s > 0 \text{ and } t > 0.$$

Anderson (1988) has shown that, wherever seller A's location, there is always a location for seller B such that no price equilibrium in pure strategies exists for the corresponding location pair.

These few examples suffice to show that no general theorem for existence in pure strategies can be obtained for the location model. To date, the most general sufficient conditions to be imposed on customer density and transportation cost functions have been derived by Champsaur and Rochet (1988). Let F be the cumulative distribution of customers over the unit interval and define

$$\alpha \equiv \max_{z \in [0, 1]} \left[\frac{f'(z)}{f(z)} - 2\frac{f(z)}{F(z)}\right],$$

$$\beta \equiv \min_{z \in [0, 1]} \left[\frac{f'(z)}{f(z)} + 2\frac{f(z)}{1 - F(z)}\right].$$

Proposition 2. *If c is three times continuously differentiable in an open interval of $\mathbb{R}$ including $[0, 1]$ and if*

$$\alpha < \frac{c'''(x)}{c''(x)} < \beta, \quad \text{for all } x \in [0, 1], \qquad (2)$$

then there exists a price equilibrium in pure strategies.

Clearly, condition (2) imposes severe restrictions on the transportation cost function $c(x)$.

The standard game-theoretic solution to the nonexistence of a noncooperative equilibrium in pure strategies is to resort to mixed strategies. When transportation costs are strictly convex, the payoff functions are continuous and, by the theorem of Glicksberg (1952), a price equilibrium in mixed strategies exists. When transportation costs are linear, the payoff functions are discontinuous, but the theorem of Dasgupta and Maskin (1986) applies and an equilibrium also exists. In this special case, Osborne and Pitchik (1987) show that for a subset of location pairs at which a pure strategy price equilibrium fails to exist, there is an equilibrium in which the firms randomize over two disjoint intervals; they show also that for *any* location pair any equilibrium in which the highest and lowest prices in the support of each firm's strategy are not too far apart must take this form. Specifically, each firm chooses its price either from an interval just below a relatively high price, or from an interval just below the price that undercuts its rival's highest price. In view of the complexity of Osborne and Pitchik's analysis, characterizing the equilibrium in mixed strategies for more general models of spatial price competition seems to be a formidable task.[5,6]

2.1.2. The outside location game

In the outside location game, firms are no longer established in the residential area, still represented by the interval $[0, 1]$, but on the right-hand side of this interval at locations a and $b \in [1, +\infty[$ with $a \le b$.

Let us denote again by $m(p_1, p_2)$ the customer $y \in [0, 1]$ satisfying

$$p_1 + c(a - y) = p_2 + c(y - b),$$

whenever firm A (resp. B) quotes price p_1 (resp. p_2).

First, we consider the Hotelling case of a linear transportation cost function $c(x) = tx$. If $p_1 < p_2$, the whole market is served by seller A. Clearly, such a situation generates a series of price cuts that result in a price tie with seller B quoting a zero price. We assume that price ties are broken in favor of the nearer firm. This is so because this firm can always price ϵ-below the price

---

[5]Another approach, which has considerable intuitive appeal, is to assume that the products sold by the firms are not homogeneous. As a result, consumers also take variables other than full price into account. Because of the (frequent) nonobservability of these variables, firms can at best determine the shopping behavior of a particular consumer up to a probability function. Describing the purchasing probabilities by the logit model [see, for example, McFadden (1984)], de Palma et al. (1986) show that a price equilibrium in pure strategies exists when products are *sufficiently heterogeneous*.

[6]Gabszewicz and Garella (1986) give up the assumption that consumers are perfectly informed about firms' prices. Instead, they suppose that consumers have subjective probabilistic beliefs about these prices. Gabszewicz and Garella then show that a price equilibrium exists provided that the two firms lie sufficiently far apart. This is reminiscent of Proposition 1.

quoted by its rival. Then, it is straightforward that the unique price equilibrium in pure strategies obtains at the pair of prices $(p_1^*, p_2^*) = (t(b - a), 0)$: any higher price p_1 of seller A could be advantageously undercut by seller B, who would then attract all the customers.

We now study the problem of existence of a price equilibrium in pure strategies for the linear-quadratic transportation cost case. The market boundary $m(p_1, p_2)$ easily obtains as the solution to the equation

$$p_1 + s(a - x)^2 + t(a - x) = p_2 + s(b - x)^2 + t(b - x) ,$$

i.e.

$$m(p_1, p_2) = \frac{p_2 - p_1 + s(b^2 - a^2) + t(b - a)}{2s(b - a)} .$$

It is readily verified that, given linear-quadratic costs, there exists a unique price equilibrium in pure strategies given by

$$p_1^* = (b - a) \frac{s(a + b + 2) + t}{3} , \qquad p_2^* = (b - a) \frac{s(4 - a - b) - t}{3} , \qquad (3)$$

when $t/s < 4 - a - b$, and by

$$p_1^* = (b - a)[s(a + b - 2) + t] , \qquad p_2^* = 0 , \qquad (4)$$

when $t/s \geq 4 - a - b$ [see Gabszewicz and Thisse (1986) for more details].

In contrast with the inside location game, we see that for both linear and linear-quadratic cost functions, a price equilibrium exists for all location pairs (a, b) when firms are located outside the segment in which consumers are located. Consequently, it seems that more stability in noncooperative price behavior is to be expected when one of the two players is endowed with a strict exogenous advantage over the other one, as in the outside location game. The fact that, in this game, seller A's location is viewed as strictly better by all consumers than seller B's location, prevents the latter from using price strategies that would attract the whole market to him. This privilege is reserved for firm A. This asymmetry between sellers no longer exists when firms are located inside the consumers' area. In that case, both firms may use "undercutting" price strategies, leading possibly to price instability. In the product differentiation context, all of this suggests that *more stability in price competition may be expected with vertically than with horizontally differentiated products*.

The foregoing analysis about the existence of a noncooperative price equilibrium is rather disappointing. Our inquiry has shown that, even in the simplest case of two firms and a uniform density of consumers located along the line, a price equilibrium in pure strategies may fail to exist for some reasonable transportation cost functions in the inside location game. Furthermore, solving the price game by using mixed strategies proves to be very difficult from a technical standpoint. On the other hand, no counterexample has been found for the outside location game (on the contrary, existence has been proved for the linear and linear-quadratic cost functions). Nevertheless, it is our feeling that *no general results can be expected to hold in spatial price competition under mill pricing*.

2.2. Variable prices and locations

2.2.1. The simultaneous game

In the foregoing subsection, firms (players) were assumed to control their *price* strategy. Now we consider the more general case where firms are allowed to choose simultaneously both price and location. While in the above two games locations a and b were viewed as *parameters* in the payoff functions π_1 and π_2, they are now considered, as well as prices p_1 and p_2, as strategic variables available to the players. Specifically, a strategy is a *pair* (p_1, a) [resp. (p_2, b)] for seller A (resp. seller B) with $p_1 \in [0, \infty[$ and $a \in S_1 = [0, 1]$ in the inside location game or $a \in S_1 = [1, \infty[$ in the outside location game (resp. $p_2 \in [0, \infty[$ and $b \in S_2 \equiv S_1$). The payoff function $\tilde{\pi}_1$ in the *simultaneous* game is given by $\tilde{\pi}_1((p_1, a), (p_2, b)) = \pi_1(p_1, p_2; a, b)$ with π_1 as defined in Subsection 2.1; and similarly for $\tilde{\pi}_2$.

It is well known that, regardless of the transportation cost function $c(x)$, *no simultaneous price–location equilibrium in pure strategies can exist in the inside location game*. To see this, assume that such an equilibrium $[(p_1^*, a^*), (p_2^*, b^*)]$ exists. Clearly, at this equilibrium both firms must have strictly positive payoffs (profits), which implies that $p_1^* > 0$ and $p_2^* > 0$. Two cases may then arise. In the first one, we have $a^* \neq 1 - b^*$. Without loss of generality, we may assume that seller B's payoffs exceed or equal seller A's payoffs. Then, firm A can increase its profits by locating at $\tilde{a} = 1 - b^*$ and by charging a price $\tilde{p}_1 = p_2^* - \epsilon$, with $\epsilon > 0$ arbitrarily small. Indeed

$$\tilde{\pi}_1((\tilde{p}_1, \tilde{a}), (p_2^*, b^*)) > \tilde{\pi}_1((p_1^*, a^*), (p_2^*, b^*)),$$

since firm A now captures the whole market at price $p_2^* - \epsilon$. Since we have assumed $\tilde{\pi}_2((p_1^*, a^*), (p_2^*, b^*)) \geq \tilde{\pi}_1((p_1^*, a^*), (p_2^*, b^*))$, we get the desired contradiction. In the second case, $a^* = 1 - b^*$. But then each player has an incentive to undercut its competitor and, as in Bertrand, to capture the whole market, again a contradiction.[7,8]

By contrast, in the outside location game there always exists a simultaneous price–location equilibrium in pure strategies. Indeed, it is readily verified that $a^* = b^* = 1$ and $p_1^* = p_2^* = 0$ is the unique equilibrium of the game. Furthermore, this solution is *socially optimal*.

2.2.2. The sequential game

We now turn to the alternative formulation in terms of a *sequential* game introduced by Hotelling himself. There, price and location strategies are assumed to be played one at a time in a two-stage process. The choice about location is viewed as prior to the decision on price, so that locations are chosen in the first stage of the sequential game while prices are decided in the second stage. Assuming that prices p_1 and p_2 are chosen at a noncooperative price equilibrium (in pure strategies) in the subgame consisting of the second stage, the corresponding equilibrium payoffs are well defined whenever this price equilibrium exists and is unique. Furthermore, they depend only upon the location choice made in the first stage. Accordingly these payoffs can be used as payoff functions in the first-stage game in which strategies are locations a and b.

We now proceed to a formal definition of a subgame-perfect equilibrium for this sequential game setting. A *subgame-perfect price–location equilibrium* is a pair of locations $(a^*, b^*) \in S_1 \times S_2$ and a pair of price functions $[p_1^*(a, b), p_2^*(a, b)]$ such that

(i) for any $(a, b) \in S_1 \times S_2$, $\pi_i[p_i^*(a, b), p_j^*(a, b); a, b] \geq \pi[p_i, p_j^*(a, b); a, b]$, $\forall p_i \geq 0$, $i = 1, 2$ and $i \neq j$, and

(ii) $\pi_1[p_1^*(a^*, b^*), p_2^*(a^*, b^*); a^*, b^*] \geq \pi_1[p_1^*(a, b^*), p_2^*(a^*, b^*); a, b^*]$, $\forall a \in S_1$; and similarly for π_2.

The concept of a subgame-perfect equilibrium captures the idea that, when firms choose their locations, they both anticipate the consequences of their choice on price competition. In particular, they are aware that this competition will be more severe if they locate close to each other, rather than far apart.

---

[7]To cope with the non-existence of a simultaneous price–location equilibrium, Lerner and Singer (1937) have proposed modifying the concept of noncooperative equilibrium. It is assumed that firms, anticipating their competitors' reaction, do not consider strategies that would eliminate these competitors from the market. This amounts to restricting the strategy sets to prices and locations yielding positive payoffs for all players. In the case of linear transportation costs, such a "modified" simultaneous equilibrium does indeed exist under reasonable assumptions [see Eaton (1972), Novshek (1980), Kohlberg and Novshek (1982)]. However, a sufficient departure from the linear case may invalidate equilibrium [see MacLeod (1985), Gabszewicz and Thisse (1986)].

[8]Note that a simultaneous price–location equilibrium can be shown to exist if the product is heterogeneous enough [see de Palma et al. (1985)].

Unfortunately, this concept is operational only if, for any location choices by firms, there exists one, and only one, corresponding price equilibrium; otherwise, either payoffs would be undefined or multivalued. From Subsection 2.1 we know how demanding these existence and, a fortiori, uniqueness conditions are.[9,10]

To illustrate, let us first consider the inside location game with $c(x) = sx^2$. Then, we know that $[p_1^*(a, b), p_2^*(a, b)]$ exists and is unique for all pairs $(a, b) \in [0, 1]^2$. Substituting $p_1^*(a, b)$ and $p_2^*(a, b)$, given by (1), in π_1 and π_2, routine calculations show that regardless of the location of the other player, the payoffs of firm A decrease when a increases, whereas the payoffs of firm B are a decreasing function of b. Consequently, each firm gains by moving away as far as possible from its competitor. Hence, the equilibrium of the first stage is given by $(0, 0)$ and the resulting prices are $p_1^*(0, 0) = p_1^*(0, 0) = s$. Clearly, these locations differ from the socially optimal locations that minimize total transportation costs, i.e. $a = b = \frac{1}{4}$.[11]

It is also interesting to characterize the subgame-perfect price–location equilibrium in the case of an outside location game with $c(x) = sx^2 + tx$ (linear-quadratic transportation costs). For this case the existence of a unique price equilibrium has been established above, for any location pair (a, b). Furthermore, the corresponding equilibrium pair of prices $[p_1^*(a, b), p_2^*(a, b)]$ is given by (3) whenever $t/s < 4 - a - b$, and by (4) if $t/s \geq 4 - a - b$. Clearly, if $2s < t$, the payoffs $\pi_2[p_1^*(a, b), p_2^*(a, b); a, b]$ of player B are necessarily equal to zero, while the payoffs $\pi_1[p_1^*(a, b), p_2^*(a, b); a, b]$ of player A

---

[9]An alternative approach is taken up by Anderson (1987). Within the original Hotelling model, he assumes that firms A and B enter sequentially and choose their prices following the rules of a *Stackelberg game*. The outcome is such that the first firm to enter the market locates at the market center and the second close to one of the market endpoints. The second firm to enter will prefer to be the price leader and the first one the price follower. It is worth noting that the introduction of space into a price duopoly allows one to endogenize the price leadership. However, it does not follow that this solution will be reached in a noncooperative setting. These profits must also be compared to the simultaneous move payoffs.

[10]Bester (1989) tackles the spatial competition problem in a completely different way. He assumes that prices are no longer set by firms but determined by a noncooperative bargaining game between sellers and buyers. More precisely, the bargaining procedure is taken to be a modified version of Rubinstein's (1982) model in which an outside option (purchasing from a competitor) is introduced. Bester shows that there exists a unique perfect equilibrium which, in turn, leads to a unique price system (given firms' locations). Introducing these prices into the payoff functions, he then proves the existence of a noncooperative Nash equilibrium in location. At this solution, firms locate symmetrically but not coincidentally.

[11]In the case of linear transportation costs, Osborne and Pitchick's (1987) calculations suggest strongly that there is a subgame-perfect equilibrium in which the firms locate symmetrically inside the first and third quartiles, i.e. in the region when they randomize their pricing decision. On the other hand, when the space is given by a *circle*, Kats (1989) proves that firms choose to set up in the subregion for which a price equilibrium in pure strategies exists. This suggests that results about the randomization of pricing are very sensitive to the particular specification of the location space.

decreases when a increases, whatever $b \in [1, \infty[$. In consequence, for any b in this interval, seller A locates at $a^* = 1$. By contrast, when $2s > t$, seller B can always choose a location b in $[1, \infty[$ so as to verify the condition $t/s < 4 - a - b$, guaranteeing himself strictly positive payoffs. Furthermore, seller A's payoffs still decrease with a for any b in $[1, \infty[$, so that seller A still locates at $a^* = 1$. The corresponding value of b which maximizes seller B's payoffs then obtains from the first-order condition

$$\frac{\mathrm{d}}{\mathrm{d}b} \, \pi_2[p_1^*(1, b), \, p_2^*(1, b); 1, b] = 0 \, ,$$

i.e. $b^* = (5s - t)/3s > 1$ since $2s > t$. Hence, if $2s > t$, the equilibrium locations are unique and such that $a^* = 1$ and $b^* = (5s - t)/3s$.

The equilibrium analysis of games with both prices and locations as strategic variables is almost as disappointing as the approach with variable prices but parametric locations. For inside location games (horizontal product differentiation), simultaneous equilibrium never exists; on the other hand, for outside location games (vertical product differentiation), such an equilibrium always exists. The existence of a subgame-perfect equilibrium relies heavily on the existence and uniqueness of a price equilibrium in second-stage subgames; these conditions are hardly met for inside location games. Hence *it appears that outside games have more stability than inside games.* Furthermore, the sequential game approach sheds some light on an important issue in the economics of imperfect competition: Do firms selling substitute products prefer to "copy" each other when selecting their products or, on the contrary, do they differentiate them in some optimal manner? It was Hotelling's belief that "buyers are confronted everywhere with an excessive sameness" [Hotelling (1929, p. 547)], a conjecture which has come to be known in the literature as the "Principle of Minimum Differentiation". However, we must conclude from our analysis that both in the inside and outside location games *firms tend to relax price competition* at the subgame perfect price–location equilibrium *by locating apart from each other.*[12]

3. Location under discriminatory price competition

3.1. Variable prices and parametric locations

From now on we limit ourselves to the inside location game.

Let us consider a model similar to the one described in Subsection 2.1.1, but

[12]See d'Aspremont et al. (1983) for a general argument. However, firms may want to agglomerate when the product is sufficiently heterogeneous [see de Palma et al. (1985)].

in which firms are no longer constrained to sell at the *same* mill prices. Instead, we suppose that firms deliver the product to the customers and can, therefore, exercise price discrimination. Since firms observe the customer's locations, they can charge location-specific prices. In general, the difference between delivered prices quoted by the same firm at two distinct locations does not equal the transportation cost between these points. Thus, firms price discriminate, setting *different* mill prices at the firms' door. As in Hotelling, we also assume that transportation costs are linear in distance.

Given sellers A and B located respectively at distances a and b from the extremities of the segment $[0, 1]$, a strategy for seller A is a *price schedule* $p_1(\cdot)$ that specifies for each location $y \in [0, 1]$ the *delivered price* at which A is willing to sell its product to the customers at y. Formally, we suppose that $p_1(\cdot)$ belongs to the class $\mathcal{P}_1$ of measurable functions defined over $[0, 1]$ which satisfy a.e. the inequality $p_1(y) \geq t|a - y|$. If this latter condition were not verified, then seller A could do at least as well, for any given price of B at y, by pricing at cost $t|y - a|$. Similarly, a strategy for seller B is a measurable function $p_2(\cdot)$ defined on $[0, 1]$ for which $p_2(y) \geq t|1 - b - y|$ holds a.e. We denote by $\mathcal{P}_2$ the strategy set of seller B.[13]

Since the product is homogeneous, customers buy from the seller quoting the lower delivered price. In the event of a price tie, we suppose that customers choose to buy from the nearer firm. This can be justified by the fact that this firm can always price ϵ-below its rival. When customers are equidistant from both firms, any allocation of the local demand is acceptable. Indeed, as will be seen, in equilibrium no seller makes positive profit at such a point. Hence, the payoff function of seller A is

$$\pi_1[p_1(\cdot), p_2(\cdot); a, b] = \int_{M_1} [p_1(y) - t|y - a|] \, dy$$

where $M_1 = \{y \in [0, 1]; \ p_1(y) < p_2(y) \ \text{or} \ (p_1(y) = p_2(y) \ \text{and} \ |y - a| < |1 - b - y|)\}$.

A *noncooperative price schedule equilibrium* in pure strategies of the above game is a pair $[p_1^*(\cdot), p_2^*(\cdot)]$ of price schedules such that

$$\pi_i[p_i^*(\cdot), p_j^*(\cdot); a, b] \geq \pi_i[p_i(\cdot), p_j^*(\cdot); a, b], \quad \forall p_i(\cdot) \in \mathcal{P}_i, i = 1, 2$$

and $i \neq j$.

---

[13]This assumption is far from being innocuous. Indeed, allowing the firms to use dominated strategies, i.e. firms can charge delivered prices below unit cost over a non-negligible set of locations, yields additional price equilibria; see, for example, Thisse and Vives (1992).

Since transportation costs to a point are unaffected by transport to other points and since marginal production costs are constant (zero), there is a separate Bertrand game at every point y. (Of course, arbitrage could link "local" markets through possible resales among consumers located at different points. However, we will see that arbitrage is not binding in equilibrium.) A standard Bertrand-like argument then runs as follows. Assume that for customers at y, A is the nearer firm. Despite the assumption of zero marginal production costs, seller A has a (transport) cost advantage which allows him to undercut any price set by seller B. The price undercutting process will stop when B can no longer reduce its price, i.e. when price is equal to $t|1 - b - y|$, the transportation cost incurred by the second-nearer firm. Returning to the allocation rule introduced above, customers at y buy from the nearer firm, i.e. seller A. The set of customers equidistant from sellers A and B has a zero measure provided only that the two firms are not coincidentally located $(a \neq 1 - b)$. Thus, we have:

Proposition 3.   *There exists a unique price schedule equilibrium; it is given by*

$$p_1^*(y) = p_2^*(y) = \max\{t|y - a|, t|1 - b - y|\}$$

for almost all $y \in [0, 1]$.

It is readily verified that the market is segmented at the point where customers are equidistant from both sellers: $m = \frac{1}{2}(1 - b + a)$. Furthermore, arbitrage is never profitable since the difference between two delivered prices is smaller than or equal to the corresponding transportation cost. This equilibrium was first identified by Hoover (1937) and formally investigated by Lederer and Hurter (1986).

Two remarks are in order. First, Proposition 3 guarantees the existence of an equilibrium for *any* location pair (a, b). This is to be contrasted with the mill pricing case where an equilibrium exists only when sellers A and B are sufficiently far apart (see Proposition 1).[14] Second, the existence property is *general* and extends to the cases of: (i) multi-dimensional space; (ii) non-uniform or atomic distributions of customers; (iii) continuous, decreasing and location-specific demand functions; and (iv) increasing and firm-specific transportation cost functions in distance [see Thisse and Vives (1988)]. Essentially, the argument is similar to that used in the above example. But $p_i^*(y)$ may now

---

[14]Kats (1987) considers price discrimination with m tiers in which each firm charges the same mill price per tier. When in the first stage a firm chooses m mill prices and, in the second stage, m scalars describing the size of each tier, he shows that $m \geq 2$ is already *sufficient* to restore existence. Furthermore, when $m \to \infty$, the m tier equilibrium converges to the equilibrium identified in Proposition 3. See also Kats (1990) for further developments.

differ from (6), thus reflecting the properties of the local demand and the sellers' costs. The key assumptions are the constant marginal production costs and constant returns w.r.t. the volume hauled.

3.2. Variable prices and locations

We concentrate on the sequential equilibria only. Because of the lack of space, we will limit ourselves to the inside location game.

Consider the model described in Subsection 3.1. A diagrammatic argument will be sufficient to prove the existence of a location equilibrium. Assume first that seller B, located at distance b from the right endpoint of the unit interval, is the only firm on the market. He then supplies all the customers and the corresponding total transport costs are given by the area of the triangles $BC0$ and $BD1$ in Figure 2. Now let seller A be located at distance a from the left endpoint. Given the resulting equilibrium price schedules (see Proposition 3), seller A supplies the customers located in $[0, \bar{m}]$ and receives a payoff equal to the area of the quadrilateral shaded horizontally. Then, it is readily verified that this area is precisely the difference between the total transportation costs borne by seller B when he is alone on the market and the total transportation costs borne by sellers A and B when they are both on the market. Hence, in order to maximize his profits, A must choose to locate at a point generating the largest decrease in total transportation costs. Consequently, if both firms locate

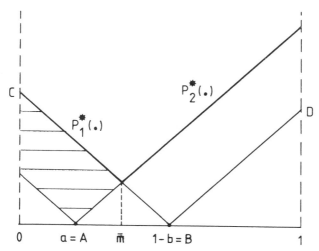

Figure 2. Diagrammatic determination of the best reply location.

at the transportation cost-minimizing points, i.e. $a = b = \frac{1}{4}$, no firm can increase its profits by unilaterally changing its location. Moreover, it is easily checked that any other pair of locations is not an equilibrium. Thus, we have:

Proposition 4. *The socially optimal location pair is the equilibrium of the first-stage game.*

The above argument, developed formally by Lederer and Hurter (1986), can be generalized to the case of: (i) multidimensional space; (ii) nonuniform or atomic distributions of customers; and (iii) increasing and firm-specific transportation cost functions.[15] The critical assumptions for the proof turn out to be the constant marginal production and transportation costs and the (perfectly) inelastic demand. For example, Gupta (1991) shows that increasing marginal production cost induces firms to choose locations outside the first and third quartiles, while Hamilton et al. (1989) demonstrate that using price-sensitive local demands leads firms to locate inside the quartiles.

The conclusions of the analysis of competition under discriminatory pricing are more encouraging than those derived under mill pricing. There exists a price schedule equilibrium for a wide class of problems. *The fact that each firm has more flexibility in its response to its rivals helps in restoring existence.*[16] To gain the customers at one point, a firm has only to change its local price. By contrast, under mill pricing a price cut affects the whole set of the firm's customers, thus generating more potential instability in the competitive process. Furthermore, the existence of a subgame–perfect price schedule–location equilibrium has been established for a significant class of problems. In particular, provided that firms have access to the same transportation technology, they never locate coincidentally in equilibrium. The reason is identical to that found in the mill pricing case: firms want to avoid the damage of price competition by separating from each other in space.

4. Location under nonprice competition

In some industries firms do not exert any control over their price because of either cartel agreements or price regulation by public authorities. Hence, competition among firms must take alternative forms. In particular, firms may

---

[15]The case of a heterogeneous product is dealt with by Anderson and de Palma (1988).

[16]In some sense, price discrimination operates with respect to mill pricing as mixed strategies operate with respect to pure strategies by enlarging the space of strategies.

compete by choosing location in such a way that they obtain the largest possible sales (which amounts here to profit maximization since prices are parametric and marginal production costs are zero).[17] The locational process may imply that either all firms locate simultaneously (Subsection 4.1) or sequentially (Subsection 4.2).

4.1. Simultaneous locations

As in the previous sections, let us assume that customers are distributed uniformly over the segment [0, 1] and that each consumer buys exactly one unit of the product. Since the product is homogeneous, we know that each consumer wants to purchase from the firm with the lowest full price. In this section the assumption is made that the mill price is given and equal for all firms. Consequently, consumers will choose to patronize the nearest firm. (When several firms are equidistant from a customer, we assume that each has an equal probability to sell.) Finally, it is assumed that transport is under the customers' control and that the cost of carrying one unit of the product is a continuous and increasing function of the distance.

In the present class of games, firms' strategies are given by locations only. Furthermore, it is readily verified that a firm's payoff is given by the measure of the set of consumers for whom this firm is the nearest one. (If several firms are located at the same point, they equally share the corresponding market segment.) To start with, let us consider the case of two firms. If firms A and B are located, respectively, at distances a and b from the extremities of [0, 1], their payoffs are given by

$$S_1(a, b) = \frac{1 + a - b}{2} \quad \text{and} \quad S_2(a, b) = \frac{1 - a + b}{2},$$

if $a \neq b$ and $a < 1 - b$,

$$S_1(a, b) = S_2(a, b) = \tfrac{1}{2}, \quad \text{if } a = b.$$

Clearly, the payoff functions exhibit a discontinuity when the two firms cross each other outside the market center.

Interestingly, in spite of the discontinuity of the payoffs, a single equilibrium can be shown to exist. Indeed, let firm 1, say, be located outside the center. In this case there is no location where firm 2 can maximize its sales since firm 2's

---

[17]Hotelling has suggested reinterpreting that model to explain the choice of political platforms in party competition, when parties aim at maximizing their constituency. This idea has been elaborated by Downs (1957) and developed further by many others. A recent survey of this literature is provided by Enelow and Hinich (1984).

sales exhibit a downward discontinuity at firm 1's location where the former approaches the latter on the larger side of the market. This prevents any pair of noncentral locations to be an equilibrium. Assume now that both firms are placed at the center. Then, each of them gets half of the market and any unilateral move of a firm away from the center leads to a decrease in its sales. In other words, *the clustering of the two firms at the market center is the only location equilibrium in pure strategies*.

The case of n firms, with $n \geqslant 3$, has been studied by Lerner and Singer (1937) and, more recently, by Eaton and Lipsey (1975) and Denzau et al. (1985).

For $n = 3$, no location equilibrium exists. The argument runs as follows. Assume that an equilibrium exists where the three firms are separated. Then the two peripheral firms have an incentive to sandwich the interior firm which finds itself with an infinitesimal volume of sales. As a result, this firm wants to leapfrog one of its rivals in order to obtain a positive market share, thus generating instability. Suppose, now, that two firms are clustered and the third isolated. Then the latter can increase its sales by selecting a location next to the clustering. Finally, if the three firms are agglomerated, each one gets one-third of the market. By choosing a location close to the clustering, any firm can gain a larger volume of sales.

Somewhat surprisingly, existence is restored for $n \geqslant 4$. Let us briefly describe the main results [see Eaton and Lipsey (1975) and Denzau et al. (1985) for more details]. When $n = 4$, there exists a unique equilibrium for which two firms are located at the first quartile and the two others at the third one. For $n = 5$, the equilibrium is unique and such that two firms are located at the first sextile, two others at the fifth one, and one firm is isolated at the market center. If $n \geqslant 6$, there exists continuum of equilibrium configurations, characterized as follows:

   (i) no more that two firms are at the same location;
   (ii) peripheral firms are paired with their neighbors;
   (iii) paired firms have equal sales; and
   (iv) isolated firms have sales which are at least as large as those of paired firms but not more than twice as great.[18]

At first glance it seems that we have obtained for the inside location game more positive results than those derived in Section 2. However, as noticed by Eaton and Lipsey (1975), they are not very robust to the specification of the model. In particular, they turn out to be very sensitive to the assumption of a

---

[18]One conclusion of the foregoing analysis is that Hotelling's Principle of Minimum Differentiation is valid only for $n = 2$. Nevertheless, de Palma et al. (1985) show that the Principle holds for n firms when the products are heterogeneous enough.

uniform customer distribution. To show this, let us assume that consumers are continuously distributed over [0, 1] according to the cumulative function $F(x)$. Then, in the two-firm case, we have:

Proposition 5. *If* $n = 2$, *there exists a unique location equilibrium in pure strategies for which the two firms are located at the median of the cumulative function F.*

In contrast, there are no equilibria in pure strategies when $n \geqslant 3$ and when the customer density is strictly convex or strictly concave, however close it is to the uniform one. This has led Osborne and Pitchik (1986) to investigate the existence problem for arbitrary distributions by resorting to mixed strategies. Here also, the Dasgupta–Maskin theorem applies and a location equilibrium in mixed strategies does exist. Osborne and Pitchik then show that, for $n \geqslant 3$, the game has a symmetric equilibrium $(M, \ldots, M)$, where M is the equilibrium mixed strategy. As observed by the authors themselves, an explicit characterization of M appears to be impossible. Yet, when n becomes large, M approaches the customer distribution F. In this case, one can say that firm's location choices mirror the customer distribution.

Finally, returning to the three-firm case with a uniform distribution, Shaked (1982) has shown that firms randomize uniformly over $[\frac{1}{4}, \frac{3}{4}]$, which suggests some tendency towards agglomeration. Osborne and Pitchik have identified an asymmetric equilibrium for the same problem in which two firms randomize, putting most weight near the first and third quartiles, while the third firm locates at the market center with probability one.[19,20]

---

[19]Palfrey (1984) has studied an interesting game in which two established firms compete in location to maximize sales but, at the same time, strive to reduce the market share of an entrant. More specifically, the incumbents are engaged in a noncooperative Nash game with each other, whereas both are Stackelberg leaders with respect to the entrant who behaves like the follower. The result is that the incumbents choose sharply differentiated, but not extreme, locations (in the special case of a uniform distribution, they set up at the first and third quartiles). The third firm always gets less than the two others.

[20]In contrast to the standard assumption of a fixed, given distribution of consumers, Fujita and Thisse (1986) introduce the possibility of consumers' relocation in response to firms' location decisions. Thus, the spatial distribution of consumers is treated as *endogenous*, and a *land market* is introduced on which consumers compete for land use. The game can be described as follows. Given a configuration of firms, consumers choose their location at the corresponding residential equilibrium, which is of the competitive type. With respect to firms, consumers are the followers of a Stackelberg game in which firms are the leaders. Finally, firms choose their location at the Nash equilibrium of a noncooperative game the players of which are the firms. The results obtained within this more general framework prove to be very different from the standard ones. For example, in the two- and three-firm case, the optimal configuration can be sustained as a location equilibrium if the transport costs are high enough or if the amount of vacant land is large enough.

4.2. Sequential locations

In practice, it is probably quite realistic to think of firms entering the market sequentially according to some dynamic process. If firms are perfectly mobile, then the problem associated with the entry of a new firm is equivalent to the one treated in Subsection 4.1 since the incumbents can freely make new location decisions. However, one often observes that location decisions are not easily modified. At the limit, they can be considered as irrevocable.

When entry is sequential and when location decisions are made once and for all, it seems reasonable to expect that an entrant also anticipates subsequent entry by future competitors. Accordingly, at each stage of the entry process the entrant must consider as given the locations of firms entered at earlier stages, but can treat the locations of firms entering at later stages as *conditional* upon his own choice. In other words, the entrant is a follower with respect to the incumbents, and a leader with respect to future competitors. The location chosen by each firm is then obtained by backward induction from the optimal solution of the location problem faced by the ultimate entrant, to the firm itself. This is the essence of the solution concept proposed by Prescott and Visscher (1977).

To illustrate, assume a uniform distribution of consumers along $[0, 1]$. For $n = 2$, the two firms locate at the market center as in the above. When $n = 3$, we have seen that no pure strategy equilibrium exists in the case of simultaneous choice of locations but an equilibrium with foresighted sequential entry does. Indeed, it can be shown that the first firm locates at $\frac{1}{4}$ (or at $\frac{3}{4}$), the second at $\frac{3}{4}$ (or at $\frac{1}{4}$) and the third anywhere between them.[21] For larger values of n, characterizing the equilibrium becomes very cumbersome (see, however, Prescott and Visscher for such a characterization when the number of potential entrants is infinite).[22]

5. Concluding remarks

Spatial competition is an expanding field lying at the interface of game theory, economics, and regional science. It is still in its infancy but attracts more and more scholars' interest because the competitive location problem emerges as a prototype of many economic situations involving interacting decision-makers.

---

[21] See Dewatripont (1987) for a possible selection where the third firm uses its indifference optimality in order to influence the other two firms' location choice.

[22] Although the sequential location models discussed here have been developed in the case of parametric prices, the approach can be extended to deal with price competition too [see, for example, Neven (1987)].

In this chapter we have restricted ourselves to the most game-theoretic elements of location theory. In so doing, we hope to have conveyed the message that space can be used as a "label" to deal with various problems encountered in industrial organization. The situations considered in this chapter do not exhaust the list of possible applications in that domain. Such a list would include intertemporal price discrimination and the supply of storage, competition between multiproduct firms, the incentive to innovate for imperfectly informed firms, the techniques of vertical restraints, the role of advertising, and incomplete markets due to spatial trading frictions. Most probably, Hotelling was not aware that game theory would so successfully promote the ingenious idea he had in 1929.

References

Anderson, S.P. (1987) 'Spatial competition and price leadership', *International Journal of Industrial Organization*, 5: 369–398.

Anderson, S.P. (1988) 'Equilibrium existence in a linear model of spatial competition', *Economica*, 55: 479–491.

Anderson, S.P. and A. de Palma (1988) 'Spatial price discrimination with heterogeneous products', *Review of Economic Studies*, 55: 573–592.

Bester, H. (1989) 'Noncooperative bargaining and spatial competition', *Econometrica*, 57: 97–119.

Champsaur, P. and J.-Ch. Rochet (1988) 'Existence of a price equilibrium in a differentiated industry', INSEE, Working Paper 8801.

Dasgupta, P. and E. Maskin (1986) 'The existence of equilibrium in discontinuous economic games: Theory and applications', *Review of Economic Studies*, 53: 1–41.

d'Aspremont, C., J.J. Gabszewicz and J.-F. Thisse (1979) 'On Hotelling's "Stability in Competition"', *Econometrica*, 47: 1145–1150.

d'Aspremont, C., J.J. Gabszewicz and J.-F. Thisse (1983) 'Product differences and prices', *Economics Letters*, 11: 19–23.

Denzau, A., A. Kats and S. Slutsky, (1985) 'Multi-agent equilibria with market share and ranking objectives', *Social Choice and Welfare*, 2: 95–117.

de Palma, A., V. Ginsburgh, Y.Y. Papageorgiou and J.-F. Thisse (1985) 'The principle of minimum differentiation holds under sufficient heterogeneity', *Econometrica*, 53: 767–781.

de Palma, A., M. Labbé and J.-F. Thisse, (1986) 'On the existence of price equilibria under mill and uniform delivered price policies', in: G. Norman, ed., *Spatial pricing and differentiated markets*. London: Pion, 30–42.

Dewatripont, M., (1987) 'The role of indifference in sequential models of spatial competition', *Economics Letters*, 23: 323–328.

Downs, A., (1957), *An economic theory of democracy*. New York: Harper and Row.

Eaton, B.C., (1972) 'Spatial competition revisited', *Canadian Journal of Economics*, 5: 268–278.

Eaton, B.C. and R.G. Lipsey, (1975) 'The principle of minimum differentiation reconsidered some new developments in the theory of spatial competition', *Review of Economic Studies*, 42: 27–49.

Economides, N., (1984), 'The principle of minimum differentiation revisited', *European Economic Review*, 24: 345–368.

Economides, N., (1986) 'Nash equilibrium in duopoly with products defined by two characteristics', *Rand Journal of Economics*, 17: 431–439.

Enelow, J.M. and M.J. Hinich, (1984) *The spatial theory of voting. An introduction*. Cambridge: Cambridge University Press.

Fujita, M. and J.-F. Thisse, (1986) 'Spatial competition with a land market: Hotelling and Von Thunen unified', *Review of Economic Studies*, 53: 819–841.

Gabszewicz, J.J. and P. Garella, (1986), ''Subjective' price search and price competition', *Industrial Journal of Industrial Organization*, 4: 305–316.

Gabszewicz, J.J. and J.-F. Thisse, (1986) 'Spatial competition and the location of firms', *Fundamentals of Pure and Applied Economics*, 5: 1–71.

Garella, P.G. and X. Martinez-Giralt, (1989) 'Price competition in markets for dichotomous substitutes', *International Journal of Industrial Organization*, 7: 357–367.

Glicksberg, I.L., (1952) 'A further generalization of the Kakutani fixed point theorem with applications to Nash equilibrium points', *Proceedings of the American Mathematical Society*, 38: 170–174.

Gupta, B., (1991) 'Competitive spatial price discrimination with nonlinear production cost', University of Florida, Department of Economics, mimeo.

Hamilton, J.H., J.-F. Thisse, and A. Weskamp, (1989) 'Spatial discrimination: Bertrand vs. Cournot in a model of location choice', *Regional Science and Urban Economics*, 19: 87–102.

Hoover, E.M., (1937) 'Spatial price discrimination', *Review of Economic Studies*, 4: 182–191.

Hotelling, H., (1929) 'Stability in competition', *Economics Journal*, 39: 41–57.

Kats, A., (1987) 'Location-price equilibria in a spatial model of discriminatory pricing', *Economics Letters*, 25: 105–109. Erratum: personal communication.

Kats, A., (1989) 'Equilibria in a circular spatial oligopoly', Virginia Polytechnic Institute, Department of Economics, mimeo.

Kats, A., (1990) 'Discriminatory pricing in spatial oligopolies', Virginia Polytechnic Institute, Department of Economics, Working Paper E-90-04-02.

Kohlberg, E. and W. Novshek, (1982) 'Equilibrium in a simple price-location model', *Economics Letters*, 9: 7–15.

Lederer, P.J. and A.P. Hurter, (1986) 'Competition of firms: discriminatory pricing and location', *Econometrica*, 54: 623–640.

Lerner, A. and H.W. Singer, (1937) 'Some notes on duopoly and spatial competition', *Journal of Political Economy*, 45: 145–186.

MacLeod, W.B., (1985) 'On the non-existence of equilibria in differentiated product models', *Regional Science and Urban Economics*, 15: 245–262.

McFadden, D.L., (1984) 'Econometric analysis of qualitative response models', in: Z. Griliches and M.D. Intriligator, eds., *Handbook of econometrics*, Vol. II. Amsterdam: North-Holland, pp. 1395–1457.

Neven, D., (1987) 'Endogenous sequential entry in a spatial model', *International Journal of Industrial Organization*, 4: 419–434.

Novshek, W., (1980) 'Equilibrium in simple spatial (or differentiated product) models', *Journal of Economic Theory*, 22: 313–326.

Osborne, M.J. and C. Pitchik, (1986) 'The nature of equilibrium in a location model', *International Economic Review*, 27: 223–237.

Osborne, M.J. and C. Pitchik, (1987) 'Equilibrium in Hotelling's model of spatial competition', *Econometrica*, 55: 911–923.

Palfrey, T.S., (1984) 'Spatial equilibrium with entry', *Review of Economic Studies*, 51: 139–156.

Ponsard, C., (1983) *A history of spatial economic theory*. Berlin: Springer-Verlag.

Prescott, E.C. and M. Visscher, (1977) 'Sequential location among firms with foresight', *Bell Journal of Economics*, 8: 378–393.

Rubinstein, A., (1982) 'Perfect equilibrium in a bargaining model', *Econometrica*, 50: 97–108.

Shaked, A., (1982) 'Existence and computation of mixed strategy Nash equilibrium for 3-firms location problem', *Journal of Industrial Economics*, 31: 93–96.

Shilony, Y., (1981) 'Hotelling's competition with general customer distributions', *Economics Letters*, 8: 39–45.

Thisse, J.-F., and X. Vives, (1988) 'On the strategic choice of spatial price policy', *American Economic Review*, 78: 122–137.

Thisse, J.-F. and X. Vives, (1992) 'Basing point pricing: competition versus collusion', *Journal of Industrial Economics*, to appear.

Chapter 10

STRATEGIC MODELS OF ENTRY DETERRENCE

ROBERT WILSON*

Stanford Business School

Contents

*Assistance provided by NSF grant SES8908269.

Handbook of Game Theory, Volume 1, Edited by R.J. Aumann and S. Hart

1. Introduction

In the 1980s the literature of economics and law concerning industry structure bloomed with articles on strategic aspects of entry deterrence and competition for market shares. These articles criticized and amended theories that incompletely or inconsistently accounted for strategic behavior. The aftermath is that game-theoretic models and methods are standard tools of the subject – although not always to the satisfaction of those concerned with empirical and policy issues; cf. Fisher (1989) for a critique and Shapiro (1989) for a rebuttal.

This chapter reviews briefly the popular formulations of the era and some interesting results, but without substantive discussion of economic and legal issues. The standard examination of the issues is Scherer (1980) and game-theoretic texts are Tirole (1988) and Fudenberg and Tirole (1991); see also Salop (1981). Issue-oriented expositions are the chapters by Gilbert (1989a) and Ordover and Saloner (1989) in *The Handbook of Industrial Organization*. Others emphasizing game-theoretic aspects are Wilson (1985, 1989a, 1989b), Fudenberg and Tirole (1986c), Milgrom (1987), Milgrom and Roberts (1987, 1990), Roberts (1987), Gilbert (1989b) and Fudenberg (1990). The combined length of these surveys matches the original articles, so this chapter collects many models into a few categories and focuses on the insights offered by game-theoretic approaches.

The motives for these studies are the presumptions, first, that for an incumbent (unregulated) firm one path to profits is to acquire or maintain monopoly power, which requires exclusion of entrants and expulsion, absorption, intimidation, or cartelization of competitors; and second, that monopoly power has adverse effects on efficiency and distribution, possibly justifying government intervention via antitrust and other legal measures. We examine here only the possibilities to exclude or expel entrants.

A single issue motivates most game-theoretic studies: when could an incumbent profitably deter entry or survival in a market via a strategy that is credible – in the sense that it is part of an equilibrium satisfying selection criteria that exclude incredible threats of dire consequences? This issue arises because non-equilibrium theories often presume implicitly that deterrence is easy or impossible. To address the matter of credibility, all studies assume some form of perfection as the equilibrium selection criterion: subgame perfection, sequential equilibrium, etc. in increasing selectivity. The models fall into three categories.

- *Preemption*. These models explain how a firm claims and preserves a monopoly position. The incumbent obtains a dominant position by arriving first in a natural monopoly; or more generally, by early investments in research and product design, or durable equipment and other cost reduction. The hallmark is commitment, in the form of (usually costly) actions that irreversibly strengthen the incumbent's options to exclude competitors.

• *Signaling.* These models explain how an incumbent firm reliably conveys information that discourages unprofitable entry or survival of competitors. They indicate that an incumbent's behavior can be affected by private information about costs or demand either prior to entry (limit pricing) or afterwards (attrition). The hallmark is credible communication, in the form of others' inferences from observations of costly actions.

• *Predation.* These models explain how an incumbent firm profits from battling a current entrant to deter subsequent potential entrants. In these models, a "predatory" price war advertises that later entrants might also meet aggressive responses; its cost is an investment whose payoff is intimidation of subsequent entrants. The hallmark is reputation: the incumbent battles to maintain other's perception of its readiness to fight entry.

Most models of preemption do not involve private information; they focus exclusively on means of commitment. Signaling and predation models usually require private information, but the effects are opposite. Signaling models typically produce "separating" equilibria in which observations of the incumbent's actions allow immediate inferences by entrants; in contrast, predation models produce "pooling" equilibria (or separating equilibria that unravel slowly) in which inferences by entrants are prevented or delayed.[1]

These three categories are described in the following sections. We avoid mathematical exposition of the preemption models but specify some signaling and predation models. As mentioned, all models assume some form of perfection.

2. Preemption

A standard example of preemption studied by Eaton and Lipsey (1977), Schmalansee (1978) and Bonanno (1987), is an incumbent's strategy of offering a large product line positioned to leave no profitable niche for an entrant. A critique by Judd (1985) observes, however, that if the incumbent can withdraw products cheaply, then an entrant is motivated to introduce a product by anticipating the incumbent's incentive to withdraw close substitutes in order to avoid depressed prices for its other products.

---

[1]The distinction between signaling models and those predation models based on reputational effects is admittedly tenuous, as for example in the cases that a signaling model has a pooling equilibrium or an attrition model unravels slowly.

A second example invokes switching costs incurred by customers, which if large might deter entry. Klemperer (1987a, 1987b) uses a two-period model, and in Klemperer (1989) a four-period model, to study price wars to capture customers: monopoly power over customers provides later profits that can be substantially dissipated in the initial competition to acquire them – possibly with the motive of excluding opportunities for a later entrant. Farrell and Shapiro (1988) consider an infinite-horizon example with overlapping generations of myopic customers who live two periods; two firms alternate roles in naming prices sequentially. The net result is that the firms rotate: each captures periodically all the customers and then profits from them in the interim until they expire and it re-enters the market to capture another cohort. However, these conclusions are altered substantially by Beggs and Klemperer (1992) in an infinite-horizon model with continual arrival of new (non-myopic) customers having diverse tastes, continual attrition of old customers, and two firms with differentiated products. For a class of Markovian strategies, price wars occur initially when both firms have few captive customers, but when the population is stationary (as in an established market) the competitive process converges monotonically over time to a stationary configuration of prices and market shares. In particular, an incumbent's monopoly can be invaded by an entrant who eventually achieves a large share. This model casts doubt on interpreting switching costs as barriers to entry in stable markets: switching costs induce an incumbent to price high to exploit its captive market, enabling an entrant to capture new arrivals at lower but still profitable prices.

This is an instance of the general effect that [in the colorful terminology of Fudenberg and Tirole (1986c)] a "fat cat" incumbent with a large stock of "goodwill" with customers (due to switching costs or perhaps advertising) prefers to exploit its existing stock rather than countering an entrant. The incumbent may choose its prior investment in goodwill to take this effect into account, either investing in goodwill and conceding entry, or not investing and deterring entry.

Farrell and Saloner (1986) illustrate that switching costs can have appreciable effects in situations with growing demand affected by network externalities; that is, each customer's valuation of a product grows with the number of others adopting the product. In this case an incumbent can profit from aggressive pricing to prevent entry, because the losses are recouped later as profits from more numerous captive customers, especially if the prevention of entry encourages standardization on the incumbent's product and thereby lessens subsequent risks of entry.

On the supply side, Bernheim (1984) studies a model in which incumbents expend resources (e.g., advertising) to raise an entrant's sunk costs of entry; cf. Salop and Scheffman (1983, 1987) and Krattenmaker and Salop (1986, 1987) for an elaboration of the basic concept of "raising rivals' costs" as a competi-

strategy in other contexts than entry deterrence.[2] From each initial configuration, entry proceeds to the next larger equilibrium number of firms. He notes that official measures designed to facilitate entry can have ambiguous effects because intermediate entrants may be deterred by prospects of numerous arrivals later. Waldman (1987) re-analyzes this model allowing for uncertainty about the magnitude of the sunk cost incurred by entrants; in this case, entry deterrence is muted by each incumbent's incentive to "free ride" on others' entry-deterring actions. This result is not general: he shows also that an analogous variant of a model in Gilbert and Vives (1986) retains the opposite property that there is no free-rider effect.

Another example, studied by Ordover, Saloner and Salop (1990), refers to "vertical foreclosure". In the simplest case, one of two competing firms integrates vertically with one of two suppliers of inputs, enabling the remaining supplier to raise prices to the integrated firm's downstream competitor, thereby imposing a disadvantage in the market for final products. The authors examine a four-stage game, including an initial stage at which the two downstream firms bid to acquire one upstream supplier, and a later opportunity for the losing bidder to acquire the other supplier. Particular assumptions are used but the main conclusion is that foreclosure occurs if the residual supplier's gain exceeds the loss suffered by the unintegrated downstream firm. This circumstance precludes a successful offer from the latter to merge and thereby counter its competitor's vertical integration. Strategic complements [Bulow, Geanakoplos and Klemperer (1985a)], in the form of Bertrand price competition at both levels, implies this condition and therefore also implies that foreclosure occurs; but it is false in the case of strategic substitutes. As usually modeled, Cournot quantity competition implies strategic substitutes, but foreclosure can still occur in a duopoly. The particular forms of pricing and contracting (including commitment to exclusive dealing by the integrated firm) assumed in this model are relaxed in the more elaborate analysis by Hart and Tirole (1990) allowing arbitrary contractual arrangements.

Vertical integration is a particular instance of long-term contracting between a seller and a buyer, which has been studied by Aghion and Bolton (1987) and Rasmusen, Ramseyer and Wiley (1991) in the context of entry deterrence. They observe that an incumbent seller and buyer can use an exclusive-dealing contract to exercise their joint monopoly power over an entrant: penalties payable by the buyer to the seller if the buyer deals with the entrant are in effect an entry fee that extracts the profit the entrant might otherwise obtain.

---

[2]Entry costs are sunk if they cannot be recovered by exit; e.g., investments in equipment are not sunk if there is a resale market, but they are sunk to the degree the equipment's usefulness is specific to the firm or the product. Coate and Kleit (1990) argue from an analysis of two cases that the requirements of the theory of "raising rivals' costs" are rarely met in practice. See also Kleit and Coate (1991).

In particular, contractually created entry fees can prevent or delay (until expiration of the contract) entry by firms more efficient than the incumbent seller.

In a natural monopoly the first firm to install ample (durable) capacity obtains incumbency and deters entrants on a similar scale, provided all economies of scale are captured. Several critiques and extensions of this view have been developed. Learning effects (i.e., production costs decline as cumulative output increases) can engender a race among initial rivals. An incumbent can benefit from raising its own opportunity cost of exit: the standard example is a railroad whose immovable durable tracks ensure that it would remain a formidable competitor against truck, barge, or air carriers whose capacity can be moved to other routes. Eaton and Lipsey (1980) note that if capacity has a finite lifetime, then the incumbent must renew it prematurely to avoid preemptive investment by an entrant that would eliminate the incumbent's incentive to continue.

Gelman and Salop (1983) observe that entry on a small scale can still be profitable: there exists a scale and price small enough that the incumbent prefers to sell the residual demand at the monopoly price rather than match the entrant's price. They observe further that the entrant can extort the incumbent's profit by selling discount coupons that the incumbent has an incentive to honor if the discounted price exceeds its marginal cost. In the United States, the airlines' coupon war of the early 1980s is an evident example.

Even in an oligopoly, incumbent firms have incentives to install more capacity (or alter the positioning of their product designs) when entry is possible; cf. Spence (1977, 1979), Dixit (1979, 1980), Eaton and Lipsey (1981), Ware (1984) and, for models with sequential entry, Prescott and Visscher (1977)[3], and Eaton and Ware (1987). Profitable entry is prevented by capacities (and product designs) that prevent an additional firm from recovering its sunk costs of entry and fixed costs of operation. Conceivably, extra unused capacity might be held in reserve for price wars against entrants, and indeed Bulow, Geanakoplos and Klemperer (1985b) provide an example in the case of strategic complements. However, in the case of strategic substitutes (the usual case when considering capacities as strategic variables) capacity is fully used for production, as demonstrated in the model of Eaton and Ware (1987). The Stackleberg model of Basu and Singh (1990), however, allows a role for an incumbent to use inventories strategically.

The thrust of these models is to develop the proposition [e.g., Spence (1979) and Dixit (1980)] that incumbency provides an inherent advantage to move first to commit to irreversible investments in durable capacity that restrict the

---

[3]An additional feature is added by Spence (1984): investments in capacity are fully appropriable by the firm but other cost-reducing investments in process and product design are not fully appropriable; moreover, if these spillover effects strengthen competitors, then each firm's incentive to make such investments is inhibited.

opportunities available to entrants. Ware (1984), and Arvan (1986) for models with incomplete information, show that this advantage is preserved even when the entrant has a subsequent opportunity to make a comparable commitment. Bagwell and Ramey (1990) examine this proposition in more detail in a model in which the incumbent has the option to avoid its fixed cost by shutting down when sharing the market is unprofitable; also see Maskin and Tirole (1988). They observe that the entrant can install capacity large enough to induce exit by the incumbent; indeed, to avoid this the incumbent restricts its capacity to curtail its fixed cost and thereby to sustain its profitability in a shared market. This argument invokes the logic of forward induction: in a subgame-perfect equilibrium that survives elimination of weakly dominated strategies, the incumbent either restricts its capacity to maintain viability after large-scale entry, or if fixed costs are too high, cedes the market to the entrant. This strategy is akin to the one in Gelman and Salop (1983), but applied to the incumbent rather than the entrant.

When capacity can be incremented smoothly and firms have competing opportunities, an incumbent's profits might be dissipated in too-early preemptive investments to deter entrants. Gilbert and Harris (1984) study a game of competition over the timing of increments, and identify a subgame-perfect equilibrium in which all profits are eliminated.[4] Similar conclusions are derived by Fudenberg and Tirole (1985) for the case of timing of adoptions of a cost-reducing innovation in a symmetric duopoly,[5] and this is extended to the asymmetric case of an incumbent and an entrant by Fudenberg and Tirole (1986c): if Bertrand competition prevails in the product market, then the incumbent adopts just before the entrant would, and thereby maintains its advantage at the cost of some dissipation of potential profit. This result is similar to the role of preemptive patenting in maintaining a monopolist's advantage, as analyzed by Gilbert and Newbery (1982).

Examining an issue raised by Spence (1979), Fudenberg and Tirole (1983) suppose that firms build capacity smoothly at bounded rates over time, which allows multiple equilibria. In one equilibrium the firms accumulate capacity to reach the Cournot equilibrium (or perhaps a Stackleberg equilibrium if one has a head start) but in other equilibria they stop with smaller final capacities: each firm expands farther only if another does. Indeed, they can stop at the monopoly total capacity and split the profits. In this view, an incumbent may be interested less in exploiting its head start by racing to build capacity, than in an accommodation with an entrant to ensure that both refrain from large capacities. Continual arrival of new entrants may therefore be necessary to ensure socially efficient capacities.

---

[4]Mills (1988) notes, however, that sufficiently lumpy capacity increments allow an incumbent with a first-mover advantage a substantial portion of the monopoly profit.

[5]Three or more firms yields different results.

Another example is a market without durable capacity but with high fixed costs; e.g., capacity is rented. Each period each active firm incurs a fixed cost so high that the market is a natural monopoly. Maskin and Tirole (1988) assume that an active firm is committed to its output level for two periods and two firms have alternating opportunities to choose whether to be active or not. In the symmetric equilibrium with Markovian strategies, the first firm (if the other is not active) chooses an output level large enough to deter entry by the other next period, and this continues indefinitely. In particular, suppose the profit of firm 1 in a period with outputs (q_1, q_2) is $\pi(q_1, q_2)$ and symmetrically for firm 2; also, the maximum monopoly profit $(q_2 = 0)$ covers the fixed cost c of one firm but not two. Then the optimal entry-deterring output is the minimum value of q for which

$$\pi(q, q) - c + \frac{\delta}{1 - \delta} [\pi(q, 0) - c] \leq 0,$$

if the discount factor δ is not too small. That is, if q is the optimal output and next period the other firm were to incur the fixed cost c enabling it to choose a positive output, then it too would choose q; therefore, this output must be sufficiently large to ensure that the present value of successful expulsion of the incumbent is not positive. If the period length is short ($\delta \approx 1$), however, then such a market is easily "contestable" since the commitment period is negligible; in particular, the entry-deterring output grows and the incumbent's profit shrinks as the period length is shortened.

There can also be asymmetric Markovian equilibria if the fixed cost and the discount factor are large enough; e.g., the first firm merely uses its two-period reaction function and then the second never enters. And via the Folk Theorem, there are many symmetric subgame-perfect equilibria that are not Markovian and that yield higher profits.

In general, ease of entry need not ensure low prices, due to the Folk Theorem. For instance, using a model of a market for a durable good, Ausubel and Deneckere (1987) observe that an incumbent monopolist can persist in charging nearly the monopoly price without incurring entry. The entrant is deterred from entering by the prospect of marginal-cost pricing thereafter, and the incumbent is deterred from offering lower prices by the prospect of entry and even lower prices thereafter that are still high enough to justify entry. The feature enabling this result is the Coase property of durable good pricing:[6] if the period length is short and marginal cost is constant, then in any subgame-perfect equilibrium with stationary strategies for the customers, the price of a

---

[6]The Coase property is stated here for the case of a continuous demand function intersecting the seller's supply function; cf. Gül, Sonnenschein and Wilson (1986) for other technical assumptions.

monopolist (not threatened with entry) is close to marginal cost. As shown also by Gül (1987) in greater generality, this relatively unprofitable prospect can be used as a punishment to construct subgame-perfect equilibria of duopolies that sustain the incumbent's punishments required above.

The overall theme of preemption models is that costly irreversible investments that enhance incumbents' competitive strength (or burden entrants) provide genuine commitment that can deter entry. The models in the next sections, in contrast, suppose that incumbents cannot make commitments.

3. Signaling

Signaling models examine costly credible "communication" that selects the firms to enter or survive in a market. Typically some aspect of each firm's profitability is private information, such as its marginal or fixed cost. Moreover, the only credible signal of a firm's competitive strength is the demonstration itself, via endurance of lower profits longer than it would tolerate if its cost were higher. We mention two prominant examples. Among firms currently active in a market, the battle for survival is modeled as a war of attrition. In the case of incumbents threatened with entry, their current prices signal costs or demand and thereby affect the potential entrant's decision about proceeding with entry.

3.1. Attrition

Attrition models study markets with excess numbers and examine the process that selects survivors. The formulation of Fudenberg and Tirole (1986a, 1986c) is representative. Consider a symmetric market in which at each time $t \geq 0$ each of N firms $i = 1, \ldots, N$ initially active in the market obtains net profit at the rate $\pi_n(t) - c_i$ if it is one of n firms remaining active, and zero if it has irrevocably exited earlier. For instance, if i is the solve survivor when its last remaining competitor exits at time t, then its present value of continuation is, using $n = 1$,

$$V_n(t, c_i) = \int_t^\infty [\pi_n(s) - c_i] e^{-r[s-t]} \, ds$$

when the interest rate is r. Suppose the profit functions π_n are monotone and continuous as functions of time, and uniformly decreasing in the number of active firms. Furthermore, the firms' fixed operating costs c_i are privately known, each drawn independently according to the differentiable distribution function F having an interval support including both extreme possibilities: the cost might be so high that the firm is unprofitable as a monopolist, or so low that it could forever profit as one of N firms.[7]

In the end-game between two firms $(n = 2)$, firm i prefers at time t to continue in the duopoly if

$$\pi_2(t) - c_i + h(t)V_1(t, c_i) \geq 0 ,$$

where $h(t)$ is the hazard rate at which the other firm exits, leaving i with a perpetual monopoly. Representing each firm's strategy in a symmetric equilibrium as the lowest cost $C(t)$ inducing it to exit at time t if the other has not exited previously, this hazard rate is

$$h(t) = [1 - F(C(t))]' / F(C(t)) ,$$

based on the inference that the other's cost is less than $C(t)$ if it has not exited previously. An equilibrium requires, therefore, that the inequality above is actually an equality at $c_i = C(t)$. This condition yields a differential equation characterizing the equilibrium strategy; moreover, its boundary condition is given by the initial condition $V_1(0, C(0)) = 0$ indicating that a firm unable to profit as a monopolist exits immediately. If profits increase with time, then $C(t)$ and $\pi_2(t)$ may intersect at some time after which a duopoly is viable; hence, the actual strategy is to continue as long as one's cost is less than the greater of $C(t)$ and $\pi_2(t)$. If each firm is initially viable as a monopolist, then the net result is that the higher-cost firm eventually exits; or if both have sufficiently low costs, then a duopoly persists forever.

The basic theory of attrition games is developed by Nalebuff and Riley (1985) and Riley (1980). Milgrom and Weber (1985) study the symmetric equilibria of symmetric attrition games in which, as above, each party has a privately known cost of delay that increases linearly with time. They provide an analysis in terms of distributional strategies; the hazard rate of exit is shown to decline with time; and mixed strategy equilibria are characterized.[8] Additional applications to markets with declining demand are discussed by Ghemawat and Nalebuff (1985, 1990) using models with complete information, and by Fishman (1990) who includes an initial entry phase. One model supposes identical cost structures but firms differ in their capacities, which impose fixed costs in

---

[7]As described by Fudenberg and Tirole (1986a), this support assumption ensures a unique equilibrium in the example below.

[8]Section 4 describes an alternative version of attrition, called "chicken", derived from models of reputational effects, in which the hazard rate increases. See Ordover and Rubinstein (1986) for a related model in a different context.

proportion to capacity: the subgame-perfect equilibrium has larger firms exiting earlier. The second allows firms to shrink their capacities: again, larger firms contract earlier.

3.2. Limit pricing

Studies of limit pricing examine the incentives of incumbent firms to signal their private information about costs or demands to deter misguided entry. The motive is clearest in the case of an incumbent monopolist with a privately known marginal cost who anticipates that a potential entrant will enter if it perceives that its profits would exceed its privately known sunk cost of entering. Suppose that profits are lower for the entrant (and higher for the incumbent) if the incumbent's cost is lower, and lower for the incumbent after entry. Moreover, prior to entry the entrant can observe the price chosen by the incumbent but not its marginal cost. Suppose first that the incumbent antici-pates naive inferences by the entrant; for instance, the entrant infers that the marginal cost is the one for which the price is the myopically optimal monopoly price. Then the incumbent prefers to cut its price somewhat to reduce the entrant's assessment of its cost, and therefore to reduce the chance of entry. In reverse, suppose the incumbent anticipates sophisticated inferences by the entrant; then the incumbent cannot charge the higher myopically optimal monopoly price without inducing false hopes in the entrant and thereby encouraging entry. Thus, one anticipates an equilibrium in which the incum-bent shaves its price before entry: this provides an accurate signal to the entrant, who then enters only if the (correctly) anticipated profit exceeds its sunk cost.

This logic is formalized in a model developed by Milgrom and Roberts (1982a). Knowing its marginal cost c, the incumbent chooses its pre-entry price p to maximize the expected present value of its pre-entry profit $\pi(p, c)$ and post-entry profit $\pi_n(c)$ with $n = 1$ or 2 firms:

$$\pi_1(p, c) + \delta\{\pi_1(c) - h(p)[\pi_1(c) - \pi_2(c)]\},$$

where $h(p)$ is the probability of entry. If $P(c)$ is this optimal price, and supposing P is invertible (i.e., the equilibrium is separating, so the signal is "accurate"), then entry occurs if the entrant's sunk cost is less than its anticipated profit $\pi_2^e(\hat{c})$ based on the inferred cost $\hat{c} = P^{-1}(p)$. Thus, if the entrant's sunk cost is drawn independently according to the distribution function F, then from the incumbent's perspective:

$$h(p) = F(\pi_2^e(P^{-1}(p))).$$

Combining this equilibrium condition with the previous one yields a differential equation that determines the incumbent's strategy. If there is an upper bound on the incumbent's cost, then the boundary condition is simply the requirement that for this highest cost the price is the myopically optimal price: this reflects the usual property of separating signaling equilibria that (in this model) the highest cost incumbent has no fear of being mistaken as having a higher cost.

Milgrom and Roberts actually assume that the entrant's private information consists of its own marginal cost, which is distributed independently of the incumbent's marginal cost, but the analysis is similar. Regularity conditions that ensure the existence of a separating equilibrium are provided by Milgrom and Roberts. In some cases a unique separating equilibria is obtained by eliminating weakly dominated strategies, but usually partial pooling and full pooling equilibria exist too. Mailath (1987) establishes general results about signaling games that, in the context of the Milgrom–Roberts model, imply (subject to a parameter restriction) existence and uniqueness of a separating equilibrium. Ramey (1987) demonstrates that a pooling equilibrium is necessary if the gain from entry deterrence is sufficiently large. In particular, if costs are independent, then the incumbent's gain from reducing the likelihood of entry can be so great that no amount of price reduction can credibly signal low costs; in such cases there are no separating equilibria. More generally, Cho (1990a, 1990b) establishes that for a large (and relevant) domain of parameters the stable equilibria must be partially pooling; thus, the incumbent's action leaves the entrant with some residual uncertainty.

Matthews and Mirman (1983) extend the model to the case that the entrant's observation of the price is affected by noise, which in some cases assures a unique equilibrium. Saloner (1982) studies a multiperiod model in which an entrant has repeated opportunities to enter; in this case, one effect of noisy signaling is that there can be more (i.e., mistaken) entry than with complete information. Bagwell and Ramey (1988) adapt the model to the case that the incumbent uses both its price and another expenditure (such as advertising) to signal.[9] Elimination of weakly dominated strategies yields a unique separating equilibrium in which the incumbent acts as it would if the entrant were informed and its cost were lower than it is; analogous results are obtained for cases with pooling equilibria if an additional "intuitive" equilibrium selection criterion due to Cho and Kreps (1987) is used. Presumably the effects of these selection criteria apply also to the Milgrom–Roberts model.

A variety of other specifications of the incumbent's private information have been suggested. Roberts (1985) supposes that in an initial phase *after* entry the incumbent has superior information about demand (and its ouput is not

---

[9]Matthews and Fertig (1990) analyze an alternative motive for advertising by the incumbent, based on counteracting false advertising by the entrant.

observable by the entrant). Thus, as in the predation models reviewed in Section 4, the incumbent drives the price down to influence the entrant's decision to exit. As in other signaling models, however, the entrant makes the correct inferences (in equilibrium) so the exit decision is not actually biased, but entry is discouraged by the entrant's anticipation that this behavior by the incumbent reduces expected profits in the initial phase. As in limit pricing, moreover, the incumbent is forced to lower prices in the initial phase lest he encourage the entrant to stay when it is unprofitable. Fudenberg and Tirole (1986b) note that the incumbent's superior demand information is unnecessary for this result: if both firms are uncertain about demand conditions, then the incumbent prefers to encourage exit by lowering the price observed by the entrant – again, provided the entrant cannot observe the incumbent's action.

A variant of limit pricing is studied by Saloner (1987) in the context that two incumbents negotiate a merger. If the firms have private information about their costs, then the bargaining process encourages each to expand output or cut prices to signal to the other that it will be a formidable competitor if the merger fails. This motive is strengthened if there is also a threat of entry, especially if deterring entry is vital to the success of the merger. Thus, limit pricing could deter new entry and simultaneously facilitate the incumbent's merger.

As described above, the threat of entry lowers prices. Harrington (1986) notes that the effect is reversed if the entrant's marginal cost is highly correlated with the incumbent's. The reason is that entry is deterred differently: in the case of independent costs, by a low cost for the incumbent; but in the case of similar costs, by a *high* cost for the incumbent, because that indicates low profits for the entrant too. With highly correlated costs, therefore, the incumbent's limit price generally *exceeds* the myopically optimal monopoly price. The case that the entrant's information is strictly inferior to the incumbent's produces a "pooling" equilibrium: for all costs of the incumbent in a middle range the incumbent charges the same price, which is below or above the monopoly price as the correlation between their costs is low or high. This price is the same as the myopically optimal price for the least cost of the incumbent that, if known to the entrant, would make entry unprofitable (in expectation). Harrington (1987) extends this analysis to the case of multiple incumbents having the same cost of producing products that are perfect substitutes. He retains the key assumption that the entrant has inferior information and observes only a single signal, say the price on the presumption that the incumbents' separate outputs are unobservable. In this case the equilibrium is again pooling except for costs so low that deterrence cannot be avoided and costs so high that the entrant is deterred surely: in the middle range the price is constant at the price for the least cost in the high range of entry-deterring costs. Neither of these models is developed for the case that the entrant's information is not inferior (e.g., the entrant's sunk cost is privately known), which might alter or eliminate the pooling equilibrium, as in Milgrom and Robert's model where the probability $h(p)$ of entry varies smoothly.

Bagwell and Ramey (1991) find a dramatically different equilibrium when the incumbents' products are differentiated and the entrant observes their individual pre-entry prices. In this case there are equilibria in which the incumbents charge their myopic pre-entry prices; i.e., the equilibrium prices for the associated Bertrand game without threat of entry. Each incumbent anticipates that entry is unaffected by its own price because the entrant can still infer the cost parameter from others' prices. Moreover, equilibrium refinements derived from stability arguments suffice to eliminate all other equilibria.

In sum, signaling models interpret battles for survival among incumbent firms, as well as incumbents' limit pricing to deter new entrants, as communication motivated by implicit bargaining over who shall retain or acquire market shares. The language is restricted to choices of prices, outputs, and other significant decisions that, because they are costly, credibly convey information by averting speculative inferences that survival or entry might be more profitable than it actually is. In some limit-pricing models the net effect is to induce entry when and only when it is profitable for the entrant. Cho's application of stability criteria, and Harrington's analysis of oligopolistic incumbents with a common cost and a single price signal, are sufficient however to indicate that communication can be imperfect (even in the absence of noise) due to pooling equilibria that prevent exact inferences by an entrant. Attrition and limit prices below the myopically optimal prices confer benefits on customers, but in the case of common costs, higher limit prices injure customers until entry occurs.

The next section examines a complementary hypothesis about the effects of an incumbent's private information.

4. Predation

Predation models aim to explain why an incumbent might willingly incur losses battling an entrant, as in a price war. The hypothesized motive is that the cost of the battle is an investment that pays off later, either by expelling the entrant or by deterring later entrants. To introduce this hypothesis, we mention two examples indicating that private information might be an important ingredient; however, we do not repeat here the analyses of basic issues in the references cited in the introduction.

A prominent view of predation is that it is irrational; e.g., the incumbent could obtain the same result at less cost by buying the entrant; cf. McGee (1958). This poses a bargaining problem and, as described in Section 3, Saloner

(1987) assumes the incumbent has private information: it prices aggressively in the product market to signal its competitive strength were the entrant to refuse the terms offered for purchase. Burns (1986) provides some empirical data.

An opposing rationalist view sees predation as the punishment phase of a subgame-perfect equilibrium of a repeated game between the incumbent and the entrant; cf. Milgrom and Roberts (1982b, appendix A). Like other Folk Theorem arguments, this one is usually deemed inadequate because it presumes an equilibrium selection in favor of the incumbent. An alternative view examines asymmetries favoring the incumbent, such as capital-market imperfections and related features that prevent competition on equal terms; cf. Telser (1966) and Poitevin (1989). Benoit (1984) shows that moderate asymmetries can produce severely asymmetric outcomes. Suppose the entrant has limited financial resources, in the sense that the entrant can survive at most n periods battling the incumbent before it is forced to exit. Assume also that for the incumbent, battling to expel the entrant is profitable if $n \leq m$, where $m \geq 1$. Then an induction argument implies that the incumbent is willing to battle for any value of n: when $n \leq m + 1$, battling for *one* period reduces the entrant's remaining resources, allowing continuation for at most m periods, whereupon the entrant knows that it will lose the ensuing battle and therefore prefers to exit immediately with its remaining resources intact. Anticipating this, the incumbent is willing to battle for the one period required to reach this situation; and anticipating this, the entrant prefers to forgo entry or to exit immediately when $n \leq m + 1$.

However, this view of predation encounters an argument examined by Selten (1978) and Rosenthal (1981). Suppose the market terminates after a finite number of periods and the entrant can enter (without sunk costs) in any period; actually, Selten assumes a series of different entrants, but this is immaterial. Then a costly battle in the past period is useless, and by backwards induction the incumbent is unwilling to battle in any period. Thus, if the duration of the market is finite, then in effect one must suppose that $m = 0$ in Benoit's construction, and therefore his induction argument fails.[10]

Several models rely on private information to resurrect predatory battles to expel entrants. Benoit's analysis considers a version in which the entrant's financial resources are privately known. To explain inequalities in the parties' resources, Poiteven (1989) examines a model in which the two firms' financial obligations differ because the entrant must obtain capital via debt that credibly signals to lenders its private information about profitability. Sharfstein and Bolton (1990) study the optimal design of a contract between the entrant and its financiers, noting that a contract that naively terminates funding if profits

---

[10]The tone of Selten's and Rosenthal's expositions is actually to argue against the plausibility of results that depend on long chains of backward induction from a known fixed finite terminus.

are low encourages the incumbent to meet entry with aggressive pricing. Judd and Peterson (1986) apply analogous ideas to limit-pricing contexts.

Milgrom and Roberts (1982b, appendix B) suggest an elegant version in which absence of common knowledge about the incumbent's information eliminates Selten's backward induction. Consider a finite sequence of different entrants, all of whom know that battling any entrant is unprofitable for the incumbent; however, the incumbent is unsure whether they know this, ascribing positive probability to the event that some (at least those late in the sequence) are unsure whether a battle is costly or profitable – and sufficiently unsure to be unwilling to take the risk. In this case the incumbent battles early entrants (who would therefore be reckless to enter) in the mistaken belief that this might deter later entrants by preserving their uncertainty. The incumbent thinks that failure to battle any entrant might reveal that battles are unprofitable and induce a flood of subsequent entrants. Even if the early entrants are well informed, they are deterred by the incumbent's readiness to battle and so the incumbent's mistaken beliefs are not challenged until later.

Kreps and Wilson (1982) and Milgrom and Roberts (1982b) study other versions of this "demonstration effect" derived from the entrant's uncertainty about the incumbent's payoffs or feasible actions. The former studies an N-period market with an incumbent facing a single entrant (or a sequence of entrants). In each period the entrant enters or not, and if it enters the incumbent concedes or fights. The incumbent knows privately that it is permanently weak or strong, determined initially with probabilities $1 - p$ and p; and similarly the entrant is weak or strong with probabilities $1 - q$ and q. Normalized per-period payoffs are shown in Table 1: assume $a > 0$ and $0 < b < 1 < B$ so that fighting is unprofitable for weak types but not for strong types. Assume that each party's payoff is the sum of its per-period payoffs, although similar results are valid for any discount factor close to 1.

If both parties are surely weak, then Selten's (subgame-perfect) equilibrium applies: the entrant always enters and the incumbent always concedes. A similar sequential equilibrium applies if only the incumbent is surely weak. Now suppose the entrant is surely weak but the incumbent might be strong:

Table 1
Per-period payoffs

Actions		Payoffs			
Entrant	Incumbent	Entrant		Incumbent	
		Weak	Strong	Weak	Strong
No entry		0	0	a	a
Entry	Concedes	b	B	0	-1
	Fights	$b-1$	$B-1$	-1	0

$q = 0$ but $p > 0$. In this case there is a sequential equilibrium described as follows. With n periods remaining and a current belief that the incumbent is strong with probability $\hat{p}$, the entrant enters if $\hat{p} < b^n$, enters with probability $1/a$ if $\hat{p} = b^n$, and stays out otherwise. Following entry, the strong incumbent surely fights, and the weak incumbent fights if $\hat{p} \geqslant b^{n-1}$ and otherwise fights with a probability such that the entrant's belief next period is b^{n-1} using Bayes' Rule. If the incumbent ever concedes, then the entrant believes thereafter that it is surely weak.

The analogous model adopted by Milgrom and Roberts (1982b) amends the formulation as follows. First, each period's entrant has a privately known type affecting its payoff from forgoing entry; similarly, the incumbent has a privately known type (fixed for the entire game) affecting its per-period payoff from fighting entry. These type parameters have independent non-atomic distributions. Second, the incumbent has private information about whether it is forced always to concede or always to fight, or it can choose each period. The second ensures that the sequential equilibrium is unique, and the first allows pure strategies.[11] Easley, Masson and Reynolds (1985) use an alternative specification in which the incumbent's private information is knowledge of demand in multiple markets: demand is high in all markets or so low that entry is unprofitable. In the analogous equilibrium, the incumbent responds to early entrants with secret price cutting that mimics the effect of low demand.

In all these formulations the equilibrium produces the intended result. In the model above, if the duration N of the market is so long that $p > b^N$, then even the weak incumbent initially fights entry, and anticipating this behavior the entrant stays out. The entrant's belief remains fixed at p until the last few periods (independent of N) when it first ventures to enter. This equilibrium illustrates the weak incumbent's incentive to maintain a reputation for possibly being strong: maintenance of the reputation (preserving $\hat{p} = p$ early in the game) is expensive when the entrant recklessly challenges the incumbent too early, but the incumbent perceives benefits from deferral of further entry. The notion that reputational effects could motivate predatory responses to entry was proposed by Yamey (1972).

This equilibrium extends to the case that each party has private information about whether it is weak or strong. To illustrate the close connection with attrition models, consider the version obtained in the limit as the period length shrinks, although preserving the assumption that the market has a finite duration T. Of course a strong entrant enters and a strong incumbent fights at every time, so failure to enter reveals a weak entrant and failure to fight reveals a weak incumbent. Using the limit of the above equilibrium, a

---

[11] In the Kreps–Wilson model, other sequential equilibria can be ruled out by stability arguments; cf. Cho and Kreps (1987).

revealed-weak entrant stays out thereafter (if the incumbent might be strong), and a revealed-weak incumbent encounters entry and concedes thereafter. Thus, after revelation of the weakness of the entrant the incumbent accrues profits at the rate a, or after revelation of the weakness of the incumbent the entrant accrues profits at the rate b if weak or B if strong. If neither is revealed, then their beliefs when a duration t remains are represented by a pair (p_t, q_t), where $(p_T, q_T) = (p, q)$. In the state space of these beliefs, the equilibrium assigns a special role to a locus along which the two weak parties have expected values of continuation that are each zero. Along this locus each weak party selects a stopping time at which it first reveals its weakness by not entering or by conceding. As long as neither reveals weakness their beliefs evolve along the locus, reaching $(1, 1)$ at a time (*before* expiration at $t = 0$) when each concludes the other is strong. In particular, letting $\alpha = b/[1 - b]$, this locus satisfies $p_t^{1/a} = q_t^{1/\alpha}$, or in a time parameterization, $p_t = k_1 t^{-1/\alpha}$ and $q_t = k_2 t^{-1/a}$, where the constants k_i depend on initial conditions. The behavioral strategies that in combination with Bayes' Rule generate this locus can be represented in terms of the weak parties' hazard rates of revealing actions: the weak incumbent's hazard rate of conceding is $[\alpha t(1 - p_t)]^{-1}$ and the weak entrant's hazard rate of not continuing entry is $[at(1 - q_t)]^{-1}$. At the first instant $t = T$, however, the beliefs are not on this locus, so depending on which side of the locus the beliefs lie, one or the other randomizes as to whether to adopt the revealing action, with probabilities such that application of Bayes' Rule yields a posterior on the locus if the revealing action is not taken. Consequently, k_1 and k_2 are determined by the two requirements that $(p_\tau, q_\tau) = (1, 1)$ for some remaining duration $\tau > 0$, and $p_T = p$ or $q_T = q$ depending on which initial belief is unchanged after the initial randomization.

Two-sided reputational equilibria of this sort are akin to attrition: each weak party continues the costly battle in the hope that the other will concede defeat first if it is also weak. Fudenberg and Kreps (1987) address cases in which the incumbent faces several entrants simultaneously or in succession, and depending on whether entrants who have exited can re-enter if the incumbent is revealed weak. Reputational effects persist but depend on the ability of entrants to re-enter. If they can re-enter, then the behavior with many entrants faced sequentially is similar to the behavior with many entrants faced simultaneously. That is, the reputation of the incumbent predominates. Fudenberg and Kreps also develop a point made in the Milgrom–Roberts model; namely, the incumbent, even if his reputation predominates, may prefer that each contest is played behind a veil, isolated from others. This happens when the incumbent has a very high prior probability of being strong, and also the entrants each have a high probability of being strong. The incumbent's reputation causes all weak entrants to concede immediately, but to defend those gains the incumbent must fight many strong entrants. If the contests were isolated, the incumbent would do nearly as well against his weak opponents and better against the strong.

The structure of such games is a variation on the infinitely repeated games addressed by the Folk Theorem, but where only one of the parties in the stage game plays repeatedly. In such cases, results analogous to the Folk Theorem obtain although often they are interpreted in terms of reputation effects. Fudenberg and Levine (1989a, 1989c) present general analyses for the case that the long-lived player has private information about his type, including formulations in which his actions are imperfectly observed by others.[12] The key result, stated for simultaneous-move stage games, is that in any Nash equilibrium the long-run player's payoff is no less in the limit as the interest rate shrinks than what he would achieve from the pure strategy to which he would most like to commit himself – provided the prior probability is positive of being of a type that would optimally play this "Stackleberg" strategy were his type known. The lower bound derives from the fact that the short-run players adopt best responses to the Stackleberg strategy whenever they attach high probability to the long-run player using this strategy; consequently, if the long-run player uses the Stackleberg strategy consistently, then the short-run players eventually infer that this strategy is likely and respond optimally.[13]

This result establishes the essential principle that explains reputational effects. Moreover, the thrust of models based on reputational effects is, in effect, to select among the equilibria allowed by Folk Theorems: such arguments would not be compelling if the resulting equilibrium (in which the incumbent deters entry) were sensitive to the prior distribution of its possible types, but in fact Fudenberg and Levine's results include a robustness property – entry deterrence occurs for a wide class of prior distributions in both finite and infinite-horizon models.

5. Concluding remarks

Previous theories of entry deterrence and market structure sorely needed amendment to account for strategic features. The formulations and analytical methods of game theory helped clarify the issues and suggest revisions of

[12]This work is reviewed by Fudenberg (1990) and portions are included in Fudenberg and Tirole (1991). See also Fudenberg, Kreps and Maskin (1990) and Fudenberg and Levine (1989b) for related results in settings without private information, as well as Fudenberg and Maskin (1986) for the case that both players are long-lived.

[13]That is, for each $\epsilon > 0$ there exists a number K such that with probability $1 - \epsilon$ the short-run players play best responses to the Stackleberg strategy in all but K periods; moreover, there exists an upper bound on K that is independent of the interest rate and the equilibrium under consideration. See the appendix of Fudenberg and Levine (1989c) for a general statement of this lemma.

long-standing theoretical constructs. A principal contribution of the game-theoretic approach is the precise modeling it enables of timing and informational conditions. In addition, it provides a systematic means of excluding incredible threats by imposing perfection criteria; e.g., subgame-perfect, sequential, or stable equilibria. Applications of these tools provide "toy" models that illustrate features discussed in informal accounts of entry deterrence. The requirements of precise modeling can also be a limitation of game theory when general conclusions are sought. In particular, the difficulties of analyzing complex models render this approach more a means of criticism than a foundation for construction of general theories of market structure.

The plethora of predictions obtainable from various formulations indicate that empirical and experimental studies are needed to select among hypotheses. Many models present econometric difficulties that impede empirical work, but this is realistic: the models reveal that strategic behavior can depend crucially on private information inaccessible to outside observers. Estimation of structural models is likely to be difficult, therefore, but it may be possible to predict correlations in the data. Experimental studies may be more effective; cf. Isaac and Smith (1985), Camerer and Weigelt (1988), Jung et al. (1989), and Neral and Ochs (1989).

Bibliography

Aghion, Phillippe and Patrick Bolton (1987) 'Contracts as a barrier to entry', *American Economic Review*, 77: 388–401.

Arvan, Lanny (1986) 'Sunk capacity costs, long-run fixed costs, and entry deterrence under complete and incomplete information', *Rand Journal of Economics*, 17: 105–121.

d'Aspremont, Claude and Phillipe Michel (1990) 'Credible entry-deterrence by delegation', Discussion Paper 9056, Center for Operations Research and Econometrics. Louvain-la-Nueve, Belgium: Université Catholique de Louvain.

Ausubel, Lawrence M. and Raymond J. Deneckere (1987) 'One is almost enough for monopoly', *Rand Journal of Economics*, 18: 255–274.

Bagwell, Kyle and Garey Ramey (1988) 'Advertising and limit pricing', *Rand Journal of Economics*, 19: 59–71.

Bagwell, Kyle and Garey Ramey (1991) 'Oligopoly limit pricing', *Rand Journal of Economics*, 22: 155–172.

Bagwell, Kyle and Garey Ramey (1990) 'Capacity, entry, and forward induction', D.P. #888, Northwestern University, mimeo.

Bagwell, Kyle and Michael H. Riordan (1991) 'High and declining prices signal product quality', *American Economic Review*, 81: 224–239.

Basu, Kaushik and Nirvikar Singh (1990) 'Entry deterrence in Stackleberg perfect equilibria', *International Economic Review*, 31: 61–71.

Beggs, Alan and Paul Klemperer (1992) 'Multi-period competition with switching costs', *Econometrica*, 60: to appear.

Benoit, Jean-Pierre (1984) 'Financially constrained entry in a game with incomplete information', *Rand Journal of Economics*, 15: 490–499.

Bernheim, B. Douglas (1984) 'Strategic deterrence of sequential entry into an industry', *Rand Journal of Economics*, 15: 1–11.

Bonanno, Giacomo (1987) 'Location choice, product proliferation, and entry deterrence', *The Review of Economic Studies*, 54: 37–45.

Bulow, Jeremy, John Geanakoplos and Paul Klemperer (1985a), 'Multimarket oligopoly: Strategic substitutes and complements', *Journal of Political Economy*, 93: 488–511.

Bulow, Jeremy, John Geanakoplos and Paul Klemperer (1985b) 'Holding idle capacity to deter entry', *Economic Journal*, 95: 178–182.

Burns, M.R. (1986) 'Predatory pricing and the acquisition cost of competitors', *Journal of Political Economy*, 94: 266–296.

Camerer, Colin and Keith Weigelt (1988) 'Experimental tests of a sequential equilibrium reputation model', *Econometrica*, 56: 1–36.

Cho, In-Koo (1990a) 'Strategic stability in repeated signaling games', University of Chicago, mimeo.

Cho, In-Koo (1990b) 'Separation or not: A critique of "appearance based" selection criteria', University of Chicago, mimeo.

Cho, In-Koo, and David Kreps (1987) 'Signaling games and stable equilibria', *Quarterly Journal of Economics*, 102: 179–221.

Coate, Malcolm B. and Andrew N. Kleit (1990) 'Exclusion, collusion and confusion: The limits of raising rivals' costs', Working paper #179. Washington DC: Federal Trade Commission.

Dixit, Avinash (1979) 'A model of duopoly suggesting a theory of entry barriers', *Bell Journal of Economics*, 10: 20–32.

Dixit, Avinash (1980) 'The role of investment in entry deterrence', *Economic Journal*, 90: 95–106.

Easley, David, R.T. Masson and R.J. Reynolds (1985) 'Preying for time', *Journal of Industrial Organization*, 33: 445–460.

Eaton, B. Curtis, and R.G. Lipsey (1977) 'The theory of market preemption: The persistence of excess capacity and monopoly in growing spatial markets', *Economica*, 46: 149–158.

Eaton, B. Curtis, and R.G. Lipsey (1980) 'Exit barriers are entry barriers: The durability of capital as a barrier to entry', *The Bell Journal of Economics*, 11: 721–729.

Eaton, B. Curtis and R.G. Lipsey (1981) 'Capital, commitment and entry equilibrium', *Bell Journal of Economics*, 12: 593–604.

Eaton, B. Curtis and Roger Ware (1987) 'A theory of market structure with sequential entry', *Rand Journal of Economics*, 18: 1–16.

Farrell, Joseph and Garth Saloner (1986) 'Installed base and compatibility: Innovation, product pre-announcements, and predation', *American Economic Review*, 76: 940–955.

Farrell, Joseph and Carl Shapiro (1988) 'Dynamic competition with switching costs', *Rand Journal of Economics*, 19: 123–137.

Fisher, Franklin (1989) 'Games economists play: A noncooperative view', *Rand Journal of Economics*, 20: 113–124.

Fishman, Arthur (1990) 'Entry deterrence in a finitely-lived industry', *Rand Journal of Economics*, 21: 63–71.

Fudenberg, Drew (1990) 'Explaining cooperation and commitment in repeated games', VIth World Congress of the Econometric Society, Barcelona.

Fudenberg, Drew and David Kreps (1987) 'Reputation in the simultaneous play of multiple opponents', *Review of Economic Studies*, 54: 541–568.

Fudenberg, Drew and David Levine (1989a) 'Reputation and equilibrium selection in games with a patient player', *Econometrica*, 57: 759–778.

Fudenberg, Drew and David Levine (1989b) 'Equilibrium payoffs with long-run and short-run players and imperfect public information', MIT and UCLA, mimeo.

Fudenberg, Drew and David Levine (1989c) 'Reputation, unobserved strategies, and active supermartingales', MIT and UCLA, mimeo.

Fudenberg, Drew and Eric Maskin (1986) 'The folk theorem in repeated games with discounting or with incomplete information', *Econometrica*, 54: 533–554.

Fudenberg, Drew and Jean Tirole (1983) 'Capital as a commitment: strategic investment to deter mobility', *Journal of Economic Theory*, 31: 227–250.

Fudenberg, Drew and Jean Tirole (1985) 'Preemption and rent equalization in the adoption of new technology', *Review of Economic Studies*, 52: 383–401.

Fudenberg, Drew and Jean Tirole (1986a) 'A theory of exit in oligopoly', *Econometrica*, 54: 943–960.

Fudenberg, Drew and Jean Tirole (1986b) 'A "signal jamming" theory of predation', *Rand Journal of Economics*, 17: 366–377.

Fudenberg, Drew and Jean Tirole (1986c) *Dynamic models of oligopoly*. Chur: Harwood Academic Publishers.

Fudenberg, Drew and Jean Tirole (1991), *Game theory*. Cambridge, Mass.: MIT Press.

Fudenberg, Drew, David Kreps and Eric Maskin (1990) 'Repeated games with long-run and short-run players', *Review of Economic Studies*, 57: 555–573.

Fudenberg, Drew, Richard Gilbert, Joseph Stiglitz and Jean Tirole (1983) 'Preemption, leapfrogging and competition in patent races', *European Economic Review*, 22: 3–31.

Gabszwicz, Jean, L. Pepall, and J. Thisse (1990), 'Sequential entry, experience goods, and brand loyalty', CORE Discussion Paper #9063, Universite Catholique de Louvain, Belgium.

Gelman, Judith, and Steven Salop (1983) 'Judo economics: Capacity limitation and coupon competition', *Bell Journal of Economics*, 14: 315–325.

Ghemawat, Pankaj, and Barry Nalebuff (1985) 'Exit', *Rand Journal of Economics*, 16: 184–194.

Ghemawat, Pankaj and Barry Nalebuff (1990) 'The devolution of declining industries', *Quarterly Journal of Economics*, 105: 167–186.

Gilbert, Richard (1989a) 'Mobility barriers and the value of incumbency', in: R. Schmalmansee and R. Willig, eds., *Handbook of industrial organization*, Vol. 1. Amsterdam: North-Holland/Elsevier Science Publishers, pp. 475–536.

Gilbert, Richard (1989b) 'The role of potential competition in industrial organization', *Journal of Economic Perspectives*, 3: 107–127.

Gilbert, Richard and Richard Harris (1984) 'Competition with lumpy investment', *Rand Journal of Economics*, 15: 197–212.

Gilbert, Richard and David Newbery (1982) 'Preemptive patenting and the persistence of monopoly', *American Economic Review*, 72: 514–526.

Gilbert, Richard and Xavier Vives (1986) 'Entry deterrence and the free rider problem', *Review of Economic Studies*, 53: 71–83.

Gül, Faruk (1987) 'Noncooperative collusion in durable goods oligopoly', *Rand Journal of Economics* 18: 248–254.

Gül, Faruk, Hugo Sonnenschein and Robert Wilson (1986) 'Foundations of dynamic monopoly and the Coase conjecture', *Journal of Economic Theory*, 39: 155–90.

Harrington, Joseph E. (1984) 'Noncooperative behavior by a cartel as an entry-deterring signal', *Rand Journal of Economics*, 15: 416–433.

Harrington, Joseph E. (1986) 'Limit pricing when the potential entrant is uncertain of its cost function', *Econometrica*, 54: 429–437.

Harrington, Joseph E. (1987) 'Oligopolistic entry deterrence under incomplete information', *Rand Journal of Economics*, 18: 211–231.

Hart, Oliver and Jean Tirole (1990) 'Vertical integration and market foreclosure', *Brookings Papers on Economic Activity: Microeconomics*, 2: 205–286.

Huang, Chi-fu and Lode Li (1991) 'Entry and exit: Subgame perfect equilibria in continuous-time stopping games', Working Paper (Series D) #54, School of Organization and Management, Yale University, mimeo.

Isaac, R. Marc and Vernon Smith (1985) 'In search of predatory pricing', *Journal of Political Economy*, 93: 320–345.

Judd, Kenneth (1985) 'Credible spatial preemption', *Rand Journal of Economics*, 16: 153–166.

Judd, Kenneth and Bruce Peterson (1986) 'Dynamic limit pricing and internal finance', *Journal of Economic Theory*, 39: 368–399.

Jung, Yun Joo, John Kagel and Dan Levin (1989) 'On the existence of predatory pricing in the laboratory: An experimental study of reputation and entry deterrence in the chain-store game', University of Pittsburgh, mimeo.

Kleit, Andrew and Malcolm Coate (1991) 'Are judges smarter than economists? Sunk costs, the threat of entry, and the competitive process', Bureau of Economics Working Paper 190. Washington, DC: Federal Trade Commission.

Klemperer, Paul (1987a) 'Markets with consumer switching costs', *Quarterly Journal of Economics*, 102: 375–394.

Klemperer, Paul (1987b) 'Entry deterrence in markets with consumer switching costs', *Economic Journal*, 97: 99–117.

Klemperer, Paul (1989) 'Price wars caused by switching costs', *Review of Economic Studies*, 56: 405–420.

Krattenmaker, Thomas G. and Steven C. Salop (1986) 'Competition and cooperation in the market for exclusionary rights', *American Economic Review*, 76(2): 109–113.

Krattenmaker, Thomas G. and Steven C. Salop (1987) 'Exclusion and antitrust', *Regulation*, 11: 29–33.

Kreps, David and Robert Wilson (1982) 'Reputation and incomplete information', *Journal of Economic Theory*, 27: 253–279.

Mailath, George (1987) 'Incentive compatibility in signaling games with a continuum of types', *Econometrica*, 55: 1349–1365.

Maskin, Eric and Jean Tirole (1988) 'A theory of dynamic oligopoly', Parts I and II, *Econometrica*, 56: 549–599.

Matthews, Steven and Doron Fertig (1990), 'Advertising signals of product quality', CMSEMS #881, Northwestern University, mimeo.

Matthews, Steven and Leonard Mirman (1983) 'Equilibrium limit pricing: The effects of private information and stochastic demand', *Econometrica*, 51: 981–995.

McGee, John S. (1958) 'Predatory price cutting: The Standard Oil (N.J.) Case', *Journal of Law and Economics*, 1: 137–169.

McLean, R. and Michael Riordan (1989) 'Industry structure with sequential technology choice', *Journal of Economic Theory*, 47: 1–21.

Milgrom, Paul (1987) 'Predatory pricing' in: J. Eatwell, M. Milgate, and P. Newman, eds., *The New Palgrave: A dictionary of economics*, Vol. 3. London: Macmillan Press, pp. 937–938.

Milgrom, Paul and John Roberts (1982a) 'Limit pricing and entry under incomplete information: An equilibrium analysis', *Econometrica*, 50: 443–459.

Milgrom, Paul and John Roberts (1982b) 'Predation, reputation, and entry deterrence', *Journal of Economic Theory*, 27: 280–312.

Milgrom, Paul and John Roberts (1987) 'Informational asymmetries, strategic behavior, and industrial organization', *American Economic Review*, 77: 184–193.

Milgrom, Paul and John Roberts (1990) 'New theories of predatory pricing', in: G. Bonanno and D. Brandolini, eds., *Industrial structure in the new industrial economics*. Oxford: Oxford University Press.

Milgrom, Paul R. and Robert J. Weber (1985) 'Distributional strategies for games with incomplete information', *Mathematics of Operations Research*, 10: 619–632.

Mills, David (1988) 'Preemptive investment timing', *Rand Journal of Economics*, 19: 114–122.

Nalebuff, Barry and John G. Riley (1985) 'Asymmetric equilibria in the war of attrition', *Journal of Theoretical Biology*, 113: 517–527.

Neral, John and Jack Ochs (1989) 'The sequential equilibrium theory of reputation building: A further test', University of Pittsburg, mimeo.

Nti, K. and M. Shubik (1981) 'Noncooperative oligopoly with entry', *Journal of Economic Theory*, 24: 187–204.

Ordover, Janusz and Garth Saloner (1989) 'Predation, monopolization, and antitrust', in: R. Schmalansee and R. Willig, eds., *Handbook of industrial organization*, Vol. 1. Amsterdam: North-Holland/Elsevier Science Publishers, pp. 537–596.

Ordover, Janusz and Ariel Rubinstein (1986) 'A sequential concession game with asymmetric information', *Quarterly Journal of Economics*, 101: 879–888.

Ordover, Janusz, Garth Saloner and Steven Salop (1990) 'Equilibrium vertical foreclosure', *American Economic Review*, 80: 127–142.

Peltzman, Sam (1991), '*The Handbook of Industrial Organization*: A review article', *Journal of Political Economy*, 99: 201–217.

Poitevin, Michel (1989) 'Financial signalling and the "deep pocket" argument', *Rand Journal of Economics*, 20: 26–40.

Prescott, Edward C. and Michael Visscher (1977) 'Sequential location among firms with perfect foresight', *Bell Journal of Economics*, 8: 373–393.

Ramey, Garey (1987) 'Limit pricing and sequential capacity choice', University of California at San Diego #87–30, mimeo.

Rasmusen, Eric, J. Mark Ramseyer and John S. Wiley (1991) 'Naked exclusion', *American Economic Review*, 81: 1137–1145.

Riley, John G. (1980) 'Strong evolutionary equilibria in the war of attrition', *Journal of Theoretical Biology*, 82: 383–400.

Roberts, John (1985) 'A signaling model of predatory pricing', *Oxford Economic Papers, Supplement*, 38: 75–93.

Roberts, John (1987) 'Battles for market share: Incomplete information, aggressive strategic pricing and competitive dynamics', in: T. Bewley, ed., *Advances in economic theory*. Cambridge: Cambridge University Press, chapter 4, pp. 157–195.

Rosenthal, Robert (1981) 'Games of perfect information, predatory pricing and the chain store paradox', *Journal of Economic Theory*, 25: 92–100.

Saloner, Garth (1982) *Essays on information transmission under uncertainty*, chapter 2, PhD Dissertation, Stanford Business School, Stanford CA.

Saloner, Garth (1987) 'Predation, mergers and incomplete information', *Rand Journal of Economics*, 18: 165–186.

Salop, Steven (1979) 'Strategic entry deterrence', *American Economic Review*, 69: 335–338.

Salop, Steven, ed. (1981) *Strategy, predation, and antitrust analysis*. Washington: Federal Trade Commission.

Salop, Steven and David Scheffman (1983) 'Raising rivals' costs', *American Economic Review*, 73: 267–271.

Salop, Steven and David Scheffman (1987) 'Cost-raising strategies', *Journal of Industrial Economics*, 36: 19–34.

Schary, Martha (1991) 'The probability of exit', *Rand Journal of Economics*, 22: 339–353.

Scherer, Frederic (1980) *Industrial market structure and economic performance*, 2nd edn. Chicago: Rand McNally.

Schmalansee, Richard (1978) 'Entry deterrence in the ready-to-eat breakfast cereal industry', *Rand Journal of Economics*, 9: 305–327.

Schmalansee, Richard (1981) 'Economies of scale and barriers to entry', *Journal of Political Economy*, 89: 1228–1232.

Schmidt, Klaus (1991) 'Reputation and equilibrium characterization in repeated games of conflicting interests', Discussion Paper A-333, Rheinische Friedrich-Wilhelms-Universität, Bonn.

Schwartz, M., and Earl A. Thompson (1986) 'Divisionalization and entry deterrence', *Quarterly Journal of Economics*, 101: 307–321.

Seabright, Paul (1990) 'Can small entry barriers have large effects on competition?', Discussion Paper 145, University of Cambridge, UK.

Selten, Reinhard (1978) 'The chain store paradox', *Theory and Decision*, 9: 127–159.

Shapiro, Carl (1989) 'The theory of business strategy', *Rand Journal of Economics*, 20: 125–137.

Sharfstein, David (1984) 'A policy to prevent rational test-market predation', *Rand Journal of Economics*, 15: 229–243.

Sharfstein, David and Patrick Bolton (1990) 'A theory of predation based on agency problems in financial contracting', *American Economic Review*, 80: 93–106.

Spence, A. Michael (1977) 'Entry, capacity, investment and oligopolistic pricing', *Bell Journal of Economics*, 8: 534–544.

Spence, A. Michael (1979) 'Investment, strategy and growth in a new market', *Bell Journal of Economics*, 10: 1–19.

Spence, A. Michael (1984) 'Cost reduction, competition and industry performance', *Econometrica*, 52: 101–122.

Sutton, John (1991) *Sunk costs and market structure*. Cambridge, MA: MIT Press.

Telser, Lester (1966) 'Cutthroat competition and the long purse', *Journal of Law and Economics*, 9: 259–277.

Tirole, Jean (1988) *The theory of industrial organization*. Cambridge, Mass.: MIT Press.

Veendorp, E.C.H. (1991) 'Entry deterrence and divisionalization', *Quarterly Journal of Economics*, 106: 297–307.

Waldman, Michael (1987) 'Noncooperative entry deterrence, uncertainty, and the free rider problem', *Review of Economic Studies*, 54: 301–310.

Waldman, Michael (1991) 'The role of multiple potential entrants/sequential entry in noncooperative entry deterrence', *Rand Journal of Economics*, 22: 446–453.

Ware, Roger (1984) 'Sunk costs and strategic commitment: A proposed three-stage equilibrium', *Economic Journal*, 94: 370–378.

Whinston, Michael D. (1988) 'Exit with multiplant firms', *Rand Journal of Economics*, 19: 568–588.

Whinston, Michael D. (1990) 'Tying, foreclosure and exclusion', *American Economic Review*, 80: 837–859.

Wilson, Robert (1985) 'Reputations in games and markets', in: Alvin Roth, ed., *Game theoretic models of bargaining with incomplete information*. Cambridge: Cambridge University Press, chapter 3, pp. 27–62.

Wilson, Robert (1989a) 'Entry and exit', in: G. Feiwel, *The economics of imperfect competition and employment*. London: Macmillan Press, chapter 8, pp. 260–304.

Wilson, Robert (1989b) 'Deterrence in oligopolistic competition', in: P. Stern, R. Axelrod, R. Jervis and R. Radner, eds., *Perspectives on deterrence*. New York: Oxford University Press, chapter 8, pp. 157–190.

Yamey, B.S. (1972) 'Predatory price cutting: Notes and comments', *Journal of Law and Economics*, 15: 129–142.

Chapter 11

PATENT LICENSING

MORTON I. KAMIEN*

Northwestern University

Contents

*I wish to acknowledge the referees' very useful suggestions. Support for this work was provided by the Heizer Research Center on Entrepreneurship.

Handbook of Game Theory, Volume 1, Edited by R.J. Aumann and S. Hart

1. Introduction

Patents were first granted by the Republic of Venice in 1474. A patent is meant to serve as an incentive for invention by providing a patentee a certain period of time, usually between 16 and 20 years, during which he also can attempt to profit from it. In return for the granting of a patent, society receives disclosure of information that might be kept secret otherwise, as well as technological advances.

One source of profit for the inventor is through licensing of patent. The other, of course, is through his own working of the patent. The common modes of patent licensing are a royalty, possibly nonuniform, per unit of output produced with the patented technology, a fixed fee that is independent of the quantity produced with the patented technology, or a combination of a fixed fee plus a royalty. The patentee can choose which of these modes of licensing to employ and how to implement them. That is, he can decide on whether to set a royalty rate and/or a fixed fee for which any firm can purchase a license or auction a fixed number of licenses. He may also devise other licensing mechanisms. Obviously he will choose, short of any legal or institutional constraints, the licensing mechanism that maximizes his profits. According to Rostoker (1984), royalty plus fixed fee licensing was used 46 percent of the time, royalty alone 39 percent, and fixed fee alone 13 percent of the time, among the firms surveyed. Actual patent licensing practices are also described by Taylor and Silberston (1973), and Caves, Crookell and Killing (1983). The requirements for obtaining a patent and its duration in leading industrialized countries is summarized in Kitti and Trozzo (1976).

Formal analysis of the profits a patentee can realize from licensing can be traced back to Arrow (1962) for inventions that reduce production costs: to Usher (1964) for new product innovations; and McGee (1966) who considered both. Arrow was concerned with the question of whether a purely competitive or monopolistic industry had a greater incentive to innovate. In a certain sense this was an attempt, on a formal level, to test Schumpeter's (1942) argument that monopolistic industries, those in which individual firms have a measure of control over their products price, provide a more hospitable atmosphere for innovation than purely competitive ones. Arrow addressed this question by comparing the profits a patentee could realize from licensing his invention by means of a uniform royalty per unit of output to a purely competitive industry with the profitability of the identical invention to a monopolist. He showed that the inventor's licensing profits to a perfectly competitive industry exceeds the profitability of the same invention to a monopolist, regardless of whether or not it is drastic. (A "drastic" invention is one for which the post-invention monopoly price is below the pre-invention competitive price.) The intuitive reason for this conclusion is that the standard of comparison of the profitability of an invention for a monopolist is against the positive pre-invention monopoly profit, while for a perfectly competitive industry it is against the zero pre-invention profits. Arrow acknowledged that this conclusion regarding the comparative profitability of the identical invention to a monopolist and a perfectly competitive industry could be reversed if the appropriability of profits from an invention were greater for a monopolist than for a perfectly competitive industry. It was Schumpeter's contention that this was precisely the case. There were a series of challenges and modifications of Arrow's conclusions by Demsetz (1969) and others, a discussion of which can be found in Kamien and Schwartz (1982).

McGee independently addressed the question of patent licensing, but did not attempt to draw any inferences regarding the relative attractiveness of an invention to a monopolist and a perfectly competitive industry. However, he introduced the concept of a derived demand for a license, a concept that has been emphasized since, and suggested that licenses might be auctioned.

Beginning in the late 1960s, papers by Scherer (1967), Barzel (1968), and Kamien and Schwartz (1972, 1976), set the stage for papers by Loury (1979), Dasgupta and Stiglitz (1980a, 1980b), Lee and Wilde (1980), and Reinganum (1981, 1982), which have come to define the theory of patent races. A comprehensive review of this literature is provided by Reinganum (1989) and Baldwin and Scott (1987). The analysis of optimal patent licensing may be regarded as complementary to the work on patent races, as in the latter work the reward for being the first to obtain a patent is supposed to be given.

Meanwhile, theoretical work on patent licensing languished. Kamien and Schwartz (1982) attempted to extend Arrow's work to licensing of a patent by means of a royalty to a Cournot oligopoly. They found the optimal fixed fee plus unit royalty the patentee should employ under the supposition that the licensee's profits remain the same as before the invention. Thus, their analysis did not allow for the patentee's ability to exploit the licensee's competition for a license. The employment of a game-theoretic framework for the analysis of the patentee's licensing strategies was introduced independently by Kamien and Tauman (1984, 1986) and Katz and Shapiro (1985, 1986). It is this work and work flowing from it that will be the primary focus of this survey.

The interaction between a patentee and licensees is described in terms of a three-stage noncooperative game. The patentee plays the role of the Stackelberg leader to the licensees, who are the followers. The patentee exercises the role of a leader in the sense that he maximizes his licensing profit against the followers' demand function (reaction or best response function) for licenses. The licensees are assumed to be members of an n-firm oligopoly, producing an identical product. Entry into the industry is assumed to be unprofitable, i.e., the cost of entry exceeds the profits an entrant could realize. The firms in the oligopoly can compete either through quantities or prices. The industry's aggregate output and product price is determined by the Cournot equilibrium in the former case and the Bertrand equilibrium in the latter. In the simplest version of the game, the oligopoly faces a linear demand function for its product. The patented invention reduces the cost of production, i.e., it is a process innovation. Licensing of a product innovation can also be analyzed in this game-theoretic framework.

In the game's first stage the patentee announces either the price of a license at which any firm can purchase one or the number of licenses he will auction. In its second stage, the firms decide independently and simultaneously whether or not to purchase a license, or how much to bid for a license. In the game's third stage each firm, licensed and unlicensed, decides independently and simultaneously either how much to produce or charge for its product. The subgame-perfect Nash equilibrium (SPNE) in pure strategies is the solution concept employed. Thus, the analysis of the game is conducted backward from its last stage to its first. That is, each firm calculates its operating profit in the game's third stage equilibrium if it were or were not a licensee given the number of other licensees. This calculation defines the value of a license to a firm, the most it would pay for a license in the game's second stage, for each number of other licensed firms. These values, as a function of the number of other licensees, in turn, define a firm's demand function for a license. In the game's first stage the patentee decides what price to sell licenses at or the number of licenses to auction so as to maximize the profit, taking into account the aggregate demand function for licenses. The game is only played once, there is no uncertainty, and all relevant information is common knowledge to all the players. Resale of licenses is ruled out.

Roughly speaking, the following results emerge from the analysis of the different modes of licensing under the assumption that the firms are Cournot competitors in its third stage. In general, auctioning licenses, that is, offering a fixed number of licenses to the highest bidders, yields the patentee a higher profit then offering licenses at a fixed fee or royalty rate to any firm wishing to purchase one. For modest cost-reducing inventions, licensing by means of the "chutzpah" mechanism, to be described below, provides the patentee higher profits than a license auction.

The essential reason that licensing by means of an auction enables the patentee to realize higher profits than by fixed fee licensing is that a nonlicensee's profits are lower in the former case than in the latter. This is because, in the case of an auction, if a firm does not get a license, it competes with licensed firms equal in number to the number of licenses auctioned, while in the case of a license fee, if it does not purchase a licence, it competes with one fewer licensee. That is, the firm can reduce the number of licensees by one by not purchasing a license in the case of a licensee fee, but cannot reduce the number of licenses by bidding zero in the auction case. As the most a firm will pay for a license equals the difference between its profits as a licensee versus a nonlicensee, it will pay more if licenses are auctioned than if they are sold for a fixed fee. Therefore the patentee in his role as the Stackelberg leader is able to extract higher total licensing profits by auctioning licenses than by selling them for a fixed fee. This difference in the patentee's licensing profits declines as the number of potential licensees increases and vanishes altogether in the limit as their number approaches infinity. Under either method of licensing, licensed and unlicensed firms' profits are in general below what they were before the invention's introduction. The exceptions occur if firms only realized perfectly competitive (zero) profits originally or if the invention is drastic and licensed for a fixed fee. In the last instance the single licensee is no worse off than he was originally. The patentee never licenses more firms than the number for which the Cournot equilibrium price equals the perfectly competitive price with the original inferior technology. If the invention is nondrastic the number of licensees is at least equal to one-half the number of potential licensees. Only a drastic invention is licensed exclusively to one firm. Consumers are always better off under either of these modes of licensing as total industry output increases with the introduction of the superior technology, and the product's price declines.

Licensing a nondrastic invention by means of a unit royalty is less profitable for the patentee than licensing by means of an auction or a fixed fee. The reason is that for any royalty rate below the magnitude of the unit cost reduction afforded by the invention, each firm will purchase a license. But if all firms purchase licenses, it is most profitable for the licensee to raise the royalty rate to exactly the magnitude of the unit cost reduction. He cannot raise it higher as then no firm would purchase a license because it would be more profitable for it to use the old technology. This in turn means that the most the patentee can extract from a licensee is the difference between his profits as a licensee and his original profits, which exceed a nonlicensee's profits with auction or fixed fee licensing. In other words, a nonlicensee can guarantee himself a higher profit if licenses are sold by means of royalty than either for a fixed fee or auctioned. The patentee's total royalty licensing profits approach those under auction or fixed fee licensing as the number of potential licensees approaches infinity regardless of the type of invention. If the invention is nondrastic, then under royalty licensing the licensees are no worse off than they were originally, while consumers are no better off, as there is no expansion in total output accompanying the introduction of the new technology. For a drastic invention, royalty licensing causes the single licensee and the nonlicensees to be worse off than originally, unless their profits were already zero, while consumers enjoy a lower product price and expanded output.

The above discussion provides the flavor of the game-theoretic approach to patent licensing and the types of results obtainable.

Section 2 deals with licensing by means of an auction. This is followed in Section 3 by an analysis of fixed fee licensing of a cost-reducing innovation and then of a new product (Section 4). Licensing by means of a royalty is taken up in Section 5 and is followed by fixed fee plus royalty licensing (Section 6). An optimal licensing mechanism, the "chutzpah" mechanism, is described in Section 7. All of the above analyses assume that the firms that are the potential licensees engage in Cournot competition. In Section 8, patent licensing in the presence of Bertrand competition is analyzed. This is followed by a brief summary.

It should be noted that, throughout the analyses of the different licensing modes, licensees' and nonlicensees' profit functions, as well as the patentee's, are denoted by the same symbols but different arguments in the different sections. However, the appropriate arguments of these functions should be clear from the context.

2. The license auction game

This is essentially the game introduced by Katz and Shapiro except that in their original version there is no specification of the structure of the industry, the firms of which are the potential licensees. We posit an industry consisting of $n \geq 2$ identical firms producing the same good with a linear cost function $f(q) = cq$, where q is the quantity produced by a firm, and $c > 0$ is the constant marginal cost of production. The inverse demand function for this good is given by $P = a - Q$, where $a > c$ and Q is the aggregate quantity demanded and produced. In addition to the n firms there is an inventor with a patent for a technology that reduces the marginal cost of production from c to $c - \varepsilon$, $\varepsilon > 0$. He seeks to maximize his profit by licensing his invention rather than using it himself to compete with the existing firms directly. The firms seek to maximize their production profits less licensing costs.

The game is noncooperative and consists of three stages. In its first stage the patentee decides how many licenses, k, to auction. All the firms decide independently and simultaneously how much to bid for a license in its second stage. Finally, in the game's third stage each firm, licensed and unlicensed, determines its profit-maximizing level of output. Thus, the patentee's strategies consist of choosing an integral number, $k \in \{0, 1, \ldots, n\}$, of licenses to auction. The ith firm's strategy consists of choosing how much to bid for a license, $b_i(k)$, which is a function of the number k of licenses auctioned in the game's second stage. Licenses are sold to the highest bidders at their bid price and in the event of a tie, licensees are chosen arbitrarily.

At the end of the game's second stage the original n firms divide into two groups, a subset S of k licensees and its complement N/S, the subset of $n - k$ nonlicensees. The members of S can produce with the superior cost function $f(q) = (c - \varepsilon)q$ and those in N/S with the inferior cost function $f(q) = cq$. The ith firm's strategy in the game's third stage consists of choosing it profit-maximizing level of output, $q_i(k, S)$, which depends on k and on whether or not it is among the licensees, S. Let $\pi_i = \pi_i(k, (b_1(k), q_1(k, S)), \ldots, (b_n(k), q_n(k, S)))$ be the ith firm's profit under the $(n + 1)$-tuple of strategies $(k, (b_1(k), q_1(k, S)), \ldots, (b_n(k), q_n(k, S)))$. Then the ith firm's payoff is

$$\pi_i(k, (b_1(k), q_1(k, S)), \ldots, (b_i(k), q_i(k, S)), \ldots, (b_n(k), q_n(k, S)))$$
$$= \begin{cases} (p - c + \varepsilon)q_i - b_i, & i \in S, \\ (p - c)q_i, & i \notin S, \end{cases} \tag{1}$$

where $P = a - \Sigma_1^n q_j$. The patentee's profit is

$$\hat{\pi}(k, (b_1(k), q_1(k, S)), \ldots, (b_n(k), q_n(k, S))) = \sum_{j \in S} b_j(k). \tag{2}$$

The payoffs (1) and (2), together with the patentee's and firms' strategy sets described above, define a strategic form game.

For this game the SPNE in pure strategies is the solution concept employed. The $(n + 1)$-tuple $(k^*, (b_1^*, q_1^*), \ldots, (b_n^*, q_n^*))$ with the corresponding set S^* of licensees, is a SPNE in pure strategies if

(i) k^* is the patentee's best reply strategy to the n firms' strategies $(b_1^*(k), q_1^*(k, S)), \ldots, (b_n^*(k), q_n^*(k, S))$;

(ii) for each k, $b_i^*(k)$ is the ith firm's best reply, given $b_j^*(k)$, for $i \neq j$;

(iii) for each k and S, $q_i^*(k, S)$ is the ith firm's best reply, given $q_j^*(k, S)$, $i \neq j$.

Note that the difference between the requirement for a SPNE and a Nash equilibrium in this game is that (ii) hold for any k, not just for $k = k^*$, and (iii) hold for any k and S, not just for $k = k^*$ and $S = S^*$.

As is customary for the development of the SPNE in pure strategies of a staged game, we work backwards from its last stage to its first. It can be shown [Kamien and Tauman (1984)] that the third stage Cournot equilibrium outputs q_i^* of the licensed and unlicensed firms, respectively, are

$$q_i^* = \begin{cases} (a - c - k\varepsilon)/(n + 1) + \varepsilon, & i \in S, \\ (a - c - k\varepsilon)/(n + 1), & i \notin S, \end{cases} \tag{3a}$$

provided the number of licensees, $k \leq (a - c)/\varepsilon$, and

$$q_i^* = \begin{cases} (a - c + \varepsilon)/(k+1), & i \in S, \\ 0, & i \notin S, \end{cases} \tag{3b}$$

if the number of licensees $k \geqslant (a - c)/\varepsilon$.

The zero output level of unlicensed firms if $k \geqslant (a - c)/\varepsilon$ follows from the requirement that a firm's output be non-negative. Note that in the first case, (3a), both licensed and unlicensed firms produce positive quantities in equilibrium, a licensed firm producing exactly ε more than an unlicensed one. Since in equilibrium all the licensed firms produce the identical quantity, let $q_i^* = \bar{q}$, $i \in S$, and similarly for unlicensed firms, let $q_i^* = \underline{q}$, $i \notin S$. The firms' third stage Cournot equilibrium profits are, for $k \leqslant (a - \bar{c})/\varepsilon$,

$$\pi_i^* = \begin{cases} \bar{q}^2 - b_i, & i \in S, \\ \underline{q}^2, & i \notin S, \end{cases} \tag{4a}$$

and, for $k \geqslant (a - c)/\varepsilon$,

$$\pi_i^* = \begin{cases} \bar{q}^2 - b_i, & i \in S, \\ 0, & i \notin S. \end{cases} \tag{4b}$$

The ratio $(a - c)/\varepsilon = K$ is the number of identical firms producing with marginal cost $c - \varepsilon$, such that the Cournot equilibrium price equals c.

This can be seen by observing that if $k \leqslant (a - c)/\varepsilon$, then from (3a) the Cournot equilibrium price is

$$P = a - k(\underline{q} + \varepsilon) - (n - k)\underline{q} = a - k\varepsilon - n\underline{q} = (a + nc - k\varepsilon)/(n + 1). \tag{5}$$

Setting $P = c$ in (5) yields

$$K = (a - c)/\varepsilon. \tag{6}$$

Similarly, for $k \geqslant (a - c)/\varepsilon$, the equilibrium price

$$P = a - k(a - c + \varepsilon)/(k + 1) = [a + k(c - \varepsilon)]/(k + 1), \tag{7}$$

and substituting $P = c$ into (7) gives

$$K = (a - c)/\varepsilon. \tag{8}$$

From this we can state:

Definition 1. An invention is *drastic* if $K \leqslant 1$. In other words, a drastic invention is one for which a monopolist operating with the superior technology would set a price at or below the marginal cost of the inferior technology.

Knowing the licensed and unlicensed firms' third stage Cournot equilibrium profits, (4a) and (4b), we can turn to the analysis of its second stage. In this stage the firms independently and simultaneously decide on how much to bid for a license. Each firm takes all the other firms' bids as given in deciding its own. The difference between a licensee's and nonlicensee's profit defines the most a firm will pay for a license. Thus, letting $\bar{\pi}$ refer to a licensee's profit and $\underline{\pi}$ to a nonlicensee's, and substituting K for $(a - c)/\varepsilon$ in (4a), yields

$$\bar{\pi}(k) = [\varepsilon(1 + (K - k)/(n + 1))]^2 - b_i, \tag{9a}$$

$$\underline{\pi}(k) = [\varepsilon(K - k)/(n + 1)]^2. \tag{9b}$$

Thus, the most a firm will bid for a license is

$$b_i = \bar{\pi}(k) - \underline{\pi}(k) = \varepsilon^2[1 + 2(K - k)/(n + 1)], \quad k \leqslant K. \tag{10}$$

Since the right-hand side is identical for every firm, their bids are identical, $b_i = b$. Obviously, no unlicensed firm will bid more than b to become a licensee because its net profit would decline. Similarly, a licensee would bid neither more than b for a license, nor less and become a nonlicensee. Thus, b is each firm's Nash equilibrium bid. Expression (10) represents the demand function for licenses in the region $k \leqslant K$.

In the region $k \geqslant K$, the demand function for licenses can be determined in a similar way by using the profits of a licensee and a nonlicensee as defined in (4b). Specifically, the demand function for licenses is

$$b = [\varepsilon(K + 1)/(k + 1)]^2, \quad k \geqslant K. \tag{11}$$

The patentee's objective in the game's first stage is to choose k, the number of licenses to auction, so as to maximize his licensing profits. Namely, to

$$\max_{1 \leqslant k \leqslant n} bk$$

$$\text{s.t. } b = \begin{cases} \varepsilon^2[1 + 2(K - k)/(n + 1)], & 1 \leqslant k \leqslant K, \\ [\varepsilon(K + 1)/(k + 1)]^2, & k \geqslant K \geqslant 1. \end{cases} \tag{12}$$

Now, assuming that k is continuous, then for $1 \leqslant k \leqslant K$,

$$d(bk)/dk = \varepsilon^2[(n + 1 + 2K - 4k)/(n + 1)] \tag{13}$$

and

$$d^2(bk)/d^2k = -4\varepsilon^2/(n + 1) < 0. \tag{14}$$

Thus, the global maximum of the patentee's profit occurs at

$$k^* = (n + 1 + 2K)/4 , \quad \text{for } 1 \leq k \leq K , \tag{15}$$

provided $k^* \leq \min(K, n)$, because the number of licenses cannot exceed n or K. That is, (15) gives the k^* that is the interior maximum number of licenses, provided it can be achieved. Thus, corner solutions have to be checked for as well.

Now, the right-hand side of (15) equals or exceeds K if $n + 1 + 2K \geq 4K$. Thus, $k^* = K$ if and only if $n + 1 \geq 2K$ as $n > K$ in this case, unless $n = K = 1$. Similarly, the right-hand side of (15) equals or exceeds n if $n + 1 + 2K \geq 4n$. Thus, $k^* = n$ if and only if $2K \geq 3n - 1$, as $K > n$ in this case, unless $n = K = 1$.

Turning next to the case of $k \geq K \geq 1$, the demand function for licenses is given by (11) and the patentee seeks to

$$\max_{k \geq K} kb . \tag{16}$$

Now, from (11), $db/dk = -2b/(k + 1)$, and the elasticity of the license demand function, $-(b/k) \, dk/db = (k + 1)/2k \leq 1$, as $k \geq 1$. Thus, as the demand function for licenses is inelastic in the region $k \geq K \geq 1$, it follows that the patentee's profit increases as the number of licenses offered declines, and so he sets $k^* = K$, its lower bound. Finally, if the invention is drastic, $1 \geq K \geq 0$, the patentee auctions a single license, since $k < 1$ is meaningless. All the above can be summarized as:

Proposition 1. *The license auction game has a unique SPNE in pure strategies in which the patentee never auctions more than K licenses, and if the invention is drastic, only a single license. Specifically, if the invention is not drastic and $1 \leq k \leq K$, then*

$$k^* = \begin{cases} n , & 2K \geq 3n - 1 , \\ (n + 1 + 2K)/4 , & 3n - 1 \geq 2K \geq n + 1 , \\ K , & n + 1 \geq 2K , \end{cases} \tag{17a}$$

and if $k \geq K$,

$$k^* = \begin{cases} K , & K \geq 1 , \\ 1 , & K \leq 1 . \end{cases} \tag{17b}$$

The intuition of Proposition 1 is that if the invention affords only a modest reduction in unit production costs, i.e., $2(a - c)/(3n - 1) \geq \varepsilon$, then it is optimal for the patentee to license all the firms in the industry. However, in doing so, he must, along with his announcement that n licenses will be auctioned, state a reservation price slightly below the magnitude $b(n)$, the benefit to a firm if all are licensed, below which he will not sell a license. The necessity for the license reservation price is to prevent a firm from offering nothing for a license, because it knows it will get one anyway. Thus, for a modest cost-reducing invention the industry's structure, in terms of the number of operating firms, remains unchanged, but all of them operate with the new technology. On the other hand, if the magnitude of the unit cost reduction permitted by the innovation is somewhat greater, i.e., $\varepsilon \geqslant 2(a-c)/(3n-1)$, then not all the firms obtain licenses, but all of them, licensed and unlicensed, continue to operate. The industry now becomes one with a mixed technology, some firms operating with the superior technology and others with the inferior one. Finally, if the cost reduction afforded by the invention is sufficiently large, i.e., $\varepsilon \geqslant 2(a-c)/(n+1)$, then the number of firms in the industry is reduced to the number of licensees, K; the remaining firms cease operating. With a nondrastic invention at least one-half of the industry's firms are licensed. In the special case of a drastic innovation, i.e., $\varepsilon \geqslant a-c$, the number of operating firms in the industry is reduced to one.

The patentee's equilibrium licensing profits, $\hat{\pi}$, can be calculated by substitution for k^* from (17a) and (17b) into (10) and (11) to determine the equilibrium bids b^*, and then multiplying by k^*, i.e.,

$$\hat{\pi} = \begin{cases} \varepsilon^2 n(2K+1-n)/(n+1), & 2K \geqslant 3n-1, & 1 \leqslant k^* = n \leqslant K, \\ \varepsilon^2(2K+1+n)^2/8(n+1), & 3n-1 \geqslant 2K \geqslant n+1, & 1 \leqslant k^* \leqslant \min(K,n), \\ \varepsilon^2 K, & n+1 \geqslant 2K, & 1 \leqslant k^* = K \leqslant n, \\ \varepsilon^2(K+1)^2/4, & K \leqslant 1, & 1 = k^*. \end{cases}$$

(18)

The firms' equilibrium profits can be calculated by recalling that in equilbrium licensees' net profits must equal nonlicensees' (if this were not so an incentive to revise bids for licenses would exist), $\bar{\pi}(k) = \underline{\pi}(k)$, and substituting for k^* from (17a) and (17b) into (9b) to obtain:

$$\pi^* = \begin{cases} \varepsilon^2(K-n)^2/(n+1)^2, & 2K \geqslant 3n-1, \\ \varepsilon^2[(K-n)+(3n-1-2K)/4]^2/(n+1)^2, & 3n-1 \geqslant 2K \geqslant n+1, \\ 0, & n+1 \geqslant 2K, \\ 0, & 0K \leqslant 1, \end{cases}$$

(19)

The licensees' profits all decline relative to their original profits, which equal $[\varepsilon K/(n + 1)]^2$. Were they able to credibly collude, the industry's firms would resist the introduction of the superior technology, or at least the patentee's means of licensing it. Without the means for credible collusion, the patentee is able to exploit their noncooperative behavior to his advantage. These calculations also reveal that the aggregate quantity produced in the third-stage equilibrium exceeds the aggregate quantity produced before the introduction of the cost-reducing technology and therefore that the market price of good declines, to the consumer's benefit.

The above conclusions regarding the auction game as summarized in Proposition 1 have been extended by Kamien, Oren and Tauman (1988) to the class of product demand functions that are downward sloping, differentiable, and for which the total revenue function is strictly concave in the quantity sold. In this generalization, the price elasticity of demand, along with the magnitude of the cost reduction from the invention, determine the equilibrium number of licensees, k^*.

3. The fixed fee licensing game

In the model, introduced by Kamien and Tauman (1984, 1986), the patentee sets a price at which any firm wishing to can buy a license. The license price is independent of the number of units produced with the superior technology and therefore is a fixed cost to the firm just as in the license auction case. Fixed fee licensing may be a more practical alternative than licensing by means of an auction if the cost of organizing the auction is taken into account.

The fixed fee licensing game is similar to the license auction game in that it, too, is a three-stage game. It is in the derivation of the demand function for licenses that the major difference between the two games arises. In determining how much to bid for a license in the auction game a potential buyer looks at the difference in his profits if he is or is not a licensee, with the knowledge that the number of licensees will be the same regardless of whether of not he purchases one (unless, of course, the number of licenses auctioned equals the total number of firms). However, in the fixed fee game a potential buyer, while also comparing his profits as a licensee versus a nonlicensee, knows that if he does not purchase a license that there will be one fewer licensee. That is, in the fixed fee game a potential licensee regards $\pi_i(n, k) - \pi_j(n, k - 1)$, $i \in S$, $j \not\in S$, as the value of a license, where $\pi_i(n, k)$ is the profit he will realize at the Cournot equilibrium of a n-firm oligopoly in which k firms have licenses and $\pi_j(n, k - 1)$ his profit if he does not purchase a license and there are only $k - 1$ licensees. In the license auction game the similar comparison is between $\pi_i(n, k)$, $i \in S$, and $\pi_j(n, k)$, $j \not\in S$, because the potential licensee knows that even if he does not purchase a license, someone else will, because k licenses are being auctioned. Thus, the opportunity cost of not being a licensee is higher in the license auction game than in the fixed fee game. Therefore the patentee can in general profit more by auctioning licenses than by selling them at a fixed fee.

However, if the patentee does employ a fixed fee as a licensing device, then in the game's first stage he deduces that the most a firm will pay for a license, call it α, equals $\pi_i(n, k) - \pi_j(n, k - 1)$, $i \in S$, $j \notin S$. From this he can, as in the license auction game, derive a demand function for licenses. This can be done by first noting that the counterparts to the equilibrium quantities of a licensee and nonlicensee in (3a) are, for $k \leq K$,

$$q_i^* = \begin{cases} \varepsilon(K - k)/(n + 1) + \varepsilon, & i \in S, \\ [\varepsilon(K - k) + \varepsilon]/(n + 1), & i \notin S. \end{cases} \tag{20}$$

Thus, under license auctioning, licensees' and nonlicensees' equilibrium outputs differ by ε, while here they differ by $n\varepsilon/(n + 1)$. Obviously this difference disappears in the limit as $n \to \infty$, i.e., the industry becomes perfectly competitive. For $k \geq K$, the equilibrium outputs of licensees and nonlicensees are the same as in (3b). The respective profits of a licensee and a nonlicensee, the counterpart to (4a), are, for $k \leq K$,

$$\pi_i^* = \begin{cases} \bar{q}^2 - \alpha, & i \in S, \\ \underline{q}^2, & i \notin S, \end{cases} \tag{21}$$

where $\bar{q}$ refers to a licensee's third-stage equilibrium output and $\underline{q}$ to a nonlicensee's. The resulting second-stage demand function for licenses is

$$\alpha = \varepsilon^2[n + 2 + 2(K - k)]/(n + 1)^2, \quad k \leq K. \tag{22}$$

For $k \geq K$, the demand function for licenses is the same as (11), with b replaced by α. The patentee chooses the fixed fee α so as to maximize his licensing profits, given the demand function for licenses. This leads to the counterpart to Proposition 1, namely:

Proposition 2. *The fixed fee licensing game has a SPNE in pure strategies in which never more than K licenses are sold, and only one license is sold if the invention is drastic. Specifically, if the invention is not drastic and $k \leq K$, then*

$$k^* = \begin{cases} n, & 2K \geq 3n - 2, \\ (n + 2 + 2K)/4, & 3n - 2 \geq 2K \geq n + 2, \\ K, & n + 2 \geq 2K, \end{cases} \tag{23a}$$

and, if $k \geq K$, then

$$k^* = \begin{cases} K, & K \geq 1, \\ 1, & K \leq 1. \end{cases} \qquad (23b)$$

The patentee's licensing profits are, in general, lower under fixed fee licensing than under license auctioning. The difference is illustrated most dramatically when the invention is drastic. If a single license is auctioned, the patentee realizes the entire monopoly profit of the single licensee as a consequence of the bidding among the firms because an unlicensed firm's profit is zero. On the other hand, the price he can set to sell only one license cannot exceed the difference between the monopoly profit of a single licensee and the Cournot equilibrium profit of a firm in the original n-firm oligopoly operating with the inferior technology, i.e., $[\varepsilon K/(n+1)]^2$. This difference in the patentee's profit declines as the number of firms in the original oligopoly increases and vanishes completely in the limit as n goes to infinity. This is true in general regarding the patentee's profit under the two licensing schemes and not just for the case of a drastic invention. It also follows from (17a) and (23a) that a license auction leads to at least $(n+1)/2$ firms being licensed, while for fixed fee licensing at least $(n+2)/2$ are. Thus, fixed fee licensing is slightly better for consumers because the market price declines with the number of licensees.

As the license auction game, the fixed fee licensing game has been extended to more general product demand functions by Kamien, Oren and Tauman (1988).

4. Fixed fee licensing of a product innovation

Another application of the fixed fee licensing game has been to licensing production of a new product [Kamien, Tauman and Zang (1988)]. In this model the existence of two goods, a "superior" good, the demand for which is denoted by Y, and an "inferior" good, the demand for which is denoted by X, is posited. The two goods are substitutes, their degree of substitutability is denoted by δ, and one good is superior to the other in the sense that if its price is less than or equal to the others, then only the superior good is purchased. The respective inverse demand functions for the inferior and superior goods are

$$X = P_2 - P_1, \quad P_2 \geq P_1 \qquad (24)$$

and

$$Y = \begin{cases} a - P_2 + \delta P_1, & P_2 \geq P_1, \\ a - (1 - \delta)P_2, & P_2 \leq P_1, \end{cases} \qquad (25)$$

with P_1 and P_2 denoting their respective prices and, $0 \leq \delta < 1$. From (24) and (25) it follows that the inverse demand function for the inferior good, if the superior good is not produced at all, $Y = 0$, is

$$X = a - (1 - \delta)P_1 . \tag{26}$$

The unit cost of producing the inferior good is assumed to be constant and is denoted by c_1. It is supposed that the invention reduces the constant unit cost of producing the superior good from $\bar{c}_2$, a cost at which its production is unprofitable, to c_2, at which it becomes profitable. The interaction among the firms, who are the potential licensees, and the patentee is again described as a three-stage game. Also, there exists a number, $K = [a - (1 - \delta)c_1]/(1 - \delta)(c_1 - c_2)$, of producers of the superior product such that its Cournot equilibrium price equals the inferior good's unit cost of production. A "drastically superior" product is one for which $K \leq 1$. Finally, as in the case of a simple cost-reducing invention, it is possible to derive a demand function for licenses in the game's second stage, which the patentee employs in the first stage to set his profit-maximizing license fee.

Among the results obtained in this game are that if the new product is drastically superior, then only it is sold. Both the inferior product and the superior one are sold (in the limit as $n \to \infty$) if their unit production costs are sufficiently close, and only the superior product if not. For sufficiently large n, the inferior product's price declines as a result of the superior product's introduction. The distinctive feature of the patentee's new product licensing behavior is that his license fee must balance the effect of increasing the number of licensees, which depresses the demand for the inferior good and increases the demand for the superior one, against the increase in competition among the superior good's producers. Finally, it is also possible to use this framework to characterize when an inventor would be better off attempting to provide a higher quality product versus reducing the unit cost of an existing one, supposing that the costs of the two possibilities are the same. In particular, a product-quality-improving invention will be more profitable than a cost-reducing invention if the demand for the superior product is sufficiently high. On the other hand, an inventor can always realize a higher profit from a drastic cost-reducing invention than from a quality-improving one.

5. Royalty licensing

Licensing by means of royalty is the form most commonly employed in the real world. The reason is that the demand for the product that will be produced with the new cost-reducing technology is typically uncertain, especially future demand, for a variety of reasons. For example, new substitute products may come on to the market or there may be a general downturn in the economy, or new uses may be discovered for this product. The licensee may be reluctant to pay a fixed fee for a license out of fear that the demand for the product will decline, while the patentee may be reluctant to sell a license for a fixed fee out of expectation that the demand for the product will rise. The same concerns apply to the licensing of the manufacture of a new product. A royalty alleviates a licensee's fear of overpaying for a license and the patentee's of undercharging. The reward to the patentee and the payments of the licensee are directly related to the demand for the product through time. Indeed, in the case of licensing the sale of a new product, the patentee often offers a lower royalty rate if sales exceed a certain prespecified level. This is done to provide an incentive for the seller of the licensed product to push its sales above the sale of other products that he may also be selling.

In the game-theoretic analysis of patent licensing by means of a royalty these factors have not yet been considered. Instead, as in the analysis of licensing by means of an auction or a fixed fee, it is assumed that the values of all the relevant variables are known with certainty. Thus, again, the analysis takes the form of a three-stage game involving the n-licensee and the patentee, who is the Stackelberg leader. Since the patentee's licensing profit increases linearly with the total quantity produced with his lower cost technology it is in his interest to license as many of the firms as possible. In fact, if the invention is nondrastic, he can license all of the industry's n firms by setting the royalty rate, r, equal to the magnitude, ε, of the reduction in unit cost permitted by the new technology. At this royalty rate each firm is indifferent between being a licensee and a nonlicensee. Thus, no firm has an incentive to deviate and become a nonlicensee. The patentee has no incentive to raise the royalty rate, as this will cause no licenses to be purchased, for the firms can do better by operating with the original technology, and no incentive to lower it, as all the firms will still be licensees but his profit will decline. Because at a royalty rate $r = \varepsilon$ the licensees are indifferent between being and not being a licensee, it might appear that the situation in which no firm purchases a license could be an equilibrium of the game as well. But this is not so, because the patentee's offer of licenses at any royalty rate below ε will cause all the firms to purchase them. Thus, a royalty rate of $r = \varepsilon$, and all the firms purchasing licenses, constitute a unique SPNE equilibrium of this game.

If the invention is not drastic, then in equilibrium the patentee's profit equals the per unit royalty times the number of units produced. But since the royalty rate $r = \varepsilon$, the equilibrium level of output after licensing is exactly the same as before licensing, namely the Cournot output of an n-firm oligopoly, $Q = n(a - c)/(n + 1)$. Thus, the patentee's licensing profit is $rn(a - c)/(n + 1)$. On the other hand, if he were to auction licenses, say, of an invention that reduces units costs sufficiently so that $k^* = K$, then his profit would be

$\varepsilon(a - c)$, which clearly exceeds his licensing profits by means of a royalty, as $r = \varepsilon$. These results obtain for all nondrastic inventions, not just for one where $k^* = K$. The exception occurs in the limit as $n \to \infty$, the industry becomes perfectly competitive, and the patentee's profits are the same under either mode of licensing.

If the invention is drastic, the patentee sets the royalty rate $r = (a - c + \varepsilon)/2$, the monopoly price that the firm employing the superior technology would charge. His licensing profit equals the corresponding monopoly profit.

Formally, the patentee seeks a royalty rate r to maximize his licensing profits, i.e.,

$$\max_{0 \leqslant r \leqslant \varepsilon} nr(a - c + \varepsilon - r)/(n + 1), \tag{27}$$

as all firms will be licensed, and the third-stage Cournot equilibrium output of a license is $q_i = (a - c + \varepsilon - r)/(n + 1)$, $i = 1, \ldots, n$. The first-order condition for one interior maximum yields

$$r^* = (a - c + \varepsilon)/2 = \varepsilon(K + 1)/2. \tag{28}$$

It is obvious that (28) can hold only if $K \leqslant 1$, the invention is drastic, because r cannot exceed ε. Thus, if $K \geqslant 1$, $r^* = \varepsilon$. The patentee's licensing profits, $\hat{\pi}$, are

$$\hat{\pi} = \begin{cases} n\varepsilon^2 K/(n + 1), & K \geqslant 1, \\ \varepsilon^2(K + 1)^2/4, & K \leqslant 1. \end{cases} \tag{29}$$

It is evident from a comparison of (28) with (18) that the patentee's licensing profits are higher if he auctions licenses than if he employs a royalty, except if the invention is drastic or the industry is perfectly competitive ($n \to \infty$). All of this can be summarized as:

Proposition 3. *The royalty licensing game has a SPNE in pure strategies in which all the firms are licensed at a royalty rate equal to the magnitude of the reduction in the unit production costs, if the invention is nondrastic, and only one firm is licensed at a royalty rate equal to the monopoly price set by the firm operating with the superior technology, if the invention is drastic. All the licensees' profits remain the same as they were originally if the invention is nondrastic, and the single licensee's profit falls to zero, if it is drastic. The product's market price declines only if the invention is drastic. The patentee's licensing profits are less than in the auction or fixed fee licensing games unless the invention is drastic or the industry is perfectly competitive.*

6. Fixed fee plus royalty licensing

Licensing by means of a fixed fee plus royalty is the form most commonly employed. Analysis of this form of licensing was conducted by Kamien and Tauman (1984). As in the previous analyses, the game between the patentee and the n potential licensees, who engage in Cournot competition amongst themselves, involves three stages. The patentee maximizes his licensing profits in this case by choosing both a fixed fee and a royalty rate, given the potential licensees' demand function for licenses. The analysis of this game is technically more complicated than those involving the patentee choosing only one strategic variable. The main conceptual difference is that the patentee faces a conflict between choosing a fixed fee that limits the number of licensees and a royalty that expands their number. However, as in the auction license game or fixed fee game, no more than K firms are licensed if $1 \leq K \leq n$ and only one if $K \leq 1$. Also, as in the fixed fee game, no fewer than $(n + 2)/2$ firms are licensed.

7. An optimal licensing mechanism: The "chutzpah" mechanism

In general, among the three licensing modes considered thus far, a license auction provides the patentee the highest profit. However, is it the best the patentee can do or is there a superior mechanism? Kamien, Oren and Tauman have proposed a licensing mechanism, the "chutzpah" mechanism, based on the work of Kamien, Tauman and Zamir (1990), dealing with the value of information that in general yields the patentee a higher profit than a license auction. In particular, this mechanism is more profitable for the patentee than a license auction for modest cost-reducing inventions.

To understand the "chutzpah" mechanism, recall that in a license auction the most a firm will bid for a license is the difference between its Cournot equilibrium profits as a licensee and as a nonlicensee, for a given number, k, of licenses auctioned. However, the firm's Cournot equilibrium profit, if it is a nonlicensee and there are k licensees, is at least as high as its profit, if there are $n - 1$ licensees. That is, a firm's lowest equilibrium profit occurs if it alone is unlicensed. In the "chutzpah" mechanism the patentee exploits this fact by setting a fixed fee that extracts essentially the entire difference between the profits of a licensee and the profits of a single nonlicensee. Moreover, the patentee also extracts from each nonlicensee essentially the difference between his equilibrium profit in the presence of k licensees and his profit as the only nonlicensee. Thus, both licensees and nonlicensees are compelled to yield almost all their profit to the patentee – as in the game to be described, it is their dominant strategy to do so after dominated strategies are eliminated. In this mechanism, licensees pay for a license and nonlicensees pay to prevent additional licensing.

Formally, licensing by means of the "chutzpah" mechanism is a three-stage game. In the first stage the patentee determines the configuration of licensees and nonlicensees that maximizes total industry profit and how much to "charge" licensees and nonlicensees. In its second stage the firms simultaneously and independently decide on whether or not to accept his offer. In the third stage, each firm determines its Cournot equilibrium output. This last stage is identical to the third stage of the previously described licensing games. Thus, (4a) and (4b), excluding the bid b_i, can be employed to indicate licensees' and nonlicensees' profits for every number k of licensees.

The patentee's first objective in the "chutzpah" mechanism is to determine the number of licenses that maximizes total industry profit, Π, i.e., to

$$\max_{1 \leq k \leq n} \Pi(k) = \max_{1 \leq k \leq n} k \hat{\pi}(k) + (n - k) \underline{\pi}(k), \tag{30}$$

where $\hat{\pi}(= \bar{\pi} + b)$ and $\underline{\pi}(k)$ refer to licensees' and nonlicensees' operating profits, respectively. Now let $\bar{k}$ be the maximizer of (30) and $\bar{\Pi} = \Pi(\bar{k})$ be the corresponding maximum total industry profit. Let $\underline{\pi}(n - 1)$ refer to a single nonlicensee's Cournot equilibrium profit, i.e., if all but this firm are licensed. Now the greatest licensing profit, G, the patentee can realize is

$$G = \bar{\Pi} - n\underline{\pi}(n - 1) \tag{31}$$

because $\bar{\Pi}$ is the maximum total industry profit under licensing and $\underline{\pi}(n - 1)$ is what each firm can guarantee itself.

Definition 2. A licensing mechanism is optimal if it achieves G.

From (31) and Definition 2 it follows that

Proposition 4. *A license auction is an optimal licensing mechanism for a sufficiently large, $\varepsilon \geq 2(a - c)/(n + 1)$, unit cost-reducing invention and a linear product demand function.*

Proof. Consider first $1 \leq k \leq K$. Then, from (30), and suppressing the argument k of Π, $\hat{\pi}$ and $\underline{\pi}$,

$$\Pi = k\hat{\pi} + (n - k)\underline{\pi} = k(\hat{\pi} - \underline{\pi}) + n\underline{\pi}$$
$$= k[(\varepsilon(K - k)/(n + 1) + \varepsilon)^2 - (\varepsilon(K - k)/(n + 1))^2]$$
$$\quad + n(\varepsilon(K - k)/(n + 1))^2$$
$$= k[2\varepsilon^2(K - k)/(n + 1) + \varepsilon^2] + n[\varepsilon(K - k)/(n + 1)]^2$$

upon substitution for $\hat{\pi}$, licensees' operating profits, and $\underline{\pi}$, nonlicensees' operating profits from (9a) and (9b), and the collection of terms. Now, after some algebra,

$$\partial \Pi / \partial k = [2(K - k) + (n + 1)(n + 1 - 2k)][\varepsilon /(n + 1)]^2 .$$

However, $\varepsilon \geq 2(a - c)/(n + 1)$ is equivalent to $n + 1 \geq 2K$, and since $k \leq K$, it follows that $n + 1 - 2k \geq 0$. Thus, $\partial \Pi / \partial k \geq 0$ for all $k \leq K$, and $\bar{k} = K$ for a maximum of Π. But when K firms are licensed, the remaining unlicensed firms cease operating and $\underline{\pi}(n - 1) = 0$.

Now for $k \geq K \geq 1$ it follows from (3b) and (4b) that $\Pi = k[\varepsilon(K + 1)/(k + 1)]^2$, as $\underline{\pi} = 0$, and it is easy to show that $\partial \Pi / \partial k \leq 0$. Thus, again $\bar{k} = K$ and $\underline{\pi}(n - 1) = 0$. Finally, for $K \leq 1$ only one license is auctioned and $\underline{\pi}(n - 1) = 0$. Therefore, a license auction is an optimal licensing mechanism because it yields the patentee $G = \bar{\Pi}$. $\square$

The "chutzpah" mechanism is, therefore, relevant if the unit cost-reducing invention is more modest, i.e., $\varepsilon \leq 2(a - c)/(n + 1)$. After having determined the total industry profit-maximizing number of licenses, $\bar{k}$, the patentee asks each firm for a fee

$$\beta_0 = \begin{cases} \hat{\pi}_i(\bar{k}) - (\underline{\pi}(n - 1) + \rho/n) , & i \in S , \\ \underline{\pi}_i(\bar{k}) - (\underline{\pi}(n - 1) + \rho/n) , & i \notin S , \end{cases} \tag{32}$$

where S refers to the subset of licensees, and $\rho > 0$ is an arbitrarily small number. If each firm agrees to pay its fee, then $\bar{k}$ are licensed and all n engage in the third-stage game. However, if a nonempty subset R, $|R| = r$, of them refuse to pay their fees, then those who agreed to pay are offered licenses at the price

$$\beta_1 = \hat{\pi}_i(n - r) - \pi_i(0) - \rho/n , \quad i \notin R , \tag{33}$$

where $\hat{\pi}_i(n - r)$ refers to the third-stage Cournot equilibrium operating profits of a licensee if $n - r$ firms are licensed, and $\pi_i(0)$ each firm's profit if none is licensed, i.e., its original preinvention equilibrium profit. From (32) it is clear that if every firm agrees to pay the fee, β_0, proposed by the patentee, then each licensee and nonlicensee will ultimately realize a profit slightly above what it can unilaterally guarantee itself. On the other hand, if there are $r > 0$ of them who refuse to pay, then they will each earn $\underline{\pi}(n - r)$, the profits of a nonlicensee in the presence of $n - r$ licensees, which is below their original profit, $\pi(0)$. The remaining $n - r$ firms who did agree to pay β_0 will only have to pay β_1 for a license, and will ultimately each realize a profit slightly above their original profit, $\pi(0)$. From this, we can then state:

Proposition 5. *By eliminating dominated strategies, it is a dominant strategy for each of the n firms to accept the initial offer in the "chutzpah" mechanism.*

Proof. Let r be the number of firms who rejected the initial offer (the "refuseniks") and firm i be one of $n - r$ firms who accepted it. Now i has the option of buying a license at the price β_1. If he purchases a license his net profit will be $\hat{\pi}(t) - \beta_1$, when t is the total number of firms, including firm i, out of the $n - r$ who had the option to buy a license at price β_1, that chose to exercise it. On the other hand, if he chooses not to purchase a license, then his profit will be $\pi(t - 1)$, the profit of a nonlicensee in the presence of $t - 1$ licensees. But by (3a), (4a), and (33), it follows that

$$\hat{\pi}(t) - \beta_1 \geq \pi(0) + \rho/n > \underline{\pi}(t - 1) , \tag{34}$$

since $t \leq n - r$. Thus, if $r > 0$, and i has accepted the patentee's initial offer, then regardless of what the other firms do he should purchase a license at the price β_1.

What remains to be shown is that i should always accept the initial offer. Suppose i rejects the initial offer. Then his profit will be $\pi(t)$, the profit of a nonlicensee in the presence of some $t \leq n - r$ licensees. But again by (3a) and (4a), $\pi(t) \leq \pi(0) + \rho/n$. Thus, if some firms have rejected the initial offer, then regardless of the actions of other firms it is best for i to accept the initial offer and then purchase a license. So firm i, by accepting the initial offer and purchasing a license, will be better off than it was originally, i.e., when it earned $\pi(0)$. But this was due to some other firms refusing to agree to the patentee's initial offer. Suppose now that all the other firms except i agree to the patentee's initial offer. Then, based on the above argument, it is a dominant strategy for each of them to purchase a license. Firm i's profit will then be $\underline{\pi}(n - 1)$, the profit of a nonlicensee in the presence of $n - 1$ licensees. This is clearly below $\pi(0)$, his original profit and, even worse, below $\underline{\pi}(n - 1) + \rho/n$, which is his net profit if he accepts the initial offer. Finally, then, it is in i's best interest to accept the initial offer. $\square$

Thus,

Proposition 6. *The "chutzpah" mechanism is almost optimal in that it enables the patentee to realize licensing profits of $\bar{\Pi} - n\underline{\pi}(n - 1) - \rho$.*

The patentee would prefer to employ the "chutzpah" mechanism to an auction for licensing modest unit cost-reducing inventions, $\varepsilon \leq 2(a - c)/(n + 1)$. For less modest inventions he could do as well by auctioning licenses. The counterparts of Propositions 5 and 6 have been established for more general product demand functions in Kamien, Oren and Tauman (1988).

Note that the "chutzpah" mechanism relies on both a carrot and a stick. The stick part arises from the threat of putting the firm in its least profitable position by licensing all the other firms. The carrot part comes from providing the firm with a small reward, ρ, for paying the fee. Both the carrot and stick are necessary for the mechanism to work. If the reward is eliminated, $\rho = 0$, then each firm is indifferent between paying and not paying the fee because it will realize its lowest possible profit in either case. On the other hand, if the stick is relaxed, then the patentee cannot extract as much from licensing his invention. Also, the "chutzpah" mechanism relies on the possibility of having two opportunities for a firm to purchase a license for its implementation. Were there only one opportunity to purchase a license, this mechanism would appear not to be implementable. How the mechanism would have to be modified to accommodate more opportunities to purchase a license or in the limit as $\rho \to 0$, remain open questions. From a real world standpoint it is difficult to cite an example of a counterpart to the "chutzpah" mechanism. The fact that it applies for modest cost-reducing innovations suggests that when its implementation costs are taken into account, a patentee might well resort to one of the more traditional licensing modes. Thus, at present the "chutzpah" mechanism should be regarded as a theoretical standard towards which any practical licensing mode might aspire.

8. Licensing Bertrand competitors

Thus far the analysis of alternative means of licensing has been conducted under the supposition that the potential licensees compete through selection of quantities. However, if they engage in price competition, Bertrand competition, and each firm's original unit production cost is constant, c, then the analysis of licensing schemes is far simpler. As is well known, price competition among firms with constant unit cost drives their profits to zero. Thus, the firm with the lower cost, $c - \varepsilon$, will drive the others out of business and realize a profit of $\varepsilon Q(c)$, where $Q(c)$ refers to the total quantity demanded at price c. The single firm with the superior technology could then become operative and it certainly will not charge a lower price if the intention is nondrastic. On the other hand, if the invention is drastic, the single firm with the superior technology will set a monopoly price $P_m \leq c$ and realize a profit of $\varepsilon Q(P_m)$, where $Q(P_m)$ is the quantity demand at the price P_m. In either event, there will be only one licensee and nonlicensees' profits, $\pi = 0$. It follows, therefore, that the patentee can extract the licensee's entire profit by auctioning a single license, setting a fixed license fee equal to the licensee's profit, or setting a royalty equal to ε, for a nondrastic invention, or the monopoly price P_m, for a drastic invention. The patentee's licensing profits are the same under any of these alternatives.

9. Concluding remarks

Game-theoretic methods have made it possible to address questions with regard to patent licensing that could not be analyzed seriously otherwise. Obviously much remains to be done in bringing the models of patent licensing closer to reality. For example, introducing uncertainty regarding the magnitude of the cost reduction provided by an invention or the commercial success of a new product into the analysis of patent licensing. Jensen (1989) has begun analysis in this direction. Another obvious topic is the licensing of competing inventions, i.e., those that achieve the same end by different means. Still another is the question of licensing inventions in the absence of complete patent protection. Muto (1987, 1990), and Nakayama and Quintas (1991) have begun analysis of licensing when the original licensee cannot prevent his immediate licensees from relicensing to others. They introduce a solution concept called "resale-proofness" and employ it to analyze the scope of relicensing. The basic idea is that relicensing may be limited to a subset of the firms in an industry, because the relicensing profit realizable by a licensee is below the decline in profits he will suffer as a result of having one more firm with the superior technology to compete with. An extreme case of this negative externality effect occurs when a firm that is a member of the industry invents a superior technology and only employs it alone.

It is often the case that a survey of a line of research is a signal of it having peaked. This is certainly not true for game-theoretic analysis of patent licensing.

References

Arrow, K.J. (1962) 'Economic welfare and the allocation of resources for invention', in: R.R. Nelson, ed., *The rate and direction of incentive activity.* Princeton: Princeton University Press.

Baldwin W.L. and T.J. Scott (1987) *Market structure and technological change.* New York: Harwood Academic Press.

Barzel, Y. (1968) 'Optimal timing of innovations', *Review of Economics and Statistics,* 50: 348–355.

Caves, R.E., H. Crookell and J.P. Killing (1983) 'The imperfect market for technology licenses', *Oxford Bulletin for Economics and Statistics,* 45: 249–268.

Dasgupta P. and J. Stiglitz (1980a) 'Industrial structure and the nature of innovation activity', *Economic Journal,* 90: 266–293.

Dasgupta P. and J. Stiglitz (1980b) 'Uncertainty, industrial structure, and the speed of R&D', *Bell Journal of Economics,* 11: 1–28.

Demsetz, H. (1969) 'Information and efficiency: Another viewpoint', *Journal of Law and Economics,* 12: 1–22.

Jensen, R. (1989) 'Reputational spillovers, innovation, licensing and entry', *International Journal of Industrial Organization,* to appear.

Kamien M.I. and N.L. Schwartz (1972) 'Timing of innovation under rivalry', *Econometrica,* 40: 43–60.

Kamien M.I. and N.L. Schwartz (1976) 'On the degree of rivalry for maximum innovative activity', *Quarterly Journal of Economics,* 90: 245–260.

Kamien M.I. and N.L. Schwartz (1982) *Market structure and innovation*. Cambridge: Cambridge University Press.

Kamien M.I. and Y. Tauman (1984) 'The private value of a patent: A game theoretic analysis', *Journal of Economics* (*Supplement*), 4: 93–118.

Kamien M.I. and Y. Tauman (1986) 'Fees versus royalties and the private value of a patent', *Quarterly Journal of Economics*, 101: 471–491.

Kamien, M.I., S. Oren and Y. Tauman (1988) 'Optimal licensing of cost-reducing innovation', *Journal of Mathematical Economics*, to appear.

Kamien, M.I., Y. Tauman and I. Zang (1988) 'Optimal license fees for a new product', *Mathematical Social Sciences*, 16: 77–106.

Kamien, M.I., Y. Tauman and S. Zamir (1990) 'The value of information in a strategic conflict', *Games and Economic Behavior*, 2: 129–153.

Katz M.L. and C. Shapiro (1985) 'On the licensing of innovation', *Rand Journal of Economics*, 16: 504–520.

Katz M.L. and C. Shapiro (1986) 'How to license intangible property', *Quarterly Journal of Economics*, 101: 567–589.

Kitti C. and C.L. Trozzo (1976) *The effects of patents and antitrust laws, regulations and practices on innovation*. Vol. 1. Arlington, Virginia: Institute for Defense Analysis.

Lee T. and L. Wilde (1980) 'Market structure and innovation: A reformulation', *Quarterly Journal of Economics*, 94: 429–436.

Loury, G.C. (1979) 'Market structure and innovation', *Quarterly Journal of Economics*, 93: 395–410.

McGee, J.S. (1966) 'Patent exploitation: Some economic and legal problems', *Journal of Law and Economics*, 9: 135–162.

Muto, S. (1987) 'Possibility of relicensing and patent protection', *European Economic Review*, 31: 927–945.

Muto, S. (1990) 'Resale proofness and coalition-proof Nash equilibria', *Games and Economic Behavior*, 2: 337–361.

Nakayama, M. and L. Quintas (1991) 'Stable payoffs in resale-proof trades of information', *Games and Economic Behavior*, 3: 339–349.

Reinganum, J.F. (1981) 'Dynamic games of innovation', *Journal of Economic Theory*, 25: 21–41.

Reinganum, J.F. (1982) 'A dynamic game of R&D: Patent protection and competitive behavior', *Econometrica*, 50: 671–688.

Reinganum, J.F. (1989) 'The timing of innovation: Research, development and diffusion', in: R. Willig and R. Schmalensee, eds., *Handbook of Industrial Organization*. Amsterdam: North-Holland.

Rostoker, M. (1984) 'A survey of corporate licensing', *IDEA*, 24: 59–92.

Scherer, F.M. (1967) 'Research and development resource allocation under rivalry', *Quarterly Journal of Economics*, 81: 359–394.

Schumpeter, J.A. (1942), *Capitalism, socialism and democracy*. Harper Calophon, ed., New York: Harper and Row (1975).

Taylor C.T. and Z.A. Silberston (1973), *The economic impact of the patent system*. Cambridge: Cambridge University Press.

Usher, D. (1964) 'The welfare economics of invention', *Economica*, 31: 279–287.

Chapter 12

THE CORE AND BALANCEDNESS

YAKAR KANNAI*

The Weizmann Institute of Science

Contents

*Erica and Ludwig Jesselson Professor of Theoretical Mathematics. I am very much indebted to T. Ichiishi, M. Wooders, and to the editors of this Handbook for some very helpful remarks concerning this survey, and to R. Holzman for a very careful reading of the manuscript.

Handbook of Game Theory, Volume 1, Edited by R.J. Aumann and S. Hart

0. Introduction

Of all solution concepts of cooperative games, the core is probably the easiest to understand. It is the set of all feasible outcomes (payoffs) that no player (participant) or group of participants (coalition) can improve upon by acting for themselves. Put differently, once an agreement in the core has been reached, no individual and no group could gain by regrouping. It stands to reason that in a free market outcomes should be in the core; economic activities should be advantageous to all parties involved. Indeed, the concept (though not the term) appeared already in the writings of Edgeworth (1881) (who used the term "contract curve"), and in the deliberations concerning allocation of the costs involved in the Tennessee Valley Project [Straffin and Heaney (1981)].

Unfortunately, for many games, feasible outcomes which cannot be improved upon may not exist – the cake may not be big enough. In such cases one possibility is to ask that no group could gain much by recontracting. It is as if communications and coalition formations are costly. The minimum size of the set of feasible outcomes required for non-emptiness of the core is given by the so-called balancedness condition. The sets containing outcomes upon which nobody could improve by much are called ε-cores.

This chapter is organized as follows. In part I we survey the theory of cores in the case of transferable utility games – i.e., games in which the worth of a coalition S [the characteristic function $v(S)$] is a single number, and a feasible outcome is an assigment of numbers (payoffs) to the individual players such that the total payoff to the grand coalition N is no larger then $v(N)$. In Section 1 we discuss the case of a game with finitely many players. In particular we prove the criterion [due to Bondareva (1963) and Shapley (1967)] for non-emptiness of the core [how big should $v(N)$ be for that?]. The important concepts of balanced collections of coalitions (a suitable generalization of the concept of a partition $S_1, \ldots, S_k$ of N) and of balanced inequalities [an appropriate generalization of the super-additivity condition $v(N) \geq v(S_1) + \cdots + v(S_k)$ – a condition which is obviously necessary for non-emptiness of the core] are introduced. In Sections 2 and 3 we consider games with infinitely many players – in Section 2 we discuss the case where the set of players is countable, and in Section 3 the case of an uncountable set is considered. Already in the countable case there is a difficulty in the definition of a payoff – should we restrict ourselves to countably additive measures or should finitely additive ones be allowed as well? In the uncountable case one encounters additional problems with the proper definition of a coalition, and measure-theoretic and point-set-topologic considerations enter. The contributions of Schmeidler (1967) and Kannai (1969) are surveyed. Results on convex games and on other special classes of games, due mostly to Shapley (1971),

Schmeidler (1972a) and Delbaen (1974), are discussed in Section 4, as well as the determination of the extreme rays of certain cones of games [Rosenmüller (1977)].

In Part II we survey the theory of cores of games with non-transferable utility. For such games one has to specify, for every coalition S, a set $V(S)$ of feasible payoff vectors x (meaning that the ith component x_i is the utility level for the ith player, $i \in S$). In Section 5 we consider games with a finite set of players, and we prove the fundamental theorem, due to Scarf (1967), on the non-emptiness of the core of a balanced game, by a variant of the proof given by Shapley (1973). We also survey a certain generalization of the concept of a balanced collection of coalitions due to Billera (1970), and quote a characterization, also due to Billera (1970), of games with non-empty cores, valid if all sets $V(S)$ are convex. In Section 6 we quote results on non-transferable utility games with an infinite set of players. There are substantial topological difficulties here, and many problems are still open. We survey a non-emptiness theorem for the countable case due to Kannai (1969), and quote an example by Weber (1981) showing that this theorem cannot be improved easily. An existence theorem due to Weber (1981) for a somewhat weaker core is formulated.

In Part III we survey some economic applications of the theory. In Section 7 we present a simple model of an exchange economy with a finite set of players (traders). We follow Scarf (1967) in constructing a balanced game with non-transferable utility from this economy. We survey the theory, due to Shapley and Shubik (1969), of market games with transferable utility and identify these games with totally balanced games. We also quote the Billera and Bixby (1974) results on non-transferable utility games derived from economies with concave utility functions. We conclude Section 7 by explaining how one might obtain a proof of the existence of a competitive equilibrium from the existence of the core, and by mentioning other economic setups leading to games with non-empty cores. We do not deal with assignment games and their various extensions owing to lack of space. Section 8 is devoted to a (very brief) survey of the subject of ε-cores for large (but finite) market games. The classical definitions and results of Shapley and Shubik (1966) for replicas of market games with transferable utility are stated. The far-reaching theory, initiated by Wooders (1979) and extended further in many directions, is indicated. We mention various notions of ε-cores of economies and of non-transferable utility games. We conclude by a remark on the continuity properties of ε-cores. The initiated reader will note the omission of market games with an infinite set of players. The reasons for this omission – besides the usual one of lack of space – are that perfectly (and imperfectly) competitive economies are treated fully elsewhere in this Handbook (Chapters 14 and 15 and the chapter on 'values of perfectly competitive economies' in a forthcoming volume of this Handbook), and that this theory has very little to do with the theory of balanced games, as treated in Sections 2, 3 and 6 of the present chapter. (We also did not include a detailed discussion of non-exchange economies, externalities, etc.)

I. GAMES WITH TRANSFERABLE UTILITY

1. Finite set of players

Let $N = \{1, 2, \ldots, n\}$ be the set of all players. A subset of N is called a *coalition*. The *characteristic function* (or the worth function) is a real-valued function v defined on the coalitions, such that

$$v(\emptyset) = 0 .$$ (1.1)

An outcome of the game (a *payoff vector*) is simply an n-dimensional vector $x = (x_1, \ldots, x_n)$; the intuitive meaning is that the ith player "receives" x_i. Usually one requires that the payoff vector satisfies (at least) the following conditions:

$$\sum_{i=1}^{n} x_i = v(N)$$ (1.2)

(*feasibility* and *Pareto-optimality*), and

$$x_i \geqslant v(\{i\}) , \quad i = 1, \ldots, n$$ (1.3)

(*individual rationality*). Condition (1.2) incorporates both the requirement that the members of the grand coalition N can actually achieve the outcome x ($\Sigma_{i=1}^{n} x_i \leqslant v(N)$ – feasibility) and cannot achieve more ($\Sigma_{i=1}^{n} x_i \geqslant v(N)$ – Pareto optimality). Condition (1.3) means that no individual can achieve more than the amount allocated to him as a payoff. Note that individual rationality and feasibility are not necessarily compatible; clearly

$$\sum_{i=1}^{n} v(\{i\}) \leqslant v(N)$$ (1.4)

is needed. We will assume that the set of payoff vectors satisfying (1.2) and (1.3) is non-empty. If equality holds in (1.4) we are left with the trivial case $x_i \equiv v(\{i\})$. Hence we will assume that in (1.4) the inequality is strict, so that we deal with an $(n - 1)$ dimensional simplex of individually-rational, Pareto-optimal outcomes.

If $\Sigma_{i \in S} x_i < v(S)$ for a coalition S, then the members of S can improve their payoffs by their own efforts. The *core* is the set of all feasible payoffs upon which no individual and no group can improve, i.e., for all $S \subset N$,

$$\sum_{i \in S} x_i \geq v(S) .$$ (1.5)

[Note that individual rationality and Pareto optimality are special cases of (1.5) – when we take S to be the singletons or N, respectively – while feasibility requires an inequality in the other direction. Thus $v(N)$ plays a dual role in the theory.]

It is clear that additional super-additivity conditions, besides (1.4), are necessary for the existence of elements in the core. Let $S_1, \ldots, S_k$ be a partition of N (i.e., $S_i \cap S_j = \emptyset$ if $i \neq j$, $S_i \subset N$ for $1 \leq i \leq k$ and $N = \bigcup_{i=1}^{k} S_i$). It follows from (1.2) and (1.5) that

$$\sum_{i=1}^{k} v(S_i) \leq v(N)$$ (1.6)

has to be satisfied for the core to be non-empty. Condition (1.6) is, unfortunately, far from being sufficient, as the following example shows.

Example 1.1. $n = 3$, $v(S) = 1$ for all coalitions with two or three members, $v(\{i\}) = 0$ for $i = 1, 2, 3$. Then (1.6) is satisfied. However, writing conditions (1.5) explicitly for all two-person coalitions and summing them up, we obtain the inequality

$$2 \sum_{i=1}^{3} x_i \geq 3$$

or

$$\sum_{i=1}^{3} x_i \geq 1.5 .$$

Hence x is not feasible when $v(N) = 1$, and becomes feasible (and the core becomes non-empty) only if $v(N) \geq 1.5$.

The proper generalization of the concept of a partition is that of a balanced collection of coalitions, defined as follows. The collection $\{S_1, \ldots, S_k\}$ of coalitions of N is called *balanced* if there exist positive numbers $\lambda_1, \ldots, \lambda_k$ such that for every $i \in N$, $\Sigma_{j : S_j \ni i} \lambda_j = 1$. The numbers $\lambda_1, \ldots, \lambda_k$ are called *balancing weights*.

Every partition is a balanced collection, with weights equal to 1. For every positive integer j, set $S_j = N \setminus \{ j \}$. Then $\{ S_j \}$ is a balanced collection with $\lambda_j \equiv 1 / (n-1)$. Note that it is possible to write the balancedness condition as

$$\sum_{j=1}^{k} \lambda_j I_{S_j}(i) \equiv I_N(i) , \tag{1.7}$$

where $I_S(i)$ is the *indicator function* of S [$I_S(i) = 1$ if $i \in S$, $I_S(i) = 0$ otherwise].
Games with non-empty cores are characterized by

Theorem 1.1 [Bondareva (1963) and Shapley (1967)]. *The core of the game v is non-empty iff for every balanced collection $\{ S_1, \ldots, S_k \}$ with balancing weights $\lambda_1, \ldots, \lambda_k$, the inequality*

$$\sum_{j=1}^{k} \lambda_j v(S_j) \leq v(N) \tag{1.8}$$

holds.

Note that (1.8) is a generalization of (1.6). Note also that in Example 1.1 the inequality (1.8) implies in particular that $v(N) \geq 1.5$ if one considers the balanced collection $\{ \{2, 3\}, \{1, 3\}, \{1, 2\} \}$ (with weights $1/2$). A game satisfying the inequalities (1.8) for all balanced collections is called *balanced*.

Proof. (i) *Necessity.* Let $\{ S_j \}_{j=1}^{k}$ be a balanced collection with balancing weights $\lambda_1, \ldots, \lambda_k$. If the core is non-empty and x is a payoff vector in the core, then by (1.5)

$$\sum_{i \in S_j} x_i \geq v(S_j) , \quad j = 1, \ldots, k . \tag{1.9}$$

Multiplying both sides of (1.9) by λ_j and summing from 1 to k, we obtain:

$$\sum_{j=1}^{k} \lambda_j \sum_{i \in S_j} x_i \geq \sum_{j=1}^{k} \lambda_j v(S_j) . \tag{1.10}$$

By balancedness, the left-hand side of (1.10) is equal to $\sum_{i=1}^{n} x_i$. Hence by (1.2) the left-hand side of (1.10) is equal to $v(N)$ and (1.8) follows.

(ii) *Sufficiency.* The statement, "the validity of (1.8) for all balanced collections implies that the system (1.5) of linear inequalities is compatible with (1.2) (i.e., that the core is non-empty)", is a statement in the duality theory of linear inequalities. In fact, the validity of (1.8) for all balanced collections is equivalent to the statement that the value v_p of the linear program

$$\text{maximize} \sum_{S \subset N} v(S) y_S \tag{1.11}$$

subject to

$$\sum_{S \subset N} I_S(i) y_S = 1, \quad i = 1, \dots, n, \tag{1.12}$$

$$y_S \geq 0, \quad S \subset N, \tag{1.13}$$

satisfies $v_p = v(N)$. [Clearly $v_p \geq v(N)$.] But then the value v_d of the dual program

$$\text{minimize} \sum_{i=1}^{n} x_i \tag{1.14}$$

subject to

$$\sum_{i=1}^{n} I_S(i) x_i \geq v(S), \quad S \subset N, \tag{1.15}$$

satisfies $v_d = v(N)$ as well, i.e., there exists a vector $(x_1, \dots, x_n)$ satisfying the inequalities (1.15) [the same as the inequalities (1.5)] such that $\sum_{i=1}^{n} x_i = v(N)$. $\square$

Note that a different formulation of duality theory is needed for games with infinitely many players (see Theorem 2.1 and the proof of Theorem 2.2).

For certain applications of Theorem 1.1 the set of all balanced collections of subsets of N is much too large. It turns out that a substantially smaller subset suffices.

We say that the balanced collection $\{S_1, \dots, S_k\}$ is a *minimal balanced collection* if no proper subcollection is balanced. It is easy to see that if a balanced collection is minimal, then $k \leq n$, the balancing weights are unique, strictly positive, and rational, and that any balanced collection is the union of the minimal balanced collections that it contains. Moreover, the balancing weights for a balanced collection C are convex combinations of the balancing weights of the minimal balanced collection contained in C [Shapley (1967), Owen (1982)]. From these facts it is not difficult to derive the following theorem, also due to Bondareva (1963) and Shapley (1967).

Theorem 1.2. *The core of the game is non-empty iff for every minimal balanced collection $\{S_1, \dots, S_k\}$ with balancing weights $\lambda_1, \dots, \lambda_k$, the inequality (1.8) holds.*

Table 1
Balanced sets for $n = 4$

	Weights
$\{12\}, \{34\}$	1, 1
$\{123\}, \{4\}$	1, 1
$\{12\}, \{3\}, \{4\}$	1, 1, 1
$\{123\}, \{124\}, \{34\}$	1/2, 1/2, 1/2
$\{1\}, \{2\}, \{3\}, \{4\}$	1, 1, 1, 1
$\{12\}, \{13\}, \{23\}, \{4\}$	1/2, 1/2, 1/2, 1
$\{123\}, \{14\}, \{24\}, \{3\}$	1/2, 1/2, 1/2, 1/2
$\{123\}, \{14\}, \{24\}, \{34\}$	2/3, 1/3, 1/3, 1/3
$\{123\}, \{124\}, \{134\}, \{234\}$	1/3, 1/3, 1/3, 1/3

The determination of all minimal balanced collections in N is not easy for large n. An algorithm is given in Peleg (1965). Table 1 of all minimal balanced collections (up to symmetries) for $n = 4$, is taken from Shapley (1967).

In general, the core is a compact convex polyhedron, and determination of the payoffs in the core involves solving the linear system (1.2), (1.5). For the special class of convex games, introduced by Shapley (1971) and described in Section 4, one can write down explicitly the extreme points of the core.

We close this section with an example of a balanced game with a single point in the core; some feel uneasy about the intuitive meaning of this payoff.

Example 1.2. $n = 3$, $v(S) = 0$ unless $S = N$, $S = \{1, 2\}$ or $S = \{1, 3\}$; for those S, $v(S) = 1$. The only payoff in the core is $x_1 = 1$, $x_2 = x_3 = 0$. The coalition $\{2, 3\}$ cannot improve upon $x_2 + x_3 = 0$; yet this coalition could block the payoff by disagreeing to cooperate with 1.

This example underlines the meaning of (1.5) as requiring that no coalition S could improve upon $\Sigma_{i \in S} x_i$, rather than that no coalition S could "object" or "block the payoff" [Shapley (1972)].

2. Countable set of players

In this section we assume that the characteristic function v is defined on the subsets of a countable set N of players [and (1.1) is satisfied]. Without loss of generality N is the set of positive integers. We may look for outcomes of the form $x = (x_1, x_2, \ldots, x_n, \ldots)$, where x_i is the amount "paid" to player i, $i \in N$, and restrict ourselves to vectors x such that (1.3) is satisfied for all i and (1.2) is replaced by

$$\sum_{i=1}^{\infty} x_i = v(N) \, . \tag{2.1}$$

For technical reasons it will be convenient to assume here and in the next section that $v(S) \geq 0$ for all coalitions S. In particular $v(\{i\}) \geq 0$ for all $i \in N$ [as a matter of fact, one usually makes the stronger assumption that $v(\{i\}) = 0$ for all $i \in N$]. It then follows from (2.1) and (1.3) that the series $\sum_{i=1}^{\infty} x_i$ converges (absolutely), or that $x \in l^1$. We can now define the core as the set of l^1 vectors satisfying (2.1) and (1.5) for *all* (finite or infinite) subsets S of N. The concept of a balanced collection of subsets of N carries over verbatim from the finite case – condition (1.7) makes perfectly good sense, and it is proved exactly as in the finite case that balancedness of the game is necessary for non-emptiness of the core.

Unfortunately, the analog of the sufficiency part of Theorem 1.1 does not carry over, as the following example shows.

Example 2.1. [Kannai (1969)]. Let $v(S)$ vanish for all $S \subset N$ except when S contains an infinite segment, i.e. $\exists k \in N$ such that $S \supset \{i \in N : i \geq k\}$, and for those S, $v(S) = 1$. Then the inequalities (1.8) clearly hold for all balanced collections, but if (1.5) is valid, then $\sum_{i=k}^{\infty} x_i = 1$ for all k, so that $x \notin l^1$.

Clearly, a version of the duality theorem, valid for infinite systems, is required. We quote the relevant theorem in a form due to Ky Fan, which, while perhaps not the simplest to apply in the finite case, is the most transparent in the infinite case.

The following is Theorem 13 in Fan (1956):

Theorem 2.1. *Let $\{x_\nu\}_{\nu \in I}$ be a family of elements, not all 0, in a real normed linear space B, and let $\{\alpha_\nu\}_{\nu \in I}$ be a corresponding family of real numbers. Let*

$$\sigma = \sup \sum \lambda_j \alpha_{\nu_j} \tag{2.2}$$

when $k = 1, 2, 3, \ldots, \nu_j \in I$ and λ_j vary under the conditions

$$\lambda_j > 0 \, (1 \leq j \leq k) \, ; \quad \left\| \sum \lambda_j x_{\nu_j} \right\| = 1 \, . \tag{2.3}$$

Then
(i) The system

$$f(x_\nu) \geq \alpha_\nu \quad (\nu \in I) \tag{2.4}$$

of linear inequalities has a solution $f \in B^*$ if and only if σ is finite.

(ii) *If the system* (2.4) *has solutions* $f \in B^*$, *and if the zero functional is not a solution of* (2.4), *then* σ *is equal to the minimum of the norms of all solutions of* (2.4).

(Here B^* denotes the conjugate space of the normed linear space B.)

Inspecting Theorem 2.1, one realizes that in order to obtain a payoff vector in l^1, l^1 has to be regarded as the conjugate space B^* of a Banach space B. (Note that this condition is essential for the validity of the compactness argument needed for passing from finite to infinite set of inequalities.) But then $B = c_0$ – the subspace of l^∞ consisting of all sequences $(y_1, \ldots, y_n, \ldots)$ such that y_n tends to zero. Interpreting the inequalities (1.5) as inequalities of the type (2.4) implies that the indicator functions I_S for the relevant coalitions S have to be elements of c_0. But $I_S \in c_0$ iff S is finite. We are thus led to the following theorem.

Theorem 2.2. *Let* v *satisfy*

$$v(S) = 0, \quad \text{if } S \text{ is infinite and } S \neq N, \tag{2.5}$$

and let v *be balanced. Then there exists a vector* $x \in l^1$ *such that* x *is in the core of* v.

Proof. To apply Theorem 2.1, consider $B = c_0$ with the l^∞ (maximum) norm (then $B^* = l^1$), I is the set consisting of all finite subsets of N, x_ν stands for I_S, the indicator function of the coalition S, and $\alpha_\nu = v(S)$. Then the system (2.4) is just the system (1.5), and the condition (2.3) reads

$$-1 \leq \sum_{j=1}^{k} \lambda_j I_{S_j}(i) \leq 1 \tag{2.6}$$

(in fact the sum is always non-negative).

There exists a collection $\{T_l\}, 1 \leq l \leq m$, and positive numbers μ_l such that the collection $\{S_1, \ldots, S_k\} \cup \{T_1, \ldots, T_m\}$ is balanced, with weights $\lambda_1, \ldots, \lambda_k, \mu_1, \ldots, \mu_m$. By (1.8)

$$\sum_{j=1}^{k} \lambda_j v(S_j) + \sum_{l=1}^{m} \mu_l v(T_l) \leq v(N). \tag{2.7}$$

Here $\sum_{j=1}^{k} \lambda_j v(S_j) \leq v(N)$ so that σ [defined by (2.2)] is finite. $\square$

The indicator functions of infinite sets belong to l^∞. The conjugate space of l^∞

is *ba* [Dunford and Schwartz (1958)] – the space of finitely additive measures on N. Accordingly, we may define the concept of a payoff to mean a (not necessarily countably additive) measure μ defined on the subsets of N.

We replace (2.1) (feasibility and Pareto optimality) by

$$\mu(N) = v(N) , \tag{2.8}$$

and individual rationality (1.3) by

$$\mu(\{i\}) \geqslant v(\{i\}) , \quad i \in N . \tag{2.9}$$

The core is defined as the set of measures μ satisfying, besides (2.8) and (2.9), the "group rationality" conditions [replacing (1.5)]:

$$\mu(S) \geqslant v(S) , \quad \text{for all } S \subset N . \tag{2.10}$$

Exactly as in the proof of Theorem 2.2, we can prove the following theorem [a special case of a theorem due to Schmeidler (1967)]:

Theorem 2.3. *The core of v is non-empty if and only if v is balanced.*

A measure μ can be decomposed into a sum of a countably-additive measure μ_1 (an element of l^1) and a purely finitely additive measure μ_2. If μ is non-negative, then μ is purely finitely additive iff $\mu(\{i\}) = 0$ for all $i \in N$ [or $\mu(S) = 0$ for all finite sets S]. In Example 2.1 all elements in the core are purely finitely additive. For every ultra-filter F [Dunford and Schwartz (1958)] which refines the filter of all sets with finite complements there is a purely finitely additive measure μ_F such that $\mu_F(S) = 1$ iff $S \in F$. Each such μ_F is in the core of the game given in Example 2.1, and the core is the set of all infinite convex combinations of measures μ_F. One might regard F as an ideal player, since for every player $i \in N$ there corresponds the ultra-filter $F(i)$ consisting of all $S \subset N$ such that $i \in S$. The ideal players stand for "crowds" or "multitudes" which are stronger than the combined strengths of the individuals they contain.

Natural questions that arise are (assuming, of course, that the game is balanced):

(i) When does there exist a countably-additive μ in the core?

(ii) When are all elements in the core countably-additive? Rephrasing, we ask: (i) when do the ordinary players have some power and the ideal players do not have all the power (in the core), and (ii) when do the ideal players have no power? The following examples show that the condition (2.5) is far from being necessary for a positive answer to question (i), or even for question (ii).

Example 2.2. $v(S) = 1$ if $1 \in S$ and S is infinite, otherwise $v(S) = 0$. In the only element in the core player 1 gets 1, all other individuals and all sets not containing 1 get zero.

A more complicated situation is exhibited in the next example.

Example 2.3. Set $(k, \infty) = \{i \in N : k < i\}$ and define the game v by $v(\{1\} \cup (k, \infty)) = 1$ for all $k \geq 2$, $v(\{2\} \cup (k, \infty)) = 1$ for all $k \geq 2$, $v(S) = 0$ for all other $S \subset N$, $S \neq N$. This game will be balanced if $v(N) \geq 1$, but an l^1 vector x in the core must satisfy $x_1, x_2 \geq 1$. Hence the minimal value which has to be set for $v(N)$ so that there exist l^1 elements in the core is 2.

Theorem 2.1 and the examples lead one to look for "relatives" of the given game v, in the class of games that satisfy (2.5). Thus, one feels that the game described in Example 2.3 is related to the game defined by $v(\{1\}) = v(\{2\}) = v(N) = 1$, $v(S) = 0$ otherwise. Similarly, the game of Example 2.2 is related to the game $v(\{1\}) = v(N) = 1$, $v(S) = 0$ otherwise, whereas the game of Example 2.1 is not related to any such game. A precise concept of "relatedness" of games will now be formulated.

Definition 2.1. Let v_1 and v_2 be balanced games defined on the subsets of N. The game v_2 is called an *extension* of v_1 if $v_2(S) \geq v_1(S)$ for all $S \subset N$ and $v_2(N) = v_1(N)$.

Definition 2.2. The game v is said to be generated by the finite subsets of N if $v(S) = 0$ if S is infinite and $S \neq N$, v is balanced, and

$$v(N) = \sup \sum_{i=1}^{k} \lambda_i v(S_i) , \tag{2.11}$$

where the supremum is extended over all finite sequences λ_i and S_i, $\lambda_i > 0$ and $S_i \subset N$, $S_i \neq N$ and $\sum_{i=1}^{k} \lambda_i I_{S_i} \leq I_N$.

Following Kannai (1969) we can now state a solution to question (i).

Theorem 2.4. *A balanced game v defined on the subsets of N has a countably additive measure in its core iff there exists a common extension of both v and a game that is generated by the finite subsets of N.*

For the (not difficult) proof we refer to Kannai (1969).
We know of no simple general answer to question (ii) (see also Section 4).

Example 2.2 seems to indicate that there exists no such simple answer. We note the following simple remark.

Remark 2.1. If v is balanced and $v(\{1, \ldots, n\}) \to v(N)$, then every element of the core is countably additive.

3. Uncountable set of players

In the general case the notion of an individual player is not always very meaningful. Instead, a game is just a set function v defined on a field Σ of subsets of a set X, such that (1.1) is satisfied. We assume in this section that v is non-negative. A feasible payoff is a (finitely additive) non-negative measure μ, defined on Σ, such that

$$\mu(X) = v(X) \tag{3.1}$$

[compare (2.1)]. The subsets $S \in \Sigma$ are called coalitions. A coalition S cannot improve upon the payoff μ if

$$\mu(S) \geq v(S) . \tag{3.2}$$

The payoff μ is in the core of v if (3.2) is satisfied for all S in Σ.
  Schmeidler (1967) called a game v *balanced* if

$$\sup \sum_{i=1}^{k} \lambda_i v(S_i) \leq v(X) , \tag{3.3}$$

where the sup is taken over all finite sequences λ_i and S_i, the λ_i are non-negative numbers, the S_i are elements of Σ, and

$$\sum_{i=1}^{k} \lambda_i I_{S_i}(x) \leq I_X(x) , \tag{3.4}$$

for all $x \in X$. [Here, as in Sections 1 and 2, $I_S(x)$ is the indicator function of S.] The following theorem (of which Theorem 2.3 is a special case) is proved in Schmeidler (1967).

Theorem 3.1. *The game v has a non-empty core iff v is balanced.*

  It is proved, exactly as in Section 1, that if the core is non-empty, then v is balanced. For the converse, consider the Banach space $B = B(X, \Sigma)$ – the space of all uniform limits of finite linear combinations of indicator functions of sets in Σ. It is well known [Dunford and Schwartz (1958)] that the conjugate space B^* is (isometrically isomorphic to) the space of all bounded additive set functions defined on Σ, normed by the total variation. Hence one can deduce Theorem 3.1 from Fan's theorem (Theorem 2.1). $\square$

The set of all countably additive set functions on Σ is in general a proper subset of $B(X, \Sigma)^*$, which is not closed in the w^*-topology on $B(X, \Sigma)^*$. Hence one cannot apply Ky Fan's method directly to find out whether there exists a countably additive measure in the core of μ [problem (i) in Section 2]. One can nevertheless proceed indirectly and prove

Theorem 3.2.   *A game v has a countably additive measure in its core iff there exists a non-negative set function $w(S)$ defined on Σ such that $w(\emptyset) = 0$ and such that for each decreasing sequence $\{S_i\}$ of elements of Σ with empty intersection we have $w(S_i) \to 0$, and*

$$\sup\left(\sum_{i=1}^{n} \lambda_i v(S_i) - \sum_{j=1}^{m} \mu_j w(T_j) \right) \leq v(X) , \tag{3.5}$$

where the supremum is taken over all finite sequences $\{\lambda_i\}_{i=1}^{n}$, $\{\mu_j\}_{j=1}^{m}$ of positive numbers, and $\{S_i\}_{i=1}^{n}$, $\{T_j\}_{j=1}^{m}$ of elements of Σ, such that

$$\left| \sum_{i=1}^{n} \lambda_i I_{S_i}(x) - \sum_{j=1}^{m} \mu_j I_{T_j}(x) \right| \leq 1 , \tag{3.6}$$

for all $x \in X$.

For a proof, see Kannai (1969).

In Section 2 we made use of the facts that $l^1 = (c_0)^*$ and that the indicator functions of finite sets are in c_0, to prove the existence of countably additive measures in cores of certain classes of games. A Banach space whose dual consists of countably additive measures is the space $C(K)$ of continuous functions on a compact Hausdorff space K. In analogy to Section 2 we make the following definitions.

Definition 3.1.   Let v_1 and v_2 be balanced games defined on a field Σ of subsets of X. The game v_2 is called an extension of v, if $v_2(S) \geq v_1(S)$ for all $S \in \Sigma$, and $v_2(X) = v_1(X)$.

Definition 3.2.   Let $\mathscr{F}$ be a subfamily of Σ. The game v is said to be generated by $\mathscr{F}$ if

$$v(S) = 0 , \quad \text{if } S \not\subseteq \mathscr{F} \text{ and } S \neq X , \tag{3.7}$$

and

$$v(X) = \sup \sum_{i=1}^{k} \lambda_i v(S_i) , \tag{3.8}$$

where the supremum is extended over all finite sequences of λ_i and S_i, $\lambda_i > 0$ and $S_i \in \Sigma$, $S_i \neq X$, and

$$\sum_{i=1}^{k} \lambda_i I_{S_i}(x) \leq I_X(x) , \quad \text{for all } x \in X .$$

Unlike the countable case we also need a stronger concept of extension. For this we set for any game v and for any subset $S \subset X$,

$$b_v(S) = \sup \sum_{i=1}^{k} \lambda_i v(S_i) , \tag{3.9}$$

where the supremum is extended over all finite sequences λ_i and S_i, $\lambda_i > 0$ and $S_i \in \Sigma$, $S_i \neq X$, and $\sum_{i=1}^{k} \lambda_i I_{S_i}(x) \leq I_S(x)$ for all $x \in X$. [Thus, a game is balanced iff $v(X) \geq b_v(X)$.]

Definition 3.3. The extension w of v is said to be restricted if $w(S) \leq b_v(S)$ for all $S \in \Sigma$.

We can now state a theorem on countably additive measures in cores of games defined on the Borel subsets of a compact Hausdorff space.

Theorem 3.3. *Let X be a compact Hausdorff space and let Σ be the Borel field of X. The balanced game v (defined on Σ) has in its core a regular countably additive measure iff there exists a game which is both an extension of v and a restricted extension of a game generated by the closed subsets of X.*

(We recall that the Borel field of a topological space is the σ-field generated by the closed subsets of X.)

Outline of proof. (i) Assume that the regular countably additive measure μ is in the core of the game v. Define a game u by $u(S) = \mu(S)$ if S is closed and $u(S) = 0$ otherwise. Then u is clearly generated by the closed subsets of X, and by regularity u is a restricted extension of u (obviously, μ is an extension of u). (ii) Let u be a game generated by the closed subsets of X. We cannot apply

Theorem 2.1 as in the previous sections, since the indicator function I_S of a closed subset S of X is not, in general, a continuous function. In order to translate the inequalities (3.2) into inequalities of the form (2.4) with $x_\nu \in B = C(X)$ (and $f \in B^*$), we consider the system of linear inequalities

$$L(g) \geqslant u(S), \quad \text{if } g \geqslant I_S, \tag{3.10}$$

for all Borel subsets $S \subset X$ and $g \in C(X)$ (such that $g \geqslant I_S$). It follows from Theorem 2.1 that there exists a functional $L \in B^*$ satisfying all inequalities (3.10) and $\|L\| = u(X) = v(X)$. By the Riesz representation theorem, there exists a regular countably additive measure μ defined on Σ such that

$$L(g) = \int_X g \, d\mu, \quad \text{for all } g \in C(X). \tag{3.11}$$

By Urysohn's lemma [Dunford and Schwartz (1958)] and regularity, (3.10) and (3.11) imply that (3.2) is satisfied (for the game u) for all closed subsets S of X. Let w be a restricted extension of u and an extension of v. Then μ is in the core of w, and thus also in the core of v. Further details can be found in Kannai (1969). □

It is possible to combine Theorems 2.4 and 3.3. This can be done by noting that N is a dense subset of a compact Hausdorff space βN – the Stone–Čech compactification of N. (In fact, the elements of $\beta N \setminus N$ are the ultrafilters that support the purely finitely additive measures on N.) The following theorem is proved in Kannai (1969):

Theorem 3.4. *Let X be a completely regular Hausdorff space and let Σ be the Borel field of X. The balanced game v (defined on Σ) has in its core a regular countably additive measure concentrated on a countable union of compact sets iff there exists a game which is both an extension of v and a restricted extension of a game generated by the compact subsets of X.*

Recall that a measure μ is said to be concentrated on a set Y if $S \cap Y = \emptyset$ implies $\mu(S) = 0$. Note that the compact subsets of N with the discrete topology are precisely the finite sets. Thus Theorem 2.4 is contained in Theorem 3.4.

4. Special classes of games

In this section we consider games v defined on a field Σ of subsets of a set X such that (1.1) is satisfied. (This includes all cases discussed in the previous sections.) We assume also that v is non-negative. An interesting class of games is the following:

Definition 4.1. A game v is called *convex* if for all coalitions S, T,

$$v(S) + v(T) \leq v(S \cup T) + v(S \cap T). \qquad (4.1)$$

Consider first the case of a finite set of players N. Let π be a permutation of N, i.e., $\pi: N \to N$ is one-to-one, and set

$$T(\pi, k) = \{ i \in N: \pi(i) \leq k \}, \quad k = 0, 1, \ldots, n, \qquad (4.2)$$

and

$$x_i(\pi) = v(T(\pi, \pi(i))) - v(T(\pi, \pi(i) - 1)), \quad i = 1, \ldots, n. \qquad (4.3)$$

Theorem 4.1 (Shapley). *The vertices of the core of a convex game are the payoffs* $(x_1(\pi), \ldots, x_n(\pi))$ *for all permutations* π *of* N.

Shapley (1971) also noted that the average of all $n!$ payoff vectors (4.3) is the value (see the chapter on 'the Shapley value' in a forthcoming volume of this Handbook). Ichiishi (1981) observed that the converse of Theorem 4.1 is also true, i.e., if all $n!$ payoff vectors $(x_1(\pi), \ldots, x_n(\pi))$ are contained in the core of a game, then the game is convex.
Another class of games is the following:

Definition 4.2. A game v is called *exact* if for every $S \in \Sigma$ there exists a μ in the core of v such that $\mu(S) = v(S)$.

It was proved by Shapley (1971) in the finite case and by Schmeidler (1972a) in general that convex games are exact. Moreover, Schmeidler (1972a) proved also that for exact games one can solve completely question (ii) of Section 2.

Theorem 4.2. *Let* v *be an exact game. Every element in the core of* v *is countably additive iff for any monotone increasing sequence* S_n *in* Σ *with* $\bigcup_{i=1}^{\infty} S_n = X$, $v(S_n) \to v(X)$.

If v is not exact but $b_v(X)$ is finite, one can define a new game $\bar{\bar{v}}$, called the *exact cover* of v, by

$$\bar{\bar{v}}(S) = \sup \left[\sum_{i=1}^{k} \lambda_i v(S_i) - \lambda b_v(X) \right], \qquad (4.4)$$

where the supremum is taken over all finite sequences λ_i, S_i and numbers λ such that $\lambda_i > 0$, $\lambda > 0$, $S_i \in \Sigma$, and $\Sigma_{i=1}^{k} \lambda_i I_{S_i} - \lambda \leq I_S$. Schmeidler (1972a) proved that

$$\bar{\bar{v}}(S) = \inf\{\lambda(S): \lambda \text{ in the core of } v\}. \tag{4.5}$$

It follows that every element in the core of the balanced game v is countably additive iff for any monotone increasing sequence S_n in Σ with $\bigcup_{n=1}^{\infty} S_n = X$, $\bar{\bar{v}}(S_n) \to v(X)$. We call this condition "continuity at X" for $\bar{\bar{v}}$.

Schmeidler conjectured that if v is exact, continuous at X, and Σ is not a σ-field, then there exists an exact game w defined on the σ-field generated by Σ such that $w|_{\Sigma} = v$ and w is continuous at X. Delbaen (1974) disproved this conjecture. Delbaen also determined the exposed points of the core of some convex games, and studied various continuity properties of the core and elements in it.

Another conjecture of Schmeidler is still open: an exact game v such that, for any monotone decreasing sequence S_n in Σ such that $\bigcap_{n=1}^{\infty} S_n = \emptyset$, $v(S_n) \to 0$, has a countably additive measure in its core.

Shapley (1971) noted that the set of all convex games defined on the same field Σ of subsets of X is a convex cone, and raised the question of determining its extreme rays. Rosenmüller (1977, and references quoted there) has answered this question, as well as that of characterizing the extreme rays of the cone of super-additive games. An essential step is a representation of convex games as envelopes of affine games (analogously to the representation of ordinary convex functions as envelopes of affine functions). Rosenmüller also relates the structure of extreme convex games to the structure of extreme elements in the core.

Rabie (1981) has exhibited an exact game for which the value is not an element of the core.

It is well known that the core is contained in every Von Neumann–Morgenstern stable set (Chapter 17 of this Handbook). For convex games, the core is the unique stable set [Shapley (1971)].

II. GAMES WITH NON-TRANSFERABLE UTILITY

5. Finite set of players

As in Section 1, the set of players is the set $N = \{1, \ldots, n\}$, and a coalition is a subset of N. Since utility cannot be transferred between different players (even if they are members of the same coalition), we always have to specify all components of a vector $x = (x_1, \ldots, x_n) \in R^N (= R^n)$. Here x_i is the amount paid to the *i*th player. We denote by R^S the subspace of R^N defined by $x_i = 0$ for $i \notin S$. A coalition S controls the projection x^S of x on R^S given by the restriction of x to the coordinates indexed by the elements of S.

Formally, it is convenient to define a non-transferable utility *n*-person game (sometimes called an *n*-person game without side payments) as a set-valued function V (a correspondence) defined on the coalitions, such that:

$$V(\emptyset) = \emptyset, \tag{5.1}$$

$$\text{for all } S \neq \emptyset, V(S) \text{ is a non-empty closed subset of } R^N, \tag{5.2}$$

$$\text{if } x \in V(S) \text{ and } y_i \leq x_i \text{ for all } i \in S, \text{ then } y \in V(S). \tag{5.3}$$

The meaning of (5.3) is that $V(S)$ is a "cylinder", that is, the Cartesian product of a subset of R^S with $R^{N \setminus S}$ (this is done only for technical convenience), and that a coalition S can achieve, along with every vector, all vectors paying less to every member of S (this is a more substantive assumption). A transferable utility game v can be translated into a non-transferable utility game V by setting

$$V(S) = \left\{ x \in R^N : \sum_{i \in S} x_i \leq v(S) \right\} \tag{5.4}$$

for all non-empty coalitions S. This example suggests the following condition, which we will always assume.

There exists a closed set $F \subset R^N$ such that

$$V(N) = \{ x \in R^N : \exists y \in F \text{ with } x_i \leq y_i \text{ for all } i \in N \}. \tag{5.5}$$

Thus, a payoff x is feasible if there exists $y \in F$ with $x_i \leq y_i$ for all $i \in N$. It is individually rational if for no $i \in N$ there exists $y \in V(\{i\})$ such that $x_i < y_i$. To simplify matters, we will assume that

$$V(\{i\}) = \{ x \in R^N : x_i \leq 0 \}. \tag{5.6}$$

Feasible, individually rational payoff vectors exist only if F contains at least some vectors with non-negative components. We will assume that

$$F \cap \{ x \in R^N : x_i \geq 0 \text{ for all } i \in N \} \text{ is a non-empty compact set}. \tag{5.7}$$

The coalition S can improve upon the vector x if there exists $y \in V(S)$ with $x_i < y_i$ for all $i \in S$. By (5.3) S can improve upon x iff $x \in \text{int } V(S)$. Hence the core of the game V, defined as the set of all feasible payoff vectors that cannot be improved upon by any coalition, coincides with $V(N) \backslash \bigcup_{S \subset N} \text{int } V(S)$.

It is clear that $V(N)$ has to be sufficiently large for the core to be non-empty. In analogy to the terminology used in the case of transferable utility, Scarf (1967) defined a *balanced game* to be a game V in which the relation

$$\bigcap_{i=1}^{k} V(S_i) \subset V(N) \tag{5.8}$$

holds for every balanced collection $S_1, \ldots, S_k$ of subsets of N [compare (1.7) and the definition following the statement of Theorem 1.1]. [Note that if V is obtained from the transferable utility game v by (5.4) and v is balanced in the sense that all inequalities (1.8) hold, then (5.8) holds for V.] Scarf proved the following

Theorem 5.1. *Every balanced game has a non-empty core.*

Our proof of Theorem 5.1 follows mostly Shapley (1973), and incorporates some ideas due to Kannai (1970a).

The first step of the proof consists of the establishment of topological lemmas which generalize Sperner's lemma and the Knaster, Kuratowski and Mazurkiewicz theorem [Burger (1963)]. We need some more notation. Let $e^1, \ldots, e^n$ be the unit vectors in R^N. For every non-empty coalition S, let A^S be the convex hull of $\{e^i : i \in S\}$, and let m_S denote the barycenter of A^S ($m_S = \Sigma_{i \in S} e^i / |S|$). Let Σ be a simplicial subdivision of A^N. Let $V(\Sigma)$ denote the set of vertices of Σ (i.e., the set of vertices of simplices in Σ). A *labelling* of $V(\Sigma)$ is a function f from the vertices of Σ into the non-empty subsets of N ($f(q) \subset N, f(q) \neq \emptyset$). (Recall that conventionally – but not here – a labelling is a map $f : V(\Sigma) \rightarrow N$.)

Theorem 5.2 (Generalized Sperner's lemma). *Let f be a labelling of $V(\Sigma)$ such that for every $S \subset N, S \neq \emptyset$,*

$$f(q) \subseteq S, \quad \text{if } q \in V(\Sigma) \cap A^S. \tag{5.9}$$

Then there exists a (*not necessarily fully dimensional*) *simplex σ in Σ such that the collection $\{f(q): q \in \sigma\}$ is balanced.*

Proof. Let B denote the relative boundary of A^N (i.e., $B = \bigcup_{S \neq N} A^S$). Set $s = |S|$. There exists a map $g : B \rightarrow B$ such that

$$\text{if } S \neq N \text{ and } x \in A^S, \text{ then } g(x) \notin A^S. \tag{5.10}$$

One possible choice of g is the antipodal map (with m_N as the origin). Let h denote the usual radial deformation of the punctured simplex $A^N \backslash \{m_N\}$ onto B. Define a map $\tilde{f} : V(\Sigma) \to A^N$ by

$$\tilde{f}(q) = \left(\sum_{i \in f(q)} e^i \right) \Big/ |f(q)| \tag{5.11}$$

and extend $\tilde{f}$ linearly on every simplex of Σ, obtaining a piecewise linear map $\tilde{f} : A^N \to A^N$. We claim that $m_N \in \tilde{f}(A^N)$. Otherwise, the map $\varphi = g \circ h \circ \tilde{f} : A^N \to A^N$ is well defined and continuous (since the range of $\tilde{f}$ is contained in the punctured simplex) and $\varphi(A^N) \subset B$. By the Brouwer fixed point theorem there exists a point $x \in A^N$ such that $\varphi(x) = x$. Hence $x \in B$, and there exists a coalition $S \neq N$ such that $x \in A^S$. Set $s = |S|$. There exists a simplex $\sigma = \{q_1, \ldots, q_s\} \in \Sigma$ such that $q_i \in A^S$ for all $1 \leqslant i \leqslant s$ and x is in the convex hull of $q_1, \ldots, q_s$. By (5.9) $f(q_i) \subset S$ for all i, and by (5.11) $\tilde{f}(q_i) \in A^S$ for all i. By construction $\tilde{f}(x)$ is contained in the convex hull of $\{\tilde{f}(q_1), \ldots, \tilde{f}(q_s)\}$. Hence $\tilde{f}(x) \in A^S$. But the restriction of h to B is the identity map. Hence $\varphi(x) = g(\tilde{f}(x))$. By (5.10) $\varphi(x) \notin A^S$, contradicting $x = \varphi(x)$. Let now $x \in A^N$ be such that $\tilde{f}(x) = m_N$, and let $\sigma \in \Sigma$ be a $(k-1)$ dimensional simplex containing x in its interior, $\sigma = \{q_1, \ldots, q_k\}$, $(k \leqslant n)$. Then there exist positive numbers $\alpha_1, \ldots, \alpha_k$ such that $\sum_{j=1}^k \alpha_j = 1$ and $x = \sum_{j=1}^k \alpha_j q_j$. Set $f(q_j) = S_j$, $1 \leqslant j \leqslant k$. By (5.11) and the construction of $\tilde{f}$,

$$m_N = \sum_{j=1}^k \alpha_j \left(\sum_{i \in S_j} e^i \right) \Big/ |S_j| . \tag{5.12}$$

The ith component of the vector equation (5.12) reads

$$\frac{1}{n} = \sum_{\{j : i \in S_j\}} \frac{\alpha_j}{|S_j|} . \tag{5.13}$$

Setting $\lambda_j = n\alpha_j / |S_j|$, we see from (5.13) that the collection $\{S_1, \ldots, S_k\}$ is balanced, with the balancing weights $\lambda_1, \ldots, \lambda_k$. $\quad\square$

For an elementary proof (independent of Sperner's lemma) of the Brouwer fixed point theorem, see, for example, Kannai (1981).

As in Shapley (1973) we deduce from Theorem 5.2 a generalized Knaster–Kuratowski–Mazurkiewicz theorem.

Theorem 5.3 (K–K–M–S theorem). *Let $\{C_S\}$ be a family of closed subsets of A^N indexed by the non-empty coalitions such that for every $T \subset N$,*

$$\bigcup_{S \subset T} C_S \supset A^T \,. \tag{5.14}$$

Then there exists a balanced collection $\{S_1, \ldots, S_k\}$ such that

$$\bigcap_{i=1}^{k} C_{S_i} \neq \emptyset \,. \tag{5.15}$$

Proof. Let $\Sigma^{(m)}$ be a sequence of simplicial partitions of A^N such that the maximal diameter of the simplices in $\Sigma^{(m)}$ tends to zero as $m \to \infty$. For each $q \in V(\Sigma^{(m)})$ let $T(q)$ be the set of indices ($\subset N$) such that q is contained in the relative interior of the simplex spanned by $\{e^i\}_{i \in T(q)}$; thus $q \in A^{T(q)}$. By (5.14) there exists $S \subset T(q)$ such that $q \in C_S$. Let $f^{(m)}(q)$ be such a set S. Then the labelling $f^{(m)}(q)$ satisfies (5.9), and by Theorem 5.2 there exists a simplex $\sigma^{(m)} \in \Sigma^{(m)}$ such that the collection $\{f^{(m)}(q): q \in \sigma^{(m)}\}$ is balanced, for every m. But the number of balanced collections (of subsets of N) is finite and A^N is compact. Hence we can choose a subsequence $\{m_l\}_{l=1}^{\infty}$ such that $\sigma^{(m_l)} \to q_0 \in A^N$ and the collections $\{f^{(m_l)}(q): q \in \sigma^{(m_l)}\}$ are all identical to a fixed balanced collection $S_1, \ldots, S_k$. For each j, $1 \leq j \leq k$, q_0 is the limit of a sequence of vertices $q_j^{(m_l)}$ with $f^{(m_l)}(q_j^{(m_l)}) = S_j$. Thus $q_j^{(m_l)} \in C_{S_j}$. But the sets C_{S_j} are closed. Hence $q_0 \in \bigcap_{j=1}^{k} C_{S_j}$. $\square$

Proof of Theorem 5.1. Assume first that there exists a constant M such that for each $S \subset N$ and $x \in V(S)$ the estimate

$$x_i \leq M \,, \quad \text{for all } i \in S \,, \tag{5.16}$$

holds. Set $\tilde{e}_i = -nMe^i$ for $i \in N$, and define the simplex $\tilde{A}^S$ to be the convex hull of $\{\tilde{e}_i\}_{i \in S}$. (Clearly, Theorem 5.3 continues to hold for suitable closed coverings of $\tilde{A}^N$.) Set

$$t(x) = \sup\{t: x + t(1, \ldots, 1) \in \bigcup_{S \subset N} V(S)\} \,. \tag{5.17}$$

By (5.2) and (5.16) the supremum in (5.17) is finite and is actually a maximum, and defines a continuous function of $x \in R^N$. Set now

$$C_S = \{x \in \tilde{A}^N: x + t(x)(1, \ldots, 1) \in V(S)\} \,. \tag{5.18}$$

The sets C_S are closed by continuity of t and (5.2). We want to show that (5.14) is satisfied (for $\tilde{A}^T$). Let $x \in C_S \cap \tilde{A}^T$. We will show that $S \subset T$. [For all $x \in \tilde{A}^T$ there exists at least one $S \subset N$ such that $x \in C_S$. Hence (5.14) follows.] If $T = N$ there is nothing to prove; we may assume therefore that $T \neq N$. Since

$x \in \tilde{A}^T$ we have $x_j \leqslant -nM/|T| < -M$ for at least one $j \in T$. Taking $S = \{j\}$ in (5.17) we obtain:

$$t(x) > M .$$ (5.19)

Combining (5.19) with (5.16) we find that $x_i < 0$ for all $i \in S$. On the other hand, $x \in \tilde{A}^T$ implies $x_i = 0$ for $i \not\in T$. Hence $S \subset T$.

It follows from Theorem 5.3 that there exists a balanced collection $\{S_1, \ldots, S_k\}$ and a point $x \in \tilde{A}^N$ such that $x \in \bigcap_{i=1}^k C_{S_i}$. The point $y = x + t(x)(1, \ldots, 1)$ therefore belongs to $\bigcap_{i=1}^k V(S_i)$, but not to $\cdot\bigcup_{S \subset N} \operatorname{int} V(S)$. By (5.8) $y \in V(N)$. Hence y is in the core, and the core is not empty if (5.16) is satisfied.

For the general case, note that by (5.7) the set of non-negative elements of $V(N)$ is bounded from above. For all large positive M, consider the game v_M defined by

$$V_M(S) = V(S) \cap \{x \in R^N; x_i \leqslant M \text{ for } i \in S\} .$$ (5.20)

Then $\exists y^{(M)} \in V_M(N) \subset V(N)$ such that no coalition S can improve upon $y^{(M)}$ in the game V_M. By (5.6) $y_i^{(M)} \geqslant 0$ for all $i \in N$. Hence there exists a converging subsequence $y^{(M_l)} \to \bar{y} \in V(N)$. If $\bar{y}$ is not in the core of V, then there exists a coalition S and a point $z \in V(S)$ such that $z_i > \bar{y}_i$ for all $i \in N$. But $z \in V_{M_l}(S)$ for all large l and $z_i > y_i^{(M_l)}$ for all $i \in N$ and l large, a contradiction. $\square$

Remark 5.1. While the proof given here for the non-emptiness of the core appears to be non-constructive, it is possible to modify the argument to obtain a computational procedure (such was, in fact, the original proof). For details, see for example, Scarf (1967), Shapley (1973), and Scarf (1973).

Unfortunately, unlike the transferable utility case, balancedness of the game is not necessary for non-emptiness of the core. The following simple example is due to Billera (1970).

Example 5.1. Define a three-person game by

$$V(123) = \{(x_1, x_2, x_3); x_1 \leqslant 0.5, x_2 \leqslant 0.5, x_3 \leqslant 0\} ,$$ (5.21)

$$V(12) = \{(x_1, x_2, x_3); x_1 + x_2 \leqslant 1\}$$ (5.22)

and

$$V(S) = \{(x_1, x_2, x_3); x_i \leqslant 0 \text{ for all } i \in S\}, \quad \text{for all other } S \subset N .$$ (5.23)

This game has a non-empty core consisting of the point $(0.5, 0.5, 0)$, but is not balanced [the vector $y = (1, 0, 0)$ is not contained in $V(N)$, even though $y \in V(\{1, 2\}) \cap V(\{3\})$ (and the collection $\{\{1, 2\}, \{3\}\}$ is balanced)].

Billera (1970) introduced certain extensions of the concepts of balanced collections and balanced games. Here we follow the slightly less general approach due to Shapley (1973).

Let there be given an array of non-negative numbers $\pi = \{\pi_{S,i}: S \subset N, i \in N, \pi_{S,i} = 0$ for $i \notin S, \Sigma_{i \in S} \pi_{S,i} > 0$ for $S \neq \emptyset\}$. The collection $\{S_1, \ldots, S_k\}$ of subsets of N is called π-balanced if there exist positive numbers $\lambda_1, \ldots, \lambda_k$ such that

$$\sum_{j=1}^{k} \lambda_j \pi_{S_j,i} = 1, \quad \text{all } i \in N. \tag{5.24}$$

An n-person game v is said to be π-balanced if the relation (5.8) holds whenever $S_1, \ldots, S_k$ is a π-balanced collection.

Billera (1970) proved

Theorem 5.4. *A π-balanced game has a non-empty core.*

The proof in Billera (1970) follows Scarf's (1967) original proof. Shapley (1973) proved Theorem 5.4 in the case that the numbers $\pi_{S,i}$ are strictly positive (for $i \in S$) by noting that Theorems 5.2 and 5.3 continue to hold (with the same proof) if balanced is replaced by π-balanced.

Billera (1970) did obtain a necessary *and* sufficient condition for the non-emptiness of the core if the sets $V(S)$ are all convex. [This generalizes the case of games derived from games with transferable utility, since in that case the sets $V(S)$ are half-spaces, by (5.4).] Recall that the *support function h_C* of a convex set $C \subset R^N$ is defined by

$$h_C(x) = \sup_{y \in C} x \cdot y. \tag{5.25}$$

Theorem 5.5. (Billera). *Let V be a game in which $V(S)$ is convex for all $S \subset N$ and let h_S be the support function of $V(S)$. Then V has a non-empty core iff for every $S \subset N, S \neq N$, there exists a non-zero vector $\pi^{(S)} \in R^N$ with $\pi_i^{(S)} \geq 0$ for all $i \in N, \pi_i^{(S)} = 0$ for all $i \notin S$ and $h_S(\pi^{(S)}) < \infty$, such that for each $x \in E^N$,*

$$h_N(x) = \max\{\Sigma \lambda_S h_S(\pi^{(S)})\}, \tag{5.26}$$

where the maximum is extended over all non-negative λ_S such that $\Sigma \lambda_S \pi^{(S)} = x$.

6. Infinite set of players

If the set of players is infinite, much less is known about the non-emptiness of the core than in the finite case, difficulties appearing already in the countable case. In analogy to the discussions in Sections 3 and 4, we wish to define a game on the elements of a field Σ of subsets of a certain set X. Indeed, no further structure is required if $X = N$ is the set of all positive integers and Σ is the set of all subsets of N. For the uncountable case we need also a σ-finite measure μ defined on Σ [i.e., $X = \bigcup_{i=1}^{\infty} E_i$, where $E_i \in \Sigma$ and $\mu(E_i) < \infty$ for all $i = 1, 2, \ldots$]. (If $X = N$ take μ to be the counting measure.) X is interpreted as the set of players, and Σ is the set of permissible coalitions. Two coalitions are identified if their symmetric difference has μ measure zero. In what follows we will not distinguish between identified coalitions. An outcome of the game (a "payoff vector") is a function $u \in L_\infty = L_\infty(X, \Sigma, \mu)$, interpreted as giving (μ a.e.) player $p \in X$ the utility level $u(p)$. We associate with every $S \in \Sigma$ a set $V(S) \subset L_\infty$ (interpreted as the set of possible utility levels obtainable by S). We assume that (5.1) is satisfied. A natural replacement of (5.3) is the assumption that for each S,

$$\text{if } u \in V(S) \text{ and } v \le u \text{ a.e. in } S, \text{ then } v \in V(S). \tag{6.1}$$

For technical convenience we assume, without loss of generality, that each $V(S)$ contains the origin of L_∞. Then it suffices to consider the intersection of $V(S)$ with the non-negative orthant of L_∞. We shall use the notation $V(S)$ to denote this intersection.

Attempting to find a replacement for (5.2), we are confronted with a substantial difficulty, since it is not at all clear which topology to use on the set of outcomes. On the one hand, the improving upon (or blocking) relation is not continuous in the L_∞ (norm) topology (and of course not in coarser topologies); the set $\{v \in L_\infty : v(p) > u(p) \text{ a.e.}\}$ is not open in the L_∞ topology. On the other hand, compactness of $V(X)$ is required for the validity of several arguments, and the finer the topology, the less compact sets one obtains. We are thus led to consider three different topologies on L_∞:
   (i) the usual norm topology;
   (ii) a topology P defined by requiring that for all pairs $u_1, u_2 \in L_\infty$ with $u_1 > u_2$, the sets $\{u \in L_\infty : u_1 > u > u_2\}$ are open; and
   (iii) the w^* topology on L_∞.
   A coalition S can improve upon an outcome u if $u(S) > 0$ and there exists $v \in V(S)$ such that $v > u$ (a.e.) in S. The core is the set of all outcomes in $V(N)$ which cannot be improved upon by any coalition.
   Consider first the countable case $X = N$, and assume that

for all $S \in \Sigma$, $V(S)$ is closed in the P topology ,   (6.2)

and that, in addition,

the set $V(N)$ is compact in the w^* topology .   (6.3)

(*Note that the w^* topology coincides with the Tychonoff product topology on bounded subsets of l_∞.*) We define the balanced cover of V with respect to S, $B_V(S)$, to be the closure (in P) of the set of all vectors $u \in l_\infty$ such that $u \in V(S_i)$, $1 \leqslant i \leqslant k$, for all collections $S_1, \ldots, S_k$ of subsets of S (with $S_i \neq N$) for which there exist positive constants $\lambda_1, \ldots, \lambda_k$ such that $\Sigma^k_{i=1} \lambda_i I_{S_i} = I_S$ [compare (3.9)]. We say that the game V is *balanced* if

$$B_V(N) \subset V(N) .   (6.4)$$

Analogously to Definition 3.1, we say that a balanced game W is an extension of the balanced game V if for all $S \subset N$, $V(S) \subset W(S)$, and $V(N) = W(N)$. Similarly to Definition 3.2, we say that V is generated by the finite subsets of N if for all infinite subsets S of N other than N, $u = 0$ in S for all $u \in V(S)$, and $V(N)$ is the w^* closure of $B_V(N)$. As in Definition 3.3, a game W is called a restricted extension of V if W is an extension of V and $W(S) \subset B_V(S)$ for all $S \subset N$, $S \neq N$.

Kannai (1969) proved the following theorem.

Theorem 6.1.   *A sufficient condition for the non-emptiness of the core of a balanced game V defined on N and satisfying (5.1), (6.1), (6.2) and (6.3), is that there exists a restricted extension W' of a game W generated by the finite subsets of N, such that W' is also an extension of V.*

We know of no analog of this theorem for the uncountable case. Moreover, an assumption about the relation between V and a game generated by the finite subsets is apparently needed in the countable case, since games with empty cores exist [even if one strengthens (6.2), (6.3) and the concept of balanced-ness, as we will do in the sequel].

In fact, consider now for the general case, the following strengthening of (6.2):

for all $S \in \Sigma$, $V(S)$ is closed in the norm topology ,   (6.5)

and the stronger version of (6.3);

there exists a norm-compact set $F \subset L_\infty$ such that

$$V(X) = \{u \in L_\infty : u \geqslant 0 \text{ and there exists } v \in F \text{ with } v \geqslant u\} .   (6.6)$$

We say that the game V is *B-balanced* [see Billera and Bixby (1973)] if for every collection $S_1, \ldots, S_k$ of coalitions, and non-negative numbers $\lambda_1, \ldots, \lambda_k$ such that $\sum_{i=1}^{k} \lambda_i I_{S_i} \leqslant I_X$, it is true that

$$\sum_{i=1}^{k} \lambda_i V(S_i) \subset V(X) . \tag{6.7}$$

The assumption of B-balancedness is, unfortunately, insufficient for the non-emptiness of the core, as follows from the following theorem by Weber (1981):

Theorem 6.2. *There exists a B-balanced game V defined on the subsets of N and satisfying* (5.1), (6.1), (6.5) *and* (6.6), *such that the core of V is empty.*

A positive result was obtained by Weber (1981) for the so-called weak core. We say that an outcome u can be strongly improved upon by a coalition S if $\mu(S) > 0$ and there exists a positive number c and an outcome $v \in V(S)$ such that $v - u \geqslant c$ a.e. in S. An outcome $u \in V(X)$ is in the weak core if no coalition S can strongly improve upon u. Note that the concept of a weak core coincides with the concept of a core in the case of finitely many players. Note also that the weak core is related to the concept of an ε-core due to Shapley and Shubik (1966) in the finite case and defined for markets with a continuum of traders by Kannai (1970b). (See also Section 8 of the present chapter.)

Theorem 6.3. *Let V be a B-balanced game satisfying* (5.1), (6.1), (6.5) *and* (6.6). *Then the weak-core of the game V is non-empty.*

For a proof, see Weber (1981).

A different approach to the question of non-emptiness of the core, based on applying Fan's theorem on linear inequalities (Theorem 2.1 here) and generalized versions of this theorem to the non-transferable utility case is due to Ichiishi and Weber (1978). Unfortunately, their conditions are rather involved.

III. ECONOMIC APPLICATIONS

7. Market games with a finite set of players

Intuition suggests that free economic activity should be advantageous to all parties involved. Technically this means that for a cooperative game to serve reasonably well as a model of a free market, the core of this game should not be empty.

Consider an exchange economy. Here the set of players (traders) is $N = \{1, 2, \ldots, n\}$, and every trader $i \in N$ is characterized by means of an initial endowment vector $a^{(i)}$ and a preference ordering $\geq_i$. We assume (for simplicity) that for all $i \in N$, $a^{(i)} \in \Omega$, where Ω is the non-negative orthant of R^m (the commodity space) and $\geq_i$ is a complete, continuous and convex preference ordering defined on Ω. Let u_i be a continuous utility function representing the order $\geq_i$. A non-transferable utility game (see Section 5) V may now be defined, following Scarf (1967), as follows. For every non-empty coalition S, set

$$V(S) = \left\{ x \in R^N : \exists y^{(i)}, i \in S, \text{ such that } \sum_{i \in S} y^{(i)} = \sum_{i \in S} a^{(i)} \right.$$

$$\left. \text{and } x_i \leq u_i(y^{(i)}) \text{ for } i \in S \right\}, \tag{7.1}$$

while for $S = \emptyset$ set (5.1). Then (5.2) and (5.3) are satisfied, and adding suitable constants to the functions u_i we can assume that (5.6) holds. Then (5.7) follows. The game V thus defined is called a *market game*. The convexity of the preferences $\geq_i$ implies that the balancedness condition (5.8) is satisfied. Hence:

Theorem 7.1.   *The core of a market game is non-empty.*

For details of the proof, see Scarf (1967).

Remark 7.1.   It is possible to define the core of an exchange economy directly, without passing through the market game. We say that the coalition S can *improve upon* the allocation $x^{(1)}, \ldots, x^{(n)}$ (an *allocation* is an n-tuple of vectors $x^{(1)}, \ldots, x^{(n)}$, $x^{(i)} \in \Omega$ for all $i \in N$, such that $\Sigma_{i=1}^n x^{(i)} = \Sigma_{i=1}^n a^{(i)}$) if there exists an allocation $y^{(1)}, \ldots, y^{(n)}$ with $\Sigma_{i \in S} y^{(i)} = \Sigma_{i \in S} a^{(i)}$ and such that $y^{(i)}$ is strictly preferred by trader i to $x^{(i)}$ (i.e., $y^{(i)} \geq_i x^{(i)}$ but not $x^{(i)} \geq_i y^{(i)}$) for all $i \in S$. The *core of the economy* is the set of all allocations upon which no coalition can improve. Clearly, $x^{(1)}, \ldots, x^{(n)}$ is in the core of the economy iff the vector $(u_1(x^{(1)}), \ldots, u_n(x^{(n)}))$ is in the core of the market game V.

Given a game V and a non-empty subset $T \subset N$, we define the *subgame* of V on T to be the restriction of V to the set of subsets of T. For an arbitrary game V, the subgame of V on T may or may not be balanced. If, however, all subgames of V are balanced, we say that the game V is *totally balanced*. Any subgame of a market game is obviously a market game. Hence we have.

Theorem 7.2.   *Every market game is totally balanced.*

The problem of characterizing all market games is still open. Theorems 7.4 and 7.6 in the sequel lead one to conjecture that the converse of Theorem 7.2 is true, i.e., that for every totally balanced game V defined on the coalitions of N there exists an exchange economy such that V is given by (7.1).

The first case that was completely settled was that of market games with transferable utility, analyzed by Shapley and Shubik (1969). For any non-empty coalition S we define the worth $v(S)$ (the characteristic function) by

$$v(S) = \sup_{x \in V(S)} \sum_{i \in S} x_i, \tag{7.2}$$

where $V(S)$ is given by (7.1). It is easy to see that the supremum in (7.2) is actually a maximum, and can also be defined by

$$v(S) = \max \sum_{i \in S} u_i(y^{(i)}), \tag{7.3}$$

where the maximum is extended over all allocations $y^{(1)}, \ldots, y^{(n)}$ such that $\sum_{i \in S} y^{(i)} = \sum_{i \in S} a^{(i)}$. Similarly to Theorem 7.2, we have (with the obvious modifications in the definitions of a subgame and of a totally balanced game):

Theorem 7.3. *Every market game with transferable utility is totally balanced.*

Shapley and Shubik (1969) proved that the converse of Theorem 7.3 is also true.

Theorem 7.4. *Every totally balanced game with transferable utility is a market game, i.e., the game can be written in the form (7.3) for a certain collection of $a^{(i)}, u_i, i \in N$.*

Shapley and Shubik take Ω to be the non-negative orthant in R^N, $a^{(i)} = e_i$ (the ith unit vector in R^N) for all $i \in N$, and

$$u_i(x) = u(x) = \max \sum_{S \subset N} \alpha_S v(S), \tag{7.4}$$

where the maximum is extended over all sets of non-negative numbers α_S satisfying

$$\sum_{S;\, S \ni i} \alpha_S = x_i, \quad \text{for all } i \in N. \tag{7.5}$$

for details and an intuitive interpretation of this economy, see Shapley and Shubik (1969).

Billera and Bixby, in a series of papers, have considered games with non-transferable utility where all sets $V(S)$ are convex, replaced the concept of balancedness by that of B-balancedness [see (6.7)], and modified the concept of an exchange economy – see Billera and Bixby (1973, 1974) and Billera (1974) – and succeeded in characterizing, in this framework, all market games. Thus, assume that for every $i \in N$ the preference relation $\geqslant_i$ is defined on a convex set Ω_i and is representable in Ω_i by a concave, upper-semicontinuous utility function u_i such that u_i is bounded below on Ω_i. Define $V(S)$ by (7.1) (where $y^{(i)}$ is restricted to be in Ω_i for all $i \in S$). Then the sets $V(S)$ are convex and the B-balancedness condition (6.7) is satisfied. Such a game V is called a *modified market game*. (Note that the set of modified market games does not contain, and is not contained in, the set of market games, even though the intersection is non-empty.) The same argument as the one leading to Theorem 7.2 now yields:

Theorem 7.5. *Every modified market game is totally B-balanced.*

The converse was also proved by Billera and Bixby.

Theorem 7.6. *If the sets $V(S)$ are convex for all $S \subset N$ and V is totally B-balanced, then V is a modified market game.*

Billera and Bixby (1974) proved that if, for every coalition S, the Pareto surface P_S of $V(S)$ [i.e., the set of maximal elements in $V(S)$ with respect to the normal partial order $\leqslant$ on R^N] is closed, then the functions u_i can be taken to be continuous. They also construct representations of the games by means of economies with production and by means of economies for which the commodity spaces are infinite dimensional.

Remark 7.2. The difference between market games and modified market games should not be underestimated. While it is true that every convex preference ordering $\geqslant_i$ can be approximated arbitrarily closely by preference orderings induced by concave utility functions, in a market game the sets $V(S)$ may be far from convex and cannot, in general, be transformed into convex sets by means of suitable choices of utility functions. An example may be found in Kannai and Mantel (1978).

Closely related to the concept of the core of a market game is the concept of equilibrium allocations (of an exchange economy). An allocation $x^{(1)}, \ldots, x^{(n)}$ is called an *equilibrium allocation* if there exists a non-zero vector $p \in R^m$ (called a *price vector*) such that

$$x^{(i)} \text{ is maximal with respect to } \geqslant_i \text{ in the set } \{x : p \cdot x \leqslant p \cdot a(i)\},$$
$$\text{for all } i \in N . \tag{7.6}$$

It is not difficult to show (see Chapter 14) that an equilibrium allocation is in the core of the economy (defined in Remark 7.1). Hence the well-known existence theorems for equilibrium allocations [e.g., Debreu (1959)] imply Theorem 7.1 (without using balancedness). It is not true that every allocation in the core is an equilibrium allocation. A large body of literature, reviewed in Chapters 14 and 15 of this Handbook, is devoted to the study of the relations between the core and the set of competitive equilibria for large (and infinite) sets of traders. We note here that if the preference ordering $\geq_i$ is strictly convex for all $i \in N$, then if i and j have identical characteristics ($\geq_i = \geq_j$ and $a^{(i)} = a^{(j)}$), then $x^{(i)} = x^{(j)}$ for all core allocations $x^{(1)}, \ldots, x^{(n)}$ ("equal treatment in the core"). It follows that a core allocation for a replica economy (where we have nk traders, k traders of each "type" $i, i \in N$) induces a core allocation for the original economy. Debreu and Scarf (1963) proved that the intersection of all these cores is an equilibrium. Hence Theorem 7.1 and the compactness of the core imply the existence of a competitive equilibrium.

For the transferable utility market game, defined by (7.2) or (7.3), a *competitive payoff* is a vector $(\alpha_1, \ldots, \alpha_n)$ defined by

$$\alpha_i = u_i(x^{(i)}) - p \cdot (x^{(i)} - a^{(i)}), \quad i \in N, \tag{7.7}$$

where p is a vector in R^m and $x^{(1)}, \ldots, x^{(n)}$ an allocation for which

$$u_i(x^{(i)}) - p \cdot (x^{(i)} - a^{(i)}) \geq u_i(y) - p \cdot (y - a^{(i)}), \quad \text{for all } y \in \Omega, i \in N \tag{7.8}$$

[see Shapley and Shubik (1976) for explanations]. Shapley and Shubik (1976) prove that every payoff in the core of the totally balanced game with transferable utility v is a competitive payoff for the economy described after the statement of Theorem 7.4.

As stated in the Introduction of this chapter, concepts such as the core arose in the deliberations about the allocations of costs of cooperative projects such as those carried out by the Tennessee Valley Authority. Sorenson, Tschirhart and Whinston (1978) proved that a transferable utility game modelling a producer and a set of potential consumers (under decreasing costs) yields a convex game [(4.1)]. By Theorem 4.1 the core of this game is non-empty. Similarly, it was proved by Dinar, Yaron and Kannai (1986) that the game describing a water purification plant, where the city and the farms are the players, is a convex game.

8. Approximate cores for games and markets with a large set of players

Convexity of preference orderings was assumed in all the existence theorems stated in the previous section. Without this assumption the core might very well be empty. Shapley and Shubik (1966) showed that certain sets of payoff vectors, defined by "slightly" modifying the definition of the core, will be non-empty in many instances without convexity. These results initiated a substantial body of research on approximate cores.

Consider first a game v with transferable utility (and a finite set of players). Let a positive ε be given. The *strong ε-core* is the set of outcomes x satisfying (1.2) and

$$\sum_{i \in S} x_i \geq v(S) - \varepsilon , \quad \text{for all } S \subset N , \tag{8.1}$$

and the *weak ε-core* is the set of outcomes satisfying (1.2) and

$$\sum_{i \in S} x_i \geq v(S) - |S|\varepsilon , \quad \text{for all } S \subset N . \tag{8.2}$$

Note that an element of an ε-core is not necessarily individually rational. We will thus deal also with the *individually rational ε-core*, which is the intersection of the ε-core with the set of individually rational outcomes. It is easy to see that

$$\text{weak } \varepsilon\text{-core} \supset \text{strong } \varepsilon\text{-core} \supset \text{weak } \frac{\varepsilon}{n} - \text{core} \supset \text{core} . \tag{8.3}$$

These ε-cores (and others, to be defined in what follows) are not merely technical devices. They provide a way of taking into account the costs of forming a coalition (such as communication costs). Alternatively, we might view ε or $|S|\varepsilon$ as a threshold, below which a coalition might not consider the improvement upon x worth the trouble.

Shapley and Shubik (1966) considered replicas of a market game with transferable utility such that all traders have the same utility function. In general, replicas are obtained by considering an economy composed of n types of traders with k traders of each type. For two consumers to be of the same type, we require them to have precisely the same preference $\geq$ and the same endowment a, and in the case of transferable utility they should have the same utility function u. The economy therefore consists of nk traders, whom we index by the pair (i, j), with $i = 1, \ldots, n$ and $j = 1, \ldots, k$; we denote the set of traders by $N(k)$. The corresponding replica game with transferable utility $v^{(k)}(S)$ is defined by (7.3), where S is a subset of $N(k)$. For the existence of weak ε-cores, Shapley and Shubik (1966) proved the following theorem:

Theorem 8.1. *Let $\geq_i \equiv \geq$, $u_i \equiv U$ for all $i \in N$, and let there exist a linear function $L_0(x)$ and a continuous function $K_0(x)$, defined on Ω, such that*

$$K_0(x) \leqslant U(x) \leqslant L_0(x) , \quad \textit{for all } x \in \Omega . \tag{8.4}$$

Then for every $\varepsilon > 0$ there exists a constant k_0 such that the market games $v^{(k)}$ with $k \geqslant k_0$ possess non-empty individually rational weak ε-cores.

For the existence of strong ε-cores the following was proved by Shapley and Shubik (1966).

Theorem 8.2. *Let $\geqslant_i \equiv \geqslant$, $u_i \equiv U$ for all $i \in N$. Let U be differentiable along all rays in Ω emanating from the origin. Let there exist a concave function $C(x)$ defined on Ω such that*

$$C(x) \geqslant U(x) , \quad \textit{for all } x \in \Omega , \tag{8.5}$$

and such that for each $x \in \Omega$ there are $(m + 1)$ points $y^h \in \Omega$ (not necessarily all distinct) and non-negative numbers λ_h such that

$$\sum_{h=1}^{m+1} \lambda_h = 1 , \quad \sum_{h=1}^{m+1} \lambda_h y^h = x \quad \textit{and} \quad \sum_{h=1}^{m+1} \lambda_h U(y^h) = C(x) . \tag{8.6}$$

Then for every $\varepsilon > 0$ there exists a constant k_0 such that the market games $v^{(k)}$ with $k \geqslant k_0$ possess non-empty individually rational strong ε-cores.

Remark 8.1. Non-emptiness of the weak ε-core can be easily characterized by means of "ε-balancedness". We say that v is ε-balanced if the inequalities

$$\sum_{i=1}^{k} \lambda_i v(S_i) \leqslant v(N) + \varepsilon n \tag{8.7}$$

hold for all balanced collections $\{S_1, \ldots, S_k\}$ with balancing weights $\lambda_1, \ldots, \lambda_k$. Wooders (1979) noted that v has a non-empty ε-core iff v is ε-balanced.

A general framework for the study of market games (and others kinds of games) was introduced by Wooders (1979). As above, the replica case is considered, with nk players, k of each type. The characteristics (of players) determining a type (such as preference, endowment, etc.) are not specified explicitly. Nor is the economic activity (exchange, production, etc.) spelled out in detail. Rather, the basic datum of the problem is a real-valued function $v(s_1, \ldots, s_n)$, defined on the set of n-dimensional vectors $(s_1, \ldots, s_n)$ for which the components s_i are non-negative integers, and expressing for each coalition S containing s_1 players of type $1, \ldots, s_n$ players of type n, the total payoff achievable by (economic) action of all members of S. For any such coalition S and any positive integer k, denote by $S(k)$ a coalition consisting of ks_i players of type i, for each $i = 1, \ldots, n$. Thus we may denote the set of players of the kth replica game by $N(k)$. The characteristic function $v^{(k)}$ of the kth replica game is defined [for $S \subset N(k)$] by

$$v^{(k)}(S) = v(s_1, \ldots, s_n) , \tag{8.8}$$

where s_i is the number of players of type i contained in S, $i = 1, \ldots, n$ (clearly $s_i \leq k$). The function v is called a *pre-game* or a *technology*.

We say that v is superadditive if for all vectors s, s', such that s_i, s_i' are non-negative integers for $i = 1, \ldots, n$, we have

$$v(s + s') \geq v(s) + v(s') . \tag{8.9}$$

We say that the per-capita payoff is bounded if there exists a constant K such that

$$\frac{v^{(k)}(N(k))}{k} = \frac{v(k, \ldots, k)}{k} \leq K \tag{8.10}$$

for all positive integers k.

Obviously, the technology v is superadditive if and only if all replica games $v^{(k)}$ are superadditive. Remarkably, superadditivity and bounded per-capita payoff suffice for ε-balancedness, as the following theorem of Wooders (1979) shows.

Theorem 8.3. *If the technology v satisfies (8.9) and (8.10), then for every $\varepsilon > 0$ there exists a constant k_0 such that the games $v^{(k)}$ with $k \geq k_0$ possess non-empty weak ε-cores.*

Outline of proof. The superadditivity and per-capita boundedness assumptions imply that $v^{(k)}(N(k))$ is asymptotically homogeneous in k or, more precisely, that the limit

$$\lim_{k \to \infty} \frac{v^{(k)}(N(k))}{k}$$

exists. [In fact, (8.10) implies that the ratio $v^{(k)}(N(k))/k$ is bounded from above, (8.9) implies that the ratio is bounded from below, so that the sequence $v^{(k)}(N(k))/k$ has limit points. By superadditivity different subsequences cannot tend to distinct limits.] Theorem 1.2 and the discussion preceding it imply that it suffices, when considering balancedness, to assume that the balancing weights are rational. Let $\{S_1, \ldots, S_r\}$ be a balanced collection of subsets of $N(k)$ with rational weights $\lambda_i = p_i/p$, $i = 1, \ldots, r$. One may embed the sets $S_1(p_1), \ldots, S_r(p_r)$ in $N(kp)$ in such a manner that the collection $\{S_1(p_1), \ldots, S_r(p_r)\}$ is a *partition of* $N(kp)$. By the superadditivity of v,

$$\sum_{i=1}^{r} \lambda_i v^{(k)}(S_i) = \sum_{i=1}^{r} \frac{p_i}{p} v^{(k)}(S_i) \leq \sum_{i=1}^{r} \frac{v^{(kp)}(S_i(p_i))}{p} \leq \frac{v^{(kp)}(N(kp))}{p}. \qquad (8.11)$$

By asymptotic homogeneity, the right-hand side of (8.11) does not differ much from $v^{(k)}(N(k))$ if k is sufficiently large (in fact, the difference is estimable by εk). Hence $v^{(k)}$ is ε-balanced. $\quad\square$

Thus, superadditivity and asymptotic homogeneity are not very far from balancedness, because in the replica case a balanced collection can be regarded as a partition of a replica, quite unlike the situation in the case of a fixed game. This framework was generalized further (so as to cover the non-replica case as well) by Wooders and Zame (1984) [see also Wooders and Zame (1987a)]. We no longer have a fixed set of types. Instead, an abstract notion generalizing the charateristics of a trader (player, agent, etc.), called an *attribute*, is the primary ingredient. The attributes of the players are assumed to be elements of a compact metric space Y. As for types, the pair $(a, \succcurlyeq)$ or the pair (a, u) may serve as examples of attributes; a metric topology on spaces of preference orderings was exhibited in Kannai (1970b). In the replica case, an important role is played by n-dimensional vectors $(s_1, \ldots, s_n)$ expressing the number of players of various types. For the non-replica case, the corresponding concept is that of a *profile*, defined as a function f from Y to the set of non-negative integers for which $\{\eta \in Y: f(\eta) \neq 0\}$ is finite. We now define the technology (or the pre-game) to be a pair (Y, Λ), where Y is a compact metric space and Λ is a non-negative function defined on the set of profiles on Y, such that the following hold:

(i) $\Lambda(0) = 0$; $\hfill (8.12)$

(ii) $\Lambda(f + g) \geq \Lambda(f) + \Lambda(g)$ (superadditivity); $\hfill (8.13)$

(iii) there is a constant M such that

$\Lambda(f + I_{\{\eta\}}) \leq \Lambda(f) + M$ for each attribute $\eta \in Y$ and each profile f (M is called an *individual marginal bound*); $\hfill (8.14)$

(iv) for every $\varepsilon > 0$ there is $\delta > 0$ such that

$|\Lambda(f + I_{\{\eta_1\}}) - \Lambda(f + I_{\{\eta_2\}})| < \varepsilon$ for all profiles f and $\eta_1, \eta_2 \in Y$
with $\mathrm{dist}(\eta_1, \eta_2) < \delta$ (continuity) .   (8.15)

[Note that the per-capita boundedness assumption (8.10) has to be strengthened to (8.14) in the general case.]

To derive an n-person game with transferable utility from the technology (Y, Λ), we specify a function $\alpha: N \to Y$ (an attribute function). (This function determines the attribute of each player.) We associate with each coalition S a profile $\mathrm{prof}(\alpha|S)$ given by

$$\mathrm{prof}(\alpha|S)(\eta) = |\alpha^{-1}(\eta) \cap S| .$$ (8.16)

The characteristic function v_α is then defined by

$$v_\alpha(S) = \Lambda(\mathrm{prof}(\alpha|S)) .$$ (8.17)

The following theorems were proved in Wooders and Zame (1984). (The method of proof involves, inter alia, a construction of ε-balanced games approximating v_α.)

Theorem 8.4. *Let (Y, Λ) be a technology and let $\varepsilon > 0$ be given. There exists an integer $n(\varepsilon)$ such that if $n \geq n(\varepsilon)$ and $\alpha: N \to Y$ is any attribute function, then the game v_α, defined by (8.17) [and (8.16)] has a non-empty weak ε-core.*

Theorem 8.5. *Let (Y, Λ) be a technology and let $\varepsilon > 0$ be given. Then there exists an integer $n(\varepsilon)$ and a positive number $\delta(\varepsilon)$ such that: if N is any finite set and $\alpha: N \to Y$ is any attribute function with the property that for each $i \in N$ there exist $n(\varepsilon)$ distinct players $j_1, \ldots, j_{n(\varepsilon)}$ such that $\mathrm{dist}(\alpha(i), \alpha(j_k)) < \delta(\varepsilon)$ for all $k = 1, \ldots, n(\varepsilon)$, then the game v_α, defined by (8.17) [and (8.16)] has a non-empty individually rational weak ε-core.*

Remark 8.2. It can be shown that Theorem 8.5 includes Theorem 8.1 as a special case. As stated earlier, Theorems 8.3, 8.4 and 8.5 include much more general exchange economies, as well as other economic models.

In Wooders and Zame (1987a) it is proved that with a suitable choice of $n(\varepsilon)$ and $\delta(\varepsilon)$, the Shapley value of (N, v_α) is an element of the individually rational ε-core of (N, v_α). The intuition underlying this result (as well as Theorems 8.3–8.5) is that for large games, the power of improvement is concentrated in small coalitions, i.e., that if an allocation can be improved upon at all, then it can be improved upon by a small coalition. Similar observations were made in analogous settings by many authors, e.g., Schmeidler (1972b), Mas-Colell

(1979) and Hammond, Kaneko and Wooders (1985) (see also p. 393). Wooders and Zame (1987a) observe further that small coalitions cannot affect the Shapley value very much.

As in Remark 7.1, it is possible to define a concept of ε-core directly for the exchange economy (without passing through any game form). If $a, b \in R^m$, denote by $a \ominus b$ the vector whose kth component is $\max(a_k - b_k, 0)$ for $k = 1, \ldots, m$ and by e the vector $(1, \ldots, 1) \in R^m$. We say that the allocation $x^{(1)}, \ldots, x^{(n)}$ is in the (strong) ε-core of the economy if for all coalitions $S \subset N$ and allocations $y^{(1)}, \ldots, y^{(n)}$ such that $y^{(i)}$ is strictly preferred by the trader i to $x^{(i)}$ for all $i \in S$, it is *not* true that

$$\sum_{i \in S} y^{(i)} \leq \sum_{i \in S} a^{(i)} \ominus \varepsilon e . \tag{8.18}$$

The weak ε-core is obtained if (8.18) is replaced by

$$\sum_{i \in S} y^{(i)} \leq \sum_{i \in S} a^{(i)} \ominus |S| \varepsilon e , \tag{8.19}$$

and the *fat* ε-core upon replacing (8.20) by

$$\sum_{i \in S} y^{(i)} \leq \sum_{i \in S} a^{(i)} \ominus n \varepsilon e . \tag{8.20}$$

Here

$$\text{fat } \varepsilon\text{-core} \supset \text{weak } \varepsilon\text{-core} \supset \text{strong } \varepsilon\text{-core} \supset \text{core} . \tag{8.21}$$

Non-emptiness of ε-cores was demonstrated (in various contexts) by, for example, Kannai (1970b), Kannai (1972), Hildenbrand, Schmeidler and Zamir (1973) and Grodal (1976). For example, it was shown in Kannai (1972) that the weak ε-core is non-empty for the replica economy, composed of nk traders, k of each type, if k is sufficiently large. In Kannai (1972) and Hildenbrand, Schmeidler and Zamir (1973) the economies grow in a much more general way (but the characteristics or attributes are restricted to stay in a compact set). In Grodal (1976) the assumptions are further weakened. The question of the approximate validity of (7.6) (approximate equilibrium) is also treated in the literature. Grodal, Trockel and Weber (1984) showed that ε can be taken to be of the order $1/n$.

For a game V with non-transferable utility, we say that a feasible outcome is in the strong ε-core if for no non-empty coalition S does there exist a vector $y \in V(S)$ such that

$$y_i - \varepsilon > x_i , \quad \text{for all } i \in S . \tag{8.22}$$

The weak ε-core is obtained if (8.22) is replaced by

$$y_i - |S|\varepsilon > x_i, \quad \text{for all } i \in S. \tag{8.23}$$

(Note that the nontransferable utility strong ε-core corresponds, roughly, to the transferable utility weak ε-core.)

Wooders (1983) defined a concept of *non-transferable utility replica games* by modifying the concept of technology (introduced earlier for analyzing transferable utility replica games without any underlying market structure) and proved the non-emptiness of ε-cores for large non-transferable utility replica games. Informally, the characteristic function $V^{(k)}$ of the kth replica game is such that the set $V^{(k)}(S)$ depends only on the vector $(s_1, \ldots, s_n)$ ($s_i \leqslant k$ is the number of players of the ith type, $i = 1, \ldots, n$) and on k, players of the same type being substitutes for each other. Moreover, the set of possible payoff vectors for members of S [i.e., the projection of $V^{(k)}(S)$ onto R^S] is assumed to be non-decreasing in k. The per-capita boundedness assumption [the analog of (8.10)] is that the components of equal-treatment payoff vectors in $V^{(k)}(N(k))$ (i.e., payoff vectors for which players of the same type get the same amount) are bounded independently of k. These assumptions suffice for proving non-emptiness of ε-cores for large k. As in the transferable utility case, no balancedness is required, the assumptions listed above yielding a sort of approximate balancedness.

Wooders and Zame (1987b) have found a condition under which the non-transferable utility value is in the ε-core, in the general framework of non-replica, non-transferable utility technologies. This result, however, does not yield a simple theorem about non-emptiness of the ε-core. The intuition underlying these results is similar to that underlying the corresponding results in the transferable utility case; the proofs, however, are much heavier.

One advantage of ε-cores is their continuity properties. Thus, while the core of a game varies with the game in an upper semi-continuous manner, it is not a lower semi-continuous function of the game. By contrast, the various ε-cores depend in a lower semi-continuous manner on the data of the problem. It is easy to see, for example, that if the outcome x is in the ε-core of the game with transferable utility v, ($\varepsilon \geqslant 0$), then for every $\varepsilon' > \varepsilon$ and every $\eta > 0$ there exists a $\delta > 0$ such that if $|v(S) - \tilde{v}(S)| < \delta$ for all coalitions S, then there exists an outcome $\tilde{x}$ in the ε'-core of the game $\tilde{v}$ and

$$\tilde{x}_i = 0, \quad \text{if } x_i = 0, \qquad \left|\frac{\tilde{x}_i}{x_i} - 1\right| < \eta, \quad \text{if } x_i > 0. \tag{8.24}$$

Continuity properties of ε-cores of economies are established in Kannai (1970b, 1972).

It follows that, in general, the ε-core is not necessarily close to the core, even if ε is small. It is plausible that the ε-core is close to the core, for "almost all" large economies. See Anderson (1985), where it is shown that "almost always" elements of the fat ε-core are close to the demanded set for some price p [see (7.6)].

Kaneko and Wooders (1986) and Hammond, Kaneko and Wooders (1985, 1989) have developed a model in which the set of all players – the "population" – is represented by a continuum, and coalitions are represented by finite subsets of this continuum (sets with a finite numbers of points). This models a situation in which almost all gains from coalition formation can be achieved by coalitions that are small relative to the population (as is the case, for example, in classical exchange economies; see Chapter 14 in this Handbook and the discussion on pp. 390–391). It is thus a continuum analogue of the "asymptotic" approach to large games discussed above, via "technologies": there, too, almost all gains from coalition formation can be achieved by coalitions that are small relative to the population (cf. the asymptotic homogeneity condition derived in the proof of Theorem 8.3).

For the continuum with finite coalitions the definition of core is not straightforward, as the worth of the all-player set need not be defined. Instead, one defines an object called the f-core (f for finite); see Chapter 14, Section 8, in this Handbook. Once the f-core is defined, the continuum yields – as usual – "cleaner" results than the asymptotic approach: Instead of non-empty ε-cores for sufficiently large k, one gets simply that the f-core is non-empty.

In the case of exchange economies, the f-core, the ordinary core, and the Walrasian allocations are all equivalent.

Another application of this model is to economies with widespread externalities. This means that an agent's utility depends only on his own consumption and on that of the population as a whole, not on the consumptions of other individuals (as with fashions). The model has also been applied to assignment games [see Gretsky, Ostroy and Zame (1990) for a related assignment model].

For details of the formulations and proofs in the continuum case, the reader is referred to the original articles.

Quite apart from continuum models, there is a large body of literature concerning cores of assignment games (see Chapter 16 in this Handbook), as well as other special classes of (nontransferable utility) games such as convex games, etc. (Note that transferable utility convex games are treated in Section 4.) Unfortunately, lack of space prevents us from describing this important literature.

References

Anderson, R.M. (1985) 'Strong core theorems with nonconvex preferences', *Econometrica*, 53: 1283–1294.

Billera, L.J. (1970) 'Some theorems on the core of an *n*-person game without side payments', *SIAM Journal on Applied Mathematics*, 18: 567–579.

Billera, L.J. (1974) 'On games without side payments arising from a general class of markets', *Journal of Mathematical Economics*, 1: 129–139.

Billera, L.J. and R.E. Bixby (1973) 'A characterization of polyhedral market games', *International Journal of Game Theory*, 2: 254–261.

Billera, L.J. and R.E. Bixby (1974) 'Market representations of *n*-person games', *Bulletin of the American Mathematical Society*, 80: 522–526.

Bondareva, O.N. (1963) 'Some applications of linear programming methods to the theory of cooperative games', *Problemy Kybernetiki*, 10: 119–139 [in Russian].

Burger, E. (1963) *Introduction to the theory of games*. Englewood Cliffs: Prentice-Hall.

Debreu, G. (1959) *Theory of value*. New Haven: Yale University Press.

Debreu, G. and H.E. Scarf (1963) 'A limit theorem on the core of a market', *International Economic Review*, 4: 235–246.

Delbaen, F. (1974) 'Convex games and extreme points', *Journal of Mathematical Analysis and its Applications*, 45: 210–233.

Dinar, A., D. Yaron and Y. Kannai (1986) 'Sharing regional cooperative games from reusing effluent for irrigation', *Water Resources Research*. 22: 339–344.

Dunford, N. and J.T. Schwartz (1958) *Linear operators, Part I*. New York: Interscience.

Edgeworth, F.Y. (1881) *Mathematical psychics*. London: Kegan Paul.

Fan, Ky (1956) 'On systems of linear inequalities', in: *Linear inequalities and related systems, Annals of Mathematics Studies*, 38: 99–156.

Gretsky, Neil E., J.M. Ostroy and W.R. Zame (1990) 'The nonatomic assignment model', Johns Hopkins working paper No. 256.

Grodal, B. (1976), 'Existence of approximate cores with incomplete preferences', *Econometrica*, 44: 829–830.

Grodal, B., W. Trockel and S. Weber (1984) 'On approximate cores of non-convex economies', *Economic Letters*, 15, 197–202.

Hammond, P.J., M. Kaneko and M.H. Wooders (1985) 'Mass-economies with vital small coalitions; the f-core approach', Cowles Foundation Discussion Paper No. 752.

Hammond, P.J., M. Kaneko and M.H. Wooders (1989) 'Continuum economies with finite coalitions: Core, equilibria, and widespread externalities', *Journal of Economic Theory*, 49: 113–134.

Hildenbrand, W., D. Schmeidler and S. Zamir (1973) 'Existence of approximate equilibria and cores', *Econometrica*, 41: 1159–1166.

Ichiishi, T. and S. Weber (1978) 'Some theorems on the core of a non-side payment game with a measure space of players', *International Journal of Game Theory*, 7: 95–112.

Ichiishi, T. (1981) 'Super-modularity: Applications to convex games and to the greedy algorithm for LP', *Journal of Economic Theory*, 25: 283–286.

Ichiishi, T. (1988) 'Alternative version of Shapley's theorem on closed coverings of a simplex', *Proceedings of the American Mathematical Society*, 104: 759–763.

Kaneko, M. and M.H. Wooders (1982) 'Cores of partitioning games', *Mathematical Social Sciences*, 3: 313–327.

Kaneko, M. and M.H. Wooders (1986) 'The core of a game with a continuum of players and finite coalitions: The model and some results', *Mathematical Social Science*, 12: 105–137.

Kannai, Y. (1969) 'Countably additive measures in cores of games', *Journal of Mathematical Analysis and its Applications*, 27: 227–240.

Kannai, Y. (1970a) 'On closed coverings of simplexes', *SIAM Journal on Applied Mathematics*, 19: 459–461.

Kannai, Y. (1970b) 'Continuity properties of the core of a market', *Econometrica*, 38: 791–815.

Kannai, Y. (1972) 'Continuity properties of the core of a market: A correction', *Econometrica*, 40: 955–958.

Kannai, Y. (1981) 'An elementary proof of the no-retraction theorem', *American Mathematical Monthly*, 88: 262–268.

Kannai, Y. and R. Mantel (1978) 'Non-convexifiable Pareto sets', *Econometrica*, 46: 571–575.

Mas-Colell, A. (1979) 'A refinement of the core equivalence theorem', *Economics Letters*, 3: 307–310.

Owen, G. (1982) *Game theory*, second edition. New York: Academic Press.

Peleg, B. (1965) 'An inductive method for constructing minimal balanced collections of finite sets', *Naval Research Logistics Quarterly*, 12: 155–162.

Rabie, M.A. (1981) 'A note on the exact games', *International Journal of Game Theory*, 10: 131–132.

Rosenmüller, J. (1977) *Extreme games and their solutions*. Lecture Notes in Economics and Mathematical Systems No. 145. Berlin: Springer.

Scarf, H.E. (1967) 'The core of *n*-person game', *Econometrica*, 35: 50–67.

Scarf, H.E. (1973) *The computation of economic equilibria*. New Haven: Yale University Press.

Schmeidler, D. (1967) 'On balanced games with infinitely many players', Mimeographed, RM-28, Department of Mathematics ,The Hebrew University, Jerusalem.

Schmeidler, D. (1972a) 'Cores of exact games I', *Journal of Mathematical Analysis and its Applications* 40: 214–225.

Schmeidler, D. (1972b) 'A remark on the core of an atomless economy', *Econometrica*, 40: 579–580.

Shapley, L.S. (1967) 'On balanced sets and cores', *Navel Research Logistics Quarterly*, 14: 453–460.

Shapley, L.S. (1971) 'Cores of convex games', *International Journal of Game Theory*, 1: 11–26.

Shapley, L.S. (1972) 'Let's block "block"', Mimeographed P-4779, The Rand Corporation, Santa Monica, California.

Shapley, L.S. (1973) 'On balanced games without side payments', in: T.C. Hu and S.M. Robinson, eds., *Mathematical programming*. New York: Academic Press, pp. 261–290.

Shapley, L.S. and M. Shubik (1966) 'Quasi-cores in a monetary economy with nonconvex preferences', *Econometrica* 34: 805–828.

Shapley, L.S. and M. Shubik (1969) 'On market games', *Journal of Economic Theory*, 1: 9–25.

Shapley, L.S. and M. Shubik (1976) 'Competitive outcomes in the cores of market games', *International Journal of Game Theory*, 4: 229–237.

Shubik, M. (1982) *Game theory in the social sciences: Concepts and solutions*. Cambridge, Mass.: MIT Press.

Shubik, M. (1984) *A game-theoretic approach to political economy: Volume 2 of Shubik* (1982). Cambridge, Mass: MIT Press.

Sorenson, J., J. Tschirhart and A. Whinston (1978) 'A theory of pricing under decreasing costs', *American Economic Review*, 68: 614–624.

Straffin, P.D. and J.P. Heaney (1981) 'Game theory and the Tennessee Valley Authority', *Interational Journal of Game Theory*, 10: 35–43.

Weber, S. (1981) 'Some results on the weak core of a non-sidepayment game with infinitely many players', *Journal of Mathematical Economics*, 8: 101–111.

Wilson, R. (1978) 'Information, efficiency and the core of an economy', *Econometrica*, 46: 807–816.

Wooders, M.H. (1979) 'Asymptotic cores and asymptotic balancedness of large replica games', Stony Brook Working Paper No. 215.

Wooders, M.H. (1983) 'The epsilon core of a large replica game', *Journal of Mathematical Economics*, 11: 277–300.

Wooders, M.H. and W.R. Zame (1984) 'Approximate cores of large games', *Econometrica*, 52: 1327–1350.

Wooders, M.H. and W.R. Zame (1987a) 'Large games: Fair and stable outcomes', *Journal of Economic Theory*, 42: 59–63.

Wooders, M.H. and W.R. Zame (1987b) 'NTU values of large games', IMSSS Technical Report No. 503, Stanford University.

Yannelis, N.C. (1991) 'The core of an economy with differential information', *Economic Theory*, 1: 183–198.

Chapter 13

AXIOMATIZATIONS OF THE CORE

BEZALEL PELEG*

The Hebrew University of Jerusalem

Contents

*Partially supported by the S.A. Schonbrunn Chair in Mathematical Economics at The Hebrew University of Jerusalem.

Handbook of Game Theory, Volume 1, Edited by R.J. Aumann and S. Hart

1. Introduction

The core is, perhaps, the most intuitive solution concept in cooperative game theory. Nevertheless, quite frequently it is pointed out that it has several shortcomings, some of which are given below:

(1) The core of many games is empty, e.g. the core of every essential constant-sum game is empty.

(2) In many cases the core is too big, e.g. the core of a unanimity game is equal to the set of all imputations.

(3) In some examples the core is small but yields counter-intuitive results. For example, the core of a symmetric market game with m sellers and n buyers, $m < n$, consists of a unique point where the sellers get all the profit [see Shapley (1959)].

For further counter-intuitive examples, see, for example, Maschler (1976) and Aumann (1985b, 1987).

In view of the foregoing remarks it may be argued that an intuitively acceptable axiom system for the core might reinforce its position as the most "natural" solution (provided, of course, that it is not empty). But in our opinion an axiomatization of the core may serve two other, more important goals:

(1) By obtaining axioms for the core, we single out those important properties of solutions that determine the most stable solution in the theory of cooperative games. Thus, in Subsections 2.2 and 2.4 we shall see that the core of TU games is determined by individual rationality (IR), superadditivity (SUPA), and the reduced game property (RGP). Also, the core of NTU games is characterized by IR and RGP (see Subsection 3.2). Furthermore, the converse reduced game property (CRGP) is essential for the axiomatization of the core of (TU) market games (see Subsection 2.3). Therefore we may conclude that four properties, IR, SUPA, RGP, and CRGP, play an important role in the characterization of the core on some important families of games.

(2) Once we have an axiom system for the core we may compare it with systems of other solutions the definitions of which are not simple or "natural". Indeed, we may claim that a solution is "acceptable" if its axiomatization is similar to that of the core. There are some important examples of this kind: (a) The prenucleolus is characterized by RGP together with the two standard assumptions of symmetry and covariance [see Sobolev (1975)]. (b) The Shapley value is characterized by SUPA and three more "weaker" axioms [see Shapley (1981)]. (c) The prekernel is determined by RGP, CRGP, and three more standard assumptions [see Peleg (1986a)].

We now review briefly the contents of this chapter. Section 2 is devoted to

TU games. In Subsection 2.1 we discuss several properties of solutions to coalitional games. An axiomatization of the core of balanced games is given in Subsection 2.2. The core of market games is characterized in Subsection 2.3, and the results of Subsection 2.2 are generalized to games with coalition structures in Subsection 2.4. In Section 3 we present the results for NTU games. First, in Subsection 3.1 we introduce reduced games of NTU games. Then, an axiom system for the core of NTU games is presented in Subsection 3.2. Finally, we review Keiding's axiomatization of the core of NTU games in Subsection 3.3.

2. Coalitional games with transferable utility

2.1. *Properties of solutions of coalitional games*

Let U be a (nonempty) set of *players*. U may be finite or infinite. A *coalition* is a nonempty and *finite* subset of U. A *coalitional game with transferable utility* (a TU game) is a pair (N, v), where N is a coalition and v is a function that associates a real number $v(S)$ with each subset S of N. We always assume that $v(\emptyset) = 0$. Let N be a coalition. A *payoff vector* for N is a function $x : N \to R$ (here R denotes the real numbers). Thus, R^N is the set of all payoff vectors for N. If $x \in R^N$ and $S \subset N$, then we denote $x(S) = \Sigma_{i \in S} x^i$. (Clearly, $x(\emptyset) = 0$.)

Let (N, v) be a game. We denote

$$X^*(N, v) = \{x \mid x \in R^N \text{ and } x(N) \leqslant v(N)\} .$$

$X^*(N, v)$ is the set of feasible payoff vectors for the game (N, v).

Now we are ready for the following definition.

Definition 2.1.1. Let Γ be a set of games. A *solution* on Γ is a function σ which associates with each game $(N, v) \in \Gamma$ a subset $\sigma(N, v)$ of $X^*(N, v)$.

Intuitively, a solution is determined by a system of "reasonable" restrictions on $X^*(N, v)$. We may, for example, impose certain inequalities that guarantee the "stability" of the members of $\sigma(N, v)$ in some sense. Alternatively, σ may be characterized by a set of axioms.

We shall be interested in the following solution.

Definition 2.1.2. Let (N, v) be a game. The *core* of (N, v), $C(N, v)$, is defined by

$$C(N, v) = \{x \mid x \in X^*(N, v) \text{ and } x(S) \geqslant v(S) \text{ for all } S \subset N\} .$$

We remark that $x \in C(N, v)$ iff no coalition can improve upon x. Thus, each member of the core consists of a highly stable payoff distribution.

We shall now define some properties of solutions that are satisfied by the core (on appropriate domains). This will enable us to axiomatize the core of two important families of games in the following two subsections.

Let Γ be a set of games and let σ be a solution on Γ.

Definition 2.1.3. σ is *individually rational* (IR) if for all $(N, v) \in \Gamma$ and all $x \in \sigma(N, v)$, $x^i \geq v(\{i\})$ for all $i \in N$.

IR says that every player i gets, at every point of σ, at least his solo value $v(\{i\})$. If, indeed, all the singletons $\{i\}$, $i \in N$, may be formed, then IR follows from the usual assumption of utility maximization [see Luce and Raiffa (1957, section 8.6)]. We remark that the core satisfies IR.

For our second property we need the following notation. Let (N, v) be a game.

$$X(N, v) = \{x \mid x \in R^N \text{ and } x(N) = v(N)\} .$$

Definition 2.1.4. The solution σ satisfies *Pareto optimality* (PO) if $\sigma(N, v) \subset X(N, v)$ for every $(N, v) \in \Gamma$.

PO is equivalent to the following condition: if $x, y \in X^*(N, v)$ and $x^i > y^i$ for all $i \in N$, then $y \notin \sigma(N, v)$. This formulation seems quite plausible, and similar versions to it are used in social choice [Arrow (1951)] and bargaining theory [Nash (1950)]. Nevertheless, it is actually quite a strong condition in the context of cooperative game theory. Indeed, the players may fail to agree on a choice of a Pareto-optimal point [i.e. a member of $X(N, v)$], because different players have different preferences over the Pareto-optimal set.

Clearly, the core satisfies PO. However, PO does not appear explicitly in our axiomatization of the core.

The following notation is needed for the next definition. If N is a coalition and $A, B \subset R^N$, then

$$A + B = \{a + b \mid a \in A \text{ and } b \in B\} .$$

Definition 2.1.5. The solution σ is *superadditive* (SUPA) if

$$\sigma(N, v_1) + \sigma(N, v_2) \subset \sigma(N, v_1 + v_2)$$

when (N, v_1), (N, v_2), and $(N, v_1 + v_2)$ are in Γ.

Clearly, SUPA is closely related to additivity. Indeed, for one-point solutions it is equivalent to additivity. The additivity condition is usually one of the axioms in the theory of the Shapley value [see Shapley (1981)]. Most writers accept it as a natural condition. Shapley himself writes:

Plausibility arguments (for additivity) can be based on games that consist of two games played separately by the same players (e.g., at different times, or simultaneously using agents) or, better, by considering how the value should act on probability combinations of games [Shapley (1981, p. 59)].

Only in Luce and Raiffa (1957, p. 248) did we find some objections to the additivity axiom. They disagree with the foregoing arguments proposed by Shapley. However, they emphasize that, so far as the Shapley value is concerned, additivity must be accepted.

Intuitively, SUPA is somewhat weaker than additivity (for set-valued functions). Fortunately, the core satisfies SUPA.

The last two properties pertain to restrictions of solutions to subcoalitions.

Definition 2.1.6. Let (N, v) be a game, let $S \subset N$, $S \neq \emptyset$, and let $x \in X^*(N, v)$. The *reduced game* with respect to S and x is the game $(S, v_{x,S})$, where

$$
v_{x,S}(T) = \begin{cases} 0, & T = \emptyset, \\ v(N) - x(N - T), & T = S, \\ \max\{v(T \cup Q) - x(Q) \mid Q \subset N - S\}, & \text{otherwise}. \end{cases}
$$

(Here, $N - T = \{i \in N \mid i \notin T\}$.)

Remark 2.1.7. For $x \in X(N, v)$ Definition 2.1.6 coincides with the definition of reduced games in Davis and Maschler (1965).

Let M be a coalition and let $x \in R^M$. If T is a coalition, $T \subset M$, then we denote by x^T the restriction of x to T.

Remark 2.1.8. The reduced game $(S, v_{x,S})$ describes the following situation. Assume that *all* the members of N agree that the members of $N - S$ will get x^{N-S}. Then, the members of S may get $v(N) - x(N - S)$. Furthermore, suppose that the members of $N - S$ continue to cooperate with the members of S (subject to the foregoing agreement). Then, for every $T \subset S$, $T \neq S, \emptyset$, $v_{x,S}(T)$ is the total payoff that the members of T expect to get. However, we notice that the expectations of different coalitions may not be compatible because they may require the cooperation of the *same* subset of $N - S$ (see Example

2.1.9). Thus, $(S, v_{x,S})$ is not a game in the ordinary sense; it serves only to determine the distribution of $v_{x,S}(S)$ to the members of S.

Example 2.1.9. Let (N, v) be the simple majority three-person game[1] $[2; 1, 1, 1]$. Furthermore, let $x = (1/2, 1/2, 0)$ and $S = \{1, 2\}$. The reduced game $(S, v_{x,S})$ is given by $v_{x,S}(\{1\}) = v_{x,S}(\{2\}) = v_{x,S}(\{1, 2\}) = 1$. Notice that player i, $i = 1, 2$, needs the cooperation of player 3 in order to obtain $v_{x,S}(\{i\})$.

Let Γ be a set of games.

Definition 2.1.10. A solution σ on Γ has the *reduced game property* (RGP) if it satisfies the following condition. If $(N, v) \in \Gamma$, $S \subset N$, $S \neq \emptyset$, and $x \in \sigma(N, v)$, then $(S, v_{x,S}) \in \Gamma$ and $x^S \in \sigma(S, v_{x,S})$.

Definition 2.1.10 is due to Sobolev (1975) who used it in his axiomatic characterization of the prenucleolus.

Remark 2.1.11. RGP is a condition of *self-consistency*. If (N, v) is a game and $x \in \sigma(N, v)$, that is, x is a solution to (N, v), then for every $S \subset N$, $S \neq \emptyset$, x^S is consistent with the expectations of the members of S as reflected by the game $(S, v_{x,S})$. The reader may also find discussions of RGP in Aumann and Maschler (1985, Sections 3 and 6) and in Thomson (1985, Section 5).

We remark that the core satisfies RGP on the class of all games [see Peleg (1986a)]. The following weaker version of RGP is very useful. First, we introduce the following notation.

Notation 2.1.12. If D is a finite set, then we denote by $|D|$ the number of members of D.

Definition 2.1.13. A solution σ on a set Γ of games has the *weak reduced game property* (WRGP) if it satisfies the following condition: if $(N, v) \in \Gamma$, $S \subset N$, $1 \leqslant |S| \leqslant 2$, and $x \in \sigma(N, v)$, then $(S, v_{x,S}) \in \Gamma$ and $x^S \in \sigma(S, v_{x,S})$.

Clearly, RGP implies WRGP. The converse is generally not true. Thus, WRGP may be used in axiomatizations of the core when RGP is not satisfied.

From a practical (or, at least, computational) point of view the following problem may be interesting. Let σ be a solution, let (N, v) be a game, and let $x \in X(N, v)$. Furthermore, let π be a set of nonempty subsets of N. Then we ask whether or not σ satisfies

---

[1]Thus, $v(\emptyset) = v(\{1\}) = v(\{2\}) = v(\{3\}) = 0$ and $v(S) = 1$ otherwise.

$$[x^S \in \sigma(S, v_{x,S}) \quad \text{for all} \quad S \in \pi] \Rightarrow x \in \sigma(N, v).$$

The foregoing remark motivates the following definition. If N is a coalition then we denote

$$\pi = \pi(N) = \{S \subset N \,|\, |S| = 2\}.$$

Definition 2.1.14. A solution σ on a set Γ of games has the *converse reduced game property* (CRGP) if the following condition is satisfied. If $(N, v) \in \Gamma$, $x \in X(N, v)$, and for every $S \in \pi(N)$, $(S, v_{x,S}) \in \Gamma$ and $x^S \in \sigma(S, v_{x,S})$, then $x \in \sigma(N, v)$.

CRGP has the following simple interpretation. Let x be a Pareto-optimal payoff vector [i.e. $x(N) = v(N)$]. Then x is an "equilibrium" payoff if every pair of players is in "equilibrium".

We remark that CRGP was first used in Harsanyi (1959) as the basis for the extension of Nash's solution to multi-person pure bargaining games. Also, it has been used in the axiomatization of the prekernel [see Peleg (1986a)]. The core satisfies CRGP [Peleg (1986a)].

We close this subsection with a simple result and definitions.

Lemma 2.1.15. *Let σ be a solution on a set of games. If σ satisfies IR and WRGP, then it also satisfies PO.*

Definition 2.1.16. A solution σ on a set Γ of games satisfies *nonemptiness* (NE) if $\sigma(N, v) \neq \emptyset$ for every $(N, v) \in \Gamma$.

Definition 2.1.17. Let σ be a solution on a set Γ of games. σ is *weakly symmetric* (WS) if the following condition is satisfied. If $N = \{i, j\}$, $(N, v) \in \Gamma$, $v(\{i\}) = v(\{j\})$, and $(x^i, x^j) \in \sigma(N, v)$, then $(x^j, x^i) \in \sigma(N, v)$.

Clearly the core satisfies WS. Actually, the core satisfies a much stronger version of symmetry, namely anonymity (i.e. it is independent of the names of the players).

2.2. An axiomatization of the core

Let U be a set of players. In this subsection we assume that U contains at least three members. We denote $\Gamma_c = \{(N, v) \,|\, C(N, v) \neq \emptyset\}$. Γ_c is the set of all balanced games [see Owen (1982, Chapter VIII)].

Theorem 2.2.1.   *There is a unique solution on Γ_c that satisfies NE, IR, SUPA, and WRGP, and it is the core.*

Outline of the proof.   Clearly, we need only to prove the uniqueness part of the theorem. This part follows from the following lemmata and corollary.

Lemma 2.2.2.   *Let σ be a solution on a set Γ of games. If σ satisfies IR and WRGP, then $\sigma(N, v) \subset C(N, v)$ for every $(N, v) \in \Gamma$.*

Corollary 2.2.3.   *Let σ be a solution on Γ_c that satisfies NE, IR, and WRGP. If the core of a game (N, v) consists of a unique point, then $\sigma(N, v) = C(N, v)$.*

Lemma 2.2.4.   *Let (N, v) be a game and let $x \in C(N, v)$. If $|N| \geqslant 3$, then there exist coalitional functions w and u on N such that: (i) $C(N, w) = \{x\}$; (ii) $C(N, u) = \{0\}$, and (iii) $v = u + w$.*

Now let σ be a solution on Γ_c that satisfies NE, IR, SUPA, and WRGP. By Lemma 2.2.2, $\sigma(N, v) \subset C(N, v)$ for every $(N, v) \in \Gamma_c$. Also, by Corollary 2.2.3 and Lemma 2.2.4, if $(N, v) \in \Gamma_c$ and $|N| \geqslant 3$, then $C(N, v) \subset \sigma(N, v)$. To complete the proof we use WRGP to show that $C(N, v) \subset \sigma(N, v)$ when $(N, v) \in \Gamma_c$ and $|N| \leqslant 2$.

Remark 2.2.5.   The axioms NE, IR, SUPA, and WRGP are independent [see Peleg (1986a, Examples 5.8–5.11)].

2.3. An axiomatization of the core of market games

We start this subsection with a definition of market games. Let U be a set of players. A *market* is a quadruple (N, E_+^m, A, W). Here N is a coalition (the set of traders); E_+^m is the non-negative orthant of the m-dimensional Euclidean space (the commodity space); $A = \{a^i \mid i \in N\}$ is an indexed collection of points in E_+^m (the initial endowments); and $W = \{w^i \mid i \in N\}$ is an indexed collection of continuous concave real functions on E_+^m (the utility functions). Let S be a coalition, $S \subset N$. A *feasible S-allocation* is an indexed collection $x^S = \{x^i \mid i \in S\}$ such that $x^i \in E_+^m$ for all $i \in S$ and $\Sigma_{i \in S} x^i = \Sigma_{i \in S} a^i$. We denote by X^S the set of all feasible S-allocations.

Definition 2.3.1.   A game (N, v) is a *market game* if there exists a market (N, E_+^m, A, W) such that

$$v(S) = \max\left\{ \sum_{i \in S} w^i(x^i) \mid x^S \in X^S \right\}$$

for every $S \subset N$.

Definition 2.3.1 is due to Shapley and Shubik (1969).

Let (N, v) be a game. A *subgame of* (N, v) is a game (T, v^T), where $T \subset N$ and $v^T(S) = v(S)$ for all $S \subset T$.

Definition 2.3.2. A game (N, v) is *totally balanced*[2] if $C(T, v^T) \neq \emptyset$ for every $T \subset N$, $T \neq \emptyset$.

By Theorem 5 of Shapley and Shubik (1969) a game is a market game if and only if it is totally balanced. We denote by Γ_t the set of all market games. Also, we assume that U contains at least four players.

Theorem 2.3.3. *There is a unique solution on Γ_t that satisfies NE, IR, WS, SUPA, WRGP, and CRGP, and it is the core.*

For a proof of Theorem 2.3.3 see Peleg (1985b).

Remark 2.3.4. It may be shown that each of the axioms NE, IR, SUPA, WRGP, and CRGP is independent of the other four axioms and WS. We do not know whether or not WS is independent of the rest of the axioms. However, we can prove that WS is independent of NE, IR, SUPA, and WRGP.[3]

Remark 2.3.5. A comparison of Theorems 2.2.1 and 2.3.3 is instructive. The core satisfies NE, IR, SUPA, and WRGP on Γ_t. Because Γ_t is a proper subset of Γ_c, there may be additional solutions on Γ_t that satisfy the foregoing four axioms. Indeed, Example 5.5 of Peleg (1989) gives us a solution σ on Γ_t with the following properties: (a) σ is different from the core, and (b) σ satisfies NE, IR, SUPA, WS, and WRGP. We conclude from that example that CRGP is essential for the characterization of the core on Γ_t. If we examine the proof of Theorem 2.2.1 we find that the analogue of Lemma 2.2.4 on Γ_t is not true. The failure of that analogue explains why the two theorems are different. Also, the proof of Theorem 2.3.3 is more difficult than that of Theorem 2.2.1.

---

[2]A game (N, u) is *balanced* if $C(N, u) \neq \emptyset$ [see Shapley (1967) for the origin of this terminology].
[3]Theorem 3.15 in Peleg (1989) implies that WS is redundant. However, its proof is incorrect. E.J. Balder and A. van Breukelen have found an error in the proof of Lemma 4.7 of Peleg (1989).

2.4. Games with coalition structures

Let U be a set of players with at least three members and let N be a coalition. A *coalition structure* (c.s.) for N is a partition of N. A game with c.s. is a triple (N, v, b), where (N, v) is a game and b is a c.s. for N (see the chapter on 'coalition structures' in a forthcoming volume of this Handbook for discussion of games with coalition structures). In this subsection, Theorem 2.2.1 is generalized to games with coalition structures. We denote by Δ the set of all games with coalition structures.

Let $(N, v, b) \in \Delta$. We use the following notation:

$$X^*(N, v, b) = \{x \mid x \in R^N \text{ and } x(B) \leq v(B) \text{ for every } B \in b\} \,.$$

Let $\Delta_0 \subset \Delta$.

Definition 2.4.1. A *solution* on Δ_0 is a function σ which associates with each game with c.s. $(N, v, b) \in \Delta_0$ a subset $\sigma(N, v, b)$ of $X^*(N, v, b)$.

Definition 2.4.2. Let $(N, v, b) \in \Delta$. The core of (N, v, b), $C(N, v, b)$, is defined by

$$C(N, v, b) = \{x \mid x \in X^*(N, v, b) \text{ and } x(S) \geq v(S) \text{ for all } S \subset N\} \,.$$

Let N be a coalition, let b be a c.s. for N, and let $S \subset N$, $S \neq \emptyset$. We use the following notation.

$$b \mid S = \{B \cap S \mid B \in b \text{ and } B \cap S \neq \emptyset\} \,.$$

Clearly, $b \mid S$ is a c.s. for S. Now we are ready for the following definitions.

Definition 2.4.3. Let $(N, v, b) \in \Delta$, let $S \subset N$, $S \neq \emptyset$, and let $x \in X^*(N, v, b)$. The *reduced game* with respect to S and x is the game with c.s. $(S, v_{x,S}, b \mid S)$ where

$$v_{x,S}(T) =
\begin{cases}
0, & T = \emptyset, \\
v(B) - x(B - T), & T \in b \mid S, T \subset B \text{ and } B \in b, \\
\max\{v(T \cup Q) - x(Q) \mid Q \subset N - S\}, & \text{otherwise} \,.
\end{cases}$$

Definition 2.4.4. Let $\Delta_0 \subset \Delta$. A solution σ on Δ_0 has the *weak reduced game property* (WRGP) if it satisfies the following condition. If $(N, v, b) \in \Delta_0$,

$S \subset N$, $1 \leq |S| \leq 2$, and $x \in \sigma(N, v, b)$, then $(S, v_{x.S}, b \,|\, S) \in \Delta_0$ and $x^S \in \sigma(S, v_{x.S}, b \,|\, S)$.

Let $\Delta_c = \{(N, v, b) \,|\, C(N, v, b) \neq \emptyset\}$.

Theorem 2.4.5. *There is a unique solution on Δ_c that satisfies NE, IR, SUPA and WRGP, and it is the core.*

The proof of Theorem 2.4.5 is similar to that of Theorem 2.2.1.

3. Coalitional games without side payments

3.1. Reduced games of NTU games

Let U be a set of players and let N be a coalition. If x, $y \in R^N$, then $x \geq y$ if $x^i \geq y^i$ for all $i \in N$, and $x \gg y$ if $x^i > y^i$ for all $i \in N$. We denote $R^N_+ = \{x \in R^N \,|\, x \geq 0\}$. Let $A \subset R^N$. A is *comprehensive* if $x \in A$ and $x \geq y$ imply that $y \in A$. The boundary of A is denoted by ∂A. Finally, cl A denotes the closure of A.

Definition 3.1.1. A *nontransferable utility* (*NTU*) *game* is a pair (N, V), where N is a coalition and V is a function that assigns to each coalition $S \subset N$ a subset $V(S)$ of R^S, such that

$V(S)$ is nonempty and comprehensive ; (1)

$V(S) \cap (x^S + R^S_+)$ is bounded for every $x^S \in R^S$; (2)

$V(S)$ is closed ; (3)

if x^S, $y^S \in \partial V(S)$ and $x^S \geq y^S$, then $x^S = y^S$. (4)

Conditions (1) and (3) are standard. (2) guarantees that $V(S)$ is a proper subset of R^S. It is a very weak requirement of boundedness. Condition (4) is the familiar nonlevelness property [see, for example, Aumann (1985a)].

Definition 3.1.2. Let (N, V) be an NTU game, let $x \in V(N)$, and let $S \subset N$, $S \neq \emptyset$. The *reduced game* with respect to S and x is the game $(S, V_{x.S})$, where

$$V_{x.S}(S) = \{y^S \,|\, (y^S, x^{N-S}) \in V(N)\} ,$$ (5)

$$V_{x,S}(T) = \bigcup_{Q \subset N-S} \{y^T \mid (y^T, x^Q) \in V(T \cup Q)\} \quad \text{if } T \subset S, \ T \neq S. \tag{6}$$

In the reduced game the players of S are allowed to choose only payoff vectors y^S that are compatible with x^{N-S}, the fixed payoff distribution to the members of $N - S$. On the other hand, proper subcoalitions T of S may count on the cooperation of subsets Q of $N - S$ provided that in the resulting payoff vectors for $T \cup Q$ each member i of Q gets exactly x^i.

Remark 3.1.3. Reduced games of NTU games were first used in Greenberg (1985). The present definition is due to Peleg (1986b). However, it is interesting to notice that the idea of considering reduced games of NTU games may be traced back to Harsanyi (1959). Also, Lensberg (1988) and Thomson (1984) have recently used reduced games in their axiomatization of various solutions of pure bargaining games.

We close this subsection with the following lemma.

Lemma 3.1.4. *Let (N, V) be an NTU game, let $x \in V(N)$, and let $S \subset N$, $S \neq \emptyset$. Then the reduced game $(S, V_{x,S})$ is a game [i.e. it satisfies (1)–(4)].*

3.2. An axiomatization of the core of NTU games

Let U be a set of players with at least three members.

Definition 3.2.1. Let (N, V) be an NTU game and let $x \in V(N)$. A coalition $S \subset N$ can *improve upon* x if there exists $y^S \in V(S)$ such that $y^S \gg x^S$. x is in the *core* of (N, V), $C(N, V)$, if no coalition can improve upon x.

We denote $\Gamma = \{(N, V) \mid C(N, V) \neq \emptyset\}$. A *solution* on Γ is a function σ that assigns to each NTU game $(N, V) \in \Gamma$ a subset $\sigma(N, V)$ of $V(N)$. We shall consider the following properties of solutions.

Definition 3.2.2. A solution σ on Γ satisfies *nonemptiness* (NE) if $\sigma(N, V) \neq \emptyset$ for every $(N, V) \in \Gamma$.

Let (N, V) be an NTU game and let $i \in N$. We denote

$$v^i = \sup\{x^i \mid x^i \in V(\{i\})\} .$$

By (1) and (2) v^i is well defined.

Definition 3.2.3. A solution σ on Γ satisfies *individual rationality* (IR) if for every $(N, V) \in \Gamma$ and every $x \in \sigma(N, V)$, $x^i \geq v^i$ for all $i \in N$.

Obviously, the core satisfies IR.

Definition 3.2.4. A solution σ on Γ has the *reduced game property* (RGP) if it satisfies the following condition. If $(N, V) \in \Gamma$, $S \subset N$, $S \neq \emptyset$, and $x \in \sigma(N, V)$, then $(S, V_{x, S}) \in \Gamma$ and $x^S \in \sigma(S, V_{x, S})$.

RGP has been used in the axiomatization of the Nash solution [Lensberg (1988)] and the egalitarian solution [Thomson (1984)] of pure bargaining games. The core satisfies RGP.

Definition 3.2.5. A solution σ on Γ has the *converse reduced game property* (CRGP) if it satisfies the following condition. If $(N, V) \in \Gamma$, $x \in V(N)$, and for every two-player coalition S, $S \subset N$, $(S, V_{x,S}) \in \Gamma$ and $x^S \in \sigma(S, V_{x,S})$, then $x \in \sigma(N, V)$.

We remark that the core satisfies CRGP. Now we are ready for the following theorems.

Theorem 3.2.6. *Assume that U is infinite. Then there is a unique solution on Γ that satisfies NE, IR, and RGP, and it is the core. Furthermore, NE, IR, and RGP are independent.*

Theorem 3.2.7. *Assume that U is finite. Then there is a unique solution on Γ that satisfies NE, IR, RGP, and CRGP, and it is the core. Furthermore, NE, IR, RGP, and CRGP are independent.*

Proofs of Theorems 3.2.6 and 3.2.7 are given in Peleg (1985a).

3.3. A review of "An axiomatization of the core of a cooperative game" by H. Keiding

Keiding (1986) considers a larger class of NTU games. More precisely, he uses the following definition.

Definition 3.3.1. An NTU *game* is a pair (N, V), where N is a coalition and V is a function that assigns to each coalition $S \subset N$ a subset $V(S)$ of R^S such that (1) is satisfied for all S, and (4) is satisfied only for $S = N$.

Let Γ be the set of all games. A *solution* is a function σ that assigns to each $(N, V) \in \Gamma$ a subset $\sigma(N, V)$ of $V(N)$.

Keiding is interested in the following properties of solutions. A game (N, V) is *trivial* if $V(S) = \{x^S \in R^S \mid x^S \leq 0\}$ for all $S \neq \emptyset$, N. Let σ be a solution.

Axiom 3.3.2 (triviality). If (N, V) is trivial and $V(N) \cap R_+^N = \{0\}$, then $\sigma(N, V) = \{0\}$.

Axiom 3.3.3 (covariance). For all $a \in R^N$, if $(N, V + \{a\})$ is defined by $(V + \{a\})(S) = V(S) + a^S$, $S \subset N$, then $\sigma(N, V + \{a\}) = \sigma(N, V) + a$.

Axiom 3.3.4 (antimonotonicity). If (N, V) and (N, W) are games such that $V(S) \subset W(S)$ for all $S \subset N$, and $V(N) = W(N)$, then $\sigma(N, W) \subset \sigma(N, V)$.

Keiding writes: "This axiom seems quite reasonable: large sets $W(S)$ mean that coalitions are powerful so that fewer of the feasible payoff vectors in $W(N) = V(N)$ may qualify as final outcomes."

Axiom 3.3.5 (continuity). If (N, V) and (N, W) are games such that cl $V(S) =$ cl $W(S)$ for all $S \subset N$ and $V(N) = W(N)$, then $\sigma(N, V) = \sigma(N, W)$.

Clearly, Axiom 3.3.5 is a very weak technical assumption.

For a solution σ and a game (N, V), we define $(D_\sigma V)(S)$, $S \subset N$, $S \neq \emptyset$, N inductively as follows. For each S with $|S| = 1$ we put $(D_\sigma V)(S) = $ cl $V(S)$. Suppose that $(D_\sigma V)(S^*)$ is defined for all $S^* \subset S$, $S^* \neq \emptyset$, S; if $(D_\sigma V)(S^*) = \emptyset$ for some S^* put $(D_\sigma V)(S) = \emptyset$; otherwise let $(D_\sigma V)(S)$ be the closed comprehensive hull of $\sigma(S, V^*)$, where (S, V^*) is the game defined by $V^*(S^*) = (D_\sigma V)(S^*)$ for $S^* \neq S$, and $V^*(S) = V(S)$. If $(D_\sigma V)(S) \neq \emptyset$ for all $S \subset N$, we define the σ-*derived* game $(N, D_\sigma V)$ of (N, V) by $D_\sigma V(S) = (D_\sigma V)(S)$ for all $S \subset N$ $S \neq \emptyset$, N, and $D_\sigma V(N) = V(N)$.

Axiom 3.3.6 (independence of σ-irrelevant alternatives). If (N, V) is a game such that the σ-derived game $(N, D_\sigma V)$ is defined, then $\sigma(N, V) = \sigma(N, D_\sigma V)$.

Keiding interprets the last axiom in the following way:

To get an understanding of Axiom 3.3.6, it is helpful to think of outcomes as results of a bargaining procedure, where coalitions S may object to payoffs x by reference to some y in $V(S)$ which they can enforce by themselves. In order for such an objection to be credible, it must be "really" enforceable by S, that is it must not in its turn be objected against by some subcoalition of

S. By the logic of our approach σ should be used repeatedly to decide which of the elements *y* of *V*(*S*) are "really" enforceable in the above sense. The construction of the σ-derived game keeps exactly such elements and excludes the non-enforceable. Thus, the axiom says that the non-enforceable options of *V*(*S*) do not count when $\sigma(N, V)$ is determined.

Keiding proves the following theorem.

Theorem 3.3.7. *Let σ be a solution on Γ with the following properties*:
(a) *σ satisfies Axioms 3.3.2–3.3.6*;
(b) *if τ is a solution satisfying Axioms 3.3.2–3.3.6, then $\sigma(N, V) \subset \tau(N, V)$ for all $(N, V) \in \Gamma$.*
Then $\sigma(N, V) = C(N, V)$ for every game (N, V).

References

Arrow, K.J. (1951) *Social choice and individual values*. New York: Wiley.

Aumann, R.J. (1985a) 'An axiomatization of the non-transferable utility value', *Econometrica*, 53: 599–612.

Aumann, R.J. (1985b) 'On the non-transferable utility value: A comment on the Roth–Shafer examples', *Econometrica*, 53: 667–677.

Aumann, R.J. (1987) 'Value, symmetry, and equal treatment: A comment on Scafuri and Yannelis', *Econometrica*, 55: 1461–1464.

Aumann, R.J. and M. Maschler (1985) 'Game theoretic analysis of a bankruptcy problem from the Talmud', *Journal of Economic Theory*, 36: 195–213.

Davis, M. and M. Maschler (1965) 'The kernel of a cooperative game', *Naval Research Logistics Quarterly*, 12: 223–259.

Greenberg, J. (1985) 'Cores of convex games without side payments', *Mathematics of Operations Research*, 10: 523–525.

Harsanyi, J.C. (1959) 'A bargaining model for the cooperative *n*-person game', in: A.W. Tucker and R.D. Luce, eds., *Contributions to the theory of games, IV*, Annals of Mathematics Studies 40. Princeton: Princeton University Press, pp. 325–335.

Keiding, H. (1986) 'An axiomatization of the core of a cooperative game', *Economics Letters*, 20: 111–115.

Lensberg, T. (1988) 'The stability of the Nash solution', *Journal of Economic Theory*, 45: 330–341.

Luce, R.D. and H. Raiffa (1957) *Games and decisions*. New York: Wiley.

Maschler, M. (1976) 'An advantage of the bargaining set over the core', *Journal of Economic Theory*, 13: 184–192.

Nash, J.F. (1950) 'The bargaining problem', *Econometrica*, 18: 155–162.

Owen, G. (1982) *Game theory*, 2nd edn. New York: Academic Press.

Peleg, B. (1985a) 'An axiomatization of the core of cooperative games without side payments', *Journal of Mathematical Economics*, 14: 203–214.

Peleg, B. (1985b) 'An axiomatization of the core of market games', Center for Research in Mathematical Economics and Game Theory, RM 68, The Hebrew University of Jerusalem.

Peleg, B. (1986a) 'On the reduced game property and its converse', *International Journal of Game Theory*, 15: 187–200.

Peleg, B. (1986b) 'A proof that the core of an ordinal convex game is a von Neumann–Morgenstern solution', *Mathematical Social Sciences*, 11: 83–87.

Peleg, B. (1989) 'An axiomatization of the core of market games', *Mathematics of Operations Research*, 14: 448–456.

Shapley, L.S. (1959) 'The solutions of a symmetric market game', in: A.W. Tucker and R.D. Luce, eds., *Contribution to the theory of games IV*, Annals of Mathematics 40. Princeton: Princeton University Press, pp. 145–162.

Shapley, L.S. (1967) 'On balanced sets and cores', *Naval Research Logistics Quarterly*, 14: 453–460.

Shapley, L.S. (1981) 'Valuation of games', in: W.F. Lucas, ed., *Game theory and its applications*. Providence: American Mathematical Society, pp. 55–67.

Shapley, L.S. and M. Shubik (1969) 'On market games', *Journal of Economic Theory*, 1: 9–25.

Sobolev, A.I. (1975) 'The characterization of optimality principles in cooperative games by functional equations', *Mathematical Methods in the Social Sciences*, 6: 150–165 [in Russian].

Thomson, W. (1984) 'Monotonicity, stability, and egalitarianism', *Mathematical Social Sciences*, 8: 15–28.

Thomson, W. (1985) 'Axiomatic theory of bargaining with a variable population: A survey of recent results', in: A.E. Roth, ed., *Game-theoretic models of bargaining*. Cambridge: Cambridge University Press, pp. 233–258.

Chapter 14

THE CORE IN PERFECTLY COMPETITIVE ECONOMIES

ROBERT M. ANDERSON*

University of California at Berkeley

Contents

*This research was supported in part by grants from the National Science Foundation. The author is grateful for the helpful comments of Bob Aumann, Don Brown, Harrison Cheng, Birgit Grodal, Sergiu Hart, Werner Hildenbrand, Ali Khan, Alejandro Manelli, Andreu Mas-Colell, Salim Rashid, Martin Shubik, Rajiv Vohra, Nicholas Yannelis and Bill Zame.

Handbook of Game Theory, Volume 1, Edited by R.J. Aumann and S. Hart

1. Introduction

In this chapter we survey results on the cores of perfectly competitive exchange economies, i.e. economies in which the endowment of each agent is negligible on the scale of the whole economy. The subject began with the pioneering work of Edgeworth (1881). Edgeworth gave a geometrical proof, in the case of two commodities and two traders, that as one replicated the economy the core collapsed to the set of Walrasian equilibria. Edgeworth claimed in passing that his proof generalized to arbitrary numbers of commodities and arbitrary numbers of agents in the base economy being replicated.

The subject lay dormant for nearly a century until Shubik (1959) recognized the importance of Edgeworth's contribution. Debreu and Scarf (1963) gave the first proof of the theorem that Edgeworth had claimed: that, in replica sequences of economies with strongly convex preferences, the intersection of the cores of the replications coincides exactly with the set of Walrasian equilibria. Their proof is quite different from Edgeworth's.

Aumann (1964) formulated a model of a large economy with a measure space of agents. In this model, he showed that the core coincided with the set of Walrasian equilibria. Moreover, Aumann required only minimal assumptions; for example, neither convexity nor monotonicity of the preferences nor boundedness of the endowments is required. Aumann's proof makes use of some of the key ideas in the Debreu and Scarf proof.

In the contributions of Edgeworth, Debreu and Scarf, and Aumann, the conclusion is clean and neat: the core (in Aumann's case) or the intersection of the cores of all replicas (in the other cases) *coincides* with the set of Walrasian equilibria. Moreover, the Debreu and Scarf paper is completely elementary, with a proof that is a model of simplicity and elegance. The Aumann paper, of course, uses more sophisticated mathematics. However, since Aumann found the proper mathematical formulation for the problem, the proof is (modulo the mathematical prerequisites) simple, and the conclusion is very strong.

Following these pioneering contributions, core theory became one of the principal focuses of mathematical economics in the 1960s and 1970s. The study turned primarily in the direction of limit theorems for sequences of large finite economies. Here, the simplicity that had been found in the replica and continuum cases disappeared. One of the key elements of the Debreu and Scarf argument, the equal treatment property which permitted one to collapse the cores of all the different replicas into the same space, does not generalize even to sequences with different numbers of traders of the various types. Moreover, the strong statement that the core (in Aumann's continuum setting) or the intersection of the cores (in the Debreu and Scarf replica setting) coincides with the set of Walrasian equilibria is simply not true in the case of general sequences of finite economies. Instead, one must substitute weaker forms of convergence. Convexity of preferences, which plays no role whatever in Aumann's theorem, is seen to make a crucial difference in the form of convergence in large finite economies.

In short, the quest in this literature was to find the appropriate way to come back from the limit case considered by Aumann to characterize the limiting behavior of large finite economies more general than those considered by Debreu and Scarf. A wide array of mathematical tools was employed, including the theory of weak convergence of probability measures [introduced by Hildenbrand (1970)], non-standard analysis [introduced by Brown and Robinson (1975)] and differential topology [introduced by Debreu (1975)]. Critical (and under-appreciated) contributions were made by Kannai (1970) and Bewley (1973a), the latter making use of an early core result of Vind (1965). Much of the work of this period is reported in Hildenbrand's classic book [Hildenbrand (1974)], the standard reference for the measure-theoretic and weak convergence approach to the study of the core and, indeed, the standard reference for much of the mathematics in current use in economic theory.

In the second half of the 1970s, independent work by E. Dierker (1975), Keiding (1974), and Anderson (1978) permitted elementary and shorter proofs of the main convergence results. In addition, the statements of the theorems could be given without referring to the weak convergence machinery. The author's paper came directly out of work using nonstandard analysis. These advances make it possible to communicate the main convergence theorems (including the subtle interplay between the assumptions and the variations in convergence forms that they produce) to a wider audience.

In the second half of the 1980s there was renewed research on the core. There were significant improvements in the results on the rate of core convergence. Counterexamples [especially those in Manelli (1991)] highlighted the extent to which core convergence results for large finite economies are dependent on strong monotonicity and convexity assumptions, even though these assumptions play no role in Aumann's continuum formulation; in addition, monotonicity plays no role in the Debreu–Scarf replica formulation. The introduction of the f-core [Kaneko and Wooders (1986, 1989) and Hammond, Kaneko and Wooders (1989)] for the first time permitted the proof of an equivalence theorem in the presence of externalities which result in failure of the First Welfare Theorem.

The literature exhibits a great variety of assumptions and conclusions; in many cases, assumptions that appear quite different in different papers are actually closely related. In order to bring some order to the literature, and to explain clearly (we hope) how the form of core convergence depends on the assumptions made, we shall first develop a taxonomy of assumptions and conclusions. The survey will then consist of tables listing the various results, with their assumptions and conclusions described in terms of the taxonomy. In this way it is hoped that the reader can come to appreciate the essential differences and similarities of the various results. However, since the assumptions and conclusions of the papers vary somewhat from those given in the taxonomy, the reader is cautioned to consult the original sources for exact statements of the theorems.

We should close by noting that there are many topics relating to the core that space does not permit us to discuss. In particular, we do not discuss here the following topics:

(1) nonemptiness of the core: see Chapter 12 of this Handbook;

(2) approximate cores: see Kannai (1970), Hildenbrand, Schmeidler and Zamir (1973), Grodal and Hildenbrand (1974), Khan (1974), Grodal (1976), Grodal, Trockel and Weber (1984), Wooders and Zame (1984) and Anderson (1985), and the references contained there;

(3) economies with imperfect competition, i.e. a single agent possesses a non-negligible fraction of the social endowment: see Chapter 15 of this Handbook;

(4) cores of production economies, because there is no generally accepted definition of the core in this case, and because the assumption of perfect competition which is reasonable in consumption is not reasonable in the case of production: see Hildenbrand (1968, 1974), Champsaur (1974), and Oddou (1976);

(5) cores of games other than economies (see Chapters 12 and 13 of this Handbook);

(6) cores with transactions costs [Khan and Rashid (1976)].

2. Basics

In this section we give some basic definitions and blanket assumptions. Not all of the assumptions are required for every theorem, but we regard them to be more technical in nature than those assumptions we have chosen to highlight in the taxonomy created in the next two sections. For this reason, we shall make these assumptions throughout, and not try to indicate which theorems can be proven without them.

Suppose $x, y \in \mathbf{R}^k$, $A \subset \mathbf{R}^k$. x^i denotes the ith component of x; $x \geqslant y$ means $x^i \geqslant y^i$ for all i; $x > y$ means $x \geqslant y$ and $x \neq y$; $x \gg y$ means $x^i > y^i$ for all i; $\|x\|_\infty = \max_{1 \leqslant i \leqslant k} |x^i|$; $\|x\|_1 = \Sigma_{i=1}^k = 1^k |x^i|$; $\mathbf{R}_+^k = \{x \in \mathbf{R}^k : x \geqslant 0\}$.

A preference is a binary relation $>$ on $\mathbf{R}_+^k$ satisfying the following conditions: (i) continuity: $\{(x, y): x > y\}$ is relatively open in $\mathbf{R}_+^k \times \mathbf{R}_+^k$; and (ii) irreflexivity: $x \not> x$. Let $\mathscr{P}$ denote the set of preferences. We write $x \sim y$ if $x \not> y$ and $y \not> x$.

An exchange economy is a map $\chi: A \to \mathscr{P} \times \mathbf{R}_+^k$, where A is a finite set. For $a \in A$, let $>_a$ denote the preference of a [i.e. the projection of $\chi(a)$ onto $\mathscr{P}$] and $e(a)$ the initial endowment of a [i.e. the projection of $\chi(a)$ onto $\mathbf{R}_+^k$]. An allocation is a map $f: A \to \mathbf{R}_+^k$ such that $\Sigma_{a \in A} f(a) = \Sigma_{a \in A} e(a)$. A coalition is a non-empty subset of A. A coalition S can improve on an allocation f if there exists $g: S \to \mathbf{R}_+^k$, $g(a) >_a f(a)$ for all $a \in S$, and $\Sigma_{a \in S} g(a) = \Sigma_{a \in S} e(a)$. The core of χ, $\mathscr{C}(\chi)$, is the set of all allocations which cannot be improved on by any coalition.

A price p is an element of $\mathbf{R}^k$ with $\|p\|_1 = 1$. Δ denotes the set of prices, $\Delta_+ = \{p \in \Delta: p \geq 0\}$, $\Delta_{++} = \{p \in \Delta: p \gg 0\}$. The demand set for $(>, e)$, given $p \in \Delta$, is $D(p, (>, e)) = \{x \in \mathbf{R}_+^k: p \cdot x \leq p \cdot e, \ y > x \Rightarrow p \cdot y > p \cdot e\}$. The quasidemand set for $(>, e)$, given $p \in \Delta$, is $Q(p, (>, e)) = \{x \in \mathbf{R}_+^k: p \cdot x \leq p \cdot e, \ y > x \Rightarrow p \cdot y \geq p \cdot e\}$. By abuse of notation, we let $D(p, a) = D(p, (>_a, e(a)))$ and $Q(p, a) = Q(p, (>_a, e(a)))$ if $a \in A$.

An income transfer is a function $t: A \to \mathbf{R}$. By abuse of notation, we write $D(p, a, t) = \{x \in \mathbf{R}_+^k: p \cdot x \leq p \cdot e + t(a), \quad y > x \Rightarrow p \cdot y > p \cdot e + t(a)\}$ and $Q(p, a, t) = \{x \in \mathbf{R}_+^k: p \cdot x \leq p \cdot e + t(a), \ y > x \Rightarrow p \cdot y \geq p \cdot e + t(a)\}$.

A Walrasian equilibrium is a pair (f, p), where f is an allocation, $p \in \Delta$, and $f(a) \in D(p, a)$ for all $a \in A$. If (f, p) is a Walrasian equilibrium, then f is called a Walrasian allocation and p is called a Walrasian equilibrium price. Let $\mathscr{W}(\chi)$ denote the set of Walrasian equilibrium prices. A Walrasian quasiequilibrium is a pair (f, p), where f is an allocation, $p \in \Delta$, and $f(a) \in Q(p, a)$ for all $a \in A$. If (f, p) is a Walrasian quasiequilibrium, then f is called a quasi-Walrasian allocation and p is called a Walrasian quasiequilibrium price. Let $\mathscr{Q}(\chi)$ denote the set of Walrasian quasiequilibrium prices.

The following theorem, which asserts that the set of Walrasian allocations is contained in the core, is an important strengthening of the First Welfare Theorem. It provides a means of demonstrating non-emptiness of the core in situations where one can prove the existence of Walrasian equilibrium; see Debreu (1982).

Theorem 2.1. *Suppose $\chi: A \to \mathscr{P} \times \mathbf{R}_+^k$ is an exchange economy, where A is a finite set. If f is a Walrasian allocation, then $f \in \mathscr{C}(\chi)$.*

Proof. Suppose (f, p) is a Walrasian equilibrium. Suppose a coalition $S \neq \emptyset$ can improve on f by means of g, so that $g(a) >_a f(a)$ for $a \in S$ and $\Sigma_{a \in S} g(a) = \Sigma_{a \in S} e(a)$. Since $f(a) \in D(p, a)$, $p \cdot g(a) > p \cdot e(a)$ for $a \in S$. Then

$$p \cdot \sum_{a \in S} g(a) = \sum_{a \in S} p \cdot g(a) > \sum_{a \in S} p \cdot e(a) = p \cdot \sum_{a \in S} e(a), \tag{1}$$

which contradicts $\Sigma_{a \in S} g(a) = \Sigma_{a \in S} e(a)$. Since f cannot be improved on by any coalition, $f \in \mathscr{C}(\chi)$. $\square$

3. Assumptions on preferences and endowments

In this section we define the various assumptions on preferences and endowments that are used in the core theorems that we shall discuss. Within each
section, the assumptions are numbered from weakest to strongest. For example, Assumption B4 implies Assumption B3, which implies B2, which implies
B1. Where appropriate, brief discussions of the economic significance will be
given. In each case, we consider a sequence of exchange economies
$\chi_n: A_n \to \mathcal{P} \times R_+^k$.

1. Convexity

(a) **C1** (bounded non-convexity). Assumption C1 is that preferences
    exhibit bounded non-convexity in the sense that

$$\frac{1}{|A_n|} \max_{a \in A_n} \sup_{x \in R_+^k} \gamma(\{y \in R_+^k : y > x\}) \to 0 , \tag{2}$$

where $\gamma(B)$ is the Hausdorff distance between the set B and its
convex hull, i.e. $\gamma(B) = \sup_{c \in \text{con } B} \inf_{b \in B} |b - c|$.

(b) **C2** (Convexity). Assumption C2 is that preferences are convex; in
    other words, $\{y \in R_+^k : y > x\}$ is a convex set for every $x \in R_+^k$.

(c) **C3** (Strong convexity). Assumption C3 is that preferences are strong-
    ly convex; in other words, if $x \neq y$, then either $(x + y)/2 > x$ or
    $(x + y)/2 > y$.

2. Smoothness

In the following list, Assumption SB neither implies nor is implied by S.

(a) **S** (Smoothness). Assumption S is that preferences are smooth, in
    other words, $\{(x, y) \in R_{++}^k \times R_{++}^k : x \sim y\}$ is a C^2 manifold; see Mas-
    Colell (1985).

   Comment. When we list Assumptions C3 and S together, we will
   assume that preferences are differentiably strictly convex; in other
   words, the indifference surfaces have non-vanishing Gaussian curvature [Debreu (1975)]. This is the condition required to make the
   demand function differentiable, as long as the demand stays in the
   interior of R_+^k. Giving a complete definition of Gaussian curvature
   would take us too far afield, but the idea is simple. The distance
   between the indifference surface and the tangent plane to the surface
   at a point x can be approximated by a Taylor polynomial. The linear
   terms are zero (that is the definition of the tangent plane); non-
   vanishing Gaussian curvature says that the quadratic terms are nondegenerate. Geometrically, this is saying that the indifference surface

is not flatter than a sphere. Since we do not assume that the indifference curves do not cut the boundary of R_+^k, the demand functions may have kinks where consumption of a commodity falls to 0.

(b) **SB** (Smoothness at the boundary). Assumption SB is that indifference surfaces with a point in the interior of R_+^k do not intersect the boundary of R_+^k [Debreu (1975)].

Comment. This is a strong assumption; it implies that all consumers consume strictly positive amounts of all commodities at every Walrasian equilibrium. S, SB and C3 together imply that the demand function is differentiable. SB is inconsistent with M4; when we list M4 and SB together as assumptions, we will assume that M4 holds only on the interior of the consumption set.

3. Transitivity and completeness

T (Transitivity and completeness). Assumption T is that preferences are transitive and complete; in other words, preferences satisfy (i) transitivity: if $x > y$ and $y > z$, then $x > z$; and (ii) negative transitivity: if $x \not> y$ and $y \not> z$, then $x \not> z$.

Comment. The rather strange-looking condition (ii) guarantees that the indifference relation induced by $>$ is transitive.

4. Monotonicity

(a) **M1** (Local non-satiation). Assumption M1 is that preferences are locally non-satiated; in other words, for every x and every $\delta > 0$, there is some y with $y > x$ and $|y - x| < \delta$.

(b) **M2** (Uniform properness). Assumption M2 is that preferences are uniformly proper; in other words, there is an open cone $V \subset R_+^k$ such that if $y \in x + V$, then $y > x$.

(c) **M3** (Weak monotonicity). Assumption M3 is that preferences are weakly monotone; in other words, if $y \gg x$, then $y > x$.

(d) **M4** (Monotonicity). Assumption M4 is that preferences are monotone; in other words, if $x > y$, then $x > y$.

Comment. Note that M4 plus continuity will imply that if $x^i > 0$, then the individual would be willing to give up a positive amount of the ith commodity in order to get a unit of the jth commodity. This has the flavor of assuming that marginal rates of substitution are bounded away from zero and infinity, and is used for the same purpose: to show that prices are bounded away from zero. Note, however, that preferences can be monotone and have the tangent to the indifference curve be vertical at a point.

5. Positivity

(a) **P1** (Positivity of social endowment). Assumption P1 is that the social endowment of every commodity is strictly positive; in other words, $\Sigma_{a \in A}\, e(a) \gg 0$.

   Comment. This is a fairly innocuous assumption; if the social endowment of a commodity is zero, then the commodity can be excluded from all considerations of exchange. One considers the economy with this commodity excluded from the commodity space, and with preferences induced from the original commodity space. The core of the new economy corresponds exactly with the core of the original economy under the obvious identification. Note, however, that the set of Walrasian equilibria *is* changed by this exclusion; with zero social endowment of a desirable commodity, there may well be no Walrasian equilibrium. Hence, the core equivalence theorem may fail without assuming P1.

(b) **P2** (Positivity of individual endowments). Assumption P2 is that each individual has a strictly positive endowment of each commodity; in other words, $\forall a \in A \; e(a) \gg 0$.

   Comment. This is a very strong assumption. Casual empiricism indicates that most individuals are endowed with their own labor and a very limited number of other commodities. We believe this assumption should be avoided if at all possible.

6. Boundedness

We shall assume throughout that the per capita endowment is bounded, i.e.

$$\sup_n \frac{1}{|A_n|}\left| \sum_{a \in A_n} e(a) \right| < \infty . \tag{3}$$

(a) **B1** (No large individual). Assumption B1 is satisfied if

$$\max\left\{ \frac{|e(a)|}{|A_n|} : a \in A_n \right\} \to 0 \text{ as } n \to \infty . \tag{4}$$

   Comment. Assumption B1 does not rule out the possibility that, in the limit, a negligible fraction of the agents will possess a significant fraction (or even all) of the social endowment. It does say that no *one* individual can possess a non-negligible fraction of the social endowment.

(b) **B2**. Assumption B2 is satisfied if

$$\max\left\{ \frac{|e(a)|}{\sqrt{|A_n|}} : a \in A_n \right\} \to 0 \text{ as } n \to \infty . \tag{5}$$

(c) **B3** (Uniform integrability). Assumption B3 is satisfied if the sequence of endowment maps is uniformly integrable. In other words, for any sequence of sets of individuals $E_n \subset A_n$ with $|E_n|/|A_n| \to 0$,

$$\frac{|\Sigma_{a \in E_n} e(a)|}{|A_n|} \to 0 \text{ as } n \to \infty. \tag{6}$$

Comment. Uniform integrability has a natural economic interpretation. It says in the limit that no group composed of a negligible fraction of the agents in the economy can possess a non-negligible fraction of the social endowment. It is clearly stronger than Assumption B1. Assumption B3 is needed in approaches to limit theorems based on weak convergence methods to guarantee that the continuum limit of a sequence of economies reflects the behavior of the sequence. In elementary approaches, one can dispense with it (although the conclusion is weakened somewhat by doing so). It is probably easier to appreciate the significance of the assumption by considering the following example of a sequence of *tenant farmer* economies in which the assumption fails. We consider a sequence $\chi_n: A_n \to (\mathcal{P} \times R_+^k)$, where $A_n = \{1, \dots, n^2\}$. For all $a \in A_n$, the preference of a is given by a utility function $u(x, y) = 2\sqrt{2}x^{1/2} + y$. The endowment is given by

$$e_n(a) = \begin{cases} (n+1, 1) & \text{if } a = 1, \dots, n, \\ (1, 1) & \text{if } a = n+1, \dots, n^2. \end{cases} \tag{7}$$

Think of the first commodity as land, while the second commodity is food. The holdings of land are heavily concentrated among the agents $1, \dots, n$, a small fraction of the total population. Land is useful as an input to the production of food; however, the marginal product of land diminishes rapidly as the size of the plot worked by a given individual increases. The sequence χ_n satisfies B1 since $\max|e_n(a)|/|A_n| < (n+2)/n^2 \to 0$. However, if we let $E_n = \{1, \dots, n\}$, then

$$\frac{|E_n|}{|A_n|} = \frac{n}{n^2} \to 0, \tag{8}$$

but

$$\frac{1}{|A_n|} \left| \sum_{a \in E_n} e_n(a) \right| > \frac{n^2 + n}{n^2} \nrightarrow 0, \tag{9}$$

so χ_n fails to satisfy B3.

(d) **B4** (Uniform boundedness). Assumption B4 is satisfied if there exists $M \in R$ such that $\max\{|e(a)|: a \in A_n\} \leq M$ for all $n \in N$.

   Comment. Assumption B4 clearly implies Assumptions B1–B3. It is a strong assumption. If one needs to assume it in a given theorem, it indicates that the applicability of the conclusion to a given large economy depends in a subtle way on the relationship of the largest endowment to the number of agents.

7. Distributional assumptions

Here, we are using the word "distribution" in its probabilistic sense; we look at the measure on $\mathscr{P} \times R_+^k$ induced by the economy. There is a complete separable metric on $\mathscr{P}$ [Hildenbrand (1974)]. When we use the term "compact", we shall mean compact with respect to this metric. The economic implication of compactness is to make any monotonicity or convexity condition apply in a uniform way. For example, a compact set K of monotone preferences is equimonotone, i.e. for any compact set X contained in the interior of R_+^k, there exists $\delta > 0$ such that $x + e^j - \delta e^i > x$ for all $x \in X$ and all $> \in K$ [Grodal (1976)]. Similarly, a compact set of strongly convex preferences is equiconvex [see Anderson (1981a) for the definition]. Indeed, although the compactness assumptions are needed to use the weak convergence machinery, they can be replaced by equimonotonicity or equiconvexity assumptions in elementary approaches to core convergence theorems.

   (a) **D1** (Tightness). Assumption D1 is satisfied if the sequence of distributions induced on $\mathscr{P} \times R_+^k$ is tight. In other works, given any $\delta > 0$, there exists a compact set $K \subset \mathscr{P} \times R_+^k$ such that

$$\frac{|\{a \in A_n: (>_a, e(a)) \in K\}|}{|A_n|} > 1 - \delta . \tag{10}$$

   (b) **D2** (Compactness). Assumption D2 holds if there is a compact set $K \subset \mathscr{P} \times R_+^k$ such that $(>_a, e(a)) \in K$ for all $a \in A_n$ and every n.

   (c) **D3** (Type). The sequence of economies is called a type sequence of economies if there is a finite set T (the set of types) such that $(>_a, e(a)) \in T$ for all $a \in A_n$ and every n.

   Comment. The assumption of a finite number of types is obviously restrictive, since it will require a large number of identical individuals in the economies. On the other hand, this assumption makes the analysis much easier. Theorems for type sequences have often pointed the way to more general theorems. However, the proofs do not generalize; new methods are typically needed, and the conclusions in the general case are usually weaker. Occasionally (as when dispersion is needed), type sequences are *less* well-behaved than general sequences. Thus it is dangerous to assume that behavior in the type case reflects fully the behavior of general economies. Note that Assumption D3 implies Assumption B4 (uniform boundedness of endowments).

(d) **D4** (Replica). The sequence of economies is called a replica sequence if it is a type sequence, and the economy χ_n has exactly n individuals of each type.

   Comment. The comment in Assumption D3 applies here, but more strongly. Great caution is required in inferring general behavior from replica results.

8. Support assumption

(a) **DI1** (No isolated individuals, usual metric). Assumption DI1 is satisfied if, for every $\delta > 0$,

$$\inf_n \min_{a \in A_n} \frac{|\{b \in A_n : d_1((>_a, e(a)), (>_b, e(b))) < \delta\}|}{|A_n|} > 0, \qquad (11)$$

where d_1 is the usual metric on $\mathscr{P} \times \boldsymbol{R}_+^k$ [Hildenbrand (1974)].

(b) **DI2** (No isolated individuals, Hausdorff metric). Assumption DI2 is satisfied if, for every $\delta > 0$,

$$\inf_n \min_{a \in A_n} \frac{|\{b \in A_n : d_2((>_a, e(a)), (>_b, e(b))) < \delta\}|}{|A_n|} > 0, \qquad (12)$$

where d_2 is constructed in the following way. If a preference is continuous, then $\{(x, y) \in \boldsymbol{R}_+^k \times \boldsymbol{R}_+^k : x \not\succ y\}$ is closed. Thus, the Hausdorff metric on closed sets induces a metric d_2' on the space of preferences; let d_2 be the product of d_2' and the Euclidean metric on $\boldsymbol{R}_+^k$.

   Comment. DI1 and DI2 say that there are no "isolated" individuals whose characteristics persist throughout the sequence but "disappear" in the limit. DI2 is implied by D4; however, DI1 and DI2 neither imply nor are implied by any of D1–D3. d_2' is much finer than the topology on preferences associated with the d_1 metric, which considers two preferences close if their restrictions to bounded sets are close. Because the space of preferences with the d_2' metric is not separable, DI2 is considerably stronger than DI1.

9. Purely competitive sequences

Since the space of agents' characteristics is a complete separable metric space, there is a metric (called the Prohorov metric) which metrizes the topology of weak convergence on the space of distributions on $\mathscr{P} \times R_+^k$ [Billingsley (1968), Hildenbrand (1974)]. Hildenbrand (1970) introduced weak convergence into the study of the core. In Hildenbrand (1974), he defined a purely competitive sequence of economies to be one the distributions of which converge in the topology of weak convergence, and moreover the average social endowments of which converge to the social endowment of the limit. Any purely competitive sequence satisfies B3 (uniform integrability of endowments) and D1 (tightness). Conversely, if a sequence satisfies B3 and D1, then every sub-sequence contains a further subsequence which is purely competitive. Thus, limit theorems for purely competitive sequences are essentially equivalent to theorems for sequences satisfying B3 and D1.

4. Types of convergence

The type of convergence that holds depends greatly on the assumptions on the sequence of economies. The various possibilities can best be thought of as lying on four largely (but not completely) independent axes: the type of convergence of individual consumptions to demands, the equilibrium nature of the price at which the demands are calculated, the degree to which the convergence is uniform over individuals, and the rate at which convergence occurs.

1. Individual convergence conclusions

In what follows we shall suppose that f is a core allocation in an economy $\chi: A \to \mathscr{P} \times R_+^k$, and that $a \in A$. We describe two sets of conclusions: those beginning with ID relate the core allocation to the demand, while those beginning with IQ relate it to the quasidemand. Since a demand vector is always a quasidemand vector, conclusion IDi is stronger than conclusion IQi for each i. Let us say that one conclusion is "informally stronger" than another if it more closely conforms to the motivation for studying core convergence as described at the beginning of Section 5. Then ID5 is informally stronger than ID4T or ID4N, which are informally stronger than ID3U or ID3N, which are informally stronger than ID2, which is informally stronger than IQ1. Indeed, under certain standard (but not innocuous) assumptions, one can show that ID5 $\Rightarrow$ {ID4T, ID4N} $\Rightarrow$ {ID3U, ID3N} $\Rightarrow$ ID2 $\Rightarrow$ IQ1. However, Manelli (1990b) has constructed an example with a sequence of core allocations satisfying ID4N (and E3, which is described below), where IQ1 nonetheless fails; in the example, preferences are not monotone.

    (a) **IQ1** (Demand-like). Conclusion IQ1 is that the consumption of the individual a is quasidemand-like, but not necessarily close to a's quasidemand set. Specifically, we define

$$\rho_{Q1}(f, a, p) = |p \cdot (f(a) - e(a))|$$
$$+ \inf\{\delta \geq 0: y >_a f(a) \Rightarrow p \cdot y > p \cdot e(a) - \delta\}. \quad (13)$$

Conclusion IQ1 is that there exists $p \in \Delta$ such that $\rho_{Q1}(f, a, p)$ is small.

Comment. This is a δ-satisficing notion: the consumption is as good as anything that costs δ less than the endowment. Note that if $\rho_{Q1}(f, a, p) = 0$, then $f(a) \in Q(p, a)$.

(b) **IQ2, ID2** (Near demand in utility). Conclusion IQ2 (ID2) is that there is a price vector p such that the utility of the consumption of individual a is close to the utility of consuming a's quasidemand (demand). Specifically, we assume that the specification of the economy includes a specification of particular utility functions representing the preferences of the individuals. We then define

$$\rho_{Q2}(f, a, p) = \inf_{x \in Q(p, a)} |u_a(f(a)) - u_a(x)| ,$$
$$\rho_{D2}(f, a, p) = \inf_{x \in D(p, a)} |u_a(f(a)) - u_a(x)| . \quad (14)$$

Conclusion IQ2 (ID2) is that there exists $p \in \Delta$ such that $\rho_{Q2}(f, a, p)$ $(\rho_{D2}(f, a, p))$ is small.

(c) Conclusion IQ3U neither implies nor is implied by conclusion IQ3N; conclusion ID3U neither implies nor is implied by conclusion ID3N.

 (i) **IQ3U, ID3U** (Indifferent to demand with income transfer). Conclusion IQ3U (ID3U) is that there is a price vector p and an income transfer t such that individual a is indifferent between consuming his/her assigned bundle and consuming $Q(p, a, t)$ $(D(p, a, t))$. Specifically, we define

$$\rho_{Q3U}(f, a, p) = \inf\{|t(a)|: \exists x \ f(a) \sim x, x \in Q(p, a, t)\} ,$$
$$\rho_{D3U}(f, a, p) = \inf\{|t(a)|: \exists x \ f(a) \sim x, x \in D(p, a, t)\} . \quad (15)$$

Conclusion IQ3U (ID3U) is that there exists $p \in \Delta$ such that $\rho_{Q3}(f, a, p)$ $(\rho_{D3}(f, a, p))$ is small.

 (ii) **IQ3N, ID3N** (Near demand with an income transfer). Conclusion IQ3N (ID3N) is that there is a price vector p and an income transfer t such that individual a's consumption bundle is near $Q(p, a, t)$ $(D(p, a, t))$. Specifically, we define

$$\rho_{\mathrm{Q3N}}(f, a, p) = \inf\{|f(a) - x|: x \in Q(p, a, t)\},$$
$$\rho_{\mathrm{D3N}}(f, a, p) = \inf\{|f(a) - x|: x \in D(p, a, t)\}. \tag{16}$$

Conclusion IQ3N (ID3N) is that there exists $p \in \Delta$ such that $\rho_{\mathrm{Q3N}}(f, a, p)$ ($\rho_{\mathrm{D3N}}(f, a, p)$) is small.

(d) Conclusion IQ4T neither implies nor is implied by conclusion IQ4N; conclusion ID4T neither implies nor is implied by conclusion ID4N.

  (i) **IQ4T, ID4T** (Demand with an income transfer). Conclusion IQ4T (ID4T) is that there is a price vector p and an income transfer t such that individual a's consumption bundle is an element of $Q(p, a, t)$ ($D(p, a, t)$). Specifically, we define

$$\rho_{\mathrm{Q4T}}(f, a, p) = \inf\{|t(a)|: f(a) \in Q(p, a, t)\},$$
$$\rho_{\mathrm{D4T}}(f, a, p) = \inf\{|t(a)|: f(a) \in D(p, a, t)\}. \tag{17}$$

  Conclusion IQ4T (ID4T) is that there exists $p \in \Delta$ such that $\rho_{\mathrm{Q4T}}(f, a, p)$ ($\rho_{\mathrm{D4T}}(f, a, p)$) is small.

  Comment. If p is a supporting price (see conclusion E2S, below), then $f(a) \in Q(p, a, t)$ ($D(p, a, t)$) with $t(a) = p \cdot f(a) - p \cdot e(a)$.

  (ii) **IQ4N, ID4N** (Near demand). Conclusion IQ4N (ID4N) is that there is a price vector p such that the consumption of individual a is near a's demand set. Specifically, we define

$$\rho_{\mathrm{Q4N}}(f, a, p) = \inf\{|x - f(a)|: x \in Q(p, a)\},$$
$$\rho_{\mathrm{D4N}}(f, a, p) = \inf\{|x - f(a)|: x \in D(p, a)\}. \tag{18}$$

  Conclusion IQ4N (ID4N) is that there exists $p \in \Delta$ such that $\rho_{\mathrm{Q4N}}(f, a, p)$ ($\rho_{\mathrm{D4N}}(f, a, p)$) is small.

(e) **IQ5, ID5** (In demand set). Conclusion IQ5 (ID5) is that there is a price vector p such that $f(a) \in Q(p, a)$ ($D(p, a)$).

2. Equilibrium conclusions on price

These conclusions concern a price vector p. If the individual convergence conclusion is of the form IQi, the equilibrium conclusion on price refers to Walrasian quasiequilibrium; if the individual convergence conclusion is of the form IDi, the equilibrium conclusion on price refers to Walrasian equilibrium.

(a) **E1** (Any price). Conclusion E1 is that the price p is an arbitrary member of Δ.

(b) Conclusion E2A neither implies, nor is implied by, conclusion E2S.

(i) **E2A** (Approximate equilibrium price). Conclusion E2A is that the price p is an approximate equilibrium price. Specifically, define

$$\mathcal{Q}_\delta(\chi) = \left\{ p: \exists g(a) \in Q(p, a), \left| \sum_{a \in A} g(a) - e(a) \right| \leq \delta \right\},$$

$$\mathcal{W}_\delta(\chi) = \left\{ p: \exists g(a) \in D(p, a), \left| \sum_{a \in A} g(a) - e(a) \right| \leq \delta \right\}. \tag{19}$$

$\mathcal{Q}_\delta(\chi)$ $(\mathcal{W}_\delta(\chi))$ is the set of δ-Walrasian quasiequilibrium (δ-Walrasian equilibrium) prices.

(ii) **E2S** (Supporting price). Conclusion E2S is that the price p is a supporting price. In other words, if $y > f(a)$, then

$$p \cdot y \geq p \cdot f(a) \tag{20}$$

if the individual convergence conclusion is of the form IQi and

$$p \cdot y > p \cdot f(a) \tag{21}$$

if the individual convergence conclusion is of the form IDi. Let $\mathcal{SQ}(f)$ denote the set of supporting prices for f in the sense of equation (20) and $\mathcal{SD}(f)$ denote the set of supporting prices for f in the sense of equation (21).

Comment. The use of a supporting price plays a critical role in rate of convergence results [Debreu (1975), Grodal (1975), Cheng (1981, 1982, 1983a), Anderson (1987), Geller (1987) and Kim (1988)] and other applications of differentiable methods [see Mas-Colell (1985)].

(c) **E3** (Equilibrium price). If the individual convergence conclusion is of the form IQi, conclusion E3 is that the price p is a Walrasian quasiequilibrium price, i.e. $p \in Q(\chi)$. If the individual convergence conclusion is of the form IDi, conclusion E3 is that the price p is a Walrasian equilibrium price, i.e. $p \in \mathcal{W}(\chi)$.

3. Uniformity conclusions

The uniformity conclusions operate jointly with the individual convergence conclusions and the equilibrium conclusions on price. The conclusion triple (Ii, Ej, Um) holds if the following is true: given any $\epsilon > 0$ and any $\delta > 0$, there exists $n_0 \in N$ such that for $n > n_0$ and $f \in \mathcal{C}(\chi_n)$, there exists

$$p \in \begin{cases} \Delta & \text{if } j = 1, \\ \mathcal{Q}_\delta(\chi) & \text{if } j = 2A \text{ and } i = Q. \ldots, \\ \mathcal{W}_\delta(\chi) & \text{if } j = 2A \text{ and } i = D. \ldots, \\ \mathcal{SQ}(f) & \text{if } j = 2S \text{ and } i = Q. \ldots, \\ \mathcal{SD}(f) & \text{if } j = 2S \text{ and } i = D. \ldots, \\ \mathcal{Q}(\chi) & \text{if } j = 3 \text{ and } i = Q. \ldots, \\ \mathcal{W}(\chi) & \text{if } j = 3 \text{ and } i = D. \ldots \end{cases} \tag{22}$$

such that

(a) **U1** (Convergence in measure).

$$\frac{|\{a \in A_n : \rho_i(f, a, p) > \epsilon\}|}{|A_n|} < \epsilon. \tag{23}$$

Comment. Convergence in measure says that, at a core allocation in a large economy, *most* agents have consumption vectors that are close (in the sense specified by the individual convergence conclusion) to demand.

(b) **U2** (Convergence in mean).

$$\frac{\Sigma_{a \in A_n} \rho_i(f, a, p)}{|A_n|} < \epsilon. \tag{24}$$

Comments. Convergence in mean is stronger than convergence in measure. It asserts that the *average* deviation (in the sense specified by the individual convergence conclusion) is small. The difference between convergence in measure and mean is closely connected to the difference between conclusion pairs (ID4N, E1) (near demands for some price) and (ID4N, E2) (near demands for an approximate equilibrium price): U2 and ID4N imply E2, but U1 and ID4N need not imply E2.

(c) **U3** (Uniform convergence).

$$\rho_i(f, a, p) < \epsilon \quad \text{for all } a \in A_n. \tag{25}$$

Comment. Uniform convergence is stronger than convergence in measure and convergence in mean. It is the only conclusion that asserts that *all* individuals are close (in the sense specified by the individual convergence conclusion) to demand.

4. Rate of convergence conclusions

An entry of "$1/n$" in the tables indicates that the rate at which convergence (in the sense specified by the individual convergence conclusion and the uniformity conclusion) occurs is $O(1/|A_n|)$. Similarly, an entry of "$1/n^2$" indicates the rate is $O(1/|A_n|^2)$. An entry "$1/n^{2-\epsilon}$" indicates that, for every $\epsilon > 0$, the rate is $O(1/|A_n|^{2-\epsilon})$.

5. Most economies

These conclusions describe various formulations of the notion that convergence holds for most sequences of economies. The three formulations of this notion (probability one in replica sequences, probability one in random economies, and topological) are incomparable.

   (a) (Probability one in replica sequences). Consider a finite economy, with fixed preferences, but with the endowments allowed to vary. We will replicate this economy.

      (i) **MR1** (Weak law of large numbers). Fix the social endowment in the unreplicated economy, and consider possible reallocations of the social endowment among the types. Conclusion MR1 holds if, for all $\epsilon > 0$, the measure of the set of endowment reallocations for which the uniformity conclusion fails in the n-fold replica tends to zero as $n \to \infty$. For example, the conclusion triple (ID4N, E3, U3) holds in conjunction with MR1 if, for all $\epsilon > 0$, the measure of the set of endowment reallocations such that for some f in the core of the n-fold replica, for every $p \in \mathcal{W}(\chi)$, $\rho_{D4N}(f, a, p) > \epsilon$ for some $a \in A_n$, tends to 0 as $n \to \infty$.

      (ii) **MR2** (Strong law of large numbers). Conclusion MR2 holds if, except for a set of endowments of Lebesgue measure zero, the resulting replica sequence converges in the sense specified by the other convergence conclusions.

   (b) **MP** (Probability one in random economies). Conclusion MP holds if the conclusion is true with probability one with respect to a certain distribution over sequences of economies. Specifically, we consider an arbitrary measure μ on the space of agents' characteristics $\mathcal{P} \times R_+^k$ such that $0 \ll \int e \, d\mu \ll \infty$. We then form a (random) sequence of economies by sampling with replacement from this measure. Let ω_n be the nth sample. Now let $A_n = \{1, \ldots, n\}$ and $\chi_n: A_n \to \mathcal{P} \times R_+^k$ be defined by $\chi_n(i) = \omega_i$. Conclusion MP holds if convergence (in the sense specified by the other conclusions) holds with probability one in the space of sample sequences.

     Comment. The formulation of this assumption implies that assumptions B3 (uniform integrability) and D1 (tightness) hold with probability one.

   (c) **MT1** (Residual). Conclusion MT1 holds if the form of convergence specified by the other conclusions holds for all sequences of economies converging to limit economies in a residual set (i.e. the complement of a countable union of nowhere dense sets; see Royden (1968).

(d) **MT2** (Open dense). The space of distributions of characteristics can be given a metric topology, as discussed in the subsection on Distributional Assumptions above. Conclusion MT2 holds if the form of convergence specified by the other conclusions holds for all sequences of economies converging to limit economies in an open dense set.

Comment. The topological and probabilistic notions of "most" economies are not comparable. The topological notion makes sense on spaces (including the space of preferences $\mathscr{P}$) on which there are no natural candidates for a canonical measure. The justification for this as an appropriate notion of "most" economies comes from the Baire Category Theorem [Royden (1968)] and from an argument about stability under perturbations. Note, however, that open dense sets in R^n may have arbitrarily small (though positive) Lebesgue measure. The notion of a residual set is weaker than that of an open dense set; its justification as a notion of "most" economies also comes from the Baire Category Theorem.

5. Survey of convergence results

In assessing convergence results for cores of large finite economies, we should keep in mind three motivations for the study of the core. The first two relate to what the core convergence results tell us about Walrasian equilibrium, and are normative in character. The fact that Walrasian allocations lie in the core is an important strengthening of the first theorem of welfare economics, which asserts that Walrasian allocations are Pareto optimal. This is a strong stability property of Walrasian equilibrium: no group of individuals would choose to upset the equilibrium by recontracting among themselves. It has a further normative significance. If we are satisfied that the distribution of initial endowments has been done in an equitable manner, no group can object that it is treated unfairly at a core allocation. Since Walrasian allocations lie in the core, they possess this desirable group fairness property. Remarkably, this strengthening of the first welfare theorem requires no assumptions on the economy: it follows directly from the definition of Walrasian equilibrium.

The second motivation concerns the relationship of the core convergence theorems to the second welfare theorem. The second welfare theorem asserts, under appropriate hypotheses, the any Pareto-optimal allocation is a Walrasian equilibrium for some redistribution of endowments. The core convergence theorems assert that core allocations of large economies are nearly Walrasian (in the senses discussed in the previous section), *without* any necessity for redistribution of endowments. This is a strong "unbiasedness" property of Walrasian equilibrium: if a social planner were to insist that only Walrasian outcomes were to be permitted, that insistence by itself would not substantially narrow the range of possible outcomes beyond the narrowing that occurs in the core. The insistence would have no hidden implications for the welfare of different groups beyond whatever equity issues arise in the initial endowment distribution. Indeed, assuming that the distribution of endowments is equitable, any allocation that is far from being Walrasian will not be in the core, and hence will treat some group unfairly.

The extent to which this unbiasedness property is compelling depends largely on which of the individual convergence conclusions IQ1–ID5 and equilibrium conclusions on price E1–E3 hold. The unbiasedness property as stated above in words corresponds to conclusions ID2 or ID4N (individual allocations are near to demands, either in utility or in consumption), and E3 (the demands are taken with respect to a Walrasian equilibrium price). We shall see that the combination of E3 with ID2 or ID4N occurs only under rather strong assumptions. Conclusion E2 (the price is one in which excess demand *almost* equals 0) is must easier to obtain, and appears to the author to be nearly as strong. It is quite plausible that markets never exactly clear; rather, at any given time, excess demand (viewed as a flow) is close to 0. The excess demand flow is accommodated by inventory adjustments for a time, until such adjustments can no longer be made. At that point, the market will switch to a new approximate equilibrium price. To the extent that this story captures what really happens in a market economy, conclusion E2 is sufficient (in combination with ID2 or ID4N) to justify the unbiasedness claim for Walrasian equilibrium. However, in situations in which only E1 is provable, the unbiasedness claim cannot be justified by the formal result.

One should be cautious about interpreting the support for Walrasian equilibrium provided by the two arguments as supporting the desirability of allowing the "free market" to operate. Implicit in the definition of Walrasian equilibrium is the notion that economic agents act as price-takers. If this assumption were false, then the theoretical advantages of Walrasian allocations would shed little light on the policy issue of whether market or planned economies produce more desirable outcomes. The fact that prices are used to equate supply and demand does not guarantee that the result is Walrasian: an agent possessing market power may choose to supply quantities different from the competitive supply for the prevailing price, thereby altering that price and leading to an outcome that is not Pareto optimal. This positive issue, whether we expect the allocations produced by the market mechanism to exhibit price-taking behavior, provides the third motivation for the core convergence results.

Edgeworth (1881), criticizing Walras (1874), took the view that the core, rather than the set of Walrasian equilibria, was the best description of the possible allocations that the market mechanism could produce. In particular, the definition of the core does not impose the assumption of price-taking behavior made by Walras. Furthermore, if any allocation not in the core arose, some group would find it in its interests to recontract. Edgeworth thus argues that the core is the significant positive equilibrium concept.

Taking Edgeworth's point of view, a core convergence theorem with either of the individual convergence conclusions ID2 and ID4N can be viewed as a justification of the price-taking assumption. Any allocation produced by the market mechanism will lie in the core. Consequently, the utility level (with ID2) or the consumption (with ID4N) will be close to that afforded by the competitive demands at the market-clearing price. In short, the exploitation of market power gives rise to little change in the outcome. Furthermore, the incentive to depart from price-taking behavior is sufficiently small that it may well be overwhelmed by transactions costs or costs of acquiring information; this question, though, is best studied in the context of non-cooperative game theory.

The core convergence theorem thus provides a positive argument in favor of the price-taking assumption. Note, however, that the boundedness assumptions B1–B4 enter into this is an important way. Whether the core convergence theorems can be viewed as providing support for the price-taking assumption in a given real economy depends in a subtle way on the relationship of the distribution of endowments to the number of agents; furthermore, this conclusion becomes more delicate as the boundedness assumption is strengthened. Edgeworth's view was that the presence of firms, unions, and other large economic units makes the core substantially large than the set of Walrasian equilibria, a view the author shares.

We now present in tabular form the principal results on core convergence. In the tables, we describe the assumptions required and the conclusions of the theorems by indicating which assumptions and conclusions in the taxonomy developed in the previous two sections most closely approximate the statement of the theorem. For a full statement of any given theorem, the reader should refer to the reference given.

Simplicity would require giving only the best results, eliminating those that represented important stages in the line of discovery, but have now been superseded. Fairness to the many contributors would require listing all the intermediate results which led to the later discoveries. We have chosen a middle route, which lists some of the intermediate results that were most important in the evolution. Each table lists a group of theorems of similar type; within each table, earlier results are listed first. We also provide a table of known counterexamples.

5.1. Non-convex preferences: Demand-like theorems

Historically, the study of convergence properties of the core in sequences of finite economies began with the study of sequences satisfying very special

Table 1
Demand-like theorems with non-convex preferences

Assumptions			Conclusions		
Preference endowment	Sequence	Individual	Uniform	Methodology and references
				U1	Vind (1965) (elementary). Vind's Lemma does not fit well in our taxonomy. The individual convergence conclusion is weaker than IQ1, and only applies to individuals with bounded endowments.
C1 T				U2	Arrow and Hahn (1971); this theorem does not fit well in our taxonomy. A particularly strong version of bounded non-convexity is required. See also Nishino (1970).
M4 P2	D2 B4				Brown and Robinson (1974), under the endogenous assumption that core allocations are uniformly bounded; Brown and Khan (1980) removed this endogenous assumption.
M4 T P1	D2 B4	IQ1 E1	U1	Khan (1974) (non-standard analysis); Khan also gives convergence theorems for approximate cores.
	D1 B3				Grodal and Hildenbrand (1973); for a published version, see Hildenbrand (1974, Theorem 3, p. 202) (weak convergence).
M3	B1		U2	E. Dierker (1975) and Anderson (1978) (elementary); see also Keiding (1974). Anderson's proof arose from a nonstandard proof, after Khan and Rashid solved a key technical problem. In hindsight, Anderson's proof is connected to the proof of Arrow and Hahn (1971). Manelli (1991) has shown that M3 cannot be weakened to M2.
M4 T P1	D2 DI1 B4		U3	Cheng (1983b) (elementary); the non-isolated condition is stronger than DI1 and the conclusion is weaker than IQ1.

Note: See also Table 3.

properties, and proceeded to consider increasingly general sequences. In the light of developments of the late 1970s, it is more efficient to proceed in the opposite direction. Theorems that hold for very general economies are presented in Table 1. The conclusions of these theorems are quite weak; however, these theorems can be used to give efficient derivations of the stronger convergence conclusions that follow from stronger assumptions.

We begin with the statement of a result due to E. Dierker (1975) and Anderson (1978).

Theorem 5.1. *Suppose* $\chi: A \to \mathcal{P} \times \mathbf{R}_+^k$ *is an exchange economy satisfying M3 and P1, and such that, for each* a, $>_a$ *satisfies the following free disposal condition:*

$$x \gg y, \ y > z \Rightarrow x > z . \tag{26}$$

If $f \in \mathcal{C}(\chi)$, *then there exists* $p \in \Delta$ *such that*

$$\sum_{a \in A} \rho_{Q1}(f, a, p) \leqslant 4k \max\{\|e(a)\|: a \in A\} . \tag{27}$$

The proof involves the following main steps [see Anderson (1978) for details]:

(1) Suppose $f \in \mathcal{C}(\chi)$. Define $\gamma(a) = \{x - e(a): x >_a f(a)\} \cup \{0\}$, $\Gamma = \Sigma_{a \in A} \gamma(a)$, $-\Omega = \{x \in \mathbf{R}^k: x \ll 0\}$. It is easy to check that $f \in \mathcal{C}(\chi) \Rightarrow \Gamma \cap (-\Omega) = \emptyset$.

(2) Let $z = (\max\|e(a)\|_\infty, \ldots, \max\|e(a)\|_\infty)$. Use the Shapley–Folkman Theorem to show that

$$(\mathrm{con} \ \Gamma) \cap (-z - \Omega) = \emptyset . \tag{28}$$

(3) Use Minkowski's Theorem to find a price $p \neq 0$ separating Γ from $-z - \Omega$.

(4) Verify that $p \geqslant 0$ and p satisfies the conclusion of the theorem.

5.2. Strongly convex preferences

Theorem 5.1 can be used as a first step is proving stronger conclusions for sequences of economies satisfying stronger hypotheses, notably strong convexity. The results in Table 2 can be proved using the following argument [see Anderson (1981a) for details]:

Table 2
General sequences with strongly convex preferences

Assumptions			Conclusions		
Preference endowment	Sequence	Individual	Uniform	Methodology and references	
C3 M4 T P1	D2 B3	ID4N E2A	U2	Bewley (1973a), building on Kannai (1970) and Vind (1965) (measure theory).	
	D2 DI1 B4		U3	Bewley (1973a) (measure theory).	
C3 M4 P1	D1 B3	ID4N E1	U1	Grodal and Hildenbrand (1973); for a published version, see Hildenbrand (1974, Theorem 1, p. 179) (weak convergence).	
		ID4N E2A	U2	Anderson (1977) (non-standard analysis); a key technical problem has been resolved by Khan and Rashid (1976).	
				Anderson (1981a) (elementary).	
	D1 B1	ID4N E1	U1	Anderson (1981a) (elementary); results without assuming B1 were obtained by Khan (1976) and Trockel (1976); these involve a rescaling of preferences that is hard to place in our taxonomy.	

Note: See also Table 3.

(1) Consider a sequence of economies $\chi_n : A_n \to \mathscr{P} \times \boldsymbol{R}_+^k$ satisfying the hypotheses in Table 2. Suppose $f_n \in \mathscr{C}(\chi_n)$. Verify that the preferences exhibit equimonotonicity and equiconvexity conditions, as discussed under Distributional Assumptions (item 7 of Section 3) above.

(2) Let p_n be the price associated with f_n by Theorem 5.1.

(3) Use the equimonotonicity condition to show that $\{p_n\}$ is contained in a compact subset of Δ^0, i.e. prices of all goods are uniformly bounded away from 0.

(4) Use the boundedness of the prices and the fact that $\rho_{Q1}(f_n, a, p_n)$ is small for most agents to show that there is a compact set which contains $f_n(a)$ for most agents a.

(5) Use the equiconvexity of preferences, the boundedness of $f_n(a)$ for most a, and the fact that $\rho_{Q1}(f_n, a, p_n)$ is small for most agents a to show that $f_n(a)$ is near $D(p_n, a)$ for most agents a.

5.3. Rate of convergence

In assessing the significance of core convergence results for particular economic situations, it is important to know the rate at which convergence occurs, in other words, how many agents are needed to ensure that core allocations are a given distance (in an appropriate metric) from being competitive. Results on the rate of convergence are presented in Table 3.

Debreu (1975) measured the convergence rate in terms of the ID4N-E3 metric, i.e. he measured the distance in the commodity space to the nearest Walrasian equilibrium. Debreu proved a convergence rate of $1/n$ for generic replica sequences; Grodal (1975) extended this result to generic non-replica sequences. It is easy to see from Debreu's proof that this rate is best possible for generic replica sequences with two goods and two types of agents; indeed, if an equal treatment allocation can be improved on by any coalition, it can be improved on by the coalition Debreu considers. Debreu's proof consists of the following main steps:

(1) Consider a sequence of allocations f_n, where f_n is in the core of the n-fold replica of an economy. Let p_n denote the supporting price at f_n. Using the smoothness of the preferences, show that $p_n \cdot (f_n(a) - e(a)) = O(1/n)$, and so $\rho_{Q1}(f_n, a, p_n) = O(1/n)$.

(2) Since p_n is a supporting price, $f_n(a) = D(p_n, a, t_n)$, where $t_n(a) = p_n \cdot (f_n(a) - e(a))$. The non-vanishing Gaussian curvature condition implies that demand is C^1, so $|f_n(a) - D(p_n, a)| = O(|t_n|) = O(1/n)$. Since f_n is an allocation, market excess demand at p_n (in the unreplicated economy) is $O(1/n)$.

(3) For a set of probability one in the space of endowments, the unreplicated economy is regular, i.e. the Jacobian of market demand has full rank at each Walrasian equilibrium. For such endowments, we can find a Walrasian equilibrium price q_n and Walrasian allocation $g_n(a) = D(q_n, a)$ such that $|p_n - q_n|$ is of the order of magnitude of the market excess demand at p_n, so $|p_n - q_n| = O(1/n)$. Using once more the fact that demand is C^1, we have $|f_n(a) - g_n(a)| = O(1/n)$.

There has been considerable progress on the rate of convergence. Debreu's proof shows that the rate of convergence, measured by ρ_{Q1} at the supporting price, is $O(1/n)$. However, Anderson (1987) showed there exist prices for which the rate of convergence (measured by ρ_{Q1}) is $1/n^2$. The main ideas of the proof are as follows:

(1) Consider a sequence of core allocations $f_n \in \mathscr{C}(\chi_n)$, where $\chi_n: A_n \to \mathscr{P} \times R_+^k$ is a sequence of exchange economies, and $|A_n| = n$; let γ_n and Γ_n be derived from f_n is the same way that γ and Γ are derived from f in the proof of Theorem 5.1. Let p_n be the price vector which minimizes $|\inf p_n \cdot \Gamma_n|$; this is called the gap-minimizing price. Let $g_n(a) = \operatorname{argmin}(p_n \cdot \gamma_n(a))$. Notice that p_n is a supporting price at $\gamma_n(a)$, *not* at f_n.

Table 3
Rate of convergence

Assumptions		Conclusions				Methodology and References
Preference endowment	Sequence	Individual	Uniform	Most	Rate	
C3 S SB M4 T P2	D4 B4	ID4N E3	U3	MR2	$\frac{1}{n}$	Debreu (1975) (differential topology); this is the best possible generic rate for replica sequences with two goods and two types of agents; for (non-generic) examples with slow convergence, see Shapley (1975) and Aumann (1979).
	D2 B4			MT2		Grodal (1975) (differential topology, measure theory).
M3 P1	B4	IQ1 E1	U2	All		E. Dierker (1975) and Anderson (1978) (elementary).
C3 S M4 T P2	D3 B4	ID4N E3	U3	MR2		Cheng (1981) (differential topology); a mild "indecomposability condition" is also assumed.
C3 S M4 P1	D2 B4	ID4N E2A	U2	All	$\frac{1}{\sqrt{n}}$	Anderson (1981a) and Cheng (1983a) (elementary); preferences may have kinks in the interior of the consumption set. Cheng (1983a) has shown this rate is best possible if there are kinks in the preferences or if the indecomposability condition is violated.
S SB M4 T P2	D1 B4				$\frac{1}{n^{2}}$	Anderson (1987) (elementary); S, SB, M4, T and P2 are only required for a positive fraction of agents; all must satisfy M3.
S M4 T P1		IQ1 E1		MR1	$\frac{1}{n^{4-\epsilon}}$	Kim (1988) (elementary); a mild "indecomposability condition" is also assumed.
C3 S SB M4 T P2	D4 B4	ID4N E3	U3		$\frac{1}{n^{2-\epsilon}}$	Geller (1987) (number theory, differential topology); there are two goods and three or more agent types.

(2) Use the Shapley–Folkman Theorem to verify that one may find a coalition S_n such that $|S_n|/|A_n|$ and $|\Sigma_{a \in S_n} p_n \cdot g_n(a)|/|\inf p_n \cdot \Gamma_n|$ are bounded away from 0 and $\Sigma_{a \in S_n} g_n(a)$ is bounded.

(3) Use the fact that sums of smooth sets become flatter as the number of sets grows, and the fact that $|S_n|$ grows linearly with n, to show that $|\Sigma_{a \in S_n} p_n \cdot g_n(a)| = O(1/n)$, and hence $|\inf p_n \cdot \Gamma_n| = O(1/n)$, then proceed as in the proof of Theorem 5.1.

Geller (1987) provided the first result in which the rate of convergence measured in the sense of Debreu (1975) (the ID4N–E3 metric) is faster than $O(1/n)$. His theorem is of the weak law of large numbers (MR1) form, with a rate $O(1/n^{2-\epsilon})$, provided there are two goods and at least three types of agents. The argument is quite delicate, but the following gives a hint of the main steps.

(1) In a replica sequence with two types of agents, the net trade of one type is the negative of the net trade of the other type. Thus, a candidate improving coalition can be characterized by subtracting the number of agents of the second type from the number of the first type. There are thus $2n + 1$ essentially distinct candidate-improving coalitions in the n-fold replica. The net trades of these coalitions are equally spaced along a line segment of length $O(n)$; in particular, they do not become more closely crowded as $n \to \infty$. As we noted above, with two types of agents, Debreu's $O(1/n)$ rate is the best possible.

(2) Now suppose there are three types of agents and two commodities. The number of essentially distinct candidate-improving coalitions is of order n^2, and these are arranged near a line segment of length $O(n)$. Thus, the average distance between the net trades of adjacent candidate-improving coalitions is $O(n^{-1})$. Using number-theoretic results on lattices with two generators in the real line, one can show that the maximum distance between the net trades of adjacent candidate improving coalitions is $O((\log n)/n)$ with high probability in the space of endowments.

(3) (a) Using Debreu's result, one can show that with high probability, every core allocation is $O(1/n)$ from a Walrasian allocation. Using the lattice results, one can show that with high probability, every allocation within $O(1/n)$ of some Walrasian allocation has the maximum distance between net trades of adjacent candidate improving coalitions of order $O((\log n)/n^{1/2})$.

(b) Use the flattening property of the sums of smooth sets, as in item 3 in the outline of the proof of Anderson (1987), to show that $|\inf p_n \cdot \Gamma_n| = O((\log n))^2/n^2)$. Now, proceed as in Theorem 5.1 and use the equal treatment property to show that

$$\max_{a \in A_n} \rho_{Q1}(f, a, p) = O\left(\frac{(\log n)^2}{n^3}\right), \tag{29}$$

with high probability. Since the distance in the commodity space to the Walrasian equilibrium in generically the square root of the gap measured by ρ_{Q1}, the rate of convergence in Debreu's sense is $O(\log n/n^{3/2})$.

(4) Now iterate items (3a) and (3b), as follows. We know that, with high probability, every core allocation is $O((\log n)/n^{3/2})$ from a Walrasian equilibrium. Use the lattice argument to show that, with high probability, every allocation within $O((\log n)/n^{3/2})$ of a Walrasian allocation has the maximum distance between net trades of adjacent candidate improving coalitions of order $O((\log n)^2/n^{3/4})$. By the argument in item (3b), we find that with high probability, every core allocation is within $O((\log n)^2/n^{7/4})$ of a Walrasian allocation, completing the second iteration. On the mth iteration, we find that with high probability, every core allocation is within $O((\log n)^m/n^{2-2^{-m}})$. Thus, given $\epsilon > 0$, we find the rate is $O(1/n^{2-\epsilon})$ within a finite number of iterations.

5.4. Decentralization by an equilibrium price

An example due to Bewley (1973a) shows that core allocations need not be close to Walrasian equilibria for all sequences of economies, even under strong assumptions on the preferences and endowments. Given smoothness assumptions, however, core allocations are close to Walrasian equilibria generically: for a set of endowment of probability 1 (MR2) in replica sequences [Debreu (1975)] and for an open and dense set of characteristics (MT2) in general sequences [Grodal (1975)]. H. Dierker (1975) showed this conclusion holds even without smooth preferences, at the cost of weakening the notion of genericity to MT1. These results are presented in Table 4.

Table 4
Decentralization by an equilibrium price

Assumptions		Conclusions			
Preference endowment	Sequence	Individual	Uniform	Most	Methodology and references
C3 S SB M4 T P2	D4 B4	ID4N E3	U3	MR2	Debreu (1975) (differential topology)
	D2 B4			MT2	Grodal (1975) (differential topology, measure theory)
C3 SB M4 T P2	D2 DI1 B4			MT1	H. Dierker (1975) (weak convergence, differential topology)
	D2 B4		U1		

Note: See also Table 7.

5.5. Non-convex preferences: Stronger conclusions

While arguments about diminishing marginal utility suffice to indicate that the preference over two goods (the consumption of other goods held constant) should usually be convex, there are nonetheless compelling examples in which convexity fails. For example, having two small apartments (one at location A, the second at location B) may not be as useful as one large apartment at either A or B. In light of the motivation for the study of core convergence given at the beginning of this section it is highly desirable to prove results with individual convergence conclusions stronger than IQ1 and E1 for finite economies with non-convex preferences. These are presented in Tables 5, 6 and 7.

Table 5 presents results about the utility levels achieved by agents at core allocations. The first two entries in the table (which depend on convexity) are included to provide a comparison to the last two entries, which have no convexity requirement. The essential idea is to show that the conclusion of Theorem 5.1, which measures budget deviations in monetary terms, implies convergence of utilities as long as the utility representations are chosen in a reasonable way. Thus, the results in Table 5 give conditions in which it is possible to convert an ϵ expressed in terms of income in the IQ1 convergence conclusion into an ϵ expressed in terms of utility.

Table 6 explores the extent to which it is possible to duplicate the utility levels of a core allocation by those of a Walrasian equilibrium or quasiequilibrium if one first makes small income transfers. Thus, these results place the ϵ solely in the incomes of the agents. These results provide the clearest relationship between core convergence and the Second Welfare Theorem. The Second Welfare Theorem asserts that Pareto optima are Walrasian equilibria or quasiequilibria after income transfers; the results in Table 6 show that, in the case of core allocations, these transfers can be made small. The proofs are elementary, and depend on studying the gap-minimizing price as described in item 1 in the outline of the proof of Anderson (1987) in Subsection 5.3.

Table 5
Utility

Assumptions			Conclusions		
Preference endowment	Sequence		Individual	Uniform	Methodology and references
C2 M4 T P1	D3 B4		ID2 E1	U1	Follows from Hildenbrand and Kirman (1973) (elementary).
	D3 DI1 B4			U3	
M4 T P1	D1 B1			U1	Anderson (1990a) (elementary).
	D2 B4			U2	

Table 6
Income transfers

Assumptions		Conclusions			Methodology and references
Preference endowment	Sequence	Individual	Uniform	Rate	
M3 T		IQ3U E3	U1		
	B2	IQ3U E1	U2		
		ID3U E3	U1		Anderson (1986) (elementary).
M4 T P1		ID3U E1			
	D1 B3		U2		
S SB M4 T P2	D1 B4	ID3U E3		$\frac{1}{n}$	

Table 7
Near demand with non-convex preferences

Assumptions		Conclusions			Methodology and references
Preference endowment	Sequence	Individual	Uniform	Most	
C3 M4 T P1		ID4N E1	U1		Hildenbrand (1974, Theorem 1, p. 179; Example 3, p. 138) (weak convergence); see also Bewley (1973a).
M4 T P1	D1 B3	ID4N E2A	U2	MP	Anderson (1985) (non-standard analysis and simple measure theory).
					Hoover (1989) (simple measure theory).
				MT1	Combine Hildenbrand (1974, Proposition 4, p. 200 and condition (*), p. 201) and Mas-Colell and Neuefeind (1977) (weak convergence and differential topology); for a statement and non-standard proof, see Anderson (1981b, 1985).

Table 7 provides results showing that, for most economies with non-convex preferences, agents' consumptions are close to their demand sets. Thus, these results place the ϵ in the commodity space distance to the demand set (a finer metric than utility or income) and in the space of economies. The first entry (which does require convexity) is included for comparison purposes.

5.6. Non-monotonic preferences

While it has long been recognized that convexity assumptions fundamentally alter the form of core convergence, relatively little attention has been paid to monotonicity. The results of Debreu and Scarf (1963) for replica sequences and Aumann (1964) for non-atomic economies require only local non-satiation (M1) of preferences. While essentially all the known convergence results for non-replica sequences assumed weak monotonicity (M3) or monotonicity (M4), most researchers appear to have thought that the assumption was inessential, and could be removed by a modification of the proofs. Manelli (1991) gave two examples which showed that this is not the case (see Table 10).

In Manelli's first example, we consider a sequence of finite exchange economies $\chi_n : A_n \to \mathscr{P} \times \mathbf{R}_+^2$ with $A_n = \{1, \ldots, n+2\}$. The endowment map is $e(1) = e(2) = 0$, $e(a) = (1, 1)$ $(a = 3, \ldots, n+2)$. Let V denote the cone $\{0\} \cup \{x \in \mathbf{R}_{++}^2 : 0.5 < x^1/x^2 < 2\}$. Consider the allocation

$$f(1) = (n, 0), \qquad f(2) = \left(0, \frac{n}{2}\right), \qquad f(a) = (0, \tfrac{1}{2}) \quad (a = 3, \ldots, n).$$

(30)

The preferences have the property that

$$x >_a f(a) \Leftrightarrow x - f(a) \in V.$$

(31)

It is not hard to see that there are complete, transitive (T) uniformly proper (M2) preferences that satisfy equation (31). It is not hard to verify that $f \in \mathscr{C}(\chi)$. Given $p \in \Delta_+$,

$$\frac{1}{n+2} \sum_{a \in A} |p \cdot (f(a) - e(a))|$$

$$= \frac{n|p^1| + (n/2)|p^2| + n|p^1| + (p^2/2)|}{n+2} \geqslant \frac{n}{2(n+2)} \nrightarrow 0.$$

(32)

Manelli's second example shows that even a uniformly bounded sequence of core allocations may fail to converge in the IQ1–E1 sense unless preferences are convex.

Manelli's examples have forced a reassessment of the core convergence literature. Unless reasonable sufficient conditions to guarantee core convergence in the absence of monotonicity can be found, the economic interpretations discussed in the beginning of this section are open to question. Manelli has provided a number of sufficient conditions, two of which are presented in Table 8.

Table 8
Theorems with non-monotonic preferences

Assumptions		Conclusions		
Preference endowment	Sequence	Individual	Uniform	Methodology and references
C3 M1 T	D4 B4	ID4N	U3	Debreu and Scarf (1963) elementary.
C1 M2 T	B1 DI2	IQ1 E1	U2	Manelli (1990a) elementary.
C3 M2 T	B1 D2 DI2	ID4N E1	U1	

Table 9
Type (including replica) sequences

Assumptions		Conclusions		
Preference endowment	Sequence	Individual	Uniform	Methodology and references
C3 M1 T P2	D4 B4	ID4N E3	U3	Debreu and Scarf (1963) (elementary); see also Debreu and Scarf (1972), Johansen (1978) and Schweizer (1982).
C3 M4 T P1				Hildenbrand and Kirman (1976, Theorem 5.1) (elementary).
	D3 DI1 B4	ID4N E1		Hildenbrand and Kirman (1973, 1976, Theorem 5.2) (elementary).
M4 P1	D4 B4	ID4N E3		Brown and Robinson (1974) (non-standard analysis) and Hildenbrand (1974, Corollary 1, p. 201) (weak convergence). *Note:* The theorem only applies to core allocations with the equal treatment property.
M4 P2	D3 B4	1D4N E1	U1	Follows easily from Brown and Robinson (1974, Theorem 2) (non-standard analysis).
M4 T P1				Hildenbrand (1974, Proposition 4, p. 200) (weak convergence).
M4 P1		ID4N E2A	U2	Anderson (1981b) (elementary).

Note: See also Table 3.

Table 10
Counterexamples

Assumptions		Conclusions				Methodology and references
Preference endowment	Sequence	Individual	Uniform	Most	Rate	
C3 M4 T P1	D2 DI1 B4	IQ4N E3	U1	All		Bewley (1973a).
M4 T P2	D2 DI1	IQ3N E1				Anderson and Mass-Colell (1988).
C3 S SB M4 T P2		IQ4N E3	U3	MR2	$\frac{1}{n^{1+\epsilon}}$	Debreu (1975) if there are two goods and two types of agents.
C3 S M4 P1	D4 B4				$\frac{1}{n^{1/2+\epsilon}}$	Cheng (1983a).
C3 SB M4 T P1			U1		Any rate	Shapley (1975); preferences are C^1.
C3 S SB M4 T P2						Aumann (1979); demands are C^1.
C2 M2 T P1	D2 B4		U2	All		Manelli (1991); core allocations not uniformly integrable.
	D3 B4	IQ1				Manelli (1990b).
M2 T P1	D2 B4	E1				Manelli (1991); core allocations uniformly bounded.
			U1			Manelli (1990b).

5.7. Replica and type sequences

We now turn to results for type sequences of economies, including replica sequences. These are the oldest results on core convergence, and the easiest to prove directly. However, the assumption of a finite number of types is extremely strong. Neither the very strong conclusions nor the original proofs generalize to non-type sequences. Note in particular that, in Debreu and Scarf (1963), only local non-satiation (M1), not weak monotonicity (M3), is required. However, Manelli (1991) has shown that even the weakest forms of core convergence may fail in general sequences of finite economies in the absence of M3.[1] Similarly, the equilibrium conclusion E3 (decentralization by an equilibrium price of the given economy) in Debreu and Scarf (1963) does not readily generalize; the only known theorems giving the conclusion E3 outside the replica context are given in Table 14.2. The results are presented in Table 14.9; results with strongly convex preferences are presented in the top half of the table, while results without convexity assumptions are presented in the bottom half.

5.8. Counterexamples

There are in the literature a large number of counterexamples indicating that results in the preceding tables cannot be further strengthened. These are summarized in Table 10. Each line presents a *false* statement, as demonstrated by a counterexample. The counterexamples of Shapley (1975) and Aumann (1979) show that the rate of convergence can be arbitrarily slow. The demand functions are differentiable in Aumann's example but not in Shapley's.

6. Economies with a continuum of agents

Economies with a continuum of agents were introduced by Aumann (1964) as an idealization of the notion of an economy with a "large" number of agents, much as continuum models are used in physics to describe the properties of large systems of interacting molecules or particles. Instead of a finite set, we take the set of traders A to be an atomless probability space, such as the unit interval $[0, 1]$ endowed with the Lebesgue measure structure. The required techniques from measure theory are described in Hildenbrand (1974), Kirman (1982), and Mas-Colell (1985).

Definition 6.1. (1) A *non-atomic exchange economy* is a function χ: $A \to \mathcal{P} \times \mathbf{R}_+^k$ where
  (a) $(A, \mathcal{A}, \mu)$ is an atomless probability space;
  (b) χ is measurable, where $\mathcal{P}$ is given the metric associated with the topology of closed convergence [Hildenbrand (1974) or Mas-Colell (1985)];
  (c) $0 \ll \int_A e(a)\, d\mu \ll (\infty, \ldots, \infty)$.

---

[1]Indeed, Manelli (1990b) has even constructed a *replica* sequence of economies (with non-convex, non-monotone preferences) where convergence in the weak IQ1–E1 sense fails. In this example, however, the core allocations do converge in the commodity space to the Walrasian equilibrium allocations (i.e. convergence is in the ID4N–E3 sense).

(2) An *allocation* is an integrable function $f: A \rightarrow R^k_+$ such that $\int_A f(a)\,\mathrm{d}\mu = \int_A e(a)\,\mathrm{d}\mu$.

(3) A *coalition* is a set $S \in \mathscr{A}$ with $\mu(S) > 0$.

(4) A coalition S can *improve* on an allocation f if there exists an integrable function $g: S \rightarrow R^k_+$ such that $g(a) >_a f(a)$ for almost all $a \in S$ and $\int_S g(a)\,\mathrm{d}\mu = \int_S e(a)\,\mathrm{d}\mu$;

(5) The *core* of χ, denoted $\mathscr{C}(\chi)$, is the set of all allocations which cannot be improved on by any coalition;

(6) A *Walrasian equilibrium* is a pair (f, p) where f is an allocation, $p \in \Delta$, and $f(a) \in D(p, a)$ for almost all $a \in A$; $\mathscr{W}(\chi)$ denotes the set of Walrasian equilibrium prices;

(7) A *Walrasian quasiequilibrium* is a pair (f, p) where f is an allocation, $p \in \Delta$, and $f(a) \in Q(p, a)$ for almost all $a \in A$; $\mathscr{Q}(\chi)$ denotes the set of Walrasian quasiequilibrium prices.

It is easy to show that if (f, p) is a Walrasian equilibrium of a non-atomic exchange economy, then $f \in \mathscr{C}(\chi)$. The proof is essentially the same as that of Theorem 2.1.

The key mathematical result underlying core theory in the continuum model is Lyapunov's theorem [Hildenbrand (1974) or Mas-Colell (1985)], which asserts that the range of any measure defined on an atomless measure space and taking values in R^k is convex. As a consequence of Lyapunov's Theorem, one can show under very mild assumptions that the core of a continuum economy *coincides* with the set of Walrasian equilibria.

Theorem 6.2. *Suppose $\chi: A \rightarrow \mathscr{P} \times R^k_+$ is an exchange economy, where $>_a$ satisfies local non-satiation (M1) for almost all $a \in A$.*

(1) *If $f \in \mathscr{C}(\chi)$, then there exists $p \neq 0$ such that (f, p) is a Walrasian quasiequilibrium.*

(2) *If in addition $>_a$ satisfies monotonicity (M3) for almost all $a \in A$, then $p \geqslant 0$ and (f, p) is a Walrasian equilibrium.*

One can prove item (2) following essentially the same steps as those for Theorem 5.1, substituting Lyapunov's Theorem for the Shapley–Folkman Theorem and making use of some advanced measure-theoretic results such as Von Neumann's Measurable Selection Theorem. Item (1) follows from Aumann's original proof, which is more like the proof of the Debreu–Scarf Theorem [Debreu and Scarf (1963)] than that of Theorem 5.1.

In order to compare the results in the continuum with those in large finite economies, Table 11 places Aumann's Theorem within our taxonomy of results.

Table 11
Economies with a continuum of agents

Assumptions		Conclusions		
Preference endowment	Sequence	Individual	Uniform	Methodology and references
M1	B3	1Q5 E3	U2	Aumann (1964) (measure theory)
M3		ID5 E3		

The reader will be struck by the contrast between the simplicity of the table for the continuum case and the complexity of the tables in the asymptotic finite case. It is particularly worthwhile comparing the continuum table with the table of counterexamples for the asymptotic case; the complex relationship between the assumptions and the conclusions found in the large finite context is entirely lost in the continuum.

To understand the divergence in behavior between large finite economies and measure space economies, it is useful to examine how the purely technical assumptions implicit in the measure space formulation may in fact correspond to assumptions with economic content in sequences of finite economies.

(1) *Integrability of endowment.* The assumption that the endowment map in the measure space economy is integrable corresponds to the assumption that the endowment maps in a sequence of economies are uniformly integrable (B2). In particular, sequences like the tenant farmer economies described following the definition of condition (B2) are ruled out. While Khan (1976) and Trockel (1976) (using non-standard analysis and measure theory, respectively) weakened the uniform integrability assumption by altering the underlying measure on the set of agents, neither result encompasses the tenant farmer sequence. However, the tenant farmer sequence does satisfy the hypotheses of E. Dierker (1975) and Anderson (1978).

(2) *Measurability of preference map.* At first sight, measurability of the map which assigns preferences to agents is a purely technical assumption. However, it carries the implication that the sequence of preference maps is tight, i.e. given $\epsilon > 0$, there is a compact set K of preferences so that $\mu(\{a \in A: >_a \in K\}) > 1 - \epsilon$. Of course, the set of continuous preferences is compact in the topology of closed convergence. However, the subset consisting of monotone preferences (M3) is not compact, so the assumption that the preference map is measurable combined with the assumption that almost every agent has a monotone preference has economic content; it corresponds to an "equimonotonicity" condition on sequences of finite economies, as discussed under Distributional Assumptions in Section 3 above. Note further that the topology of closed convergence heavily discounts the behavior of preferences with respect to large consumptions. If large consumptions are important

because of large endowments or a failure of monotonicity, the topology is too coarse to permit the analysis of the core; this is a key reason for the discrepancy between conclusion (1) in theorem 6.2 (which requires only locally non-satiated preferences) and the non-convergence examples of Manelli (1991). However, strengthening the topology to avoid this discounting of large consumptions would make the topology highly non-compact, and would thus make measurability of the preference map a strong assumption.

(3) *Integrability of allocations.* If preferences are not "equimonotone" (see Distributional Assumptions in Section 3, above), then core allocations in sequences of finite economies may fail to be uniformly integrable. Such allocations do not correspond to an integrable allocation in the measure space limit economy. Thus, restricting attention to integrable allocations amounts, from the perspective of sequences of finite economies, to a strong endogenous assumption. This is the second key factor explaining the discrepancy between (1) in Theorem 6.2 and the examples of Manelli (1991).

(4) *Integrability of coalitional improvements.* Just as the integrability requirement on allocations can make the core of the measure space economy smaller than the cores of sequences of finite economies, the requirement that an improving allocation for a coalition be integrable imposes a restriction on coalitions that is not present in sequences of finite economies, potentially making the core of the measure space economy bigger than the cores of sequences of finite economies. Example 4.5.7 in Anderson (1991) provides just such an example. A sequence of finite economies χ_n is constructed, with the endowment e_n uniformly bounded. e_n is not Pareto optimal; however, any Pareto-improving allocation g_n is necessarily not uniformly integrable. In the limit non-atomic economy, the endowment map is a Walrasian equilibrium, and so in particular is in the core; no coalition can improve on it because doing so would require a non-integrable reallocation of consumption.

(5) *Failure of lower hemicontinuity of demand.* There is a sharp discrepancy between the major role played by convexity in the large finite context and its total irrelevance in non-atomic exchange economies, as can be seen by comparing Tables 1, 2, 6, 7, 8 and 9 with Table 11. If $\rho_{Q1}(f, a, p) = 0$ [see condition (I1) in Section 4 above], then $f(a) \in Q(p, a)$. In a non-atomic exchange economy, Lyapunov's Theorem asserts the exact convexity of a certain set, which then guarantees that if f is a core allocation, then $\rho_{Q1}(f, a, p) = 0$ almost surely, and hence core allocations are Walrasian quasiequilibria. In the large finite context, the Shapley–Folkman theorem asserts that the analogous set is approximately convex, leading to the conclusion that $\rho_{Q1}(f, a, p)$ is small for most agents. However, since neither the quasidemand correspondence nor the demand correspondence are lower hemicontinuous, knowing that $\rho_{Q1}(f, a, p)$ is small does not guarantee that $f(a)$ is near the demand or quasidemand of agent a. Anderson and Mas-Colell

(1988) provide an example of a sequence of economies and core allocations in which all agents' consumptions are uniformly bounded away from the agents' demand correspondences, even if one allows income transfers.

7. Non-standard exchange economies

Non-standard analysis provides an alternative formulation to Aumann's continuum model for the notion of a large economy. A *hyperfinite exchange economy* is an exchange economy in which the set of agents is hyperfinite, i.e. it is uncountable, but possesses all the formal properties of a finite set of agents. Thus, the core of such an economy can be defined exactly as in the case of a finite exchange economy.

A construction known as the Loeb measure [Loeb (1975)] permits one to convert the hyperfinite set of agents into an atomless measure space, in a way which converts summations into integrations. It is a consequence of Aumann's Theorem on economies with a continuum of agents that every core allocation of a suitable hyperfinite exchange economy is close to a Walrasian allocation of the associated Loeb measure economy.

A powerful result known as the Transfer Principle asserts that every property formalizable in a certain language which holds for hyperfinite exchange economies holds for sufficiently large finite economies. Thus, the derivation of limit results for finite exchange economies from results for continuum economies comes almost for free. Where the properties of continuum economies diverge from those of large finite economies (as discussed at the end of Second 6), the hyperfinite exchange economy will always reflect the behavior of large finite economies. Indeed, given a sequence of finite economies χ_n, let χ be the corresponding hyperfinite economy. By examining the relationship of χ to the corresponding Loeb measure economy, one can see the exact reason why the measure space limit economy fails to capture the behavior of the large finite economies χ_n. For a detailed treatment of hyperfinite exchange economies, including their use to derive limit theorems for large finite economies and a comparison with economies with a measure space of economic agents, see Anderson (1991).

8. Coalition size and the f-core

There is an extensive literature on the core where the size of coalitions is restricted. Schmeidler (1972), Grodal (1972), and Vind (1972) showed that the core of a non-atomic exchange economy does not change if one restricts coalitions in any of the following ways:

(1) considering only coalitions S with $\mu(S) < \epsilon$, where $\epsilon \in (0, 1]$;

(2) considering only coalitions S with $\mu(S) < \epsilon$, where the characteristics of the agents in S are taken from at most $k + 1$ balls of radius less than ϵ, where k is the number of commodities;

(3) considering only coalitions S with $\mu(S) = \alpha$, where $\alpha \in (0, 1]$.

The proof make use of Lyapunov's Theorem.

Mas-Colell (1979) gave an asymptotic formulation of these results for sequences of large finite economies. He showed that, given $\epsilon > 0$, there exists $m \in N$ such that for sufficiently large economies, any allocation which cannot be improved on by a coalition with m or fewer members must be ϵ-competitive in the IQ1–E1 sense. Chae (1984) studied the core in overlapping generations economies where only finite coalitions are allowed.

The f-core of a non-atomic exchange economy was developed by Kaneko and Wooders (1986, 1989) and Hammond, Kaneko and Wooders (1989). It is intended to model situations in which trades are carried out only within finite groups of agents. The definition involves a delicate mixing of notions from finite and non-atomic economies, but the essential idea is as follows.

(1) An f-allocation is, roughly speaking, an allocation which can be achieved by partitioning the economy into coalitions each of which consists of only a finite number of agents and allowing trade only within the coalitions.

(2) A coalition S can f-improve on an f-allocation f if there exists an improving allocation $g: S \to R^k_+$ which is an f-allocation for the subeconomy consisting of the agents in S.

(3) The f-core consists of all those f-allocations which cannot be f-improved on.

In the presence of externalities, there is no natural definition for the core in the spirit of Aumann's definition for a nonatomic economy. Moreover, the First Welfare Theorem may fail: Walrasian allocations are typically not Pareto optimal.

The f-core provides a suitable alternative to the core for modelling situations with widespread externalities. A widespread externality occurs if the utility of each agent depends on the agent's consumption and the distribution of consumption in the economy as a whole, but not on the consumption of any other individual agent. Since the consumption of a finite coalition does not affect the distribution of consumption in the non-atomic economy, it is impossible for a finite coalition to internalize a widespread externality. Hence, allocations in the f-core are typically not Pareto optimal. Hammond, Kaneko and Wooders (1989) proved the equivalence of the f-core and the set of Walrasian allocations in the presence of widespread externalities; Kaneko and Wooders (1989) used this to derive an asymptotic convergence theorem for large finite economies.

9. Number of improving coalitions

Consider a finite exchange economy $\chi: A \to \mathscr{P} \times R_+^k$ and a Pareto-optimal allocation f. Note that it is not possible for both a coalition S and its complement $A \setminus S$ to improve on f, for then $A = S \cup (A \setminus S)$ could improve on f, so f would not be Pareto optimal. Thus, at most half the coalitions can improve on a given Pareto-optimal allocation.

Mas-Colell (1978) proved under smoothness assumptions that if f_n is a sequence of allocations for $\chi_n \to \mathscr{P} \times R_+^k$ and f_n is bounded away from being competitive in the ID4T–E2S sense, then the proportion of coalitions in A_n which can improve on f_n tends to $\frac{1}{2}$.

10. Infinite-dimensional commodity spaces

Infinite-dimensional commodity spaces arise naturally in many economic problems.

(1) The space $\mathscr{M}([0, 1])_+$ of countably additive finite non-negative Borel measures on $[0, 1]$, endowed with the topology of weak convergence, is the natural space for the study of commodity differentiation [Mas-Colell (1975), Jones (1984)].

(2) The spaces $L^p([0, 1])_+ (1 \le p \le \infty)$ consisting of non-negative measurable functions[2] $X: [0, 1] \to R$ satisfying

$$\int_{[0,1]} X(t)^p \, d\mu < \infty \ (1 \le p < \infty) \,,$$

$$\exists M \in N \ \mu(\{t: X(t) > M\}) = 0 \quad (p = \infty) \,, \tag{33}$$

where μ is Lebesgue measure; $L^\infty([0, 1])_+$ and $L^2([0, 1])_+$ are natural commodity spaces for situations involving uncertainty [Gabscewicz (1968), Zame (1986), Ostroy and Zame (1988), Mertens (1990)], with $L^2([0, 1])_+$ being particularly natural for applications in finance which use Brownian motion or normally distributed random variables.

(3) The spaces $l_+^p \ (1 \le p \le \infty)$ of non-negative real sequences satisfying

$$\sum_{i=1}^\infty (x^i)^p < \infty \quad (1 \le p < \infty) \,,$$

$$\sup x^i < \infty \quad (p = \infty) \,, \tag{34}$$

[2]More precisely, we take equivalence classes of such functions, where functions are equivalent if they are equal almost surely.

are natural spaces for studying consumption over an infinite time horizon [Bewley (1973b)].

(4) The space $C(X)_+$ of bounded continuous real-valued, non-negative functions on a compact Hausdorff space X (in particular $X = [0, 1]$) is useful because many infinite-dimensional spaces can be represented as $C(X)$ spaces for an appropriate choice of X [Gabszewicz (1968)].

The problem of existence of Walrasian equilibrium in economies with infinite-dimensional commodity spaces and a first number of agents has been extensively studied [Aliprantis, Brown and Burkinshaw (1989), Mas-Colell and Zame (1991)]. In an economy with a continuum of agents and an infinite-dimensional commodity space, the equivalence of the core and the set of competitive equilibria may fail, as shown in the following example:

Example 10.1. (1) Let $\mathcal{M}([0, 1])$ denote the space of (signed) Borel measures on the interval $[0, 1]$ endowed with the norm topology generated by

$$\|\mu\| = \sup\{|\mu(B)| + |\mu([0, 1]\backslash B)|: B \in \mathcal{B}\}, \tag{35}$$

where $\mathcal{B}$ denotes the σ-algebra of Borel subsets of $[0, 1]$. We let $\mathcal{M}([0, 1])_+$ denote the cone of non-negative measures in $\mathcal{M}([0, 1])$. We consider an exchange economy $\chi: [0, 1] \rightarrow \mathcal{P}(\mathcal{M}([0, 1])_+) \times \mathcal{M}([0, 1])_+$, where $\mathcal{P}(\mathcal{M}([0, 1])_+)$ denotes the space of continuous preferences on $\mathcal{M}([0, 1])_+$.

(2) Each agent $a \in [0, 1]$ has a preference relation given by a utility function

$$u_a(\mu) = 2\mu(\{a\}) + \mu([0, 1]\backslash\{a\}); \tag{36}$$

in other words, agent a has marginal utility of consumption 2 on his/her "birthday" a, and marginal utility 1 at all other dates. Each agent's endowment is Lebesgue measure, hereafter denoted λ.

(3) This economy has a unique Walrasian equilibrium. The price is the linear functional $p: \mathcal{M}([0, 1])_+ \rightarrow \mathbf{R}_+$ defined by $p(\mu) = \mu([0, 1])$, while the Walrasian allocation is $f(a) = \delta_a$, where δ_a denotes the point mass at a [i.e. $\delta_a(B) = 1$ if $a \in B$, $\delta_a(B) = 0$ if $a \notin B$].

(4) Consider the following allocation:

$$F(a) = \begin{cases} \delta_a + \frac{1}{4}\nu & \text{if } a \in [0, \frac{1}{2}], \\ \frac{7}{8}\delta_a & \text{if } t \in (\frac{1}{2}, 1], \end{cases} \tag{37}$$

where $\nu(B) = \lambda(B \cap (\frac{1}{2}, 1])$. Notice that this produces utility levels $u_a(F(a)) = 2\frac{1}{8}$ if $a \in [0, \frac{1}{2}]$ and $1\frac{3}{4}$ if $a \in (\frac{1}{2}, 1]$.

(5) We claim that F lies in the core of χ. Consider a potential improving coalition $S \subset [0, 1]$. If $G: S \rightarrow \mathcal{M}([0, 1])_+$ improves on F, feasibility implies that

$$\int_S u_a(G(a))\, d\lambda \le \lambda(S)[2\lambda(S) + 1(1 - \lambda(S))] = \lambda(S)[1 + \lambda(S)]. \tag{38}$$

Since we must have $u_a(G(a)) > 1\frac{3}{4}$ almost surely, it follows that $1 + \lambda(S) > 1\frac{3}{4}$, so $\lambda(S) > \frac{3}{4}$. The set of feasible utility payoffs to a coalition depends only on the measure of the coalition, not on whether the agents are drawn from $[0, \frac{1}{2}]$ or $(\frac{1}{2}, 1]$. The reservation utility needed to entice an agent from $[0, \frac{1}{2}]$ to join the coalition exceeds the reservation utility needed to entice an agent from $(\frac{1}{2}, 1]$ to join. Therefore we may assume without loss of generality that $S \supset (\frac{1}{2}, 1]$. Let $s = \lambda(S \cap [0, \frac{1}{2})) > \frac{1}{4}$.

(6) For each $a \in S$, let $c_a = \|G(a)\|$. We must provide a utility level exceeding $\frac{7}{4}$ to each of the agents $a \in (\frac{1}{2}, 1]$. This will require that $\int_{S \cap (\frac{1}{2}, 1]} c_a\, d\lambda > \frac{1}{2}(\frac{7}{8}) = \frac{7}{16}$. Therefore, $\int_{S \cap [0, \frac{1}{2}]} c_a < (s + \frac{1}{2}) - \frac{7}{16} = s + \frac{1}{16}$.

(7) It follows from item (6) that $c_a < (s + \frac{1}{16})/s = 1 + (1/16s)$ for every $a \in C$, where $C \subset S \cap [0, \frac{1}{2}]$ is a set of positive Lebesgue measure. The total endowment of the coalition S at the point a is $s + \frac{1}{2}$; hence, the utility of agent $a \in C$ is less than $2(s + \frac{1}{2}) + (1 + (1/16s) - (s + \frac{1}{2})) = \frac{3}{2} + s + (1/16s)$. Since the reservation utility needed to entice each of the agents in C to join the coalition is $\frac{17}{8}$, it follows that

$$s + \frac{1}{16s} > \frac{17}{8} - \frac{3}{2} = \frac{5}{8}, \tag{39}$$

$$\frac{d}{ds}\left(s + \frac{1}{16s}\right) = 1 - \frac{1}{16s^2}, \tag{40}$$

which is non-negative for $s \in [\frac{1}{4}, \frac{1}{2}]$. Thus, the maximum value of $s + (1/16s)$ on the interval $[\frac{1}{4}, \frac{1}{2}]$ is attained at $s = \frac{1}{2}$, where it equals $\frac{5}{8}$. Thus, equation (39) has no solution for $s \in [\frac{1}{4}, \frac{1}{2}]$, which shows that $F \in \mathscr{C}(\chi_n)$.

While there have been quite a number of papers concerning core equivalence in continuum economies with an infinite-dimensional commodity space, there is still no systematic delineation of what assumptions are crucial for obtaining equivalence. There is very little work on core convergence with a large finite number of agents and an infinite-dimensional commodity space, outside the replica context. Both problems are attractive areas for future research. Given the limitations of space, we shall limit ourselves to providing the following list of papers on core equivalence and/or convergence in exchange economies with an infinite-dimensional commodity space: Gabszewicz (1968), Bewley (1973b), Mas-Colell (1975), Jones (1984), Gretsky and Ostroy (1985), Zame (1986), Aliprantis, Brown and Burkinshaw (1987), Cheng (1987), Rustichini and Yannelis (1991), Ostroy and Zame (1988), Aliprantis and Burkinshaw (1989), Anderson (1990b), Mertens (1990).

Bibliography

Aliprantis, Charalambos D. and Owen Burkinshaw (1989) 'When is the core equivalence theorem valid?' Social Science Working Paper #707, Division of the Humanities and Social Sciences, California Institute of Technology.

Aliprantis, Charalambos D., Donald J. Brown and Owen Burkinshaw (1987) 'Edgeworth equilibria', *Econometrica*, 55: 1109–1137.

Aliprantis, Charalambos D., Donald J. Brown and Owen Burkinshaw (1989) *Existence and optimality of competitive equilibrium*. Berlin: Springer-Verlag.

Anderson, Robert M. (1977) 'Star-finite probability theory', Ph.D. Dissertation, Yale University.

Anderson, Robert M. (1978) 'An elementary core equivalence theorem', *Econometrica*, 46: 1483–1487.

Anderson, Robert M. (1981a) 'Core theory with strongly convex preferences', *Econometrica*, 49: 1457–1468.

Anderson, Robert M. (1981b) 'Strong core theorems with nonconvex preferences', Cowles Foundation Discussion Paper No. 590, Yale University.

Anderson Robert M. (1985) 'Strong core theorems with nonconvex preferences', *Econometrica*, 53: 1283–1294.

Anderson, Robert M. (1986) 'Core allocations and small income transfers', Working Paper 8621, Department of Economics, University of California at Berkeley; in: John Geanakoplos and Pradeep Dubey, eds., *Essays in honor of Martin Shubik* (to appear).

Anderson, Robert M. (1987) 'Gap-minimizing prices and quadratic core convergence', *Journal of Mathematical Economics*, 20: 599–601.

Anderson, Robert M. (1988) 'The second welfare theorem with nonconvex preferences', *Econometrica*, 56: 361–382.

Anderson, Robert M. (1990a) 'Core allocations are almost utility-maximal', *International Economic Review*, 31: 1–9.

Anderson, Robert M. (1990b) 'Large square economies: An asymptotic interpretation', preprint, Department of Economics, University of California at Berkeley, April.

Anderson, Robert M. (1991) 'Nonstandard analysis with applications to economics', in: Werner Hildenbrand and Hugo Sonnenschein, eds., *Handbook of mathematical economics*, Vol. IV. Amsterdam: North-Holland, Ch. 39.

Anderson, Robert M. and Andreu Mas-Colell (1988) 'An example of Pareto optima and core allocations far from agents' demand correspondences', Appendix to Anderson (1988), *Econometrica*, 56: 361–382.

Arrow, Kenneth J. and F.H. Hahn (1971) *General competitive analysis*. San Francisco: Holden-Day, Inc.

Aumann, Robert J. (1964) 'Markets with a continuum of traders', *Econometrica*, 32: 39–50.

Aumann, Robert J. (1979) 'On the rate of convergence of the core', *International Economic Review*, 20: 349–357.

Bewley, Truman F. (1973a) 'Edgeworth's conjecture', *Econometrica*, 41: 425–454.

Bewley, Truman F. (1973b) 'The equality of the core and the set of equilibria in economies with infinitely many commodities and a continuum of agents', *International Economic Review*, 14: 383–394.

Billingsley, Patrick (1968), *Convergence of probability measures*. New York: John Wiley.

Brown, Donald J. and M. Ali Khan (1980) 'An extension of the Brown–Robinson equivalence theorem', *Applied Mathematics and Computation*, 6: 167–175.

Brown Donald J. and Abraham Robinson (1974) 'The cores of large standard exchange economies', *Journal of Economic Theory*, 9: 245–254.

Brown, Donald J. and Abraham Robinson (1975) 'Nonstandard exchange economies', *Econometrica*, 43: 41–55.

Chae, Suchan (1984) 'Bounded core equivalence and core characterization', preprint, University of Pennsylvania.

Champsaur, Paul (1974) 'Note sur le noyau d'une economie avec production', *Econometrica*, 42: 933–946.

Cheng, Hsueh-Cheng (1981) 'What is the normal rate of convergence of the core (Part I)', *Econometrica*, 49: 73–83.

Cheng, Hsueh-Cheng (1982) 'Generic examples on the rate of convergence of the core', *International Economic Review*, 23: 309–321.

Cheng, Hsueh-Cheng (1983a) 'The best rate of convergence of the core', *International Economic Review*, 24: 629–636.

Cheng, Hsueh-Cheng (1983b) 'A uniform core convergence result for non-convex economies', *Journal of Economic Theory* 31: 269–282.

Cheng, Harrison H.-C. (1987) 'The principle of equivalence', Working Paper, University of Southern California.

Debreu, Gerard (1975) 'The rate of convergence of the core of an economy', *Journal of Mathematical Economics*, 2: 1–7.

Debreu, Gerard (1982) 'Existence of competitive equilibrium', in: Kenneth J. Arrow and Michael D. Intriligator, eds., *Handbook of mathematical economics*, Vol. II, Amsterdam: North-Holland, pp. 697–743.

Debreu, Gerard and Herbert Scarf (1963) 'A limit theorem on the core of an economy', *International Economic Review*, 4: 236–246.

Debreu, Gerard and Herbert Scarf (1972) 'The limit of the core of an economy', in: C.B. McGuire and Roy Radner, eds., *Decision and organization: A volume in honor of Jacob Marschak*. Amsterdam: North-Holland.

Dierker, Egbert (1975) 'Gains and losses at core allocations', *Journal of Mathematical Economics*, 2: 119–128.

Dierker, Hildegard (1975) 'Equilibria and core of large economies', *Journal of Mathematical Economics*, 2: 155–169.

Edgeworth, Francis Y. (1881) *Mathematical psychics*. London: Kegan Paul.

Gabszewicz, J. (1968) 'Coeurs et allocations concurrentielles dans des economies d'echange avec un continu de biens', Librairie Universitaire, Université Catholique de Louvain.

Geller, William (1987) 'Almost quartic core convergence', presentation to Econometric Society North American Summer Meeting, Berkeley, June 1987.

Gretsky, Neil and Joseph Ostroy (1985) 'Thick and thin market non-atomic exchange economies', in: Charalambos D. Aliprantis, Owen Burkinshaw and N.J. Rothman, eds., *Advances in equilibrium theory*, Springer-Verlag Lecture Notes in Economics and Mathematical Systems, 244.

Grodal, Birgit (1972) 'A second remark on the core of an atomless economy', *Econometrica*, 40: 581–583.

Grodal, Birgit (1975) 'The rate of convergence of the core for a purely competitive sequence of economies', *Journal of Mathematical Economics*, 2: 171–186.

Grodal, Birgit (1976) 'Existence of approximate cores with incomplete preferences', *Econometrica*, 44: 829–830.

Grodal, Birgit and Werner Hildenbrand (1973) unpublished notes.

Grodal, Birgit and Werner Hildenbrand (1974) 'Limit theorems for approximate cores', Working Paper IP-208, Center for Research in Management, University of California, Berkeley.

Grodal, Birgit, Walter Trockel and Shlomo Weber (1984) 'On approximate cores of non-convex economies', *Economics Letters*, 15: 197–202.

Hammond, Peter J., Mamoru Kaneko and Myrna Holtz Wooders (1989) 'Continuum economies with finite coalitions: Core, equilibria and widespread externalities', *Journal of Economic Theory*, 49: 113–134.

Hildenbrand, Werner (1968) 'The core of an economy with a measure space of economic agents', *Review of Economic Studies*, 35: 443–452.

Hildenbrand, Werner (1970) 'On economies with many agents', *Journal of Economic Theory*, 2: 161–188.

Hildenbrand, Werner (1971) 'Random preferences and equilibrium analysis', *Journal of Economic Theory*, 4: 414–429.

Hildenbrand, Werner (1974) *Core and equilibria of a large economy*. Princeton: Princeton University Press.

Hildenbrand, Werner (1982) 'Core of an economy', in Kenneth J. Arrow and Michael D. Intriligator, eds., *Handbook of mathematical ecoonomics*, Vol. II. Amsterdam: North-Holland, pp. 831–877.

Hildenbrand, Werner and Alan P. Kirman (1973) 'Size removes inequity', *Review of Economic Studies*, 40: 305–319.

Hildenband, Werner and Alan P. Kirman (1976) *Introduction to equilibrium analysis: Variations on themes by Edgeworth and Walras*. Amsterdam: North-Holland.

Hildenbrand, Werner, David Schmeidler and Shmuel Zamir (1973) 'Existence of approximate equilibria and cores', *Econometrica*, 41: 1159–1166.

Hoover, Douglas (1989) Private communication.

Johansen, Leif (1978) 'A calculus approach to the theory of the core of an exchange economy', *American Economic Review*, 68: 813–829.

Jones, Larry E. (1984) 'A competitive model of commodity differentiation', *Econometrica*, 52, 507–530.

Kaneko, Mamoru and Myrna Holtz Wooders (1986) 'The core of a game with a continuum of players and finite coalitions: The model and some results', *Mathematical Social Sciences*, 12: 105–137.

Kaneko, Mamoru and Myrna Holtz Wooders (1989) 'The core of a continuum economy with widespread exernalities and finite coalitions: From finite to continuum economies', *Journal of Economic Theory*, 49: 135–168.

Kannai, Yakar (1970) 'Continuity properties of the core of a market', *Econometrica*, 38: 791–815.

Keiding, Hans (1974) 'A limit theorem on the cores of large but finite economies', preprint, University of Copenhagen.

Khan, M. Ali (1974) 'Some equivalence theorems', *Review of Economic Studies*, 41: 549–565.

Khan, M. Ali (1976) 'Oligopoly in markets with a continuum of traders: An asymptotic interpretation', *Journal of Economic Theory*, 12: 273–297.

Khan, M. Ali and Salim Rashin (1976) 'Limit theorems on cores with costs of coalition formation', preprint, Johns Hopkins University.

Kim, Wan-Jin (1988) 'Three essays in economic theory', Ph.D. Dissertation, Department of Economics, University of California at Berkeley.

Kirman, Alan P. (1982) 'Measure theory with applications to economics', in: Kenneth J. Arrow and Michael D. Intriligator, eds., *Handbook of mathematical economics*, Vol. I. Amsterdam: North-Holland, pp. 159–209.

Loeb, Peter A. (1975) 'Conversion from nonstandard to standard measure spaces and applications in potential theory', *Transactions of the American Mathematical Society*, 211: 113–122.

Manelli, Alejandro (1990a) 'Core convergence without monotone preferences or free disposal', preprint, Department of Managerial Economics and Decision Sciences, Kellogg Graduate School of Management, Northwestern University, May.

Manelli, Alejandro (1990b), Private communication.

Manelli, Alejandro (1991) 'Monotonic preferences and core equivalence', preprint, Department of Managerial Economics and Decision Sciences, Kellogg Graduate School of Management, Northwestern University, September, 1989. *Econometrica*, 59: 123–138.

Mas-Colell, Andreu (1975) 'A model of equilibrium with differentiated commodities', *Journal of Mathematical Economics*, 2, 263–295.

Mas-Colell, Andreu (1977) 'Regular nonconvex economies', *Econometrica*, 45: 1387–1407.

Mas-Colell, Andreu (1978) 'A note on the core equivalence theorem: How many blocking coalitions are there?', *Journal of Mathematical Economics*, 5: 207–215.

Mas-Colell, Andreu (1979) 'A refinement of the core equivalence theorem', *Economics Letters*, 3: 307–310.

Mas-Colell, Andreu (1985), *The theory of general economic equilibrium: A differentiable approach*. Cambridge: Cambridge University Press.

Mas-Colell, Andreu and Wilhelm Neuefeind (1977) 'Some generic properties of aggregate excess demand and an application', *Econometrica*, 45: 591–599.

Mas-Colell, Andreu and William R. Zame (1991) 'Equilibrium theory in infinite dimensional spaces', in: Werner Hildenbrand and Hugo Sonnenschein, eds., *Handbook of mathematical economics*, Vol. IV. Amsterdam: North-Holland, Ch. 34.

Mertens, Jean-Francois (1990) 'An equivalence theorem for the core of an economy with commodity space $L_\infty - \tau(L_\infty, L_1)$', Discussion Paper 7028 (1970), Center for Operations Research and Econometrics, Université Catholique de Louvain, to appear.

Nishino, H. (1970) 'The cores of exchange economies and the limit theorems', in: M. Suzuki, ed., *Theory of games in competitive society*. Tokyo: Keiso-shobo, pp. 131–168.

Oddou, Claude (1976) 'Théoremes d'existence et d'equivalence pour des economies avec production', *Econometrica*, 44: 265–281.

Ostroy, Joseph M. and William R. Zame (1988) 'Non-atomic economies and the boundaries of perfect competition', Working Paper, Department of Economics, University of California at Los Angeles, August.

Robinson, Abraham (1970), *Non-standard analysis*. Amsterdam: North-Holland.

Royden, H.L. (1968), *Real analysis*. New York: Macmillan.

Rustichini, Aldo and Nicholas Yannelis (1991) 'Edgeworth's conjecture in economies with a continuum of agents and commodities', *Journal of Mathematical Economics*, 20: 307–326.

Scarf, Herbert E. (1967) 'The core of an *n*-person game', *Econometrica*, 35: 50–69.

Schmeidler, David (1972) 'A remark on the core of an atomless economy', *Econometrica*, 40: 579–580.

Schweizer, Urs (1982) 'A Lagrangean approach to the limit theorem on the core of an economy', *Zeitschrift für Nationalokonomie*, 42: 22–30.

Shapley, Lloyd S. (1975) 'An example of a slow-converging core', *International Economic Review*, 16: 345–351.

Shubik, Martin (1959) 'Edgeworth market games', *Contributions to the theory of games IV, Annals of Mathematics Studies*, 40: 267–278.

Shubik, Martin (1982), *Game theory in the social sciences: Concepts and solutions*. Cambridge, Mass.: MIT Press.

Trockel, Walter (1976) 'A limit theorem on the core', *Journal of Mathematical Economics*, 3: 247–264.

Trockel, Walter (1984) 'On the uniqueness of individual demand at almost every price system', *Econometrica*, 33: 397–399.

Vind, Karl (1964) 'Edgeworth allocations in the exchange economy with many traders', *International Economic Review*, 5: 165–177.

Vind, Karl (1965) 'A theorem on the core of an economy', *Review of Economic Studies*, 32: 47–48.

Vind, Karl (1972) 'A third remark on the core of an atomless economy', *Econometrica*, 40: 585–586.

Walras, Leon (1874), *Eléments d'économie politique pure*. Lausanne: L. Corbaz.

Wooders, Myrna Holtz and William R. Zame (1984) 'Approximate cores of large games', *Econometrica*, 52: 1327–1350.

Zame, William R. (1986) 'Markets with a continuum of traders and infinitely many commodities', Working Paper, Department of Mathematics, State University of New York at Buffalo, December.

Chapter 15

THE CORE IN IMPERFECTLY COMPETITIVE ECONOMIES

JEAN J. GABSZEWICZ[a] and BENYAMIN SHITOVITZ[b]

[a]*C.O.R.E., Université Catholique de Louvain and* [b]*Haifa University*

Contents

Handbook of Game Theory, Volume 1, Edited by R.J. Aumann and S. Hart

1. Introduction

In a pathbreaking paper, Aumann (1964) proved that in a pure exchange economy consisting of an "atomless" set of traders, the core of the market must coincide with the set of its competitive allocations. The introduction of atomless market models was meant to capture the traditional economic idea of "perfect competition". With a continuum of traders, the influence of each individual participant is "negligible", per se: the notion of perfect competition is "built into the model" [Aumann (1964, p. 40)]. The formal reason for this is that "integrating over a continuum, and changing the integrand at a single point does not affect the value of the integral, that is, the actions of a single individual are negligible" [Aumann (1964, p. 39)]. The equivalence result referred to above proves, indeed, that this model adequately captures the notion of perfect competition: in atomless economies, competitive equilibria are the sole possible outcomes of the group decision mechanism underlying the concept of core.

Nevertheless, the idea of "perfect competition" has traditionally been viewed by economists as representing an "ideal state", which essentially serves as a reference point for contrasting real market phenomena: real market competition is recognized as being far from perfect! First, even if markets often embody an "ocean" of small anonymous traders, individual merchants who are *not* anonymous may also be present. This is often the case because their endowment of some commodities is large compared with the endowments of the entire market. The most extreme case corresponds to a *monopoly*, when the whole market endowment of a good is concentrated in the hands of a single merchant. Intermediate forms arise when, although spread over a few competitors, initial ownership of resources is still "concentrated" when compared with the total endowment in the market; such intermediate forms are known as *oligopolistic* structures. Marxian economists have seen in the concentration of capital ownership, accompanied by the dispersion of the labor force, the basis for an increase in economic exploitation in capitalistic economies. There is no doubt that, for large values of n, the bargaining position of a single capitalist owning n units of capital and facing n nonunionized workers is far stronger than the position of the same capitalist, if he owns one unit of capital and faces a single worker. The concentration of ownership and market power are intimately related.

On the other hand, even if the ownership of goods is not initially concentrated, but spread over a continuum of small economic units, the possibility is always open, to some market participants, of "combining" into a restricted number of decision centers so as to bias the collective decision outcome. Von

Neumann and Morgenstern (1944) had already perceived how such a collusive process could alter the competitive outcome:

> The classical definitions of free competition all involve further postulates besides the greatness of the number (of participants). E.g. it is clear that, if a certain great group of participants act together, then the great number of participants may not become effective; the decisive exchanges may take place directly between large "coalitions" (such as trade unions, cooperatives, . . .), and not between individuals, many in number, acting independently.

When oligopolistic structures with initially concentrated ownership are present in the market, or when collusive cartels or unions are becoming effective, the operating market conditions violate those of the "ideal state" of perfect competition. Accordingly, the continuous atomless model – designed to represent this ideal state – is no longer appropriate as such; it should be amended to handle the imperfectly competitive market ingredients described above. To the extent that the economy under consideration embodies, in particular, a very large number of participants with negligible influence, the continuous model is still the most natural one to represent this "oceanic sector" of the economy. As for the non-negligible market participants – monopolists, oligopolists, cartels, syndicates or other institutional forms of collusive agreements – their formal counterpart in the model *cannot* be simply points with null measure in the continuum; such a formal representation would entail per se that the actions of these participants *are* mathematically negligible when clearly they are not. We submit that the most appropriate formal model consists in representing them as *atoms*, i.e. subsets with strictly positive mass containing no proper subset of strictly positive mass. In this alternative model, changing the integrand to an atom *does* affect the value of the integral, so that the actions of the economic unit represented by the atom are *not* mathematically negligible.

Let us illustrate this for the case of collusive agreements organized between traders in an atomless exchange economy. To this end, assume that $(T, \mathcal{T}, \mu)$ is an atomless measure space, where the set T is to be interpreted as the set of traders, and $\mathcal{T}$ as the set of possible coalitions of traders in the same economy. Imagine that, for any reason whatsoever, all traders in some *non-null* subset A in T decide to act only "in unison", for instance by delegating to a single decision unit the task of representing their economic interests in the trade. Whenever effective, such a binding agreement definitely prevents the formation of any coalition of traders including a *proper* subset of A: while such coalitions were allowed before the collusive agreement, they are henceforth forbidden. Formally, the σ-field $\mathcal{T}$ no longer represents the class of acceptable coalitions; this class is now reduced to the σ-field $\mathcal{T}_A = \{S \in \mathcal{T} \mid \text{either } S \cap A = \emptyset, \text{ or } S \cap A = A\}$, i.e. the subset A now constitutes an *atom* in $\mathcal{T}_A$. The actions of the "syndicate" of traders in A are no longer mathematically negligible.

When the "atomic" representation of "noncompetitive" traders is adopted, we end up with a *mixed* model, in which some of the traders – the "small" ones – are represented by the atomless sector, while the others – the "large" ones – are represented by atoms. It turns out that a considerable amount of research work was devoted in and around the 1970s to theorizing about the core of such a mixed general equilibrium exchange model, embodying both atoms and a continuum of traders. The object of this chapter is to give an account of this research work.

In exchange situations involving both non-negligible and negligible agents, one should not generally expect the equivalence between the core and the set of competitive allocations to hold. For instance, when a collusive agreement is signed between the traders in a non-null subset of an atomless economy, it eliminates all coalitions, including a proper subset of this set. As a consequence, the core may be enlarged to those exchange allocations which could have been improved upon otherwise via some of these excluded coalitions, thereby destroying the equivalence of the core and the set of competitive allocations. A first question of interest for the theory is thus: To what extent can the existence of atoms destroy the equivalence principle? As we shall see (Section 4), it is sometimes possible to extend Aumann's equivalence theorem to markets in which some of the traders are "large". This is true, in particular, when these traders are similar to each other, or when to each such large trader there corresponds a set of small traders which are similar to him. Such conditions suggest that when large traders find competitors similar to them in the economy, it may well imply the dilution of their market power; this is reminiscent of Bertrand price competition in a noncooperative context.

By the equivalence theorem, all core allocations can be decentralized through the competitive price mechanism when the economy is atomless; when there are atoms, this is no longer guaranteed. Does this mean that core allocations can no longer be characterized via *some* price mechanism? It is one of the most important results of this theory that core allocations can still be sustained by a price system which, although not carrying all the properties of a competitive price system, shares with it some interesting features (Sections 3 and 5). Moreover, at this price system, no small trader (in the atomless part) can, with the value of his initial endowment, buy a commodity bundle that he prefers to his part of that core allocation. *But this restriction does not apply to atoms.* Accordingly, "small" traders are "budgetarily exploited" at those prices that sustain a core allocation. "Budgetary" exploitation does not imply, however, "utility" exploitation, and analyzing when the former implies the latter (Section 6) is also an important topic.

Finally, we have already evoked the interesting interpretation of an atom as a "syndicate" of traders in an atomless economy. Accordingly the mixed model of a market is well suited to investigate the effectiveness and the stability of binding agreements among traders in the context of a pure exchange economy. This is done in Section 7. In our conclusion, we examine the question of the adequacy of the core concept for capturing the idea of imperfect competition; alternative approaches, founded on the Shapley value, the bargaining set or the von Neumann–Morgenstern solution are briefly discussed. There we examine also the problem of approximating mixed markets by finite exchange economies.

The mathematical model is presented in Section 2.

2. The mathematical model

We are interested in an exchange economy with n commodities. Following standard practice, for x and y in R^n we write $x \gg y$ to mean $x^i > y^i$ for all i; we use $x \geq y$ to mean $x^i \geq y^i$ for all i; and we use $x > y$ to mean $x \geq y$ but not $x = y$.

Let $(T, \mathcal{F}, \mu)$ be a measure space of economic agents, i.e. T denotes a set (the traders), $\mathcal{F}$ denotes a σ-field of subsets of T (the family of coalitions), and μ denotes a totally finite complete positive σ-additive measure on $\mathcal{F}$, which represents the respective "weights" of agents or groups of agents in the economy. An atom of the measure space $(T, \mathcal{F}, \mu)$ is a coalition S with $\mu(S) > 0$ such that for each coalition $R \subseteq S$ we have either $\mu(R) = 0$ or $\mu(S \backslash R) = 0$. The set T can be divided into a countable union of atoms T_1 and an atomless sector T_0.

A *commodity bundle* x is a point in the non-negative orthant Ω of R^n. An *assignment* (of commodity bundles to traders) is an integrable function x from T to Ω. In an integral we will omit the symbol $d\mu(t)$ and the indication of the dependence of the integrand on t (which stands for trader).

There is a fixed initial (density) assignment w. We assume that $\int_T w \gg 0$. This asserts that no commodity is totally absent from the market. For each trader t a relation $>_t$ is defined on Ω, which is called the *preference relation* of trader t and satisfies the standard assumptions such as strong desirability, continuity and measurability. We also assume

(H.1) Convexity assumption on the large traders. For each atom $t \in T_1$, $>_t$ is convex, i.e. $y \in \Omega$ implies $\{x \in \Omega : x >_t y\}$ is a convex set.

Note specifically that $>_t$ is not assumed to be complete, nor even transitive. In some parts of the chapter we shall assume

(H.2) Quasi-order assumption on the large traders. For each atom $t \in T_1$, $>_t$

is derived from a preference-or-indifference relation $\gtrsim_t$ on Ω, which is assumed to be a quasi-order, i.e. a reflexive, transitive and complete binary relation.

Note that $>_t$ and $\gtrsim_t$ can be derived from a measurable, continuous quasi-concave utility function $u_t(x)$, for all $t \in T_1$.

An *allocation* (or "final assignment" or "trade") is an assignment x for which $\int_T x = \int_T w$. An assignment y *dominates* an allocation x *via* a coalition S (S is then said to *improve upon* or to *block* x) if $y(t) >_t x(t)$ for almost each $t \in S$, $\mu(S) > 0$, and S is *effective* for y, i.e $\int_S y = \int_S w$. The *core* is the set of all allocations that cannot be improved upon by any (non-null) coalition.

A *price vector* p is a vector $p \in \Omega$, $p \neq 0$. A *competitive equilibrium* is a pair (p, x) consisting of a price vector p and an allocation x such that for almost all traders t, $x(t)$ is maximal with respect to $>_t$ in t's *budget set* $B_p(t) = \{x \in \Omega : p \cdot x \leq p \cdot w(t)\}$. A *competitive allocation* x is an allocation for which there exists a price vector p such that (p, x) is a competitive equilibrium. Similarly, an *efficiency equilibrium* (e.e.) is a pair (p, x) consisting of a price vector p and an allocation x such that for almost all traders t, $x(t)$ is maximal with respect to $>_t$ in t's *efficiency budget set* $E_p(t) = \{x \in \Omega : p \cdot x \leq p \cdot x(t)\}$. Note that every competitive equilibrium is an e.e., but not every e.e. is necessarily a competitive equilibrium. Furthermore, under (H.1), an allocation that is Pareto optimal is an e.e. with a suitable price system p.

Two traders s and t are said to be of the same *type* or *similar* if $w(s) = w(t)$ and, for all $x, y \in \Omega$, $x >_s y$ if and only if $x >_t y$. Note that when $s \in T_1$ is of the same type as t (not necessarily in T_1), then both s and t have the same utility function $u_s(x) = u_t(x)$ [under (H.2)]. Finally, two large traders are said to be of the same *kind* if they are of the same type and have the same measure.

3. Budgetary exploitation: A general price property of core allocations in mixed markets

We start our investigation of core allocations in mixed markets by stating a general price property of such allocations (Theorem 3.1). Aumann's equivalence theorem appears as an immediate corollary of this property when the market has no atoms. In most situations, however, the equivalence of the core and the set of competitive allocations should not be expected to hold in mixed markets. An example with no equivalence is provided in Subsection 3.2. From this example and Theorem 3.1, we shall see that small traders are necessarily "budgetarily exploited" at a core allocation; at efficiency prices, the value of their part in that core allocation cannot exceed the value of their initial endowment.

3.1. The "budgetary exploitation" theorem

Let x be a core allocation. Then in particular x is Pareto optimal and we may consequently associate with it a price vector p such that (p, x) is an efficiency equilibrium. Theorem 3.1 below states that the efficiency prices p can be chosen in such a way that, whenever an agent t is a small trader, the "value" $p \cdot x(t)$ of the bundle $x(t)$ assigned by x to him does not exceed the value $p \cdot w(t)$ of his initial bundle. In terms of value, therefore, the small traders lose, or at best they come out even, i.e. they are "budgetarily exploited". As for the large traders, considered as a group their budgetary gain is exactly equal to the sum of the losses of the small traders. About an individual large trader, however, we can say nothing – he may either gain or lose. Formally we state:

Theorem 3.1 [Shitovitz (1973)]. *Assume* (H.1) *and let x be in the core. Then there exists a price vector p such that* (i) *(p, x) is an efficiency equilibrium and* (ii) *$p \cdot x(t) \leqslant p \cdot w(t)$ for almost all $t \in T_0$.*

The formal proof of Theorem 3.1, which is omitted, is based on the notion of the integral of a set valued function (correspondence). Defining $G(t) = \{x \in \Omega : x >_t x(t)\}$ and $\int_T G \equiv \{\int_T g : g$ is integrable and $g(t) \in G(t)$ a.e.$\}$ we note that p are efficiency prices for x if and only if the hyperplane $\{x : p \cdot x = p \cdot \int_T x\}$ supports the convex set $\int_T G$ at $\int_T x$ [$\int_T G$ is convex by (H.1)]. Set

$$F(t) = \begin{cases} G(t) \cup \{w(t)\} & t \in T_0, \\ G(t) & t \in T_1. \end{cases}$$

Then, because x is a core allocation, $\int_T w$ (which equals $\int_T x$) is not an interior point of $\int_T F$, by strong desirability. Thus the convex set $\int_T F$ can be supported by the hyperplane $\{x \in R^n : p \cdot x = p \cdot \int_T x\}$ at $\int_T x$. Immediate calculations imply the theorem. Note that Aumann's Equivalence Theorem (1964) is a special case of Theorem 3.1. If there are no large traders, then the total loss of the small traders is 0, and since no small trader can gain, each one loses nothing, so we have a competitive equilibrium.

3.2. An example of a monopolistic market with no equivalence

Let us consider the following exchange economy with $T = [0, 1] \cup \{2\}$, where $T_0 = [0, 1]$ is taken with Lebesgue measure, $T_1 = \{2\}$ is an atom with $\mu(T_1) = 1$, and the number of commodities is 2.

The initial assignment is defined by

$$w(t) = \begin{cases} (4, 0), & t \in T_0, \\ (0, 4), & t \in T_1, \end{cases}$$

while the utility of the traders t is $u_t(x_1, x_2) = \sqrt{x_1} + \sqrt{x_2}$, a homogeneous utility, the same for all traders.

There is a unique competitive allocation, namely the allocation that assigns $(2, 2)$ to all traders. On the other hand, the core consists of all allocations x of the form $x(t) = (\alpha(t), \alpha(t))$ for almost all t, where $1 \leq \alpha(t) \leq 2$ for almost all $t \in T_0$, and $2 \leq \alpha(2) \leq 3$ are such that $\int_T \alpha = 4$. In particular,

$$x_0(t) = \begin{cases} (1, 1), & t \in T_0, \\ (3, 3), & t \in T_1, \end{cases}$$

is in the core and is obviously different from the competitive allocation. At x_0, the small traders are "budgetarily exploited", i.e. we have

$$p \cdot x_0(t) = p \cdot (1, 1) < p \cdot (4, 0) = p \cdot w(t),$$

where the unique efficiency prices for all points in the core are $p = (1, 1)$. Moreover, the utility of every small trader at x_0 is exactly the same as that of his initial bundle. Note that the allocation x_1, where

$$x_1(t) = \begin{cases} (3, 3), & t \in T_0, \\ (1, 1), & t \in T_1, \end{cases}$$

is not in the core; the large trader must receive at least $(2, 2)$ in the core, i.e. at least as much as at the competitive allocation (see Figure 1, in which the heavy

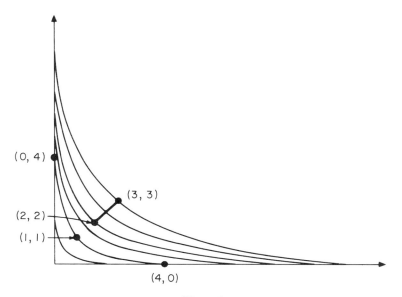

Figure 1

line indicates the set of possible bundles assigned to the large trader by core allocations).

In this example we notice that, at any core allocation, the small traders are not only "budgetarily exploited", but also exploited in utility: their utility level at a core allocation, compared with their utility level at the competitive allocation, is always smaller. As we shall see in Section 6, this property is not always satisfied; budgetary exploitation does not necessarily imply utility exploitation.

4. Competitive allocations and the core of mixed markets

We continue our investigation of cores in mixed markets by identifying some situations in which the equivalence theorem holds. Interestingly enough, these situations reveal that, in spite of their "size", large traders can be engaged in an intense competition because other traders are similar to them in the economy. These can either be other large traders or small traders in sufficiently large number. In Subsection 4.1 the case of similar large traders is considered, while in Subsection 4.2 we consider the core when to large traders correspond similar small traders.

4.1. The core when large traders are similar

A significant result in Shitovitz (1973) is his "Theorem B", which extends the Equivalence Theorem in Aumann (1964) to oligopolistic mixed markets. It states that in a market in which there are at least two large traders, and all the large traders are similar (i.e. are of the same "type"), all core allocations are competitive. Thus such a market is essentially indistinguishable from a perfectly competitive market; if all the large traders were to split into a continuum of small traders of the same "type" as the original large traders, there would be no change in the core. In this case, therefore, the presence of several large traders engenders such intense competition among them that the effect of the larger traders' size is nullified.

Theorem 4.1 [Shitovitz (1973)]. *Assume that there are at least two large traders, and that all large traders are similar. Then, under* (H.1) *and* (H.2), *the core coincides with the set of competitive allocations.*

A straightforward proof of this equivalence theorem can be found in Shitovitz (1973). Another proof was also given in Greenberg and Shitovitz (1986), using Vind (1972) and Aumann (1964).

The formal proof of Theorem 4.1 is omitted. Here, we give a heuristic argument. For simplicity in this description, let us assume that there are just two large traders W_1 and W_2, and they are of the same type with equal measure, and let x be an allocation in the core. Obviously, both large traders are indifferent between $x(W_1)$ and $x(W_2)$; this phenomenon is well known. In fact, we may go further; assume that both traders actually get the same bundle, i.e. that $x(W_1) = x(W_2)$. Therefore, by Lyapunov's theorem there exist two disjoint coalitions S_1 and S_2 of small traders trading with W_1 and W_2, respectively, i.e. $\int_R x = \int w$ for $R = S_1 \cup W_1$ and $R = S_2 \cup W_2$. Intuitively, also, it is not unreasonable to assume that two identical large traders will split the market evenly between them. This means that the market is actually composed of two monopolistic submarkets whose traders are $S_1 \cup W_1$ and $S_2 \cup W_2$, respectively. Let $p(t) = p \cdot x(t) - p \cdot w(t)$ be the budgetary profit of trader t at the "market prices" p. By Theorem 3.1 we have that each small trader t has a nonpositive profit. Suppose now that one of the large traders, say W_1, has a positive profit. Then, since the total profit of each submarket is zero, there are small traders in S_1 who have been budgetarily exploited. Therefore, by adding a sufficiently small part of these traders to the other submarket, and by distributing the excess $w(t) - x(t)$ of the part (whose value at the "market prices" is positive) among themselves and the traders of the other submarket, we obtain a new submarket whose traders t receive a new bundle in the neighborhood of $x(t)$ whose value at the "market prices" is more than the value of $x(t)$. Therefore, the traders of this new submarket can improve upon x, in contradiction to the assumption that x is in the core.

In very simple terms, what is happening is that if some of the "customers" of W_1 are "losing money", then it is worthwhile for W_2 to "steal" at least a small number of these customers from W_1 (while keeping his own customers). Therefore, none of W_1's customers can lose money, and so, by the symmetry of the situation, nobody does.

4.2. The core when to large traders correspond similar small traders

Theorem 4.1 asserts that when all large traders are competitors of the same type, the core is equivalent to the set of competitive allocations. Our next theorem (Theorem 4.2 below) asserts that the same must hold whenever to each large trader there corresponds a set of *small* traders of the same type *and* a constraint on the size of the atoms is satisfied. Let $A_1, A_2, \ldots, A_k, \ldots, \ldots$ denote the set of atoms and, for each atom A_k consider the class

$$T_k \overset{\text{def}}{=} \{t \in T \mid w(t) = w(A_k); \forall(x, y) \in \Omega \times \Omega: x \succsim_t y \Leftrightarrow x \succsim_{A_k} y\};$$

the class T_k is the set of all traders who are of the same type as the atom A_k. Denote by $\succsim_k$ the common preferences of traders in T_k and by A_{hk} the hth atom of type k, $h \in N^*$. We then obtain a partition of the set of agents into at most countably many classes T_k, and an atomless part $T \backslash \bigcup_{k \in N^*} T_k$.

Theorem 4.2 [Gabszewicz and Mertens (1971)]. *If*

$$\sum_k \left(\sum_h \frac{\mu(A_{hk})}{\mu(T_k)} \right) < 1 , \tag{$*$}$$

then, under (H.1) *and* (H.2), *the core coincides with the set of competitive allocations.*

The inequality ($*$) says that the sum over all types of the atomic proportions of the types should be less than one. This result implies in particular that if there is only a single atom in the economy, *any* non-null set of "small" traders similar to the atom "nullifies" the effect of the large trader's size. The proof of Theorem 4.2 is too long to be reported in full in the present survey. Let us however give an idea of the proof, which the reader can find in Gabszewicz and Mertens (1971, p. 714). This proof essentially rests on a lemma which states that, under condition ($*$), *all* traders in T_k – the "large" and the "small" ones – must get in the core a consumption bundle which is in the same indifference class relative to their common preferences $\succsim_k$. Indeed, the equivalence theorem is an immediate corollary of this lemma when combined with Theorem 3.1. As for the lemma, the idea of the proof is as follows. Let x be in the core. Suppose that traders in each type are represented on the unit interval, with Lebesgue measure λ, atoms of that type being subintervals. Figure 2 provides a representation of the economy where, for all pairs of traders in a given type, one trader is "below" another if, and only if, under x he prefers the other's consumption to his own. The set D represents the atomless part $T \backslash \bigcup_k T_k$.

If, contrary to the lemma, all traders in some type are not in the same indifference class, then the condition of Theorem 4.2 implies that there exists a number α, $\alpha \in]0, 1[$, such that if a horizontal straight line is drawn in Figure 2 at level α through the types, no atom is "split" by this line. Then the agents below this line are worse off in all types, and they will, supplemented by some subcoalition P in D, form a blocking coalition. The idea is to choose, by Lyapunov's theorem, a subset P of D such that the agents below α together with the subset form an α-reduction of the initial economy. Of course, the agents of a given type are not originally defined as supposed in the above reasoning, and the main difficulty of the proof of the lemma consists in "rearranging" traders in such a way that the above reasoning can be applied.

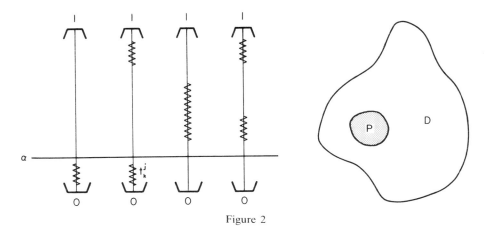

Figure 2

5. Restricted competitive allocations and the core of mixed markets

The "budgetary exploitation theorem" asserts that core allocations in mixed markets are sustained by "efficiency prices" such that, for any trader, a consumption preferred to what he gets under that allocation would also be more expensive; furthermore the values, at these prices, of the consumption received in that allocation by any *small* trader cannot exceed the value of his initial endowment. As the example in Subsection 3.2 shows, one should generally expect the latter to be *strictly* larger than the former, implying strict budgetary exploitation [i.e. $p \cdot x(t) < p \cdot w(t), t \in T_0$]. Nonetheless, under the additional assumptions introduced in Section 4, the equivalence property is restored: efficiency prices are also competitive prices.

Without requiring as much as the equivalence property for core allocations in mixed markets, one could be interested in a weaker property of price decentralization for such allocations, namely, that all small traders to be, at that allocation, in competitive equilibrium with respect to the efficiency price system p. More precisely, define a *restricted allocation* $x|_{T_0}$ to be *competitive* if there exists a price system p such that, for almost all $t \in T_0$, $p \cdot x(t) = p \cdot w(t)$ and $y >_t x(t) \Rightarrow p \cdot y > p \cdot w(t)$. That is, a competitive restricted allocation carries the property that all small traders are as in a competitive allocation w.r.t. a price vector $p : x(t)$ is a maximal element for $\succsim_t$ of the budget set $\{y \mid p \cdot y \le p \cdot w(t)\}$. But it does not follow that $x|_{T_0}$ is the restriction to T_0 of a competitive allocation, for the large traders need not be in competitive equilibrium with respect to p; and it does not follow that $x|_{T_0}$ is an allocation for the subeconomy consisting of the atomless sector T_0 alone, since the equality of supply and demand (over T_0) may be violated (it is *not* required that $\int_{T_0} x = \int_{T_0} w$). However, if at a particular core allocation x no *strict* budgetary exploitation against the small traders is observed at efficiency prices p, i.e. if $p \cdot x(t) = p \cdot w(t)$ for $t \in T_0$, the restriction $x|_{T_0}$ would then be competitive. Moreover, notice that small traders of the same type must get, at a restricted competitive allocation, equivalent bundles.

In this section we study sufficient conditions under which an allocation in the core of a mixed market has a competitive restriction on the atomless sector. To this end it is useful to introduce the notion of a split market. The market is said to be *split* with respect to (w.r.t.) a core allocation x if there exists a coalition S (called the *splitting coalition*) such that $\int_S x = \int_S w$ and $0 < \mu(S) < \mu(T)$. For the following theorem we assume that core allocations are in the interior of the commodity space and that indifference curves generated by $\succsim_t$ are C_1.

Theorem 5.1 [Shitovitz (1982a)]. *If the market is split with respect to a core allocation x, $x|_{T_0}$ is a restricted competitive allocation.*

In the literature on mixed markets, several conditions have been identified under which the market can be split w.r.t. to each core allocation, implying, with Theorem 5.1, that such core allocations are restricted competitive allocations. The first condition stated below involves only the set of atoms and has been introduced by Shitovitz (1973). Define two large traders (atoms) to be of the same *kind* if they are of the same type and have the same measure. Thus every market may be represented by $(T_0; A_1, A_2, \ldots)$, where T_0 is the atomless sector and $A_1, A_2, \ldots$ is a partition of the set of all atoms such that two atoms belong to the same A_k, iff they are of the same kind. Denote the number of atoms in A_k by $|A_k|$.

Theorem 5.2 [Shitovitz (1973)]. *Given a market $(T_0; A_1, A_2, \ldots)$, let m denote the g.c.d. (greatest common divisor) of $|A_k|$, $k = 1, 2, \ldots$; if $m \geq 2$, all core allocations are restricted competitive allocations.*

An alternative condition, implying an identical result, has been introduced in Drèze, Gabszewicz, Schmeidler and Vind (1972); this condition involves only the atomless sector. It states that, for each commodity, there exists a non-null subset of the atomless sector, the initial endowment of which is made only of that commodity. Assuming also that indifference surfaces generated by $\succsim_t$ are C_1, one can prove

Theorem 5.3 [Drèze, Gabszewicz, Schmeidler and Vind (1972)]. *If, for each commodity j, $j = 1, \ldots, n$, there exists a non-null coalition S_j, $S_j \subset T_0$ for which $\int_{S_j} w^i = 0$ and $\int_{S_j} w^j > 0$, then the market can be split and all core allocations are restricted competitive allocations.*

Finally, the following condition concerns both the set of atoms and the atomless part of the economy. Assume that there is a finite number of atoms A_h, $h = 1, \ldots, m$, and that for all h, $w(A_h) = w$. Further, assume that, for all h, $\succsim_{A_h}$ is derived from a homogeneous utility function and that there exists S_h, $S_h \subset T_0$, $\mu(S_h) > 0$, with for all $t \in S_h$, $\succsim_t = \succsim_{A_h}$ and $w(t) = w$.

Theorem 5.4 [Gabszewicz (1975)]. *Under the preceding assumptions, any allocation x in the core is a restricted competitive allocation.*

Not only can a restricted competitive allocation which is in the core be fully decentralized by efficiency prices on the atomless sector, but such a core allocation can also be transformed in a competitive allocation for the same prices under an appropriate redistribution of the initial resources of *the atoms among themselves* ($p \cdot \int_{T_0} x = p \cdot \int_{T_0} w \Rightarrow p \cdot \int_{T_1} x = p \cdot \int_{T_1} w$). Accordingly any discrimination among traders introduced by an allocation in the core which is restricted competitive – as compared with a competitive allocation – is a phenomenon affecting the atoms only; within the atomless sector, no discrimination takes place.[1] Finally, it is worthwhile to point out two additional properties of core allocations when they are also restricted competitive. Define an allocation x to be *coalitionally fair* (c-fair) relative to disjoint coalitions S_1 and S_2, if there exists no y and no i, $i = 1, 2$, such that for all $t \in S_i$, $y(t) >_t x(t)$ and $\int_{S_i} (y - w) = \int_{S_j} (x - w)$. In other words, an allocation is c-fair relative to S_1 and S_2 if neither of these coalitions could benefit from achieving the net trade of the other. The following theorem establishes a link between c-fair and restricted competitive allocations.[2]

Theorem 5.5 [Gabszewicz (1975)]. *If x is a restricted competitive allocation in the core, x is c-fair relative to all S_1 and S_2, $S_1 \subseteq T_0$, $T_1 \subseteq S_2$.*

Secondly, knowing that core allocations are also restricted competitive sometimes allows us to strengthen existing results in the literature of mixed markets. In particular, when the core of the economy defined in Subsection 4.2 consists only of restricted competitive allocations, these restricted competitive allocations are also competitive, implying therefore the equivalence theorem. A sufficient condition to that effect is $\mu(A_h)/\mu(T_h) + \mu(A_k)/\mu(T_k) < 1$, for all h, k $h, k = 1, \ldots, m$ [see Gabszewicz and Drèze (1971, Proposition 5, p. 413)], which is a considerable strengthening of Theorem 4.2.

---

[1] It may still be true, however, that the efficiency prices discriminate against the whole atomless sector when compared with the price system corresponding to a fully competitive allocation for the same economy; on this subject, see Gabszewicz and Drèze (1971).

[2] For further results on c-fair allocations, see Shitovitz (1987b).

6. Budgetary exploitation versus utility exploitation

As promised at the end of Section 3, we now treat the idea of "exploitation" of the small traders, which is fundamental in our analysis of oligopoly in mixed markets. We have expressed this idea in terms of a "value" criterion, using Pareto prices. Actually, however, each trader is concerned with his preferences rather than with any budgetary criterion. There exist classes of markets in which budgetary profit expresses the relative situations of some traders in terms of their preferences (see Subsection 6.2). But in general this is not the case, and there exist markets (even monopolistic markets) in which, although budgetarily exploited in the sense of Theorem 3.1, *all* small traders are actually better off than at any competitive equilibrium. We now examine this question in the narrow sense of a monopoly, from the viewpoint of the atom.

6.1. "Advantageous" and "disadvantageous" monopolies

In his "Disadvantageous monopolies", Aumann (1973) presents a series of examples (Examples A, B and C) showing that budgetary exploitation does not necessarily imply utility exploitation. In these examples there is a single atom $\{a\}$ and a nonatomic part (the "ocean"). All examples are two-commodity markets, with all of one commodity initially concentrated in the hands of the atom and all of the other commodity initially held by the ocean. Thus *a* is a "monopolist" both in the sense of being an atom and in the sense that he initially holds a "corner" on one of the two commodities. We omit the other details of the examples.

In Example A, the core is quite large, there is a unique competitive allocation, and from the monopolist's viewpoint, the competitive allocation is approximately in the "middle" of the core [see Figure 3(a)].

Example B is a variant of Example A. Here, the core is again quite large, there is a unique competitive allocation, and from the monopolist's viewpoint, the competitive allocation is the *best* in the core [see Figure 3(b)]. Thus the monopoly is "disadvantageous" and the monopolist would do well to "go competitive", i.e. split itself into many competing small traders.

Perhaps the most disturbing aspect of these examples is their utter lack of pathology. One is almost forced to the conclusion that monopolies which are not particularly advantageous are probably the rule rather than the exception. The conclusion is rather counterintuitive since one would conjecture that the monopoly outcome should be advantageous for the monopolist when compared with its competitive outcome. Although Aumann's examples disprove this conjecture, Greenberg and Shitovitz (1977) and Postlewaite (unpublished) have been able to show

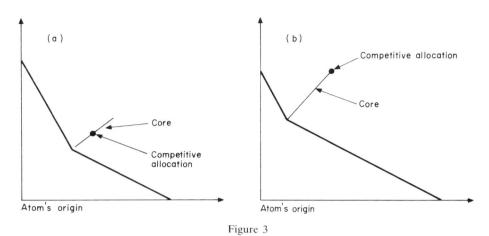

Figure 3

Theorem 6.1 [Greenberg and Shitovitz (1977)]. *In an exchange economy with one atom, and one type of small traders, for each core allocation x there is a competitive allocation y whose utility to the atom is smaller than that of x, whenever either x is an equal treatment allocation, or all small traders have the same homogeneous preferences.*

Examples have been constructed to show that the conditions of Theorem 6.1 are indispensable for the result to hold [see Greenberg and Shitovitz (1977)].

6.2. When budgetary exploitation implies utility exploitation

That budgetary exploitation implies utility exploitation for the small traders has been established for two particular classes of markets, namely "homogeneous" and "monetary" markets. A *homogeneous market* is a market in which all traders t have the same homogeneous preference relation, namely a relation derived from a concave utility function $u(x)$ that is homogeneous of degree 1 and has continuous derivatives in the neighborhood of $\int_T w$. It is easy to see that a Pareto-optimal allocation x can be written as $x(t) = \alpha(t) \cdot \int_T w$, with $\int_T \alpha(t) = 1$. Hence, the efficiency prices p are the same for all Pareto-optimal allocations. There is also a unique competitive allocation x^*. Let $p(t) = p \cdot x(t) - p \cdot w(t)$ be the "budgetary profit" of trader t. Then we have $p(t) \leq 0$ if and only if $u(x(t)) \leq u(x^*(t))$. For x in the core, therefore, it follows that small traders are at most as "satisfied" (in the sense of preferences) as they are at the competitive equilibrium, and in split markets the small traders receive in the core the same bundle that they receive at the competitive equilibrium.

In a *monetary market*, the consumption set of each trader is $R \times \Omega$, i.e. a commodity bundle (ξ, x) consists of an amount ξ of "money" and a vector $x \in \Omega$ of commodities. The utility function $U_t(\xi, x)$ of trader t is assumed to be linear in money, i.e.

$$U_t(\xi, x) = \xi + V_t(x) \, .$$

Furthermore, the initial amount of money of each trader in T_1 is supposed to be zero, while trader t holds commodities $w(t) \geqslant 0$. Note that a price vector for a monetary economy which constitutes an efficiency equilibrium with some allocation x is of the form $(1, p)$ where, for each t, $V_t(x(t) - p \cdot x(t)) = \max_{x \in \Omega}(V_t(x) - p \cdot x)$. Using a generalization of Theorem 3.1,[3] it can be proved that for each core allocation x, there exists $p \geqslant 0$ in R^n which constitutes an efficiency equilibrium with $(1, p)$ and $\xi(t) + p \cdot x(t) \leqslant p \cdot w(t) \ \forall t \in T_0$. But, defining $\xi^*(t)$ in such a way that, for almost all $t \in T$

$$\xi^*(t) + p \cdot x(t) = p \cdot w(t) \, ,$$

we obtain that $(1, p), (\xi^*, x)$ is a competitive equilibrium. Hence it follows that, for almost all $t \in T_0$, $\xi(t) \leqslant \xi^*(t)$ and therefore we obtain $U_t(\xi(t), x(t)) \leqslant U_t(\xi^*(t), x(t))$. Thus, budgetary exploitation implies utility exploitation (Samet and Shitovitz, unpublished). Note that these monetary markets can be viewed as a transferable utility side-payment game T^*, where the characteristic function v is defined as [see Aumann and Shapley (1974)]

$$v(S) = \max\left\{ \int_S u_t(x(t)) \, d\mu(t) : \int_S x(t) \, d\mu = \int_S \alpha(t) \right\} \, .$$

Using the equivalence principle and the standard definition of the core in side-payments games, we obtain:

Theorem 6.2 (Samet and Shitovitz, unpublished). *For each core allocation α of the game $(T, \mathcal{F}, \mu)$, there is an allocation α^* in the core of the game T^* such that $\alpha(t) \leqslant \alpha^*(t)$ for a.e. $t \in T_0$.*

Note that α^*, like all other core allocations of T^*, is a transferable utility competitive equilibrium.

---

[3]See Shitovitz (1982a) and Champsaur and Laroque (1974).

6.3. Disadvantageous monopolies and disadvantageous endowments

In his Example B, Aumann (1973) used the term "disadvantageous monopoly" for an exchange economy with a single atom and a nonatomic sector, whenever the core of the economy consists of a single competitive allocation *y* and a set of noncompetitive allocations, all of which are less favorable for the atom than *y*. An analogous definition can be used to characterize disadvantageous transfers among traders. In an exchange economy, a non-negative transfer of initial resources (a "gift") from traders in some subset $\hat{T}$ to traders in another subset $\bar{T}$, say, generates a *disadvantageous endowment* whenever, after the gift, there exists a competitive allocation assigning to traders in $\bar{T}$ a consumption less preferred than every consumption assigned to the same traders that was competitive before the gift.

Using Theorem 3.1, it can be shown:

Theorem 6.3 [Drèze, Gabszewicz and Postlewaite (1977)]. *Whenever the measure space of agents consists of a single atom {a} and a nonatomic part, and the atom a is a disadvantageous monopoly, there exists a gift from the nonatomic part to the atom generating a disadvantageous endowment for a.*

Thus this proposition asserts that disadvantageous endowments are no more unusual than disadvantageous monopolies!

7. Syndicates

In our Introduction we considered the possibility of interpreting an atom as a "syndicate" of traders formed in an initially atomless economy. In this section we elaborate on this interpretation by investigating the effectiveness and the stability of binding agreements among traders in the context of a pure exchange economy. To this end, it is convenient to work with a simplified version of our general exchange model. Consider an atomless exchange economy with a continuum T of traders falling into r "types", T_k, $k = 1, \ldots, r$, i.e. all traders t in T_k have the same preferences ($\succsim_t \overset{\text{def}}{=} \succsim_k; t \in T_k, k = 1, \ldots, r$) and the same initial assignment ($w(t) \overset{\text{def}}{=} w_k; t \in T_k; k = 1, \ldots, r$). For simplicity let T_k be the right open interval $[k - 1, k]$ and T_r the closed interval $[r - 1, r]$ so that $T = [0, r]$. Also let $\mathcal{T}$ denote the class of Lebesgue measurable subsets of T and μ the Lebesgue measure defined on $\mathcal{T}$. Syndicates of traders are defined informally as follows: Let A_{hk}, $h = 1, \ldots, j_k$, be a measurable subset of T_k and imagine that all traders in A_{hk} agree that (i) no *proper* subset of them will form a coalition with traders outside A_{hk}, so that only the group as a whole will enter into broader coalitions; (ii) only those allocations which assign an identical consumption vector to all $t \in A_{hk}$ will be accepted by the group.[4] Denote by A the set $\bigcup_{k=1}^{r} [\bigcup_{h=1}^{j_k} A_{hk}]$; the set A contains all agents in A who are members of a syndicate; the set A will also be referred to as a *syndicate structure*. The formal consequences of this definition of the syndicates A_{hk} are the following:

(1) the set of potential coalitions is now reduced to a subclass of the original σ-field $\mathcal{T}$, namely the subclass $\mathcal{T}_A$ defined by

$$\mathcal{T}_A = \{ S \in \mathcal{T} \mid \forall k = 1, \ldots, r, \forall h = 1, \ldots, j_k,$$

$$S \cap A_{hk} = \emptyset \text{ or } S \cap A_{hk} = A_{hk} \};$$

(2) the set of admissible allocations in the economy is now restricted to those allocations which are constant-valued on $A_{hk}, k = 1, \ldots, r; h = 1, \ldots, j_k$, i.e. which are $\mathcal{T}_A$-measurable;

(3) the core must be redefined in terms of the new class of coalitions and the new set of admissible allocations; the $\mathcal{T}_A$ − core $\mathscr{C}(\mathcal{T}_A)$ is the set of all $\mathcal{T}_A$-measurable allocations that are not dominated, via any non-null coalition in $\mathcal{T}_A$, by some $\mathcal{T}_A$-measurable allocation;

(4) on the other hand, it is readily verified that the set $\mathscr{E}$ of competitive allocations is invariant with respect to the process of syndicate formation.

The formal analogy between a syndicate in an atomless measure space and an atom in an atomic measure space is thus complete. Accordingly, general results established for an exchange economy with atoms are directly applicable to our special case. This is true in particular for Theorems 4.1, 4.2 and 5.2, which now become, respectively:

Proposition 7.1. *Assume that for all $k' \neq k$, there is no syndicate included in $T_{k'}$, while there are at least two non-null syndicates included in T_k. Then $\mathscr{C}(\mathcal{T}_A) = \mathscr{E}$.*

Proposition 7.2. $\mu(A) < 1 \Rightarrow \mathscr{C}(\mathcal{T}_A) = \mathscr{E}$.

Proposition 7.3. *Assume that the set of syndicates can be divided into at least two disjoint subsets S_1 and S_2 such that, to each syndicate in S_1 included in T_k, there corresponds a syndicate in S_2, of the same measure, also included in T_k. Then all core allocations are restricted competitive allocations.*

---

[4]This rule of uniform imputation is justified by the fact that all traders in A_{hk} are identical. This justification would no longer be valid should syndicates include traders of *different* types. We shall consider this alternative assumption below.

Proposition 7.1 states that if all existing syndicates are "of the same type", the equivalence theorem must hold. Proposition 7.2 shows that the process of syndicate formation must involve a sufficiently "broad" class of traders before it can become effective at enlarging the core of the economy.

In the absence of syndicates, the core of our (atomless) exchange economy coincides with $\mathscr{E}$; this set consists of $\mathscr{T}_A$-measurable allocations only, and thus belongs to the $\mathscr{T}_A$-core. If syndicates are to be *effective*, they must bring about an allocation that is not competitive; assuming that only allocations in the $\mathscr{T}_A$-core can emerge, we are led to investigate allocations in $\mathscr{C}(\mathscr{T}_A)\backslash\mathscr{E}$, i.e. noncompetitive core allocations. Any such allocation could be blocked by some coalition involving a *proper* subset of at least one syndicate; that is, it could be blocked if some syndicate members could be persuaded to break the agreement which binds them to the syndicate. Under a noncompetitive allocation one must accordingly reckon with a permanent temptation for some traders either to leave their syndicate or to break its rules (by secretly recontracting with outsiders). How can the syndicates achieve stability in the face of such temptations? To simplify the discussion, we shall assume that there is only one syndicate in each type k, and denote it by A_k. One can think of two types of economic considerations that may preserve the stability of syndicates, namely:

(i) comparison of the consumption of syndicate members (t in A_k) with that of "unorganized" traders of the same type (t in $T_k\backslash A_k$), and

(ii) comparison of the consumption of syndicate members with what they would receive under competitive allocations. A natural requirement for the stability of a syndicate A_k under the first type of comparison, given an allocation y, would be the existence of a non-null set of unorganized traders of the same type who are not better off than the members of syndicate A_k. This leads to the following concept of *marginal stability*.

Definition 7.1.   Given a syndicate structure $A = (A_1, \ldots, A_r)$ with $\mu(A_k) < 1$, $\forall k$, an allocation $y \in \mathscr{C}(\mathscr{T}_A)$ is *marginally stable* iff, for all k with $\mu(A_k) > 0$, there exists S_k, $S_k \subseteq T_k\backslash A_k$, $\mu(S_k) > 0$ and $y(t) \succsim_k y(\tau)$, $\tau \in A_k$, $t \in S_k$.

Clearly any competitive allocation satisfies marginal stability, but no other allocations in $\mathscr{C}(\mathscr{T}_A)$ do, as indicated in the following

Proposition 7.4 [Gabszewicz and Drèze (1971)].   *Let A be a syndicate structure with $\mu(A_k) < 1$, $k = 1, \ldots, r$; then no allocation in $\mathscr{C}(\mathscr{T}_A)\backslash\mathscr{E}$ is marginally stable.*

Having reached this negative conclusion, we turn to the other type of comparison, in which an allocation $y \in \mathscr{C}(\mathscr{T}_A)\backslash\mathscr{E}$ is compared with allocations in $\mathscr{E}$. A natural requirement for the stability of a syndicate A_k under this type of comparison, given an allocation y, $y \in \mathscr{C}(\mathscr{T}_A)\backslash\mathscr{E}$, would be that the members of A_k are better off under y than they would be under every competitive allocation.

Definition 7.2. Given a syndicate structure A, an allocation $y \in \mathscr{C}(\mathscr{T}_A)\backslash\mathscr{E}$ carries the property of total stability whenever, for all $t \in A$ and all $x \in \mathscr{E}$,

$$y(t) \gtrsim_t x(t) .$$

Proposition 7.5 [Gabszewicz and Drèze (1971)]. *There exist exchange economies for which the class of syndicate structures A, with $\mu(A_k) < 1$, $k = 1, \ldots , r$, such that $\mathscr{C}(\mathscr{T}_A)\backslash\mathscr{E}$ contains allocations that satisfy total stability, is nonempty.*

Propositions 7.3 and 7.4 raise several interesting questions insofar as the stability of syndicates is concerned. Clearly syndicates are relevant only to the extent that they may enforce noncompetitive core allocations, i.e. allocations in $\mathscr{C}(\mathscr{T}_A)\backslash\mathscr{E}$. Such allocations appear at first sight unstable because they never satisfy marginal stability (Proposition 7.3). We must infer from this that the relevant stability concept is one of total stability. Proposition 7.4 tells us that total stability may well hold for all syndicates simultaneously. Combining the two propositions, we are led to recognize the possibility that a syndicate structure may be both relevant and stable.

We have so far considered that all traders belonging to a particular syndicate are of the same type; this assumption allowed us to restrict our attention to allocations which were uniform on each syndicate ($\mathscr{T}_A$-measurable allocations). In many circumstances, however, more complex forms of syndicates should be envisaged; thus, a large corporation may be regarded as a syndicate of employees and stockholders; then all members are not of the same type and no rule of uniform imputation would be justified. The question then arises whether the above results on syndicate structures would still hold when syndicates may include traders of different types. Unfortunately, Champsaur and Laroque (1976) have shown that the analogs of Propositions 7.1 and 7.3 are no longer valid with such syndicate structures. Clearly much remains to be done on this topic, in which further research would be welcome.

8. Conclusions

The work we have reported above is only part of the full panoply of research devoted to the study of mixed markets. Other paths have been followed by researchers, and most of our concluding remarks aim at briefly describing the results they have discovered. These include in particular the study of alternative cooperative game-theoretic solution concepts in the framework of mixed markets and the problem of "approximating" mixed markets by finite exchange economies. We start, however, with a brief comment about some surprising features of the core's behavior in such markets. To this end let us return to the example presented in Section 3.2. In this example we have assumed that there is a single atom with measure equal to 1. Now consider the same exchange economy, but in which the atom has been replaced by *two* atoms of measure β and $1 - \beta$, respectively, $\beta \in]0, 1[$. While the core with a single atom consisted of all allocations x of the form $(\alpha(t), \alpha(t))$, where $1 \leq \alpha(t) \leq 2$, for almost all $t \in T_0$ and $2 \leq \alpha(2) \leq 3$ for the atom, the core now consists of the sole competitive allocation, irrespective of the value of β in $]0, 1[$; this follows immediately from Theorem 4.1. We conclude from this example that, in the abstract setup of a measure space of traders with atoms and an atomless part, the core correspondence is not lower hemicontinuous. Consider indeed the sequence of economies E_r obtained by replacing the single atom of measure 1 by two atoms of measures $1/r$ and $(r - 1)/r$, respectively. This sequence of economies converges to the economy E_∞ containing a single atom of measure 1. According to Theorem 4.1, the core of E_r consists, for all r, of the sole competitive allocation. However, for $r = \infty$, the core suddenly enlarges so as to contain a whole "segment" of allocations which are no longer competitive. Starting from these difficulties, some have questioned the abstract representation of oligopolistic markets via mixed markets. Perhaps the abstract representation itself would be responsible for those disturbing results. The question was whether the representation of a mixture of small and large traders via finite, possibly large economies, would dispose of this pathology. It turns out, however, that this is not the case: the same disturbing phenomena may well occur even if one considers *finite* exchange economies. Building a sequence of *finite* duopolistic markets $\mathcal{E}_r$ converging to an economy containing a single monopoly, $\mathcal{E}_\infty$, Gabszewicz (1977) has shown that while the cores of the duopolistic markets $\mathcal{E}_r$ converge to the set of competitive allocations, the core of the limit monopolistic market $\mathcal{E}_\infty$ is much larger and includes many noncompetitive allocations; also, in this asymptotic version the core does not behave lower hemicontinuously. Most probably it is the concept of the core itself which is responsible for the discontinuity. Some small coalitions, including the "small" large trader, are endowed with the possibility of depriving the "big" large trader of all its power, while the intuition would suggest that such coalitions will not act to do so. This seems to indicate that the concept of the core must be used with care: core analysis gives qualitative insights rather than exact predictions of market behavior.

The model of mixed markets has also been used to examine the behavior of other cooperative concepts, like the von Neumann and Morgenstern (v-N.M.)

solutions, the bargaining set, and the transferable Shapley value. Since this survey is mainly concerned with the concept of core we shall limit ourselves to a succinct presentation of these approaches. The v-N.M. solutions in large markets were analyzed by Hart (1974). He considered a nonatomic market consisting of a finite number of different types of traders, initially owning disjoint sets of goods. It is proved there that if all traders of each type form a cartel and behave like a single (atomic) trader (their "representative"), then one gets solutions (i.e. v-N.M. stable sets of allocations) of the original nonatomic market from those of the finite market made of the "representatives"; furthermore, *all* symmetric solutions are obtained in this manner. Another solution concept is the bargaining set, i.e. the set of all allocations with no "justified objection". Recently, Mas-Colell (1986) gave a uniform modified definition of the bargaining set for atomless economies, and proved its equivalence with the set of competitive allocations. Applying this concept to mixed markets, Shitovitz (1987a) has exhibited an example involving a duopolistic market (with two atoms) where the equivalence theorem between the core and the bargaining set fails to hold (compare with our Theorem 4.1 about the core). This suggests that, in some contexts, the bargaining set would lead to drastically different outcomes than the core, reflecting more adequately the relationship of market power to the size of the agents. Nevertheless, it can be shown that, when each atom has a corner on some commodities and hence is a veto player, the core and the bargaining set do coincide [Shitovitz (1987)]. As we have seen in Subsection 6.1, syndicates can be advantageous as well as disadvantageous. The existence of disadvantageous syndicates led Aumann (1973) to the conclusion that "the game theoretic notion of core is not the proper vehicle for the explanation... (of the monopolistic advantage)". He conjectured that perhaps the ideas underlying the concept of the Shapley value could better capture the bargaining power engendered by the harm the monopolist can cause by refusing to trade. Starting from this conjecture and using the asymptotic approach, Guesnerie (1977) has examined the stability of a monopoly from the viewpoint of Shapley's NTU value. He exhibits examples with no pathological features where the monopoly is unstable from this viewpoint as well, thus disproving Aumann's conjecture. Hence the Shapley value does not seem at first sight a more appropriate concept than the core for capturing the bargaining power of syndicates or monopolies. Finally, a recent paper by Legros (1987) examined the stability of syndicates with respect to the nucleolus concept. In particular, he found an example where the monopoly is unstable from the viewpoint of the nucleolus. The example seems, however, highly pathological since the author uses a utility function which is not differentiable. On the other hand, if we consider *any* homogeneous differentiable market (where the transferable utility competitive equilibrium is unique), we obtain stability of the nucleolus since the core contains the nucleolus.

Finally, we come to the problem of approximating mixed markets via large, but finite, economies, in the tradition introduced by Debreu and Scarf (1963) and Hildenbrand (1974) for the approximation of atomless economies. The first finite version of a mixed market was proposed by Drèze, Gepts and Gabszewicz (1969) and provides an asymptotic analog of Gabszewicz and Mertens (1971). It bears a relation to the latter paper that is similar to the relation of the Debreu and Scarf (1963) paper to the work by Aumann (1964). Similarly, Gabszewicz (1977) provides an asymptotic analog of Shitovitz's Theorem B (1973). Using the methods of non-standard analysis – a branch of model theory – Khan (1976) has shown that Shitovitz's results hold as well for large but finite economies in which traders, except those belonging to a finite set, are "negligible". His work also sheds some light on the interpretation and modeling of large traders via mixed markets. The next interesting contribution in this area of research is due to Trockel (1976), who uses Hildenbrand's (1974) framework, and proves the upper hemicontinuity in distribution of the core correspondence in mixed markets. Finally, using "approximate" efficiency prices, Greenberg and Shitovitz (1981) have given explicit bounds on the size of the budgetary exploitation of the small traders in a finite version of a mixed market. To this end, they make use of the novel proof of the equivalence theorem provided by Anderson (1978).

It is now common knowledge that core theory applied to atomless market models has proved to be fruitful for analyzing perfect competition. It is our personal feeling that this theory fits as well for analyzing *imperfect* competition as when the mixed market model is utilized for representing the asymmetric size of the economic agents. At this point we hope that the reader has come to share at least part of our conviction.

References

Anderson, R.M. (1978) 'An elementary core equivalence theorem', *Econometrica*, 46: 1483–1487.

Aumann, R.J. (1964) 'Markets with a continuum of traders', *Econometrica*, 32: 39–50.

Aumann, R.J. (1973) 'Disadvantageous monopolies', *Journal of Economic Theory*, 6: 1–11.

Aumann, R.J. and L.S. Shapley (1974) *Values of nonatomic games*. Princeton: Princeton University Press.

Champsaur, P. and G. Laroque (1974) 'Une nouvelle démonstration de l'équivalence entre le noyau et l'ensemble des équilibres concurrenciels', *Cahiers du Séminaire d'Econométrie*, 14: 59–71.

Champsaur, P. and G. Laroque (1976) 'A note on the core of economies with atoms or syndicates', *Journal of Economic Theory*, 13: 458–471.

Debreu, G. and H. Scarf (1963) 'A limit theorem on the core of an economy', *International Economic Review*, 4: 235–246.

Drèze, J.H., S. Gepts and J.J. Gabszewicz (1969) 'On cores and competitive equilibria', in: *La décision, agrégation et dynamique des ordres de préférence. Colloques Internationaux du C.N.R.S.*, 171: 91–114.

Drèze, J., J.J. Gabszewicz, D. Schmeidler and K. Vind (1972) 'Cores and prices in an exchange economy with an atomless sector', *Econometrica*, 40: 1091–1108.

Drèze, J.H., J.J. Gabszewicz and A. Postlewaite (1977) 'Disadvantageous monopolies and disadvantageous endowments', *Journal of Economic Theory*, 16: 116–121.

Gabszewicz, J.J. (1975) 'Coalitional fairness of allocations in pure exchange economies', *Econometrica*, 43: 661–668.

Gabszewicz, J.J. (1977) 'Asymmetric duopoly and the core', *Journal of Economic Theory*, 14: 172–179.

Gabszewicz, J.J. and J.H. Drèze (1971), 'Syndicates of traders in an exchange economy', in: G. Szegö, ed., *Differential games and related topics*. Amsterdam: North-Holland.

Gabszewicz, J.J. and J.-F. Mertens (1971) 'An equivalence theorem for the core of an economy whose atoms are not "too" big', *Econometrica*, 39: 713–721.

Greenberg, J. and B. Shitovitz (1977) 'Advantageous monopolies', *Journal of Economic Theory*, 16: 394–402.

Greenberg, J. and B. Shitovitz (1981) 'Cores of finite oligopolistic markets with non-convex preferences for small traders', in: O. Moeschlin and D. Pallaschke, eds., *Game theory and mathematical economics*. Amsterdam: North-Holland.

Greenberg, J. and B. Shitovitz (1986) 'A simple proof of the equivalence theorem for oligopolistic mixed markets', *Journal of Mathematical Economics*, 15: 79–83.

Guesnerie, R. (1977) 'Monopoly, syndicate and Shapley value: About some conjectures', *Journal of Economic Theory*, 15: 235–251.

Hart, S. (1974) 'The formation of cartels in large markets', *Journal of Economic Theory*, 7: 453–466.

Hildenbrand, W. (1974) *Core and equilibria in large economies*. Princeton: Princeton University Press.

Khan, M.A. (1976) 'Oligopoly in markets with a continuum of traders: An asymptotic interpretation', *Journal of Economic Theory*, 12: 273–297.

Legros, P. (1987) 'Disadvantageous syndicates and stable cartels: The case of the nucleolus, *Journal of Economic Theory*, 42: 30–49.

Mas-Colell, A. (1989) 'An equivalence theorem for a bargaining set', *Journal of Mathematical Economics*, 18: 129–139.

Shitovitz, B. (1973), 'Oligopoly in markets with a continuum of traders', *Econometrica*, 41: 467–505.

Shitovitz, B. (1974) 'On some problems arising in markets with some large traders and a continuum of small traders', *Journal of Economic Theory*, 4: 458–470.

Shitovitz, B. (1982a) 'On exploitation in a class of differentiable mixed markets, *Economic Letters*, 9: 301–304.

Shitovitz, B. (1982b) 'Some notes on the core of production economies with some large traders and a continuum of small traders', *Journal of Mathematical Economics*, 9: 99–105.

Shitovitz, B. (1987a) 'Notes on the bargaining set and the core in mixed markets', mimeo.

Shitovitz, B. (1987b), 'Coalitional fair allocations in differentiable mixed markets', mimeo.

Trockel, W. (1976) 'A limit theorem on the core', *Journal of Mathematical Economics*, 3: 247–264.

Vind, K. (1972) 'A third remark on the core of an atomless market', *Econometrica*, 40: 585–586.

von Neumann, J. and O. Morgenstern (1944) *Theory of games and economic behavior*. Princeton: Princeton University Press.

Chapter 16

TWO-SIDED MATCHING

ALVIN E. ROTH[a] and MARILDA SOTOMAYOR[b]

[a]*University of Pittsburgh and*
[b]*Pontificia Universidade Catolica do Rio de Janeiro*

Contents

Handbook of Game Theory, Volume 1, Edited by R.J. Aumann and S. Hart

1. Introduction

The games we consider in this chapter are "two-sided matching markets". The phrase "two-sided" refers to the fact that agents in such markets belong, from the outset, to one of two disjoint sets – e.g. firms or workers. The term "matching" refers to the bilateral nature of exchange in these markets – e.g. if I work for some firm, then that firm employs me. In recent years the game-theoretic analysis of these markets has proved useful in various empirically oriented studies. To emphasize the close connection between empirical and theoretical work in this area, this chapter begins by describing some of the phenomena the theory should be able to explain. Much of the available theory will be summarized in the body of the chapter, and the chapter will conclude by returning to consider how the theory addresses the empirical questions raised at the beginning.

   We will be concerned both with the core of the game, and with the dominant and equilibrium strategies under various rules about how the game might be played. Thus this material will serve to emphasize that the distinction between "cooperative" and "noncooperative" game theory is often somewhat artificial, since the tools of both kinds of theory can be used to study the same phenomena.

   This chapter is adapted from our monograph, Roth and Sotomayor (1990a), in which a much more complete treatment can be found.

2. Some empirical motivation

2.1. The case of American physicians

Hospitals began offering newly-graduated medical students internship positions around the turn of the century. Not until 1945 were the relevant medical associations able to institute a single market with uniform dates at which such positions would be offered.[1] Once this was accomplished, however, both students and hospitals were dismayed by the chaotic conditions that developed between the time offers of internships were first made, and the time by which students were required to accept or reject them. The situation is described as follows in Roth (1984a).

---

[1]See Roth (1984a) for a description of the difficulties encountered in setting uniform appointment dates prior to 1945, which will not be discussed here.

Basically, the problem was that a student who was offered an internship at, say, his third choice hospital, and who was informed he was an alternate (i.e. on a waiting list) at his second choice, would be inclined to wait as long as possible before accepting the position he had been offered, in the hope of evantually being offered a preferable position. Students who were pressured into accepting offers before their alternate status was resolved were unhappy if they were ultimately offered a preferable position, and hospitals whose candidates waited until the last minute to reject them were unhappy if their preferred alternate candidates had in the meantime already accepted positions. Hospitals were unhappier still when a candidate who had indicated acceptance subsequently failed to fulfil his commitment after receiving a preferable offer. In response to pressure originating chiefly from the hospitals, a series of small procedural adjustments were made in the years 1945–51. The nature of these adjustments, described next, makes clear how these problems were perceived by the parties involved.

For 1945, it was resolved that hospitals should allow students ten days after an offer had been made to consider whether to accept or reject it. For 1946, it was resolved that there should be a uniform appointment date (July 1) on which offers should be tendered . . . , and that acceptance or rejection should not be required before July 8. By 1949, [the Association of American Medical Colleges] proposed that appointments should be made by telegram at 12:01 AM (on November 15), with applicants not being required to accept or reject them until 12:00 Noon the same day. Even this twelve-hour waiting period was rejected by the American Hospital Association as too long: the joint resolution finally agreed upon contained the phrase "no specified waiting period after 12:01 AM is obligatory," and specifically noted that telegrams could be filed in advance for delivery precisely at 12:01 AM. In 1950, the resolution again included a twelve-hour period for consideration, with the specific injunction that "Hospitals and/or students shall not follow telegrams of offers of appointment with telephone calls' until after the twelve-hour grace period." [. . . the injunction against telephone calls was two-way, in order to stem a flood of calls both from hospitals seeking to pressure students into an immediate decision, and from students seeking to convert their alternate status into a firm offer].

It was eventually recognized that these problems could not be solved by compressing still further the time allowed for the last stage of the matching process, and it was agreed to try instead a centralized matching algorithm, on a voluntary basis. Students and hospitals would continue to exchange information via applications and interviews as before, but then both students and hospitals would submit rank-orderings of their potential assignments,[2] and the algorithm would be used to suggest a matching of students to hospitals, who would then, it was hoped, sign employment contracts with their suggested assignments.

The first algorithm proposed was abandoned after a year because it was observed to give students the incentive to submit a rank-ordering different from their true preferences. The algorithm proposed in its place was used for the first time in 1951, and remains in use to this day. (This algorithm will be called the NIMP algorithm, for National Intern Matching Program, the name under which the algorithm was initially administered.)

This system of arranging matches was *voluntary* – students and hospitals were free to try to arrange their own matches outside the system, and there was no way to enforce compliance on those who did participate.[3] This makes it all the more remarkable that, in the first years of operation, over 95 percent of eligible students and hospitals participated in the system, and these high rates of participation continued until the early 1970s.

Since then, although the overall rate of participation remains high, increasing numbers of students, particularly those among the growing number of medical students who are married to other medical students, have begun to seek to arrange their own matches, without going through the centralized clearinghouse. Another aspect of this market which has caused some concern in medical circles has been the resulting distribution of physicians among hospitals, with rural hospitals getting fewer interns than they wish, and a much higher percentage of interns who are graduates of foreign medical schools.

The chief phenomena we would like a theory to explain in this case are:

• What accounted for the disorderly operation of the market between 1945 and 1951?

• Why was the centralized procedure instituted in the 1951–52 market able to achieve such high rates of voluntary participation?

• Why did these high rates start to diminish by the 1970s, particularly among the growing number of medical students who were married to other medical students?

We will also want to investigate "strategic" questions of the kind that led to the scrapping of the first algorithm.

• Does the NIMP algorithm, as claimed by the sponsoring medical associations,

---

[2]Regarding the problem of formulating a rank-ordering, note that the complete job description offered by a hospital program in a given year was customarily specified in advance. Thus the responsibilities, salary, etc. associated with a given internship, while they might be adjusted from year to year in response to a hospital's experience in the previous year's market, were not a subject of negotiation with individual job candidates.

[3]The experience prior to 1950 amply demonstrated that no amount of moral suasion was effective at preventing participants from acting in what they perceived as their own best interests.

give students and hospitals the incentive to submit rank-orderings corresponding to their true preferences?

Finally, we would like to be able to get some idea of which aspects of the market could be influenced by modifying its organization, while preserving those features that have led to high rates of voluntary participation. In this regard, we will want to know:

• Can the defection of married couples be halted?
• Can the distribution of interns to rural hospitals be changed?

A preview of the proposed explanation

When we see that, in the late 1940s, there is a lot of two-way telephone traffic between hospitals and students, who sometimes renege on previous verbal agreements, we can hypothesize that there is some systematic incentive to the parties involved to behave in this way. These incentives must be mutual: if students who called hospitals that had not extended them offers were uniformly told that no places were available, the practice would be unlikely to persist in the virulent form that was observed. Situations in which there are some students and hospitals who are not matched with each other, but who *both* prefer to be matched one to the other, will therefore be called "unstable". By the same token, if the matching suggested by the NIMP algorithm was unstable in this way – i.e. if there were students and hospitals that would prefer to be matched to one another rather than to accept the suggested match – then we would expect that these students and hospitals would continue to try to locate each other, and subsequently decline to accept the assignment suggested by the matching procedure. The very high rates of voluntary participation in the years following the introduction of the NIMP procedure suggest this was not the case, and that the set of suggested assignments produced by the NIMP procedure must be "stable", i.e. must have the property that, if some student would prefer another hospital to his suggested assignment, then that hospital does not return the favor, but prefers the students assigned to it to the student in question. In Section 6 we will see that the NIMP assignments do indeed have this property. So our explanation of why the chaotic market conditions prior to 1951 vanished following the introduction of the NIMP procedure will be that it introduced this kind of stability to the market.

In a similar vein, we will observe that, as married couples became more common in this market, the procedures used to deal with them introduced instabilities once again, so that married couples could find hospitals that they preferred to their assigned matches and that were willing to offer them jobs. This will be the basis of our explanation of the defection of married couples from the system, which became so noticeable in the mid-1970s. We will also argue that the answers to our questions about how much freedom there is to alter the organization of the market while maintaining high rates of voluntary participation also hinge on whether any given organization of the market leads to stable market outcomes.

A complementary set of ideas, having to do with the strategies of individual agents in the market, will be used to explore the question of whether, as claimed, it is always in the interest of all parties to state their true preferences. We will see it is not, and that it cannot be for any procedure that produces stable outcomes. However, it is possible to arrange things so that it is always in the best interest of *some* of the parties to state their true preferences. The development of these ideas will involve us in a number of subtle issues, not the least of which is that we will be forced to reconsider and re-evaluate our conclusions about stability. If the students and hospitals may not be stating their true preferences when they submit rank-order lists for the NIMP algorithm, is there still reason to believe that the outcome is a stable set of assignments? When we look at equilibrium behavior we will see that there is.

2.2. Bidder rings in auctions

Strategic considerations of a somewhat different sort arise in the study of auction markets. The opportunities to profitably deviate from straightforward behavior are different for buyers and sellers. The sellers (and their agent the auctioneer) would like prices to be high, and the buyers would like prices to be low. The most commonly reported "strategic" behavior on the part of auctioneers or sellers is to introduce imaginary bids into the proceedings, which when practiced by auctioneers is called by a variety of colorful names, such as "pulling bids off the chandelier". And the most commonly reported strategic behavior on the part of buyers is to form *rings* that agree to coordinate their bidding in an effort to keep down the price. Cassady (1967) reports that in antique and art auctions, the subsequent auction among members of the ring, called a "knockout" auction, serves both to determine which of the ring members will receive what the ring has bought, and what payments shall be made by ring members among themselves. (The *Oxford English Dictionary* cites nineteenth-century sources for this meaning of the word "knockout", suggesting that the organization of bidder rings in this way is not only a widespread phenomenon, but also not a new one.) Cassady remarks that buyer rings are common in many kinds of auctions all over the world, although in auctions of divisible commodities (such as fish in England, timber in the United States, and wool in Australia), rings commonly divide the purchases among themselves, rather than conducting a knockout auction. An unusually detailed description of the strategic behavior of rings and auctioneers in New Jersey machine tool auctions is given by Graham and Marshall (1984). The analysis that follows will shed some light on the strategic opportunities facing the auctioneer and individual bidders, and the opportunities for bidders to organize themselves into rings.

3. Several simple models: Stability, and the polarization of interests in the core

Most theoretical work on this topic traces its history to the papers of Gale and Shapley (1962), and Shapley and Shubik (1972). Gale and Shapley formulated a model of two-sided matching without sidepayments which they called the marriage problem, and Shapley and Shubik formulated a sidepayment game which they called the assignment game. Each paper studied the core of the game, and showed it is nonempty for any preferences of the agents. Curiously, although Gale and Shapley were unaware of the 1951 NIMP algorithm, they showed the core was nonempty by formulating what can be regarded as an equivalent algorithm [Roth (1984a)]. Both for the marriage problem and the assignment game, these early papers demonstrated that, within the core, there is a surprising coincidence of interest among players on the same side of the market, and a polarization of interest between the two sides of the market. These two models are introduced in Subsections 3.1 and 3.4.

Both models involve one-to-one matching, i.e. each agent on one side of the market can be matched to at most one agent on the other side. Gale and Shapley also discussed the case of many-to-one matching, which they called the college admissions problem, but they treated this as essentially equivalent to the marriage problem. Although they considered that agents on one side of the market (e.g. colleges) could be matched to more than one agent on the other side (e.g. students), colleges' preferences were only considered to be defined over individual students, not over groups. For a number of years thereafter, the case of many-to-one matching was regarded as equivalent to one-to-one matching.

That this is not the case was observed in Roth (1985a), where some erroneous conclusions that had been reached about many-to-one matching were considered, and where a model of many-to-one matching was reformulated as a well-defined game. That model, presented in Subsection 3.2, is a straightforward generalization of the marriage model, in that colleges continue to have preferences over individual students, and their preferences over groups of students are constrained by their preferences over individuals in a simple way. The model of Subsection 3.3, in contrast, is a further generalization in which firms' preferences over groups of workers need not reflect an underlying preference over individuals.[4] We will see that while these generalizations and alternative formulations differ in important ways from the simple marriage problem, all of these models also share a number of their most striking properties.

As in any game-theoretic analysis, it will be important to keep clearly in mind the "rules of the game" by which agents may become matched to one another, as these will influence every aspect of the analysis. We will suppose the general rules are that any pair of agents on opposite sides of the market may be matched to one another if they both agree, and any agent is free to remain unmatched. We will consider more detailed descriptions of possible rules (concerning, for example, how proposals are made, or whether a marriage broker plays a role) at various points in the discussion.

3.1. The marriage model

The two finite and disjoint sets of agents in the marriage model are the set $M = \{m_1, m_2, \ldots, m_n\}$ of men, and $W = \{w_1, w_2, \ldots, w_m\}$ of women. Each man has preferences over the women, and each woman has preferences over the men. These preferences are transitive and complete, and may be such that a man m, say, would prefer to remain single rather than marry some woman w he does not care for.

The preferences of each man m will be represented by an ordered list, $P(m)$, on the set $W \cup \{m\}$. That is, a man m's preferences might be of the form $P(m) = w_1, w_2, m, w_3, \ldots, w_m$, indicating that his first choice is to be married to woman w_1, his second choice is to be married to woman w_2, and his third choice is to remain single. Similarly, each woman w in W has an ordered list of preferences, $P(w)$, on the set $M \cup \{w\}$. (An agent may also be indifferent between several possible mates.) We will usually describe an agent's preferences by writing only the ordered set of people that the agent prefers to being single. Thus the preferences $P(m)$ described above will be abbreviated by $P(m) = w_1, w_2$.

Denote by P the set of preference lists $P = \{P(m_1), \ldots, P(m_n), P(w_1), \ldots, P(w_m)\}$, one for each man and woman. A specific marriage market is denoted by the triple $(M, W; P)$. We write $w >_m w'$ to mean m prefers w to w', and $w \geqslant_m w'$ to mean m likes w at least as well as w'. Similarly

---

[4]The model presented here is a special case of one formulated in Roth (1984c), which in turn builds upon the work of Kelso and Crawford (1982). (In each of these models we may refer to the agents as firms and workers, but in the marriage model we will also refer to them as men and women, in the reformulated college admissions model as colleges and students, and in the assignment model as buyers and sellers, in order to keep in mind the particular assumptions of those models.)

we write $m >_w m'$ and $m \geq_w m'$. Woman w is called *acceptable* to man m if he likes her at least as well as remaining single, i.e. if $w \geq_m m$. Analogously, m is *acceptable* to w if $m \geq_w w$. If an individual is not indifferent between any two acceptable alternatives, he or she has *strict preferences*.

An outcome of the marriage market is a set of marriages. In general, not everyone may be married – some people may remain single. (We will adopt the convention that a person who is not married to someone is *self-matched*.) Formally we have

Definition 1. A *matching* μ is a one-to-one correspondence from the set $M \cup W$ onto itself of order two [that is, $\mu^2(x) = x$] such that if $\mu(m) \neq m$, then $\mu(m)$ is in W and if $\mu(w) \neq w$, then $\mu(w)$ is in M. We refer to $\mu(x)$ as the *mate* of x.

Note that $\mu^2(x) = x$ means that if man m is matched to woman w [i.e. if $\mu(m) = w$], then woman w is matched to man m [i.e. $\mu(w) = m$]. The definition also requires that individuals who are not single be matched with agents of the opposite set – i.e. men are matched with women. A matching will sometimes be represented as a set of matched pairs, e.g.

$$\mu = \begin{matrix} w_4 & w_1 & w_2 & w_3 & (m_5) \\ m_1 & m_2 & m_3 & m_4 & m_5 \end{matrix},$$

has m_1 married to w_4 and m_5 remaining single, i.e. $\mu(m_1) = w_4$ and $\mu(m_5) = m_5$, etc.

We will assume that each agent's preferences over alternative matchings correspond exactly to his (her) preferences over his own mates at the two matchings. Thus man m, say, prefers matching μ to matching ν if and only if he prefers $\mu(m)$ to $\nu(m)$.

A matching μ is individually irrational if it contains a matched pair (m, w) who are not mutually acceptable, and we say such a matching can be *improved upon* by an individual, since the rules allow any agent to remain single if he or she chooses. Similarly, a matching μ can be improved upon by some pair consisting of a man m and woman w if m and w are not matched to one another at μ, but prefer each other to their assignments at μ, i.e. if $w >_m \mu(m)$ and $m >_m \mu(w)$. The motivation of this terminology should be clear. Suppose such a matching μ should be under consideration – e.g. suppose no agreements have yet been reached, but courtships are under way that if successfully concluded will result in the matching μ. This state of affairs would be unstable in the sense that man m and woman w would have good reason to disrupt it in order to marry each other, and the rules of the game allow them to do so. This leads to the following definition.

Definition 2. A matching μ is *stable* if it cannot be improved upon by any individual or any pair of agents.

Note that unstable matchings are those dominated via coalitions consisting of individuals or pairs, and so unstable matchings are not in the core of the game. But the core is the set of matchings undominated by coalitions of any size, and so the set of stable matchings might strictly contain the core. But for this model of one-to-one matching, that is not the case.

Theorem 1.   *The core of the marriage market equals the set of stable matchings.*

Proof.   If μ is not in the core, then μ is dominated by some matching μ' via a coalition A. If μ is not individually irrational, this implies $\mu'(w) \in M$ for all w in A, since every woman w in A prefers $\mu'(w)$ to $\mu(w)$, and A is effective for μ'. Let w be in A and $m = \mu'(w)$. Then m prefers w to $\mu(m)$ and the pair (m, w) can improve upon μ, so μ is unstable. $\square$

We will continue to speak of stable (rather than core) matchings since in the more general models of many-to-one matching that follow, the set of stable matchings will be a subset of the core. For the marriage model, Gale and Shapley proved the following.

Theorem 2 [Gale and Shapley (1962)].   *The set of stable matchings is always nonempty. And when all men and women have strict preferences it contains an M-optimal stable matching, which all the men like at least as well as every other stable matching, and, similarly, a W-optimal stable matching.*

We will defer discussion of the proof until the more general model of Subsection 3.3.[5] We turn next to a many-to-one generalization of the marriage model in which it continues to be meaningful to speak of firms as having preferences over individual workers.

3.2. The reformulated college admissions model

There are two finite and disjoint sets, $\mathscr{C} = \{C_1, \ldots, C_n\}$ and $S = \{s_1, \ldots, s_m\}$, of colleges and students, respectively. Each student has preferences over the

---

[5]Roth and Vande Vate (1990) construct another kind of existence proof, based on the observation that a sequence of matchings generated by allowing randomly chosen blocking pairs to form must converge with probability one to a stable matching. (The difficulty lies in the fact that cycles of unstable matching may arise.)

colleges, and each college has preferences over individual students, exactly as in the marriage model.

The first difference from the marriage model is that, associated with each college C is a positive integer q_C called its *quota*, which indicates the maximum number of positions it may fill. (That all q_C positions are identical is reflected in the fact that students' preferences are over colleges – they do not distinguish between positions.)

An outcome is a matching of students to colleges, such that each student is matched to at most one college, and each college is matched to at most its quota of students. A student who is not matched to any college will be "matched to himself" as in the marriage model, and a college that has some number of unfilled positions will be matched to itself in each of those positions. A matching is bilateral, in the sense that a student is enrolled at a given college if and only if the college enrolls that student. To give a formal definition, first define, for any set X, an *unordered family of elements of X* to be a collection of elements, not necessarily distinct. So a given element of X may appear more than once in an unordered family of elements of X.

Definition 3. A *matching* μ is a function from the set $\mathscr{C} \cup S$ into the set of unordered families of elements of $\mathscr{C} \cup S$ such that:
(1) $|\mu(s)| = 1$ for every student s and $\mu(s) = s$ if $\mu(s) \notin \mathscr{C}$;
(2) $|\mu(C)| = q_C$ for every college C, and if the number of students in $\mu(C)$, say r, is less than q_C, then $\mu(C)$ contains $q_C - r$ copies of C;
(3) $\mu(s) = C$ if and only if s is in $\mu(C)$.

So $\mu(s_1) = C$ denotes that student s_1 is enrolled at college C at the matching μ, and $\mu(C) = \{s_1, s_3, C, C\}$ denotes that college C, with quota $q_C = 4$, enrolls students s_1 and s_3 and has two positions unfilled.

At this point in our description of the marriage model we had only to say that each agent's preferences over alternative matchings correspond exactly to his preferences over his own assignments at those matchings. We can now say this about students, since at each matching a student is either unmatched or matched to a college, and we have already described student's preferences over colleges. But, while we have described colleges' preferences over students, each college with a quota greater than 1 must be able to compare *groups* of students in order to compare alternative matchings, and we have yet to describe the preferences of colleges over groups. (Until we have described colleges' preferences over matchings, the model will not be a well-defined game.)

The assumption we will make connecting colleges' preferences over groups of students to their preferences over individual students is one insuring that, for example, if $\mu(C)$ assigns college C its third and fourth choice students, and

$\mu'(C)$ assigns it its second and fourth choice students, then college C prefers $\mu'(C)$ to $\mu(C)$. Specifically, let $P^{\#}(C)$ denote the preference relation of college C over all assignments $\mu(C)$ it could receive at some matching μ. A college C's preferences $P^{\#}(C)$ will be called *responsive* to its preferences $P(C)$ over individual students if, for any two assignments that differ in only one student, it prefers the assignment containing the more preferred student. That is, we assume colleges' preferences are responsive, as follows.

Definition 4. The preference relation $P^{\#}(C)$ over sets of students is *responsive* [to the preferences $P(C)$ over individual students] if, whenever $\mu'(C) = \mu(C) \cup \{s_k\} \backslash \{\sigma\}$ for σ in $\mu(C)$ and s_k not in $\mu(C)$, then C prefers $\mu'(C)$ to $\mu(C)$ [under $P^{\#}(C)$] if and only if C prefers s_k to σ [under $P(C)$].

We will write $\mu'(C) >_C \mu(C)$ to indicate that college C prefers $\mu'(C)$ to $\mu(C)$ according to its preferences $P^{\#}(C)$, and $\mu'(C) \geq_C \mu(C)$ to indicate that C likes $\mu'(C)$ at least as well as $\mu(C)$, where the fact that $\mu'(C)$ and $\mu(C)$ are not singletons will make clear that we are dealing with the preferences $P^{\#}(C)$, as distinct from statements about C's preferences over individual students. Note that C may be indifferent between distinct assignments $\mu(C)$ and $\mu'(C)$ even if C has strict preferences over individual students.

Note also that different responsive preference orderings $P^{\#}(C)$ exist for any preference $P(C)$, since, for example, responsiveness does not specify whether a college prefers to be assigned its first and fourth choice students instead of its second and third choice students. However, the preference ordering $P(C)$ over individual students can be derived from $P^{\#}(C)$ by considering a college C's preferences over assignments $\mu(C)$ containing no more than a single student (and $q_C - 1$ copies of C). The assumption that colleges have responsive preferences is essentially no more than the assumption that their preferences for sets of students are related to their ranking of individual students in a natural way. (Of course the assumption that colleges have preferences over individual students is nontrivial, and it is this assumption which is relaxed in Subsection 3.3.)

Some of the results that follow will depend on the assumption that agents have strict preferences. Surprisingly, we will only need to assume that colleges have strict preferences over individuals: it will not be necessary to assume they have strict preferences over groups of students. The reasons for this will not become completely clear until Corollary 17, which says that when colleges have strict preferences over individuals, they will not be indifferent between any groups of students assigned to them at stable matchings, even though they may be indifferent between other groups of students.

A matching μ is individually irrational if $\mu(s) = C$ for some student s and college C such that either the student is unacceptable to the college or the college is unacceptable to the student. We will say the unhappy agent can *improve upon* such a matching. Similarly, a college C and student s can improve upon a matching μ if they are not matched to one another at μ, but would both prefer to be matched to one another than to (one of) their present assignments. That is, the pair (C, s) can improve upon μ if $\mu(s) \neq C$ and if $C >_s \mu(s)$ and $s >_C \sigma$ for some σ in $\mu(C)$. [Note that σ may equal either some student s' in $\mu(C)$, or, if one or more of college C's positions is unfilled at $\mu(C)$, σ may equal C.] It should be clear that matchings blocked in this way by an individual or by a pair of agents are unstable in the sense discussed for the marriage model, since there are agents with both the incentive and the power to disrupt such matchings. So, as in the marriage model, we now define stable matchings – although we will immediately have to ask whether the set of stable matchings defined this way can serve the same role as in the marriage model.

Definition 5. A matching μ is *stable* if it cannot be improved upon by any individual agent or any college–student pair.

It is not obvious that this definition will still be adequate, since we now might need to consider coalitions consisting of colleges and several students (all of whom might be able to enroll simultaneously at the college), or even coalitions consisting of multiple colleges and students. However, when preferences are responsive, nothing is lost by concentrating on simple college–students pairs. The set of stable matchings is equal to the core defined by *weak* domination [Roth (1985b)].[6] So it is a subset of the core. To see why an outcome which is not strictly dominated might nevertheless be unstable, suppose college C with quota 2 is the first choice of students $s_1, s_2,$ and s_3, and has preferences $P(C) = s_1, s_2, s_3$. Then a matching with $\mu(C) = \{s_1, s_3\}$ can be improved upon by (C, s_2), but the resulting match, $\mu'(C) = \{s_1, s_2\}$, involves a coalition of three agents, $\{C, s_1, s_2\}$, and s_1 is indifferent between μ and μ', since he is matched to C at both matchings.

We will defer until the next section the proof that the set of stable matchings is always nonempty, and contains optimal stable matchings for each side of the market. But note that if the preferences of the colleges for groups of students are not responsive (to some set of preferences over individual students), the core may be empty.

---

[6]A matching μ dominates another matching μ' if there is a coalition A of agents which is effective for μ – i.e. whose members can achieve their parts of μ by matching among themselves, without the participation of agents not in A – and such that all members of A prefer their matches under μ to those under μ'. In contrast, a matching μ *weakly* dominates another matching μ' if only some of the members of the effective coalition A prefer μ to μ', so long as no other members of A have the reverse preference. The core is the set of matchings that are not dominated, and the core defined by weak domination is the set of matchings that are not (even) weakly dominated.

3.3. Complex preferences over groups

Let the two sets of agents be n firms $\mathcal{F} = \{F_1, \ldots, F_n\}$, and m workers $W = \{w_1, \ldots, w_m\}$. For simplicity assume all firms have the same quota, equal to m, so each firm could in principle hire all the workers. This will allow us to describe matchings a little more simply, since it will not be necessary to keep track of each firm's quota by saying, for example, that a firm that does not employ any workers is matched to m copies of itself.

Definition 6. A *matching* μ is a function from the set $F \cup W$ into the set of all subsets of $\mathcal{F} \cup W$ such that:
  (1) $|\mu(w)| = 1$ for every worker w and $\mu(w) = w$ if $\mu(w) \notin \mathcal{F}$;
  (2) $|\mu(F)| \leqslant m$ for every firm F [$\mu(F) = \emptyset$ if F is not matched to any workers];
  (3) $\mu(w) = F$ if and only if w is in $\mu(F)$.

Workers have preferences over individual firms, just as in the college admissions problem, and firms have preferences over subsets of W. For simplicity assume all preferences are strict. So a worker w's preferences can be represented by a list of acceptable firms, e.g. $P(w) = F_i, F_j, F_k, w$; and a firm's preferences by a list of acceptable subsets of workers, e.g. $P^\#(F) = S_1, S_2, \ldots, S_k, \emptyset$; where each S_i is a subset of W. Each agent compares different matchings by comparing his (its) own assignment at those matchings. The preferences of all the agents will be denoted by $\boldsymbol{P} = (P^\#(F_1), \ldots, P^\#(F_n), P(w_1), \ldots, P(w_m))$. Keep in mind that firms' preferences are over *sets* of employees.
    Faced with a set S of workers, each firm F can determine which subset of S it would most prefer to hire. Call this F's choice from S, and denote it by $\mathrm{Ch}_F(S)$. That is, for any subset S of W, F's *choice set* is $\mathrm{Ch}_F(S) = S'$ such that S' is contained in S and $S' >_F S''$ for all S'' contained in S. Since preferences are strict, there is always a single set S' that F most prefers to hire, out of any set S of available workers. (Of course S' could equal S, or it could be empty.)
    We will assume that firms regard workers as substitutes rather than complements, as follows.[7]

Definition 7. A firm F's preferences over sets of workers has the property of *substitutability* if, for any set S that contains workers w and w', if w is in $\mathrm{Ch}_F(S)$, then w is in $\mathrm{Ch}_F(S - w')$.

That is, if F has "substitutable" preferences, then if its preferred set of

[7]This kind of condition on preferences was proposed by Kelso and Crawford (1982).

employees from S includes w, so will its preferred set of employees from any subset of S that still includes w. [By repeated application, if $w \in Ch_F(S)$, then for any S' contained in S with $w \in S'$, $w \in Ch_F(S')$.] This is the sense in which the firm regards worker w and the other workers in $Ch_F(S)$ more as substitutes than complements: it continues to want to employ w even if some of the other workers become unavailable.

So substitutability rules out the possibility that firms regard workers as complements, as might be the case of an American football team, for example, that wanted to employ a player who could throw long passes and one who could catch them, but if only one of them were available would prefer to hire a different player entirely. Note that responsive preferences have the substitutability property: in the college admissions model, the choice set from any set of students of a college with quota q is either the q most preferred acceptable students in the set, or all the acceptable students in the set, whichever is the smaller number.

A matching μ can be *improved upon* by an individual worker w if $w >_w \mu(w)$, and by an individual firm F if $\mu(F) \neq Ch_F(\mu(F))$. Note that μ may be improved upon by an individual firm F without being individually irrational, since it might still be that $\mu(F) >_F \emptyset$. This definition reflects the assumption that workers' preferences are over firms (and not over coworkers), so that F may fire some workers in $\mu(F)$ if it chooses, without affecting other members of $\mu(F)$. Similarly, μ can be improved upon by a worker–firm pair (w, F) if w and F are not matched at μ but would both prefer if F hired w: i.e. if $\mu(w) \neq F$ and if $F >_w \mu(w)$ and $w \in Ch_F(\mu(F) \cup w)$. If the firms have responsive preferences this is equivalent to the definition we used for the college admissions model. We define stable matchings the same way also.

Definition 8. A matching μ is *stable* if it cannot be improved by any individual agent or any worker–firm pair.

Since "improvement" is now defined in terms of firm's preferences over sets of workers, this definition of stability has a slightly different meaning than the same definition for the college admissions model. Nevertheless, it is still a definition of *pairwise* stability, since the largest coalitions it considers are worker–firm pairs. So we still have to consider whether something is missed by not considering larger coalitions. It turns out that pairwise stability is still sufficient: as when preferences are responsive, we can show that, for any preferences P, the set $S(P)$ of stable matchings equals the core defined by weak domination, $C_W(P)$, and is always nonempty.

Theorem 3. *When firms have substitutable preferences (and all preferences are strict)* $S(P) = C_W(P)$.

Theorem 4. *When firms have substitutable preferences, the set of stable matchings is always nonempty.*

The proof of Theorem 4 will be by means of the following algorithm:

In Step 1, each firm proposes to its most preferred set of workers, and each worker rejects all but the most preferred acceptable firm that proposes to it. In each subsequent step, each firm which received one or more rejections at the previous step proposes to its most preferred set of workers that includes all of those workers who it previously proposed to and have not yet rejected it, but does not include any workers who have previously rejected it. Each worker rejects all but the most preferred acceptable firm that has proposed so far. The algorithm stops after any step in which there are no rejections, at which point each firm is matched to the set of workers to which it has issued proposals that have not been rejected.

Proof of Theorem 4. The matching μ produced by the above algorithm is stable. The key observation is that, because firms have substitutable preferences, no firm ever regrets that it must continue to offer employment at subsequent steps of the algorithm to workers who have not rejected its earlier offers. That is, at every step in the algorithm each firm is proposing to its most preferred set of workers that does not contain any workers who have previously rejected it. So consider a firm F and a worker w not matched to F at μ such that $w \in Ch_F(\mu(F) \cup w)$. At some step of the algorithm, F proposed to w and was subsequently rejected, so w prefers $\mu(w)$ to F, and μ is not improvable by the pair (w, F). Since w and F were arbitrary, and since μ is not improvable by any individual, μ is stable. $\square$

We call this algorithm a "deferred acceptance" procedure, to emphasize that workers are able to hold the best offer they have received, without accepting it outright. For the moment we present this algorithm only to show that stable matchings always exist. That is, although the algorithm is presented as if at each step the firms and workers take certain actions, we will not consider until Section 5 whether they would be well advised to take those actions, and consequently whether it is reasonable for us to expect that they would act as described, if the rules for making and accepting proposals were as in the algorithm.

This result also establishes the nonemptiness of the set of stable matchings for the marriage and college admissions models, which are special cases of the present model. The algorithm and proof presented here are simple generalizations of those presented by Gale and Shapley (1962). And, as in the marriage and college admissions models, we can further note the surprising fact that the set of stable matchings contains elements of the following sort.

Definition 9. A stable matching is *firm optimal* if every firm likes it at least as well as any other stable matching. A stable matching is *worker optimal* if every worker likes it at least as well as any other stable matching.

Theorem 5 [Kelso and Crawford (1982)]. *When firms have substitutable preferences, and preferences are strict, the deferred acceptance algorithm with firms proposing produces a firm-optimal stable matching.*

Theorem 5 can be proved by showing that in the deferred acceptance algorithm with firms proposing, no firm is ever rejected by an *achievable* worker, where a worker w is said to be achievable for a firm F if there is some stable matching μ at which $\mu(w) = F$.

Since, unlike the marriage model and like the college admissions model, this model is not symmetric between firms and workers, it is not immediately apparent that a deferred acceptance algorithm with workers proposing will have an analogous result, but it does. In the algorithm with workers proposing, workers propose to firms in order of preference, and a firm rejects at any step all those workers who are not in the firm's choice set from those proposals it has not yet rejected. We can state the following result.

Theorem 6 [Roth (1984c)]. *When firms have substitutable preferences, and preferences are strict, the deferred acceptance algorithm with workers proposing produces a worker-optimal stable matching.*

The key observation for the proof is that, because firms have substitutable preferences, no firm ever regrets that it rejected a worker at an earlier step, when it sees who proposed at the current step. One can then show that no worker is ever rejected by an achievable firm.

These results cannot be generalized in a strightforward way to the symmetric case of many-to-many matching in which workers may take multiple jobs, even when both sides have substitutable, or even responsive, preferences.[8] The reason is not that the analogously defined pairwise stable matchings do not have similar properties in such a model, but that pairwise stable matchings are no longer always in the core.

Before moving on, an example will help clarify things.

Example 7. An example in which firms have substitutable (but nonresponsive) preferences. There are two firms and three workers, with preferences as follows.

[8]See Blair (1988) and Roth (1991).

$$P^\#(F_1) = \{w_1, w_2\}, \{w_1, w_3\}, \{w_2, w_3\}, \{w_3\}, \{w_2\}, \{w_1\},$$

$$P^\#(F_2) = \{w_3\},$$

$$P(w_1) = F_1, F_2,$$

$$P(w_2) = F_1, F_2,$$

$$P(w_3) = F_1, F_2.$$

Note that

$$\mu = \begin{matrix} F_1 & F_2 \\ \{w_1, w_2\} & \{w_3\} \end{matrix}$$

is the unique stable matching.

If we look just at single workers, we see that F_1 prefers w_3 to w_2 to w_1. But $P^\#(F_1)$ is not responsive to these preferences over single workers, since $\{w_1, w_2\} >_{F_1} \{w_1, w_3\}$ even though w_3 alone is preferred to w_2 alone. But the preferences are substitutable. Recall the earlier discussion of why the college admissions model needed to be reformulated to include colleges' preferences over groups, and observe once again that the class of many-to-one matching problems, of which this is an example, would not be well-defined games if we specified only the preferences of firms over individuals. Indeed, if we defined stability only in terms of preferences over individuals, the matching μ would be unstable with respect to the pair (F_1, w_3) since w_3 prefers F_1 to F_2 and F_1 prefers w_3 (by himself) to w_2 (by himself). But μ is not unstable in this example because F_1 does not prefer $\{w_1, w_3\}$ to $\{w_1, w_2\}$. □

3.4. The assignment model

In this model money plays an explicit role. There are two finite disjoint sets of players P and Q, containing m and n players, respectively. Members of P will sometimes be called P-agents and members of Q called Q-agents, and the letters i and j will be reserved for P and Q agents, respectively. Associated with each possible partnership $(i, j) \in P \times Q$ is a non-negative real number α_{ij}. A game in coalitional function form with sidepayments is determined by (P, Q, α), with the numbers α_{ij} being equal to the worth of the coalitions $\{i, j\}$ consisting of one P agent and one Q agent. The worth of larger coalitions is determined entirely by the worth of the pairwise combinations that the coalition members can form. That is, the coalitional function v is given by

$v(S) = \alpha_{ij}$ if $S = \{i, j\}$ for $i \in P$ and $j \in Q$;

$v(S) = 0$ if S contains only P agents or only Q agents; and

$v(S) = \max(v(i_1, j_1) + v(i_2, j_2) + \cdots + v(i_k, j_k))$ for arbitrary coalitions S, with the maximum to be taken over all arrangements of $2k$ distinct players $i_1, i_2, \ldots, i_k$ belonging to S_P and $j_1, j_2, \ldots, j_k$ belonging to S_Q, where S_P and S_Q denote the sets of P and Q agents in S (i.e. the intersection of the coalition S with P and with Q), respectively.

So the rules of the game are that any pair of agents $(i, j) \in P \times Q$ can together obtain α_{ij}, and any larger coalition is valuable only insofar as it can organize itself into such pairs. The members of any coalition may divide among themselves their collective worth in any way they like. An imputation of this game is thus a non-negative vector (u, v) in $\mathbf{R}^m \times \mathbf{R}^n$ such that $\Sigma_{i \in P} u_i + \Sigma_{j \in Q} v_j = v(P \cup Q)$. The easiest way to interpret this is to take the quantities α_{ij} to be amounts of money, and to assume that agents' preferences are concerned only with their monetary payoffs.

We might think of P as a set of potential buyers of some objects offered for sale by the set Q of potential sellers, and each seller owns and each buyer wants exactly one indivisible object. If each seller j has a reservation price c_j, and each buyer i has a reservation price r_{ij} for object j, we may take α_{ij} to be the potential gains from trade between i and j; that is, $\alpha_{ij} = \max\{0, r_{ij} - c_j\}$. If buyer i buys object j from seller j at a price p, and if no other monetary transfers are made, the utilities are $u_i = r_{ij} - p$ and $v_j = p - c_j$. So, when no other monetary transfers are made, $u_i + v_j = \alpha_{ij}$ when i buys from j. But note that transfers between agents are *not* restricted to those between buyers and sellers; e.g. buyers may make transfers among themselves as in the bidder rings of Subsection 2.2.[9]

We can also think of the P and Q agents as being firms and workers, etc. As in the marriage model, we look here at the simple case of one-to-one matching, with firms constrained to hire at most one worker.[10] In such a case, the α_{ij}'s represent some measure of the joint productivity of the firm and worker, while transfers between a matched firm and worker represent salary. Transfers can also take place between workers (as when workers form a labor union in which the dues of employed members help pay unemployment benefits to unemployed members), or between firms.

The maximization problem to determine $v(S)$ for a given matrix α is called an *assignment problem*, so games of this form are called *assignment games*. We will be particularly interested in the coalition $P \cup Q$, since $v(P \cup Q)$ is the

---

[9]A model in which it is assumed that transfers cannot be made between agents on the same side of the market is considered by Demange and Gale (1985), who show that many of the results presented here for other models can be obtained in a model of this kind allowing rather general utility functions.

[10]The case of many-to-one matching has some important differences, analogous to those found between the marriage and college admissions models: see Sotomayor (1988).

maximum total payoff available to the players, and hence determines the Pareto set and the set of imputations.

Consider the following linear programming (LP) problem P_1:

$$\text{maximize} \sum_{i,j} \alpha_{ij} \cdot x_{ij}$$

$$\text{subject to (a)} \sum_{i} x_{ij} \leq 1,$$

$$\text{(b)} \sum_{j} x_{ij} \leq 1,$$

$$\text{(c)} \ x_{ij} \geq 0.$$

We may interpret x_{ij} as, for example, the probability that a partnership (i, j) will form. Then the linear inequalities of type (a), one for each j in Q, say that the probability that j will be matched to some i cannot exceed 1. The inequalities of form (b), one for each i in P, say the same about the probability that i will be matched.

It can be shown [see Dantzig (1963, p. 318)] that there exists a solution of this LP problem which involves only values of zero and one. [The extreme points of systems of linear inequalities of the form (a), (b), and (c) have integer values of x_{ij}, i.e. each x_{ij} equals 0 or 1.] Thus the fractions artificially introduced in the LP formulation disappear in the solution and the (continuous) LP problem is equivalent to the (discrete) assignment problem for the coalition of all players, that is, the determination of $v(P \cup Q)$. Then $v(P \cup Q) = \Sigma \ \alpha_{ij} \cdot x_{ij}$, where x is an optimal solution of the LP problem.

Definition 10. A *feasible assignment* for (P, Q, α) is a matrix $x = (x_{ij})$ (of zeros and ones) that satisfies (a), (b) and (c) above. An *optimal assignment* is a feasible assignment x such that $\Sigma_{i,j} \ \alpha_{ij} \cdot x_{ij} \geq \Sigma_{i,j} \ \alpha_{ij} \cdot x'_{ij}$, for all feasible assignments x'.

So if x is a feasible assignment, $x_{ij} = 1$ if i and j form a partnership and $x_{ij} = 0$ otherwise. If $\Sigma_j \ x_{ij} = 0$, then i is *unassigned*, and if $\Sigma_i \ x_{ij} = 0$, then j is likewise unassigned. A feasible assignment x corresponds exactly to a matching μ as in Definition 1, with $\mu(i) = j$ if and only if $x_{ij} = 1$.

Definition 11. The pair of vectors (u, v), $u \in R^m$ and $v \in R^n$ is called a *feasible payoff* for (P, Q, α) if there is a feasible assignment x such that

$$\sum_{i \in P} u_i + \sum_{i \in Q} v_j = \sum_{i \in P, j \in Q} \alpha_{ij} \cdot x_{ij}.$$

In this case we say (u, v) and x are *compatible* with each other, and we call $((u, v); x)$ a *feasible outcome*. Note again that a feasible payoff vector may involve monetary transfers between agents who are not assigned to one another.

As in the earlier models, the key notion is that of stability.

Definition 12. A feasible outcome $((u, v); x)$ is *stable* [or the payoff (u, v) with an assignment x is stable] if
  (i) $u_i \geqslant 0, v_j \geqslant 0$,
  (ii) $u_i + v_j \geqslant \alpha_{ij}$ for all $(i, j) \in P \times Q$.

Condition (i) (individual rationality) reflects that a player always has the option of remaining unmatched (recall that $v(i) = v(j) = 0$ for all individual agents i and j). Condition (ii) requires that the outcome cannot be improved by any pair: if (ii) is not satisfied for some agents i and j, then it would pay them to break up their present partnership(s) (either with one another or with other agents) and form a new partnership together, because this could give them each a higher payoff.

From the definition of feasibility and stability it follows that

Lemma 8. *Let $((u, v), x)$ be a stable outcome for (P, Q, α). Then*
  (i) $u_i + v_j = \alpha_{ij}$ *for all pairs (i, j) such that $x_{ij} = 1$;*
  (ii) $u_i = 0$ *for all unassigned i, and $v_j = 0$ for all unassigned j at x.*

The lemma implies that at a stable outcome, the only monetary transfers that occur are between P and Q agents who are matched to each other. (Note that this is an implication of stability, not an assumption of the model.)

Now consider the LP problem P_1^* that is the dual of P_1, i.e. the LP problem of finding a pair of vectors (u, v), $u \in R^m$, $v \in R^n$, that minimizes the sum

$$\sum_{i \in P} u_i + \sum_{i \in Q} v_j$$

subject, for all $i \in P$ and $j \in Q$, to
  (a*) $u_i \geqslant 0, v_j > 0$,
  (b*) $u_i + v_j \geqslant \alpha_{ij}$.
Because we know that P_1 has a solution, we know also that P_1^* must have an optimal solution. A fundamental duality theorem [see Dantzig (1963, p. 129)] asserts that the objective functions of these dual LPs must attain the same value. That is, if x is an optimal assignment and (u, v) is a solution of P_1^*, we have that

$$\sum_{i \in P} u_i + \sum_{i \in Q} v_j = \sum_{P \times Q} \alpha_{ij} \cdot x_{ij} = v(P \cup Q) . \tag{1}$$

This means that $((u, v), x)$ is a feasible outcome. Moreover, $((u, v), x)$ is a stable outcome for (P, Q, α), since (a*) ensures individual rationality and $u_i + v_j \geqslant \alpha_{ij}$ for all $(i, j) \in P \times Q$ by (b*). It follows, by the definition of $v(S)$, that for any coalition $S = S_P \cup S_Q$, where S_P is contained in P and S_Q in Q,

$$\sum_{i \in S_P} u_i + \sum_{i \in S_Q} v_j \geqslant v(S) \,. \tag{2}$$

But (1) and (2) are exactly how the core of the game is determined: (1) ensures the feasibility of (u, v) and (2) ensures its nonimprovability by any coalition. Conversely, any payoff vector in the core, i.e. satisfying (1) and (2), satisfies the conditions for a solution to P_1^*. Hence we have shown

Theorem 9 [Shapley and Shubik (1972)]. *Let (P, Q, α) be an assignment game. Then*
   (a) *the set of stable outcomes and the core of (P, Q, α) are the same;*
   (b) *the core of (P, Q, α) is the (nonempty) set of solutions of the dual LP of the corresponding assignment problem.*

The following two corollaries make clear why, in contrast to the discrete models considered earlier, we can concentrate here on the payoffs to the agents rather than on the underlying assignment (matching).[11]

Corollary 10. *If x is an optimal assignment, then it is compatible with any stable payoff (u, v).*

Corollary 11. *If $((u,v), x)$ is a stable outcome, then x is an optimal assignment.*

In this model also there are optimal stable outcomes for each side of the market. Note that in view of Corollary 10, the difference between different stable outcomes in this model has to do only with the payments to each player, not to whom players are matched.

Theorem 12 [Shapley and Shubik (1972)]. *There is a P-optimal stable payoff $(\bar{u}, \underline{v})$, with the property that for any stable payoff (u, v), $\bar{u} \geqslant u$ and $\underline{v} \leqslant v$; there is a Q-optimal stable payoff $(\underline{u}, \bar{v})$ with symmetrical properties.*

---

[11]Becker (1981), who uses the assignment model to study marriage and household economics, makes use of the fact that stable outcomes all correspond to optimal assignments, and that the optimal assignment is typically unique, to study which men are matched to which women (e.g. if high wage earners marry good cooks), for different assumptions about how the assignment matrix is derived.

4. The structure of the set of stable matchings

In each of the models we have described, the set of stable matchings is nonempty.[12] In fact, for each side of the market, there exists an optimal stable matching that all agents on that side of the market like at least as well as any other stable matching.[13] That such side-optimal stable matchings always exist is more than a little surprising in models in which firms compete with one another for good workers, and workers compete with one another for desirable jobs. It turns out to be only the tip of the iceberg, in terms of the welfare comparisons which can be made between different stable matchings. In this section we describe some of these.

One question is whether the optimal stable matching for agents on one side of the market is Pareto optimal for them as well. This turns out to be one of the respects in which many-to-one matching is not equivalent to the special case of one-to-one matching. In the marriage model the optimal stable matching for each side of the market is weakly Pareto optimal for that side. (Since the market is symmetric between men and women, we consider here only the man-optimal stable matching μ_M.)

Theorem 13 [Roth (1982a)]. Weak Pareto optimality for the men. *In the marriage model there is no individually rational matching μ (stable or not) such that $\mu \geq_m \mu_M$ for all m in M.*

However, the following example shows that this result cannot be strengthened to strong Pareto optimality.

Example 14 [Roth (1982a)]. Let $M = \{m_1, m_2, m_3\}$ and $W = \{w_1, w_2, w_3\}$ with preferences over the acceptable people given by:

$$P(m_1) = w_2, w_1, w_3 ; \qquad P(w_1) = m_1, m_2, m_3 ;$$
$$P(m_2) = w_1, w_2, w_3 ; \qquad P(w_2) = m_3, m_1, m_2 ;$$
$$P(m_3) = w_1, w_2, w_3 ; \qquad P(w_3) = m_1, m_2, m_3 .$$

Then

$$\mu_M = \begin{matrix} w_1 & w_2 & w_3 \\ m_1 & m_3 & m_2 \end{matrix}$$

---

[12]This nonemptiness is related to the two-sidedness of the models: one-sided and three-sided models may have empty cores.

[13]For the discrete markets this is only the case when preferences are strict: when they are not, it is easy to see that although the set of stable matchings remains nonempty, it may not contain any such side-optimal matchings.

is the man-optimal stable matching. Nevertheless,

$$\mu = \frac{w_1 \quad w_2 \quad w_3}{m_3 \quad m_1 \quad m_2}$$

leaves m_2 no worse than under μ_M, but benefits m_1 and m_3. So there may in general be matchings that all men like at least as well as the M-optimal stable matching, and that some men prefer. We shall return to this fact in our discussion of the strategic options available to coalitions of men.

Theorem 13 cannot be generalized to both sides of the college admissions model. We can state the following result instead.

Theorem 15 [Roth (1985a)]. *When the preferences over individuals are strict, the student-optimal stable matching is weakly Pareto optimal for the students, but the college-optimal stable matching need not be even weakly Pareto optimal for the colleges.*

However, as we have already seen through the existence of optimal stable matchings for each side of the market, there are some important properties concerning welfare comparisons within the set of stable matchings that hold both for one-to-one and many-to-one matching. There are also welfare comparisons that can be made in the case of many-to-one matching that have no counterpart in the special case of one-to-one matching.

We first consider some comparisons of this latter sort, for the college admissions model, concerning how well a given college might do at different stable matchings. Theorem 16 says that for every pair of stable matchings, each college will prefer every student who is assigned to it at one of the two matchings to every student who is assigned to it in the second matching but not the first. An immediate corollary is that in a college admissions problem in which all preferences over individuals are strict (and responsive), no college will be indifferent between any two (different) groups of students that it enrolls at stable matchings. The manner in which these results are mathematically unusual can be understood by noting that this corollary, for example, can be rephrased to say that if a given matching is stable (and hence in the core), and if some college is indifferent between the entering class it is assigned at that matching and a different entering class that it is assigned at a different matching, then the second matching is *not* in the core. We thus have a way of concluding that an outcome is not in the core, based on the direct examination of the preferences of only *one* agent (the college). Since the definition of the core involves preferences of coalitions of agents, this is rather unusual.

Theorem 16 [Roth and Sotomayor (1989)]. *Let preferences over individuals be strict, and let μ and μ' be stable matchings for a college admissions problem $(S, \mathscr{C}, P)$). If $\mu(C) >_C \mu'(C)$ for some college C, then $s >_C s'$ for all $s \in \mu(C)$ and $s' \in \mu'(C) - \mu(C)$. That is, C prefers every student in its entering class at μ to every student who is in its entering class at μ' but not at μ.*

Given that colleges have responsive preferences, the following corollary is immediate.

Corollary 17 [Roth and Sotomayor (1989)]. *If colleges and students have strict preferences over individuals, then colleges have strict preferences over those groups of students that they may be assigned at stable matchings. That is, if μ and μ' are stable matchings, then a college C is indifferent between $\mu(C)$ and $\mu'(C)$ only if $\mu(C) = \mu'(C)$.*

And since the set of stable matchings depends only on the preferences over individuals, and not on the preferences over groups (so long as these are responsive to the preferences over individuals) the following result is also immediate.

Corollary 18 [Roth and Sotomayor (1989)]. *Consider a college C with preferences $P(C)$ over individual students, and let $P^{\#}(C)$ and $P^*(C)$ be preferences over groups of students that are responsive to $P(C)$ (but are otherwise arbitrary). Then for every pair of stable matchings μ and μ', $\mu(C)$ is preferred to $\mu'(C)$ under the preferences $P^{\#}(C)$ if and only if $\mu(C)$ is preferred to $\mu'(C)$ under $P^*(C)$.*

An example will illustrate Theorem 16 and Corollaries 17 and 18.

Let the preferences over individuals be given by

$$P(s)_1 = C_5, C_1 ; \qquad\qquad P(C)_1 = s_1, s_2, s_3, s_4, s_5, s_6, s_7 ;$$
$$P(s)_2 = C_2, C_5, C_1 ; \qquad\qquad P(C)_2 = s_5, s_2 ;$$
$$P(s)_3 = C_3, C_1 ; \qquad\qquad P(C)_3 = s_6, s_7, s_3 ;$$
$$P(s)_4 = C_4, C_1 ; \qquad\qquad P(C)_4 = s_7, s_4 ;$$
$$P(s)_5 = C_1, C_2 ; \qquad\qquad P(C)_5 = s_2, s_1 ;$$
$$P(s)_6 = C_1, C_3 ;$$
$$P(s)_7 = C_1, C_3, C_4 ,$$

and let the quotas be $q_{C_1} = 3$, $q_{C_j} = 1$ for $j = 2, \ldots, 5$. Then the set of stable outcomes is $\{\mu_1, \mu_2, \mu_3, \mu_4\}$ where

$$
\begin{array}{cccccc}
 & C_1 & C_2 & C_3 & C_4 & C_5 \\
\mu_1 = & s_1 s_3 s_4 & s_5 & s_6 & s_7 & s_2 \\
\mu_2 = & s_3 s_4 s_5 & s_2 & s_6 & s_7 & s_1 \\
\mu_3 = & s_3 s_5 s_6 & s_2 & s_7 & s_4 & s_1 \\
\mu_4 = & s_5 s_6 s_7 & s_2 & s_3 & s_4 & s_1
\end{array}
$$

Note that these are the only stable matchings, and

$$\mu_1(C_1) >_{C_1} \mu_2(C_1) >_{C_1} \mu_3(C_1) >_{C_1} \mu_4(C_1),$$

for any responsive preferences.

We turn next to consider welfare comparisons involving more than one agent, on the set of stable matchings. Again, we concentrate primarily on the college admissions model. (The proofs all involve some version of Theorem 16.) However, these results [which are proved in Roth and Sotomayor (1990a)] all have parallels in the case of one-to-one matching, where they were first discovered.

We begin with a result which says that if an agent prefers one stable matching to another, then any agents on the other side of the market who are matched to that agent at either matching have the opposite preferences.[14]

Theorem 19. *If μ and μ' are two stable matchings for $(S, \mathcal{C}, P)$ and $C = \mu(s)$ or $C = \mu'(s)$, with $C \in \mathcal{C}$ and $s \in S$, then if $\mu(C) \geq_C \mu'(C)$ then $\mu'(s) \geq_s \mu(s)$ [and if $\mu'(s) >_s \mu(s)$ then $\mu(C) \geq_C \mu'(C)$].*

The equivalent result for the assignment model, which is easy to prove, says that if i prefers a stable payoff (u, v) to another stable payoff (u', v'), his mate(s) will prefer (u', v').

Theorem 20. *Let $((u, v), x)$ and $((u', v'), x')$ be stable outcomes for (P, Q, α). Then if $x'_{ij} = 1$, $u'_i > u_i$ implies $v'_j < v_j$.*

Proof. Suppose $v'_j \geq v_j$. Then $\alpha_{ij} = u'_i + v'_j > u_i + v_j \geq \alpha_{ij}$, which is a contradiction.

---

[14]The case of the marriage model was shown by Knuth (1976), and an extended version of this result was given by Gale and Sotomayor (1985a), who show its usefulness as a lemma in a number of other proofs.

The next result concerns the common preferences of agents on the same side of the market. Stated here for the college admissions model, it also holds for the assignment model. We write $\mu >_{\mathscr{C}} \mu'$ to mean that every college likes μ at least as well as μ', and some college strictly prefers μ, i.e. $\mu(C) \geqslant_C \mu'(C)$ for all $C \in \mathscr{C}$ and $\mu(C) >_C \mu'(C)$ for some $C \in \mathscr{C}$. So the relation $>_{\mathscr{C}}$ represents the common preferences of the colleges, and we define the relation $>_S$ analogously, to represent the common preferences of the students. The relations $>_{\mathscr{C}}$ and $>_S$ are only partial orders on the set of stable matchings, which is to say that there may be stable matchings μ and μ' such that neither $\mu >_S \mu'$ nor $\mu' >_S \mu$. An additional definition will help us summarize the state of affairs.

Definition 13. A *lattice* is a partially ordered set L any two of whose elements x and y have a "sup", denoted by $x \vee y$ and an "inf", denoted by $x \wedge y$. A lattice L is *complete* when each of its subsets X has a "sup" and an "inf" in L.

Hence, any nonempty complete lattice P has a least element and a greatest element. The next result therefore accounts for the existence of optimal stable matchings for each side of the market.

Theorem 21. *When all preferences over individuals are strict, the set of stable matchings in the college admissions model is a lattice under the partial orders $>_{\mathscr{C}}$ and $>_S$. Furthermore, these two partial orders are duals: if μ and μ' are stable matchings for $(S, \mathscr{C}, P)$, then $\mu >_{\mathscr{C}} \mu'$ if and only if $\mu' >_S \mu$.*

This theorem provides a more complete description of those structural properties of the set of stable matchings that account for the existence of optimal stable matchings for each side of the market. And the theorem shows that the optimal stable matching for one side of the market is the worst stable matching for the other side. Knuth (1976) attributes the lattice result for the marriage model to J.H. Conway. Shapley and Shubik (1972) established the same result for the assignment model.

4.1. Size of the core

Knuth (1976) examined the computational efficiency of the deferred acceptance procedure for the marriage model, and observed that the task of computing a single stable matching is not computationally onerous (it can be completed in polynomial time). However, even in the marriage model, the task of computing *all* the stable matchings can quickly become intractable as the size of the problem grows, for the simple reason that the number of stable matchings can grow exponentially. The next result, which follows a construction found in Knuth (1976), describes the case of a marriage problem in which there are n men and n women, which we will speak of as a problem of size n.

Theorem 22 [Irving and Leather (1986)]. *For each $i \geq 0$ there exists a stable marriage problem (M, W, P) of size $n = 2^i$ with at least 2^{n-1} stable matchings.*

However, because of the special structure of the core in these games, we can answer some questions about the core without computing all its elements. For example, suppose we simply wish to know which pairs of agents may be matched to one another at some stable matching, i.e. which pairs of agents are achievable for one another. The following result says that these can be found by following any path through the lattice from the man-optimal stable matching μ_M to the woman optimal stable matching μ_W.

Theorem 23 [Gusfield (1985)]. *Let $\mu_M = \mu_0 >_M \mu_1 >_M \mu_2 >_M \cdots >_M \mu_t = \mu_W$ be a sequence of stable matchings encountered on any path through the lattice of stable matchings of a marriage problem. Then every achievable pair appears in at least one of the matchings in the sequence.*

For the assignment model, since the core is a convex polyhedron we cannot ask how many elements it contains, but we can ask how many extreme points it might have. We can state the following result.

Theorem 24 [Balinski and Gale (1990)]. *In the assignment game, the core has at most $\binom{2m}{m}$ extreme points, where $m = \min\{|P|, |Q|\}$.*

4.2. The linear structure of the set of stable matchings in the marriage model

That the marriage and assignment models share so many properties has been a long-standing puzzle, since many of these results (e.g. the existence of optimal stable outcomes for each side of the market, and the lattice structure throughout the set of stable outcomes) require the assumption of strict preferences in the marriage model, while in the assignment model all admissible preferences must allow agents to be indifferent between different matches if prices are adjusted accordingly.[15] However, a structural similarity between the two models is seen in the rather remarkable result of Vande Vate (1989) that

---

[15] Roth and Sotomayor (1990b) show, however, that the two sets of results can be derived under common assumptions if one requires merely that the core defined by weak domination coincides with the core.

finding the stable matchings in the marriage model can also be represented as a linear programming problem.[16] The argument proceeds by first showing that the problem can be phrased as an integer program, and then observing that when the integer constraints are relaxed, the problem nevertheless has integer solutions.

For simplicity consider the special case in which $|M| = |W|$ and every pair (m, w) is mutually acceptable, and all preferences are strict. Thus, every man is matched to some woman and vice versa, under any stable matching. Let the *configuration* of a matching μ be a matrix x of zeros and ones such that $x_{mw} = 1$ if $\mu(m) = w$ and $x_{mw} = 0$ otherwise.

We will also consider matrices x of dimension $|M| \times |W|$ the elements of which may not be integers, i.e. matrices which may not be the configuration of any matching. Let $\Sigma_i x_{iw}$ denote the sum over all i in M, $\Sigma_j x_{mj}$ denote the sum over all j in W, $\Sigma_{j_m > w} x_{mj}$ denote the sum over all those j in W that man m prefers to woman w, and $\Sigma_{i_w > m} x_{iw}$ denote the sum over all those i in M that woman w prefers to man m.

We can characterize the set of stable matchings by their configurations:

Theorem 25 (Vande Vate).   *A matching is stable if and only if its configuration x is an* integer *matrix of dimension $|M| \times |W|$ satisfying the following set of constraints*:
  (1) $\Sigma_j x_{mj} = 1$ *for all m in M*,
  (2) $\Sigma_i x_{iw} = 1$ *for all w in W*,
  (3) $\Sigma_{j_m > w} x_{mj} + \Sigma_{i_w > m} x_{iw} + x_{mw} \geq 1$ *for all m in M and w in W*,
and
  (4) $x_{mw} \geq 0$ *for all m in M and w in W*.

Constraints (1), (2) and (4) require that if x is integer it is the configuration of a matching, i.e. its elements are 0's and 1's and every agent on one side is matched to some agent on the opposite side. It is easy to check that constraint (3) is equivalent to the nonexistence of blocking pairs. [To see this, note that if x is a matching, i.e. a matrix of 0's and 1's satisfying (1), (2), and (4), then (3) is not satisfied for some m and w only if $\Sigma_{j_m > w} x_{mj} = \Sigma_{i_w > m} x_{iw} = x_{mw} = 0$, in which case m and w form a blocking pair.]

Thus, an integer $|M| \times |W|$ matrix x is the configuration of a stable matching if and only if x satisfies (1)–(4). Of course there will in general be an infinite set of noninteger solutions of (1)–(4) also, and these are not matchings. However, we may think of them as corresponding to "fractional matchings", in which x_{mw} denotes something like the fraction of the time man m and woman w are matched, or the probability that they will be matched.

---

[16]Subsequent, simpler proofs are found in Rothblum (1992) and Roth, Rothblum and Vande Vate (1992).

The surprising result is that the integer solutions of (1)–(4), i.e. the stable matchings, are precisely the extreme points of the convex polyhedron defined by the linear constraints (1)–(4). That is, we have the following result.

Theorem 26 (Vande Vate). *Let C be the convex polyhedron of solutions to the linear constaints* (1)–(4). *Then the integer points of C are precisely its extreme points. That is, the extreme points of the linear constraints* (1)–(4) *correspond precisely to the stable matchings.*

4.3. Comparative statics: New entrants

The results of this subsection concern the effect of adding a new agent to the market. Following Kelso and Crawford (1982), who established the following result for a class of models including the assignment model, a number of authors have examined the effect on the optimal stable matchings for each side of the market of adding an agent on one side of the market. Briefly, the results are that, measured in this way, agents on opposite sides of the market are complements, and agents on the same side of the market are substitutes.[17] This result seems to be robust, with a recent paper by Crawford (1988) establishing the result for a general class of models with substitutable preferences introduced in Roth (1984c). As it applies to the simple model with substitutable preferences described in Subsection 3.3, his result is the following.

Theorem 27 [Crawford (1991)]. *Suppose $\mathcal{F}$ is contained in $\mathcal{F}^*$ and μ_W and μ_F are the W-optimal and F-optimal stable matchings, respectively, for a market with substitutable preferences* $(W, \mathcal{F}, P)$ *and let μ_W^* and μ_F^* be the W- and F-optimal stable matchings for $(W, \mathcal{F}^*, P^*)$, where P^* agrees with P on $\mathcal{F}$. Then*

$$\mu_W^* \geq_W \mu_W \text{ and } \mu_F^* \geq_W \mu_F \text{ under } P^*; \text{ and } \mu_W \geq_F \mu_W^*, \text{ and } \mu_F \geq_F \mu_F^*.$$

Symmetrical results are obtained if S is contained in S^*.

The next result, which we state for the assignment model, shows that when a new agent enters the market there will be some P and Q agents for whom we can unambiguously compare *all* stable outcomes of the two markets.[18] Suppose some P agent i^* enters the market $M = (P, Q, \alpha)$. The new market is then $M^{i*} = (P \cup \{i^*\}, Q, \alpha')$, where $\alpha'_{ij} = \alpha_{ij}$ for all $i \in P$ and $j \in Q$.

---

[17]Cf. Shapley (1962) for a related linear programming result.
[18]A similar result for the marriage market is given in Roth and Sotomayor (1990a).

Theorem 28. Strong dominance [Mo (1988)]. *If i^* is matched under some optimal assignment for M^{i*}, then there is a nonempty set A of agents in $P \cup Q$ such that every Q agent in A is better off and every P agent in A is worse off at any stable outcome of the new market than at any stable outcome of the old market. That is, for all (u', v') and (u, v) stable for M^{i*} and M, respectively, we have*

(a) *if a P agent i is in A, then $u_i \geqslant u'_i$;*
(b) *if a Q agent j is in A, then $v_j \leqslant v'_j$.*

The final result of this subsection can be thought of as describing how much the entry of an agent i^* in the assignment model can move the core of the game. There will be some agents whose worst core payoff in one of the two games (with and without i^*) is exactly equal to their best core payoff in the other.

Corollary 29 [Mo (1988)]. *Let $(\bar{u}', \underline{v}')$ be the P-optimal stable payoff for M^{i*}. Let $(\underline{u}, \bar{v})$ be the Q-optimal stable payoff for M. If i^* is matched under some optimal assignment for M^{i*}, there exists a nonempty set A of agents in $P \cup Q$ such that*

(a) *if a P agent i is in A, then $\bar{u}'_i = \underline{u}_i$;*
(b) *if a Q agent j is in A, then $\underline{v}'_j = \bar{v}_j$.*

5. Strategic results

We now turn to a different class of questions, motivated by the claim made in the literature distributed to participants in the hospital-intern market that the NIMP algorithm makes it unprofitable for either students or hospitals to state anything other than their true preferences. While we will defer consideration of the NIMP algorithm itself until Section 6, we consider here the extent to which it is possible to minimize the strategic complexity of matching, and what can be said about the strategic properties of procedures which lead to stable matchings.

To set the stage, consider the procedure by which graduating students at the United States Naval Academy obtain their first posts as Naval officers. The following description is taken from the *New York Times* (30 January 1986, p. 8).

Midshipmen who will graduate from the Naval Academy in June decided this week whether they wanted to be aviators or nuclear submariners, destroyermen or engineers, marines or oceanographers.... From late Thursday afternoon through the wee hours of Friday morning, the first classmen, or seniors, lined up according to their standing in the class, walked up to a long table lined with officers from each specialty, and made their choices on a first-come, first-served basis . . . .

It is easy to see that each agent in this procedure has a dominant strategy, since a student can do no better than to select his first choice of those specialties remaining when his turn comes, and since the various Naval specialties have no choices of any sort to make. And if the preferences of each specialty over the students correspond exactly to students' class standings, then the matching which results from this procedure is stable. Of course if any of the specialties have different preferences, the matching may not be stable, but the rules by which the Navy is run do not permit specialties to refuse positions to some students and offer positions to students they prefer but who have lower class standings.[19]

However, in markets that allow the agents on the two sides of the market to freely negotiate with one another, the empirical evidence suggests that the stability or instability of the final matching is important. So we will want to consider whether any procedures exist which yield stable matchings for all preferences, and which give each agent a dominant strategy. It will be sufficient for this purpose to confine our attention to the special class of "revelation mechanisms" which are functions from the stated preferences of the agents to the set of matchings. We will call a revelation mechanism which always chooses a matching that is stable with respect to the stated preferences a *stable matching mechanism*.[20] If any procedures with the desired properties exist, then there will exist a revelation mechanism which is a stable mechanism and which makes it a dominant strategy for each agent to state his true preferences.[21]

The next theorem states that no such mechanism exists for the marriage model. Since the marriage model is a special case of the college admissions and substitutable preferences models, the theorem implies that no such mechanism exists for those models either.[22]

Theorem 30. Impossibility Theorem [Roth (1982a)]. *No stable matching mechanism for the marriage model exists for which stating the true preferences is a dominant strategy for every agent.*

---

[19]That is, in the Navy's game, the outcome of this procedure is in the core, even if it can be improved upon by some student–specialty pair, since the rules do not permit the specialties to be active players.

[20]Note that the Naval Academy procedure just described can be thought of as a revelation mechanism, albeit one in which the preferences of the specialties are ignored. However, it is not a stable matching mechanism, since although it produces a stable matching for some preferences, there are (many) preferences for which the matching it produces is unstable.

[21]Various formalizations of this observation go under the name of the *revelation principle*, and are widely used in game-theoretic proofs.

[22]Notice that impossibility theorems are strongest when stated on the narrowest domain, since if no mechanism exists which works for all examples of the narrow domain, then certainly no mechanism exists which works for all examples of wider domains.

Proof*.   Since a matching mechanism is a function that produces a matching for *any* stated preferences, to prove the theorem it is sufficient to demonstrate some particular marriage market such that, for any stable matching mechanism, truthtelling will not be a dominant strategy for all agents. So consider a market with two men and two women, with preferences P given by $P(m_1) = w_1, w_2$; $P(m_2) = w_2, w_1$; $P(w_1) = m_2, m_1$; $P(w_2) = m_1, m_2$. Then there are two stable matchings, μ and ν, given by $\mu(m_i) = w_i$ for $i \in \{1, 2\}$, and $\nu(m_i) = w_j$ for $i, j \in \{1, 2\}$, $j \neq i$. So any stable mechanism must choose one of μ or ν when preferences P are stated: suppose the mechanism chooses μ. Observe that if w_2, say, changes her stated preference from $P(w_2)$ to $Q(w_2) = m_1$ while everyone else states their true preferences, then ν is the only stable matching with respect to the stated preferences $P' = (P(m_1), P(m_2), P(w_1), Q(w_2))$, and so any stable mechanism must select ν when the stated preferences are P'. So it is not a dominant strategy for all agents to state their true preferences, since w_2 does better to state $Q(w_2)$. Similarly, if the mechanism chooses ν when the preferences P are stated, then m_2 can profitably mis-state his preferences. $\square$

The same result can be stated for the assignment model.

Since we have defined a matching mechanism as a procedure which can be applied to any marriage market (i.e. as a function defined for all marriage markets), the Impossibility Theorem says we cannot find a stable mechanism that will not *sometimes* give some agent an incentive to mis-state his or her preferences. But we might hope to find a stable matching mechanism that only seldom gave agents such incentives, in which case the problem of incentives might not be very important. The following result, which can be thought of as a corollary of the proof of the Impossibility Theorem, and which strengthens it, states that no such mechanism can be found. Instead, that at least one agent will have incentive to behave strategically seems to be the usual case.

Corollary 31 [Roth and Sotomayor (1990a)].   *When any stable mechanism is applied to a marriage market in which preferences are strict and there is more than one stable matching, then at least one agent can profitably misrepresent his or her preferences, assuming the others tell the truth. (This agent can misrepresent in such a way as to be matched to his or her most preferred achievable mate under the true preferences at every stable matching under the mis-stated preferences.)*

---

*Alcade and Barbera (1991) have strengthened the impossibility theorem by observing that there exists no efficient and individually rational matching mechanism for the marriage model, for which stating the true preferences is a dominant strategy for every agent.

The proof of Corollary 31 depends on demonstrating the following. Suppose a stable mechanism selects a point different from the W-optimal stable matching μ_W, say. Then a woman w who prefers μ_W can profitably misrepresent her preferences by removing from her stated preference list of acceptable men all men who rank below $\mu_W(w)$ (as in the proof of the Impossibility Theorem). Similarly, if the mechanism selects a point different from the M-optimal stable matching, some man can profitably misrepresent his preferences.

The Impossibility Theorem and the parallel result for the assignment model tell us that in each of the models considered here there will be no way to organize the market so as to achieve a stable matching without sometimes presenting at least some of the agents with nontrivial strategic decisions. And Corollary 31 shows that only in rare cases will it be an equilibrium for all agents to state their true preferences. So we turn next to investigating which agents may have incentives to misrepresent their preferences, and what equilibrium behavior looks like, as a function of how the market is organized.

It was observed in Roth (1985a) that the answers to these questions differ in important ways depending on whether we are considering one-to-one or many-to-one matching, and so we shall deal with these two cases separately. We begin with our models of one-to-one matching, namely the marriage and assignment models.

5.1. Strategic behavior in models of one-to-one matching

The first result for the marriage model states that the incentive to state other than true preferences can be confined to the agents on one or the other side of the market.

Theorem 32 [Dubins and Freedman (1981), Roth (1982a)]. *In the marriage model, the mechanism that yields the M-optimal stable matching (in terms of the stated preferences) makes it a dominant strategy for each man to state his true preferences. (Similarly, the mechanism that yields the W-optimal stable matching makes it a dominant strategy for every woman to state her true preferences.)*

To place in context the parallel result for the assignment model, it will be helpful to first consider the case in which there is only a single agent on one side of the market. This can be thought of as a market consisting of a single seller, who owns one unit of an indivisible object, and n buyers, each of whom is interested in purchasing it. Each buyer b places a monetary value $\$r_b$ on the object, which is the maximum amount he is willing to pay, and the seller similarly places a value $\$r_s$ on the object, which is the price below which he will not sell. We will call these monetary values the *reservation prices* of the agents.

An example of this market is given by the vector of reservation prices $r = (r_1, \ldots, r_n, r_{n+1})$. In order to characterize the core and the set of stable payoff vectors, it will be convenient to define a *reordering* of the players, $1^*, 2^*, \ldots, n+1^*$ so that $r_{1^*} \geq r_{2^*} \geq \cdots \geq r_{n+1^*}$. That is, under this alternative ordering, player 1^* is that player in N who has the highest reservation price (or one of the highest, if there is a tie), and $n+1^*$ is the player with the lowest reservation price.

It is straightforward to verify that the core of this game (which by Theorem 9 equals the set of stable outcomes) corresponds to those transactions in which the object is sold to the agent with the highest reservation price, at a price between the highest and second highest reservation prices.[23] So at the seller-optimal stable outcome the price equals r_{1^*} and at the buyer-optimal stable outcome it equals r_{2^*}. It is easy to see why the Impossibility Theorem applies to this model (and therefore to the general assignment model as well), since if the seller does not have the highest reservation price, he can raise his payoff by stating a reservation price equal to the highest stated reservation price whenever the seller optimal outcome is not chosen, and similarly, if the buyer optimal outcome is not chosen, the buyer with the highest reservation price can profit by lowering his stated price to just above the second highest stated price.

As in the marriage model, however, it is possible to make it a dominant strategy for either side of the market to state true reservation prices, by using the mechanism that always selects the optimal core outcome for that side. We concentrate here on the mechanism that, for any stated reservation prices, chooses the buyer optimal core outcome. This is a well-known mechanism, variants of which are used in the auction of some U.S. government securities, for example. It is called the *sealed-bid, second-price auction* mechanism, and can be thought of as follows: each buyer writes down a number (his bid, or stated reservation price) in an envelope, without knowing what number will be written down by any other buyer. The seller also writes down a number. All the envelopes are opened, and placed in order $r_{1^*} \geq \cdots \geq r_{n+1^*}$, with the seller being player 1^* only if his number is strictly greater than all the others, in which case there is no sale. Otherwise, buyer 1^* receives the object, and pays the seller the price $p = r_{2^*}$. This mechanism is sometimes also called a *Vickrey auction*, after the economist who first observed the following result in a celebrated paper.

Theorem 33 [Vickrey (1961)]. *In a second-price, sealed bid auction (which always yields the buyer optimal core outcome in terms of the stated reservation prices), it is a dominant strategy for every buyer to state his true reservation price.*

---

[23]Unless the seller has the highest reservation price, in which case there is no sale. By Lemma 8 no monetary transfers other than the transfer of the selling price can take place in the core.

Proof.   Consider a buyer b who states his true reservation price r_b, resulting in a vector r of stated reservation prices. Given the stated reservation prices of the others, b could not have helped himself, and could have hurt himself, if he had instead stated some reservation price different from the true one. If $b = 1^*$ with respect to these stated prices, i.e. if his true reservation price is the highest stated price, then he gets the object at price $p = r_{2^*}$, which gives him a positive profit whenever r_{2^*} is strictly less than $r_b = r_{1^*}$. If he had stated a different reservation price, the outcome would not change at all so long as his stated price remains above r_{2^*}. But if he states a reservation price $r_b' < r_{2^*}$ (where by 2^* we still mean the player with the second highest of the original reservation prices), buyer b will forgo his profit, and receive a payoff of 0. (What happens when $r_b = r_{2^*}$ depends on what tie-breaking rule is used, but does not change the argument.) Now suppose that $b \neq 1^*$. Then b receives a payoff of 0, and would continue to do so for any stated preference $r_b' \leq r_{1^*}$. The only way b can change his payoff is by stating a reservation price $r_b' > r_{1^*}$, but in this case he buys the object at a price greater than his true reservation price, which gives him a negative profit. So it is a dominant strategy for each buyer to state his true reservation price.[24]   □

This brings us back to the case of the general assignment model. The following lemma shows a critical way in which the Vickrey second price auction is generalized by the mechanism which gives P agents their optimal stable outcome $(\bar{u}, \underline{v})$. Just as the second price auction gives the winning buyer his marginal contribution $r_{1^*} - r_{2^*}$ (and gives each other buyer his marginal contribution, which is 0), the P-optimal stable mechanism gives each P agent his marginal contribution.

Lemma 34 [Demange (1982), Leonard (1983)].   *For all i in P, $\bar{u}_i = v(P, Q) - v(P - \{i\}, Q)$.*

This permits the following parallel to Theorem 32.

Theorem 35 [Demange (1982), Leonard (1983)].   *The mechanism that yields the P-optimal stable outcome $(\bar{u}, \underline{v})$ makes truthtelling a dominant strategy for each P agent.*

---

[24]Note that an important feature of this mechanism is that the price stated by a bidder determines if he is the winner, but does not determine the price he pays (as it would in a conventional first-price sealed bid auction in which the high bidder 1^* pays r_{1^*}). Of course, this is not the whole argument: a useful exercise for the reader to check that he has understood is to consider why a *third-price* sealed bid auction, i.e. one at which buyer 1^* receives the object but pays price r_{3^*}, does not make it a dominant strategy for each buyer to state his true reservation price.

Returning to the marriage model, we state the following two theorems which strengthen and amplify Theorem 32.

Theorem 36 [Dubins and Freedman (1981)]. *Let P be the true preferences of the agents, and let $\bar{P}$ differ from P in that some coalition $\bar{M}$ of the men mis-state their preferences. Then there is no matching μ, stable for $\bar{P}$, which is preferred to μ_M by all members of $\bar{M}$.*

The original proofs of Theorems 32 and 36 in Roth (1982a) and Dubins and Freedman (1981) were rather lengthy. A short proof of the following result, gives a much shorter proof of those two theorems.

Theorem 37. Limits on successful manipulation [Demange, Gale and Sotomayor (1987)]. *Let P be the true preferences (not necessarily strict) of the agents, and let $\bar{P}$ differ from P in that some coalition C of men and women mis-state their preferences. Then there is no matching μ, stable for $\bar{P}$, which is preferred to every stable matching under the true preferences P by all members of C.*

To see that Theorem 37 will provide a proof of Theorems 32 and 36, consider the special case where all the coalition members are men. Then Theorem 37 implies that no matter which stable matching with respect to $\bar{P}$ is chosen, at least one of the liars is no better off than he would be at the M-optimal matching under P.[25]

Note also that Theorem 36 implies Theorem 32. Initially Theorem 36 was sometimes further interpreted as stating that no *coalition* of men could profitably misrepresent their preferences in one-to-one matching situations of the kind modelled by the marriage model, when an M-optimal stable mechanism was employed. That this is not a robust interpretation can be seen by re-examining Example 14, and observing that if man m_2 in that example were to misrepresent his preferences by listing w_3 as his first choice, then the M-optimal stable matching with respect to the stated preferences P' in which all agents but m_2 state their true preferences is equal to μ. That is, if m_2 misrepresents his preferences in this way under an M-optimal stable matching mechanism, then the resulting matching is $\mu'_M = \mu$ instead of μ_M. So m_2 is able to help the other men at no cost to himself. Note, however, that if there were any way at all in which the other men could pay m_2 for his services, then it

---

[25]When preferences are not strict, there may of course not be an M-optimal stable matching, and so we have to rephrase Theorem 32 to avoid speaking of *the* M-optimal stable mechanism. Instead, we can consider the deferred acceptance procedure with men proposing, and with a tie-breaking procedure.

would be possible for a coalition of men to form and collectively profit from this misrepresentation. Since m_2 receives the same mate at both matchings, presumably even a very small payment would make it worth his while to become part of a coalition to change the final outome from μ_M to μ, and since the gains to the other men in this coalition might be substantial, there would be ample motivation for such a coalition to form. Thus the negative implications of Theorem 36 (and also of Theorem 37) for strategic behavior by coalitions depend on the fact that, in the model of the marriage market that we are working with, we have assumed that no possibility whatsoever exists for such "sidepayments" between agents.[26] If this assumption is relaxed even a little, we see that coalitions of men can profitably manipulate even the M-optimal stable mechanism. We turn next to consider this in detail for the case of one seller and many buyers considered in connection with Theorem 33.

It is clear in that model that a *coalition* of bidders may be able, by suppressing some bids, to lower the price at which the object is sold in a second-price, sealed bid auction, or an ascending bid auction[27] (or for that matter in virtually any kind of auction). We will concentrate here on the second-price, sealed bid auction. Consider a vector r of reservation prices for which the seller's reservation price is strictly less than the second highest, so that the sale price, $p = r_{2^*}$, is greater than the seller's (auctioneer's) reservation price. Suppose the seller has the $(k + 1)$st highest reservation price, i.e. the seller is player $(k + 1)^*$. Then the coalition consisting of bidders 1^* through k^* can, by suppressing $k - 1$ bids (or submitting only one bid greater than the seller's reservation price), obtain the object at price $p' = r_{k+1^*} < r_{2^*}$.

Of course, if this was the end of the matter, the buyer who took possession of the object would benefit, but his co-conspirators would not. However, there is money in this model, so the k members of the coalition can share the wealth, for example by having a subsequent auction among themselves, with the proceeds distributed among the coalition members. Thus it is possible for a

---

[26]We have also assumed that each agent is concerned only with his own mate at any matching, and not with the mates of any other agents, and that the game is played only once, so that there is no possibility of a coalition forming to trade favors over time. In Subsection 5.3 we will also see how this result breaks down if we relax the assumption of complete information.

[27]One reason the second-price, sealed bid auction is of great interest is because of the relationship it has to the more commonly observed *ascending bid* (also called "English") auctions, in which the auctioneer keeps raising the price so long as two or more bidders indicate that they are still interested, and stops as soon as the next-to-last bidder drops out of the bidding. At that point the sale is made to the remaining bidder at the price at which the next to last bidder dropped out. (If the price at which the next to last bidder drops out is lower than the auctioneer's reservation price, the auctioneer acts as if there were a bidder who continued bidding until the auctioneer's reservation price is reached.) Suppose for simplicity that the bidders cannot see which other bidders are still bidding: then the problem facing a bidder b in this auction is simply to decide at what price to drop out of the auction. So in this case these two auctions are *strategically equivalent*, and the incentives facing the players are the same.

coalition of bidders acting together (a "bidder ring") to profit from understating their bids and sharing the benefits among themselves, even though it is not possible for a single bidder acting alone to do better than to state his true reservation price.

Note how this compares with the results for the marriage model. In both models it is a dominant strategy for an individual agent to state his true preferences when his choice consists of what preferences to state to the stable mechanism that chooses the optimal stable outcome for his side. In both models, no coalition of these agents may, by mis-stating their preferences, arrange so that they all do better under such a mechanism than when they all state their true preferences, *unless they are able to make sidepayments within the coalition.* That is, the conclusion of Theorem 36 is true in this model as well: if some coalition of bidders mis-states its reservation prices so that the vector of reservation prices is $\bar{r}$ instead of r, then there is no outcome *in the core with respect to* $\bar{r}$ that all members of the ring prefer to the result of truthful revelation. This is because no money other than the purchase price is transferred at core outcomes. But, as we have just seen, a coalition can profit by understating its preferences and then making sidepayments among its members.

Having gotten some idea of what can be said about dominant strategies and the limits on how much an individual agent can manipulate a stable mechanism, and what possibilities are open to coalitions, we now turn to the questions associated with equilibrium behavior.

5.1.1. Equilibrium behavior

The first result suggests that we may see matchings that are stable with respect to the true preferences even when agents do not state their true preferences.

Theorem 38 [Roth (1984b)]. *Suppose each man chooses his dominant strategy and states his true preferences, and the women choose any set of strategies (preference lists) $P'(w)$ that form an equilibrium for the matching game induced by the M-optimal stable mechanism. Then the corresponding M-optimal stable matching for (M, W, P') is one of the stable matchings of (M, W, P).*

Theorem 38 states that *any* equilibrium in which the men state their true preferences produces a matching that is stable with respect to the *true* preferences. Note that the conclusion would not hold if we did not restrict our attention to equilibria in which the men play undominated strategies. For example, when every agent states that no other agent is acceptable, the result is an equilibrium at which all agents remain single.

When preferences are strict, the next result presents a sort of converse to

Theorem 38, since it says that any matching μ which is stable under the true preferences can be obtained by an equilibrium set of strategies.

Theorem 39 [Gale and Sotomayor (1985b)]. *When all preferences are strict, let μ be any stable matching for (M, W, P). Suppose each woman w in $\mu(M)$ chooses the strategy of listing only $\mu(w)$ on her stated preference list of acceptable men (and each man states his true preferences). This is an equilibrium in the game induced by the M-optimal matching mechanism (and μ is the matching that results).*

The next theorem describes an equilibrium even for the case when preferences need not be strict. Furthermore, this equilibrium is a "strong equilibrium for the women", in that no coalition of women can achieve a better outcome for all of its members by having its members change their strategies.

Theorem 40 [Gale and Sotomayor (1985b)]. *Let P' be a set of preferences in which each man states his true preferences, and each woman states a preference list which ranks the men in the same order as her true preferences, but ranks as unacceptable all men who ranked below $\mu_W(w)$. These preferences P' are a strong equilibrium for the women in the game induced by an M-optimal stable matching mechanism (and μ_W is the matching that results).*

Note that these last two theorems describes strategies which put a great burden on the amount of information the women must have in order to implement them. In Subsection 5.3 we will relax the assumption that agents know one another's preferences. In the meantime, it should be clear that advising a woman to play the strategy of Theorem 40, for example, will be singularly unhelpful in most of the practical situations to which we might want to apply a theory of matching, since the strategy requires each woman w to know $\mu_W(w)$. This leads us to consider what advice we can give in environments in which information about other players' preferences may not be readily available to the players.

5.1.2. Good and bad strategies

The problems of coordination and information that may arise in implementing equilibria do not arise in the same way for players who have a dominant strategy. In particular, Theorem 32 implies that when an M-optimal stable matching procedure is used, a man may confidently state his true preferences, without regard to what the preferences of the other men and women may be. So this is a good strategy for the men, and other strategies are, in comparison, bad. Although we have seen that stating the true preferences is not a good strategy in the same way for the women, we turn now to considering what classes of strategies might be bad, in the sense of being dominated by other available strategies.

The first result states that, although it may not be wise for a woman to state her true preferences when the M-optimal stable matching mechanism is used, it can never help her to state preferences in which her first choice mate according to her stated preferences is different from her true first choice.

Theorem 41 [Roth (1982a)]. *Any strategy $P'(w)$ in which w does not list her true first choice at the head of her list is strictly dominated, in the game induced by the M-optimal stable mechanism.*

Theorem 42 states that Theorem 41 describes essentially all the dominated strategies.

Theorem 42 [Gale and Sotomayor (1985b)]. *Let $P'(w)$ be any strategy for w in which w's true first choice is listed first, and the acceptable men in $P'(w)$ are also acceptable men in w's true preference list $P(w)$. Then $P'(w)$ is not a dominated strategy when the M-optimal stable mechanism is used.*

5.2. Many-to-one matching: The college admissions model

We return now to the case of many-to-one matching, and the kind of strategic question that caused the initial 1950 algorithm in the hospital-intern labor market to be abandoned in favor of the NIMP algorithm: Is it always in agents' best interest to state their true preferences? From the Impossibility Theorem for the special case of the marriage market (Theorem 30) we know that no stable matching mechanism can have this property for all agents. But in the marriage market we observed that a mechanism that produced the optimal stable matching for one side of the market made it a dominant strategy for agents on that side to state their true preferences (Theorem 32). We might therefore hope that the parallel result holds for the college admissions model. However, this is not the case: as the next theorem shows, Theorem 32 is one of those results that does not generalize from the case of one-to-one matching to the case of many-to-one matching.

Theorem 43 [Roth (1985a)]. *No stable matching mechanism exists which makes it a dominant strategy for all colleges to state their true preferences.*

An immediate corollary of the proof of Theorem 43 is that Theorem 37 is another of the results which does not generalize from the special case of the marriage model. That is, we have

Corollary 44 [Roth and Sotomayor (1990a)].   *In the college admissions model, the conclusions of Theorem 37 for the marriage model do not hold. A coalition of agents (in fact even a single agent) may be able to misrepresent its preferences so that it does better than at any stable matching.*

Although Theorem 43 shows that no stable matching mechanism gives colleges a dominant strategy, the situation of students is as in the marriage problem. That is, we have the following result.

Theorem 45 [Roth (1985a)].   *A stable matching mechanism for the college admissions model which yields the student-optimal stable matching makes it a dominant strategy for all students to state their true preferences.*

As in the case of the marriage model, these results do little to help us identify "good" strategies for either the students or the colleges when the college-optimal stable mechanism is used. No agents have dominant strategies under that mechanism, so they all face potentially complex decision problems. And we cannot even say as much about equilibria as we could for the marriage market, since there are lots of Nash equilibria, and no easy way to distinguish among them, since the lack of dominant strategies prevents us from eliminating unreasonable equilibria as in Theorem 38. However, since Theorem 45 establishes that the student-optimal stable mechanism makes it a dominant strategy for students to state their true preferences, we might hope to have at least a one-sided generalization of Theorem 38, which would say that every equilibrium of the student-optimal stable mechanism at which students state their true preferences is stable with respect to the true preferences. But this is another result which fails to generalize, even in this partial way, from the special case of the marriage model. Again, the result is a corollary of the proof of Theorem 43.

Corollary 46 [Roth and Sotomayor (1990a)].   *In the college admissions model, the conclusions of Theorem 38 for the marriage model do not hold, even for the student-optimal stable mechanism. When all students state their true preferences, there may be equilibria of the student-optimal stable mechanism which are not stable with respect to the true preferences.*

In general, although there are equilibrium misrepresentations that yield stable matchings with respect to the true preferences, there are also equilibrium misrepresentations that yield any individually rational matching, stable or not.

Theorem 47 [Roth (1985a)]. *There exist Nash equilibrium misrepresentations under any stable matching mechanism that produce any individually rational matching with respect to the true preferences.*

But the equilibria referred to in this theorem may require a great deal of both information and coordination, since, for example, an individually rational matching μ may be achieved at equilibrium if each agent x states that $\mu(x)$ is his or her only acceptable mate.

5.3. Incomplete information

As we have seen, the (implicit) assumption of complete information makes its presence felt in a burdensome way in some of the equilibrium strategies which arise (cf. Theorems 39, 40, and 47). In this subsection we consider which of the results we have discussed so far are robust to a relaxation of the complete information assumption, and which are not.

A one-to-one marriage game with incomplete information about others' preferences will be given by a collection

$$\Gamma = (N = M \cup W, \{D_i\}_{i \in N}, g, U = X_{i \in N}U_i, F) .$$

The set N of players consists of the men and women to be matched. The sets D_i describe the decisions facing each player in the course of any play of the game (i.e. an element d_i of D_i specifies the action of player i at each point in the game at which he has decisions to make). The function g describes how the actions taken by all the agents correspond to matchings and lotteries over matchings, i.e. $g: X_{i \in N}D_i \to L[\mathcal{M}]$, where $\mathcal{M}$ is the set of all matchings between the sets M and W, and $L[\mathcal{M}]$ is the set of all probability distributions (lotteries) over $\mathcal{M}$. The set U_i is the set of all expected utility functions defined over the possibile mates for player i and the possibility of remaining single, and F is a probability distribution over n-tuples of utility functions $u = \{u_i\}_{i \in N}$, for u_i in U_i. The interpretation is that a player's "type" is given by his utility function, and at the time players must choose their strategies each player knows his own type, and the probability distribution F over vectors u is common knowledge. The special case of a game of complete information occurs when the distribution F gives a probability of one to some vector u of utilities. We will typically be concerned with games in which only a countable subset of U has positive probability. In any event, since each player i knows his own utility function u_i, he can compute a conditional probability $p_i(u_{-i} | u_i)$ for each vector of other players' utilities u_{-i} in $U_{-i} \equiv X_{j \neq i}U_j$, by applying Bayes' rule to F.

This is not the most general kind of incomplete information model we might consider. The only unknown information is the other players' utilities. In particular, players know their own utilities for being matched with one another even though they do not know what "type" the other is. Each player's utility payoff depends on his own type, and on the actions of all the players (through the matching that results), but not on the types of the other players, i.e. players' types do not effect their desirability, only their desires. This seems like a natural assumption for elite professional markets for entry level positions. For example, in the hospital-intern market, after the usual interviewing has been completed, top students are able to rank prestigious programs, and vice versa. But agents do not know how their top choices rank *them*. (Note the difference between this kind of model and one in which the interviewing process itself is modelled, in which agents would in effect be uncertain about their own preferences.)

A *strategy* for player i is a function σ_i from his type (which in this case is his utility function) to his decisions, i.e. $\sigma_i: U_i \rightarrow D_i$. If $\sigma = \{\sigma_i\}_{i \in N}$ denotes the strategy chosen by each player, then for each vector u of players' utility functions, $\sigma(u) = \{d_i \in D_i\}_{i \in N}$ describes the decisions made by the players, which result in the matching (or lottery over matchings) $g(\sigma(u))$. Consequently, a set of strategy choices σ results in a lottery over matchings, the probabilities of which are determined by the probability distribution F over vectors u, and by the function g. The expected utility to player i who is of type u_i is given by

$$u_i(\sigma) = \sum_{u_{-i} \in U_{-i}} p_i(u_{-i} \mid u_i) u_i[g(\sigma(u_{-i}, u_i))] \, .$$

A *Bayesian equilibrium*[28] is a σ^* such that, for all players i in N and all utility functions u_i in U_i, $u_i(\sigma^*) \geq u_i(\sigma^*_{-i}, \sigma_i)$ for all other strategies σ_i for player i. That is, when player i's utility is u_i the strategy σ^*_i determines player i's decision $d^*_i = \sigma^*_i(u_i)$, and the equilibrium condition requires that for all players i and all types u_i which occur with positive probability, player i cannot profitably substitute another decision $d_i = \sigma_i(u_i)$.

Recall that a general matching game with incomplete information about others' preferences is given by $\Gamma = (N = M \cup W, \{D_i\}_{i \in N}, g, U = X_{i \in N} U_i, F)$. We may call $[\{D_i\}_{i \in N}, g]$ the *mechanism*, and $[U, F]$ the *state of information* of the game. Then a game Γ is specified by a set of players, a mechanism, and a state of information. Note that we are here considering much more general kinds of mechanisms than the simple "revelation mechanisms" of the kind observed in the NIMP algorithm, for example, in which

---

[28]See Chapter 5 in this Handbook on incomplete information.

agents are just asked to state their preferences. Since we will be stating an impossibility theorem, we want to consider very general mechanisms.

The first result is an impossibility theorem that provides a strong negation to the conclusions of Theorem 38 about equilibria in the complete information case when the M-optimal stable mechanism is employed. It says that, in the incomplete information case, no equilibrium of any mechanism can have the stability properties that every equilibrium[29] of the M-optimal stable mechanism has in the complete information case. (The strategy of the proof is to observe that, by the revelation principle, if any such mechanism existed then there would be a stable revelation mechanism with truthtelling as an equilibrium, and then to show that no such revelation mechanism exists.)

Theorem 48 [Roth (1989)].   *If there are at least two agents on each side of the market, then for any general matching mechanism $[\{D_i\}_{i \in N}, g]$ there exist states of information $[U, F]$ for which every equilibrium σ of the resulting game Γ has the property that $g(\sigma(u)) \not\in L[S(u)]$ for some $u \in U$. (And the set of such u with $g(\sigma(u)) \not\in L[S(u)]$ has positive probability under F.) That is, there exists no mechanism with the property that at least one of its equilibria is always stable with respect to the true preferences at every realization of a game.*

The next theorem states that the conclusion of Theorem 36 also does not generalize to the case of incomplete information. It is possible for coalitions of men, by mis-stating their preferences, to obtain a preferable matching (even) from the M-optimal stable mechanism. This is so even though, as we will briefly discuss, it remains a dominant strategy for each man to state his true preferences.

Theorem 49 [Roth (1989)].   *In games of incomplete information about preferences, the M-optimal stable mechanism may be group manipulable by the men.*

As discussed earlier, the fact that, even in the case of complete information it is possible for a coalition of men to mis-state their preferences in a way that does not hurt any of them and helps some of them, means that the conclusion from Theorem 36 that coalitions of men cannot collectively manipulate the M-optimal mechanism to their advantage cannot be expected to be very robust. Once there is any possibility that the men can make any sort of sidepayments among themselves, this conclusion is no longer justified. The proof of Theorem 49 depends on observing that uncertainty about the preferences of other agents allows some transfers in an expected utility sense, with

---

[29] In undominated strategies.

men able to trade a gain in one realization for a gain in another. Building on Example 14, it is not hard to show that this can occur even when there is arbitrarily little uncertainty about the preferences.

In contrast to the results for equilibria, the results concerning dominant strategies in the complete information case do generalize to the case of incomplete information. This can be seen by a pointwise argument on realizations of the types of the players. In this way Roth (1989) observed that the conclusions of Theorem 32 and Theorem 41 generalize to the present case: when an M-optimal stable mechanism is used, it is a dominant strategy for each man to state his true preferences, and any strategy for a woman is dominated if her stated first choice is not her true first choice for each of her possible types.

6. Empirical overview

We return now to see what the theory described here can tell us about the principal example which we used to motivate our consideration of stability in two-sided matching markets, namely the hospital-intern labor market. We begin with the formal statement of the result promised in the preview given in Subsection 2.1.1.

Theorem 50 [Roth (1984a)]. *The NIMP algorithm is a stable matching mechanism, i.e. it produces a stable matching with respect to any stated preferences. (In fact, it produces the hospital-optimal stable matching.)*

This result lends support to the conjecture offered in the first part of Subsection 2.1.1 that the difference between the chaotic markets of the late 1940s and the orderly operation of the market with such high rates of voluntary participation starting in the early 1950s can be attributed to the stability of the matchings produced by the centralized procedure.[30]

However in Subsection 2.1 we also referred to the fact that, at least as early as 1973, significant numbers of married couples declined to take part in the NIMP procedure, or to accept the jobs assigned to them by that procedure. If it is the stability of the matching which contributes to voluntary participation in a centralized matching procedure, this should make us suspect that something about the presence of couples introduced instabilities into the market. In fact, the NIMP program included a specific procedure for handling couples that will

---

[30]The theorem also explains the way in which the NIMP algorithm is equivalent to the deferred acceptance procedure with hospitals proposing, since it also produces the hospital-optimal stable matching. However the internal working of the two algorithms differ in ways that are important for their implementation – see Roth (1984a) and Roth and Sotomayor (1990a).

make it fairly clear how these instabilities arose (and why they were so prevalent), at least until 1983, when the procedure for married couples was modified.

Briefly, the situation prior to 1983 was this. Couples graduating from medical school at the same time, and wishing to obtain two positions in the same community, had two options. One option was to stay outside of the NIMP program and negotiate directly with hospital programs. Alternatively, they could (after being certified by the Dean of their medical school as a legitimate couple) enter the NIMP program together to be matched by a special "couples algorithm".

This couples algorithm can be described roughly as follows. The couple was required to specify one of its members as the "leading member", and to submit a rank ordering of positions for each member of the couple, i.e. a couple submitted two preference lists, one for each member. The leading member of the couple was then matched to a position in the usual way, the preference list of the other member of the couple was edited to remove distant positions, and the second member was then matched if possible to a position in the same vicinity as the leading member.

It is easy to see why instabilities often result. Consider a couple $\{s_1, s_2\}$ whose first choice is to have two particular jobs in Boston, and whose second choice is to have two particular jobs in New York. Under the couples algorithm, the designated "leading member" might be matched to his or her first choice job in Boston, while the other member might be matched to some relatively undesirable job in Boston. If s_1 and s_2 were ranked by their preferred New York jobs higher than students matched to those jobs, an instability would now exist, since the couple would prefer to take the two New York jobs, and the New York hospitals would prefer to have s_1 and s_2.

Notice that, to describe this kind of instability, we are implicitly proposing a modification of the basic model of agents in the market. A couple consists of a pair of students who have a single preference ordering over pairs of positions. Part of the problem with the couples algorithm just described is that it did not permit couples to state their preferences over pairs of positions. Starting with the 1983 match, modifications were made so that couples could for the first time express such preferences within the framework of the centralized matching scheme. However, the following theorem shows that the problem goes deeper than that.

Theorem 51 [Roth (1984a)].[31] *In the hospital-intern problem with couples, the set of stable matchings may be empty.*

---

[31] This result was independently proved by Sotomayor in an unpublished note.

In view of the evidence in favor of the proposition that high voluntary rates of participation are associated with the stability of the matching mechanism, this suggests that the problem with married couples may be a persistent one. In a similar way, the next theorem suggests that the distribution of interns to rural hospitals discussed in Subsection 2.1 also is not likely to respond to any changes in procedures which achieve high degrees of voluntary compliance.

Theorem 52 [Roth (1986)].   *When all preferences over individuals are strict, the set of interns employed and positions filled is the same at every stable matching. Furthermore, any hospital that does not fill its full quota at some stable matching is matched with exactly the same set of interns at every stable matching.*

6.1. Some further remarks on empirical matters

There are several reasons why we have devoted some attention, in a survey largely concerned with mathematical theory, to the way that American physicians get their first jobs. One reason is to suggest why we think that the body of theory developed here has empirical content. Another reason is simply to give readers an idea of what empirical work connected with theory of this kind might look like. And a third reason is because it seems likely that the lessons learned from the rather special market for American medical interns may generalize to a much wider variety of entry level labor markets and other matching processes.

Regarding the empirical content of the theory, we have laid great weight in our explanation of the history of the medical market on the fact that the centralized market mechanism introduced in 1951 is a stable matching mechanism, and on the fact that the growing numbers of married couples in the market introduce instabilities. It might be objected that these are coincidental features of the market, and that the true explanations of, for example, the rates of participation lie elsewhere. For example, it might be postulated that *any* centralized market organization would have solved the problems experienced prior to 1951, and that the difficulties with having married couples in the market have less to do with instabilities of the kind dealt with here than with the difficulties that young couples have in making decisions.

Ideally, we would like to be able to conduct carefully controlled experiments designed to distinguish between any such alternative hypotheses.[32] But for theories involving the histories of complex natural organizations, we often have to settle for finding "natural experiments" which let us distinguish as well as we

---

[32]And laboratory experimentation is indeed becoming more common: see the chapter by Shubik on that subject in a forthcoming volume of this Handbook, or see *Handbook of experimental economics* [Kagel and Roth (1992)].

can between competing hypotheses. A very nice natural experiment involving these matters can be found when we look across the Atlantic ocean and examine how new physicians in the United Kingdom obtain their first jobs. The following very brief description is taken from Roth (1991).

Around the middle of the 1960s, the entry level market for physicians in England, Scotland, and Wales began to suffer from some of the same acute problems that had arisen in the American market in the 1940s and 1950s. Chief among these was that the date of appointment for "preregistration" positions (comparable to American internships, and required of new medical school graduates) had crept back in many cases to years before the date a student would graduate from medical school. The market for these positions is regional rather than national, and this problem occurred more or less in the same way in many of the regional markets. (These regional markets have roughly 200 positions each, so they are two full orders of magnitude smaller than the American market.)

The British medical authorities were aware of the experience of the American market, and in many of the regional markets it was decided to introduce a centralized market mechanism using a computerized algorithm to process preference lists obtained from students and hospitals, modelled loosely after the American system, but adapted to local conditions. Most of these algorithms were not stable matching mechanisms, and it appears that a substantial majority of those that were not failed to solve the problems they were designed to address, and were eventually abandoned. [Before being abandoned at least some experienced serious incentive problems, the evidence being a lack of voluntary participation, or a variety of unstraightforward strategic behavior. Some of the ways in which mechanisms failed, and the kind of strategic behavior they elicited, are extremely instructive; see Roth (1991) or Roth and Sotomayor (1990a) for details.] As far as can so far be determined, only two stable matching mechanisms were introduced. Both were largely successful and remain in use to this day. The similarity of the British experience in markets with unstable mechanisms to the American situation prior to 1951, and the similarity of the British experience in the markets with stable mechanisms to the American experience after 1951, support the argument that stability plays at least something like the role we have attributed to it.

The nature of this kind of empirical investigation is of course very different from the purely mathematical investigation of abstract cases. Particular models adapted to the institutional details of the markets in question must be considered (just as considering instabilities involving married couples required us to extend the basic hospital intern model). To give a bit of the flavor of this, one example comes to mind.

One of the stable matching procedures was introduced in a region of Scotland where, in keeping with previous custom, certain kinds of hospital programs were permitted to specify that they did not wish to employ more than one female physician at any time. A program taking advantage of this option might submit a preference list on which several women graduates were highly ranked, but nevertheless stipulate that no more than one of these should be assigned to it. In analyzing such a model, it is of course necessary to consider whether the introduction of such "discriminatory quotas" influences the existence of stable matchings. We leave as an exercise for the reader to show that the model of many-to-one matching with substitutable preferences can be used to address this question, and to prove the following proposition.

Proposition 53 [Roth (1991)].   *In the hospital-intern model with discriminatory quotas, the set of stable matchings is always nonempty.*

Regarding directions for future empirical work, we remark that the two studies discussed here [Roth (1984a, 1991)] are both part of a line of work that seeks to identify markets in which it is possible to establish a particularly close connection between the observed market outcome and the set of stable outcomes. This connection can be made so closely because the markets in question used computerized matching procedures which can be examined to determine the precise relationship between the submitted preferences and the market outcome. But the kind of theory developed here is by no means limited to such markets, and as more becomes known about the behavior of other entry level labor markets, for example, we should be better able to associate certain phenomena with markets that achieve stable outcomes, and other phenomena with markets that achieve unstable outcomes. In this way it should be possible to extend the empirical investigation of the predictions of this kind of theory to two-sided matching markets which are operated in a completely decentralized manner.

An interesting intermediate case, which has been described in Mongell (1987) and Mongell and Roth (1991), concerns the procedures by which the social organizations known as sororities, which operate on many American college campuses, are matched each year with new members. A centralized procedure is employed which in general would not lead to a stable matching, but because the agents in that market respond to the incentives which the procedure gives them not to state their full true preferences, much of the actual matching in that market is done in a decentralized after-market. In the data examined by Mongell and Roth, the strategic behavior of the agents led to stable matches. (This study reaffirms the importance of examining systems of rules from the point of view of how they will behave when participants respond strategically to the incentives which the rules create.)

Finally, what more general conclusions can be draw from the empirical observations we have so far been able to make of two-sided matching markets?

While some of these have been widely interpreted as evidence that "game theory works", our own view is that a somewhat more cautious interpretation is called for. First, while there is a wide variety of game-theoretic work concerning a diversity of environments, there has so far been very much less empirical work that provides tests of game-theoretic predictions. This is no doubt due to the difficulty of gathering the kind of detailed information about institutions and agents that game-theoretic theories employ, and for this reason much of the most interesting empirical work has involved controlled experiments under laboratory conditions.[33] What has made the empirical work on two-sided matching markets different is that it has proved possible to identify naturally occurring markets for which the necessary information can be found. Which brings us to the question: How does the theory fare when tested on the markets observed to date?

Even here, the answer is a little complex. We certainly cannot claim that the evidence supports the simple hypothesis that the outcome of two-sided matching markets will always be stable, since we have observed markets that employ unstable procedures and produce unstable matchings at least some of the time. And even those markets that eventually developed procedures to produce stable matchings operated for many years without such procedures before the problems they encountered in doing so led them to develop the rules they successfully use today.

However the evidence is much clearer when we turn from simple predictions to conditional predictions. The available evidence strongly supports the hypothesis that if matching markets are organized in ways that produce unstable matchings, then they are prone to a variety of related problems and market failures that can largely be avoided if the markets are organized in ways that produce stable matchings. So the kinds of empirical work described here go a long way towards supporting the contention that (at least parts of) game theory may reasonably be thought of as a source of useful theories about complex natural phenomena, and not merely of idealized or metaphorical descriptions of the behavior of perfectly rational agents.

Bibliography

Alcalde, José and Salvador Barbera (1991) 'Top dominance and the possibility of strategy proof stable solutions to the marriage problem', Universitat Antònoma de Barcelona, mimeo.
Alkan, Ahmet (1988a) 'Existence and computation of matching equilibria', Bogazici University, mimeo.
Alkan, Ahmet (1988b) 'Auctioning several objects simultaneously', Bogazici University, mimeo.

---

[33]For discussions of experiments, see a forthcoming volume of this Handbook, the surveys in Roth (1987a, 1987b), or *Handbook of experimental economics* (Kagel and Roth (1992)].

Alkan, Ahmet (1988c) 'Nonexistence of stable threesome matchings', *Mathematical Social Sciences*, 16: 207–209.

Alkan, Ahmet and David Gale (1990) 'A constructive proof of non-emptiness of the core of the matching game', *Games and Economic Behavior*, 2: 203–212.

Allision, Lloyd (1983) 'Stable marriages by coroutines', *Information Processing Letters*, 16: 61–65.

Balinski, M.L. and David Gale (1990) 'On the core of the assignment game', in: L.J. Leifman, ed., *Functional analysis, optimization and mathematical economics*. Oxford: Oxford University Press, pp. 274–289.

Bartholdi, John J. Ill and Michael A. Trick (1986) 'Stable matching with preferences derived from a psychological model', *Operations Research Letters*, 5: 165–169.

Becker, Gary S. (1981) *A treatise on the family*. Cambridge, Mass.: Harvard University Press.

Bennett, Elaine (1988) 'Consistent bargaining conjectures in marriage and matching', *Journal of Economic Theory*, 45: 392–407.

Bergstrom, Theodore and Richard Manning (1982) 'Can courtship be cheatproof?' (personal communication).

Bird, Charles G. (1984) 'Group incentive compatibility in a market with indivisible goods', *Economic Letters*, 14: 309–313.

Blair, Charles (1984) 'Every finite distributive lattice is a set of stable matchings', *Journal of Combinatorial Theory* (Series A), 37: 353–356.

Blair, Charles (1988) 'The lattice structure of the set of stable matchings with multiple partners', *Mathematics of Operations Research*, 13: 619–628.

Brams, Steven J. and Philip D. Straffin, Jr. (1979) 'Prisoners' dilemma and professional sports drafts', *American Mathematical Monthly*, 86: 80–88.

Brissenden, T.H.F. (1974) 'Some derivations from the marriage bureau problem', *The Mathematical Gazette*, 58: 250–257.

Cassady, Ralph Jr. (1967) *Auctions and auctioneering*. Berkeley: University of California Press.

Checker, Armand (1973) 'The national intern and resident matching program, 1966–72', *Journal of Medical Education*, 48: 106–109.

Crawford, Vincent P. (1991), 'Comparative statics in matching markets', *Journal of Economic Theory*, 54: 389–400.

Crawford, Vincent P. and Elsie Marie Knoer (1981) 'Job matching with heterogeneous firms and workers', *Econometrica*, 49: 437–450.

Crawford, Vincent P. and Sharon C. Rochford (1986) 'Bargaining and competition in matching markets', *International Economic Review*, 27: 329–348.

Curiel, Imma J. (1988) *Cooperative game theory and applications*, Doctoral dissertation. Katholieke Universiteit van Nijmegen.

Curiel, Imma J. and Stef H. Tijs (1985) 'Assignment games and permutation games', *Methods of Operations Research*, 54: 323–334.

Dantzig, George B. (1963) *Linear programming and extensions*. Princeton: Princeton University Press.

Demange, Gabrielle (1982) 'Strategyproofness in the assignment market game', mimeo, Laboratoire d'Econometrie de l'Ecole Polytechnique, Paris.

Demange, Gabrielle (1987) 'Nonmanipulable cores', *Econometrica*, 55: 1057–1074.

Demange, Gabrielle and David Gale (1985) 'The strategy structure of two-sided matching markets', *Econometrica*, 53: 873–888.

Demange, Gabrielle, David Gale and Marilda Sotomayor (1986) 'Multi-item auctions', *Journal of Political Economy*, 94: 863–872.

Demange, Gabrielle, David Gale and Marilda Sotomayor (1987) 'A further note on the stable matching problem', *Discrete Applied Mathematics*, 16: 217–222.

Diamond, Peter and Eric Maskin (1979) 'An equilibrium analysis of search and breach of contract, I: Steady states', *Bell Journal of Economics*, 10: 282–316.

Diamond, Peter and Eric Maskin (1982) 'An equilibrium analysis of search and breach of contract, II: A non-steady state example', *Journal of Economic Theory*, 25: 165–195.

Dubins, L.E. and D.A. Freedman (1981) 'Machiavelli and the Gale–Shapley algorithm', *American Mathematical Monthly*, 88: 485–494.

Francis, N.D. and D.I. Fleming (1985) 'Optimum allocation of places to students in a national university system', *BIT*, 25: 307–317.

Gale, David (1968) 'Optimal assignments in an ordered set: An application of matroid theory', *Journal of Combinatorial Theory*, 4: 176–180.

Gale, David (1984) 'Equilibrium in a discrete exchange economy with money', *International Journal of Game Theory*, 13: 61–64.

Gale, David and Lloyd Shapley (1962) 'College admissions and the stability of marriage', *American Mathematical Monthly*, 69: 9–15.

Gale, David and Marilda Sotomayor (1985a) 'Some remarks on the stable matching problem', *Discrete Applied Mathematics*, 11: 223–232.

Gale, David and Marilda Sotomayor (1985b) 'Ms Machiavelli and the stable matching problem', *American Mathematical Monthly*, 92: 261–268.

Gardenfors, Peter (1973) 'Assignment problem based on ordinal preferences', *Management Science*, 20: 331–340.

Gardenfors, Peter (1975) 'Match making: Assignments based on bilateral preferences', *Behavioral Science*, 20: 166–173.

Graham, Daniel A. and Robert C. Marshall (1984) 'Bidder coalitions at auctions', Duke University Department of Economics, mimeo.

Graham, Daniel A. and Robert C. Marshall (1987) 'Collusive bidder behavior at single object second price and English auctions', *Journal of Political Economy*, 95: 1217–1239.

Graham, Daniel A., Robert C. Marshall and Jean-Francois Richard (1987) 'Auctioneer's behavior at a single object English auction with heterogeneous non-cooperative bidders', Working paper #87–01, Duke University Institute of Statistics and Decision Sciences.

Graham, Daniel A., Robert C. Marshall and Jean-Francois Richard (1990) 'Differential payments within a bidder coalition and the Shapley value', *American Economic Review*, 80: 493–510.

Granot, Daniel (1984) 'A note on the room-mates problem and a related revenue allocation problem', *Management Science*, 30: 633–643.

Gusfield, Dan (1987) 'Three fast algorithms for four problems in stable marriage', *SIAM Journal on Computing*, 16: 111–128.

Gusfield, Dan (1988) 'The structure of the stable roommate problem: Efficient representation and enumeration of all stable assignments', *SIAM Journal on Computing*, 17: 742–769.

Gusfield, Dan and Robert W. Irving (1989) *The stable marriage problem: Structure and algorithms.* Cambridge, Mass.: MIT Press.

Gusfield, Dan, Robert W. Irving, Paul Leather and M. Saks (1987) 'Every finite distributive lattice is a set of stable matchings for a *small* stable marriage instance', *Journal of Combinatorial Theory* A, 44: 304–309.

Harrison, Glenn W. and Kevin A. McCabe (1988) Stability and preference distortion in resource matching: An experimental study of the marriage market', mimeo, Department of Economics, University of New Mexico.

Hull, M. Elizabeth C. (1984) 'A parallel view of stable marriages', *Information Processing Letters*, 18: 63–66.

Hwang, J.S. (1978) 'Complete unisexual stable marriages', *Soochow Journal of Mathematics*, 4: 149–151.

Hwang, J.S. (1986) 'The algebra of stable marriages', *International Journal of Computer Mathematics*, 20: 227–243.

Hwang, J.S. (undated) Modelling on college admissions in terms of stable marriages', mimeo.

Hwang, J.S. and H.J. Shyr (1977) 'Complete stable marriages', *Soochow Journal of Mathematical and Natural Sciences*, 3: 41–51.

Hylland, Aanund and Richard Zeckhauser (1979) 'The efficient allocation of individuals to positions', *Journal of Political Economy*, 87: 293–314.

Irving, Robert W. (1985) 'An efficient algorithm for the stable room-mates problem', *Journal of Algorithms*, 6: 577–595.

Irving, Robert W. (1986) 'On the stable room-mates problem', mimeo, Department of Computing Science, University of Glasgow.

Irving, Robert W. and Paul Leather (1986) 'The complexity of counting stable marriages', *SIAM Journal of Computing*, 15: 655–667.

Irving, Robert W., Paul Leather and Dan Gusfield (1987) 'An efficient algorithm for the "optimal" stable marriage', *Journal of the ACM* 34: 532–543.

Itoga, Stephen Y. (1978) 'The upper bound for the stable marriage problem', _Journal of the Operational Research Society_, 29: 811–814.

Itoga, Stephen Y. (1981) 'A generalization of the stable marriage problem', _Journal of the Operational Research Society_, 32: 1069–1074.

Itoga, Stephen Y. (1983) 'A probabilistic version of the stable marriage problem', _BIT_, 23: 161–169.

Jones, Philip C. (1983) 'A polynomial time market mechanism', _Journal of Information and Optimization Sciences_, 4: 193–203.

Kagel, John and Alvin E. Roth, eds. (1992) _Handbook of experimental economics_. Princeton, NJ: Princeton University Press.

Kamecke, Ulrich (1987) 'A generalization of the Gale–Shapley algorithm for monogamous stable matchings to the case of continuous transfers', Discussion paper, Rheinische Friedrich-Wilhelms Universitat, Bonn.

Kamecke, Ulrich (1989) 'Non-cooperative matching games', _International Journal of Game Theory_, 18: 423–431.

Kamecke, Ulrich (1992) 'On the uniqueness of the solution to a large linear assignment problem', _Journal of Mathematical Economics_, forthcoming.

Kaneko, Mamoru (1976) 'On the core and competitive equilibria of a market with indivisible goods', _Naval Research Logistics Quarterly_, 23: 321–337.

Kaneko, Mamoru (1982) 'The central assignment game and the assignment markets', _Journal of Mathematical Economics_, 10: 205–232.

Kaneko, Mamoru (1983) 'Housing markets with indivisibilities', _Journal of Urban Economics_, 13: 22–50.

Kaneko, Mamoru and Myrna Holtz Wooders (1982) 'Cores of partitioning games', _Mathematical Social Sciences_, 3: 313–327.

Kaneko, Mamoru and Myrna Holtz Wooders (1985) 'The core of a game with a continuum of players and finite coalitions: Nonemptiness with bounded sizes of coalitions', mimeo, Institute for Mathematics and its Applications, University of Minnesota.

Kaneko, Mamoru and Myrna Holtz Wooders (1986) 'The core of a game with a continuum of players and finite coalitions: The model and some results', _Mathematical Social Sciences_, 12: 105–137.

Kaneko, Mamoru and Yoshitsugu Yamamoto (1986) 'The existence and computation of competitive equilibria in markets with an indivisible commodity', _Journal of Economic Theory_, 38: 118–136.

Kapur, Deepak and Mukkai S. Krishnamoorthy (1985) 'Worst-case choice for the stable marriage problem', _Information Processing Letters_, 21: 27–30.

Kelso, Alexander S., Jr. and Vincent P. Crawford (1982) 'Job matching, coalition formation, and gross substitutes', _Econometrica_, 50: 1483–1504.

Knuth, Donald E. (1976) _Marriages stables_. Montreal: Les Presses de l'Universite de Montreal.

Leonard, Herman B. (1983) 'Elicitation of honest preferences for the assignment of individuals to positions', _Journal of Political Economy_, 91: 461–479.

Masarani, F. and S.S. Gokturk (1988) 'On the probabilities of the mutual agreement match', _Journal of Economic Theory_, 44: 192–201.

McVitie, D.G. and L.B. Wilson (1970a) 'Stable marriage assignments for unequal sets', _BIT_, 10: 295–309.

McVitie, D.G. and L.B. Wilson (1970b) 'The application of the stable marriage assignment to university admissions', _Operational Research Quarterly_, 21: 425–433.

McVitie, D.G. and L.B. Wilson (1971) 'The stable marriage problem', _Communications of the ACM_, 14: 486–492.

Mo, Jie-ping (1988) 'Entry and structures of interest groups in assignment games', _Journal of Economic Theory_, 46: 66–96.

Moldovanu, Benny (1990) 'Bargained equilibria for assignment games without side payments', _International Journal of Game Theory_, 18: 471–477.

Mongell, Susan J. (1987) _Sorority rush as a two-sided matching mechanism: A game-theoretic analysis_, Ph.D. dissertation. Department of Economics, University of Pittsburgh.

Mongell, Susan J. and Alvin E. Roth (1986) 'A note on job matching with budget constraints', _Economics Letters_, 21: 135–138.

Mongell, Susan J. and Alvin E. Roth (1991) 'Sorority rush as a two-sided matching mechanism', *American Economic Review*, 81: 441–464.

Mortensen, Dale T. (1982) 'The matching process as a Noncooperative bargaining game', in: J. McCall, ed., *The Economics of Information and Uncertainty*. Chicago: University of Chicago Press, pp. 233–258.

Owen, Guillermo (1975) 'On the core of linear production games', *Mathematical Programming*, 9: 358–370.

Prasad, Kislaya (1987) 'The complexity of games II: Assignment games and indices of power', mimeo, Department of Economics, Syracuse University.

Proll, L.G. (1972) 'A simple method of assigning projects to students', *Operational Research Quarterly*, 23: 195–201.

Quinn, Michael J. (1985) 'A note on two parallel algorithms to solve the stable marriage problem', *Bit*, 25: 473–476.

Quint, Thomas (1987a) 'Elongation of the core in an assignment game', Technical report, IMSSS, Stanford.

Quint, Thomas (1987b) 'A proof of the nonemptiness of the core of two sided matching markets', CAM report #87–29, Department of Mathematics, UCLA.

Quint, Thomas (1988) 'An algorithm to find a core point for a two-sided matching model', CAM report #88–03, Department of Mathematics, UCLA.

Quint, Thomas (1991) 'The core of an m-sided assignment game', *Games and Economic Behavior*, 3: 487–503.

Quinzii, Martine (1984) 'Core and competitive equilibria with indivisibilities,' *International Journal of Game Theory*, 13: 41–60.

Rochford, Sharon C. (1984) 'Symmetrically pairwise-bargained allocations in an assignment market', *Journal of Economic Theory* 34: 262–281.

Ronn, Eytan (1986) *On the complexity of stable matchings with and without ties*, Ph.D. dissertation, Yale University.

Ronn, Eytan (1987) 'NP-complete stable matching problems', *Journal of Algorithms*, forthcoming.

Roth, Alvin E. (1982a) 'The economics of matching: Stability and incentives', *Mathematics of Operations Research*, 7: 617–628.

Roth, Alvin E. (1982b) 'Incentive compatibility in a market with indivisible goods', *Economics Letters*, 9: 127–132.

Roth, Alvin E. (1984a) 'The evolution of the labor market for medical interns and residents: A case study in game theory', *Journal of Political Economy*, 92: 991–1016.

Roth, Alvin E. (1984b) 'Misrepresentation and stability in the marriage problem', *Journal of Economic Theory*, 34: 383–387.

Roth, Alvin E. (1984c) 'Stability and polarization of interests in job matching', *Econometrica*, 52: 47–57.

Roth, Alvin E. (1985a) 'The college admissions problem is not equivalent to the marriage problem', *Journal of Economic Theory*, 36: 277–288.

Roth, Alvin E. (1985b) 'Common and conflicting interests in two-sided matching markets', *European Economic Review* (Special issue on Market Competition, Conflict, and Collusion), 27: 75–96.

Roth, Alvin E. (1985c) 'Conflict and coincidence of interest in job matching: Some new results and open questions', *Mathematics of Operations Research*, 10: 379–389.

Roth, Alvin E. (1986) 'On the allocation of residents to rural hospitals: A general property of two-sided matching markets', *Econometrica* 54: 425–427.

Roth, Alvin E. (1987a) 'Laboratory experimentation in Economics', in: Truman Bewley, ed., *Advances in economic theory, Fifth World Congress*. Cambridge University Press, pp 269–299. (Preprinted in *Economics and Philosophy*, Vol. 2, 1986, 245–273.)

Roth, Alvin E., ed. (1987b) *Laboratory experimentation in economics: Six points of view*. Cambridge: Cambridge University Press.

Roth, Alvin E. (1988) 'Laboratory experimentation in economics: A methodological overview', *Economic Journal*, 98: 974–1031.

Roth, Alvin E. (1989) 'Two sided matching with incomplete information about others' preferences', *Games and Economic Behavior*, 1: 191–209.

Roth, Alvin E. (1991) 'A natural experiment in the organization of entry level labor markets:

Regional markets for new physicians and surgeons in the U.K.', *American Economic Review*, 81: 415–440.

Roth, Alvin E. and Andrew Postlewaite (1977) 'Weak versus strong domination in a market with indivisible goods', *Journal of Mathematical Economics*, 4: 131–137.

Roth, Alvin E. and Marilda Sotomayor (1988a) 'Interior points in the core of two-sided matching problems', *Journal of Economic Theory*, 45: 85–101.

Roth, Alvin E. and Marilda Sotomayor (1989) 'The college admissions problem revisited', *Econometrica*, 57: 559–570.

Roth, Alvin E. and Marilda Sotomayor (1990a) *Two-sided matching: A study in game-theoretic modelling and analysis*, Econometric Society Monograph Series. Cambridge: Cambridge University Press.

Roth, Alvin E. and Marilda Sotomayor (1990b) 'Stable outcomes in discrete and continuous models of two-sided matching: A unified treatment', University of Pittsburgh, mimeo.

Roth, Alvin E. and John H. Vande Vate (1990) 'Random paths to stability in two-sided matching', *Econometrica*, 58: 1475–1480.

Roth, Alvin E., Uriel G. Rothblum and John H. Vande Vate (1992) 'Stable matchings, optimal assignments, and linear programming', *Mathematics of Operations Research*, forthcoming.

Rothblum, Uriel G. (1992) 'Characterization of stable matchings as extreme points of a polytope', *Mathematical Programming*, forthcoming.

Samet, Dov and Eitan Zemel (1984) 'On the core and dual set of linear programming games', *Mathematics of Operations Research*, 9: 309–316.

Sasaki, Hiroo (1988) 'Axiomatization of the core for two-sided matching problems', Economics Discussion paper #86, Faculty of Economics, Nagoya City University, Nagoya, Japan.

Sasaki, Hiroo and Manabu Toda (1986) 'Marriage problem reconsidered: Externalities and stability', mimeo, Department of Economics, University of Rochester.

Satterthwaite, Mark A. (1975) 'Strategy-proofness and Arrow's conditions: Existence and correspondence theorems for voting procedures and social welfare functions', *Journal of Economic Theory*, 10: 187–217.

Scotchmer, Suzanne and Myrna Holtz Wooders (1988) 'Monotonicity in Games that Exhaust Gains to Scale', mimeo, University of California, Berkeley.

Shapley, Lloyd S. (1962) 'Complements and substitutes in the optimal assignment problem', *Naval Research Logistics Quarterly*, 9: 45–48.

Shapley, Lloyd S. and Herbert Scarf (1974) 'On cores and indivisibility', *Journal of Mathematical Economics*, 1: 23–28.

Shapley, Lloyd S. and Martin Shubik (1972) 'The assignment game I: The core', *International Journal of Game Theory*, 1: 111–130.

Sondak, Harris and Max H. Bazerman (1987) 'Matching and negotiation processes in quasi-markets', *Organizational Behavior and Human Decision Processes*, forthcoming.

Sotomayor, Marilda (1986a) 'On incentives in a two-sided matching market', Working paper, Department of Mathematics, Pontificia Universidade Catolica do Rio de Janeiro.

Sotomayor, Marilda (1986b) 'The simple assignment game versus a multiple assignment game', Working paper, Department of Mathematics, Pontificia Universidade Catolica do Rio de Janeiro.

Sotomayor, Marilda (1987) 'Further results on the core of the generalized assignment game', Working paper, Department of Mathematics, Pontificia Universidade Catolica do Rio de Janeiro.

Sotomayor, Marilda (1988) 'The multiple partners game', William Brock and Mukul Majumdar, eds., in: *Equilibrium and dynamics: Essays in honor of David Gale*, in preparation.

Stalnaker, John M. (1953) 'The matching program for intern placement: The second year of operation', *Journal of Medical Education*, 28: 13–19.

Thompson, Gerald L. (1980) 'Computing the core of a market game', in: A.V. Fiacco and K.O. Kortanek, eds., *Extremal methods and systems analysis*, Lecture Notes in Economics and Mathematical Systems #174. Berlin: Springer, pp. 312–334.

Thompson, William (1986) 'Reversal of asymmetries of allocation mechanisms under manipulation', *Economics Letters*, 21: 227–230.

Toda, Manabu (1988) 'The consistency of solutions for marriage problems', Department of Economics, University of Rochester, mimeo.

Tseng, S.S. and R.C.T. Lee (1984) 'A parallel algorithm to solve the stable marriage problem', *Bit*, 24: 308–316.

Vande Vate, John H. (1989) 'Linear programming brings marital bliss', *Operations Research Letters*, 8: 147–153.

Vickrey, W. (1961), 'Counterspeculation, auctions, and competitive sealed tenders', *Journal of Finance*, 16: 8–37.

Wilson, L.B. (1972) 'An analysis of the stable marriage assignment algorithm', *BIT*, 12: 569–575.

Wilson, L.B. (1977) 'Assignment using choice lists', *Operational Research Quarterly*, 28: 569–578.

Wood, Robert O. (1984) 'A note on incentives in the college admissions market', mimeo, Stanford University.

Chapter 17

VON NEUMANN–MORGENSTERN STABLE SETS

WILLIAM F. LUCAS

The Claremont Graduate School

Contents

Handbook of Game Theory, Volume 1, Edited by R.J. Aumann and S. Hart

1. Introduction

Most approaches to multiperson game theory divide into either the non-cooperative methods involving equilibrium points or else the various cooperative models. In the cooperative case one assumes that the participants can communicate, form coalitions, and make binding agreements. These games are primarily concerned with which coalitions will form and how the resulting gains (or losses) will be allocated among the participants. The cooperative models are usually described in terms of a characteristic function which assigns a real number (or else a set of realizable outcomes) to each potential subset (coalition) of the set of players. The possible outcomes are represented as payoff vectors corresponding to the distribution of utility to the players. Some of these outcomes will be preferred by the players over others, and certain final distributions are more likely to occur. Many different models have been proposed over the past fifty years to analyze these cooperative interactions, and alternate notions of a solution have been proposed. The first such general model was presented by von Neumann and Morgenstern (1944) in *Theory of Games and Economic Behavior*. The solution concept that they proposed is now referred to as a "stable set" or a "(von Neumann–Morgenstern) solution."

  In this chapter we will describe their original model for the coalitional games, provide some illustrations, analyze the three-person case in detail, and discuss some of the mathematical properties of stable sets. Special classes of games such as the simple or symmetric cases, as well as particular types of solutions such as the finite, discriminatory, and symmetric ones are of particular interest from mathematical as well as empirical behavior viewpoints.

2. Abstract games and stable sets

In general a multiperson cooperative game involves a set of realizable outcomes and some preference relation between these outcomes. Some outcomes will be more desired, more likely to occur, or more fair than others. We thus define an *abstract game* (U, d) to consist of a set U of elements called *imputations* and a binary irreflexive relation d on U referred to as *domination*. *Irreflexive* means that no element in U can dominate itself. If the set U is a subset of *n*-space $\boldsymbol{R}^n$, then (U, d) is called an *n-person abstract game* and we refer to $N = \{1, 2, \ldots, n\}$ as the set of *n players* $1, 2, \ldots, n$. Figures 1 and 2 describe two three-person abstract games where U consists of the five vector outcomes in $\boldsymbol{R}^3$, corresponding to the five partitions of $N = \{1, 2, 3\}$, and the

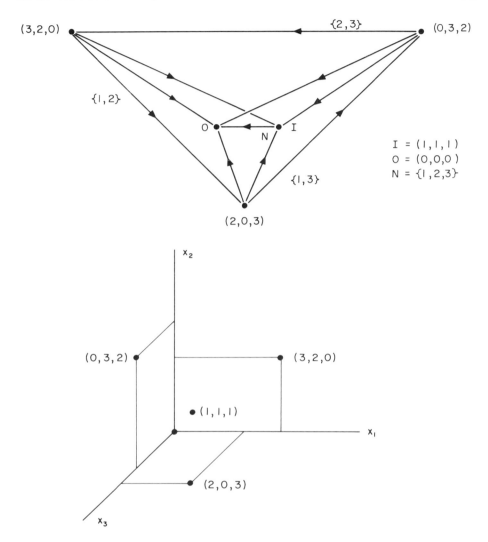

Figure 1. A three-person spatial game.

dominance relation d is indicated by the arrows. Any such directed graph can be so interpreted as an abstract game.

The *core C* of an abstract game consists of the set of elements in U which are maximal with respect to the dominance relation d. No element in U can dominate any element in the core. The vector $y^W = (3, 2.2, 0)$ in Figure 2 is the only element in the core for this game.

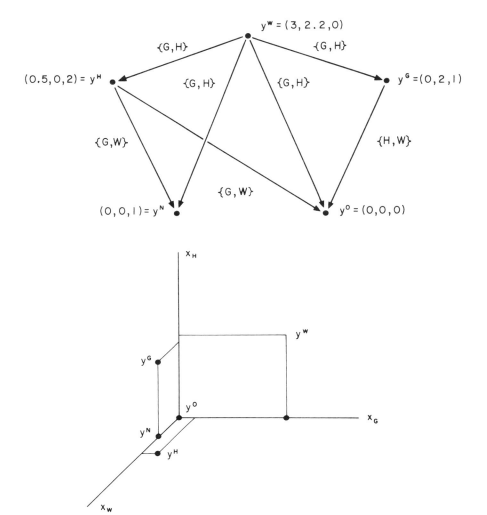

Figure 2. The satellite game.

No element is undominated in Figure 1, and thus the core for the corresponding game is the empty set $\emptyset$. Even when the core C of a game (U, d) is a nonempty set, it might be "too small" to serve as a reasonable solution concept for the game, as will be illustrated in Example 3. It is not always the case that every element in $U - C$ is dominated by an element in C as occurred in Figure 2. These considerations lead naturally to the consideration of other solution notions, and von Neumann and Morgenstern (1944) introduced the concept of stable set for cooperative games. (Originally stable sets were called "solutions", but it is more common today to use the term "solution" or "solution concept" for any one of the many solution ideas which have been proposed for the cooperative games.)

A subset V of U is called a *stable set* (or a von Neumann–Morgenstern *solution*) for an abstract game (U, d) whenever

$$V \cap D(V) = \emptyset$$

and

$$V \cup D(V) = U ,$$

where the *dominion* function D is defined for any subset X of U by

$$D(X) = \{ y \in U : y \text{ is dominated by some } x \in X \} .$$

These two conditions are called *internal stability* and *external stability*, respectively. They state that no element in a stable set V can dominate another element in V, and any element in $U - V$ is dominated by at least one element in V. In other words, the set V is "domination free", and "setwise dominates" all elements not in V. This definition for a stable set V can be expressed by the one equation,

$$V = U - D(V) ,$$

which describes V as a fixed subset under the mapping $f(X) = U - D(X)$, where $X \subset U$. The single outcome $y^{W} = (3, 2.2, 0)$ is the unique stable set V for the game in Figure 2, since y^{W} dominates the other four outcomes. The game in Figure 1 has no stable set, due essentially to the odd cycle of domination between $(3, 2, 0)$, $(2, 0, 3)$, and $(0, 3, 2)$.

The core C of any game (U, d) can be expressed by the equation

$$C = U - D(U) .$$

The core of a given game is a unique set, although in many cases it is the empty set $\emptyset$. A stable set V is never empty (unless $U = \emptyset$). However, there are games for which no stable set exists, as in Figure 1. A game typically does not have only one stable set. It follows from our definitions that

$$C \subset V \quad \text{and} \quad V \cap D(C) = \emptyset$$

for any stable set V. C will be the unique stable set whenever $D(C) = U - C$. When C is not a stable set by itself, then one attempts to enlarge C by adding elements from $U - (C \cup D(C))$ to reach a stable set V. That is, elements are added to C in such a manner as to maintain internal stability at each step in hope of eventually obtaining external stability as well.

Because of some theoretical and practical difficulties with stable sets, several variations, extensions, and generalizations of this notion have been proposed. The stationary sets of Weber (1974), the subsolutions and supercore of Roth (1976), and the absorbing sets of Chang (1985), are a few examples of these. A traditional view of abstract games and its relation to graph theory notions is indicated in Berge (1957) and Richardson (1955). A recent abstract approach to stable set theory and its connections to other solution concepts in game theory is given in Greenberg (1989, 1990).

Example 1. Three players denoted by 1, 2, and 3 can partition themselves into a coalition structure in five ways:

$$\{\{1\}, \{2\}, \{3\}\}, \{\{1,2\}, \{3\}\}, \{\{1,3\}, \{2\}\}, \{\{1\}, \{2,3\}\} \quad \text{and}$$
$$\{\{1,2,3\}\}\,.$$

Assume that the five corresponding outcomes in three-space $\boldsymbol{R}^3$ are

$$(0,0,0), (3,2,0), (2,0,3), (0,3,2) \quad \text{and} \quad (1,1,1)\,,$$

where a vector (x_1, x_2, x_3) assigns a payoff of x_1 to player 1, x_2 to 2, and x_3 to 3, respectively. These five points are pictured both as a graph and as they are located in $\boldsymbol{R}^3$ in Figure 1. The best outcome for players 1 and 2 as a group, as well as 1 individually, is $(3,2,0)$ which is achieved when 1 and 2 form the coalition $\{1,2\}$ which excludes player 3. Players 1 and 3 together, and 3 individually, prefer the outcome $(2,0,3)$ realized by the coalition $\{1,3\}$. Similarly, $\{2,3\}$ and 2 in particular would rather have the outcome $(0,3,2)$. The coalitional preferences between these five outcomes are indicated by the arrows in Figure 1. Which outcome would occur in a play of this game, assuming that only one coalition structure and final payoff vector is allowed?

The outcome $(0,0,0)$, which results from coalition structure $\{\{1\}, \{2\}, \{3\}\}$ of all singletons, is clearly inferior to each of the other four possibilities. The outcome $(1,1,1)$ realized by the grand coalition $N = \{1,2,3\}$ is less desirable to the pair of players in each of the three two-person coalitions. It appears as though one of these two-person coalitions may ultimately form with individual payoffs of 3 and 2 units to its players. If "side payments" are allowed then, for example, player 1 may offer a side payment of $\frac{1}{2}$ unit to player 2 to realize the outcome $(\frac{5}{2}, \frac{5}{2}, 0)$ in the coalition $\{1,2\}$.

Table 1

Coalition structures	Outcomes (x_G, x_H, x_W)	Normalized outcomes
$P^0 = \{\{G\}, \{H\}, \{W\}\}$	$x^0 = (1, 2, 3)$	$y^0 = (0, 0, 0)$
$P^G = \{\{G\}, \{H, W\}\}$	$x^G = (1, 4, 4)$	$y^G = (0, 2, 1)$
$P^H = \{\{H\}, \{G, W\}\}$	$x^H = (1.5, 2, 5)$	$y^H = (0.5, 0, 2)$
$P^W = \{\{W\}, \{H, G\}\}$	$x^W = (4, 4.2, 3)$	$y^W = (3, 2.2, 0)$
$P^N = \{\{G, H, W\}\}$	$x^N = (1, 2, 4)$	$y^N = (0, 0, 1)$

Example 2. In chapter 11 of the book *The Game of Business*, John McDonald (1975) described a ten-person communication satellite game played out in the United States in the early 1970s. In particular, he focused on a three-person subgame played by the three corporations: General Telephone and Electronics Corporations (G), Hughes Aircraft Company (H), and Western Union Telegraph Company (W). The estimated benefits to the companies depended upon which coalitions were to form. There were substantial gains for those forming a two-person coalition, but only one player gained in the full three-person coalition. The expected outcomes for G, H, and W can be expressed as a three-tuple (x_G, x_H, x_W) as indicated in Table 1. The normalized outcomes (y_G, y_H, y_W) subtract off $x^0 = (1, 2, 3)$ from the initial outcomes in the previous column, and measure only the additional gains obtained when nonsingleton coalitions form. These latter points are shown both as a graph and as located in R^3 in Figure 2.

The coalition $\{G, H\}$ would clearly prefer the outcome $y^W = (3, 2.2, 0)$ over any of the other four normalized outcomes, and they have it in their power to effect this result. Furthermore, $\{H, W\}$ would prefer $y^G = (0, 2, 1)$ to $y^0 = (0, 0, 0)$, and $\{G, W\}$ would prefer $y^H = (0.5, 0, 2)$ to $y^N = (0, 0, 1)$ and to y^0. These preferences are indicated by the directed graph in Figure 2. The outcome vector $y^W = (3, 2.2, 0)$ seems to be the natural resolution to this three-person cooperative game. One would expect G and H to enter into a joint undertaking and for W to go it alone. This is in fact what happened at the time. (We will return to Example 2 in Section 4 where the possibility of side payments is considered.)

3. The classical model

The first general approach for the multiperson coalitional games was proposed in the monumental book by von Neumann and Morgenstern (1944). Their model consists of four basic concepts: a characteristic function v, a set of imputations A, a dominance relation dom, and a solution notion V called stable set. An *n-person game in characteristic function form* (with side payments) is a pair (N, v), where $N = \{1, 2, \ldots, n\}$ is a set of *players* and where v

is a real valued *characteristic function* on 2^N, the set of all subsets of N. The
function v assigns a real number $v(S)$ to each subset S of N and $v(\emptyset) = 0$ for the
empty set $\emptyset$. Intuitively, the value $v(S)$ indicates the wealth, worth, or power
which the *coalition* S can achieve when its members act together. In practice
the number $v(S)$ may be derived from a game in normal (strategic) form, but in
many applications this value arises in a more direct or natural way from the
situation being modeled. One often writes (n, v) or just v for the game (N, v).

It is often assumed that v is *superadditive*, i.e.,

$$v(S \cup T) \geqslant v(S) + v(T)$$

whenever $S \cap T = \emptyset$. Much of their theory holds without this condition.
However, we will assume that the subsequent games in this chapter are
superadditive unless stated otherwise.

A vector $x = (x_1, x_2, \ldots, x_n)$ with real components is an *imputation* for the
game (N, v) if

$$x_i \geqslant v(\{i\}) \quad \forall i \in N$$

and

$$x_1 + x_2 + \cdots + x_n = v(N) .$$

Let $A = A(v)$ be the set of all imputations. These two constraints are referred
to as *individual rationality* and *Pareto optimality* or *efficiency*, respectively. An
imputation x represents a realizable way for the n players to distribute the total
amount $v(N)$, with x_i going to player i who is unlikely to accept anything less
than his own value $v(\{i\})$.

If x and y are imputations and S is a *nonempty* subset of N, then x *dominates*
y via S, denoted $x \operatorname{dom}_S y$, if

$$x_i > y_i \quad \forall i \in S \tag{1}$$

and

$$\sum_{i \in S} x_i \leqslant v(S) . \tag{2}$$

One also says that x *dominates* y, denoted $x \operatorname{dom} y$, if there exists some
(nonempty) S such that $x \operatorname{dom}_S y$. In other words, the coalition S prefers the
distribution x over y if each member of S obtains more, and if S has it within its
power to achieve this allocation. This latter condition (2) is referred to as

effectiveness, i.e., either as S is *effective at* x or as x is *effective for* S. It is convenient to set

$$x(S) = \sum_{i \in S} x_i \,.$$

We also introduce the following notation for $x \in A$ and $B \subset A$:

$$\text{Dom}_S \, x = \{ y \in A : x \, \text{dom}_S \, y \} \,,$$

$$\text{Dom} \, x = \bigcup_{S \subset N} \text{Dom}_S \, x \,,$$

$$\text{Dom}_S \, B = \bigcup_{x \in B} \text{Dom}_S \, x \,,$$

$$\text{Dom} \, B = \bigcup_{S \subset N} \text{Dom}_S \, B \,.$$

as well as the inverse domination regions

$$\text{Dom}^{-1} B = \bigcup_{x \in B} \{ y \in A : y \, \text{dom} \, x \} \,.$$

Note that the binary relation "dom_S" is irreflexive, asymmetric, and transitive for a given coalition S. On the other hand, the relation "dom" may not in general have these latter two properties, and this is the source of many of the mathematical difficulties that arise in the von Neumann–Morgenstern theory of solutions.

A subset V or $V(v)$ of A is a von Neumann–Morgenstern *solution* or *stable set* if no x in V dominates any y in V, and if every z not in V is dominated by some $x \in V$. These two conditions can be expressed as

$$V \cap \text{Dom} \, V = \emptyset \tag{3}$$

and

$$V \cup \text{Dom} \, V = A \,, \tag{4}$$

or by one expression

$$A - \text{Dom} \, V = V \,.$$

We also say that V' is a *solution for* $B \subset A$ whenever

$$V' \cap \mathrm{Dom}\, V' = \emptyset$$

and

$$V' \cup \mathrm{Dom}\, V' \supset B\,.$$

An n-person game (N, v) in characteristic function form is an abstract game (U, d) where $U = A$ and $d = \mathrm{dom}$. One is also interested in the idea of a core $C = C(v)$ for a game v in characteristic function form. C is the set of imputations x in A that are maximal with respect to the dom relation. When v is superadditive, this is equivalent to

$$C = \{x \in A\colon x(S) \geqslant v(S)\ \forall S \subset N\}\,.$$

However, von Neumann and Morgenstern did not dwell on the notion of the core, because the first class of games they studied were the "constant-sum essential" games (defined below) which always have empty cores. The core of a game, on the other hand, is the primary topic of the five previous chapters in this Handbook.

A game (N, v) is said to be *constant-sum* if

$$v(S) + v(N - S) = v(N)\quad \forall S \subset N\,.$$

Such games are called *zero-sum* when $v(N) = 0$. This case arises if one were to derive the characteristic function values $v(S)$ from a constant-sum normal form game by solving a (two-person, zero-sum) matrix game played between each coalition S and its complement $N - S$. It turns out that the distinction between zero-sum and nonconstant-sum games is not a major one for most multiperson cooperative models, in contrast to the theory of two-person games.

A game (N, v) is *essential* if

$$\sum_{i \in N} v(\{i\}) < v(N)$$

and *inessential* when

$$\sum_{i \in N} v(\{i\}) \geqslant v(N)\,.$$

A superadditive inessential game has an additive v. In this case A, C, and the unique V consists of the single imputation $(v(1), v(2), \ldots, v(n))$, where we write $v(i)$ for $v(\{i\})$. So there is no need to study coalition formation or solution concepts in this latter case, and one typically restricts the analysis to only the class of essential games.

We can now show that the core C for any essential constant-sum game (N, v) is empty. Since any $x \in C \subset A$ must satisfy

$$\sum_{j \in N-i} x_j \geqslant v(N - i) = v(N) - v(i) \quad \forall i \in N ,$$

where we write $N - i$ for $N - \{i\}$. Summing these n relations for each $i \in N$ gives

$$(n - 1) \sum_{j \in N} x_j \geqslant nv(N) - \sum_{i \in N} v(i)$$

or

$$(n - 1)v(N) \geqslant (n - 1)v(N) + \left[v(N) - \sum_{i \in N} v(i) \right],$$

which contradicts the definition of essential.

There is no loss in generality regarding most n-person game solution concepts if we assume

$$v(i) = 0 \quad \forall i \in N .$$

One can translate any game (N, u) to this 0-*normalized* form by letting

$$v(S) = u(S) - \sum_{i \in S} u(i) \quad \forall S \subset N .$$

It is also common to assume that $v(N) = 1$. A game (N, v) with $v(i) = 0$ for all $i \in N$ and $v(N) = 1$ is said to be in $(0, 1)$-*normalized* (or *normal*) form. Any essential game (N, w) can be mapped into this form by

$$v(S) = \frac{w(S) - \Sigma_{i \in S} w(i)}{w(N) - \Sigma_{i \in N} w(i)} \quad \forall S \subset N .$$

This linear transformation on the 2^n-dimensional game spaces preserves the domination relation in A defined by (1) and (2) and hence stable sets and cores, as well as most other solution concepts for the cooperative games with side payments.

There have been many generalizations and variations made in the original model of von Neumann and Morgenstern. Alternate definitions have been given for the characteristic function v, the imputation set A, and the domination relation dom. Some of the important extensions are the games without side payments in generalized characteristic function form [see Aumann (1967)], the games in partition function form [see Thrall and Lucas (1963)], and the games in discrete partition function form [see Lucas and Maceli (1978)].

4. Stable sets for three-person games

The structure of all stable sets and the core for all three-person games in characteristic function form can be seen from the following four examples. More details and complete proofs for the general case appear in von Neumann and Morgenstern (1944).

Example 3. The *three-person veto-power* game (N, v) has $N = \{1, 2, 3\}$, $v(N) = 1 = v(12) = v(23)$, and $v(13) = 0 = v(1) = v(2) = v(3)$. [Expressions such as $v(\{1, 3\})$ and $\text{Dom}_{\{1,2\}}$ are written as $v(13)$ and Dom_{12}, respectively.] The set of imputations is

$$A = \{x = (x_1, x_2, x_3): x_1 + x_2 + x_3 = 1 \text{ and } x_1, x_2, x_3 \geq 0\} .$$

It is easy to show that the core C for this example consists of the one imputation $(0, 1, 0)$ in which the veto-power player 2 obtains the full payoff of 1. However, this is a game in which an outcome in the core may not be realized in practice, because player 2 is not a dictator. He needs the cooperation of at least one other player who will likely demand some positive payoff. One can also view player 2 as a seller of some item and players 1 and 3 as potential buyers in this three-person "market game".

Figure 3 shows the set A as equilateral triangles. The top triangle illustrates the imputations which are dominated by, or which will dominate, a typical imputation x in A. One can easily show that for any game

$$\text{Dom}_S A = \emptyset \text{ whenever } S = N \text{ or } \{i\} \text{ for } i \in N .$$

Note that the regions $\text{Dom}_S x$ and $\text{Dom}_S^{-1} x$ are relatively open sets, whereas the core and any stable set are closed sets.

One can prove that the only stable set for this game that is "symmetric" in the players 1 and 3 is

$$V^S = \{x \in A: x_1 = x_3\} .$$

This is illustrated in the lower left triangle in Figure 3 by the heavy vertical line between the core point $(0, 1, 0)$ and the midpoint $(\frac{1}{2}, 0, \frac{1}{2})$ of the opposite side of A. This set V^S reflects the fact that the coalition $\{1, 3\}$ also has veto power

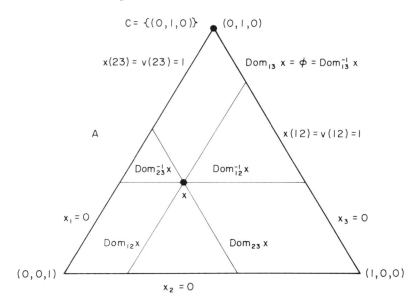

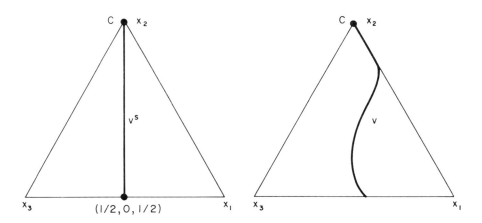

Figure 3. The three-person veto-power game.

when it acts in unison, and that this game is then a pure bargaining game between the coalitions $\{2\}$ and $\{1, 3\}$. If the union between 1 and 3 does not hold firm, then player 2 can play them off against each other and move ever closer to the core point $(0, 1, 0)$. Any possible stable set V for this game must be a continuous curve from the point $(0, 1, 0)$ to the opposite side $(x_2 = 0)$ of A

which satisfies the following Lipschitz condition: $y_2 < x_2$ implies that $y_1 \geq x_1$ and $y_3 \geq x_3$ for every x and y in V. This is illustrated in the lower right corner of Figure 3. These rather arbitrary curves V are called "bargaining curves" in von Neumann and Morgenstern (1944) where it is argued that they correspond to possible social norms or standards of behavior in a society.

In particular the two stable sets

$$V_3^0 = \{x \in A : x_1 + x_2 = 1\} \quad \text{and} \quad V_1^0 = \{x \in A : x_2 + x_3 = 1\}$$

correspond to the minimal winning coalitions $\{1, 2\}$ and $\{2, 3\}$, respectively, where either such coalitions can form and then divide the total gain among themselves in any manner.

Example 4. The *three-person constant-sum* game, or *simple majority* game, has $N = \{1, 2, 3\}$, $v(N) = 1 = v(12) = v(13) = v(23)$ and $v(1) = v(2) = v(3) = 0$. The set A is the same as in the previous example, and the core C is the empty set. The $\mathrm{Dom}_S x$ and $\mathrm{Dom}_S^{-1} x$ patterns for this case are illustrated in Figure 4. The only symmetric stable set, as well as the only finite stable set, for this game is

$$V^S = \{(\tfrac{1}{2}, \tfrac{1}{2}, 0), (\tfrac{1}{2}, 0, \tfrac{1}{2}), (0, \tfrac{1}{2}, \tfrac{1}{2})\} \,.$$

This is pictured in the lower left of Figure 4. That is, a minimal winning (or minimal veto-power) coalition of two players splits evenly and excludes the third player.

There is another class of stable sets V_i^d for this game, where $i \in N$ and $0 \leq d < \tfrac{1}{2}$, given by

$$V_i^d = \{x \in A : x_i = d\} = \{x \in A : x(N - i) = 1 - d\} \,.$$

These are called *discriminatory* stable sets: player i (the "agent") receives the amount d and the other two players bargain over how to divide the remaining amount $1 - d$. It can be proved that any stable set to this game is either V^S or of the form V_i^d. The set V_2^d is pictured in the lower right of Figure 4.

Example 5. Consider the three-person game with $N = \{1, 2, 3\}$ and $v(N) = 5$, $v(12) = 4$, $v(13) = 3$, $v(23) = 1$ and $v(1) = v(2) = v(3) = 0$. A superadditive three-person game with $v(i) = 0$ for $i = 1, 2$, and 3 will have a nonempty core if and only if $v(12) + v(13) + v(23) \leq 2v(123)$; and thus $C \neq \emptyset$ for this example. The core of this game is the inverted triangle in Figure 5 with vertices

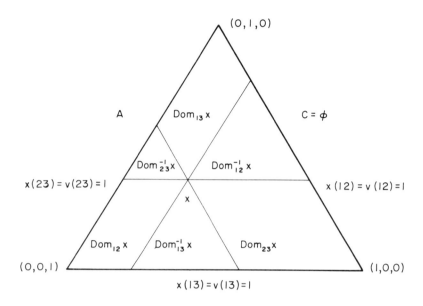

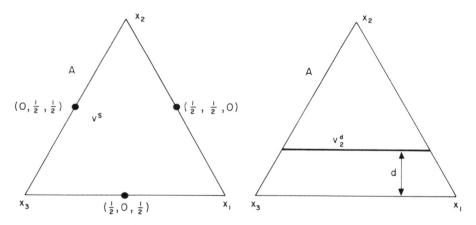

Figure 4. The three-person constant-sum game.

$(2, 2, 1), (4, 0, 1)$ and $(4, 2, -1)$ intersected with the imputation set A. Note that the three (relatively open) "corner" regions in Figure 5 are in Dom C and thus cannot intersect any stable set for this game. To find a stable set V one begins with the core C and adds imputations from the triangular region $A_1 = A - (C \cup \text{Dom } C)$. However, the domination relation within A_1 is similar

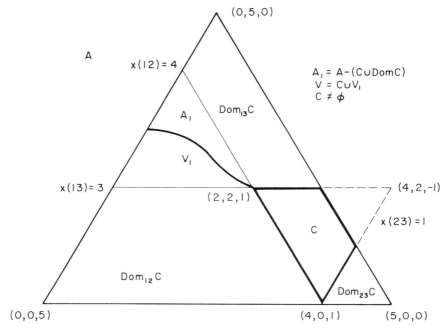

Figure 5. A game with nonempty core C.

to that for the veto-power game in Example 3, but now with player 1 as the veto-player. One can show that any stable set V for the game consists of C and a bargaining curve V_1 (like those in Figure 3) which extends from the left vertex of C to the corresponding side $x_1 = 0$ in A_1. This is illustrated in Figure 5. For an arbitrary three-person game with nonempty core there can be up to three such bargaining curves emanating from the vertices of C which are interior to A.

Example 6. Consider the same game as in Example 5 except that the one value $v(23) = 1$ is changed to $v(23) = 4$. Now the core is empty. The three lines $x_i + x_j = v(ij)$ for $\{i, j\} \subset \{1, 2, 3\}$ divide the imputation set A into seven parts as indicated in Figure 6. Only one two-person coalition $\{i, j\}$ is effective in each of the three "corner" regions A_k near the vertices $x_k = 5$, where i, j and k are the distinct members of N; and the interiors of these three regions cannot contain any solution points. The small triangle A_0 in the middle part of A has the same domination pattern as the constant-sum game in Example 4. The remaining three areas in A have domination patterns analogous to the veto-power game in Example 3. It is possible to prove that any stable set V for the

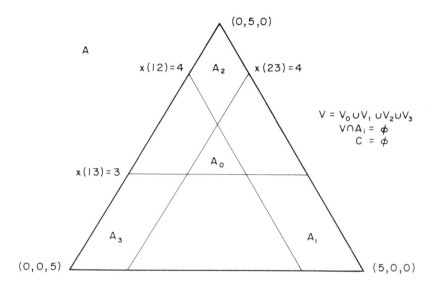

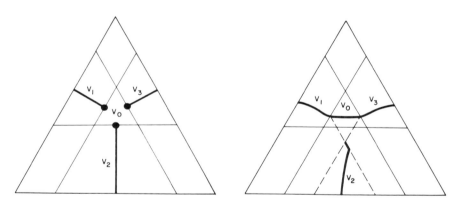

Figure 6. A game with empty core.

present game when restricted to A_0 is a set V_0 similar to the stable sets for Example 4, as indicated in Figure 4. To enlarge V_0 to V one has to add three bargaining curves V_i within the three regions $A - \text{Dom}\, V_0$ which are each similar to the stable sets given in Example 3. The nature of the two resulting types of stable sets are indicated in the lower part of Figure 6.

Examples 5 and 6 illustrate the general nature of all possible stable sets for all three-person (essential) games. Bondareva, Kulakovskaya and Naumova (1979) proved that every four-person game (and thus every five-person constant-sum game) has a stable set, but not all such solutions have been characterized. Also see Michaelis (1982) for the case $n = 4$. It is still not known whether stable sets always exist for n-person games for $n = 5$ to $n = 9$.

Example 7. GHW with side payments. Let us return to the satellite game between G, H, and W introduced in Section 2, but now we will allow for the possibility of side payments. Assume that any company can give some of its gain to another, whether they are in the same coalition or not, and that the amount transferred preserves its value. From the five normalized outcome vectors,

$$y^0 = (0, 0, 0) , \quad y^G = (0, 2, 1) , \quad y^H = (0.5, 0, 2) ,$$
$$y^W = (3, 2.2, 0) \quad \text{and} \quad y^N = (0, 0, 1) ,$$

one can naturally arrive at the characteristic function

$$v(G) = v(H) = v(W) = 0 , \quad v(HW) = 3 , \quad v(GW) = 2.5 ,$$
$$v(GH) = 5.2 = v(GHW) .$$

[One might argue that we should set $v(GHW) = v(N) = 1$ from $y^N = (0, 0, 1)$, and arrive at a nonsuperadditive game. However, $\{G, H\}$ could "pay" W some of their 5.2 to remain out of the grand coalition $N = \{G, H, W\}$, which in fact happened in the "real-world" game.] The resulting game has the set of imputations

$$A = \{(x_G, x_H, x_W) = x: x_G + x_H + x_W = 5.2, x_G \geqslant 0, x_H \geqslant 0, \text{ and } x_W \geqslant 0\} .$$

The core C is empty since

$$v(HW) + v(GW) + v(GH) = 10.7 > 10.4 = 2v(N) .$$

However, C is "just barely" empty, since a decrease of only 0.3 in the left hand side of this relation, or a similar increase in $v(N)$, would cause C to be nonempty (see Figure 7).

This game is analogous to the one in Example 6. Intuitively, one would expect the final outcome to occur in or near the small triangular region $A_0 \subset A$

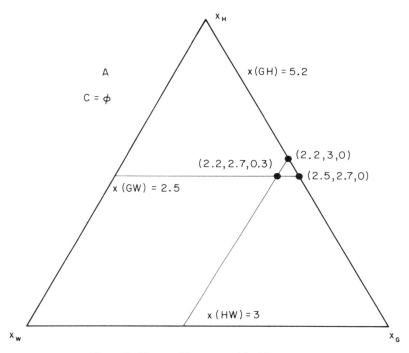

Figure 7. The satellite game with side payments.

with vertices

$$(x_G, x_H, x_W) = (2.5, 2.7, 0), (2.2, 3, 0), \quad \text{and} \quad (2.2, 2.7, 0.3).$$

In particular, one may expect G and H to form the coalition $\{G, H\}$ which realizes 5.2, to exclude W, and to settle on some point on the line segment joining $(2.5, 2,7, 0)$ and $(2.2, 3, 0)$.

In the real-world game the coalition $\{G, H\}$ did form and W was left to " go it alone". However, the U.S. Federal Communications Commission (FCC) disapproved of this proposal, mainly because it felt that W's proposal was risky due to a perceived technological weakness if W did not have assistance from H. In response, H agreed to make a free technological transfer to W to overcome FCC's objection, and the coalitions $\{G, H\}$ and $\{W\}$ each began their own projects. One might conclude that the final result was indeed in A_0, and perhaps on the line segment joining $(2.2, 3, 0)$ and $(2.2, 2.7, 0.3)$. For more details about this example, see chapter 11 in McDonald (1975).

5. Properties of stable sets

One major question regarding any solution concept concerns uniqueness. Does each game have at most one solution? We have seen for the three-person games in the previous section that stable sets are typically not unique. The only time a three-person game has a unique stable set V is when V is equal to the "rather large" core C. For the $(0, 1)$-normalized case this occurs when the three conditions $v(ij) + v(ih) \leq 1$ hold, where $\{i, j, h\} = \{1, 2, 3\}$. For the three-person constant-sum game in Example 4 we see that there is an uncountable number of stable sets V, that the union of all such V is A, and that the intersection of all V is the empty set $\emptyset$. It is quite common for an n-person game to have a plethora of stable sets and some of these may be quite "pathological" in nature. Shapley (1959) showed that for *any* closed bounded set B of n-dimensions there is an $(n + 3)$-person game with B as a disconnected component of one of the game's stable sets V. The other part of V will, of course, depend upon B. So there is a five-person game with anyone's signature (presumably a compact set) as a disconnected part of some stable set for this game. Von Neumann and Morgenstern (1944) were not particularly disturbed by the multiplicity of stable sets. They argued instead in terms of the richness of "bargaining conventions" and "standards of behavior" that could exist within a society. Although it may be a very interesting theoretical problem to characterize all stable sets for classes of games, the number is clearly excessive from the point of view of practical applications.

It is clear now that the two simple conditions (3) and (4) of internal and external stability are not in themselves sufficient to cut down the number of allowable imputation sets to serve as a suitable solution for all n-person cooperative games. This is particularly true as the number of players n increases. One must add other restrictions to narrow the number of solutions, or modify these two constraints, despite their individual desirability.

We can also observe that each stable set for any essential three-person game, except for the symmetric V^S in Example 4, has an uncountable number of imputations. Stable sets are a "global" solution concept in the sense that they provide a *set* of outcomes, and do not specify a unique result for a game. A particular stable set may correspond to a specific "standard of behavior". Various imputations within a stable set are reasonable according to this rule or standard. A change between different imputations within one stable set may be easily made, whereas a change to a different standard is more like changing the basic operating rules of this society, or the role of the individuals involved. A stable set does not indicate a specific imputation for a game, but may delineate a range of values over which the players may bargain, or suggest a smaller "game between coalitions".

Early research led to a variety of conjectures regarding the mathematical nature of stable sets. The following six statements, now known to be false, are illustrations of a few of the important ones.

(i) The intersection of all stable sets for a game (N, v) is its core C.

(ii) For every game (N, v) and any partition $P = \{S_1, S_2, \ldots, S_m\}$ of the player set N, there exists a stable set V contained in the region of A defined by

$$x(S_j) = \sum_{i \in S_j} x_i \geq v(S_j) \quad \text{for all } j = 1, 2, \ldots, m .$$

(iii) Every game has a stable set which preserves the "symmetry" of the characteristic function.

(iv) Every game has a stable set which is a finite union of "polyhedral" sets (i.e., polytopes).

(v) The union of all stable sets for a game is a connected set.

(vi) Every game has at least one stable set.

We saw in Section 4 that these six conjectures are all valid when $n = 3$. The following example shows that (i) and (ii) fail for $n = 5$. [Note that the game given in Example 8, as well as some that follow, are not superadditive. They can, however, be made into superadditive games using a technique of Gillies (1959, pp. 68–69) without altering the n, A, C or V's of the initial game. These nonsuperadditive forms greatly reduce the number of nonzero values $v(S)$.]

Example 8. Consider the game (N, v) with $N = \{1, 2, 3, 4, 5\}$ and

$$v(N) = 2, \ v(12) = v(34) = v(135) = v(245) = 1 ,$$

$$v(S) = 0 \quad \text{for all other } S \subset N .$$

It is easy to see that the core C for this example is the closed line segment joining $(1, 0, 0, 1, 0)$ and $(0, 1, 1, 0, 0)$. The unique stable set V for this game is the square

$$B = \{x \in A : x_1 + x_2 = x_3 + x_4 = 1\}$$

which has the four vertices $(1, 0, 0, 1, 0)$, $(0, 1, 1, 0, 0)$, $(1, 0, 1, 0, 0)$, and $(0, 1, 0, 1, 0)$. We can see that Dom $C \supset A - V$ because $x \in A - V$ implies $x_1 + x_2 + x_3 + x_4 + x_5 = 2$ and either $x_1 + x_2 < 1$ or $x_3 + x_4 < 1$, or both. If $x_1 + x_2 < 1$, for example, one can pick a $y \in C$ so that $y \, \text{dom}_{12} \, x$, and similarly if $x_3 + x_4 < 1$. So $V = B$ is externally stable. On the other hand, no $y \in V$ can dominate an $x \in V$ since this would require either that

$$y_1 + y_2 > 1 = v(12) \quad \text{or} \quad y_3 + y_4 > 1 = v(34) ,$$

or else $y_5 > 0$, which contradicts the assumption (2) that y is effective for $\{1, 2\}$ or $\{3, 4\}$ or else that $y \in V$, respectively. V is thus internally stable. Therefore, V is a stable set; and it is unique since no element in Dom C can be in any stable set.

Lucas (1968b, 1969a) showed that there are games with $n \geq 5$ which have unique stable sets which are *nonconvex* sets. Lucas (1968b) also provided a counterexample to (iii) with $n = 8$. This sequence of findings showed, contrary to the multiplicity of stable sets discussed above, that the set of all stable sets for a game could indeed be quite restricted. These results paved the way to disproving the major conjecture (vi). In the meantime, several generalizations of the classical model presented in Section 3 were proposed and analyzed, and the nonexistence of stable sets was demonstrated for some of these models. Stearns (1964) showed that stable sets need not exist for the n-person cooperative games without side payments (in generalized characteristic function form) for $n \geq 7$. For example, see Aumann (1967) or Lucas (1971, pp. 507–509). Lucas (1968a) proved the nonexistence of stable sets for $n \geq 11$ for the games in partition function form which had been studied by Thrall and Lucas (1963). These latter two discoveries also suggested the possibility of (vi) being false.

The primary theoretical question for any game solution concept is whether or not it always exists: Does every game have at least one solution? Although von Neumann and Morgenstern (1944) were not terribly concerned about the lack of uniqueness for stable sets, they did consider a positive response to the existence question to be crucial. On page 42 of their third edition (1953) they discuss existence and uniqueness, and state:

> There can be, of course, no concession as regards existence. If it should turn out that our requirements concerning a [stable set V] are, in any special case, unfulfillable – this would certainly necessitate a fundamental change in the theory.

Many special classes of games were known to always have stable sets, and often a great variety of different ones. It had also been known since 1953 [see Gillies (1959)] that a "positive fraction" of all n-person games had a unique stable set consisting of a large core. This will occur when all the coalition values $v(S)$ are small relative to $v(N)$, i.e., when each $v(S) \leq v(N)/(n-1)$ whenever $S \neq N$. On the other hand, some 25 years after von Neumann first conjectured (vi), Lucas (1968c) provided a negative answer to the existence question. The ten-person game which follows has no stable set. [Although this characteristic function is not superadditive, it is equivalent to a superadditive one using Gillies (1959, pp. 68–69). The superadditive form would have hundreds of nonzero values $v(S)$.]

Example 9. Consider the ten-person game:

$$v(N) = 5, \quad v(13579) = 4, \quad v(3579) = v(1579) = v(1379) = 3,$$

$$v(12) = v(34) = v(56) = v(78) = v(9, 10) = 1,$$

$$v(357) = v(157) = v(137) = v(359) = v(159) = v(139) = 2,$$

$$v(1479) = v(3679) = v(5279) = 2,$$

and

$$v(S) = 0 \quad \text{for all other } S \subset N.$$

The imputation set A for this game is a regular nine-dimensional simplex. Consider the five-dimensional hypercube

$$B = \{x \in A: x(12) = x(34) = x(56) = x(78) = x(9, 10) = 1\}.$$

The core C of this game is the five-dimensional simplex consisting of the "corner" of B, with $x(13579) \geqslant 4$. C has the six vertices $(1, 0, 1, 0, 1, 0, 1, 0, 1, 0)$ and

$$(0, 1, 1, 0, 1, 0, 1, 0, 1, 0),$$

$$(1, 0, 0, 1, 1, 0, 1, 0, 1, 0),$$

$$(1, 0, 1, 0, 0, 1, 1, 0, 1, 0),$$

$$(1, 0, 1, 0, 1, 0, 0, 1, 1, 0),$$

$$(1, 0, 1, 0, 1, 0, 1, 0, 0, 1).$$

Dom C via only the two-person coalitions $\{i, i + 1\}$ is $A - B$, similar to the five-person game in Example 8. So any stable set for this game must be contained in $B - $ Dom C, which one can show partitions into three sets $C, F,$ and E. $C \cup F$ must be in any such stable set, and $E \cap \text{Dom}(C \cup F) = \emptyset$. So any stable set for this game is made up of $C \cup F \cup V'$, where V' is a stable set for the region E. E consists of three three-dimensional triangular wedges which meet on C. There is a cyclical domination relation among these wedges via the coalitions $\{1, 4, 7, 9\}$, $\{3, 6, 7, 9\}$ and $\{5, 2, 7, 9\}$. An argument similar to that used in the nonexistence proofs by Stearns (1964) and Lucas (1968a) then shows no stable set V' exists for E. Therefore, the ten-person game in

Example 9 has no stable set. The details of this proof are provided in Lucas (1969b).

It should be observed that all of the examples introduced so far in this section are not just mathematical curiosities or mere pathologies of no interest in practical applications. Shapley and Shubik (1969) have shown that these games, which all have nonempty cores, do arise in the study of markets in economics.

It should also be noted again that von Neumann and Morgenstern (1944) were initially concerned with essential constant-sum games which always have empty cores, whereas the above examples have nonempty cores. However, Lucas and Rabie (1982) have provided a 14-person superadditive game with an empty core for which no stable set exists. On the other hand, no one has yet settled the general existence question for their original class of constant-sum games.

In light of the nonexistence of stable sets, statements (iv) and (v) should be limited to those games for which stable sets do exist. Conjecture (v) has been shown to be false by Lucas (1976) when $n = 12$. This result could also have been a stepping stone to the proof of the nonexistence of stable sets if it had been arrived at before Example 9. In the spring of 1967, Shapley (1968) discovered a 20-person game with infinitely many possible stable sets, but *each one* is highly pathological in nature. This provided a counter-example to (iv). It also ruled out the idea of a "constructive" algorithm for always determining a stable set as well as any reasonable economic interpretation for at least one stable set for every game.

6. Special classes of games

We have seen in Section 5 that stable sets fail to have many desirable properties when considering the class of *all* possible games in characteristic function form. These rather negative aspects for the general case, however, are offset by many good mathematical properties and interesting interpretations of stable sets when viewed in more restricted settings. To indicate some of the more positive results we will proceed to limit ourselves to looking at a couple of special classes of games as well as some special and fundamental types of stable sets. In Section 3 we already introduced the classes of superadditive, constant-sum, and essential games. These restrictions cut down the totality of games significantly, but do not avoid most of the problems arising in stable set theory, and we do *not* wish to restrict ourselves to just constant-sum games. In this section we will introduce two very important special classes of games: simple and symmetric. There are also several other particular classes of games for which an extensive literature exists that will not be covered in this chapter.

Names of some other classes of n-person games and some basic references are: extreme games [see Griesmer (1959) and Rosenmüller (1977)], homogeneous games [see Ostmann (1987)], quota and k-quota games [see Shapley (1953b) and Muto (1979b)], and convex games [see Shapley (1971)].

6.1. Simple games

An n-person game (N, v) is said to be a *simple game* if

$$v(S) = 0 \quad \text{or} \quad v(S) = 1 \quad \forall S \subset N \, .$$

A coalition S is called *winning* if $v(S) = 1$ and *losing* whenever $v(S) = 0$. Coalition M is *minimal winning* if M is winning and no proper subset T of M is winning. A coalition T has *veto-power* if $T \cap S \neq \emptyset$ for every winning coalition S. Simple games are also referred to as *voting games*. They provide an elementary model of voting systems in which some coalitions can pass a bill, whereas other groups of players cannot pass it. We will assume that simple games are *monotone* in the sense that

$$v(S) \geqslant v(T) \text{ whenever } S \supset T \, .$$

We will also assume that $v(N) = 1$ as well as $v(\emptyset) = 0$.

Monotone simple games arise in many other mathematical contexts besides game theory and there is a large literature on the subject. An excellent introduction is given in Shapley (1962). A survey showing the connections of n-person simple games with other mathematical subjects and a bibliography is given in Hilliard (1983). The most popular solution concepts for simple games are the values proposed by Shapley (1953a) and by Banzhaf (1968) and Coleman (1971), as well as several variations and extensions of these notions. The Shapley value is a major solution concept for general n-person games as well as for this case of simple games. It will be discussed in a subsequent volume of this Handbook. Several alternate value concepts have also appeared. Additional chapters will be devoted to limiting properties of values for games with a large number of players as well as value notions for games with a continuum of players. Extensions of the Shapley value to games without side payments exist. The Shapley value from games theory has also been employed in a large number of theoretical and practical applications. This is illustrated in chapters on the use of values to study perfectly competitive economies, other economic applications, fair cost allocation, as well as for measuring political power in voting structures (simple games).

It is easy to characterize the core for monotone simple games. Any player

$i \in N$ who is in every minimal winning coalition forms a veto-power coalition $\{i\}$. The corresponding imputation e^i which has a 1 in the ith component and 0 elsewhere is clearly in the core C for such a game. It is easy to see in this case that the core of the game is the convex hull of the points e^i where i has veto power. The core is thus empty if there are no veto-power players, as in Example 4. In Example 3, player 2 has veto power and the core of the game has the one imputation $(0, 1, 0)$.

At least one stable set exists for every simple n-person game (with finite n). If M is a minimal winning coalition in a simple game, then the set

$$V_M = \{x \in A : x(M) = v(N) = 1\}$$

is a stable set. Any imputation $y \in A - V_M$ must have $y(M) < y(N) = 1$ and can be dominated by an $x \in V_M$. Clearly V_M is also internally stable. The two stable sets $V_{12} = V_3^0$ and $V_{23} = V_1^0$ in Example 3 and the three stable sets $V_{ij} = V_k^0$ for $\{i, j, k\} = \{1, 2, 3\}$ in Example 4 illustrate this result. If there is only one minimal winning coalition M, then the stable set V_M is unique. Otherwise, there are a great number of stable sets for simple games as seen in Examples 3 and 4. The games of Shapley (1959) which have some pathological stable sets are simple games. One initial step towards characterizing all stable sets for just the four-person simple game with one veto-power player is given in Rabie (1980).

Owen (1968b, pp. 177–178) has shown that if one extends the definition of n-person simple game to the case where $N = \{1, 2, 3, \ldots\}$ has a countable infinity of players, then stable sets may not exist since minimal winning coalitions need not exist. For example, the winning coalitions S could be those subsets of $\{1, 2, 3, \ldots\}$ whose complements $N - S$ are finite. So any winning coalition has a proper subset that is also winning.

Von Neumann and Morgenstern (1944) have shown that there is a family of n-person, constant-sum simple games which has a finite stable set which they called the *main simple solution*. Assume that for such a game there is a vector $x = (x_1, x_2, \ldots, x_n)$ with each $x_i \geq 0$ such that $x(M) = 1$ whenever M is a minimal winning coalition. For each such M let $x_i^M = x_i$ if $i \in M$ and $x_i^M = 0$ when $i \notin M$. Then the set of imputations x^M forms a stable set V^{MSS}.

Example 10. Consider the six-person, constant-sum, monotone simple game which has $v(S) = 1$ for all coalitions S which have four or more players and $v(M) = 1$ for the following ten minimal winning, three-person coalitions M: $\{4, 5, 6\}$ and $\{i, j, h\}$, where $\{i, j\} \subset \{1, 2, 3\}$ and $h \in \{4, 5, 6\}$. All other coalitions with three or fewer players are losing. The vector $x = (\frac{1}{3}, \frac{1}{3}, \frac{1}{3}, \frac{1}{3}, \frac{1}{3}, \frac{1}{3})$ provides a solution for the condition above. The following ten imputations thus form a stable set for this game:

$(0, 0, 0, \frac{1}{3}, \frac{1}{3}, \frac{1}{3})$,

$(\frac{1}{3}, \frac{1}{3}, 0, \frac{1}{3}, 0, 0), (\frac{1}{3}, 0, \frac{1}{3}, \frac{1}{3}, 0, 0), (0, \frac{1}{3}, \frac{1}{3}, \frac{1}{3}, 0, 0)$,

$(\frac{1}{3}, \frac{1}{3}, 0, 0, \frac{1}{3}, 0), (\frac{1}{3}, 0, \frac{1}{3}, 0, \frac{1}{3}, 0), (0, \frac{1}{3}, \frac{1}{3}, 0, \frac{1}{3}, 0)$,

$(\frac{1}{3}, \frac{1}{3}, 0, 0, 0, \frac{1}{3}), (\frac{1}{3}, 0, \frac{1}{3}, 0, 0, \frac{1}{3}), (0, \frac{1}{3}, \frac{1}{3}, 0, 0, \frac{1}{3})$.

Another proper subclass of n-person simple games is the *weighted majority games* $[q: w_1, w_2, \ldots, w_n]$. Each player i has a positive *weight* w_i, and a coalition S of players wins if and only if

$$\sum_{i \in S} w_i \geq q .$$

The number q is called the *quota* and is usually assumed to be in the range $w \geq q > w/2$, where $w = w_1 + w_2 + \cdots + w_n$. Since these are simple games, the theory of the core and stable sets is as above. Example 10 is a constant-sum, monotone simple game which cannot be expressed as a weighted majority game. The system of ten inequalities $\sum_{i \in M} w_i \geq q$ for the minimal winning coalitions M has no feasible solution for any $q > w/2$. Applications of value theories to the weighted voting games are given in Lucas (1983) and Straffin (1983).

6.2. Symmetric games

An n-person game (N, v) is said to be *symmetric* if $v(S) = v(T)$ whenever $|S| = |T|$. Any two coalitions of the same size $s = |S|$ have the same value. In this case the characteristic function v is determined by the $n - 1$ numbers $v(s) = v(|S|) = v(S)$, where $s = 2, 3, \ldots, n$, assuming $v(1) = 0 = v(\emptyset)$.

It is easy to characterize when the core of a symmetric game is nonempty. One first observes that the core C is nonempty if and only if it contains the centroid c of A:

$$C \neq \emptyset \Leftrightarrow c = (v(n)/n, v(n)/n, \ldots, v(n)/n) \in C .$$

Note that if $x \in C$ then, using symmetry, the $n!$ permutations πx of x are also in C. Since C is a convex set, the average of these $n!$ imputations πx, which is c, is also in C. The core conditions $x(S) \geq v(S)$ applied to c state that $c(S) = sv(n)/n \geq v(s)$ for all $s \leq n$. It follows that

$$C \neq \emptyset \Leftrightarrow v(s) \leq sv(n)/n \quad \forall s \leq n .$$

For example, the three-person symmetric game has $C \neq \emptyset$ whenever $v(2) \leq 2v(3)/3$. The four-person symmetric game $v(1) = 0$, $v(2) = 0.5$, $v(3) = 0.8$, and $v(4) = 1$ has $C = \emptyset$ since $v(3) = 0.8 > 3v(4)/4$.

One can also characterize when the core C of a symmetric game is "large enough" to be the unique stable set V for the game. This makes use of the notion of the cover $\bar{v}$ of a game v. In the case of a symmetric game v the *cover* is defined to be

$$\bar{v}(s) = \max_{0 < t \leq s} sv(t)/t .$$

For the four-person game $v(1) = 0$, $v(2) = 0.5$, $v(3) = 0.8$, and $v(4) = 1$, which has an empty core, we get that $\bar{v}(4) = 4(0.8)/3 = 3.2/3 > 1 = v(4)$ for $t = 3$. For any symmetric game with a nonempty core we see that $\bar{v}(n) = v(n)$. For the four-person symmetric game $v(1) = 0$, $v(2) = 0.5$, $v(3) = 0.6$, and $v(4) = 1$ one gets $\bar{v}(s) = v(s)$ for $s = 1$, 2, and 4, but $\bar{v}(3) = 3(0.5)/2 = \frac{3}{4}$. A symmetric n-person game v with a nonempty core C will have the unique stable set $V = C$ if and only if

$$\frac{v(n) - \bar{v}(t)}{n - t} \geq \frac{v(s) - \bar{v}(t)}{s - t}$$

for all t and s which satisfy $0 \leq t < s < n$. A proof of this appears in Shapley (1976) and Menshikova (1977). Additional results on symmetric stable sets and their uniqueness for symmetric games with "large" cores are presented in Muto (1983). In particular, he gives "symmetric" stable sets for all five-person symmetric games with nonempty cores, as well as a sufficient condition for uniqueness for n-person symmetric games with nonempty cores.

6.3. Simple and symmetric games

If a monotone simple game (N, v) is also symmetric (and essential) then it is determined by one integer k with $1 < k \leq n$. In this case

$$v(S) = 1 \quad \forall S \subset N \text{ with } s \geq k ,$$

$$v(S) = 0 \quad \forall S \subset N \text{ with } s < k .$$

These games were referred to earlier as the (n, k) *games*, or the (n, k) *majority games* when $k > n/2$. These games must have $k > n/2$ in order to be superadditive (proper), and then they provide a model for voting systems in which any coalition of k or more players can pass a bill. For the case $k = n$ we get that the core $C = A = V$ is the unique stable set. For this unanimity or pure bargaining game unanimous support is needed to pass an issue, and any one player can veto a proposed bill. It is easy to check that $C = \emptyset$ when $k < n$.

The $(3, 2)$ game was analyzed in Example 4. The unique finite or symmetric stable set $V^S = \{(\frac{1}{2}, \frac{1}{2}, 0), (\frac{1}{2}, 0, \frac{1}{2}), (0, \frac{1}{2}, \frac{1}{2})\}$ can be interpreted as one of the minimal winning coalitions forming and splitting the gain evenly. It can also be interpreted as those in a minimal sized veto–power coalition (which is also winning in this game) getting the same amount while excluding the other player. We will see in the next section that it is this veto-power interpretation that is the one that generalizes to (n, k) games in general. The $(3, 2)$ game also has three "totally discriminatory" solutions $V_i^d = V_i^0 = \{x \in A: x_i = 0\} = \{x \in A: x_j + x_l = 1\}$ for $\{i, j, l\} = \{1, 2, 3\}$. These correspond to a minimal winning coalition $\{j, l\}$ forming and bargaining over how to split the one unit.

The $(4, 3)$ game has a unique "symmetric" stable set V^S composed of the three line segments:

$$[(\tfrac{1}{2}, \tfrac{1}{2}, 0, 0), (0, 0, \tfrac{1}{2}, \tfrac{1}{2})],$$

$$[(\tfrac{1}{2}, 0, \tfrac{1}{2}, 0), (0, \tfrac{1}{2}, 0, \tfrac{1}{2})],$$

$$[(\tfrac{1}{2}, 0, 0, \tfrac{1}{2}), (0, \tfrac{1}{2}, \tfrac{1}{2}, 0)].$$

This can be interpreted as any two complementary two-person coalitions pairing off against each other and playing the pure bargaining game between each other. The two players in the same minimal veto-power coalition must get the same amount. This game also has four totally discriminatory stable sets $V_i^0 = \{x \in A: x_i = 0\} = \{x \in A: x(\{j, l, h\}) = 1\}$ for $\{i, j, l, h\} = \{1, 2, 3, 4\}$. These correspond to a minimal winning coalition of three players playing the resulting three-person unanimity game. The $(4, 3)$ game also has discriminatory stable sets with $d > 0$ as well as a great number of nondiscriminatory and nonsymmetric stable sets.

The four-person (n, k) game $(4, 2)$ is nonsuperadditive (*improper*) since, for example, $v(12) + v(34) = 1 + 1 > 1 = v(1234) = v(N)$. This game does, however, have four finite, "nonsymmetric" stable sets of the form

$$V_i = \{x \in A: x_i = x_j = 0 \text{ and } x_h = x_l = \tfrac{1}{2}\},$$

where $\{j, h, l\} = N - i$. The set V_i is symmetric with respect to the three players j, h, and l. Player i is excluded first by $\{j, h, l\}$ who have veto-power. They then play the three-person, constant-sum game in Example 4. This results in a minimal winning coalition $\{h, l\}$ excluding a player j and splitting the one unit. However, the two excluded players also could form a minimal winning coalition $\{i, j\}$ in this improper game, so that this simple economic interpretation is questionable for such nonsuperadditive games and nonsymmetric stable sets. The stable sets V_i are examples of what are called "semi-simple" stable sets. This game $(4, 2)$ also has a unique finite "symmetric" stable set of the form

$$V^S = \{x \in A: x_i = 0 \text{ and } x_j = x_h = x_l = \tfrac{1}{3} \text{ for some } i \in N\} \ .$$

The (n, k) games with odd n and $k = (n + 1)/2$ are called the (n, k) *simple majority games*. There are unique finite and symmetric stable sets V^S for these games which are composed of the $n!/m!(n - m)!$ distinct imputations that are permutations of the components of

$$(\underbrace{1/m, 1/m, \dots, 1/m}_{m}, \underbrace{0, \dots, 0}_{n - m}) \ ,$$

where $m = (n + 1)/2$ is the size of a minimal winning or minimal veto-power coalition M. The m players in some M each get the same amount $1/m$ while the other $n - m$ players are "completely defeated" and get 0. This game also has "completely discriminatory" stable sets of the form $V^0_{N-M} = \{x \in A: x_i = 0 \ \forall i \in N - M\}$ as well as many other stable sets.

7. Symmetric stable sets

Many n-person games have a great number of different stable sets and many of these are of a rather bewildering nature. On the other hand, when one restricts the classes of games or the types of stable sets allowed, then a much more pleasing theory emerges, at least for the smaller values of n. In the previous section we introduced the class of symmetric games and provided a few examples of stable sets for some games in this class. We also referred to some of the stable sets described above as being "symmetric", although we have not yet given a formal definition of this latter use of the term. Symmetric stable sets very often provide useful interpretations and valuable insights into the likely dynamics of coalition formation and bargaining mechanisms in various political or economic situations. Symmetric stable sets also provide beautiful geometric structures, which extend to higher dimensions as well. In this section we provide a brief introduction to the symmetric theory of stable sets and provide a few references which lead into what is now a very extensive literature on this topic.

We defined an n-person game (N, v) to be symmetric if coalitions of the same size have the same value, i.e., $v(S) = v(T)$ whenever $s = |S| = |T| = t$.

We will now define what is meant by a "symmetric stable set". Let π be a permutation of the integers (players) $1, 2, \ldots, n$, and define $\pi x = (x_{\pi(1)}, x_{\pi(2)}, \ldots, x_{\pi(n)})$ to be the corresponding reordering of the components of the imputation $x = (x_1, x_2, \ldots, x_n)$. For $x \in A$ and $B \subset A$ we also define

$$\langle x \rangle = \{ y \in A : y = \pi x \text{ for any permutation } \pi \}$$

and

$$\langle B \rangle = \{ y \in A : y = \pi x \text{ for any } \pi \text{ and any } x \in B \} .$$

The subset B is said to be *symmetric* if $\langle B \rangle = B$. In particular, a stable set V is *symmetric* if $\langle V \rangle = V$. For example, the six permutations of $(1, 2, 3)$ are $(1, 2, 3)$, $(1, 3, 2)$, $(2, 1, 3)$, $(2, 3, 1)$, $(3, 1, 2)$, and $(3, 2, 1)$. The finite stable set in Example 4

$$V^S = \{ (\tfrac{1}{2}, \tfrac{1}{2}, 0), (\tfrac{1}{2}, 0, \tfrac{1}{2}), (0, \tfrac{1}{2}, \tfrac{1}{2}) \} = \langle (\tfrac{1}{2}, \tfrac{1}{2}, 0) \rangle$$

is symmetric in this technical sense, because these six permutations merely map each of these three imputations into the set V^S. For example, if $\pi = (3, 1, 2)$, then π maps 1 into 3, 2 into 1, and 3 into 2; and π applied to $(\tfrac{1}{2}, \tfrac{1}{2}, 0)$ is $(\tfrac{1}{2}, 0, \tfrac{1}{2})$. Thus $\langle V^S \rangle = V^S$.

The word "symmetric" is also used on occasion in a weaker sense. We say that the simple game in Example 3 is symmetric with respect to the two players 1 and 3, although this three-person game is not a symmetric game. An interchange of players 1 and 3 in the characteristic function of this game leaves it unchanged. We also refer to the one stable set $V^S = \{ x \in A : x_1 = x_3 \}$ as being symmetric in the players 1 and 3, whereas this is not a symmetric stable set. The simple game in Example 10 and the main simple stable set presented there also has several symmetries in this weaker sense. The improper simple and symmetric (n, k) game $(4, 2)$ had four stable sets V_i described above. Each V_i is symmetric in the other three players j, h, and l; but it is not a symmetric stable set.

One of the early major results on symmetric stable sets for general (n, k) games was given by Bott (1953). He proved that there is a unique symmetric stable set V^S for every (n, k) game with $k > n/2$. Recall that these games are defined by $v(S) = 1$ for $s \geq k$ and $v(S) = 0$ for $s < k$. V^S for the case where $k = (n + 1)/2$ and n is odd (i.e., the constant-sum case) was presented in the previous section, and is the only case where V^S is a finite set of imputations. Coalitions of size k are minimal winning ones, whereas coalitions of cardinality $p = n - k + 1$ are the minimal sized ones with veto power. To describe Bott's solution, let $n = qp + r$, where q and r are integers with $0 \leq r < p$. Then consider the set B of all the imputations $a \in A$ of the form

$$a = (\underbrace{a_1, \ldots, a_1}_{p}, \underbrace{a_2, \ldots, a_2}_{p}, \ldots, \underbrace{a_q, \ldots, a_q}_{p}, \underbrace{0, \ldots, 0}_{r}) \, .$$

Bott's unique stable set V^S consists of the set $\langle B \rangle$ of all permutations of all imputations of the form a. V^S for the $(3, 2), (4, 3)$, and $(n, (n + 1)/2)$ games presented before are of this form. The $(5, 4)$ game has

$$V^S = \langle \{(a_1, a_2, a_3, a_4, 0) \in A: a_1 = a_2 \text{ and } a_3 = a_4\} \rangle \, .$$

Bott's stable set has an interesting interpretation. The n players in N partition themselves into q disjoint coalitions $M_1, M_2, \ldots, M_q$ of size p, each of which has veto power, plus a set R of r "left over" players. The coalitions M_j act as players in a q-person unanimity (n, k) game with $n = k = q$ which can have any payoff $(b_1, b_2, \ldots, b_q)$ in its core $C(q) = A(q)$. Each player i in a particular blocking coalition M_j will then receive the same amount $a_i = b_j/p$. Meanwhile, any of the excluded players i in R will receive $x_i = 0$. This theme of games played between blocking coalitions M_j and then an equal split within each M_j persists for more complicated symmetric stable sets for symmetric games, as demonstrated by Heijmans (1987).

The result of Bott presented above can be extended to nonsuperadditive (improper) (n, k) games where $k \leqslant (n + 1)/2$. In this case too there is a unique symmetric stable set V^S given by

$$V^S = \langle (\underbrace{1/(n - k + 1), \ldots, 1/(n - k + 1)}_{n - k + 1 = p}, \underbrace{0, \ldots, 0}_{k - 1 = r}) \rangle \, .$$

For a proof of this, as well as extensions of the above work to "semi-symmetric" stable sets and to nonsimple symmetric games, consult Muto (1978, 1980).

The nature of all stable sets for all three-person games was exhibited in Section 4. An examination of these for the symmetric three-person games shows that there is precisely one symmetric stable set for each such game. Nering (1959) showed that symmetric stable sets exist for every four-person symmetric game and that they are not always unique. Heijmans (1987) has described all symmetric stable sets for all symmetric four-person games. His reduction techniques allow one to reduce the problem to one in fewer dimensions. He thus analyzes a handful of cases in a planar triangle, reminiscent of von Neumann and Morgenstern's analysis of all three-person games, to arrive at his results. Muto (1983) also described symmetric stable sets for all symmetric five-person games with nonempty cores, and provided sufficient

(n, k) games when $k < (n + 1)/2$. We also saw in Example 10 that the main simple stable set of von Neumann and Morgenstern for the six-person game presented there was finite but not (totally) symmetric.

Many of the finite stable sets will be symmetric with respect to some subset T of the set of players N. We thus introduce the notation $\langle x \rangle_T = \{ y \in A : y$ is obtained from x by permuting the coordinates x_i where $i \in T\}$ and for $B \subset A$

$$\langle B \rangle_T = \bigcup_{x \in B} \langle x \rangle_T .$$

Note that $\langle x \rangle_N$ and $\langle B \rangle_N$ are the same as $\langle x \rangle$ and $\langle B \rangle$, respectively, which were defined in Section 7.

Finite stable sets are known to exist for many four-person games. We will examine, in particular, the four-person, constant sum games (in $(0, 1)$ normalized form). In this case the characteristic function v is determined by only the three values,

$$b_1 = v(14) , \qquad b_2 = v(24) , \qquad b_3 = v(34) ,$$

because $v(1234) = v(N - i) = 1$ and $v(i) = 0$ for all $i \in N$, and $v(jh) = 1 - v(i4)$ for $\{i, j, h\} = \{1, 2, 3\}$. Each such game corresponds to a point $b = (b_1, b_2, b_3)$ in the unit cube U. It is sufficient, using symmetry, to consider only the games (N, v) corresponding to the points b in the cube which are in the four-sided polyhedron

$$P = \{ b \in U : b_1 \leqslant b_2 \leqslant b_3 \text{ and } b_2 + b_3 \leqslant 1 \} .$$

P has vertices $(0, 0, 0)$, $(0, \frac{1}{2}, \frac{1}{2})$, $(\frac{1}{2}, \frac{1}{2}, \frac{1}{2})$ and $(0, 0, 1)$. Von Neumann and Morgenstern (1944) and Mills (1954, 1959) determined finite stable sets for the four vertices of P, three of the six edges of P, and in a three-dimensional neighborhood in P near the center $(\frac{1}{2}, \frac{1}{2}, \frac{1}{2})$ of the cube U. Two edges and one face of P are known to possess no finite stable sets.

The three vertices $b^0 = (0, 0, 0)$, $b^1 = (0, 0, 1)$ and $b^2 = (0, \frac{1}{2}, \frac{1}{2})$ of P give rise to the following finite stable sets $V(b^i)$, $i = 0, 1, 2$, consisting of three, four, and seven imputations, respectively:

$$V(b^0) = \langle (0, \tfrac{1}{2}, \tfrac{1}{2}, 0) \rangle_{\{1,2,3\}} ,$$

$$V(b^1) = \{ (\tfrac{1}{3}, \tfrac{1}{3}, 0, \tfrac{1}{3}) \} \cup \langle (\tfrac{1}{3}, 0, \tfrac{2}{3}, 0) \rangle_{\{1,2,4\}} ,$$

$$V(b^2) = \{ (0, \tfrac{1}{2}, \tfrac{1}{2}, 0) \} \cup \langle (0, \tfrac{1}{4}, \tfrac{1}{2}, \tfrac{1}{4}) \rangle_{\{1,2,4\}} \cup \langle (0, \tfrac{1}{2}, \tfrac{1}{4}, \tfrac{1}{4}) \rangle_{\{1,3,4\}} .$$

The stable set $V(b^0)$ is merely the stable set V^s for the three-person constant-sum game, since player 4 is a "dummy" in the game corresponding to b^0. The vertex $b^3 = (\frac{1}{2}, \frac{1}{2}, \frac{1}{2})$ in P corresponds to the only four-person, constant-sum, symmetric game, and it is known to have at least the following three types of finite stable sets of 10, 13, and 13 points, respectively:

$$V_1(b^3) = \langle (\tfrac{1}{3}, \tfrac{1}{3}, \tfrac{1}{3}, 0) \rangle \cup \langle (\tfrac{1}{3}, \tfrac{1}{3}, \tfrac{1}{6}, \tfrac{1}{6}) \rangle ,$$

$$V_2(b^3(c)) = \{(\tfrac{1}{4}, \tfrac{1}{4}, \tfrac{1}{4}, \tfrac{1}{4})\} \cup \langle (\tfrac{3}{8} - c, \tfrac{3}{8} - c, 2c, \tfrac{1}{4}) \rangle, \text{ where } 0 \leqslant c \leqslant \tfrac{1}{24} ,$$

$$V_3(b^3; 1) = \{(\tfrac{1}{4}, \tfrac{1}{4}, \tfrac{1}{4}, \tfrac{1}{4})\} \cup \langle (0, \tfrac{1}{4}, \tfrac{3}{8}, \tfrac{3}{8}) \rangle_{\{2,3,4\}}$$

$$\cup \langle (\tfrac{1}{4}, 0, \tfrac{3}{8}, \tfrac{3}{8}) \rangle_{\{2,3,4\}} \cup \langle (\tfrac{1}{4}, \tfrac{1}{8}, \tfrac{1}{4}, \tfrac{3}{8}) \rangle_{\{2,3,4\}} .$$

The last one $V_3(b^3; 1)$ isolates upon player 1, and is not symmetric. Three similar stable sets $V_3(b^3; i)$ exist which likewise focus on players $i = 2, 3$, and 4, respectively. The stable sets $V(b^1), V(b^2), V_1(b^3)$ and $V_3(b^3, 1)$ are illustrated in Figure 8.

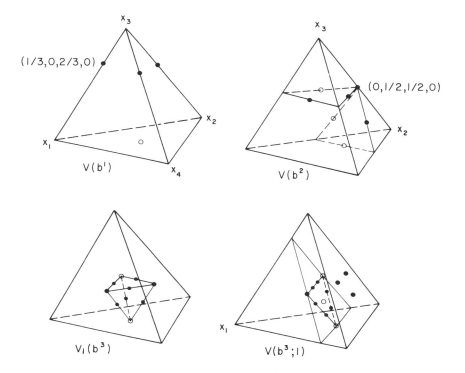

Figure 8. Some four-person finite stable sets.

Some finite stable sets are known for four of the six edges of P.

(i) The "main space" diagonal $b = (z, z, z)$, $0 \leqslant z \leqslant \frac{1}{2}$, which connects the vertices $b^0 = (0, 0, 0)$ and $b^3 = (\frac{1}{2}, \frac{1}{2}, \frac{1}{2})$ of P has the finite stable sets

$$V_z[b^0, \tfrac{4}{5}b^3] = \left\langle\left(\frac{1}{2} - \frac{z}{4}, \frac{1}{2} - \frac{z}{4}, \frac{z}{2}, 0\right)\right\rangle_{\{1,2,3\}}$$

$$\cup \left\langle\left(\frac{1}{2} - \frac{z}{4}, \frac{1}{2} - \frac{z}{4}, 0, \frac{z}{2}\right)\right\rangle_{\{1,2,3\}}$$

$$\cup \left\langle\left(\frac{1}{2} - \frac{z}{2}, \frac{1}{2} - \frac{z}{2}, \frac{z}{2}, \frac{z}{2}\right)\right\rangle_{\{1,2,3\}}$$

when $0 < z \leqslant \frac{2}{5}$; and

$$V_z[\tfrac{4}{5}b^3, b^3] = V_z[b_0, \tfrac{4}{5}b^3] \cup \left\langle\left(\frac{1}{2} - \frac{z}{4}, \frac{1}{2} - \frac{z}{2}, \frac{z}{4}, \frac{z}{2}\right)\right\rangle_{\{1,2,3\}}$$

when $\frac{2}{5} \leqslant z \leqslant \frac{1}{2}$. These stable sets have 9 and 15 imputations, respectively.

(ii) The "main face" diagonal $b = (0, z, z)$, $0 \leqslant z \leqslant \frac{1}{2}$, which connects vertices b^0 and $b^2 = (0, \frac{1}{2}, \frac{1}{2})$ of P has the seven-point finite stable set

$$V_z[b^0, b^2] = \left\{\left(0, \frac{1}{2}, \frac{1}{2}, 0\right)\right\} \cup \left\langle\left(\frac{1}{2} - \frac{z}{2}, \frac{1}{2}, 0, \frac{z}{2}\right)\right\rangle_{\{2,3\}}$$

$$\cup \left\langle\left(\frac{1}{2} - \frac{z}{2}, \frac{1}{2}, \frac{z}{2}, 0\right)\right\rangle_{\{2,3\}} \cup \left\langle\left(\frac{1}{2} - z, \frac{1}{2}, \frac{z}{2}, \frac{z}{2}\right)\right\rangle_{\{2,3\}},$$

which converges to $V(b^0)$ and $V(b^2)$ as z approaches 0 and $\frac{1}{2}$, respectively. Mills (1959) showed that this is the unique finite stable set for the *interior* of this edge.

(iii) Consider the other "space" diagonal $(\frac{1}{2} - z, \frac{1}{2} - z, \frac{1}{2} + z)$, $0 \leqslant z \leqslant \frac{1}{2}$, which joins the vertices b^3 and $b^1 = (0, 0, 1)$ of P. For $0 \leqslant z < \frac{1}{18}$, there is the stable set $V(R(e))$ given below. Von Neumann and Morgenstern (1944) stated that finite stable sets exist for the intervals $\frac{1}{18} < z \leqslant \frac{1}{6}$ and $\frac{1}{6} \leqslant z \leqslant \frac{1}{2}$, but they did not explicitly display them. It is not known whether finite ones exist or not when $z = \frac{1}{18}$.

(iv) The "interior" edge $b = (z, \frac{1}{2}, \frac{1}{2})$, $0 \leqslant z \leqslant \frac{1}{2}$, which joins the vertices b^2 and b^3 of P has the finite stable set $V(R(e))$ given below when $\frac{2}{5} < z \leqslant \frac{1}{2}$. No finite one is known for $0 < z \leqslant \frac{2}{5}$.

(v) and (vi) Mills (1954, 1959) proved that no finite stable set can exist for games corresponding to the interiors of the other two edges of P, i.e., $b = (0, 0, 2z)$ or $(0, \frac{1}{2} - z, \frac{1}{2} + z)$ for $0 < z < \frac{1}{2}$.

Mills (1959) proved that there exist no finite stable sets on the face of P with $b_1 = 0$, except for the edge $b = (0, z, z)$ where $0 \leqslant z \leqslant \frac{1}{2}$ and the vertex

$b^1 = (0, 0, 1)$, which were covered above. There are no published results about finite stable sets for the interiors of the other three faces of P, except where these faces meet the solid region discussed in the next paragraph.

Von Neumann and Morgenstern (1944, 3rd edn., pp. 321–329) showed the existence of finite stable sets in a three-dimensional region R in P located near the center point $b^3 = (\frac{1}{2}, \frac{1}{2}, \frac{1}{2})$ of the unit cube U. For any $b \in P$ let

$$u_1(b) = (-1 + b_1 + b_2 + b_3)/2,$$

$$u_2(b) = (1 - b_1 + b_2 - b_3)/2,$$

$$u_3(b) = (1 + b_1 - b_2 - b_3)/2,$$

$$u_4(b) = (1 - b_1 - b_2 + b_3)/2,$$

and then for $i \in N = \{1, 2, 3, 4\}$ define

$$\underline{u}(b) = \min_i u_i(b) \quad \text{and} \quad \bar{u}(b) = \max_i u_i(b).$$

Whenever $b \in P$ and $\frac{2}{3}\bar{u} < 2e \leq \underline{u}$ then there is the finite stable set

$$V(R(e)) = \left\{ x \in A: x_i = \begin{cases} u_i + e & \text{if } y_i = \frac{3}{8} \\ u_i & \text{if } y_i = \frac{1}{4} \\ u_i - 2e & \text{if } y_i = 0 \end{cases}, \text{ where } y \in V_2(b^3(0)) \right\},$$

which has 13 points and is somewhat similar to $V_2(b^3(c))$ presented above when $c = 0$. One can prove that the required bounds on the parameter e are satisfied when b is near b^3, e.g., one can pick the region to be

$$R = \{b \in P: 5b_1 + 5b_2 + b_3 > 5\}.$$

There are four constant-sum, weighted majority games $[q: w_1, w_2, w_3, w_4, w_5]$ with five persons that have finite stable sets. Their main simple stable sets are as follows.

(i) $[3: 1, 1, 1, 1, 1]$. This is the five-person simple-majority game $(5, 3)$ that has the unique symmetric finite stable set

$$V^S = \langle (\tfrac{1}{3}, \tfrac{1}{3}, \tfrac{1}{3}, 0, 0) \rangle$$

consisting of 10 imputations.

(ii) $[4: 1, 1, 1, 2, 2]$ has the seven-point stable set

$$V = \{(0, 0, 0, \tfrac{1}{2}, \tfrac{1}{2})\} \cup \langle (\tfrac{1}{4}, \tfrac{1}{4}, 0, \tfrac{1}{2}, 0) \rangle_{\{1,2,3\}} \cup \langle (\tfrac{1}{4}, \tfrac{1}{4}, 0, 0, \tfrac{1}{2}) \rangle_{\{1,2,3\}}.$$

(iii) $[5: 1, 1, 2, 2, 3]$ has the five-point stable set

$$V = \{(\tfrac{1}{5}, \tfrac{1}{5}, 0, 0, \tfrac{3}{5})\} \cup \langle(0, 0, \tfrac{2}{5}, 0, \tfrac{3}{5})\rangle_{\{3,4\}} \cup \langle(\tfrac{1}{5}, 0, \tfrac{2}{5}, \tfrac{2}{5}, 0)\rangle_{\{1,2\}}.$$

(iv) $[4: 1, 1, 1, 1, 3]$ has the five-point stable set

$$V = \{(\tfrac{1}{4}, \tfrac{1}{4}, \tfrac{1}{4}, \tfrac{1}{4}, 0)\} \cup \langle(\tfrac{1}{4}, 0, 0, 0, \tfrac{3}{4})\rangle_{\{1,2,3,4\}}.$$

There is rather little known about the existence and nature of finite stable sets for *n*-person games with $n > 4$. A few families of finite stable sets for infinitely many values of *n* have been discovered in addition to those previously mentioned. Only one additional and more recent result in this direction will be presented here.

McKelvey and Ordeshook (1977) described new finite nonsymmetric stable sets $V^5(a, \gamma)$ given below for the simple majority game $(5, 3)$ or $[3: 1, 1, 1, 1, 1]$ which consist of ten imputations of the form

$$(a, a, b, 0, 0) \quad (0, a, 0, b, a) \quad (a, 0, 0, a, b) \quad (b, 0, a, 0, a) \quad (a, b, 0, 0, a)$$

$$(0, 0, b, a, a) \quad (a, 0, a, b, 0) \quad (0, a, a, 0, b) \quad (b, a, 0, a, 0) \quad (0, b, a, a, 0),$$

where $2a + b = 1$ and $\tfrac{1}{4} < b < \tfrac{1}{2}$. [When $a = \tfrac{1}{4}$ (or $b = \tfrac{1}{2}$), then the set $V^5(\tfrac{1}{4}, \gamma) \cup \langle(\tfrac{1}{4}, \tfrac{1}{4}, \tfrac{1}{4}, \tfrac{1}{4}, 0)\rangle$ is a stable set.] Each set $V^5(a, \gamma)$ is a proper subset of the symmetric set $\langle(a, a, b, 0, 0)\rangle = W^5$ of 30 points. There are several such stable sets $V^5(a, \gamma)$ for each value a depending upon the particular selection γ of ten such imputations from the set W^5.

For the nine-person simple majority game $(9, 5)$ or $[5: 1, 1, 1, 1, 1, 1, 1, 1, 1]$ Michaelis (1981) found analogous types of stable sets $V^9(a, \gamma)$ of 126 imputations each of which is a subset of the symmetric set $\langle(a, a, a, a, b, 0, 0, 0, 0)\rangle = W^9$ of 630 points and where $4a + b = 1$ and $\tfrac{1}{6} < b < \tfrac{1}{3}$. [When $a = \tfrac{1}{6}$ (or $b = \tfrac{1}{3}$), then $V^9(\tfrac{1}{6}, \gamma) \cup \langle(\tfrac{1}{6}, \tfrac{1}{6}, \tfrac{1}{6}, \tfrac{1}{6}, \tfrac{1}{6}, \tfrac{1}{6}, 0, 0, 0)\rangle$ is a stable set.] He also proved that there are no such stable sets contained in $W^7 = \langle(a, a, a, b, 0, 0, 0)\rangle$ for the seven-person simple majority game.

It has also been shown that when *n* is odd and not of the form $2^p - 1$, then the simple majority game $[(n + 1)/2: 1, 1, \ldots, 1]$ has various stable sets $V^n(a, \gamma)$ of $\binom{n}{(n+1)/2}$ imputations each.

These are proper subsets of the symmetric set $\langle(a, \ldots, a, b, 0, \ldots, 0)\rangle = W^n$ of $((n + 1)/2)\binom{n}{(n+1)/2}$ points and have $(n - 1)a/2 + b = 1$ and $2/(n + 3) < b < 4/(n + 3)$ (or $2/(n + 3) < a < 2(n + 1)/(n + 3)(n - 1)$). No stable set of this form can exist when *n* is of the form $2^p - 1$, because $\binom{n}{(n+1)/2}$ is then odd and the following characterization is impossible to achieve.

One can characterize these stable sets $V^n(a, \gamma)$ as the subsets of $\langle(a, \ldots, a, b, 0, \ldots, 0)\rangle = W^n$ which are *complete* in the sense that for each

$S \subset N$ with $|S| = (n - 1)/2$ there is a unique imputation x with $x_i = 0$ for all $i \in S$, and *complementary* in the sense that if $x \in V^n(a, \gamma)$, then x' is also in this stable set where $x'_i = a$ when $x_i = 0$ and $x'_i = 0$ when $x_i = a$. The detailed proof of this characterization and of the existence of such sets appears in Lucas, Michaelis, Muto and Rabie (1981, 1982).

Many of the particular games mentioned so far belong to a special class of games known as "extreme" games. This class of games contains most of the games for which finite stable sets have been determined, and it provides a useful scheme for studying games with finite stable sets. A brief introduction to extreme games is presented in Lucas and Michaelis (1982). A more detailed exposition is given in the monograph on this topic by Rosenmüller (1977). Simple games and finite stable sets have many other connections with other discrete structures in traditional mathematics such as finite projective geometries [for example, see Richardson (1956) and Hoffman and Richardson (1961)].

It is true that an arbitrary n-person cooperative game rarely has any finite stable sets. There is, nonetheless, an extensive and very rich theory about this topic which is of great interest in its own right from both a theoretical and applied point of view. Von Neumann and Morgenstern (1944) created stable set theory as an applied subject for use in the social and behavioral sciences. Finite stable set theory often does correspond to obvious social outcomes and it has also provided new insights into nonobvious group behavior. Experimental work on multiperson group interactions often conforms to these theoretical outcomes. Furthermore, it appears as though the theory of finite stable sets should be of major interest as pure geometry and combinatorics, as well as having potential applications in other directions such as the physical sciences.

Much of traditional geometry deals with points, lines, and subspaces, and their interrelations. Many contemporary fields such as discrete optimization, however, are also concerned with nonlinear spatial notions of a more "directional" or "angular" nature, e.g., cones and polytopes. These higher-dimensional objects may display new types of geometrical or combinatorial relationships. So finite stable sets should be considered as a new subject somewhat like the existing areas of finite projective geometry, or various discrete systems of designs or schemes. Recall that the domination cones of any imputation is a finite set of open "generalized orthants". A finite stable set V gives rise to a finite number of such overlapping orthants which *covers* precisely the set $A - V$. This provides a new type of geometry of points and "space" filling cones emanating from these points.

As combinatorial objects, finite stable sets have a variety of possible applications to areas such as statistical designs and scheduling theory as suggested in Lucas, Michaelis, Muto and Rabie (1982). Different parts of a stable set can be considered as "multidimensional keys" and "locks", or codes.

There are also many physical systems, such as crystals, molecules, atoms, and nuclei, where "bodies are held in position in space". The forces between the particles would appear to be less than uniform in all directions. Finite stable sets, which often display partial symmetry as well as full symmetry in some cases, may give insights into what physical configurations can arise when noncentral force fields are involved.

10. Some conclusions

Work by Emile Borel and John von Neumann on matrix games in the 1920s eventually led to the theory of n-person noncooperative games as well as various results about equilibrium outcomes. Although individual illustrations of cooperative games appeared for some time before the famous book by von Neumann and Morgenstern (1944), they presented the first general model and solution concept for the multiperson cooperative theory. There are now several variations and extensions of their model, plus some two to three score of alternate solution concepts. One now views stable set theory as only one of several approaches for analyzing coalitional games. Although there may be some shortcomings with stable set theory from an applied point of view, it is nevertheless one of the most interesting and richest of these theories mathematically.

Stable sets are defined in terms of two simple conditions: internal and external stability; along with a rather simple preference relation called domination. These stability concepts, (3) and (4), are rather basic and fundamental mathematical notions, and presume very little about the nature or structure of social institutions and interactions. Similarly, the definition of dominance is quite simple and straightforward and arises in other contexts. It does use the relation "greater than" in (1) and it sums numbers in the effectivity condition $x(S) \leq v(S)$ given in (2). [The models for games without side payments generalize condition (2) from closed half spaces to other regions in space subject to certain natural restrictions.] So the assumptions built into the dominance relation are also very minimal. In light of the very general nature of the classical model of stable sets, it is truly amazing how many insights into social economic, and political behavior it does provide.

Stable sets often predict likely social structures and how groups will organize themselves. They show the important role of minimal winning coalitions and minimal sized veto (or blocking) coalitions. They often show how a game will decompose into subgames between critical coalitions. They exhibit a variety of standards of behavior and delineate bargaining ranges. They predict the formation of cartels and illustrate the stability of discrimination and its limits. It is quite remarkable how so few assumptions can lead to so many insights into coalition formation, competition, and distributions of wealth. There are also many situations where stable set theory matches well with experimental results.

On the other hand, we saw in Section 5 that stable set theory has some highly undesirable properties. Although some lack of uniqueness for cooperative game solution concepts seems reasonable, there are clearly too many stable sets for most games. At the same time there are some games, presumably rare, for which no stable sets at all exist. A particular stable set may also contain many imputations. This latter multiplicity is not so bothersome, however, when one views stable set theory as delineating the possible coalitions and resulting "subgames" that might be pursued in group bargaining, rather than predicting a precise outcome. One can also add additional axioms to conditions (3) and (4) in our definition of stable set and thus cut down on the multiplicity of such solutions. For example, we saw that many games had unique symmetric stable sets. Moreover, one can make changes in the definition of dominance which can restrict the number of stable sets. Lucas (1965a) showed that "almost all" four-person games in "partition function form" have a unique stable set, even though this theory includes the von Neumann–Morgenstern model as a "highly degenerate" special case. For some other possible changes in the dominance relation see Lucas and Maceli (1978).

When economic problems are formulated as n-person cooperative games they usually have nonempty cores, and the core typically serves as a satisfactory solution concept. Many of the other known solution concepts also give results "akin" to the core for such games. In fact, there are several major theorems about various solution concepts, both cooperative and noncooperative, converging to the same outcomes (often a unique "price" vector) as the number of players in the game approaches infinity. Stable set theory appears as a noticeable outlier to these important results. On the other hand, there are many games which have empty cores. These arise most often in modeling politics and voting systems. In these situations stable set theory is often more informative than "core-like" solution concepts. Even in economics, stable sets may provide significant insights if one forgoes the common assumption of "perfect competition". The theories on bargaining sets, kernels, and nucleoli discussed by Maschler in Chapter 18 of this Handbook also give interesting results for games with empty cores. Note that a game has a nucleolus point for each coalition structure (partition) of the player set N; and not just the one nucleolus outcome $v(N)$ for this grand coalition N which is always in the core when the latter is nonempty. So stable sets and bargaining set theory are still among the most insightful models if one is concerned about games with empty cores or if one has some doubt whether perfect competition is involved. It is interesting to note that the notion of the core of an n-person game arose three times in the volume by von Neumann and Morgenstern (1944), but they chose to stress the stable set instead. They were working initially with essential constant-sum games which have empty cores. Moreover, it was known that Morgenstern was a strong critic of many assumptions and approaches taken in economic theory, including the idea of perfect competition. It is of further interest to note that two of the most popular solution concepts applicable to games with empty cores, stable sets and bargaining set theory, are known to have some close connections in some cases. The "symmetric type" stable sets can be built up from the nucleoli in a natural way for some special classes of games, as discussed in Lucas (1990).

Princeton University Press, a major publisher of important mathematics and science books, recently entered the text by von Neumann and Morgenstern (1944) into a publishers' competition as the most influential book published by them. In addition to its major impact on game theory and social science, it initiated or made significant contributions to several other subjects. This includes extensive form and normal (strategic) form games, monotone simple games, the "first value theory" (the main simple stable sets), finite stable set theory, and the most popular approach to utility theory (in the appendix of their second edition in 1947). Although there are some serious flaws in their theory of stable sets, it is nevertheless a very rich and insightful contribution and is still a leading contender for explaining much of what occurs in multiperson competitions, especially in the case of games with empty cores.

The book by Shubik (1982) has an extensive bibliography on game theory, and his Appendix B provides a most helpful descriptive list of over 100 references on stable set theory through 1973.

Bibliography

Aumann, R.J. (1967) 'A survey of cooperative games without side payments', in: M. Shubik, ed., *Essays in mathematical economics: In honor of Oskar Morgenstern*. Princeton: Princeton University Press, pp. 3–27.

Banzhaf, J.F. (1968) 'One man, 3.312 votes: A mathematical analysis of the electoral college', *Villanova Law Review*, 13: 303–332.

Berge, C. (1957) *Théorie générale des jeux à n personnes*. Paris: Gauthier-Villars.

Bondareva, O.N., T.E. Kulakovskaya and N.I. Naumova (1979) 'The solution of the four-person games', *Leningrad University Vestnik*, 6: 104–105.

Bott, R. (1953) 'Symmetric solutions to majority games', in: H.W. Kuhn and A.W. Tucker, eds., *Contributions to the theory of games*, Vol. II, *Annals of Mathematics Studies* No. 28. Princeton: Princeton University Press, pp. 319–323.

Chang, Chih (1985) 'Solution concepts for *n*-person cooperative games', Ph.D. thesis, Center for Applied Mathematics, Cornell University, Ithaca.

Coleman, J.S. (1971) 'Control of collectivities and the power of collectivity to act', in: B. Lieberman, ed., *Social choice*. New York: Gordon and Breach, pp. 269–300.

Dubey, P. and L.S. Shapley (1979) 'Mathematical properties of the Banzhaf power index', *Mathematics of Operations Research*, 4: 99–131.

Gillies, D.B. (1959) 'Solutions to general non-zero-sum games', in: A.W. Tucker and R.D. Luce, eds., *Contributions to the theory of games*, Vol. IV, *Annals of Mathematics Studies* No. 40. Princeton: Princeton University Press, pp. 47–85.

Greenberg, J. (1989) 'Deriving strong and coalition-proof Nash equilibria from an abstract system', *Journal of Economic Theory*, 49: 195–202.

Greenberg, J. (1990) *The theory of social situations: An alternate game-theoretic approach.* Cambridge: Cambridge University Press.

Griesmer, J.H. (1959) 'Extreme games with three values', in: A.W. Tucker and R.D. Luce, eds., *Contributions to the theory of games*, Vol. IV, *Annals of Mathematics Studies* No. 40. Princeton: Princeton University Press, pp. 189–212.

Hart, S. (1973) 'Symmetric solutions of some production economies', *International Journal of Game Theory*, 2: 247–265.

Hart, S. (1974) 'Formation of cartels in large markets', *Journal of Economic Theory*, 7: 453–466.

Heijmans, J.G.C. (1986) 'Discriminatory and symmetric von Neumann–Morgenstern solutions for a class of symmetric games', Ph.D. thesis, School of Operations Research and Industrial Engineering, Cornell University, Ithaca.

Heijmans, J. (1987) 'Abstract stable sets, symmetric games and symmetric von Neumann–Morgenstern solutions; Application: The symmetric vN–M solutions of the symmetric (0, 1)-normalized 4-person games', Paper presented at the International Conference on Game Theory and Applications, O.S.U., Columbus, Ohio, July, 5–16 (unpublished).

Heijmans, J. (1991) 'Discriminatory von Neumann–Morgenstern solutions', *Games and Economic Behavior*, 3: 438–452.

Hilliard, M.R. (1983) 'Weighted voting theory and applications', Ph.D. thesis, School of Operations Research and Industrial Engineering, Cornell University, Ithaca.

Hoffman, A.J. and M. Richardson (1961) 'Block design games', *Canadian Journal of Mathematics*, 13: 110–128.

Kuhn, H.W., ed. (1953) 'Report of an informal conference on the theory of N-person games', Princeton University, dittoed, out of print.

Lucas, W.F. (1965a) 'Solutions for four-person games in partition function form', *SIAM Journal of Applied Mathematics*, 13: 118–128.

Lucas, W.F., ed. (1965b) 'Report of the fifth conference on game theory', Princeton University, mimeographed, out of print.

Lucas, W.F. (1966) 'n-person games with only 1, $n - 1$ and n-person coalitions', *Zeitschrift für Wahrscheinlichkeitstheorie und verwandte Gebiete*, 6: 287–292.

Lucas, W.F. (1967) 'A counterexample in game theory', *Management Science*, 13: 766–767.

Lucas, W.F. (1968a) 'A game in partition function form with no solution', *SIAM Journal of Applied Mathematics*, 16: 582–585.

Lucas, W.F. (1968b) 'On solutions for n-person games', RM-5567-PR, The RAND Corp., Santa Monica.

Lucas, W.F. (1968c) 'A game with no solution', *Bulletin of the American Mathematical Society*, 74: 237–239.

Lucas, W.F. (1969a) 'Games with unique solutions that are nonconvex', *Pacific Journal of Mathematics*, 28: 599–602.

Lucas, W.F. (1969b) 'The proof that a game may not have a solution', *Transactions of the American Mathematical Society*, 137: 219–229.

Lucas, W.F. (1971) 'Some recent developments in n-person game theory', *SIAM Review*, 13: 491–523.

Lucas, W.F. (1976) 'Disconnected solutions', *Bulletin of the American Mathematical Society*, 82: 596–598.

Lucas, W.F. (1983) 'Measuring power in weighted voting systems', in: S.J. Brams, W.F. Lucas, and P.D. Straffin, Jr., eds., *Political and related models*. New York: Springer-Verlag, pp. 183–238.

Lucas, W.F. (1990) 'Developments in stable set theory', in: T. Ichiishi, A. Neyman, and Y. Tauman, eds., *Game theory and applications*. New York: Academic Press, pp. 300–316.

Lucas, W.F. and J.C. Maceli (1978) 'Discrete partition function games', in: P.C. Ordeshook, ed., *Game theory and political science*. New York: New York University Press, pp. 191–213.

Lucas, W.F. and K. Michaelis (1982) 'Finite solution theory for coalitional games', *SIAM Journal on Algebraic and Discrete Methods*, 3: 551–565.

Lucas, W.F. and M. Rabie (1980) 'Existence theorems in game theory', School of Operations Research and Industrial Engineering, Technical Report No. 473, Cornell University, Ithaca.

Lucas, W.F. and M. Rabie (1982) 'Games with no solutions and empty cores', *Mathematics of Operations Research*, 7: 491–500.

Lucas, W.F., K. Michaelis, S. Muto and M. Rabie (1981) 'Detailed proofs for a family of finite solutions', School of Operations Research and Industrial Engineering, Technical Report No. 523, Cornell University, Ithaca.

Lucas, W.F., K. Michaelis, S. Muto and M. Rabie (1982) 'A new family of finite solutions', *International Journal of Game Theory*, 11: 117–127.

Maschler, M., ed. (1971) 'Recent advances in game theory' (A Meeting of the Princeton University Conference). Philadelphia: Ivy Curtis Press, lithograph, out of print.

McDonald, J. (1975) *The game of business*. New York: Doubleday; Anchor Books (paper, 1977).

McKelvey, R.D. and P.C. Ordeshook (1977) 'An undiscovered von Neumann–Morgenstern solution for the (5, 3) majority rule game', *International Journal of Game Theory*, 6: 33–34.

Menshikova, O.R. (1977) 'Necessary and sufficient conditions of core stability in symmetric cooperative games', Technical Report, Leningrad State University (in Russian).

Michaelis, K. (1981) 'A survey of finite stable sets for cooperative games', M.S. Thesis, Operations Research, Cornell University, Ithaca.

Michaelis, K. (1982) 'On the nature of solution concepts for cooperative games', Ph.D. thesis, Field of Regional Science, Cornell University, Ithaca.

Mills, W.H. (1954) 'The four person game-edge of the cube', *Annals of Mathematics*, 59: 367–378.

Mills, W.H. (1959) 'The four person game – finite solutions on the face of the cube', in: A.W. Tucker and R.D. Luce, eds., *Contribution to the theory of games*, Vol. IV, *Annals of Mathematics Studies* No. 40. Princeton: Princeton University Press, pp. 125–143.

Muto, S. (1978) 'Stable sets for symmetric, *n*-person cooperative games', Ph.D. thesis, School of Operations Research and Industrial Engineering, Cornell University, Ithaca.

Muto, S. (1979a) 'Symmetric solutions for symmetric, constant-sum, extreme games with four values', *International Journal of Game Theory*, 8: 115–123.

Muto, S. (1979b) 'Generalized *k*-quota solutions for (*N*, *k*) games', *International Journal of Game Theory*, 8: 165–173.

Muto, S. (1980) 'Semi-symmetric solutions for (*n*, *k*) games', *International Journal of Game Theory*, 9: 91–97.

Muto, S. (1982a) 'Symmetric solutions for (*n*, *k*) games', *International Journal of Game Theory*, 11: 195–201.

Muto, S. (1982b) 'Symmetric solutions for (*n*, *n* − 2) games with small values of $v(n-2)$', *International Journal of Game Theory*, 11: 43–52.

Muto, S. (1982c) 'On Hart production games', *Mathematics of Operations Research*, 7: 319–333.

Muto, S. (1983) 'Symmetric solutions for symmetric games with large cores', *International Journal of Game Theory*, 12: 207–223.

Nering, E.D. (1959) 'Symmetric solutions for general-sum symmetric 4-person games', in: A.W. Tucker and R.D. Luce, eds., *Contributions to the theory of games*, Vol. IV, *Annals of Mathematics Studies* No. 40. Princeton: Princeton University Press, pp. 111–123.

Ostmann A. (1987) 'On the minimal representation of homogeneous games', *International Journal of Game Theory*, 16: 69–81.

Owen, G. (1965) 'A class of discriminatory solutions to simple *n*-person games', *Duke Mathematics Journal*, 32: 545–553.

Owen, G. (1966) 'Discriminatory solutions of *n*-person games', *Proceedings of the American Mathematical Society*, 17: 653–657.

Owen, G. (1968a) '*n*-person games with only *n* − 1, and *n*-person coalitions', *Proceedings of the American Mathematical Society*, 19: 1258–1261.

Owen, G. (1968b) *Game theory*. Philadelphia: Saunders.

Owen, G. (1970) 'The four-person constant-sum games: Discriminatory solutions on the main diagonal', *Pacific Journal of Mathematics*, 34: 461–480.

Owen, G. (1982) *Game theory*, 2nd edn. New York: Academic Press.

Rabie, M. (1980) 'Properties of solution concepts for *n*-person cooperative games', Ph.D. thesis, School of Operations Research and Industrial Engineering, Cornell University, Ithaca.

Rabie, M. (1985) 'A simple game with no symmetric solution', *SIAM Journal on Algebraic and Discrete Methods*, 6: 29–31.

Richardson, M. (1955) 'Relativization and extension of solutions of irreflexive relations', *Pacific Journal of Mathematics*, 5: 551–584.

Richardson M. (1956) 'On finite projective games', *Proceedings of the American Mathematical Society*, 7: 458–465.

Rosenmüller, J. (1977) *Extreme games and their solutions*, Lecture Notes in Economics and Mathematical Systems, 145. New York: Springer-Verlag.

Roth, A.E. (1976) 'Subsolutions and the supercore of cooperative games', *Mathematics of Operations Research*, 1: 43–49.

Shapley, L.S. (1953a) 'A value for *n*-person games', in: H.W. Kuhn and A.W. Tucker, ed., *Contributions to the theory of games*, Vol. II, *Annals of Mathematics Studies* No. 28. Princeton: Princeton University Press, pp. 307–317.

Shapley, L.S. (1953b) 'Quota solutions of *n*-person games', in: H.W. Kuhn and A.W. Tucker, eds., *Contributions to the theory of games*, Vol. II, *Annals of Mathematics Studies* No. 28. Princeton: Princeton University Press, pp. 343–359.

Shapley, L.S. (1959) 'A solution with an arbitrary closed component', in: A.W. Tucker and R.D. Luce, eds., *Contribution the the theory of games*, Vol. IV, *Annals of Mathematics Studies* No. 40. Princeton: Princeton University Press, pp. 87–93.

Shapley, L.S. (1962) 'Simple games: An outline of the descriptive theory', *Behavioral Science*, 7: 59–66.

Shapley, L.S. (1967) 'On solutions that exclude one or more players', in: M. Shubik, ed., *Essays in mathematical economics: In honor of Oskar Morgenstern*. Princeton: Princeton University Press, pp. 57–61.

Shapley, L.S. (1968) 'Notes on *n*-person games VIII: A game with infinitely "flaky" solutions', unpublished manuscript.

Shapley, L.S. (1976), unpublished manuscript.

Shapley, L.S. (1971) 'Cores of convex games', *International Journal of Game Theory*, 1: 11–26.

Shapley, L.S. and M. Shubik (1969) 'On market games', *Journal of Economic Theory*, 1: 9–25.

Shubik, M. (1982) *Game theory in the social sciences: Concepts and solutions*. Cambridge, Mass.: MIT Press.

Sokolina, N.A. (1986) 'Some solutions with a special form – discriminating against a coalition of two players', *Vestnik Leningrad State University*, *Mathematics*, 19: 40–48.

Stearns, R.E. (1964) 'On the axioms for a cooperative game without side payments', *Proceedings of the American Mathematical Society*, 15: 82–86.

Straffin, P.D., Jr. (1983) 'Power indices in politics', in: S.J. Brams, W.F. Lucas and P.D. Straffin, Jr., eds., *Political and related models*. New York: Springer-Verlag, pp. 256–321.

Thrall, R.M. and W.F. Lucas (1963) '*n*-person games in partition function form', *Naval Research Logistics Quarterly*, 10: 281–298.

Von Neumann, J. and O. Morgenstern (1944) *Theory of games and economic behavior*. Princeton: Princeton University Press; 2nd edn. 1947; 3rd edn. 1953.

Weber, R.J. (1973a) 'Symmetric simple games', Technical Report No. 173, School of Operations Research and Industrial Engineering, Cornell University, Ithaca.

Weber, R.J. (1973b) 'A generalized discriminatory solution for a class of *n*-person games', Technical Report No. 174, School of Operations Research and Industrial Engineering, Cornell University, Ithaca.

Weber, R.J. (1973c) 'Discriminatory solutions for [*n*, *n* − 2]-games', Technical Report No. 175, School of Operations Research and Industrial Engineering, Cornell University, Ithaca.

Weber, R.J. (1974) 'Bargaining solutions and stationary sets in *n*-person games', Ph.D. thesis, School of Operations Research and Industrial Engineering, Cornell University, Ithaca.

Weber R.J. (1982) 'Distributive solutions for absolutely stable games', *International Journal of Game Theory*, 11: 53–56.

Wolfe, P., ed. (1955) 'Report on an informal conference on recent developments in the theory of games', Princeton University, mimeographed, out of print.

Wolfe, P., ed. (1957) 'Report of the third conference on games', Princeton University, mimeographed, out of print.

Chapter 18

THE BARGAINING SET, KERNEL, AND NUCLEOLUS

MICHAEL MASCHLER*

The Hebrew University of Jerusalem

Contents

*I wish to express my gratitude to Bezalel Peleg, who read this survey critically, and to Robert J. Aumann, with whom I had several conversations concerning the preparation of this survey. Their comments were indeed illuminating and very useful. I am also grateful to Daniel Granot for many helpful comments. In particular he enlightened me about actual events that led to the successful computation of the kernel and the nucleolus in many applications.

Handbook of Game Theory, Volume 1, Edited by R.J. Aumann and S. Hart

1. Introduction

In this survey I shall try to lead the reader along the same path I have followed since beginning to work in the area of solution concepts in cooperative game theory. For me it has been a fascinating experience, to watch how a few, almost naïve ideas, in the creation of which I was participating, developed into deep mathematical theorems, growing to an important chapter in game theory and proving useful in several fields of the social sciences. This could not have been achieved without the combined effort of many scholars, from practically all over the world. I wish to express here my deep appreciation and gratitude to those scholars, many of whom are personal friends and colleagues. Research in this field is still growing. Indeed it has recently gained new impetus, so that it is time to pause and review what has been accomplished. This is the aim of the survey.

While preparing this survey I had several ambitions. I wanted it to serve as an introduction to the interested reader who does not know the theory. At the same time, I wanted to bring the reader to the frontiers of today's research. I also wanted to add critical comments on the significance, scope, and limitations of the bargaining set and the solution concepts derived from it. Such comments in the literature are scarce, mainly because these were not known at the time of the creation of the solution concepts. I needed time and experience to get a clearer understanding of these issues.

While preparing this survey I had to decide whether to report also on unpublished results, or to stay only with the published ones. Certainly, research memoranda often undergo thorough changes before they get published. They have not been refereed and some of them will never be printed. Nevertheless, I decided to report on certain unpublished results because I felt that I could not otherwise do justice to the subject.

The bibliography includes an extensive literature on the subjects of this survey – as complete as I could make it. I covered the *Mathematical Review* citations throughout 1991. Some citations were obtained only by searching cross references, when I could not get the reprints, or was unable to read them due to language barriers. Some references contain important recent contributions that could not be surveyed because they are so recent.

The references are grouped into subjects to enable their users to find material relevant to a particular topic.[1]

I intend to keep in my computer an updated file of the references and to make it available upon request. Therefore, I shall be grateful if authors

---

[1]References which do not end with bold face letters are references cited in this survey, but are not related to the bargaining set–kernel–nucleolus theory.

continue sending me their current papers related to the bargaining set, kernel, and nucleolus, both as working papers and as reprints.

After a short section on terminology and notations (Section 2), the bargaining set, kernel, and nucleolus are introduced, together with their presolutions. Their main properties are described and ways to compute them are discussed (Sections 3, 4 and 5).

Section 6 is devoted to the axiomatic foundation of the prekernel and the prenucleolus. These axiomatizations provide a theoretical justification of these solutions concepts, which is important in view of the fact that the rationale embodied in the original definition is quite obscure. The case of the prenucleolus is particularly interesting because it enables a deep understanding of the difference between this solution and the Shapley value.

In Section 7 we provide dynamic processes which lead the participants in a cooperative game to reach the bargaining set, or the kernel, or the nucleolus, or many other bargaining sets via a sequence of steps that make good intuitive sense. The existence of such processes is a unique feature of the bargaining set theory. Similar processes for other solution concepts are hardly known at present. The development of these ideas not only introduced dynamics into game theory, it also enriched the theory of dynamic systems by introducing vector-valued Lyapunov functions to treat set-valued dynamical systems.

The ideas embodied in the bargaining set, kernel, and nucleolus spawned many other related solution concepts. Only the most related ones are reported in Section 8 as the scope of this chapter did not allow me to do full justice to the topic. Nevertheless, I have tried to indicate what directions the various modifications have taken.

Sections 9 and 10 are devoted to applications. Section 9 describes some classes of games of which the solutions, or at least their properties, are known. Section 10 discusses certain subjects in economics for which the solutions discussed in this survey were recommended. Section 11 treats some psychological aspects, namely to what extent people behave, or are willing to behave, in accordance with the recommendations of the bargaining set. It turns out that although some laboratory experiments are quite supportive, others exhibit issues that are much deeper and require further research to arrive at better normative theories. The section concludes with the analysis of actual elections in some European countries and in Israel.

In Section 12 we report on the status of the bargaining set theory for games without side payments, and in Section 13 on results concerning games with an infinite number of players, countable number as well as a continuum. The research for these classes of games has only begun, and we are far from having a rich theory with justifications from many directions as we have in the side-payment, finite-number-of-players case. However, important results have already been achieved in these cases as well, which makes me hope that a solid theory is only a matter of time.

2. Basic definitions and notations for games with side payments

We shall be concerned with cooperative games with side payments $(N; v)$, where $N = \{1, 2, \ldots, n\}$ is the set of *players* and $v: 2^N \to \mathcal{R}$ is the *characteristic function*.[2] For mathematical convenience we shall require that

$$v(\emptyset) = 0 . \tag{2.1}$$

Note that we do not require the game to be superadditive; however, we shall sometimes require that it is *zero-monotonic*;[3] namely, that if $(N; w)$ is zero-normalized and strategically equivalent to $(N; v)$ then

$$S, T \subset N, \ T \supset S \Rightarrow w(T) \geqslant w(S) . \tag{2.2}$$

This class contains the class of superadditive games.

For results concerning the bargaining set we may interpret the *worth* $v(S)$ of a coalition[4] S to be an amount *in monetary units* that the coalition S can make (in a certain time period) if it is *formed*.[5] For results concerning the kernel and the nucleolus we must require that utilities are transferable, because these solution concepts are not covariant with respect to utility transformations which merely preserve monotonicity and risk aversion.[6] To be on the safe side, we assume that when a coalition is formed, it is well known what the proceeds will be and these are *independent*[7] *of actions taken by members outside* S.

Presumably, when players face such a game they will end up forming *disjoint coalitions*[8] which form a partition of N, namely a set of nonempty and disjoint

---

[2]Also called the *coalition function*.

[3]Also called *weakly superadditive*.

[4]Subsets of N will be called *coalitions*.

[5]To this we add the (obvious) assumption that each player prefers more money to less and the (restrictive) assumption that each player is risk averse in his preferences for money. This is much less than requiring that utility for money be transferable. Indeed, if $n \geqslant 3$ and the players do have transferable utility for money, then there exists an infinitely divisible and desirable commodity – which can be called "money" – towards which the players' utilities are linear. We only require that they be concave. [See Aumann (1960, 1967) for a discussion of this issue.]

[6]However, interpreting $v(S)$ as money makes sense even for these solution concepts if the rules of the game preclude lotteries on outcomes and money has an absolute meaning (e.g., when a judge is called to prescribe outcomes in monetary units and he does not care about the players' utilities towards money).

[7]Of course, there are other interpretations of $v(S)$ which do not require this heavy restriction on the games – for example, the *highest security level* that can be achieved by joint action of members of S. Even though we also employ such interpretations from time to time, one must remember that they are open to criticism and each application which uses them should be approached cautiously.

[8]It is customary to justify the requirement of forming disjoint coalitions by saying that if, in reality, two overlapping coalitions form we would express this by saying that their union actually formed.

subsets of N the union of which is N. Such a partition will be called a *coalition structure* (c.s.).[9]

An *imputation* for a coalition structure $\mathscr{B}$ is a *payoff vector* $x = (x_1, x_2, \ldots, x_n)$ satisfying[10]

$$x(B) = v(B), \quad \text{all } B \text{ in } \mathscr{B} \text{ (group rationality)}, \tag{2.3}$$

$$x_i \geq v(\{i\}), \quad \text{all } i \text{ in } N \text{ (individual rationality)}. \tag{2.4}$$

A *preimputation* for $\mathscr{B}$ is a group rational payoff vector; i.e., (2.4) is not required. We shall denote by $X(\mathscr{B})$ and $X^0(\mathscr{B})$ the spaces of all imputations and preimputations for the c.s. $\mathscr{B}$, respectively. We shall sometimes write $X[X^0]$ instead of $X(\{N\})[X^0(\{N\})]$ and call this *the imputation [preimputation] space of the game*.

3. The bargaining set

Consider a group of players N who face a game $(N; v)$. A basic question would be: What coalitions will form and how will their members share the proceeds? In my opinion, no satisfactory answer has so far been given to this important question.[11] The theory of the bargaining set answers a more modest question: How would or should the players share the proceeds, given that a certain c.s. $\mathscr{B}$ has formed? From a normative point of view, the reason for asking such a question stems from the need to let the players know what to expect from each coalition structure so that they can then make up their mind about the coalitions they want to join, and in what configuration. From a descriptive point of view, one can reason as follows. During the course of negotiations there comes a moment when a certain coalition structure is "crystallized". The players will no longer listen to "outsiders", yet each coalition has still to adjust the final share of its proceeds. (This decision may depend on options outside

---

[9]See Aumann and Drèze (1974) for a discussion concerning the interpretations of this concept and for the study of various solution concepts for coalition structures.

[10]By $x(S)$ we mean $\Sigma_{i \in S} x_i$, if $S \neq \emptyset$, and 0 if $S = \emptyset$.

[11]True, if $v(N) > \Sigma \{v(B): B \in \mathscr{B}\}$, for every partition $\mathscr{B}$ of N, considerations involving Pareto optimality yield some ground to the claim that N should form. These arguments are not too compelling, however, because it is possible that Pareto optimal imputations will be contested by some players who can achieve more if they defect and form their own coalition. Think of the three-person zero-normalized game, where $v(\{i, j\}) = 1$ whenever $i \neq j$ and $v(\{1, 2, 3\}) = 1.2$. By the symmetry of the situation, if N forms, the players should end up with equal share; but then, every two-person coalition would prefer to defect with a $(1/2, 1/2)$ split. The players may find it too risky to share later the extra 0.2 in a three-person coalition formation. Thus, it is perhaps safer to predict that in such games a two-person coalition will form, even though the outcome will not be Pareto optimal.

the coalition, even though the chances of defection are slim.) With these ideas in mind, let us introduce the bargaining set.

Definition 3.1. Let x be an imputation in a game $(N; v)$ for a coalition structure $\mathscr{B}$. Let k and l be two distinct players in a coalition B of $\mathscr{B}$. An *objection* of k against l at x is a pair $(C; y)$, satisfying:

(i) $C \subset N$, $k \in C$, $l \notin C$;

(ii) $y \in \Re^C$, $y(C) = v(C)$; [12]

(iii) $y_i > x_i$,   all $i \in C$. [13]

It is important to understand that the purpose of claiming an objection is not actually to defect from $\mathscr{B}$. After all, we have said that $\mathscr{B}$ has been crystallized. The purpose is to indicate to l that k can get more by taking his business someplace else, and since this can be done without the consent of player l ($l \notin C$), perhaps l is getting too much and should transfer some of his share in $v(B)$ to k. Should he? Not necessarily! Player l should not yield if he can protect his share x_l; namely, if he has a counter-objection in the following sense:

Definition 3.2. Let $(C; y)$ be an objection of k against l at x, $x \in X(\mathscr{B})$, $k, l \in B \in \mathscr{B}$. A counter-objection to this objection is a pair $(D; z)$, satisfying:

(i) $D \subset N$, $l \in D$, $k \notin D$;

(ii) $z \in \Re^D$, $z(D) = v(D)$;

(iii) $z_i \geqslant y_i$,   all i in $D \cap C$;

(iv) $z_i \geqslant x_i$,   all i in $D \setminus C$.

In the counter-objection, player l claims that he can protect his share by forming D. He does not need the consent of k ($k \notin D$), he can give each member of D his original payment, and if some members of D were offered some benefits from k, he can match the offer. [14] Note that k can object against l only if they belong to the same coalition of the coalition structure.

We say that an objection is *justified* if it has no counter-objection; otherwise,

---

[12] $\mathscr{R}^C$ is the set of real $|C|$-tuples ($|C|$ being the cardinality of C) whose coordinates are indexed by the members of C. Thus, $\mathscr{R}^{\{i,j\}}$ is the set of pairs (y_i, y_j) with real components.

[13] We could have replaced the strong inequality here by a weak one for all i's, except one, and get the same bargaining set.

[14] We could even insist on a strong inequality in (iii) and still get the same bargaining set.

we say that the objection is *unjustified*. With these definitions we arrive at the bargaining set.

Definition 3.3.[15] Let $(N; v)$ be a cooperative game with side payments. The *bargaining set*[16] $\mathcal{M}_1^i(\mathcal{B})$ *for a coalition structure* $\mathcal{B}$ is

$$\mathcal{M}_1^i(\mathcal{B}) := \{x \in X(\mathcal{B}): \text{ every objection at } x \text{ can be countered}\}$$
$$= \{x \in X(\mathcal{B}): \text{ there exists no justified objection at } x\}. \quad (3.1)$$

If $X(\mathcal{B})$ is replaced by $X^0(\mathcal{B})$, the set is called the *prebargaining set* and denoted $\mathcal{P}\mathcal{M}_1^i$.

It is customary to shorten and write $\mathcal{M}_1^i$ instead of $\mathcal{M}_1^i(\{N\})$ and call it *the bargaining set of the game*. (It is also customary to talk about $\mathcal{M}_1^i$ and mean the *union* of the various $\mathcal{M}_1^i(\mathcal{B})$'s. The reader should be able to deduce the correct meaning from the context.) Similar conventions are for the prebargaining set.

One rationale for the bargaining set is this: there has been a bargaining process which has stabilized on a certain coalition structure $\mathcal{B}$ and a certain imputation x. Stability then implies that, for this $\mathcal{B}$, the conditions of the bargaining set are met.

The bargaining might take the following course. During the negotiation stage, all kinds of offers and counteroffers are made for the purpose of trying to convince potential partners to form coalitions. Then there comes a stage at which a coalition structure crystallizes. Nobody, at that stage, really wants to leave the coalition in which he is a partner, but the players still argue about the proper way to share the proceeds. At this stage, when a player expresses a justified objection, it should be interpreted as if he is saying to the other player: "I like you, and want to be with you in the coalition, but you are getting too much. In fact, not that I really want to leave you, but I can take my business elsewhere and earn more. If you try to find other partners you will find yourself losing. So why shouldn't you give me some of your share and we will both be happy?" Expressing an unjustified objection is not convincing. By expressing a counter-objection, the other player is in fact saying: "I like you too in our coalition, but I do not feel that I should compensate you. Even if you move away, I can still protect my share without you. Sometimes, I shall even destroy your ambition – which happens if our new potential partners have

---

[15]The concepts of an objection and a counter-objection, as well as another version of a bargaining set, were originally discovered by Aumann and Maschler (1964). The present bargaining set, among other variants, was implicitly hinted there but not developed. In view of Theorem 3.5, the present bargaining set turned out to be more fundamental. It was introduced in Davis and Maschler (1963, 1967).

[16]The various indices attached to $\mathcal{M}$ are a result of "historical" idiosyncrasies. They came to distinguish this bargaining set from others.

a nonempty intersection; but even if this is not the case, and we can both gain by departure, *as long as we are in the same coalition (and the same coalition structure), there is no reason for me to yield any part of my share to you.* If we move to another coalition structure, that is another story. We shall then have to look for bargaining set outcomes in that coalition structure. It may well be that we all will like that coalition structure more than the present one."

I elaborated on the above rationale in order to answer two frequently asked questions: Why stop at counter-objections and not talk about counter-counter-objections, etc.? What kind of an objection is it if both players can move somewhere else and *both* make profit? (This happens if the partners in the objection and in the counter-objection form disjoint sets.) The answer is that it is not the purpose of an objection to carry it out. Rather, it is to convince your partner to give you part of his share, and *stay in the coalition* without carrying out any threat. Nowhere is it said that one coalition structure is preferred to others. Of course, the above arguments will be greatly enhanced if we can back them up by a dynamic process that leads the players to some outcomes in the bargaining set. This aspect will be discussed in Section 7.

It should be clear from the definitions that nowhere do we claim that all points in $\mathcal{M}_1^i$ have equal merit. They do not! It is claimed only that the points *not* in $\mathcal{M}_1^i$ are unstable. The bargaining set, like the core, eliminates imputations, narrowing the predictions (or recommendations) to a smaller set of imputations.

Example 3.4. Let[17] $N = 123$, $v(i) = 0 \; \forall i \in N$, $v(12) = 20$, $v(13) = 30$, $v(23) = 40$, $v(123) = 42$. The bargaining set for each coalition structure is

$$\mathcal{M}_1^i(\{1, 2, 3\}) = \{(0, 0, 0)\},$$

$$\mathcal{M}_1^i(\{12, 3\}) = \{(5, 15, 0)\},$$

$$\mathcal{M}_1^i(\{13, 2\}) = \{(5, 0, 25)\},$$

$$\mathcal{M}_1^i(\{23, 1\}) = \{(0, 15, 25)\},$$

$$\mathcal{M}_1^i(\{123\}) = \{(4, 14, 24)\}.$$

In this particular example we see that the bargaining set for each c.s. consists of a one-element set. The bargaining set for the c.s. $\{1, 2, 3\}$ is obvious. When a two-person coalition forms, each member receives his *quota*[18] in the game. For the grand coalition, the players reduce their quotas equally.

---

[17]In order to simplify notation, we shall often ignore some braces and commas when describing coalitions and coalition structures. For example, we shall write 12 instead of $\{1, 2\}$ and $\{12, 3, 45\}$ instead of $\{\{1, 2\}, \{3\}, \{4, 5\}\}$.

[18]The quota vector $\omega = (\omega_1, \omega_2, \omega_3)$ is defined by the system of equations: $\omega_i + \omega_j = v(i, j)$, all $i, j \in N$, $i \neq j$. Here, $\omega = (5, 15, 25)$.

The figures in this example show that the bargaining set does not predict that ("rational") players will end up at one of the above outcomes. Rather, it seems that the players will look at these outcomes as a *starting point for further bargaining*, leading to outcomes that deviate from the above outcomes by a "second order of magnitude" – hence the title "bargaining set". For example, the players may reason that a two-person coalition is bound to arise, because the difference of 1 is significant enough to cause two players to join and reject the third. Then, each player will be willing to sacrifice a small amount from his quota in order to guarantee his participation in a two-person coalition. Under another variant, player 3, who has more to lose if left alone, may be willing to pay, say 1, to players 2 and 3, in order to "convince" them to form a three-person coalition. This is certainly better for him than to remain alone. For experimental purposes, two conjectures are plausible:

(i) The derivations will be small, and the *average* over many games, ending with formation of the same c.s., will be the bargaining set outcome (up to a "least noticeable difference"). The underlying assumption here is that willingness to shave one's quota will be the same for all parties concerned.

(ii) The tendency to sacrifice will be larger for players having higher quotas, because these players have more to lose if left alone. Thus, the above averages will tend to be more egalitarian than the payments in $\mathcal{M}_1^i$.

We shall discuss the results of some experiments in Section 11. It will be seen that reality exhibits facets deeper than these oversimplified conjectures.

Clearly, the bargaining set $\mathcal{M}_1^i$ contains the core for each c.s.,[19] because at core imputations there are no objections, and a fortiori no justified objections. In general, the bargaining set may contain imputations outside the core.

The core, however, is empty in many cases, so that one advantage of the bargaining set over the core is the following important result:

Theorem 3.5. *For every game* $(N; v)$, *if* $X(\mathcal{B}) \neq \emptyset$, *then* $\mathcal{M}_1^i(\mathcal{B}) \neq \emptyset$.

The original proof of this theorem was given by Davis and Maschler (1963, 1967) for $\mathcal{B} = \{N\}$ and by Peleg (1963a, 1963d, 1967) for an arbitrary c.s. A key result needed in both proofs is the fact that *the relation "a player has a justified objection against another player" is acyclic* (though not necessarily transitive). The proof of the theorem uses the K.K.M. Lemma.[20] The extension of this lemma to a Cartesian product of simplices, needed for the case of a general c.s., uses the Brouwer fixed point theorem.

Maschler and Peleg (1966) gave an algebraic proof of Theorem 3.5. That proof was subsequently simplified by Schmeidler (1969a, 1969b), who invented

---

[19]The *core* $\mathscr{C}(\mathcal{B})$ for a c.s. $\mathcal{B}$ is $\{x \in X(\mathcal{B}): x(S) \geq v(S), \text{ all } S, S \subseteq N\}$ [Aumann and Drèze (1974)].

[20]The Lemma of Knaster, Kuratowski and Mazurkiewicz (1929). See also Kuratowski (1961).

the *nucleolus*[21] in order to exhibit a unique point in the *kernel*.[22] The kernel was known to be a subset of the bargaining set.

Each of the two types of proof has its own merits. The algebraic proofs immediately introduce several related solution concepts and exhibit inclusion relations among them. The analytic proof yields ideas that help in establishing some analogous proofs for the case of games without side payments (see Section 12). It is also shorter if one wants only to prove the nonemptiness of $\mathcal{M}_1^i(\mathcal{B})$.

Another type of proof of this theorem involves dynamic systems. This will be discussed in Section 7.

Undoubtedly, the core is a very important solution concept since it yields itself easily and convincingly to many applications. Nevertheless, it would be a mistake to say that the bargaining set should be considered only if the core is empty. We refer the reader to Maschler (1976), where an economic example of games with a nonempty core is given, yet points in the bargaining set outside the core make more sense intuitively.

It is quite straightforward to show that the bargaining set is covariant with respect to strategic equivalence.[23] Thus it passes one requirement needed in order to earn the title of "a game theoretical solution concept". What can we say about its structure? Maschler (1966) has translated the definition of $\mathcal{M}_1^i$ into a system of weak linear inequalities involving v, connected by the connectives "and" and "or". This shows that $\mathcal{M}_1^i(\mathcal{B})$ *consists of a finite union of compact convex polyhedra* (i.e., *polytopes*).

The fact that the inequalities are weak also shows that *the bargaining set is an upper-semi-continuous function of v*. It *need not* be lower-semi-continuous, as has been shown by Stearns (1968) by means of an example; so the question that now comes to mind is whether the bargaining set admits, at least, a *continuous skeleton*.[24] An affirmative answer was given by Schmeidler (1969a, 1969b) and Kohlberg (1971), who proved by different methods that *the nucleolus for $\mathcal{B}$ is a continuous skeleton*.

Another consequence of the nature of the system of inequalities of Maschler (1966) is that if the characteristic function takes values from an ordered field,[25] then all the vertices of the polyhedra that constitute the bargaining set must have coordinates taken from that ordered field.

---

[21]See Section 5.

[22]See Section 4.

[23]I.e., if $(N; v)$ and $(N; w)$ are two games defined on the same set of players, and if there exists a positive number α and a vector β in $\mathfrak{R}^N$ such that $w(S) = \alpha v(S) + \beta(S)$ for all S, then, for all $\mathcal{B}$, $\mathcal{M}_1^i(N; w; \mathcal{B}) = \alpha \mathcal{M}_1^i(N; v; \mathcal{B}) + \beta$.

[24]I.e., whether a point can be chosen in each $\mathcal{M}_1^i(N; v; \mathcal{B})$, for the class of games over a fixed set of players, and a fixed coalition structure, that varies continuously with v.

[25]Say, the field of rational numbers.

Something these inequalities *do not* provide is an easy way to compute the bargaining set for arbitrary "generic" games. One reason is that these inequalities involve knowing all the minimal balanced collections[26] over $n - 1$ players, needed for merely listing all the inequalities for[27] $\mathcal{M}_1^i(\{N\})$. Another, more fundamental reason is that the amount of computations is enormous. For example, to compute by these inequalities the bargaining set $\mathcal{M}_1^i(\{N\})$, where $N = 1234$, one has to inspect 150^{12} systems, each consisting of 41 linear inequalities connected by the connective "and". True – many of these systems have no solution and others yield only imputations already given by others,[28] but at present there is no known way to tell a computer what systems could safely be ignored.[29]

It is interesting to note that it takes microseconds if one wants merely to know whether a certain imputation belongs to $\mathcal{M}_1^i(\{N\})$, $N = 1234$. There are only 197 easy inequalities to check! Thus, for any particular game with v taking rational values, one can try to determine the maximal denominator that a vertex coordinate can take and then capture the vertices of the bargaining set by a grid search.[30] Of course, one then has to determine which of the vertices (and other points captured by the grid) belongs to what polytope, and this may require additional analysis, and even brute force, if the bargaining set happens to have many polytopes.

I have occasionally heard the argument that the bargaining set is useless because it is hard to compute for a generic game. Admittedly I am biased, but nevertheless let me offer some counter-arguments.

(1) If the players really want that kind of stability which is reflected in the bargaining set, namely to be immune against objections, what is the sense in offering them other solutions? It is as if you want to buy a car and the salesman offers you *Encyclopedia Britannica* because he is unable to deliver the car.

(2) Should we discard the concept of "equilibrium" for noncooperative games simply because equilibrium points cannot be computed for any medium size generic game? With such arguments we can do away with almost any solution concept in game theory.[31]

(3) For important classes of games it is actually possible to compute the bargaining set, or at least parts of it, without referring to the inequalities described above. This is because the characteristic functions of these games

---

[26]See Shapley (1967).

[27]One needs less for other coalition structures.

[28]If $N = 123$ there are 3^6 such systems, yet $\mathcal{M}_1^i(N)$ consists of one polytope.

[29]When computing the bargaining set manually, one often sees many short cuts. The present author has computed the bargaining set of several four- and five-person games in a reasonable amount of time, but he does not know how to instruct a computer to "perceive" such short cuts.

[30]See Aumann, Peleg and Rabinowitz (1965), where such a procedure was employed in the case of the kernel.

[31]It takes $2^n - 1$ storing steps just to store a generic n-person game on a computer.

have special properties. Examples of this kind will be discussed in Sections 9 and 10.

(4) It would be nice, of course, if one could compute the bargaining set for any game. But how often does the need really arise to do so for a generic game? Was the core, for example, often computed for such games?

(5) Properties of the bargaining set frequently shed interesting insight in applications, even if the bargaining set is not computed in full, or in part.

(6) I have not lost hope that methods will eventually be found which will enable one to know exactly what polytopes of the bargaining set should be computed. This may reduced the computation time considerably. We did have such luck in connection with the kernel (Section 4).

So the players have the bargaining set at their disposal – computed and represented to them. Can we say something about the coalition structures that may form? How do coalition structures come about anyway? There may be several views on this subject. Sometimes coalition structures come about for "personal reasons" which are consciously independent of the characteristic function (yet, payoff division is still based on options outside these coalition structures). Sometimes they come about simply because players find them beneficial. We refer the reader to Aumann and Drèze (1974) for an excellent discussion of some of the issues involved. It seems to me, however, that from a normative point of view the issue of coalition formation is far from trivial. For example, here is a suggestion of Shenoy (1979).[32] A c.s. $\mathscr{B}_1$ should *not* survive if there is another c.s. $\mathscr{B}_2$, and a B in $\mathscr{B}_2$, such that for every y in $\mathscr{M}_1^i(\mathscr{B}_1)$ there is an x in $\mathscr{M}_1^i(\mathscr{B}_2)$, satisfying $x_i > y_i$ for each i in B. Under such circumstances we say that $\mathscr{B}_2$ *dominates* $\mathscr{B}_1$. Shenoy then proves that for every three-person game in which $X(\mathscr{B}) \neq \emptyset$ for every $\mathscr{B}$, there are undominated coalition structures. Whether this is true for larger games remains open. Shenoy's suggestion yields plausible surviving coalition structures for three-person games. For larger games it can be criticized, for example, on the ground that B is perhaps counting too heavily on the members of $N \backslash B$ to agree to form $\mathscr{B}_2$ and share specified proceeds in the bargaining set.

To be able to handle coalition formation normatively, one has to take into account that coalitions need not form simultaneously. Sometimes players should rush to form coalitions. In other cases it is beneficial to wait until others form coalitions. In Section 11 we shall encounter cases where real players consciously considered such possibilities. Come to think, perhaps it is beneficial for players to pay some money to other players in order to encourage them to form certain coalitions at certain stages of the process of coalition formation. These aspects of coalition formation certainly deserve careful study.

---

[32]Shenoy puts this and other suggestions in a framework of a general theory on coalition formation.

4. The kernel

The kernel was introduced as an auxiliary solution concept, the main task of which was to illuminate properties of the bargaining set and to compute at least part of this set. No intuitive meaning was attached to it.[33] Nevertheless, it was soon discovered that the kernel had many interesting mathematical properties that reflected in various ways the structure of the game. Gradually it became an important solution concept in its own right. Its intuitive meaning became clearer only at a later stage. The present section will follow this historical path.

Definition 4.1. Let x be an imputation [a preimputation] in a game $(N; v)$ for an arbitrary c.s. The *excess $e(S, x)$ of a coalition S at x* is $v(S) - x(S)$ if $S \neq \emptyset$, and 0 if $S = \emptyset$.

Thus, $e(S, x)$ represents the total gain (loss, if negative) that members of S will have if they depart from x and form their own coalition. Note that if $x \in X^0(\mathcal{B})$, then $e(B, x) = 0$ whenever $B \in \mathcal{B}$.

Definition 4.2. Let x be an imputation [a preimputation] in a game $(N; v)$. Let k and l be two distinct players in N. The *surplus of k against l at x* is

$$s_{k,l}(x) := \max_{\substack{S \ni k \\ S \not\ni l}} e(S, x). \tag{4.1}$$

Thus, $s_{k,l}(x)$ represents the most player k can hope to gain (the least to lose, if negative) if he departs from x and forms a coalition that does not need the consent of l, assuming that the other members of this coalition are happy with their payments in x.

Definition 4.3. Let $(N; v)$ be a game and let $\mathcal{B}$ be a coalition structure. The *kernel*[34] $\mathcal{K}(\mathcal{B})$ *for $\mathcal{B}$* is

$$\mathcal{K}(\mathcal{B}) := \{x \in X(\mathcal{B}): s_{k,l}(x) > s_{l,k}(x) \Rightarrow x_l = v(l),$$
$$\text{all } k, l \in B \in \mathcal{B}, k \neq l\}. \tag{4.2}$$

The *prekernel*[35] $\mathcal{PK}(\mathcal{B})$ *for $\mathcal{B}$* is

$$\mathcal{PK}(\mathcal{B}) := \{x \in X^0(\mathcal{B}): s_{k,l}(x) = s_{l,k}(x), \text{ all } k, l \in B \in \mathcal{B}, k \neq l\}. \tag{4.3}$$

---

[33]Except if one was willing to embark on the obscure notion of interpersonal comparison of utilities.
[34]Davis and Maschler (1965).
[35]Maschler, Peleg and Shapley (1972, 1979).

We shall often write $\mathcal{K}$ instead of $\mathcal{K}(\{N\})$ and call it *the kernel of the game*. A similar shortcut will be adopted for the prekernel. (Sometimes we shall use $\mathcal{K}$ to mean the *union* of the $\mathcal{K}(\mathcal{B})$'s, similarly for $\mathcal{P}\mathcal{K}$.)

Note that

$$\mathcal{P}\mathcal{K}(\mathcal{B}) \cap X(\mathcal{B}) \subseteq \mathcal{K}(\mathcal{B}) , \tag{4.4}$$

and indeed it may well happen that $\mathcal{P}\mathcal{K}$ contains payoffs which are not individually rational.

Suppose that $s_{k,l} > s_{l,k}$ at x, then player k might request player l to transfer some amount to him on the ground that, in case of departure, he hopes to gain more than [lose less than] l.[36] From the point of view of the prekernel [kernel] player l should yield [unless he is already driven to his "minimum" $v(l)$] so that such an x is not "balanced". This argument of "fair share" is reasonable only if one can assume (and make sense of it) that the utilities of all players to the same amounts of money are interpersonally the same.[37] Another way to make sense out of this reasoning is to assume that it is imposed on the players by some "big brother" who cares only about money and pays no attention to the utilities of the players towards this money.[38] Both these interpretations, as well as the decision to base every thing on "best hopes", are not too attractive.

Theorem 4.4.[39]   *For every game* $(N; v)$, $\mathcal{P}\mathcal{K}(\mathcal{B}) \neq \emptyset$. *If* $x(\mathcal{B}) \neq \emptyset$, *then also* $\mathcal{K}(\mathcal{B}) \neq \emptyset$. *The last set is a subset of* $\mathcal{M}_1^i(\mathcal{B})$.

The various nonemptiness proofs use techniques similar to the proofs of Theorem 3.5. Note that the relation $k > l$ at x, which means $s_{k,l}(x) > s_{l,k}(x)$ and $x_l > v(l)$, is transitive – not merely acyclic. The proof that the kernel is a subset of the bargaining set follows from the fact that if $l \geqslant k$ then k has no justified objection against l.

It easily follows from the definition that the kernel [prekernel] is covariant with respect to strategic equivalence. It also follows that both are finite unions of polytopes. On the face of it, it is almost as difficult to compute the kernel as to compute the bargaining set; nevertheless Aumann, Peleg and Rabinowitz (1965) and Aumann, Rabinowitz and Schmeidler (1966) succeeded in comput-

---

[36]These departures are virtual: it may well happen that to gain the surplus, both players need intersecting coalitions.

[37]This is because $s_{k,l}$ is measured in k's utility and $s_{l,k}$ is measured in l's utility.

[38]Perhaps this is not too far fetched: If you and I find a $100 bill and go to an arbitrator to decide how to split it, I am quite sure that most arbitrators will not care about our utilities for money and will suggest that we share the dollars equally.

[39]Davis and Maschler (1965), Maschler and Peleg (1966), Maschler, Peleg and Shapley (1979), Schmeidler (1969a, 1969b).

ing the kernel for many simple games.[40] Observing the results of these computations enabled Maschler and Peleg (1966, 1967) to analyze the structure of the polytopes which compose the kernel and to reduce considerably the number of systems of inequalities that need be considered in order to compute the kernel.[41] The amount of computation can further be reduced if the characteristic function possesses certain "order relations". We refer the reader to these rather technical papers, where he will also find several examples of kernels which were computed manually. From these examples we wish to report here the following interesting game:

Example 4.5. The seven-person projective game.[42] This game is given by:
$$v(124) = v(235) = v(346) = v(457) = v(561) = v(672) = v(713) = 1, \qquad v(S) = 1$$
whenever S is a superset of the above seven coalitions and $v(S) = 0$, otherwise. The kernel of this game, for the grand coalition, consists of seven straight-line segments, all emanating from the payoff: $(1/7, 1/7, \ldots, 1/7)$ and ending at a point, where a minimal winning coalition shares its value equally among its members. Thus, in this game, the kernel reflects a confrontation between two "forces" that may exist: one, in which the players say "we are all in a similar situation so let us share the proceeds equally". The other, when members of a minimal winning coalition say "the hell with the others, let us take the 1 and share it equally among ourselves".

We see from this example that the kernel may contain more than one polytope.[43] How big can the dimension of these polytopes be? The answer is given by the following:

Theorem 4.6.[44] *Let* $\mathcal{B} = \{B_1, B_2, \ldots, B_m\}$ *be a coalition structure over a set of players N. Denote* $\|\mathcal{B}\| := \max_{1 \leq j \leq m} |B_j|$, *where* $|B_j|$ *is the cardinality of* B_j. *The maximal dimension of a polytope in* $\mathcal{K}(\mathcal{B})$, *taken over the class of all games on N, is equal to*[45]

---

[40]Games for which the characteristic function takes only the values 0 and 1 are called *simple games*.

[41]Based on the above papers, Kopelowitz computed the kernel of all six- and seven-person zero-sum weighted majority games and all six-person superadditive weighted majority games [taken from Isbell's (1959) list.] The average computation time was 1 second for the six-person games and 6–7 seconds for the seven-person games. However, some seven-person games took 40–60 seconds. Based on the above papers, Beharav (1983) has constructed a computer program for finding the kernels of up to five-person games.

[42]Introduced in Von Neumann and Morgenstern (1953).

[43]It need not even be connected [see Kopelowitz (1967) and Stearns (1968)].

[44]Maschler and Peleg (1966).

[45]The term 1/2 is needed in order to get a formula that works also when $\|\mathcal{B}\| = 1$. [$\cdots$] means: the integer part of "$\cdots$".

$$n - [\log_2(\|\mathcal{B}\| - \tfrac{1}{2})] - m - 1 . \tag{4.5}$$

The above formula is sharp; namely, for every n there are games for which a polytope in their kernel attains this dimension.

This strange formula is derived by analyzing the inequalities which determine the polytopes of which the kernel is constituted.

Example 4.5 indicates that the kernel is sensitive to various symmetries that may exist in the game. The following results substantiate this claim.

Definition 4.7. A player k is said to be *at least as desirable* as a player l in a 0-normalized game $(N; v)$, if

$$v(S \cup \{k\}) \ge v(S \cup \{l\}) , \quad \text{whenever } k, l \notin S . \tag{4.6}$$

They are called *symmetric* if each one is more desirable than the other.

Theorem 4.8.[46]   *Let $(N; v)$ be a zero-normalized game. Let $x \in \mathcal{K}(\mathcal{B})$ and $k, l \in B \in \mathcal{B}$. If k is more desirable than l, then $x_k \ge x_l$. In particular, if k and l are symmetric players, then $x_k = x_l$.*

An immediate consequence of this theorem is that *each payoff vector in the kernel of a weighted majority game*[47] *(weakly) preserves the order of the weights.*

Theorem 4.9.[48]   *The kernel [prekernel] for the grand coalition is reasonable in the sense of Milnor;*[49] *i.e., if $x \in \mathcal{K}$, or $x \in \mathcal{P}\mathcal{K}$, then*

$$x_i \le \max_{S: S \ni i} [v(S) - v(S \setminus \{i\})] , \quad \text{all } i \in N . \tag{4.7}$$

In particular, a dummy[50] *i_0 receives $v(i_0)$ at each payoff of $\mathcal{K}$.*

A somewhat similar concept is that of a *pairwise reasonable* preimputation x, which means that for every pair of players i and j in N, i's payoff should not exceed j's payoff by more than the greatest amount that i's contribution to any

---

[46]Maschler and Peleg (1966).
[47]A weighted majority game $[q; w_1, w_2, \ldots, w_n]$ is defined by $v(S) = 1$ if $w(S) \ge q$ and otherwise $v(S) = 0$. In most applications one requires that $\tfrac{1}{2}w(N) < q < w(N)$.
[48]Wesley (1971). See also Maschler, Peleg and Shapley (1979).
[49]Milnor (1952). See also Luce and Raiffa (1957).
[50]I.e., a player i_0 for which $v(S) - v(S \setminus \{i_0\}) = v(i_0)$, all $S, S \ni i_0$.

coalition exceeds j's contribution to that same coalition, namely[51]

$$x_i - x_j \leq \max_{S \subseteq N \setminus \{i,j\}} [v(S \cup \{i\}) - v(S \cup \{j\})] . \tag{4.8}$$

Theorem 4.10. *The kernel is contained in the set of pairwise reasonable imputations.*[52]

Recently, there has been some interesting research whose aim was to find what parts of the kernel are contained in the set[53]

$$\mathcal{L} := \left\{ x \in X(N) : x_i \geq m_i[v] := \min_{\substack{S \ni i \\ S \neq \{i\}}} [v(S) - v(S \setminus \{i\})], \text{ all } i \in N \right\} . \tag{4.9}$$

This research started with Kikuta (1976) and culminated with the following result of Funaki (1986):

Theorem 4.11. *If $(N; v)$ is a zero-normalized monotonic game, then*

$$\mathcal{K} \subseteq \mathcal{L} \cup \mathcal{C}^0 , \tag{4.10}$$

where $\mathcal{C}^0$ is the interior of the core. In particular, if $\mathcal{L} = \emptyset$, i.e., $\Sigma_{i \in N} m_i[v] > v(N)$, then $\mathcal{K} \subset \mathcal{C}^0$, and if the core is empty, or has no interior points, then $\mathcal{K} \subseteq \mathcal{L}$.

A most useful property of the kernel (for the grand coalition) is the fact that for "ordinary" games it can be defined by means of equations instead of inequalities. This follows from

Theorem 4.12.[54] *For zero-monotonic games,*

$$\mathcal{K}(\{N\}) = \mathcal{P}\mathcal{K}(\{N\}) . \tag{4.11}$$

Inside the core there is no need to require zero-monotonicity:

[51] This definition, as well as the result that follows, is due to Shapley (private written communication at the beginning of 1980).

[52] The Shapley value is pairwise reasonable too.

[53] The requirement $S \neq \{i\}$ is needed to allow for some imputations to be outside of $\mathcal{L}$. It also makes sense intuitively.

[54] Maschler and Peleg (1967). Actually the theorem is proved there for a somewhat larger class of games. The class is further extended in Maschler, Peleg and Shapley (1979). The theorem remains correct also for the kernel for a coalition structure if the game is also *decomposable* for this c.s. [i.e., for all S, $v(S) = \Sigma_{B \in \mathcal{B}} v(S \cap B)$] [see Chang (1991)]. So far, no other conditions were found which guarantee that $\mathcal{K}(\mathcal{B}) = \mathcal{P}\mathcal{K}(\mathcal{B})$. Zero-monotonicity is not sufficient.

Theorem 4.13.[55]   *For every game,*

$$\mathcal{K}(\mathcal{B}) \cap \mathcal{C}(\mathcal{B}) = \mathcal{P}\mathcal{K}(\mathcal{B}) \cap \mathcal{C}(\mathcal{B}) . \tag{4.12}$$

These results have an interesting geometric interpretation. First take the core $\mathcal{C}(\mathcal{B})$ and suppose that it is not empty. It is a polytope in $\mathfrak{R}^n$. Through each point x in this polytope pass line segments of the form

$$R_{k,l}(x) := \{ y = x + \alpha \cdot e^k - \alpha \cdot e^l : y \in \mathcal{C}(\mathcal{B}) \} , \tag{4.13}$$

where e^t denotes the unit vector in the t direction. Pass all lines $R_{k,l}(x)$, where $k, l \in B \in \mathcal{B}$. It turns out that $\max\{\alpha : y \in \mathcal{C}(\mathcal{B})\} = -s_{l,k}(x)$ and $\min\{\alpha : y \in \mathcal{C}(\mathcal{B})\} = s_{k,l}(x)$. This brings us to the following characterization:

Theorem 4.14.   *The payoff vector x belongs to $\mathcal{K}(\mathcal{B}) \cap \mathcal{C}(\mathcal{B})$ iff all the above straight-line segments are bisected at x.*

The above result holds if one replaces "core" by "(strong) ϵ-core"[56] in the following cases:

   (i) The game is zero-monotonic and $\mathcal{B} = \{N\}$.
   (ii) The game is zero-monotonic and decomposable for $\mathcal{B}$.
   (iii) "Kernel" is replaced by "prekernel".

If the game is not of the above type, there still exists a geometric characterization of the kernel intersected with a nonempty ϵ-core, but it is somewhat more complicated.[57] We refer the reader to Maschler, Peleg and Shaply (1979) and to Chang (1991), where the above results are elaborated.

We can now provide a better intuitive interpretation for the kernel [prekernel]. The line segment $R_{k,l}(x)$ can be regarded as a *bargaining range* between k and l, at x: If player k presses player l for an amount greater than $\max\{\alpha : y \in \mathcal{C}(\mathcal{B})\}$, then l will be able to find a coalition which can block k's demand. Similarly, for the other end of $R_{k,l}(x)$. The middle point of $R_{k,l}(x)$ represents a situation in which both players are *symmetric* with respect to the bargaining range. Thus, $\mathcal{K}(\mathcal{B}) \cap \mathcal{C}(\mathcal{B})$ *is the set of payoff vectors for which every pair of players in the same coalition of the c.s. is situated symmetrically with respect to its bargaining range.* This can be regarded as an intuitive interpretation of $\mathcal{K} \cap \mathcal{C}$, an interpretation which does not directly employ interpersonal comparison of utilities and does not base its arguments on "best

---

[55]See Maschler, Peleg and Shapley (1979) for the grand coalition, and Chang (1991) for a general c.s.
  [56]I.e., $\{x \in X(\mathcal{B}) : e(S, x) \leqslant \epsilon$, all $S, S \not\subseteq \mathcal{B}, S \neq \emptyset\}$ [see Shapley and Shubik (1963, 1966)].
  [57]I.e., for an x on the boundary of $X(\mathcal{B})$, $x_k = v(k)$, $x_l > v(l)$, the part of $R_{k,l}$ outside $X(\mathcal{B})$ is allowed to be longer than the part inside.

hopes in case of virtual defections". *Points in the intersection of the kernel and the core satisfy a kind of fair division scheme.*[58]

A similar interpretation can be provided for the intersection with the (strong) ϵ-core, for the cases stated in items (i)–(iii) above. If ϵ is large enough, the intersection will contain both the kernel and the prekernel. Note, however, that the bargaining range with respect to the ϵ-core is less convincing than with respect to the core, because the ϵ-core is a less intuitive solution concept: it requires the (unrealistic) assumption that there is a cost of departure from x which is the same for all coalitions. Thus, although we have here an intuitive interpretation for the intersection of the kernel [prekernel] with the ϵ-core as a fair division scheme, we must realize that it is quite convincing if $\epsilon = 0$, but becomes less so as ϵ becomes larger.

Theorem 4.14 can be phrased in another way. For a fixed pair (k, l) and all core points x in $\mathcal{C}(\mathcal{B})$, take the middle points of $R_{k,l}(x)$. These constitute a hypersurface in $\mathcal{C}(\mathcal{B})$. The intersection of these hypersurfaces for all pairs $k, l \in B \in \mathcal{B}, k \neq l$, is precisely $\mathcal{K}(\mathcal{B}) \cap \mathcal{C}(\mathcal{B})$. This shows that *the intersection of the kernel [prekernel] with the core is a locus of the core for every coalition structure.* A similar result holds for each ϵ-core.[59] This implies that *if two games have the same imputation [preimputation] space and the same ϵ-core for a given ϵ and a given c.s., then the part of the kernel [prekernel] in the ϵ-core is identical in both games. If ϵ is big enough, then both games have the same kernel [prekernel].* This can be of help in finding the kernel of one game from the kernel of another game.[60]

Any solution concept – and in particular a solution concept like the kernel, the definition of which is not intuitively obvious – must be tested if it generates reasonable outcomes for actual games. In all games for which I know $\mathcal{K}(N)$, it yields reasonable outcomes, or at least outcomes which are not intuitively inferior to outcomes of other solution concepts. This is not the case for $\mathcal{K}(\mathcal{B})$, for c.s. other than $\{N\}$. The following is such an example:

Example 4.15. $N = 12345, v(12) = v(13) = v(14) = v(15) = v(2345) = 100, v(S) = 0$, otherwise.

This game was introduced in Davis and Maschler (1965) under the title *Me and My Aunt*, accompanied by an appropriate scenario. Its superadditive cover

---

[58]Of course, any requirement based on symmetry is enhanced if utilities are, in some sense, interpersonally compared. Similarly, any argument based on symmetry can be attacked on the ground that it does not take into account the possibility that the utility units might be interpersonally unequal.

[59]This locus, for the kernel, is somewhat more complicated when $\mathcal{K} \neq \mathcal{PK}$.

[60]See Maschler, Peleg and Shapley (1979), where this approach was used in order to extend the scope of Theorem 4.12.

is the weighted majority game $[3, 1, 1, 1, 1]$. The reader will find in that paper a lively correspondence with several distinguished game theorists concerning the question: What should the outcome be if the c.s. is $\{12, 3, 4, 5\}$? This, and more general games, called *apex games*,[61] were tested experimentally in various laboratories.[62] The kernel for the c.s. $\{12345\}$ is $(3/7, 1/7, 1/7, 1/7, 1/7)$, which is quite reasonable. The bargaining set for the c.s. $\{12, 3, 4, 5\}$ is $\{(x, 100 - x, 0, 0, 0, 0): 50 \leqslant x \leqslant 75\}$ and this is also reasonable. The kernel for this c.s., however, seems to be the wrong end of the bargaining set, namely $\{(50, 50, 0, 0, 0)\}$. Undoubtedly, this is not an outcome one would expect if 1 and 2 form a coalition. Why do we feel that player 1 is stronger? Of course it is easier for 1 to get into a coalition, because he has to convince only one out of four players, whereas the others have to get together to form their own coalition. But this is not the "fault" of the kernel because, if "time is money", communication barriers should perhaps enter the construction of the characteristic function. It seems to me, however, that player 1 has another source of strength which is not captured by the kernel. Even if coalition 2345 forms, he can control who should get more among them. If he offers, say, 40 to player 2, then the others must come out with, say, $(0, 43, 19, 19, 19)$ in order to beat this offer. In fact, precisely this behavior has often been observed in the experiments. It seems to me that this power of player 1 should be translated into profits when, say, players 1 and 2 decide to form a coalition.

5. The nucleolus

The definition of the nucleolus of a cooperative game, though mathematically attractive, is quite complicated. It is amazing, therefore, that it has already been applied successfully to several unrelated topics in the social sciences. This section will provide the definition and some of the basic properties. Applications will be discussed in other sections.

We consider again a game $(N; v)$ with side payments and a closed set X of vectors in $\mathfrak{R}^N$. For each x in X, we define a vector $\theta(x)$ to be

$$\theta(x) := (e(S_1, x), e(S_2, x), \ldots, e(S_{2^n}, x)), \tag{5.1}$$

---

[61]Namely, games that satisfy $v(1i) = q_1 + q_i$ for $i = 2, 3, \ldots, n$, $v(2, 3, \ldots, n) = q_1 + q_2 + \cdots + q_n$, and $v(S) = 0$, otherwise [Horowitz (1973)].

[62]Selten and Schuster (1968), Horowitz and Rapoport (1974), Albers (1978), Rapoport, Kahan and Wallsten (1978), Rapoport, Stein and Burkheimer (1979), Kahan and Rapoport (1979), Funk, Rapoport and Kahan (1980). A survey of these experiments can be found in Kahan and Rapoport (1984). Let us say at once that the bargaining set fares quite well in most of these experiments, but the kernel comes out poorly.

where the various excesses of all coalitions are arranged in decreasing order. Note that the order of the coalitions which appear in (5.1) depends on x and, in general, it is not uniquely determined. However, the components of $\theta(x)$ are well defined and vary continuously with x.

We say that $\theta(x)$ is *lexicographically smaller* than $\theta(y)$, and denote this by $\theta(x) < \theta(y)$ if there exists a positive integer q such that $\theta_i(x) = \theta_i(y)$ whenever $i < q$ and $\theta_q(x) < \theta_q(y)$.

Definition 5.1.[63] Let X be an arbitrary nonempty closed set in $\mathfrak{R}^N$. The *nucleolus* of X – denoted $\mathcal{N}(X)$, or $\mathcal{N}(N; v; X)$ – is the set of vectors in X whose θ's are lexicographically least; i.e.,

$$\mathcal{N}(X) := \{x \in X : \theta(x) \leqslant \theta(y), \text{ all } y \in X\} . \tag{5.2}$$

If $X = X(\{N\})$, the nucleolus is called *the nucleolus of the game*. If $X = X(\mathcal{B})$ it is called *the nucleolus of the game for the c.s.* $\mathcal{B}$. If $X = X^0(\{N\})[X^0(\mathcal{B})]$ it is called *the prenucleolus of the game* [*for the c.s.* $\mathcal{B}$]. We denote the last four cases by $\mathcal{N}$, $\mathcal{N}(\mathcal{B})$, $\mathcal{PN}$, and $\mathcal{PN}(\mathcal{B})$, respectively.

Mathematicians will certainly admire the above definition, but can it be given a convincing intuitive meaning? Here is an attempt [Maschler, Peleg and Shapley (1979)]. Consider an arbitrator, whom the players ask to decide how to share $v(N)$. The arbitrator may regard the excess of a coalition as a measure of dissatisfaction and he may be eager to decrease the excesses of the various coalitions as much as possible. This will also increase "stability".[64] He will then look for payoffs in which the highest excess is as low as possible. If there is more than one such payoff, he will tell the highest-excess coalitions: "I have helped you as much as I could, but I can still help other coalitions." He will then proceed to choose outcomes for which the second highest excess is minimal, and so on. Obviously, such "justification" raises more questions than it answers. What is more "stable", a situation in which a few coalitions of highest excess have it as low as possible, or one where such coalitions have a slightly higher excess, but the excesses of many other coalitions are substantially lowered? It is the lexicographic order that is hard to motivate. And then, there may be other objections to the goals of the above arbitrator: Why should he take into consideration only excesses and not, say, sizes of coalitions? Fortunately, the prenucleolus can be justified by much more intuitive axioms (Section 6); moreover, it is also justified by some attractive properties.

---

[63]Schmeidler (1969a, 1969b).
[64]A coalition with a high positive excess will gain a lot by departure and, even if the excess is negative, defection is still less liable if the excess is smaller.

Theorem 5.2.[65] *If X is compact, or even if it is merely closed, but $x(N) \leq$ constant for all $x \in X$, then the nucleolus $\mathcal{N}(X)$ is not empty. If, in addition, X is convex, it consists of a unique point (called the nucleolus point). For games on a fixed set of players and a fixed c.s., it varies continuously with the characteristic function v. The nucleolus point for $\mathcal{B}$ belongs to the $\mathcal{K}(\mathcal{B})$.*

Nonemptiness follows from considerations of compactness.[66] Assuming that x is not in the kernel, one can lower $\theta(x)$ lexicographically by transferring some money from one player to another, getting a contradiction thereby. Proving continuity is harder. Schmeidler proves it by giving an alternative definition for the nucleolus from which continuity follows easily and then showing that the two definitions are indeed equivalent. In addition to Schmeidler's, there is another proof by Kohlberg (1971), based on the following very interesting and important characterization of the nucleolus:

Theorem 5.3. *Let $\mathcal{B}_1$, $\mathcal{B}_2$, ... be the sets of coalitions of highest excess at x, second highest, third highest, etc. Let $\mathcal{B}_0$ be the set of single-person coalitions i, satisfying $x_i = v(i)$. Let $\mathcal{D}_t = \mathcal{B}_1 \cup \mathcal{B}_2 \cup \cdots \cup \mathcal{B}_t$. A necessary and sufficient condition for x to be the prenucleolus point for $\{N\}$ is that each $\mathcal{D}_t$ is a balanced collection. A necessary and sufficient condition for x to be the nucleolus point for N is that each $\mathcal{B}_0 \cup \mathcal{D}_t$ is weakly balanced,[67] having positive balancing coefficients on elements of $\mathcal{D}_t$.*

Owen (1977a) has generalized this result for a general c.s. Unfortunately, his characterization is hard to verify. He also provided necessary, but not sufficient, conditions as well as sufficient, but not necessary, conditions. These conditions have the merit of lending themselves more easily to verification. The proofs of Kohlberg's and Owen's results follow from extensions of Farkas' Lemma (1902).

Being a point in the kernel, the nucleolus point has all the nice properties of the kernel points. Since, obviously, it is a solution concept that does not depend on the names of the players,[68] it preserves all the symmetries of the game. The following is a neat formulation of this fact, due to Peleg.[69] It is useful in many cases for the computation of the nucleolus:

Definition 5.4. *A symmetry of a game $(N; v)$ for a c.s. $\mathcal{B}$ is a permutation $\pi: N \rightarrow N$ which keeps the members of $\mathcal{B}$ and leaves the game invariant; i.e., satisfies*

[65]Schmeidler (1969a, 1969b).
[66]If X is not bounded, one has to prove first that its nucleolus is located inside a compact subset of X.
[67]I.e., some balancing coefficients are allowed to vanish.
[68]This property is called *anonymity*.
[69]Oral communication.

$$\pi(B) = B , \qquad \text{for all } B \in \mathcal{B} , \tag{5.3}$$

$$v(\pi(S)) = v(S) , \quad \text{for all } S \subseteq N . \tag{5.4}$$

Corollary 5.5. *If π is a symmetry of a game $(N; v)$, for a given $\mathcal{B}$, and if $X(\mathcal{B}) \neq \emptyset$, then the nucleolus point ν for $\mathcal{B}$ satisfies $\nu_{\pi(i)} = \nu_i$, all $i \in N$.*

Many scholars who applied the nucleolus did so because they wanted an outcome which always belongs to the core when the core is not empty. In fact, it follows easily from its definition that *the nucleolus point [for each c.s.] belongs to the strong ϵ core [for that c.s.], whenever this set is not empty* [Schmeidler (1969a, 1969b)]. In this respect, the nucleolus has an "advantage" over the Shapley value. Since the nucleolus is not empty even if the core is empty, it can be stated picturesquely that *the nucleolus is the location of the "latent position" of the core* [Shubik (1983, p. 340)]. But where in the core does the nucleolus lie when the core is *not* empty? The answer is that, unlike the intersection of the kernel with the core, the location cannot be determined solely by the core. Maschler, Peleg and Shapley (1979) provide an example of two zero-monotonic games having the same imputation space and the same core, and yet they have different nucleolus points. Thus, the nucleolus is not a locus of the core. The following is an open problem. Can one characterize the set of all nucleolus points of all games in the class of games having a given imputation space and a given core? As a first clue to this problem, the reader is referred to the above [Maschler, Peleg and Shapley (1979)], where a sequence of geometric constructions is given which leads one to the nucleolus point. The process is too long to describe here; let us say only that its steps involve moving certain hyperplanes $x(S) = v(S) + \text{constant}$ at equal l_1-speeds. During these moves, certain hyperplanes, originally outside the core, may enter inside an ϵ-core which resulted from the movements of the hyperplanes that bounded the core. This explains why the location of the nucleolus is not a locus of the core.

We have stated above a property according to which the nucleolus has an advantage over the Shapley value. Here is another property, according to which the Shapley value has the advantage. This is *monotonicity*:

Definition 5.6. A one-point solution concept is called *monotonic* if, whenever one increases the worth of $v(N)$, without changing the worth of any other coalition, the solution specifies that *no player* gets less in the modified game.[70]

Megiddo (1974c) provided the following example of a nine-person game:

---

[70]A set-valued solution concept is called *monotonic* if, for every x in the solution of the original game, there is a y in the solution of the modified game, satisfying $y \geq x$.

Example. Let $N = \{1, 2, \ldots, 9\}$ and let $x = (1, 1, 1, 2, 2, 2, 1, 1, 1)$. Let $v(S) = 6$ for $S \in \{123, 14, 24, 34, 15, 25, 35, 789\}$, $v(S) = 9$ for $S \in \{12367, 12368, 12369, 456\}$, $v(N) = 12$, and $v(S) = \Sigma_{i \in S} x_i - 1$, otherwise. Let $w(N) = v(N) + 1$ but $w(S) = v(S)$, otherwise. The nucleolus point of $(N; v)$ is x and the nucleolus point of $(N; w)$ is $(1\frac{1}{9}, 1\frac{1}{9}, 1\frac{1}{9}, 2\frac{2}{9}, 2\frac{2}{9}, 1\frac{8}{9}, 1\frac{1}{9}, 1\frac{1}{9}, 1\frac{1}{9})$. Thus, in spite of the fact that all coalitions but N stay put, and there is more to share in $(N; w)$, player 6 gets less.[71] This is certainly an undesirable feature, and it bothered some people. One has the feeling that in any "fair" outcome *all* players should benefit if $v(N)$ increases and other coalitions stay put. For that reason, there was a suggestion [Young, Okada and Hashimoto (1982)] to use the *per-capita nucleolus*, which yields a monotonic one-point outcome in the core for games with a nonempty core.[72] This is not going to be of much help, because even the per-capita nucleolus does not satisfy a slightly stronger, but not less intuitive, *coalitional monotonicity property*.

Definition 5.7.[73] A one-point solution ϕ is called *coalitionally monotonic* if for every pair of games $(N; v)$ and $(N; w)$, satisfying

$$v(T) > w(T), \quad \text{for some subset } T \text{ of } N,$$
$$v(S) = w(S), \quad \text{for all } S, \, S \neq T, \tag{5.5}$$

it follows that

$$\phi_i[v] \geq \phi_i[w], \quad \text{for all } i \text{ in } T. \tag{5.6}$$

Surprisingly, Young (1985a) proves that *for the class of games with nonempty core there does not exist a one-point coalitionally monotonic solution concept which always lies in the core*. He proves it by exhibiting two games with a one-point core, one of which results from the other by an increase in the worth of coalitions containing a player, yet the core payment to that player decreases. There is no escape from this fact: if you want a unique outcome in the core you must face some undesirable nonmonotonicity consequences. On the other hand, if you feel that monotonicity is essential, say, because it "provides incentives" if imposed on a society [Young (1985a)], then you should sometimes discard the core, and the nucleolus is not a solution concept that you should recommend. Note that the Shapley value is a coalitionally monotonic solution concept.

Recently, Zhou (1991) proved that the nucleolus is monotonic in a weaker

---

[71] Moreover, for every payoff y in $\mathcal{M}_1^i(N; w)$, and therefore also in $\mathcal{K}(N; w)$, one of the players must get less than in x.

[72] The per-capita nucleolus is defined in the same way as the nucleolus, except that the per-capita excesses $[v(S) - x(S)]/|S|$ are taken instead of excesses for $S \neq \emptyset$. (See Section 8.)

[73] Young (1985a). He calls ϕ, in this case, "monotonic".

sense: if one increases the worth of exactly one coalition, the *total payoff* to its members does not decrease.

We shall conclude this section with a brief discussion on the possibility of computing the nucleolus. If one considers a "generic" game, the first difficulty involves simply listing the characteristic function. It must be prescribed by the $2^n - 1$ numbers $v(S)$. This limitation already restricts one to small n's. Having listed the game, one now faces the problem of computing the nucleolus. One method, suggested by Peleg [see Kopelowitz (1967)], "translates" the definition of the nucleolus into a sequence of linear programs, defined inductively as follows:

Problem k, $k = 1, 2, \ldots$:

   minimize t

   s.t. $x \in X(\mathcal{B})$,

   $\quad S \in \mathcal{A}_i \Rightarrow e(S, x) = t_i$, $\quad i \in \{1, 2, \ldots, k - 1\}$,

   $\quad S \in 2^N \setminus \{\mathcal{A}_0 \cup \mathcal{A}_1 \cup \cdots \cup \mathcal{A}_{k-1}\} \Rightarrow e(S, x) \leq t$.

Here, t_i is the optimal value of the objective function in Problem i and $\mathcal{A}_k$ is defined by $\mathcal{A}_0 = \emptyset$, and for $k \geq 1$, $\mathcal{A}_k$ is the set of coalitions attaining the excess t_k at each optimal solution (t_k, x) of Problem k. It can be shown that $\mathcal{A}_k \neq \emptyset$ as long as the previous ones do not exhaust the set of all coalitions 2^N. The process terminates when the optimal solution is a unique point, and this occurs usually long before the set of all coalitions is exhausted.

Using this process, Kopelowitz (1967) computed the nucleolus for many six- and seven-person zero-sum weighted majority games.[74] The average computation time was 10 seconds for a six-person game and 40 seconds for a seven-person game.

Kohlberg (1972) provided a single linear program the solution of which yields the nucleolus. Its disadvantage was that it involved $2^n!$ constraints – too big to compute the nucleolus of even a four-person game. Owen (1974) succeeded in reducing it to 4^n constraints at the expense of more variables and a more complicated objective function.

There have been other suggestions, based on Kohlberg's Theorem 5.3, of how to compute the nucleolus. The reader is referred to Brune (1976), Bruyneel (1979), Dragan (1981), and Wallmeier (1983, 1984).[75] I do not know of any study that compares the merits of the various proposals.

The situation may be more pleasant if one is allowed to use the special properties which a game may possess. For example, Huberman (1980), using a

---

[74]Taken from a list of Isbell (1959).

[75]Wallmeier provided a Basic program that also computes other related nucleoli (see Section 8).

slightly modified version of Peleg's algorithm, has shown that if it is known that the game has a nonempty core, then one need consider in the algorithm only *essential coalitions*, namely coalitions S, which are either singletons or for which $v(S) > \Sigma_{T \in \mathcal{S}} v(T)$, for every partition $\mathcal{S}$ of S.

The linear programs in Peleg's algorithm are huge. Even if n is moderately large, the computation of the nucleolus appears infeasible. Therefore it came as a happy surprise that Littlechild (1974a), using Peleg's procedure, found the nucleolus of the "Birmingham Airport Game" (see Section 10) involving 13 572 players of 11 different types.[76] He showed that if the players can be ordered in such a way that the worth of each coalition is equal to the worth of the least ordered player in that coalition, Peleg's programs can be solved easily.

The observation of Littlechild was further advanced by Megiddo (1978a), who gave an algorithm to find the nucleolus for cost allocation games defined over a tree (see Section 10) which requires $O(n^3)$ operations. Galil (1980) accelerated this number to $O(n \log n)$. It should be noted that their computations did not require the computation of the characteristic function in full. Thus, as has been stated explicitly by Megiddo, one can sometimes compute the nucleolus for large games without computing first the worths of all coalitions if one knows certain facts about the structure of the game. This idea was further advanced by Hallefjord, Helming and Jørnsten (1990). They presented an algorithm, based on Dragan's (1981) algorithm, for computing the nucleolus of the linear production game [see, for example, Owen (1975b)]. This is a game whose characteristic function is defined as a set of solutions of $2^n - 1$ linear programs. The authors showed that in order to compute the nucleolus, one need not solve all the linear programs. Often the number of programs that need be solved is very small indeed.

6. The reduced game property and consistency

In the previous sections some attempts were made to convince the reader that both the [pre-]kernel and the [pre-]nucleolus make little sense intuitively. In this, and in other sections, I shall do my best to convince the reader that the converse is actually true. The real question, in my opinion, is not whether a particular solution is good or bad, but rather: In what circumstances should it be recommended and what insight would it then yield? An attempt to justify a solution in this sense can be made in several ways:

(1) By examining the definition and showing that it reflects goals that some

---

[76]Thus only 11 components of the nucleolus may be different.

people, in some cases, may have. This, I believe, was done successfully for the bargaining set, but with less success for the kernel and the nucleolus.

(2) By showing that a solution concept has appealing properties that should be preserved during a "fair" bargaining, or in the verdict of an unbiased arbitrator. This has been done for the kernel and is valid a fortiori for the nucleolus.

(3) By providing a dynamic intuitive process that leads the players to the proposed solution. We shall exhibit such a process for various bargaining sets, including the kernel, in Section 7.

(4) By showing that for concrete situations the proposed solution yields results that one would otherwise expect, or want, or at least regard as plausible. We shall consider several such applications in subsequent sections.

(5) By providing an axiomatic foundation to the proposed solution and convincing the reader that people would like to obey these axioms. This is the subject of the present section.

Our first task is to study the concept of a *reduced game* and present some of its applications.

Definition 6.1.[77]   Let $(N; v)$ be any game and let x be any vector in $\Re^N$. Let S be a nonempty subset of N. *The reduced game on* S, *at* x, *denoted* $(S; v^x_{*S})$ *is defined by*[78]

$$
v^x_{*S}(T) = \begin{cases} 0, & T = \emptyset, \\ x(T), & T = S, \\ \max_{Q \subseteq S^c} [v(T \cup Q) - x(Q)], & \emptyset \neq T \subset S, T \neq S, S^c := N \backslash S. \end{cases} \tag{6.1}
$$

The idea is this: the players in N contemplate an outcome x. Then, for reasons that will be explained subsequently, each nonempty subset S examines "its own game". The members of S consider $(S; v^x_{*S})$ to be their own game on the following ground: $x(S)$ is what they got, so this should be $v^x_{*S}(S)$. Now, each other nonempty coalition T figures that it can take partners Q from S^c. Together they can make $v(T \cup Q)$, but the partners have to be paid $x(Q)$, so only the difference can be considered feasible for T. The max operation indicates that the best partner should be considered when computing $v^x_{*S}(T)$.

Remark.   It should be stressed that the worth $v^x_{*S}(T)$ is virtual: it may well

---

[77]Davis and Maschler (1965). Actually, it was defined there for $S = N \backslash \{i\}$, so that to get the present definition one should apply that definition repeatedly.

[78]In a similar fashion one defines [Peleg (1986)] a *reduced game on* S, *w.r.t. a c.s.* $\mathcal{B}$ *at* x by requiring the second line of (6.1) to hold for all T's of the form $B \cap S$ for some B in $\mathcal{B}$. It represents how members of S could perceive their own game, given that $\mathcal{B}$ was formed and x is being considered.

happen that to achieve the above maxima, two disjoint coalitions may need overlapping Q's.[79]

Now, consider a game $(N; v)$ and suppose that we live in a society whose members believe in a set-valued (or point-valued) solution concept $\mathcal{F}$. Suppose that an imputation x, for the grand coalition, belongs to $\mathcal{F}[v]$, and let S be any nonempty subset of N; then the players in S may consider their own game $(S; v_{*S}^x)$ and ask themselves whether x^S is in the solution of this game. If not, then certainly there is some instability in $\mathcal{F}$, because the players in S would want to redistribute $x(S)$, thus moving away from x. These ideas lead to a desire to adopt solution concepts that are stable, or consistent, in the following sense:

Definition 6.2. A solution concept $\mathcal{F}$, defined over a class of games Γ, is called *consistent*,[80] or *possessing the reduced game property*,[81] if

(i) Γ is rich enough in the sense that with every game $(N; v)$ in Γ, and every imputation x in $\mathcal{F}[N; v]$ for the grand coalition, and every nonempty subset S of N, the reduced game $(S; v_{*S}^x)$ belongs to Γ.

(ii) For every such x in $\mathcal{F}[N; v]$, every S, $S \subset N$, $S \neq \emptyset$,

$$x^S \in \mathcal{F}[S; v_{*S}^x] . \tag{6.2}$$

The first application of the reduced game was the observation that *the pseudo-kernel*[82] *is a consistent solution* [Davis and Maschler (1965)] *and so is*

---

[79]On this ground one can object to the credibility of the reduced game, claiming that it represents a lot of wishful thinking; namely, each subset of S hopes that the partners Q it needs will agree to cooperate. Certainly this is a valid argument and I would welcome any better way of defining how members of S should perceive their "own game". Let me point out, however, that an analogous, though not quite the same, virtual worth already exists in the concept of the characteristic function: $v(S)$ can be realized *only if overlapping coalitions do not form*. Here, $v_{*S}^x(T)$ can be realized only if other coalitions and their partners do not overlap T and its partners. Fortunately, the reduced game often has other, quite reasonable, interpretations when one considers applications. For example, if the characteristic function of a bankruptcy situation (see Section 10) is defined as in Aumann and Maschler (1985), then, as proved in that paper, for x in the core, the reduced game on a subset of the participants S is that the bankruptcy situation that results from the original game if we restrict ourselves to S, allowing its members to have *the same* claims, and letting the estate be $x(S)$. To sum up this discussion, I feel that in considering the application of the reduced game, one should check if it makes sense in the context of the application. If it does, fine. Otherwise, the interpretation of the reduced game as a way a subset of the players interprets its own game should be questioned.
[80]A term due to Hart and Mas-Colell (1989a, 1989b). It captures the spirit of our argument.
[81]A term which explains the idea more specifically.
[82]Pseudo-solutions are defined in the same way as the solutions, except that the payoff vectors are required to be non-negative instead of individually rational. The need to pass to the pseudo-kernel is due to the fact that x^S need not be individually rational in the reduced game. Note that pseudo-solutions are not covariant with strategic equivalence. Of course, if the game is zero-normalized, its pseudo-solutions coincide with the solutions. It can be shown that in another "normalization" the pseudo-kernel is equal to the prekernel [Maschler, Peleg and Shapley (1972)].

the prekernel [Maschler, Peleg and Shapley (1972)]. This is true because the $s_{i,j}(x)$'s, for i, $j \in S$, remain the same when passing from the original game to the reduced game. The above consistency properties were used in these papers and others to determine the kernel for games in which only n and $(n-1)$-person coalitions are not trivial [Davis and Maschler (1965)], to analyze the structure of the kernel [Maschler and Peleg (1967)] and to show that it consists of a unique point for the grand coalition of convex games [Maschler, Peleg and Shapley (1972)].[83] The reduced game property in these papers turned out to be an indispensable tool to get deep results on the kernel, because it made it possible to construct induction-wise proofs and compute kernels of large games from kernels of small ones.

The reduced game played a decisive role in Aumann and Drèze (1974). They defined and investigated several solution concepts for coalition structures, discovering[84] that *for* $\mathscr{C}(\mathscr{B})$, $\mathscr{M}_1^i(\mathscr{B})$, $\mathscr{K}(\mathscr{B})$ *and* $\mathscr{N}(\mathscr{B})$, *if x is in one of these solutions, then* x^B *is in the corresponding pseudo-solution of the reduced game*[85] $(B; v_{*B}^x)$, *for each B in* $\mathscr{B}$. Thus, for a c.s., the components of these solutions depend on the payoffs received by the other players, and the nature of this tie is expressed by the requirement to be in the solution of the reduced game. The proofs in Aumann and Drèze are easily modified to show that *the core, the prebargaining set, the prekernel, and the prenucleolus are consistent solution concepts*.

The rich results concerning the reduced games, as well as their intuitive appeal, raised the question whether consistency could be used to define some of the above solution concepts. The one who did this was Sobolev (1975) who gave an ingenious proof of the following:

Theorem 6.3. *The prenucleolus is the unique solution, defined over the class of all side payment games, which satisfies the following axioms:*[86]

(1) *The solution ϕ consists of a unique point for each game.*

(2) *It is Pareto optimal:* $\Sigma_{i \in N} \phi_i[N; v] = v(N)$.

(3) *It is covariant with strategic equivalence.*

(4) *The solution ϕ satisfies anonymity, i.e., it does not depend on the "names" of the players.*[87]

(5) *The solution is consistent.*

---

[83]The "profile" in Maschler and Peleg (1966) can be regarded as a "visual manifestation" of the reduced game property.

[84]The case of the nucleolus is credited there to M. Justman and the case of the kernel is credited to Maschler and Peleg (1967).

[85]Note that pseudo-core is equal to the core.

[86]Sobolev defined $v_{*S}^x(S)$ to be $v(N) - x(S^c)$ instead of the $x(S)$ of (6.1). With this definition, Pareto optimality could be deduced from the other axioms. I prefer (6.1) because it enables me to give a somewhat better interpretation of the reduced game.

[87]More precisely, it is covariant with any one-to-one and onto mapping of the set of players onto another set of players.

It is remarkable that *these same axioms also characterize the Shapley value.* The only difference is that the reduced game differs from that given by Definition 6.1. It is defined by

$$v_{*S}(T) = v(T \cup S^c) = \sum_{i \in S^c} \phi_i[v \mid T \cup S^c], \quad T \subseteq S, \tag{6.3}$$

where ϕ is the (one-point) solution concept under discussion, applied to the restriction of v to the players $T \cup S^c$. This is the result of Hart and Mas-Colell [1989a, 1989b].

Interpretation. Note that for $x = \phi[v]$ in (6.1), $v_{*S}(T)$ coincide in (6.1) and (6.3) for $T \in \{\emptyset, S\}$. For other subsets T of S, the coalitions evaluate their worth in (6.3) by asking themselves what will happen if the members of $S \setminus T$ suddenly disappear. In that case there remains a set of players, $T \cup S^c$, who will have to play together. Since they belong to a society of people who believe in ϕ, the members of S^c will ask for $\phi[v \mid T \cup S^c](S^c)$. The rest should be the worth of T in the reduced game. There are two basic differences between (6.1) and (6.3). In (6.1) T is allowed to choose partners Q from S^c. In (6.3) T is stuck with S^c. In (6.3) each player in S^c asks for his payment in the solution of the new game $(T \cup S^c; v \mid T \cup S^c)$, whereas in (6.1) each player asks for his payment x_i, which is supposed to be the solution of the original game $(N; v)$.

We see that in a deep sense the difference between the Shapley value and the prenucleolus lies in the way the subsets of N want to evaluate "their own games". Put it in a different way: if one has to choose, in any specific case, between the Shapley value and the prenucleolus, it is a good idea to examine the two types of reduced games. If one of them makes more sense for the particular case, the corresponding solution should be preferred. For example, in Aumann and Maschler (1985), bankruptcy situations originating in the Talmud are modeled as cooperative games. It turns out that for x being the prenucleolus,[88] the reduced game in the sense of (6.1) is precisely the bankruptcy game for the players in S alone, given that their estate is $x(S)$. The reduced game (6.3), when applied to the Shapley value, does not make sense in that case. Thus, if indeed the characteristic function of the paper models the situation correctly, the prenucleolus should be recommended. Hart and Mas-Colell (1989a, 1989b) provide an example of a cost allocation problem for which (6.3) makes good intuitive sense.[89]

---

[88]Which is equal to the nucleolus because the games are zero-monotonic.

[89]There are other ways to define "reduced games". Hart and Mas-Colell (1989a, 1989b) studied some of them. Sometimes they led to different solutions [see also Moulin (1985)] and sometimes they cause the axioms of Theorem 6.3 to be self-contradicting.

Note that we have axiomatized the prenucleolus – not the nucleolus itself. It would be wrong to say that these axioms characterize the nucleolus if we restrict ourselves to the class of zero-monotonic games. The reason is that even if the game is zero-monotonic, the reduced game need not be. However, Snijders (1991) showed that the same axioms axiomatize the nucleolus if, for one-person coalitions we replace $v^x_{*S}(i)$ in (6.1) by $\min\{x_i, v^x_{*S}(i)\}$.

The prekernel was axiomatized in Peleg (1986). In order to report it, we need an axiom which says that a preimputation is in the solution if its projection to every two-person coalition belongs to the solution of the reduced game for that coalition:

Definition 6.4. A solution concept $\mathcal{F}$, defined for a rich class of games Γ, is said to have *the converse reduced game property*, if for every game $(N; v)$ in the class and every preimputation x, if $x^S \in \mathcal{F}[v^x_{*S}]$ for every two-person coalition S, then $x \in \mathcal{F}[v]$.

Theorem 6.5.[90] *The prekernel is the unique solution concept defined over the class of all side payment games, which satisfies the following axioms:*
  (1) *It is never an empty set.*
  (2) *Each solution point is Pareto optimal.*
  (3) *Symmetric players receive equal payments in each solution point.*
  (4) *The solution is covariant with strategic equivalence.*
  (5) *The solution possesses the reduced game property.*
  (6) *The solution possesses the converse reduction game property.*
These axioms are independent.

It is interesting to note[91] that the last axiom can be replaced by the following:

The solution is a largest-under-inclusion set-valued solution satisfying the first five axioms.

We refer the reader to Peleg's papers for the axiomatization of the prekernel for a given coalition structure. It is interesting to note that Peleg (1985), and Chapter 13 by Peleg in this Handbook, show how the reduced game property also plays an important role in the axiomatization of the core for games with and without side payments. We also remark that the reduced game (6.1) was extended in Maschler et al. (1992) to "games with permissible coalitions and permissible imputations".

---

[90]Peleg (1986, 1987).
[91]Private communication from Bezalel Peleg.

7. The dynamic theory

In Section 3 we justified the bargaining set by presenting a dialogue between two players, k and l, facing a justified objection of k against l in which player k tries to convince player l to pass him some of his proceeds so that they can stay in their coalition of the c.s. The claim was that l was going to lose anyhow, so why not lose this way and stay in the coalition. It is not difficult to show that if k has a justified objection against l, then there is a minimal amount α, such that after its transfer k no longer has a justified objection against l.[92] Thus, every justified objection represents a *demand* of a definite size, and the bargaining set is the set of payoffs at which all demands are zero. This line of argument – static in nature – is not really convincing if we cannot show that *the willingness to settle demands brings the players to the bargaining set.*

Suppose that a transfer is made at some x to nullify a justified objection; then, after it is made, another player may have a justified objection against somebody, and so on, and we may end up with an infinite sequence of transfers. How does such a sequence behave? To make our argument solid, one has to find out if such processes always converge, and, if so, under what conditions, to the bargaining set. In other words, a dynamic backing is highly desirable.

Such a backing was supplied by Stearns (1968), who generalized his results to an even wider class of bargaining sets. That development and others will be described in this section. To simplify our presentation we shall limit ourselves to the case of the formation of the grand coalition.

We consider a game $(N; v)$. A system of functions $D \equiv \{d_{i,j}: i \in N, j \in N\}$ is called *a system of demand functions* if

(1) $d_{i,j}: X(N) \to \Re$ are lower-semicontinuous;

(2) $0 \leq d_{i,j}(x) \leq x_j - v(j)$ for all $x \in X(N)$;

(3) $d_{i,j}(x) \leq k_{i,j}(x) := \min(\frac{1}{2}[s_{i,j}(x) - s_{j,i}(x)]_+, x_j - v(j))$, for all x in $X(N)$, whenever $i \neq j$; here, $[a]_+ = a$, if $a > 0$ and 0, otherwise;

(4) $d_{i,i}(x) = 0$ for all $x \in X(N)$.

The idea is that at x, for each couple i, j, player i informs player j that he owes him $d_{i,j}(x)$. (1) is a weak continuity assumption on these depts. We are not told much about the nature of the debts, but it is assumed in (2) that after a debt is satisfied, the resulting payoff is still individually rational. It is implied by (3) that j owes nothing to i at x, if[93] $s_{i,j}(x) \leq s_{j,i}(x)$; and if the converse inequality holds, then, *after $d_{i,j}$ is passed from j to i, and the resulting payment is y, the inequality $s_{i,j}(y) \geq s_{j,i}(y)$ still holds.* (See Section 4 for the meaning of $s_{i,j}$.) In

---

[92]At most, player l would pass him $x_l - v(l)$. Then he will be able to defend himself alone.

[93]This is a mild restriction: if the inequality holds, then whatever coalition i wants to use to threaten j, j can find a coalition not containing i having a higher excess. Should this be the case, then indeed i is too weak to ask anything from j.

other words, the $s_{i,j}$'s determine a reasonable upper bounds on the demand functions, but otherwise the demand functions may be chosen quite arbitrarily.

Definition 7.1. The *bargaining set* $\mathcal{M}_D$ (for the grand coalition) is the set

$$\mathcal{M}_D := \{x \in X: d_{i,j}(x) = 0, \text{ for all } i, j \in N\}. \tag{7.1}$$

We have here many bargaining sets – one for every choice of $\{d_{i,j}\}$ – upon which only weak assumptions are made. Stearns (1968) proves that they all contain the kernel, and that the kernel itself is one of them. It is obtained by choosing $d_{i,j} = k_{i,j}$ for all i, j. This follows from the definition of $k_{i,j}$. Stearns also shows that if we define $d_{i,j}(x)$ to be the minimal amount j has to pass to i in order to nullify a justified objection of i against j, if there is one at x, and zero if there is not, then this $d_{i,j}$ has all the properties of (1)–(4) above, so that $\mathcal{M}_1^i$ is one of Stearns' bargaining sets.

Facing a positive demand $d_{i,j}(x)$, does not mean that j has to pay it immediately, or to pay all of it. It only means that, when at x, he will not pay more than this amount. Accordingly, we say that y *results from* x *by a D-bounded transfer*, if i, j and α exist, such that

$$y = x + \alpha e^i - \alpha e^j, \quad 0 \leq \alpha \leq d_{i,j}. \tag{7.2}$$

Here, e^t is the unit vector in the t-direction.

Definition 7.2. A sequence $x^1, x^2, \ldots$ is called a *D-bounded transfer sequence*, starting at x, $x \in X$, if $x^1 = x$, and every other x^t results from the previous one by a D-bounded transfer.[94] It is called a *maximal transfer sequence* if, infinitely often, "maximal transfers" are passed; i.e., if there exists γ, $0 < \gamma \leq 1$, such that, at infinite instances ν, the transfers α at x^ν are at least $\gamma \max_{i,j \in N} d_{i,j}(x^\nu)$.

Theorem 7.3.[95] *Every D-bounded transfer sequence converges. Every maximal*[96] *transfer sequence converges to a point in $\mathcal{M}_D$.*

Billera (1972a) extended Stearns' results in two ways: by asking that at each stage a fixed proportion of *all demands* be transferred at once, and by

---

[94] We allow transfers of size zero, so there is no loss of generality in assuming that all the transfer sequences are infinite.

[95] Stearns (1968).

[96] The request for a maximal transfer sequence arises from the need to prohibit smaller and smaller transfers say, among players 1, 2, 3, and meanwhile, say, player 5 owes a lot to player 6 but has no chance to reimburse him. In fact, Stearns states other similar criteria which guarantee that such circumstances will not occur. It follows from these considerations that if the players agree that nothing less than a penny should be passed, then the theorem says that every transfer sequence should reach a distance of up to a penny from the bargaining set $\mathcal{M}_D$ in a finite number of transfers.

considering also a continuous process that is defined by the differential equation

$$\frac{dx_i}{dt} = \sum_{j \in N} (d_{i,j}(x) - d_{j,i}(x)), \quad i \in N. \tag{7.3}$$

He showed that *if the demand functions are continuous, the trajectories always converge to a point in* $\mathcal{M}_D$.[97]

These results provide perhaps the strongest case for the bargaining set and its related concepts. Note, however, that they also show that not all outcomes in a bargaining set are equally plausible: Stearns already studied an example, due to Kopelowitz (1967), in which a bargaining set (that happened to be the kernel) consisted of two points. He found that one could start from a point x, arbitrarily close to the non-nucleolus point, yet the transfers would carry one to the other point. Thus, in the dynamic system that was determined by the transfer sequences, *not all points of the bargaining set were stable in the sense of Lyapunov*.

The characterization of all stable points of Stearns' dynamic process and, more generally, of all stable sets with respect to this process, is still an open problem. Only scattered results can be reported. Kalai, Maschler and Owen (1975) proved that for every bargaining set $\mathcal{M}_D$, the nucleolus is always a stable point of the process.

Since all the bargaining sets in this section contain the kernel, Stearns' transfer sequences do not necessarily converge to the nucleolus. For example, if the starting point is a kernel point other than the nucleolus, the sequence will stay put at this point and will not converge to the nucleolus. Justman (1977) has generalized Stearns' transfers by removing the restriction that transfers are made only from one player to another. He was then able to obtain a dynamic system that converges to the nucleolus. Justman did much more than this: he generalized Stearns' ideas to an *arbitrary discrete set-valued dynamic system defined over a complete metric space*, thereby extending the range of applications to many areas – not necessarily related to game theory.[98]

Based on Justman's paper, and on Kalai et al. (1975) mentioned above, Maschler and Peleg (1976) developed a theory which *characterizes closed stable sets with respect to a discrete set-valued dynamic system over a compact metric space*; the characterization being in terms of a *generalized nucleolus with respect to a vector of Lyapunov functions*. One byproduct of the theory is the

---

[97] The results are somewhat more general because Billera refers to an arbitrary polytope in $\Re^N$ inside of $X(N)$, and even to $\Re^N$ itself, in which case the pre-bargaining set obtains. Specializing in the prekernel, Wang (1974), Wang and Billera (1974) and Billera and Wu (1977) investigated the trajectories of the differential equation and showed that they can be classified into equivalence classes that depend on the "excess structure" at the limit point.

[98] Justman himself applied the theory to model an *abstract iterative process of negotiations*.

fact that *every nonempty ε-core is a stable set, in the sense of Lyapunov for Stearns' transfer sequences.* These results were extended by Yarom (1985) to the continuous case of *differential inclusions.* [See also Yarom (1982, 1983).]

8. Related solutions

The ideas behind the solution concepts, as presented in the previous sections, generated numerous other related solutions, both in the area of cooperative game theory and in its applications. In this section we shall report only the most immediate solutions. For a more comprehensive list we refer the reader to the items indexed **R** (and **RR**) in the Bibliography.

Already in Aumann and Maschler (1964), a bargaining set was introduced in which *sets of players objected against sets of players* instead of individuals against individuals. Other possibilities were also hinted at in that paper. A detailed list of various bargaining sets is given in Maschler (1963a). Dissatisfaction with the kernel of Apex games (see Section 4), and too many outcomes in $\mathcal{M}_1^i$ for these games, led Horowitz (1973, 1974) to create his *competitive bargaining set* in which a player announces a *multi-objection*, namely several simultaneous objections, regarded as offers to whomever is first to accept one of them. A counter-objection is valid only if it matches all of them. The idea of employing multi-objections was pursued in different ways by Dragan (1985, 1986, 1987a, 1987b, 1988a, 1988b). Additional variants of the bargaining set, motivated by analyses of European Governments, were created by Schofield (1976, 1978, 1980a, 1980b, 1981, 1982).

Although many of those bargaining sets are empty for some games, I do not consider that a major drawback. They represent desires for greater stability[99] which, however, cannot always be met. But in those cases where they can be met, I see no reason not to recommend their outcomes. In this connection, I would like to draw attention to Naumova (1973). The author generalized the original Aumann and Maschler (1964) paper by restricting both the objecting sets K and the sets L, against whom the objections are made, to be taken only from an a priori specified set of coalitions[100] $\mathcal{Q}$. The resulting $\mathcal{Q}$-*bargaining set* is denoted $\mathcal{M}_2^i$.[101] In this spirit she also defined a $\mathcal{Q}$-*kernel*, $\mathcal{K}_2$, which is contained in the $\mathcal{Q}$-bargaining set. Interestingly, Naumova was able to provide conditions on $\mathcal{Q}$ which guarantee that each game would have a nonempty $\mathcal{Q}$-kernel for every coalition structure, as long as $X(\mathcal{B}) \neq \emptyset$. One such condition is that each coalition of $\mathcal{Q}$ should contain a *fanatic*, namely a player who

---

[99]This does *not* mean that there is an automatic inclusion relation between some of them and $\mathcal{M}_1^i$.

[100]In another variant, the sets allowed to be used for objections and for counter-objections are also restricted.

[101]If $\mathcal{Q}$ is the set of all singletons, $\mathcal{M}_2^i = \mathcal{M}_1^i$.

does not belong to the other coalitions in $\mathcal{Q}$. The proofs show that the relation "having a justified objection" in her sense is still acyclic, which enables the use of Peleg's (1963b, 1963d, 1967) generalization of the K.K.M. Lemma.[102]

So we have a lot of bargaining sets, each interesting in itself. I am concerned, however, by the lack of known general properties of these bargaining sets and by a missing backup of a convincing dynamic theory. Further research is certainly needed in this direction.

An interesting bargaining set has recently been developed by Mas-Colell (1989). According to this definition, an objection to a payoff x in X^0 is a pair (C, y), y feasible for C, such that $y_i \geqslant x_i$ for all i in C, and at least one of these inequalities is strict. A counter-objection is a pair (D, z), z feasible for D, such that $z_i \geqslant y_i$ for all i in $C \cap D$ and $z_i \geqslant x_i$ for all i in $D \setminus C$. And again, at least one of the inequalities on z_i's must be strict. This definition has the advantage that it makes good sense also for games without side payments and games with a continuum of players. In fact, it was defined for such games in Mas-Colell (1989). He proved there that under mild assumptions on the economy, his bargaining set for such market games consists of the set of payoffs to Walrasian equilibria.

Note that the objection in this definition is not against somebody. In fact, the counter-objecting coalition can even contain the objecting one. Thus, a different scenario of claims and counter-claims must be shown to provide us with a good understanding of the significance of this bargaining set, hopefully enabling us to develop a dynamic theory analogous to Stearns' transfer schemes (Section 7).

For side payment games Mas-Colell proved that this bargaining set contains the prekernel, and so it is not empty in the space of preimputations. Recently, Vohra (1991a) proved that essentially the same bargaining set contains imputations. Thus *this bargaining set is not empty for side payment games with nonempty sets of imputation.* [See also Vohra (1991a, 1991b), Dutta, Ray, Sengupta and Vohra (1989)[103] and Grodal (1986).]

Some modifications of the kernel and the nucleolus result from modifying the excess function. They result from the feeling that the excess, as defined in Section 3, unjustifiably does not take into account the size of the coalitions. One of the most detailed studies in this direction is Wallmeier's (1980) thesis. Wallmeier defines the excess to be

$$e_f(S, x) := \begin{cases} e(S, x)/f(|S|), & \text{if } S \neq \emptyset, \\ 0, & \text{if } S = \emptyset, \end{cases} \tag{8.1}$$

---

[102]See also Naumova (1978), where more general results are obtained.

[103]Dutta et al. consider a variant in which a *sequence of counter-objections* is considered, each against the previous one, that, when taken together, enable one to decide if the original imputation is stable or not.

where $f: \{1, \ldots, n\} \to \Re_+$ is a monotonically nondecreasing function. For example, $f(|S|) = |S|$ is a case that was discussed frequently in the literature. The nucleolus based on this excess is sometimes called the *per-capita nucleolus* or the *equal division nucleolus*. [See, for example, Young, Okada and Hashimoto (1982). In this connection see also Lichtenfeld (1976) and Albers (1979b).] Wallmeier shows that concepts such as f-kernel and f-nucleolus can be defined easily in complete analogy with those based on $e(S, x)$. He also shows that many of the "classical" theorems can be generalized to these solutions.

Another interesting solution concept, called *the lexicographic kernel*, was suggested by G. Kalai[104] and studied in Yarom (1981). Its definition is similar to the nucleolus, except that the lexicographic comparisons are performed on the vectors $\psi(x)$ the coordinates of which are the $s_{ij}(x)$ arranged in decreasing order [see (4.1)]. Like the nucleolus, it is contained in every nonempty ϵ-core. It is even a locus of these sets. Unlike the nucleolus, it may consist of more than one point. Each of its points is Lyapunov-stable in Stearns' dynamic systems [Maschler and Peleg (1976)].

The nucleolus essentially results from a sequence of minimization problems; each one, after the first, has as its domain the optimal set of the previous one. This idea should have been employed in other areas. In fact, Potters and Tijs (1992) point out that the idea was already reported by Dresher (1961) for zero-sum matrix games.[105] It is suggested that a player should choose from among his optimal strategies one which maximizes his worst expected payoff, given that the opponent is allowed to mix only from pure strategies which are not active in any optimal strategy of the opponent. From this subset of his set of optimal strategies the player should choose one that maximizes his worst payoff, given that the opponent is restricted to mix only from pure strategies which were not active in any of the optimal strategies for the previous cases, etc. The process terminates when no further strategies are available to the opponent, in which case the last set of optimal strategies available to the player is the recommended set. This subset of optimal strategies is entitled by Potters and Tijs (1992) *the nucleolus of the zero-sum game, for that player*. It is aimed at exploiting the opponent's mistakes without sacrificing one's own security levels. Potters and Tijs (1992) continue to investigate this nucleolus and show that it possesses properties analogous to those of the "classical" [pre-]nucleolus. In particular, they prove an analogue of Kohlberg's criteria (Theorem 5.3). Furthermore, for each cooperative game $(N; v)$, normalized by satisfying $v(N) = 1$, they produce a matrix game the nucleolus of which is essentially the [pre-]nucleolus of $(N; v)$. Maschler et al. (1992) axiomatized this nucleolus as well as more general nucleoli defined over metric spaces.

---

[104]Oral communication.
[105]L. Shapley told us that this idea was circulated earlier among RAND game theorists and was already reported as a Research Memorandum in Brown (1950).

An interesting topic is the study of games in which the values of a characteristic function are not deterministic; i.e., when *each v(S) is a random variable having a known distribution functions*. In a series of papers, Charnes and Granot (1973, 1974, 1976, 1977) address the problem of defining a nucleolus[106] for such games. They suggest a two-step process, in which a "prior payoff vector" is "promised", one having a good chance of eventually becoming realizable. Later, if it turns out that it does not, a second stage play determines the final payoff. Granot (1977) extends these results and extends also the concepts of a "prior kernel" and a "prior bargaining set". *Nonemptiness and the various inclusion relations are valid, although the nucleolus may consist of more than one payoff.*

9. Classes of games

The bargaining set, kernel, and nucleolus and their variants were studied for several classes of games. The purpose in studying these classes was sometimes purely mathematical, motivated by the desire to better "feel" the nature of the solutions and check whether the recommendations of the theory make sense. In other cases, the motivation to study some classes of games resulted from their application, mainly to the social sciences. In the next section we shall discuss some of these applications. This section we devote to the more theoretical results.

One of the nicest results, in my opinion, is concerned with the nucleolus of *constant-sum weighted majority games*.[107] Consider, for example, the games [8, 1, 8] and [2, 2, 2]. They are, in fact, two representations of the same game, because their characteristic functions coincide. There are infinitely many other representations. The second representation, however, is more natural, because in this representation each *minimal winning* coalition carries the same weight. Such a representation is called a *homogeneous representation* and if it exists, the game is called *homogeneous*.[108] Von Neumann and Morgenstern (1953) already realized that not every weighted majority game has homogeneous weights, and some years later, Isbell (1959) expressed the desire to find for each constant-sum weighted majority game a unique (normalized) representation which will make sense intuitively and reduce to the homogeneous representation if the game is homogeneous. The question remained open for nine years until Peleg (1968) proved that *the nucleolus is always a system of weights, and these are homogeneous weights if the game is homogeneous*. Thus, if one

---

[106]Also defining a core and a Shapley value.

[107]Namely, weighted majority games in which a coalition wins iff its complement loses. In this case it is not necessary to specify the quota. It can be taken as half the sum of the weights.

[108]It is unique up to specification of the weights to the *dummy players*, and up to a multiplication of the weights by a positive constant.

agrees that the nucleolus makes sense intuitively, one finds that Isbell's desideratum has been accomplished.

Another interesting set of results deals with properties of the solutions for the *composition of games* in terms of the solutions of the components. The results are too long to reproduce here and we refer the reader to Peleg (1965a), Megiddo (1971, 1972a, 1972b, 1972c, 1973, 1974a, 1974b) and Simelis (1973a, 1973b, 1973c, 1975a, 1975b, 1976a, 1984).

Clearly, the bargaining set, the kernel and the nucleolus are known for three-person games [Davis and Maschler (1965), Grotte (1970), Brune (1983), Ostman (1984), and Ostman and Schmauch (1984)]. Their study is readily generalized to *games with only* 1-, $(n - 1)$- *and n-person permissible coalitions*. In these games, the bargaining set consists of the core if the core is not empty and it is a unique point if the core is empty. The kernel and the nucleolus coincide for these games, and the formulae that express them are simple linear formulae, each of which is valid in a region determined by the values of the characteristic function. [See Maschler (1963a), Davis and Maschler (1965), Owen (1968, 1977b), and Kikuta (1982a, 1982b, 1983).]

The "1, $(n - 1)$, n" games are particular cases of *quota games* and *m-quota games*.[109] Solutions for these games were obtained in Maschler (1964), Peleg (1964, 1965b) and in Bondareva (1965).

Considerable effort was invested to compute the kernel and the nucleolus of four-person games [see Peleg (1966b), Brune (1976), and Bitter (1982)]. These computations were intimately connected with development of general algorithms to find out *regions of linearity* of the nucleolus, namely regions in the game space in which the nucleolus is a linear function of the values of the characteristic function.[110]

An interesting and important class of games is the class of *convex games*. These games, introduced by Shapley (1971), are games the characteristic function of which satisfies for every pair of coalitions S and T:

$$v(S) + v(T) \leq v(S \cup T) + v(S \cap T) . \tag{9.1}$$

This class is interesting because all important solution concepts agree on its games: They have a unique Von Neumann–Morgenstern solution [Von Neumann and Morgenstern (1953)] coinciding with the core, and the Shapley value [Shapley (1953b)] is essentially the center of gravity of the core. They also have interesting economic applications [see Shapley (1971)]. *For convex games, the bargaining set coincides with the core and the kernel coincides with*

---

[109]*m*-quota games are defined by an *n*-tuple $(\omega_1, \omega_2, \ldots, \omega_n)$ such that $v(S) = \omega(S)$ whenever $|S| = m$, and is either equal to zero otherwise or made superadditive in the obvious way.

[110]The research in this area can be found in Kohlberg (1971, 1972), Kortanek (1973), and Brune (1983).

the nucleolus; therefore, the kernel is contained in the core, although it differs in general from the Shapley value [Maschler, Peleg and Shapley (1972)].

The above results motivated Driessen (1985a, 1985b, 1986b) to study a larger class of *k-convex games*. We shall omit the precise definition of this class, but would like to cite two important results. *For these games there is exactly one kernel point in the core, although the kernel may contain payoffs not in the core.*[111] Consequently, the bargaining set may contain points not in the core, but in any case *the core is a component of the bargaining set; namely, it is disconnected from other parts of the bargaining set.* A detailed description of these results can be found in Driessen (1985a, 1988).

The Dutch school of game theory, started by Tijs, conducted a systematic study aimed at finding several solutions to some classes of games – often classes of games with real-life applications. These studies yield insight into the actions of the various forces during the playing of the treated games. Thus, they may help people facing real conflicts to decide with better understanding which solution to adopt should they face such games. In their studies, Tijs, his colleagues and students show that *quite often the nucleolus coincides with the τ-value* – a solution concept introduced in Tijs (1981). The importance of this finding is both theoretical and practical. The fact that two solutions, based on completely different ideas, happen to coincide for some games, strengthens the reasons to adopt such solutions for those games. The practical importance lies in the fact that it is much easier to compute the τ-value. We refer the reader to Driessen and Tijs (1983, 1985), Muto, Nakayama, Potters and Tijs (1988), Potters and Tijs (1990), and Muto, Potters and Tijs (1989) for the analyses described above.[112] As an example, we shall report here the results of Potters, Poos, Tijs and Muto (1989) concerning *clan games*. These are games for which there exists a nonempty coalition C, called "the clan", such that

(i) $v(S) \geq 0$,   all S,

(ii) $M_v(i) := v(N) - v(N \setminus \{i\}) \geq 0$,   all $i, i \in N$,

(iii) $v(S) = 0$,   if $S \not\supseteq C$,

(iv) $v(N) - v(S) \geq \sum_{i \in N \setminus S} M_v(i)$,   if $S \supseteq C$.

Thus, in order to be worthy of any positive amount, a coalition S must contain the clan; however, the complement of such a coalition also has some power: it contributes towards the grand coalition at least as much as the sum of the

---

[111] The core for these games is never empty.
[112] In one class, called the class of *semiconvex games*, the kernel coincides also with the Shapley value.

contributions of all its members towards the grand coalition. Clan games occur frequently in real situations: the clan may consist of a group of people who are in possession of a technology, or it may consist of people who have copyright privileges, etc.

For such games, the authors prove that *the bargaining set (for the grand coalition) coincides with the core. Moreover, it is given by*

$$\{x \in X(N) \colon x_i \leq M_v(i) \text{ for all } i \in N \backslash C\} \; .$$

The authors also show that *the kernel of a clan game coincides with the nucleolus*, and they provide a very short algorithm to compute it.

10. Applications

Some applications of the solution concepts presented in this chapter were already discussed in previous sections, especially Section 9. Indeed, the classes of games for which solutions were computed are quite often useful for applications and have been investigated frequently in order to solve problems of an applied nature. Other applications can be found in the Bibliography, where they are denoted by either **A** or **H**, or even **F**.

Rather than surveying all of these papers, we shall attempt to concentrate on a few which bear on the basic question: *In what situations may one of the solutions of this chapter be preferred to others?* By doing so, I do not want to give the impression that either the bargaining set, or the kernel, or the nucleolus is superior to any other solution in any particular case. I would try rather to indicate *under what situations some of our solutions should be candidates for recommendation.* I myself take the position that for any "real situation" that is modelled as a game one should examine all solutions, even if one is eventually going to adopt only one of them. The reason is that each solution sheds light on one corner of the real world, so the more solutions one knows the better one really understands the issues involved.

The core is a useful solution concept which is applied in many cases. It represents a strong and often desirable stability property. But often the core is large, and then one faces the need to single out a unique outcome in the core. In such cases the nucleolus is a good candidate. Of course there are other "distinguished points", such as the center of gravity of the core. Still, the nice properties of the nucleolus (or variants of the nucleolus) render its choice quite attractive.

An interesting class, where the above desideratum holds, is that of issues concerning *cost allocations*. These issues can be cast in the following setup. A group of "players" (people, companies, states, or other organizations) is involved in a project that would provide desirable services to the members. One wants to know *how to share the costs of the project among the participants*. The project itself may be small, say building a road to serve several communities, or big, such as building a dam to serve several countries. It may involve few participants or many, such as the building of an airport, where various aircraft movements get serviced.[113]

Denote by N the set of players. Consider a coalition S. The services desired by the members of S constitute, in general, only a subset of the services desired by all the participants. Thus, one may define $c(S)$ to be the cost S would incur if its members construct a smaller project to supply only their own needs. The problem of cost allocation is to provide a definite rule for allocating $c(N)$ among the members of N, based on the above data. One principle which comes to mind is that *no subset S should be charged more than $c(S)$*, for that would mean that some groups are subsidizing others.

Looking at costs as "losses" and denoting[114] $v(S) = -c(S)$ and $x = -a$, where a is the suggested allocation of costs, it is easy to see that *a nonsubsidizing recommendation of an allocation $a = (a_1, a_2, \ldots, a_n)$ means that x belongs to the core of the game $(N; v)$*. The reader is invited to read the survey of Straffin and Heaney (1981) on the various recommendations provided by the Tennessee Valley Authority for allocating the costs of water development projects in the Tennessee River Basin. It turns out that the nonsubsidy requirement was already formulated in 1942 by Josef Ransmeier, who worked for the Authority [see also Driessen (1988)]. Several solutions were considered, in each of which the individual was first charged with his *separable costs* $SC_i := c(N) - c(N \setminus \{i\})$. The remaining *nonseparable costs* were to be divided by various methods.[115] One method – in which it was recommended to share them equally – is interesting for our purpose, because *in many cases it yields the nucleolus of the game*. [See Suzuki and Nakayama (1976), Legros (1983, 1986), and Driessen and Funaki (1991) for conditions that guarantee this result.]

With all due respect to the people who worked for the Authority, it seems to me that these methods are not sophisticated enough, because by separating the separable costs one introduces an undesirable asymmetry. Why should one consider separately the contribution of a player to the grand coalition and not, for example, the contribution of two players to the grand coalition? Thus, it seems to me that the ideas which motivated the above recommendations require a more sophisticated analysis.

The first recommendation of the nucleolus for some cost allocation games

---

[113]In this case each movement can be considered a player.

[114]Or, $v(S) = \Sigma_{i \in S} c(\{i\}) - c(S)$, if one wants a zero-normalized version, in which case $x = c(\{i\}) - a_i$.

[115]Some of which are employed also today.

was suggested by Littlechild (1974a) and Littlechild and Thompson (1977).[116] The authors consider a cost allocation game resulting from the building of an airport runway designed to serve all types of aircraft. Of course, the cost of the runway will be higher if it is going to serve larger aircraft, needing a longer runway. However, a longer runway can also serve small aircraft. Thus $c(S)$ is equal to $c(\{i\})$, where i is the largest aircraft in S. It turns out that this feature of the characteristic function makes it relatively easy to calculate the nucleolus. In fact, Littlechild (1974a) succeeded in giving a recursive formula for the nucleolus of such games.[117]

These results motivated Megiddo (1978a) to construct a "good algorithm"[118] for computing the nucleolus of cost allocation problems defined over a tree as follows: the players are the nodes of a directed tree, other than its root. $v(S)$ is the total length of arcs that belong to some path from the root to a node i, $i \in S$. Bird (1976b) and Granot and Huberman (1981, 1984) [see also Granot and Huberman (1982)] treated the *minimal cost spanning tree* cost allocation problems. The setup in these problems is a complete graph with nodes $\{0, 1, \ldots, n\}$, where 0 is a common supplier and the rest of the nodes are users. Let $c_{i,j}$ be the cost of establishing an edge between i and j. For each coalition S, we define by $c(S)$ the lowest cost to connect all nodes in $\{0\} \cup S$. (Think of constructing a cable television network.) Granot and Huberman studied the core and the nucleolus of such games. In particular, they provided ways to compute the nucleolus. *For these games, the nucleolus is the unique payoff in the intersection of the kernel and the (nonempty) core*. Other variants of cost allocation problems on a graph were analyzed by Galil (1980), and by D. Granot and F. Granot (1992).

Young, Okada and Hashimoto (1982) recommended the *per-capita nucleolus* (see Section 8) as a solution for a cost allocation problem involved in a water resources development for the Scandinavian countries, employing actual data in their analysis.

For a detailed analysis of the cost allocation problem, the reader is referred to Young (1985b), D. Granot and F. Granot (1988a), and to Driessen (1988).

Another interesting topic for which the nucleolus was sometimes recommended is *revenue allocation*. Here one is asked to distribute an amount of money in a "fair way", given the "legitimate" claims of individuals, or groups of individuals. One of the simplest examples is bankruptcy, where an estate not covering the debts has to be distributed "fairly". In other examples one wants

---

[116]These papers consider also the Shapley value and a comparison is made with the actual fees charged in the particular case of Birmingham Airport.

[117]Littlechild and Owen (1976) generalized the model by introducing into the characteristic function also the benefits which the aircraft obtain by using the airport. They prove that in some cases the nucleolus schedule for covering the costs does not depend on this revenue.

[118]Polynomial in n.

to allocate revenue resulting from a joint enterprise, where the claims result from contributions of the partners.

A related topic is *taxation*. Here, instead of claims, one considers the "taxable income" and the estate is replaced by the "budget" which has to be covered by taxes. Again, the problem is: How should one determine a tax schedule?

The literature is full of various solutions to revenue allocation problems. One way of approaching them is to determine the desired properties and see if such properties can be used as axioms in order to determine the said solutions. This approach has the advantage that it shows which (usually simple) desirable principles can bring about a preference for one solution. The reader is referred to Young (1985c, 1987), where this approach is beautifully executed. Another approach might be to convert the revenue allocation problem into a game and find out what outcomes are provided by the various game-theoretical solution concepts.

Unlike the case of cost allocation, in which a cost function is the natural thing to consider and transform into a game,[119] in a revenue allocation case one can think of various ways to convert it into a game.[120] For example, take a two-person case, where the estate is 100 and the claims are 30 and 120. Clearly, $v(1\,2) = 100$, but what should the values $v(1)$ and $v(2)$ be?

One possibility could be $v(1) = 0$ and $v(2) = 70$, where $v(i)$ is the remainder after the claim of the other person is satisfied "as much as possible". This representation seems reasonable in the case of a bankruptcy. Another representation might be $v(1) = -20$, $v(2) = 70$, which is what is left after the claim to the other partner is fully paid. This representation is better, for example, in a divorce case, where wife and husband each take home what they brought to the marriage, but are still jointly responsible for the debts acquired by the family.

The most prominent game-theoretical solutions to revenue allocation problems are the Shapley value [Shapley (1953b)] and the nucleolus. Neither of them recommends to distribute the estate in the above example in proportion to the claims, which is the practical solution in many real-life cases.

It is interesting to note that the nucleolus appeared as a suggestion for solving a bankruptcy problem 2000 yeas ago: A Mishna in the *Talmud*, written by Rabbi Nathan, describes a case of three widows of the same man, who have claims of 100, 200 and 300 units, respectively, on the estate of the late

---

[119]Even in this case, the characteristic function is not obvious if one also wants to include benefits that result from services, and take into consideration the possibility that a coalition S may decide not to put up a small project on its own, but would still agree to participate in the bigger project of N.

[120]The author is indebted to Mas-Colell, who pointed out this aspect in a lecture given in 1987 at a game theory conference held at Ohio State University.

husband. The Mishna treats three cases, where the estates are 100, 200 and 300 units. In the first case Rabbi Nathan prescribes an equal share: $(33\frac{1}{3}, 33\frac{1}{3}, 33\frac{1}{3})$. The specifications for the other cases are $(50, 75, 75)$, $(50, 100, 150)$. It turns out that if one defines

$$v(S) = \max\{\text{estate} - \text{sum of the claims of the members of } N\backslash S, 0\},$$

$$(10.1)$$

then *the above tuples are precisely the nucleoli of the resulting games (N, v).*

Of course, Rabbi Nathan did not have any knowledge of the nucleolus. How then did he arrive at his recommendations? It is proved in Aumann and Maschler (1985) that *he could have arrived at the above numbers by looking for a solution which is consistent for every two-person coalition.*[121] The (1985) paper also provides a simple algorithm that yields the nucleolus for a general bankruptcy game. *For these games, the nucleolus coincides with the kernel.*

D. Granot and F. Granot (1992c) consider revenue allocation problems resulting from a network flow. The setup is a network connecting a source and a sink. Each arc is owned by a player and different arcs are owned by different players. The problem is to allocate the revenue resulting from transferring a maximal flow from the source to the sink. They study the core, the kernel, and the nucleolus of these games in which $v(S)$ is the revenue that S can obtain by transporting maximal flow from the source to the sink. This paper is rich in results that assist in the task of computing the kernel and the nucleolus of the games considered. It also yields insight on the nature of these solution concepts. [See also D. Granot and F. Granot (1992a).]

Let me mention in passing that the nucleolus has been computed by Chetty, Dasgupta and Raghavan (1970) for a production economy involving one landowner and many landless peasants [as modeled by Shapley and Shubik (1967)], and by Legros (1987a, 1987b) for some bilateral market games with symmetric agents, where the effect of "disadvantageousness of syndication" is studied.[122] (See also the discussion at the end of this section concerning the bargaining set for some of these games.) Muto, Nakayama, Potters and Tijs (1988) considered the solutions for a wider class of games, entitled *Big Boss Games*, that contains the previous games of this paragraph. Galil (1974) studied the nucleolus of some weighted majority games with "major" players and many symmetric "minor" players.[123] He proved that under some conditions *the payoffs to the major players do no change if each minor player splits*

---

[121]See Section 6. There are many examples in the *Talmud* concerning the resolution of two-person conflicts by means of the nucleolus.

[122]In Aumann's (1973) opinion this shows that the core, which coincides with the nucleolus is *not* the right solution for the analysis of these games.

[123]Namely, carrying small and equal weights.

into k "mini-minor" symmetric players. The nucleolus of a subclass of these games in computed.

An interesting application of the kernel is provided in the analysis of Rochford (1983a, 1983b, 1984) on the Shapley and Shubik (1972c) *assignment game*. She recommends for these games the intersection of the kernel and the core and variants of it as a good solution, precisely *because it involves transactions in pairs*, for which, according to her, the equations $s_{ij} = s_{ji}$ make good sense.

Normalized assignment games are games in which the players are composed of two equal-sized groups: "consumers" and "producers". If a consumer i and a producer j form a partnership, that partnership will realize a non-negative profit $a_{i,j}$. The worth of a coalition is the maximum amount that can be realized when the members form disjoint partnerships among themselves. A collection of partnerships under which $v(N)$ is realized is called *an optimal assignment*. It is still an open problem whether the kernel of an assignment game is always contained in the core. D. Granot and F. Granot (1992c) provided some partial results on this question. They then proceeded to study the properties $\mathcal{K} \cap \mathcal{C}$. This set need not even be convex. Nevertheless, they were able to show that this set coincides with the core if and only if every consumer or producer, whose payoff in the core is not constant, appears with at least two different partners in the various optimal assignments.

D. Granot and F. Granot (1992c) also studied the nucleolus of certain assignment games and were able to provide conditions under which the computation of the nucleolus amounts to finding a core-imputation, which is lexicographically maximum. One such game is the famous Bohm-Bawerk (1923) 18-person game, the nucleolus of which is computed quite easily.

Another recent result on assignment games is due to Owen (1992), who observed that the reduced game (Section 6) of an assignment game with reservation prices, with respect to a core imputation, is itself such an assignment game, easily derived from the original game.

Among the various solutions discussed in this chapter, the bargaining set is the most intuitive one. Why is it that only a few papers study it for classes resulting from an applied field? One may imagine that the reason is the difficulty in computing the bargaining set, compared with the computation of the kernel and the nucleolus. In my opinion the reason is different: economic theory has many beautiful results connected with the core of an economy. Thus, it became natural to impose conditions on the economy which guarantee that the core is not empty. Sometimes such conditions are artificial, such as the requirement that utilities be quasi-concave. If you are dealing with a game with a nonempty core, why should you bother to search for the bargaining set, which is usually a larger set? Why should you look for outcomes having no justified objection if you can point to outcomes without objections whatsoever?

Surely one would be interested in studying the nucleolus and the intersection of the kernel with the core if one wants a solution which is *smaller* than the core, and this may be one reason why so much research effort has been dedicated to the computation of these solutions. But why care about the bargaining set?

These considerations are wrong, as I hope to convince the reader. Fortunately, applications of the bargaining set and variants of it are beginning to appear and they are in line with the following comments:

(1) It should be interesting to find the bargaining set in applied cases in which the core is empty. In many economic situations this will be the case, as indicated above. Certainly this is the case in most political conflicts.

(2) For several classes of games one can prove that *the bargaining set coincides with the core*. This, for example, is the case for *convex games* [Maschler, Peleg and Shapley (1972)] and for *big boss games* [Potters, Muto and Tijs (1990)]. In those cases the recommendation for the core is enhanced: *not only points outside the core are subject to objections – they are even subject to justified objection* and therefore "should not even be considered" reasonable outcomes.

(3) Maschler (1976) discussed an example of a five-person bilateral market game with a nonempty one-point core, due to Postlewaite and Rosenthal (1974). It is claimed that the core does not make sense, because in the core *syndication is disadvantageous* [see also Aumann (1973)].[124] The bargaining set, on the other hand, is more acceptable in that example, because it does not exhibit this phenomenon.

11. Experiments and empirical data

11.1. Laboratory experiments

There may be two basic reasons to conduct experiments designed to test game-theoretical solutions. One of them is descriptive: to find out if, and under what settings, people adopt the principles underlying the various solutions. The second is normative: to find out what goals people have, in order to construct for them solution concepts that will help them achieve their goals. It has been my experience that laymen are sometimes smarter than game theorists, coming up with ideas not previously considered by the experts. Knowledge of such ideas may enable a game theorist to improve his own solution concepts.

---

[124]The core, when it is small, is often not convincing. One disadvantage of the core is that, although it is a very stable set of outcomes, it is not always clear how the players arrive at such outcomes during a negotiation process. For example, if a player gets a low payment in the outcomes of the core, he will do his best not to cooperate with the other players in bringing about a core outcome.

The first experiment designed to test the bargaining set was conducted by the present author in 1961 [and published in Maschler (1978a)]. It was conducted in a highly uncontrolled manner but had the advantage that the subjects (high school children) had no time limit and could continue their negotiations over several sessions. They also supplied short accounts on their "strategies" and reasoning while playing each game. These turned out to be highly illuminating. Other, better controlled experiments, were conducted by Selten and Schuster (1968), Riker (1967, 1971, 1972), Albers (1975, 1978, 1981, 1986) and Albers et al. (1982, 1985), and others (see references designated by **E**). The most extensive series of experiments in this direction were carried out by Kahan and Rapoport (1974, 1977, 1979, 1980) and Rapoport and Kahan (1976, 1979, 1982, 1983). Their experiments were innovative in that they were highly controlled by a computer program. The subjects could negotiate only by using a fixed set of statements, electronically transmitted to their partners. The reader is referred to the comprehensive book of Kahan and Rapoport (1984) where their experiments, and experiments by others, are surveyed and evaluated. The book is quite supportive of the bargaining set and its modifications (see subsequent discussion) so, to balance it, I refer the reader also to Selten (1987). In that paper Selten puts forward the idea that a descriptive theory should not rely on normative solution concepts, because "ordinary people" do not reason using normative principles. To construct a descriptive theory, Selten argues, one has to look at real data and find out how people really behave. If the attempt is successful, the resulting theory may perhaps be less elegant, having to take into account many cases, each with a different formula for the possible outcomes, but it will have the advantage of yielding better predictions. Selten puts forward such a theory for the case of three-person games and claims that it indeed explains real life better.

The rest of this section will be devoted to some personal evaluations of a general nature concerning the outcomes of the various experiments, mainly from a normative point of view. It will be argued that "real life", even in a laboratory, often presents aspects deeper than the ideas encompassed by the bargaining set. This suggests that further research should be carried out if one wants to construct a normative theory which can be recommended to people facing a cooperative game.

One cannot expect people to reach the bargaining set on their own, even if they play a generic game with five or six players. How could they, if the best computers of today cannot calculate the bargaining set? People simply cannot realize the various objections and counter-objections that may be present, and will not reach the bargaining set even if they wish to achieve a solution in which for every objection there is a counter-objection.

Do people behave according to the predictions of the bargaining set if presented with a game involving few players, or a game with a simple characteristic function? If so, would it indicate that people *wish* to settle at outcomes for which every objection can be countered? If that were the case, we could suggest to them to consult an expert, when the need arises, and ask him to compute the bargaining set for them, if he can.

In many experiments the bargaining set comes out quite nicely.[125] See, for example, the experiments of Riker, Selten and Schuster, Kahan and Rapoport, and Rapoport and Kahan, cited above. In my opinion, this *does not prove* that people actually consciously wish to settle where justified objections are impossible. It could be that the outcomes of the bargaining set (for the games that were tried, and under the conditions that prevailed in the experiments) have other "stability properties" – not related to the concept of immunity against a justified objection, and these attract people. Thus, although I feel that stability as reflected by the bargaining set has an intuitive appeal, I cannot say that experiments have proven that this kind of stability is what motivates people while reaching their compromises.[126]

But suppose people do seek this kind of stability. Does it mean that they will reach an outcome in the bargaining set, given that the game is simple enough? I shall present here two examples in which the outcomes of experiments (already observed in my experiment) were different, in which I must admit that "rightly so".

Example 11.1. The subjects played many three-person zero-normalized games where $v(12) = v(13) = v(123) = 90$, $v(23) = 0$. At first they ended up with an outcome close to $(90, 0, 0)$, which is the unique outcome in the bargaining set for all nontrivial coalition structures. Soon, the weak players realized their weaknesses and decided to flip a coin so that the loser would drop from the negotiations leaving the other weak player to bargain with player 1 and settle close to an even split. Thus, for a while the players ended up (in expectation) with $(45, 22.5, 22.5)$. This did not last long. The strong player realized the threat and offered one of the weak players somewhat more than 22.5 "for sure", which was enough to detract him from tossing the coin with the other weak player. Thus the final games ended up close to $(67.5, 22.5, 0)$ or $(67.5, 0, 22.5)$. I must admit that I too will accept such a compromise. We see that *the bargaining set does not take into account threats to leave the game, split into negotiation groups*, etc. To take such aspects into account, I argued [Maschler (1963b)] that although the players were presented with a characteristic function, in view of some "standards of fairness"[127] they perceived the game

---

[125]At least in accordance with the discussion following Example 3.4.

[126]It is true that one often observes that people agree to compromise when someone presents them with a justified objection, but to the best of my knowledge such observations were never tested and documented rigorously.

[127]Such as feeling that if two players remain alone, they would split evenly.

as $w(1) = 45$, $w(2) = w(3) = 0$, $w(12) = w(13) = w(123) = 90$, $w(23) = 45$. The function w is called *the power function* of the game. The payoff $(67.5, 22.5, 0)$ is the unique outcome in the bargaining set of $(123, w)$ for the coalition structure $\{12, 3\}$.

We refer the reader to Kahan and Rapoport (1984), in which one sees that the power concepts "explained" outcomes in many other games.[128]

Example 11.2. The game was a *quota game* of four players, the quota being $\omega = [10, 20, 30, 40]$[129]. We expected that for every nontrivial c.s., the players will end up near $(10, 20, 30, 40)$, which is the nucleolus of the game and a reasonable outcome anyhow. Only half of the games ended up this way. All the others ended up with one pair of players splitting according to the quota and the other splitting differently.[130] It is easy to explain this result: as long as four players are playing, there is a pressure to share according to the quota. The moment two players remain there is also a pressure to share equally. Indeed, in all the cases of the experiment, the players who shared far from the quota were the last coalition to form![131] In two games player 1 was smart: he realized that although he is weak, *his condition would improve if he waited until another pair forms a coalition.* He therefore purposely placed extravagant demands to discourage others from proposing a coalition with him. Once another pair formed, he would approach the one left alone about an equal split. Thus, in this game, it is clear that player 1 should wait and the other players should hurry to form a coalition, so as not to remain stuck with player 1. We thus see that the bargaining set for coalition structures "assumes" that all coalitions form simultaneously, whereas in effect this is not the case in real life and there is, moreover, a strategic element concerning the questions when to form a coalition and with whom, and when to wait. This aspect is not taken care of in the theory so far.[132]

To conclude, the bargaining set, though an intuitively appealing solution

---

[128]The reader is referred to Rapoport and Kahan (1982) for more results concerning the interplay between the value and the power of a coalition.

[129]Namely, $v(ij) = \omega_i + \omega_j$, all i, j in 1234.

[130]It is remarkable to realize that such pattern of splits constitutes a von Neumann–Morgenstern solution of (essentially) this game, as discovered in Shapley (1953a).

[131]And all the share for such coalitions were between the equal share and the quota share. The players whose quota was higher than the equal share demanded and received some compensation for the past high aspiration. I believe that this phenomenon deserves to be analyzed and explained by psychologists, being one simple case where normative theories (at present) differ from descriptive ones.

[132]Even in the four-person quota game discussed above it is clear that player 1 would rather have players 2 and 3 form a coalition than, say 3 and 4, because he would rather remain with player 4 and 50 to share that with player 2 and only 30 to share. How much will he be willing to pay players 2 and 3 in order to encourage them to form a coalition first? Bezalel Peleg and I believe that we have some good answers to this question, but this research is still in progress.

concept, should be regarded only as a starting point for more elaborate theories which should be constructed in order to be able to arrive at meaningful recommendations in real cases. From the normative point of view, laboratory experiments yield important guidelines on how to proceed because they enable us to see not only what people really want, but how some smart people may grasp situations that mathematical models still oversimplify.

11.2. Evidence from empirical data

Real-life data cannot be used to check rigorously whether people really behave in accordance with any of the above solution concepts. Real-life conflicts are complicated situations, encumbered with a lot of noise, and modelling them, even as games without side payments, is usually only a crude approximation of reality, both because the factors which determine the characteristic function are not fully known and because a characteristic function itself cannot capture the full intricacies of a real situation. Obviously, representing a real situation such as a political election as a constant-sum weighted majority game with side payments, where a coalition is winning iff it has a majority, means oversimplifying reality: we are abstracting away ideologies, personal affinities, considerations about present behavior in a way that should enhance the chances of a party to get more votes in the next elections, and many other relevant issues. But even if we are going to commit ourselves to such oversimplifications as we do in this subsection, how are we going to check if a certain portfolio distribution belongs to any solution concept, if we do not know how to translate these portfolios into payments? How can we know how much each portfolio is worth to every party?[133] Any determination of such payoffs seems to rely on ad hoc principles. Clearly, any good correlation with any solution concept, based on real data, is quite unbelievable. Nevertheless, Schofield (1976, 1978, 1982) dared to examine various European governments[134] and found some interesting fits with modified bargaining sets, and even with the kernel in some cases.

Peleg (1981) undertook a less ambitious task by asking himself if the nucleolus (and other solution concepts, as well as other criteria) can be used to predict which coalition actually forms. We shall present here some of his findings:

In many weighted majority games there is a player (= party) and a winning coalition containing him, such that in this coalition he holds a strict majority.

---

[133]Not to mention other, usually secret promises to support political issues and to allocate more budget to items which interest members of the winning coalition.

[134]He proposed to estimate the payoff x_i to party i by $x_i = p(i)/p(M)$, where $p(M)$ is the total number of portfolios and $p(i)$ is the number of portfolios received by party i.

There can be at most one such player, who is called *a dominant player*. We shall restrict our attention to real-life weighted majority games that contain such a player. Furthermore, we shall restrict ourselves to real-life cases where the coalitions that actually formed often contained the dominant player. This was the case in parliaments in Denmark, Israel, Italy, The Netherlands, and Sweden, in various periods of this century.[135] This was also the case in many city councils in Israel.

Peleg (1981) examined these cases and tested them against several conjectures based on the reasonable assumption that the dominant player is given the mandate to form the coalition. Of the various conjectures, two proved successful.

(1) The dominant player will, in fact choose, a winning coalition in which he holds a strict majority. This was the case in 56 out of the 67 parliaments examined (84 percent) and in 33 out of 41 city councils (80.5 percent).

(2) The dominant player will choose that winning coalition in which *his nucleolus payoff for the resulting c.s. is highest*. This occurred in 45 parliaments (67 percent) and, even more impressively, in 35 city councils (85.4 percent).

That the first conjecture turned out successful is not surprising: common sense dictates it. That the second conjecture had relative success is more impressive: obviously, the parties did not compute any nucleolus. Can this behavior be explained on intuitive grounds?

We refer the reader to Grofman (1982) and Straffin and Grofman (1984) who introduce ideologies into the model and come up with a nontransferable utility game (see Section 13). They introduce a bargaining set appropriate to that model, which may be empty for some c.s.'s. A reasonable conjecture is that a winning c.s. will not form if the bargaining set for that c.s. is empty. They come up with the finding that their bargaining set is inferior to another solution which they propose for the situation.

12. Games without side payments

The generalization of the concepts "objection" and "counter-objection" to games without side payments[136] $(N; V)$ is straightforward. Therefore it is clear what the bargaining set $\mathcal{M}_1^i$ is. Unfortunately, as noted already by Peleg

---

[135]Exceptions in other European countries, such as France, Finland, and Norway, were usually due to the fact that the dominant player held an extreme ideological position and was therefore excluded from the cabinet.

[136]As usual, we assume (for a zero-normalized game) that $V(i) = 0 - \mathfrak{R}_+$, and that for every S, $S \subseteq N$, $V(S)$ is a comprehensive subset of $\mathfrak{R}^S$, whose intersection with $\mathfrak{R}_+^S$ is not empty and compact. Here, $\mathfrak{R}^N$ is the utility space of the set of players N. It is customary to denote the characteristic function of a game without side payments by a capital letter, to remind that its values are sets and not numbers.

(1963c), it may be empty even for three-person games. The reason is that *the relation "a player has a justified objection against another player" need not be acyclic*. The task then is to find another definition for a bargaining set that satisfies the following desiderata:

(1) It should be nonempty for every coalition structure.

(2) It should coincide with $\mathcal{M}_1^i$ if the game happens to be a side payment game.[137]

(3) It should be accepted intuitively.

Of course, the resolution of the task depends on the determination of "what is intuitively acceptable". One possibility, suggested in Billera (1968, 1970), is to decide that if there is a cycle of players $i_1, i_2, \ldots, i_t, i_1$ (belonging to the same coalition of the c.s.) such that at an imputation x (for the c.s.) each player in the cycle has a justified objection against the next one, we would still say that no player in this cycle has any claim against the next one. The idea is that there will be a *liquidation* of demands along the cycle, in which a player would tell his predecessor: "relinquish your claim against me and I in return will waive my claim against the next on the cycle on condition that he will waive his claim against the next one, etc. Eventually, your predecessor will relinquish the claim against you." The resulting bargaining set is the set of imputations x having the property that if a player k has a justified objection against a player l at x, then a chain of justified objections exists at x, going from l back to k. This bargaining set is called *the ordinal bargaining set*, and is denoted by $\mathcal{M}^0$. It was introduced in Asscher (1975b, 1976a, 1976b), where he proved, using a lemma of Billera (1968, 1970), that *it is never empty for every c.s. and coincides with $\mathcal{M}_1^i$ if the game happens to be a side payment game*. This bargaining set need not be a closed set, as was shown in Yarom (1982). In this work Yarom also proposed a nonempty subset of $\mathcal{M}^0$, called $\mathcal{M}^c$, which is a closed set and varies upper semicontinuously with the characteristic function. It also coincides with $\mathcal{M}_1^i$ for games with side payments.

Is it intuitively acceptable?

Yes, if we regard the bargaining set as a vehicle whose purpose is to *rule out* outcomes. At imputations outside $\mathcal{M}^0$ there are justified objections that cannot be liquidated. One advantage of this bargaining set is that its definition is simple and therefore its computation might not be too difficult.

No, if we want the payoffs in $\mathcal{M}^0$ to have a reasonable kind of stability. The point is that one cannot talk about liquidation without referring to the *size* of the debts being waived. A person would not agree to discard a large debt in return for a small debt against him being waived. The situation has an added complication in our case when side payments are impossible: in such games, if a player l passes some utility to player k while staying on the Pareto surface

---

[137]I.e., if $V(S) = \{x \in \Re^S : x(S) \le c_S\}$, for every coalition S, where the c_S's are constants.

(for the c.s.), and if thereby player k's utility increases by an amount α, player l's utility decreases by an amount β which is usually different from α. We therefore have to talk about *transfers* at x, $x \in X(\mathscr{B})$, represented by pairs $(\alpha_{k,l}, \beta_{k,l})$, to mean passing to $(x_1, \ldots, x_k + \alpha_{k,l}, \ldots, x_l - \beta_{k,l}, \ldots, x_n)$ which is on the same $X(\mathscr{B})$.

Let us now make the assumption[138] that for each S, $V(S)$ is convex and satisfies *nonlevelness*, namely that its Pareto hyper-surface contains no segment parallel to an axis. Under these restrictions, $\alpha_{k,l}$ is a strictly monotone function of $\beta_{k,l}$.

It is straightforward to show that *if a player k has a justified objection against l at x in $X(\mathscr{B})$, then there exists a unique transfer $(\alpha_{k,l}, \beta_{k,l})$ with $\alpha_{k,l}$ being minimal under the requirement that after the transfer, k has no longer a justified objection against l.* These minimal transfers should be considered the *claims* resulting from justified objections.

Now, let $(i_1, i_2, \ldots, i_t, i_1)$ be a cycle of players, each having a justified objection against the next one at x in $X(\mathscr{B})$. These justified objections determine claims $(\alpha_{i_1, i_2}, \beta_{i_1, i_2}), \ldots, (\alpha_{i_t, i_1}, \beta_{i_t, i_1})$. *We shall say that the players on the cycle will agree to waive their claims if for each one of them $\alpha_{i_k, i_{k+1}} \leq \beta_{i_{k-1}, i_k}$.* Here, $t + 1$ means 1. This is justified because, if each player owes more than he claims, each will be glad of a liquidation proposal. With this terminology and these definitions, Asscher (1975b, 1977) defines the *cardinal bargaining set $\mathscr{M}^c$* to be the set of outcomes x (for each c.s.), such that at x the graph of justified objections can be decomposed into cycles, and the claims can be split along these cycles[139] in such a way that all the players will agree to waive their claims. Asscher then proves that *for every coalition structure the cardinal bargaining set is a nonempty subset of the ordinal bargaining set.* Obviously, if the game has side payments, it coincides with $\mathscr{M}_1^i$.

Thus, we have a more convincing bargaining set, but the price one pays is that its definition is quite complicated. Asscher (1975a, 1975b, 1976a, 1976b, 1977) studies both bargaining sets for three-person games. One of his examples is quite striking: it is an example of a game with a lot of symmetry. Its ordinal bargaining set, as well as the cardinal, consists of four points, three of which are core points, none of which reflects the symmetry, whereas the fourth is an equal share point on the Pareto surface of $V(123)$. A slight modification of the characteristic function annihilates the core, but the symmetric payoff remains. In this example, at least, the bargaining sets seem superior to the core.

We refer the reader to the works of Mas-Colell, Vohra, Dutta et al., and Grodal, reported in Section 8, for other bargaining sets defined over the class of cooperative games without side payments (nontransferable utility cooperative games).

---

[138] In addition to the previous assumptions.
[139] A split is necessary if an edge belongs to more than one cycle.

Research concerning the extension of the kernel and the nucleolus to games without side payments is still scarce, and I shall not report on it. Let me just say that the main issue is to decide what the analogue of the excess functions should be. There have been several suggestions, starting with Kalai (1972, 1973, 1975) and continuing with Vilkov (1974, 1979), Nakayama (1983), and McLean and Postlewaite (1989). So far these attempts are not yet crystallized to a general theory similar to the one we have in the side payment case. The reader can find the relevant literature in the Bibliography marked **N**.

13. Games with an infinite number of players

13.1. Games with a countable number of players

Wesley (1971) already observed that if a game has a countable number of players, its kernel may be empty. He then used the theory of nonstandard analysis to give *conditions that guarantee that a superadditive game with a countable number of players will have a nonempty kernel for every coalition structure*. Essentially he demands that the "tails" of infinite coalitions contribute little to their worths and the $\sum_{j=1}^{\infty} \Omega(j) < \infty$, where $\Omega(j) = \sup_S[v(S) - v(S\setminus\{j\})]$. It is known that every theorem on real numbers, which can be proved by the methods of nonstandard analysis, can also be proved by standard topological methods. Nevertheless, the novelty of using nonstandard analysis lies in the fact that it helps to *find* the theorems. Theorems on games with a finite number of players can almost automatically be translated into theorems concerning games with a countable number of players, using nonstandard analysis. [In this connection see also Geanakoplos (1978) and Lewis (1983, 1985a, 1985b, 1985c).]

A different approach to games with a countable number of players can be found in Naumova (1973). She looks for limitations on the coalition structures which guarantee that the kernel, and therefore the bargaining set, will not be empty for these structures.[140] She proves that *the kernel for a c.s. $\mathcal{B}$ will not be empty if each coalition in $\mathcal{B}$ is finite and $\sum_{B \in \mathcal{B}} v(B) < \infty$.*

13.2. Limit behavior

In this subsection we report an interesting result obtained by Shapley and Shubik (1972a) [see also Shubik (1985)]. The authors consider (nk)-person games, where n is the number of a finite fixed set N of *types*, and k is a positive integer considered a variable that eventually will tend to infinity. Each type in such a game has k members. The worth of a coalition S is assumed to depend

---

[140] Her results extend also to her modification of the bargaining set (see Section 8).

only on its *profile* $\sigma(S) = \sigma = (\sigma_1, \ldots, \sigma_n)$, where $\sigma_i = \sigma_i(S)$ is the number of players of type i in S. More specifically, it is assumed that $v(S) = \phi(\sigma(S))/k$, where ϕ is a real function defined in $\Re_+^N$, concave, positively homogeneous of degree 1, and *continuously differentiable in* $\Re_+^N \setminus \{0\}$. Note that $v(kN) = \phi(1, 1, \ldots, 1)$ is independent of k.

Such games result from economic models as follows. One regards ϕ as a production function, each player-type providing a different input and all players having linear utility for the output. Denoting the above game by $k\Gamma$, Shapley and Shubik prove that *for every positive ϵ, the bargaining set $\mathcal{M}_1^i$ of $k\Gamma$ is contained in the strong ϵ-core of this game, provided that k is large enough.*[141]

Note that these conditions imply that the (nonempty) core itself converges to a single point,[142] namely the payoff generated by the competitive prices of the underlying economy [Debreu and Scarf (1963)]. The same conditions ensure the convergence of the Shapley value to this competitive payoff [Shapley (1964b)].

13.3. Games with a continuum of players

At present, there does not exist a generalization of the bargaining set $\mathcal{M}_1^i$ and the kernel to games with a continuum of players. The reason is that single players play only a small role in most models of such games, so it is not clear what should replace an objection of a single player against a single player if we have a continuum of players. One is faced with two choices.

One choice is to find definitions which will make sense also for the continuous case and generate the original solutions in the discrete case. This approach was taken by Bird (1976a) *for the case of the nucleolus.* Of course, the lexicographic minimization of $\theta(x)$ makes no sense – there are just too many coalitions – but Bird noticed that Schmeidler (1969a) used an alternative definition to prove the continuity of the nucleolus, and that definition made perfect sense for games with a continuum of players. Also, Kohlberg (1971) had a characterization which made sense in the continuous case. Bird examined the two possibilities and fortunately found that *they yielded the same set of imputations*, so it appears that we are having a perfect generalization for the concept of a nucleolus for games with a continuum of players.

But there was a price to pay. *For some games the nucleolus was an empty set and for others it consisted of more than one point.*[143] Thus, even if the nucleolus

---

[141]Actually they prove that even another bargaining set, which contains $\mathcal{M}_1^i$, is a subset of the strong ϵ-core.

[142]Players of the same type get the same payoff in the core of $k\Gamma$, $k \geq 2$, so that convergence may be taken in the sense of replacing each core point of $k\Gamma$ by a point in $\Re^N$, in which one payment is taken for each type.

[143]In many important cases the nucleolus coincided essentially with the least core.

of a game is not empty, one still would like to single out a unique point which makes sense intuitively. As an example of Bird's work we would like to sight the following result.

The game is represented as a triple $\Gamma := (v, \mathscr{B}, X)$, where v is a real function defined for all *coalitions* S, $S \in \mathscr{B}$, where $\mathscr{B}$ constitutes a Borel field on a subset X of $\mathfrak{R}$. We assume that $v(\emptyset) = 0 \leqslant v(S) \leqslant v(X) = 1$ for all sets S. The set of imputations I is considered to be the set of all countably additive non-negative measures, μ, satisfying $\mu(X) = 1$.

For two imputations μ_1 and μ_2, we say that $\mu_1 \leqslant^* \mu_2$, if

$$\sup_{\{S:\ \mu_2(S) > \mu_1(S)\}} (v(S) - \mu_1(S)) < \sup_{\{S:\ \mu_1(S) > \mu_2(S)\}} (v(S) - \mu_2(S)).$$

The nucleolus of Γ is the set of imputations that are minimal under the relation $\leqslant^*$.

A game Γ is called *an orthogonal vector measure game* if $v(S) = f(\mu_1(S), \ldots, \mu_n(S))$, where $\mu_1, \ldots, \mu_n$ are imputations, whose supports are n disjoint sets whose union is X. We normalize f by requiring that $f(0) = 0$ and $f(1) = 1$.

Theorem 13.1. *Every nonatomic[144] orthogonal vector measure game has a nonempty nucleolus.*

Another choice in trying to extend the concepts to games with a continuum of players is to look for other, say, bargaining sets, which do not employ objections of single players against single players. This approach was taken by Mas-Colell, Vohra, Dutta et al. and Grodal, whose works were reported in Section 8.

Bibliography

We have indicated at the end of each paper connected with the bargaining set–kernel–nucleolus theory the topics discussed in the paper. We use the following key:

  A – Applications of the theory.
  C – Computation of solutions.
  D – Dynamic systems leading to solutions.
  E – Experiments that test the solutions.
  F – properties and solutions to Families of games.
  G – General related subjects.
  H – Historical evidence.
  I – Infinite number of players.
  N – Non-transferable utility games.
  R – Related solution concepts.
RR – Remotely Related topics.
  T – The Theory of the solutions.

[144]The result is true also if one allows the measures to contain a finite number of atoms.

Albers, W. (1975) 'Zwei Lösungskonzepte für kooperative Mehrpersonenspiele, die auf Anspruchsniveaus der Spieler basieren', in: R. Henn et al., eds., *Operations Research Verfahren XXI*. Meisenheim am Glan: Hain, pp. 1–13. **ER**

Albers, W. (1978) 'Block forming tendencies as characteristic of the bargaining behavior in different versions of apex games', in: H. Sauermann, ed., *Beiträge zur Experimentellen Wirtschaftsforschung, Vol. VIII: Coalition forming behavior*. Tübingen: J.C.B. Mohr, pp. 172–203. **EF**

Albers W. (1979a) 'Grundzüge einiger Lösungskonzepte, die auf Forderungsniveaus der Spieler Basieren', in: W. Albers, G. Bamberg and R. Selten, eds., *Entscheidungen in kleinen Gruppen*. Meisenheim am Glan: Hain, pp. 9–37. **R**

Albers W. (1979b) 'Core-and-kernel-variants based on imputations and demand profiles', in: O. Moeschlin and D. Pallaschke, eds., *Game theory and related topics*. Amsterdam–New York–Oxford: North-Holland, pp. 3–16. **R**

Albers W. (1981) 'Some solution concepts based on power potentials', in: O. Moeschlin and D. Pallaschke, eds., *Game theory and mathematical economics*. Amsterdam–New York–Oxford: North-Holland, pp. 3–13. **ER**

Albers W. (1986) 'Reciprocal potentials in apex games', in: R.W. Scholz, ed., *Current issues in West German decision research*. Frankfurt, New York. **ER**

Albers, W., H.W. Crott and J.K. Murningham (1985) 'The formation of blocks in an experimental study of coalition formation', *Journal of Occupational Behavior*, 6: 33–48. **ER**

Albers, W., H.W. Crott and R.W. Scholz (1982) 'Equal division kernel and reference coalitions in three-person games: Results of an experiment', Working papers, Institute of Mathematical Economics, Bielefeld. **ER**

Alvarado, R. and S. Chavarria (1980) 'A game with a priori unions', *Ciencia y Tecnologia*, 4: 21–36. **A**

Ameljanczyk, A. (1980) 'Multicriterial solution of *n*-person cooperative games and their properties', *Wrocław Technical University, System Science*, 6: 279–284. **R**

Anderson, S.L. and E.A. Traynor (1962) 'An application of the Aumann–Maschler *n*-person cooperative game', in: M. Maschler, ed., *Recent advances in game theory*, Proceedings of a conference held in Princeton, in October 1961, privately printed for members of the conference. Princeton: Princeton University Conference, pp. 265–270. **A**

Asscher, N. (1975a) 'Bargaining set for 3-person games without side payments', Center for Research in Mathematical Economics and Game Theory, Inst. of Math., The Hebrew University of Jerusalem, RM11. **N**

Asscher, N. (1975b) 'Bargaining set for cooperative games without side payments', Ph.D. Thesis, The Hebrew University of Jerusalem, September [in Hebrew]. **N**

Asscher, N. (1976a) 'An ordinal bargaining set for games without side payments', *Mathematics of Operations Research*, 1: 381–389. **N**

Asscher, N. (1976b) 'Bargaining sets for games without side payments', Ph.D. Thesis, The Hebrew University of Jerusalem.

Asscher, N. (1977) 'A cardinal bargaining set for games without side payments', *International Journal of Game Theory*, 6: 87–114. **N**

Aubin, J.P. (1979) 'Mathematical methods of game and economic theory', *Studies in Mathematics and its Applications*, 7. Amsterdam–New York: North-Holland. **T**

Aumann, R.J. (1960) 'Linearity of unrestrictedly transferable utilities', *Naval Research Logistics Quarterly* 7: 281–284.

Aumann, R.J. (1967) 'A survey of cooperative games without side payments', in: M. Shubik, ed., *Essays in mathematical economics in honor of Oskar Morgenstern*. Princeton: Princeton University Press, pp. 3–27.

Aumann, R.J. (1973) 'Disadvantageous monopolies', *Journal of Economic Theory*, 6: 1–11.

Aumann, R.J. (1989) 'Lectures on game theory', *Underground classics in economics*. Boulder–San Francisco–London: Westview Press. **T**

Aumann, R.J. and J.H. Drèze (1974) 'Cooperative games with coalition structures', *International Journal of Game Theory*, 3: 217–237. **T**

Aumann, R.J. and M. Maschler (1961) 'An equilibrium theory for *n*-person cooperative games', *American Mathematical Society Notices*, 8: 261. **TR**

Aumann, R.J. and M. Maschler (1964) 'The bargaining set for cooperative games', in: M. Dresher, L.S. Shapley and A.W. Tucker, eds., *Advances in games theory*. Princeton: Princeton University Press, pp. 443–476. **TR**

Aumann, R.J. and M. Maschler (1985) 'Game theoretic analysis of a bankruptcy problem from the Talmud', *Journal of Economic Theory*, 36: 195–213. **AH**

Aumann, R.J., B. Peleg and P. Rabinowitz (1965) 'A method for computing the kernel of *n*-person games', *Mathematics of Computation*, 19: 531–551. **C**

Aumann, R.J., P. Rabinowitz and D. Schmeidler (1966) 'Kernels of superadditive simple 5-person games', RM No. 18, Research Program in Game Theory and Mathematical Economics, Dept. of Mathematics, The Hebrew University of Jerusalem. **C**

Baran-Marzak, Y. and D. Encaoua (1978) 'Determination numerique de solutions d'un jeu dans l'ensemble de negociation', RM, Sciences Economiques–Sciences Humaines–Sciences Juridiques et Politiques, Universite de Paris I, Pantheon–Sorbonne. **TC**

Baton, B. and Lemaire, J. (1981) 'The bargaining set of a reinsurance market', *Astin Bulletin*, 12: 101–114. **A**

Beharav, J. (1983) 'Adapting the formula of the kernel for cooperative games', (of Maschler/Peleg) to the computer', M.Sc. Thesis, Tel-Aviv University [in Hebrew]. **C**

Bennett, E. (1980) 'Coalition formation and payoff distribution in cooperative games', Thesis, Northerwestern University, Evanston, Illinois. **R**

Bennett, E. (1983) 'The aspiration approach to predicting coalition formation and payoff distribution *n* sidepayment games', *International Journal of Game Theory*, 12: 1–28. **R**

Bennett, E. (1984) 'A new approach to predicting coalition formation and payoff distribution in characteristic function games', in: M.J. Holler, ed., *Coalitions and collective actions*. Würzburg: Springer-Verlag. **R**

Bennett, E. (1985) 'Endogeneous vs. exogeneous coalition formation', *Economie Appliquée*, 37: 611–635. **TR**

Bennett, E. (1988) 'The aspiration core, bargaining set, kernel and nucleolus', Department of Economics, Working Paper Series, University of Kansas, Lawrence, Kansas. **R**

Bennett, E. and W. Zame (1988) 'Bargaining in cooperative games', *International Journal of Game Theory*, 17: 279–300. **RTN**

Billera, L.J. (1968) 'On cores and bargaining sets for *n*-person cooperative games without side payments', Ph.D. Thesis, The City University of New York. **N**

Billera, L.J. (1970) 'Existence of general bargaining sets for cooperative games without side payments', *Bulletin American Mathematical Society*, 76: 375–380. **N**

Billera, L.J. (1971) 'Some recent results in *n*-person game theory', *Mathematical Programing*, 1: 58–67. **T**

Billera, L.J. (1972a) 'Global stability in *n*-person games', *Transactions American Mathematical Society*, 172: 45–56. **D**

Billera, L.J. (1972b) 'A note on a kernel and the core for games without side payments', Technical Report 152, Dept. of Operations Research, Cornell University, Ithaca, New York. **N**

Billera, L.J. and L.S.-Y. Wu (1977) 'On a dynamic theory for the kernel of an *n*-person game', *International Journal of Game Theory*, 6: 65–86. **D**

Bird, C.G. (1975) 'A class of convex nuclei solution concepts from differences in coalition excesses', *SIAM Journal of Applied Mathematics*, 29: 503–510. **R**

Bird, C.G. (1976a) 'Extending the nucleolus to infinite player games', *SIAM Journal of Applied Mathematics*, 31: 474–484. **I**

Bird, C.G. (1976b) 'On cost allocation for a spanning tree: A game theoretic approach', *Networks*, 6: 335–350. **A**

Bird, C.G. and K.O. Kortanek (1974) 'Game theoretic approaches to some air pollution regulation problems', *Socio-Econ. Plan. Sci.*, 8: 141–147. **A**

Bitter, D. (1982) 'The kernel for the grand coalition of the four-person game', *International Journal of Game Theory*, 11: 215–239. **F**

Bohm-Bawerk, E. Von (1923) *Positive theory of capital* (translated by W. Smart), (original publication 1891). New York: G.E. Steckert.

Bondareva, O.N. (1965) 'Stability in *m*-quota games', *Lietuvos Matematikos Rinkinys* 5: 391–395 [Russian, English and Lithuanian summaries]. **F**

Bondareva, O.N. (1982) 'A production problem and computation of the *n*-core using coverings' (Russian, English summary), *Kibernetika (Kiev)*, 2: 101–104, 135; English translation, *Cybernetics*, 18: 260–265 (1982). **AC**

Bondareva, O.N. (1988) 'Game-theoretic analysis of a one-product "supply and demand" model', *Kibernetika (Kiev)*, 6: 118–120, 135 [Russian, English summary]. **AR**

Bondareva, O.N. (1989) 'The nucleolus of a game without side payment', Working paper No. 176, Institute of Mathematical Economics, Bielefeld University, Germany. **TN**

Bronisz, P. and L. Krůs (1986) 'Interactive system aiding decision making in multiobjective cooperative games. Mathematical background', *Systems and Modelling Simulation*, 3: 387–394. **RN**

Brown, G.W. (1950) 'A method for choosing among optimum strategies', RM–376, Rand Project. **RR**

Brune, S. (1976) 'Computation of the nucleolus for superadditive 4-person games', Working paper No. 48, Inst. of Math. Studies, Bielefeld University. **C**

Brune, S. (1983) 'On regions of linearity for the nucleolus and their computation', *International Journal of Game Theory*, 12: 47–80. **CT**

Bruyneel, G. (1978) 'On balanced sets with application in game theory', *Bull. de la Societe Mathematique de Belgique*, 30: 93–100. **C**

Bruyneel, G. (1979) 'Computation of the nucleolus of a game by means of minimal balanced sets', *Operations Research Verfahren* 34: 35–51. **C**

Buckley, J. and T.E. Westen (1976) 'Bargaining set theory and majority rule', *Journal of Conflict Resolution*, 20: 481–495. **E**

Butnariu, D. (1979) 'Two-dimensional concepts of nucleolus and kernel for an *n*-person game', in: I. Maruşciac and W.W. Breckner, eds., *Proceedings of the 3rd Colloquium on Operations Research held in Cluj-Napoca*, October 20–21, 1978. Universitatea "Babeş–Bolyai", Facultatea de Matematica, Cluj-Napoca, pp. 30–34. **R**

Cesco, J.C. and E. Marchi (1990) 'A general kernel type solution for pure exchange economies', *Revista de Matematicas-Aplicadas*, 11: 29–36. **AR**

Champsaur, P. (1975a) 'How to share the cost of a public good', *International Journal of Game Theory*, 4: 113–129. **ARR**

Champsaur, P. (1975b) 'Upper hemi continuous selection of symmetric allocations in the core of economies with public goods', RM, I.N.S.E.E., Paris. **AR**

Chang, C. (1991) 'Bisection property of the kernel', *International Journal of Game Theory*, 20: 1–11. **T**

Chang, C. (1989) 'Bargaining subsolution for *n*-person games', RM, Institute of Applied Mathematics, National Tsing Hua Univ., Hsinchu, Taiwan. **R**

C. Chang and F-C. Hsiaq (1992) 'An example on quasi zero-monotonic games', *Games and Economic Behavior* (To appear). **T**

Charnes, A. and D. Granot (1973) 'Extensions of convex nucleus solutions to chance constrained games', *Proceedings of Computer Science and Statistics: 7th Symposium of Iowa State University*, pp. 323–331. **RA**

Charnes, A. and D. Granot (1974) 'Prior solutions: Extensions of convex nucleus solutions to chance-constrained games', in: W.J. Kennedy, ed., *Proceedings of Computer Science and Statistics: 7th Annual Symposium on the Interface held at Iowa State University*, Ames, Iowa, October 18–19, 1973. Statistical Laboratory, Iowa State University. **R**

Charnes, A. and D. Granot (1976) 'Coalitional and chance-constrained solution to *n*-person games I: The satisficing probabilistic nucleolus', *SIAM Journal of Applied Mathematics*, 31: 358–367. **RT**

Charnes, A. and D. Granot (1977) 'Coalitional and chance-constrained solutions to *n*-person games II: Two stage solutions', *Operations Research*, 25: 1013–1019. **R**

Charnes, A. and M. Keane (1969) 'Convex nuclei and the Shapley value', Center for Cybernetic Studies, Research Report 12, The University of Texas. **R**

Charnes, A. and K.O. Kortanek (1969) 'On asymptotic behavior of some nuclei of *n*-person games and piecewise linearity of the nucleolus', Management Science Research Report No. 170, Graduate School of Industrial Administration, Carnegi-Mellon University, Pittsburgh, Pennsylvania. **RT**

Charnes, A. and K.O. Kortanek (1970) 'On classes of convex and preemptive nuclei for n-person games', in: H.W. Kuhn, ed., *Proceedings of the 1967 Princeton Mathematical Programming Symposium*. Princeton: Princeton University Press, pp. 377–390. **R**

Charnes, A., J. Rousseau and L. Seiford (1978) 'Complements, mollifiers and the propensity to disrupt', *International Journal of Game Theory*, 7: 37–50. **RR**

Chetty, V.K., Dasgupta, D. and T.E.S. Raghavan (1970) 'Power distribution of profits', Discussion paper No. 139, Indian Statistical Institute, Delhi Campus. **A**

Crott, H.W. and W. Albers (1981) 'The equal division kernel: An equity approach to coalition formation and payoff distribution in n-person games', *European Journal of Social Psychology*, 11: 285–306. **R**

Danilov, N.N. (1986) 'Dynamically stable principles of optimality in cooperative differential games on quick action operation', *Akademia Prikladnaya Matematika i Mekhanika*, 50: 3–16 [in Russian]. English translation: (1986), *Journal of Applied Mathematics and Mechanics*, 50: 1–11. **RRD**

D'Aspremont, C. (1973) 'The bargaining set concept for cooperative games without side payments', Ph.D. Thesis, Stanford University, Stanford, California. **N**

Davis, M. (1970) *Game theory: A non-technical introduction*. New York–London: Basic Books. **G**

Davis, M. and M. Maschler (1963) 'Existence of stable payoff configurations for cooperative games' (abstract), *Bulletin American Mathematical Society*, 69: 106–108. **T**

Davis, M. and M. Maschler (1965) 'The kernel of a cooperative game', *Naval Research Logistics Quarterly*, 12: 223–259. **TF**

Davis, M. and M. Maschler (1967) 'Existence of stable payoff configurations for cooperative games', in: M. Shubik, ed., *Essays in mathematical economics in honor of Oskar Morgenstern*. Princeton: Princeton University Press, pp. 39–52. **T**

Debreu, G. and H. Scarf (1963) 'A limit theorem on the core of an economy', *International Economic Review*, 4: 235–246.

Derks, J.J.M. (1987) 'Decomposition of games with nonempty core into veto-controlled simple games', *Operations Research Spektrum*, 9: 81–85. **RR**

Dragan, I. (1981) 'A procedure for finding the nucleolus of a cooperative n-person game', *Zeitschrift für Operations Research*, pp. 119–131. **C**

Dragan, I. (1982) 'A traffic flow model and its solution by means of the generalized nucleolus', *Libertas Mathematica*, 2: 151–158. **RCA**

Dragan, I. (1985) 'A combinatorial approach to the theory of the bargaining sets', *Libertas Mathematica*, 5: 133–150. **R**

Dragan, I. (1986) 'The bargaining set $\mathcal{M}_0$ for convex games', *Libertas Mathematica*, 6: 175–181. **RF**

Dragan, I. (1987a) 'The multicoalitional bargaining set $\mathcal{M}_0$ versus the one coalitional bargaining set $\mathcal{M}$', *Libertas Mathematica*, 7: 141–148. **R**

Dragan, I. (1987b) 'A bargaining set for games with coalition structures and thresholds', RM, University of Texas, Arlington, Texas. **R**

Dragan, I. (1988a) 'An existence theorem for the modified bargaining set of a cooperative n-person convex game', *Libertas Mathematica*, 8: 55–64. **RF**

Dragan, I. (1988b) 'The compensatory bargaining set of a cooperative n-person game with side payments', Technical report No. 256, Research Center for Advanced Study, Department of Mathematics, The University of Texas at Arlington, Texas. **R**

Dragan, I. (1990) 'The compensatory bargaining set of a big boss game', *Libertas Mathematica*, 10: 53–61. **RF**

Dresher, M. (1961) *Games of strategy*. New Jersey: Prentice Hall. **RR**

Driessen, T.S.H. (1985a) 'Contribution to the theory of cooperative games: The τ value and k-convex games', Ph.D. Thesis, Catholic University, Nijmegen, The Netherlands. **F**

Driessen, T.S.H. (1985b) 'Properties of 1-convex n-person games', *Operations Research Spektrum*, 7: 19–26. **F**

Driessen, T.S.H. (1985c) 'A new axiomatic characterization of the Shapley value', in: *IX Symposium on Operations Research*. Part II 5–8 (Osnabrück, 1984), *Methods of operations research*. Königstein/Ts: Athenäum–Hain–Hanstein, pp. 505–517. **RR**

Driessen, T.S.H. (1986a) 'Advantageous syndicates in a bilateral market game: The case of the τ-value', in: *Proceedings of the 11.SOR Conference at Darmstadt, West Germany*. **A**

Driessen, T.S.H. (1986b) 'Solution concepts of k-convex n-person games', *International Journal of Game Theory*, 15: 201–229. **F**

Driessen, T.S.H. (1987) 'The τ-value: a survey', in: *Surveys in game theory and related topics.* Amsterdam: Math. Centrum, Centrum Wisk. Inform., pp. 209–213.

Driessen, T.S.H. (1988) 'Cooperative games, solutions and applications', Theory and Decision Library, Series C: *Game theory, mathematical programming and mathematical economics.* Dordrecht–Boston–London: Kluwer Academic Press. **TAF**

Driessen, T.S.H. (1991) 'A survey of consistency properties in cooperative game theory', *SIAM Review*, 33: 43–59. **T**

Driessen, T.S.H. and Y. Funaki (1991) 'Coincidence of and linearity between game theoretic solutions', *Operations Research Spektrum*, 13: 15–30. **A**

Driessen, T., S. Muto and M. Nakayama (1989) 'A cooperative game of information trading: The core, the nucleolus and the kernel', Memorandum No. 810, Faculty of Applied Mathematics, University of Twente, The Netherlands. **A**

Driessen, T.S.H. and S.H. Tijs (1983) 'The τ-value, the nucleolus and the core for a subclass of games', *Methods of Operations Research*, 46: 395–406. **F**

Driessen, T.S.H. and S.H. Tijs (1984) 'Game-theoretic solutions for some economic stiuations', *Cahiers Centre Études Rech. Opér.*, 26: 51–58. **A**

Driessen, T.S.H. and S.H. Tijs (1985) 'The τ value, the core and semiconvex games', *International Journal of Game Theory*, 14: 229–248. **F**

Dubey, P. and A. Neyman (1984) 'Payoffs in nonatomic economies: an axiomatic approach', *Econometrica*, 52: 1129–1150.

Dubey, P. and A. Neyman (1988) 'Payoffs in nonatomic economies: an axiomatic approach', in: *The shapley value*, Cambridge: Cambridge University Press, pp. 207–216. **IF**

Dutta, B., D. Ray, K. Sengupta and R. Vohra, (1989) 'A consistent bargaining set', Working paper No. 87–21, Department of Economics, Brown University, Providence, Rhode Island, *Journal of Economic Theory*, 49: 93–112. **R**

Einy, E. (1985) 'On connected coalitions', *International Journal of Game Theory*, 14: 103–125. **A**

Farkas, J. (1902) 'Theorie der einfachen Ungleichungen', *Journal Reine Angew. Math.*, 124: 1–27.

Fischer, D. and D. Gately (1975) 'A comparison of various solution concepts for three-person cooperative games with non-empty cores', Center for Applied Economics, New York University, New York. **R**

Forman, R. and J.D. Laing (1982) 'Metastability and solid solutions of collective decisions', *Mathematical Social Sciences*, 2: 397–420. **RR**

Friedman, J.W. (1986) *Game theory with application to economics.* New York–Oxford: Oxford University Press. **T**

Funaki, Y. (1986) 'Upper and lower bounds of the kernel and nucleolus', *International Journal of Game Theory*, 15: 121–129. **T**

Funk, S.G., Am. Rapoport and J.P. Kahan (1980) 'Quota vs. positional power in four person apex games', *Journal of Experimental Psychology*, 16: 77–93. **E**

Gaidov, S.D. (1989) 'Objection and counter-objection equilibria in many-player stochastic differential games' *Serdica*, 15: 100–107. **RR**

Galil, Z. (1974) 'The nucleolus in games with major and minor players', *International Journal of Game Theory*, 3: 129–140; Also M.Sc. Thesis, Tel-Aviv University, September 1971 [in Hebrew]. **FT**

Galil, Z. (1980) 'Application of efficient megreable heaps for optimization problem on trees', *Acta Informatica*, 13: 53–58. **C**

Gately, D. (1974) 'Sharing the gains from regional cooperation: A game theoretical application to planning investment in electric power', *International Economic Review*, 15: 195–208. **R**

Geanakoplos, J. (1978) *The bargaining set and nonstandard analysis*, Publication TR-1, Harvard University: Center on Decision and Conflict in Complex Organizations. **IA**

Gerard-Varet, L.A. and S. Zamir (1987) 'Remarks on the reasonable set of outcomes in a general coalition function form game', *International Journal of Game Theory*, 16: 123–143. **T**

Granot, D. (1977) 'Cooperative games in stochastic characteristic function form', *Management Science*, 23: 621–630. **T**

Granot, D. (1984) 'A note on the room-mates problem and a related revenue allocation problem', *Management Science*, 30: 633–643. **A**

Granot, D. and F. Granot (1992a) 'A survey on cost and revenue allocation problems', 'Computational complexity of a cost allocation approach to a fixed cost spanning forest problem', *Mathematics of Operations Research* (to appear). **A**

Granot, D. (1992) 'On the reduced game of some linear production games info', RM, Faculty of Commerce and Business Administration, The University of British Columbia, Vancouver, Canada. **F**

Granot, D. and F. Granot (1992b) in: J.J. Rousseau, ed., *Systems and management science by extreme methods*. Dordrecht: Kluwer Academic Publishers, Ch. 27, pp. 427–459. **AFC**

Granot, D. and F. Granot (1992c) 'On some network flow games', *Mathematics of Operations Research*, 17: (to appear). **AF**

Granot, D. and M. Hojati (1990) 'On cost allocation in communication networks', *Networks*, 20: 209–229. **AF**

Granot, D. and G. Huberman (1981) 'Minimum cost spanning tree games', *Mathematical Programming*, 21: 1–18. **AF**

Granot, D. (1982) 'The relationship between convex games and minimum cost spanning tree games: A case for permutationally convex games', *SIAM Journal of Algebraic and Discrete Methods*, 3: 288–292. **AF**

Granot, D. and G. Huberman (1984) 'On the core and nucleolus of minimum cost spanning tree games', *Mathematical Programming*, 29: 323–347. **A**

Granot, D. and M. Maschler (1992) 'Network cost games and the reduced game property', RM, Faculty of Commerce and Business Administration, The University of British Columbia, Vancouver, Canada and Department of Mathematics, The Hebrew University, Jerusalem. **F**

Greenberg, J. (1975) 'Taxation and equilibrium in a market with public goods: In a general analysis and in a game theoretic approach', Ph.D. Thesis, The Hebrew University of Jerusalem [in Hebrew]. **RA**

Grodal, B. (1986) 'Bargaining sets and Walrasian allocations for atomless economies with incomplete preferences', RM, Mathematical Sciences Research Institute, Berkeley and University of Copenhagen. **RAIN**

Grofman, B. (1982) 'A dynamic model of protocoalition formation in ideological n-space', *Behavioral Science*, 27: 77–90. **HRN**

Grotte, J.H. (1970) 'Computation of and observations on the nucleolus, the normalized nucleolus and the central games', M.S. Thesis, Field of Applied Math., Cornell University, Ithaca, N.Y. **TCR**

Grotte, J.H. (1971/72) 'Observations on the nucleolus and the central game', *International Journal of Game Theory*, 1: 173–177. **TF**

Grotte, J.H. (1974) 'The dynamics of cooperative games', Ph.D. Thesis, Cornell University, Ithaca, New York. **DR**

Grotte, J.H. (1976) 'Dynamics of cooperative games', *International Journal of Game Theory*, 5: 27–64. **DR**

Grushko, A.N. (1987) 'A game-theoretic method for constructing an optimal hierarchical structure', in: V.N. Ligunov, ed., *Multistage, hierarchical, differential and non-cooperative games*. Kalinin: Kalinin. Gos. University, pp. 74–82 [in Russian].

Gyachene, V. (1985) 'A defended set for a simple cooperative game', *Matematicheskie Metody v Sotsial'nykh Naukakh. Akad. Nauk Litovsk. SSR, Inst. Mat. i Kibernet., Vilnius*, 18: 21–28 [in Russian, English and Lithuanian summaries]. **RF**

Gyachene, V. (1986) 'A defended set for a cooperative game', Matematicheskie Metody v Socialnykh Naukakh, 19: 20–26 [In Russian, English and Lithuanian summaries]. **R**

Hallefjord, Å., R. Helming and K. Jørnsten (1990) 'Computing the nucleolus when the characteristic functions is given implicitly: A constraint generation approach', RM Chr. Michelsen Institute, Bergen. **CAT**

Hart, S. and Mas-Colell, A. (1989a) 'Potential, value and consistency', *Econometrica*, 57: 589–614. **RR**

Hart, S. and Mas-Colell, A. (1989b) 'The potential of the Shapley value', in: A.E. Roth, ed., *The Shapley value, essays in honor of Lloyd S. Shapley*. Cambridge–New York–New Rochelle–Melbourne–Sydney: Cambridge University Press, pp. 127–137. **RR**

Heaney, J.P. (1979) 'Efficiency/equity analysis of environmental problems – a game theoretic perspective', in: S.J. Brams, A. Schotter and G. Schwödiauer, eds., *Applied game theory*. Wien–Heidelberg: Physica Verlag, pp. 352–369. **A**

Heaney, J.P. and R.E. Dickinson (1982) 'Methods for apportioning the cost of a water resource project', *Water Resources Research* 18: 476–482. **A**

Henss, R. (1986) 'Bargaining strength in three-person characteristic-function games with $v(i) > 0$, a reanalysis of Kahan and Rapoport (1977)', *Theory and Decision*, 21: 267–282. **E**

Henss, R. and M. Momper (1985) 'Neue Ansätze zur Erforschung des Verhandlungsverhaltens in experimentellen 3-Personen-spielen', Arbeiten der Fachrichtung Psychologie, Universität des Saarlandes, No. 97. **E**

Hertzmann, J. (1975) 'The nucleolus and the kernel for a class of market games', RM 12, Center for Research in Mathematical Economics and Game Theory, The Hebrew University of Jerusalem. **FA**

Horowitz, A.D. (1973) 'The competitive bargaining set for cooperative n-person games', *Journal of Mathematical Psychology*, 10: 265–289. **R**

Horowitz, A.D. (1974) Erratum, *Journal of Mathematical Psychology*, 11: 161. **R**

Horowitz, A.D. (1977) 'A test of the core, bargaining set, kernel, and Shapley models in n-person quota games with one weak player', *Theory and Decision*, 8: 49–65. **E**

Horowitz, A.D. and An. Rapoport (1974) 'Test of the kernel and two bargaining set models in four- and five-person games', in: An. Rapoport, ed., *Game theory as a theory of conflict resolution*. Dordrecht–Boston, Massachusetts: D. Reidel. **E**

Huberman, G. (1980) 'The nucleolus and the essential coalitions', in: *Analysis and optimization of systems*, Proceedings of the Fourth International Conference, Versailles, 1980, Lecture Notes in Control and Information Sciences, 28. Berlin: Springer, pp. 416–422. **TC**

Imai, H. (1983) 'Individual monotonicity and lexicographic maxmin solution', *Econometrica*, 51(2): 389–401. **NR**

'International workshop on basic problems of game theory', (1975) Held at Bad Salzuflen, September 2 to 17, 1974. Collection of abstracts, Universität Bielefeld, Arbeiten aus dem Institut für Mathematische Wirtschaftsforschung, Universität Bielefeld, Rheda. **TDA**

Isbell, J.R. (1956) 'A class of majority games', *Quarterly Journal of Mathematics, Oxford Series*, 7: 183–187.

Isbell, J.R. (1958) 'A class of simple games', *Duke Mathematics Journal*, 25: 423–439.

Isbell, J.R. (1959) 'On the enumeration of majority games', *Mathematical Tables and Other Aids to Computation*, 13: 21–28.

Justman, M. (1977) 'Iterative processes with "nucleolar" restrictions', *International Journal of Game Theory*, 6: 189–212. **RAT**

Kahan, J.P. and An. Rapoport (1974) 'Test of the bargaining set and kernel models in three-person games', in: An. Rapoport, ed., *Game theory as a theory of conflict resolution*. Dordrecht–Boston, Massachusetts: D. Reidel. **E**

Kahan, J.P. and Am. Rapoport (1977) 'When you don't need to join: The effects of guaranteed payoffs on bargaining in three-person cooperative games', *Theory of Decision*, 8: 97–126. **E**

Kahan, J.P. and Am. Rapoport (1979) 'The influence of structural relationship on coalition formation in four-person apex games', *European Journal of Social Psychology*, 9: 339–362. **E**

Kahan, J.P. and Am. Rapoport (1980) 'Coalition formation in the triad when two are weak and one is strong', *Mathematical Social Sciences* 1: 11–38. **E**

Kahan, J.P. and Am. Rapoport (1984) *Theories of coalition formation*. Hillsdale, New Jersey–London: Lawrence Erlbaum Associates. **E**

Kalai, E. (1972) 'Cooperative non-sidepayment games: Extensions of sidepayment game solutions, metrics, and representative functions', Ph.D. Thesis, Cornell University, Ithaca, New York. **N**

Kalai, E. (1973) 'Excess functions, nucleolus, kernel and ϵ-core of non-sidepayments cooperative games', Technical Report No. 12, Department of Statistics, Tel-Aviv University. **N**

Kalai, E. (1975) 'Excess functions for cooperative games without sidepayments', *SIAM Journal of Applied Mathematics*, 29: 60–71. **N**

Kalai, G., M. Maschler and G. Owen (1975) 'Asymptotic stability and other properties of trajectories and transfer sequences leading to the bargaining sets', *International Journal of Game Theory*, 4: 193–213. **D**

Kaneko, M. (1978) 'Price oligopoly as a cooperative game', *International Journal of Game Theory*, 7: 137–150. **A**

Kaufman, M. and W.H. Tack (1975) 'Koalitions-bildung und Gewinnaufteilung bei Strategisch Äquivalenten 3-Personen-spielen', *Zeitschrift für Sozialpsychologie*, 6: 227–245. **E**

Keane, M. (1969) 'Some topics in n-person game theory', Ph.D. Thesis, Northwestern University, Evanston, Illinois. **C**

Kikuta, K. (1976) 'On the contribution of a player to a game', *International Journal of Game Theory*, 5: 199–208. **T**

Kikuta, K. (1978) 'A lower bound of an imputation of a game', *Journal of Operations Research Society of Japan*, 21: 457–468. **T**

Kikuta, K. (1979) 'Some results on the contribution of a player in an *n*-person characteristic function game', Ph.D. Thesis, Osaka University. **T**

Kikuta, K. (1982a) 'The nucleolus of a game with only 1, $n - 1$ and *n*-person coalitions', Working paper No. 64, Faculty of Economics, Toyama University Gofuku, Toyama City, Japan. **F**

Kikuta, K. (1982b) 'The noncooperative games and the nucleolus for a class of n-person games', Working paper No. 66, Faculty of Economics, Toyama University, Gofuku, Toyama City, Japan. **FT**

Kikuta, K. (1983) 'On the nucleolus of $(n, n - 1)$ games', RM, Faculty of Economics, Toyama University, Gofuku, Toyama, Japan. **F**

Kikuta, K. (1986) 'A remark on the nucleolus of a cooperative game', Working paper No. 83, Faculty of Economics, Toyama University, Gofuku, Toyama City, Japan. **T**

Knastner, B., C. Kuratowski and S. Mazurkiewicz (1929) 'Ein Beweis des Fixpunktsatzes für *n*-Dimensionale Simplexe', *Fundamenta Mathematica*, 14: 132–137.

Kohlberg, E. (1971) 'On the nucleolus of a characteristic function game', *SIAM Journal of Applied Mathematics*, 20: 62–66. **T**

Kohlberg, E. (1972) 'The nucleolus as a solution of a minimization problem', *SIAM Journal of Applied Mathematics*, 23: 34–39. **CT**

Komorita, S.S. and T.P. Hamilton (1984) 'Effects of alternatives in coalition bargaining', *Journal of Experimental Social Psychology*, 20: 116–136. **E**

Komorita, S.S. and D.A. Kravitz (1979) 'The effect of alternatives in bargaining', *Journal of Experimental Social Psychology*, 15: 147–157. **E**

Kopelowitz, A. (1967) 'Computation of the kernels of simple games and the nucleolus of *n*-person games', RM 31, Research Program in Game Theory and Mathematical Economics, The Hebrew University of Jerusalem. **CF**

Kortanek, K.O. (1973) 'Piecewise linearity and uniform continuity in linear programming in *n*-person cooperative games', R.M, School of Urban and Public Affairs, Carnegie Mellon University, Pittsburgh, Pennsylvania. **T**

Kranich, L.J. (1988) 'Cooperative games with hedonic coalitions', RM, Dept. of Economics, The Pennsylvania State University, University Park, Pennsylvania. **R**

Kukushkin, L.J. (1986) '*N*-kernel stability in games with structural victory functions', *Dokladi Academii Nauk SSSR*, 290: 1045–1047 [in Russian]. **T**

Kukushkin, N.S., I.S. Menshikov, O.R. Menshikova and N.M. Moiseyev (1985) 'Stable compromises in games with structured payoff functions', *Vichislitelnoi Matematiki i Matematicheskoi Fisiki*, 25: 1761–1776, 1918 [in Russian]; English translation, *USSR Computational Mathematics and Mathematical Physics*, 25: 108–116. **TR**

Kukushkin, N.S., O.R. Menshikova and I.S. Menshikov (1986) 'Conflicts and compromises' in: *Current life, science and technology series*: "*Mathematics and cybernetics*", pp. 86–89. Moscow: Znanie [in Russian]. **T**

Kuratowski, K. (1961) *Introduction to set theory and topology*. New York: Pergamon Press.

Laffond, G. and H. Moulin (1981) 'Stability by threats and counterthreats in normal form games', in: J.P. Aubin, A. Bensoussan and I. Ekeland, eds., *Mathematical techniques of optimization, control and decision*. Boston, Mass: Birkhauser, pp. 195–212. **RR**

Laing, J.D. and R.J. Morrison (1974) 'Sequential games of status', *Behavioral Science*, 19: 177–196.

Lee, M., R.D. McKelvey and H. Rosenthal (1979) 'Game theory and the French apparentements of 1951', *International Journal of Game Theory*, 8: 27–53. **NHR**

Legros, P. (1982) 'Disadvantageous syndicates: A note on an example of Maschler', RM, Presented at the Econometric Society European Meeting, Pisa. **A**

Legros, P. (1983) 'The nucleolus and the cost allocation problem', Managerial Economics and Decision Sciences Dept., Northwestern University. **A**

Legros, P. (1984) 'Formation des coalitions et allocation des coûts: une approche par la théorie des jeux', Ph.D. Thesis, University of Paris XII, La Varenne St. Hilaire. **AF**

Legros, P. (1986) 'Allocating joint costs by means of the nucleolus', *International Journal of Game Theory*, 15: 109–119. **AF**

Legros, P. (1987a) 'Computation of the nucleolus of some bilateral market games', *International Journal of Game Theory*, 16: 1–14. **AF**

Legros, P. (1987b) 'Disadvantageous syndicates and stable cartels: the case of the nucleolus', *Journal of Economic Theory*, 42: 30–49. **A**

Lensberg, T. (1985) 'Bargaining and fair allocation', in: H.P. Young, ed., *Cost allocation: Methods, principles, applications*. Amsterdam: North Holland, pp. 101–116. **RR**

Leopold-Wildburger, U. (1987) 'The general tendency of payoff division in a 3-person characteristic function experiment', RM, Dept. of Business, University of Graz, Graz, Austria. **E**

Leopold-Wildburger, U. (1988) 'Payoff divisions on coalition formation in a three-person characteristic function experiment', RM, Dept. of Statistics, Economics and Operations Research, University of Graz, Austria and Department of Operations Research and Mathematical Economics, University of Zürich, Zürich. **E**

Levinsohn, J.R. and Am. Rapoport (1978) 'Coalition formation in multistage three-person cooperative games', in: H. Sauermann, ed., *Beiträge zur Experimentellen Wirtschaftsforschung*, Vol. VIII: *Coalition forming behavior*. Tübingen: J.C.B. Mohr. **E**

Lewis, A.A. (1983) 'The quasi-kernel of nonstandard games', RM, Cornell University, Ithaca, New York. **I**

Lewis, A.A. (1985a) 'Hyperfinite extensions of stable solutions to von Neumann cooperative games', *Southeast Asian Bulletin of Mathematics*, 9: 35–39. **I**

Lewis, A.A. (1985b) 'Hyperfinite von Neumann games', *Mathematical Social Sciences*, 9: 189–195. **I**

Lewis, A.A. (1985c) 'Loeb-measurable solutions to *finite games', *Mathematical Social Sciences*, 9: 197–247.

Lichtenberger, J. (1975) 'Drei-Personen-Spiele mit Seitenzahlungen', Ph.D. Thesis, Universität des Saarlands, Saarbrücken, B.R., Germany. **E**

Lichtenfeld, N. (1976) 'Der Durchschnittskernel. Ein Lösungskonzept für n-Personenspiele', Thesis, Münster. **R**

Lipnowski, I. (1981) 'A re-examination of Coase's theorem and the empty core: Solution concepts from the theory of cooperative games', RM, Dept. of Economics, University of Manitoba, Winnipeg, Manitoba. **A**

Littlechild, S.C. (1974a) 'A simple expression for the nucleolus in a special case', *International Journal of Game Theory*, 3: 21–29. **AFC**

Littlechild, S.C. (1974b) 'A note on the use of game theory in the reduced cost allocation problem', Working paper No. 18, Management Centre, University of Aston, Birmingham. **A**

Littlechild, S.C. and G. Owen (1976) 'A further note on the nucleolus of the airport game', *International Journal of Game Theory*, 5: 91–95. **AFC**

Littlechild, S.C. and G.F. Thompson (1977) 'Aircraft landing fees: A game theory approach', *The Bell Journal of Economics*, 8: 186–204. **AH**

Littlechild, S.C. and K.G. Vaidya (1976) 'The propensity to disrupt and the disruption nucleolus of a characteristic function game', *International Journal of Game Theory*, 5: 151–161. **R**

Lucas, W.F. (1971) 'Some recent developments in *n*-person game theory', *SIAM Reviews*, 13: 491–523. **G**

Lucas, W.F. (1972) 'An overview of the mathematical theory of games', *Management Science*, 18: 3–19. **G**

Lucas, W.F. (1981) 'Applications of cooperative games to equitable allocation', in: W.F. Lucas, ed., *Game theory and its applications*, Proceedings of Symposia in Applied Mathematics, Vol. 24. American Mathematical Society, pp. 19–36. **AG**

Lucas, W.F. (1990) 'Developments in stable set theory', in: T. Ichiishi, A. Neyman and Y. Tauman, eds., *Game theory and applications* (Columbus, Ohio, 1987). San Diego: Academic Press, pp. 300–316. **TF**

Lucchetti, R., F. Patrone, S.H. Tijs and A. Torre (1987) 'Continuity properties of solution concepts for cooperative games', *Operations Research Spektrum*, 9: 101–107. **T**

Luce, R.D. and H. Raiffa (1957) *Games and decisions*. New York: John Wiley and Sons.

Mareš, M. (1973) 'A model of the bargaining in coalition games with side payments', in: *Transactions of the Sixth Prague Conference on Information Theory, Statistical Decision Functions, Random Processes* (Technical University Prague, Prague, 1971; dedicated to the memory of Antonin Špaček), Prague: Academia, pp. 613–629. **R**

Marutian, E.S. (1983) 'On the n-core of the sum of cooperative games with a continuum of players', Akademia Nauk Armyanskoi-SSR. Doklady, 76: 61–64 [in Russian]. **I**

Mas-Colell, A. (1989) 'An equivalence theorem for a bargaining set', MSRI, Berkeley and Harvard Universities, *Journal of Mathematical Economics*, 18: 129–139. **RINA**

Maschler, M. (1962) 'An experiment in n-person games', in: M. Maschler, ed., *Recent advances in game theory*. Princeton: Princeton University Press. **E**

Maschler, M. (1963a) 'n-person games with only 1, $n - 1$, and n-person permissible coalitions', *Journal of Mathematical Analysis and Application*, 6: 230–256. **F**

Maschler, M. (1963b) 'The power of a coalition', *Management Science*, 10: 8–29. **TERR**

Maschler, M. (1964) 'Stable payoff configurations for quota games', in: M. Dresher, L.S. Shapley and A.W. Tucker, eds., *Advances in game theory*. Princeton: Princeton University Press, pp. 477–499. **F**

Maschler, M. (1966) 'The inequalities that determine the bargaining set $\mathcal{M}_1^{(i)}$', *Israel Journal of Mathematics*, 4: 127–134. **TC**

Maschler, M. (1970) 'Game theory A', Lecture notes compiled by N. Megiddo, Academon, The Hebrew University Students' Union Press, The Hebrew University of Jerusalem [in Hebrew]. **G**

Maschler, M. (1973) *Lectures on game theory*, A series of lectures given at the Seminar of the Institute for Mathematical Studies in the Social Sciences, Stanford University, Stanford, California. **G**

Maschler, M. (1976) 'An advantage of the bargaining set over the core', *Journal of Economic Theory*, 13: 184–192. **A**

Maschler, M. (1978a) 'Playing an n-person game: An experiment', in: H. Sauermann, ed., *Beiträge zur Experimentellen Wirtschaftsforschung*, Vol. VIII: *Coalition-forming behavior*. Tübingen: J.C.B. Mohr, pp. 231–328. **E**

Maschler, M. (1978b) 'Lectures on cooperative n-person game theory', Lectures given at the Institute for Advanced Studies, Vienna, Compiled by M. Winkler. **G**

Maschler, M. (1989) 'The concept and the role of consistency in cooperative games', *Proceedings of the Israel Mathematical Union Conference*, Tel Aviv (1987). Tel Aviv: Tel Aviv University. **F**

Maschler, M. (1990) 'Consistency', in: T. Ichiishi, A. Neyman and Y. Tauman, eds., *Game theory and applications*. San Diego: Academic Press, pp. 183–186. **T**

Maschler, M. and G. Owen (1989) 'The consistent Shapley value for hyperplane games', *International Journal of Game Theory*, 18: 389–407. **NTRR**

Maschler, M. and G. Owen (1992) 'The consistent Shapley value for games without side payments', in: R. Selten, ed., *Rational interaction: essays in honor of John C. Harsanyi*. Berlin: Springer-Verlag, pp. 5–12. **NTRR**

Maschler, M. and B. Peleg (1966) 'A characterization, existence proof and dimension bounds for the kernel of a game', *Pacific Journal of Mathematics*, 18: 289–328. **TC**

Maschler, M. and B. Peleg (1967) 'The structure of the kernel of a cooperative game', *SIAM Journal of Applied Mathematics*, 15: 569–604. **T**

Maschler, M. and B. Peleg (1976) 'Stable sets and stable points of set-valued dynamic systems with applications to game theory', *SIAM Journal of Control and Optimization*, 14: 985–995. **D**

Maschler, M., B. Peleg and L.S. Shapley (1972) 'The kernel and bargaining set for convex games, *International Journal of Game Theory*, 1: 73–93. **F**

Maschler, M., B. Peleg and L.S. Shapley (1979) 'Geometric properties of the kernel, nucleolus and related solution concepts', *Mathematics of Operations Research*, 4: 303–338. **T**

Maschler, M., J.A.M. Potters and S.H. Tijs (1992) 'The general nucleolus and the reduced game property', *International Journal of Game Theory* (to appear). **TR**

McKelvey, R.D., P.C. Ordeshook and M.D. Winer (1978) 'The competitive solution for n-person games without transferable utility with an application to committee games', *American Political Science Review*, 72: 599–615. **RN**

McKelvey, R.D., P.C. Ordeshook and M.D. Winer (1979) 'An experimental test of several theories of committee decision making under majority rule', in: S.J. Brams, A. Schotter and G. Schwödiauer, eds., *Applied game theory*. Würzburg: Physica-Verlag, pp. 152–167. **E**

McLean, R.P. and A. Postlewaite (1989) 'Excess functions and nucleolus allocations of pure exchange economies', *Games and Economic Behavior*, 1: 131–143. **NAR**

Medlin, S.M. (1976) 'Effects of grand coalition payoffs on coalition formation in 3-person games', *Behavioral Science*, 21: 48–61. **E**

Megiddo, N. (1971) 'The kernel and the nucleolus of a product of simple games', *Israel Journal of Mathematics*, 9: 210–221. **T**

Megiddo, N. (1972a) 'Composition of cooperative games', Ph.D. Thesis, The Hebrew University of Jerusalem [in Hebrew]. **T**

Megiddo, N. (1972b) 'Nucleoluses of compound games I: The nucleolus of the sum', RM 76, Research Program in Game Theory and Mathematical Economics, The Hebrew University of Jerusalem. **T**

Megiddo, N. (1972c) 'Nucleoluses of compound games II: General compounds with simple components', RM 77, Research Program in Game Theory and Mathematical Economics, The Hebrew University of Jerusalem. **T**

Megiddo, N. (1973) 'On the nucleolus of the composition of characteristic function games', Working paper, Department of Mathematical Sciences, Tel Aviv University, Israel. **T**

Megiddo, N. (1974a) 'Kernels of compound games with simple components', *Pacific Journal of Mathematics*, 50: 531–555. **T**

Megiddo, N. (1974b) 'Nucleoluses of compound simple games', *SIAM Journal of Applied Mathematics*, 26: 607–621. **T**

Megiddo, N. (1974c) 'On the nonmonotonicity of the bargaining set, the kernel and the nucleolus of a game', *SIAM Journal of Applied Mathematics*, 27: 355–358. **T**

Megiddo, N. (1978a) 'Computational complexity and the game theory approach to cost allocation for a tree', *Mathematics of Operations Research*, 3: 189–196. **CAF**

Megiddo, N. (1978b) 'Cost allocation for Steiner trees', *Networks*, 8: 1–6. **RR**

Menshikov, I.S. and O.R. Menshikova (1985) 'Strong equilibrium situations and the n-nucleolus in games with a hierarchical vector of interests', *Journal Vichislitelnoi Matematiki i Matematicheskoi*, 25: 1304–1313, 1437 [in Russian]; English translation, *USSR Computational Mathematics and Mathematical Physics*, 25: 14–20. **TR**

Menshikova, O.R. (1974) 'On computation of the nucleous of some classes of games', Third All-Union Conference on Game Theory, Odessa, USSR [in Russian]. **CF**

Menshikova, O.R. (1976) 'On the computation of the generalized nucleolus', *Journal Vichislitelnoi Matematiki i Matematicheskoi*, 16: 1121–1135, 1370 [in Russian]; English translation, *USSR Computational Mathematics and Mathematical Physics*, 16: 30–45 (1978). **CR**

Menshikova, O. and I. Menshikov (1983) 'The generalized nucleolus as a solution of a cost allocation problem', International Institute for Applied Systems Analysis, A-2361 Laxenburg, Austria, Collaborative paper. **A**

Michaelis, K. (1982) 'On the nature of solution concepts for cooperative games', Ph.D. Thesis, Cornell University, Ithaca, New York. **G**

Michener, H.A. and D.C. Dettman (1987) 'A test of the characteristic function and the Harsanyi function in N-person normal form sidepayment games', *Theory and Decision*, 23: 161–187. **ET**

Michener, H.A., J.A. Fleishman and J.J. Vaske (1976) 'A test of the bargaining theory of coalition formation in four-person groups', *Journal of Personality and Social Psychology*, 34: 1114–1126. **E**

Michener, H.A., J.A. Fleishman, J.J. Vaske and G.R. Statza (1975) 'Minimum resource and pivotal power theories: A competitive test in four-person coalitional situations, *Journal of Conflict Resolution*, 19: 89–107. **E**

Michener, H.A., I.J. Ginsberg and K. Yuen (1979) 'Effects of core properties in four-person games with side payments', *Behavioral Science*, 24: 263–280. **E**

Michener, H.A., W. Potter and M.M. Sakurai (1983) 'On the predictive efficiency of the core solution in side payment games', *Theory and Decision*, 15: 11–28. **E**

Michener, H.A. and M. Sakurai (1976) 'A research note on the predictive adequacy of the kernel', *Journal of Conflict Resolution*, 20: 129–141. **E**

Michener, H.A., M.M. Sakurai, K. Yuen and T.J. Kasen (1979) 'A competitive test of the $\mathcal{M}_1^{(i)}$ and $\mathcal{M}_1^{(im)}$ bargaining sets', *Journal of Conflict Resolution*, 23: 102–119. **E**

Michener, H.A. and M.S. Salzer (1989) 'Comparative accuracy of value solutions in nonsidepayment games with empty core', *Theory and Decision*, 26: 205–233. **NER**

Michener, H.A. and K. Yuen (1982) 'A competitive test of the core solution in side payment games', *Behavioral Science*, 17: 57–68. **E**

Michener, H.A., K. Yuen and I.J. Ginsberg (1977) 'A competitive test of the $\mathcal{M}_1^{(im)}$ bargaining set, kernel and equal share models', *Behavioral Science*, 22: 341–355. **E**

Michener, H.A., K. Yuen and M.M. Sakurai (1981) 'On the comparative accuracy of lex-iconographical solutions in cooperative games, *International Journal of Game Theory*, 10: 75–89. **E**

Milnor, J.W. (1952) 'Reasonable outcomes for n-person games', RM 916, The RAND Corporation, Santa Monica, California.

Moldovanu, B. (1990) 'Stable bargained equilibria for assignment games without side payments', *International Journal of Game Theory*, 19: 171–190. **NRA**

Moulin, H. (1985) 'The separability axiom and equal sharing methods', *Journal of Economic Theory*, 36: 120–148. **RR**

Moulin, H. (1988) *Axioms of cooperative decision making*. New York–New Rochelle–Melbourne–Sydney: Cambridge University Press. **TA**

Murninghan, J.K. and A.E. Roth (1977) 'The effects of communication and information availability in an experimental study of a three-person game', *Management Science*, 23: 1336–1348. **E**

Murninghan, J.K. and A.E. Roth (1978) 'Large group bargaining in a characteristic function game', *Journal of Conflict Resolution*, 22: 299–317. **E**

Murninghan, J.K. and A.E. Roth (1980) 'Effects of group size and communication availability on coalition bargaining in a veto game', *Journal of Personality and Social Psychology*, 39: 92–103. **E**

Murninghan, J.K. and E. Szwajkowski (1979) 'Coalition bargaining in four games that include a veto player', *Journal of Personality and Social Psychology*, 37: 1933–1946. **E**

Muto, S. and M. Makayama (1988) 'A cooperative game of information trading: The core and the nucleolus', Discussion paper No. 757, Dept. of Managerial Economics and Decision Science, Northwestern University. **A**

Muto, S., M. Nakayama, J. Potters and S.H. Tijs (1988) 'Big boss games', *The Economic Studies Quarterly*, 39: 303–321. **FA**

Muto, S., J. Potters and S.H. Tijs (1989) 'Information market games', *International Journal of Game Theory*, 18: 209–226. **A**

Nakayama, M. (1983) 'A note on a generalization of the nucleolus to games without sidepayments, *International Journal of Game Theory*, 12: 115–122. **N**

Nakayama, M. (1988) 'A direct proof of Aumann and Maschler's theorem of the nucleolus of a bankruptcy game', Working paper No. 81, Faculty of Economics, Toyama University, Gofuku Toyama City, Japan. **A**

Nakayama, M. and L. Quintas (1989) 'Stable payoffs in resale-proof information trades', *Games and Economic Behavior*, 3: 339–349. **A**

Naumova, N.I. (1973) 'Sufficient conditions for the nonemptiness of the bargaining set $\mathcal{M}_1^{(i)}$ in a Game with a countable number of players', in: *Advances in game theory* (Proceedings of the Second All-Union Conference on Game Theory, Vilnius, 1971), Vilnius: Izdat. "Mintis", pp. 146–149 [in Russian, English summary]. **TI**

Naumova, N.I. (1976) 'The existence of certain stable sets for games with a discrete set of players', *Vestnik Leningradskogo Universiteta, Mechanika, Astronomia* (*Leningrad*) 7: 47–54 [in Russian, English summary]; English translation in, *Vestnik Leningrad University Mathematics*, 9: 131–139 (1981). **TIR**

Naumova (1978) 'M-systems of relations and their application in cooperative games', *Vestnik Leningradskogo Universiteta, Mechanika, Astronomia* (Leningrad) 1 (1978) 60–66, 156 [in Russian, English summary]; English translation in, *Vesnik Leningrad University Mathematics* 11: 131–139 (1983). **TRI**

Niou, E.M.S. and Ordershook, P.C. (1989) 'The geographical imperatives of the balance of power in 3-country systems', *Mathematical and Computer Modelling*, 12: 519–531. **RA**

Nurmi, H. (1980) 'Game theory and power indices' *Zeitschrift für Nationalokonomie*, 40: 35–58. **E**

Oppenheimer, J. (1979) 'Outcomes of logrolling in the bargaining set and democratic theory: Some conjectures', *Public Choice*, 34: 419–434. **A**

Ordeshook, P.C. (1986) *Game theory and political theory: An introduction.* Cambridge–London–New York–New Rochelle–Melbourne–Sydney: Cambridge University Press. **FRTEN**

Orshan, G. (1987) 'The consistent Shapley value in hyperplane games from a global standpoint', M.Sc. thesis, The Hebrew University of Jerusalem. **NTRR**

Orshan, G. (1992) 'The consistent Shapley value in hyperplane games from a global stand point', *International Journal of Game Theory* (to appear). **NTRR**

Ostmann, A. (1984) 'Classifying three-person-games', Working paper No. 140, Institute of Mathematical Economics, Universität Bielefeld, 1984. **F**

Ostmann, A. (1985a) 'Für Abhängige keinen Zugewinn?', Working paper No. 141, Institute of Mathematical Economics, Universität Bielefeld. **E**

Ostmann, A. (1985b) 'Argumente und Einigungen: Klassifikation von Verteilungskonflikten', Arbeiten der Fachrichtung Psychologie, Universität des Saarlandes, No. 95. **E**

Ostmann, A. (1985c) 'Erfahrungen mit der Makler-Serie 1982/1983 (*Experimentelle Drei-Personnen-Spiele mit Seitenzahlungen*)', Arbeiten der Fachrichtung Psychologie, Universität des Saarlandes, No. 96. **E**

Ostmann, A. (1989) 'Simple games: On order and symmetry', Working paper No. 169, Institute of Mathematical Economics, Universität Bielefeld. **FRR**

Ostmann, A. and C. Schmauch (1984) 'Die Berechnung von Lösungskonzepten für Drei-Personen-Spiele mit Seitenzahlungen', Arbeiten der Fachrichtung Psychologie, Universität Des Saarlandes, No. 89. **F**

Owen, G. (1968) 'n-person games with only 1, $n - 1$ and n-person coalitions', *Proceedings of the American Mathematical Society*, 19: 1258–1261. **F**

Owen, G. (1974) 'A note on the nucleolus', *International Journal of Game Theory*, 3: 101–103. **C**

Owen, G. (1975a) 'A note on the convergence of transfer sequences in n-person games, *International Journal of Game Theory*, 4: 221–228. **D**

Owen, G. (1975b) 'On the core of linear production games', *Mathematical Programming*, 9: 358–370.

Owen, G. (1977a) 'A generalization of the Kohlberg criterion, *International Journal of Game Theory*, 6: 249–255. **T**

Owen, G. (1977b) 'Characterization of the nucleolus for a class of n-person games', *Naval Research Logistic Quarterly*, 24: 463–472. **F**

Owen, G. (1982) *Game theory*, 2nd edn. New York–London: Academic Press. **GT**

Owen, G. (1990) 'Stable outcomes in spatial voting games', *Mathematical Social Sciences*, 19: 269–279. **ARR**

Owen, G. (1992) 'The assignment game: The reduced game', *Annales d'Economie et de Statistique* (to appear). **A**

Pecherskiĭ, S.L. and A.I. Sobolev (1983) 'The problem of optimal distribution in socioeconomic problems and cooperative games', Otvetstvennii redactor E.B.Y. Yanovskaya, Nauka Leningrad Otdel, Leningrad [in Russian]. **T**

Pecherskii, S. and S. Sobolev (1986) 'Multivalued solutions of cooperative games: an axiomatic approach', *Matematicheskie Metody v Sotsialnykh Nauk*, 19: 72–86, 102 [in Russian, English and Lithuanian summaries]. **T**

Peleg, B. (1963a) 'Quota game with a continuum of players', *Israel Journal of Mathematics*, 1: 48–53. **R**

Peleg, B. (1963b) 'On bargaining sets of cooperative games', Ph.D. Thesis, The Hebrew University of Jerusalem [in Hebrew]. **T**

Peleg, B. (1963c) 'Bargaining sets of cooperative games without side payments', *Israel Journal of Mathematics*, 1: 197–200. **N**

Peleg, B. (1963d) 'Existence theorem for the bargaining set $\mathcal{M}_1^{(i)}$, (abstract), *Bulletin American Mathematical Society*, 69: 109–110. **T**

Peleg, B. (1964) 'On the bargaining set $\mathcal{M}_0$ of m-quota games', in: M. Dresher, L.S. Shapley and A.W. Tucker, eds., *Advances in game theory*. Princeton: Princeton University Press, pp. 501–512. **FR**

Peleg, B. (1965a) 'The kernel of the composition of characteristic function games', *Israel Journal of Mathematics*, 3: 127–138. **T**

Peleg, B. (1965b) 'The kernel of *m*-quota games', *Canadian Journal of Mathematics*, 17: 239–244. **F**

Peleg, B. (1966a) 'On the kernel of constant-sum simple games with homogeneous weights', *Illinois Journal of Mathematics*, 10: 39–48. **F**

Peleg, B. (1966b) 'The kernel of the general-sum four-person game', *Candian Journal of Mathematics*, 18: 673–677. **F**

Peleg, B. (1967) 'Existence theorem for the bargaining set $\mathcal{M}_1^{(i)}$', in: M. Shubik, ed., *Essays in mathematical economics in honor of Oskar Morgenstern*. Princeton: Princeton Univ. Press, pp. 53–56. **T**

Peleg, B. (1968) 'On weights of constant-sum majority games', *SIAM Journal Applied Mathematics*, 16: 527–532. **F**

Peleg, B. (1969) 'The extended bargaining set for cooperative games without side payments', Center for Game Theory and Mathematical Economics, The Hebrew University of Jerusalem, RM44. **NFT**

Peleg, B. (1979) 'Verhandlungsbereich', in: M.J. Beckmann, G. Menges and R. Selten, eds., *Handwörterbuch der Mathematischen Wirtschaftswissenschaften, Wirtschaftstheorie*. Wiesbaden: Gabler, pp. 431–436. **T**

Peleg, B. (1981) 'Coalition formation in simple games with dominant players', *International Journal of Game Theory*, 10: 11–33. **H**

Peleg, B. (1985) 'An axiomatization of the core of cooperative games without side payments', *Journal of Mathematical Economics*, 14: 203–214. **RR**

Peleg, B. (1986) 'On the reduced game property and its converse', *International Journal of Game Theory*, 15: 187–200. **T**

Peleg, B. (1987) 'On the reduced game property and its converse. A correction', *International Journal of Game Theory*, 16: 290. **T**

Peleg, B. (1988) 'Introduction to the theory of cooperative games', in preparation, RM 80–85, Center for Research in Mathematical Economics and Game Theory, The Hebrew University of Jerusalem. **TFA**

Peleg, B. (1990) 'Axiomatization of the core, the nucleolus and the prekernel', in: T. Ichiishi, A. Neyman and Y. Tauman, eds., *Game theory and applications*, San Diego: Academic Press, pp. 176–182. **T**

Peleg, B. (1992) 'Axiomatization of the core', This volume. **RR**

Peleg, B. (1992) 'Voting by count and account', in: R. Selten, ed., *Rational interaction, essays in honor of John C. Harsanyi*, Berlin: Springer-Verlag, pp. 45–51. **F**

Peleg, B. and J. Rosenmüller (1992) 'The least core, nucleolus and kernel of homogeneous weighted majority games' *Games and Economic Behavior* (to appear). **F**

Peleg, B., J. Rosenmüller and P. Sudhölter (1992) 'The kernel of homogeneous games with steps', Working Paper No. 209, Institute of Mathematical Economics, University of Bielefeld. **F**

Pérez, B.S. (1978) 'Solutions of *n*-person games', *Gaceta Mat. (1) (Madrid)*, 30: 142–150 [in Spanish]. **RT**

Postlewaite, A. and R.W. Rosenthal (1974) 'Disadvantageous syndicates', *Journal of Economic Theory*, 9: 324–326.

Potters, J.A.M. (1991) 'An axiomatization of the nucleolus' *International Journal of Game Theory*, 19: 365–373. **T**

Potters, J.A.M. and S. Tijs (1990) 'Information market games with more than one informed player', *Methods of Operations Research*, 63: 313–324. **F**

Potters, J.A.M. and S.H. Tijs (1992) 'The nucleolus of a matrix game and other nucleoli', *Mathematics of Operations Research*, 17: 164–174. **R**

Potters, J.A.M., S. Muto and S.H. Tijs (1990) 'Bargaining set and kernel for big boss games', *Methods of Operations Research*, 60: 329–335. **F**

Potters, J.A.M., R. Poos, S. Tijs and S. Muto (1989) 'Clan games', *Games and Economic Behavior*, 1: 275–293. **F**

Pressacco, F., (1979) 'The constrained competitive bargaining set for homogeneous weighted majority games', in: *Proceedings of the First AMASES Meeting (Pisa, 1977)*. Turin: Giappichelli, pp. 333–347. **RF**

Rapoport Am. (1985) 'A note on the equal division kernel and the α-power mode', *Journal of the Mathematical Sociology*, 11: 65–76. **RF**

Rapoport, Am. (1990) 'Experimental studies of interactive decisions', *Theory and decision library, series C: game theory, mathematical programming and mathematical economics*. Dordrecht: Kluwer Academic Publishers. **E**

Rapoport, Am. and J.P. Kahan (1976) 'Where three isn't always two against one: Coalitions in experimental three-person games', *Journal of Experimental Social Psychology*, 12: 253–273. **E**

Rapoport, Am. and J.P. Kahan (1979) 'Standards of fairness in the 4-person monopolistic cooperative games', in: S.J. Brams, A. Schotter and G. Schwödiauer, eds., *Applied game theory*, Proceedings of a conference at the Institute for Advanced Studies, Vienna, June 13–16, 1 1978. Würzburg: Physica Verlag. **E**

Rapoport, Am. and J.P. Kahan (1982) 'The power of a coalition and payoff disbursement in 3-person negotiable conflicts', *Journal of Mathematical Sociology*, 8: 193–224. **E**

Rapoport, Am. and J.P. Kahan (1983) 'Coalition formation in a five-person market game', *Management Science*. **E**

Rapoport, Am., J.P. Kahan and T.S. Wallsten (1978) 'Sources of power in 4-person apex games', in: H. Sauermann, ed., *Beiträge zur Experimentellen Wirtschaftsforschung*, Vol. VIII: *Coalition forming behavior*. Tübingen: J.C.B. Mohr. **E**

Rapoport, Am., J.P. Kahan, S.G. Funk and A.D. Horowitz (1979) 'Coalition formation by sophisticated players', Lecture notes in economic and mathematical systems, *Mathematical Economics*, 169. Berlin–Heidelberg–New York: Springer-Verlag. **E**

Rapoport, Am., W.E. Stein and G.J. Burkheimer (1979) *Response models for detection of change*. Dordrecht–Boston, Massachusetts: D. Reidel. **E**

Rapoport, An. (1970) *N-person game theory, concepts and applications*. Ann Arbor Science Library, Ann Arbor: The University of Michigan Press. **TE**

Riker, W.H. (1967) 'Bargaining in a three-person game', *American Political Review*, 61: 642–656. **E**

Riker, W.H. (1971) 'An experimental examination of formal and informal rules of a three-person game', in: B. Lieberman, ed., *Social choice*. New York: Gordon and Breach. **E**

Riker, W.H. (1972) 'Three-person coalitions in three-person games: Experimental verification of the theory of games', in: J.F. Herndon and J.L. Bernd, eds., *Mathematical applications in political sciences VI*. Charlottesville: University of Virginia Press. **E**

Rochford, Sh.C. (1983a) 'Pairwise-bargaining allocations in an assignment game: A new approach to the kernel', Dept. of Economics, University of Georgia, Athens, Georgia. **A**

Rochford, Sh.C. (1983b) 'Pairwise-bargained allocations in a marriage market', Ph.D. Thesis, The University of Iowa, Iowa City, Iowa. **A**

Rochford, Sh.C. (1984) 'Symmetrically pairwise-bargained allocation in assignment market', *Journal of Economic Theory*, 34: 262–281. **AN**

Rosenmüller, J. (1979) 'Konvexe Spieler–Balancierte Spiele', G. Menges, R. Selten, eds., *Handwörterbuch der Mathematischen Wirtschaftswissenschaften*, Band 1: *Wirtschaftstheorie*. Wiesbaden: Gabler, pp. 167–172. **F**

Rosenmüller, J. (1981) *The theory of games and markets*. Amsterdam–New York–Oxford: North-Holland. **TG**

Rosenmüller, J. and P. Sudhölter (1992) 'The nucleolus of homogeneous games with steps' *Discrete Applied Mathematics* (to appear). **F**

Rosenthal, R.W. (1970) 'Stability analysis of cooperative games in effectiveness form', Thesis, Stanford University, also Technical Report No. 70–11, Operations Research House, Stanford House, Stanford, California. **R**

Rosenthal, R.W. (1972) 'Cooperative games in effectiveness form', *Journal of Economic Theory*, 5: 88–101. **R**

Roth, A.E. (1973) 'Subsolutions of cooperative games', Technical report No. 118, Institute for Mathematical Studies in the Social Sciences, Stanford University, Stanford, California. **R**

Roth, A.E. (1974) 'Topics in cooperative game theory', Ph.D. Thesis, Stanford University, Stanford, California; also, Technical report SOL 74-8, Department of Operations Research, Stanford University, Stanford, Calfornia. **TR**

Sakawa, M. and I. Nishizaki (1984) 'A lexicographical solution concept in an n-person cooperative fuzzy game', *Systems and Control*, 28: 461–468 [in Japanese, English summary]. **R**

Sankaran, J.K. (1991) 'On finding the nucleolus of an *n*-person cooperative game', *International Journal of Game Theory*, 19: 329–338. **C**

Schjødt, U. and B. Sloth (1991) 'Bargaining sets with small coalitions', RM, Institute of Economics, University of Copenhagen. **RN**

Schmeidler, D. (1969a) 'The nucleolus of a characteristic function game', *SIAM Journal of Applied Mathematics*, 17: 1163–1170. **T**

Schmeidler, D. (1969b) 'Games with a continuum of players', Ph.D. Thesis, The Hebrew University of Jerusalem [in Hebrew]. **T**

Schofield, N. (1976) 'The kernel and payoffs in European government coalitions', *Public Choice*, 26: 29–49. **H**

Schofield, N. (1978) 'Generalised bargaining sets for cooperative games', *International Journal of Game Theory*, 7: 183–199. **R**

Schofield, N. (1980) 'The bargaining set in voting games', *Behavioral Science*, 25: 120–129. **R**

Schofield, N. (1981) 'Generic instability of simple majority rule', Essex Economic paper No. 174. **RR**

Schofield, N. (1982) 'Bargaining set theory and stability in coalition governments', *Mathematical Social Sciences*, 3: 9–31. **RH**

Selten, R. (1987) 'Equity and coalition bargaining in experimental three person games', in: A.E. Roth, ed., *Laboratory experimentation in economics, six points of view*. Cambridge–New York–New Rochelle–Melbourne–Sydney: Cambridge University Press. **E**

Selten R. (1988) 'Coalition probabilities in a non-cooperative model of three-person quota game bargaining', in: R. Selten, ed., *Models of strategic rationality*. Dordrecht–Boston–London: Kluwer Academic Pub. **ERR**

Selten, R. and K.G. Schuster (1968) 'Psychological variables and coalition forming behavior', in: K. Borch and J. Mossin, eds., *Risk and uncertainty*. London: Macmillan. **E**

Seo, F. and M. Sakawa (1990) 'A game theoretic approach with risk assessment for international conflict solving', *Institute of Electrical and Electronic Engineers, Transactions on Systems, Man and Cybernetics*, 20: 141–148. **A**

Shapley, L.S. (1953a) 'Quota solutions of *n*-person games', in: H. Kuhn and A.W. Tucker, eds., *Contributions to the theory of games*, Vol. II. Princeton: Princeton University Press, pp. 343–359.

Shapley, L.S. (1953b) 'A value for *n*-person games', in: H. Kuhn and A.W. Tucker, eds., *Contributions to the theory of games*, Vol. II. Princeton: Princeton University Press, pp. 307–317.

Shapley, L.S. (1964a) 'Solutions of compound simple games', in: M. Dresher, L.S. Shapley and A.W. Tucker, eds., *Advances in game theory*. Princeton: Princeton University Press, pp. 267–307.

Shapley, L.S. (1964b) 'Values of large games VII: A general exchange economy with money', RM–4248, The Rand Corporation, Santa Monica, California.

Shapley, L.S. (1967) 'On balanced sets and cores', *Naval Research Logistics Quarterly*, 14: 453–460.

Shapley, L.S. (1971) 'Cores of convex games', *International Journal of Game Theory*, 1: 11–26, 199.

Shapley, L.S. (1973) 'An example of a nonconverging kernel in a replicated market game, RAND Pub. IN-22419-NSF. **TAI**

Shapley, L.S. and M. Shubik (1963) 'The core of an economy with nonconvex preferences', RM3518, The RAND Corporation, Santa Monica, CA.

Shapley, L.S. and M. Shubik (1966) 'Quasi-cores in a monetary economy with nonconvex preferences', *Econometrica*, 34: 805–827.

Shapley, L.S. and M. Shubik (1967) 'Ownership and production function', *Quarterly Journal of Economics*, 81: 88–111.

Shapley, L.S. and M. Shubik (1972a) 'Convergence of the bargaining set for differentiable market games', Internal note, the RAND Corporation, Santa Monica, California. **ATI**

Shapley, L.S. and M. Shubik (1972b) 'The kernel and bargaining set for market games', Working notes. **A**

Shapley, L.S. and M. Shubik (1972c) 'The assignment game I: The core', *International Journal of Game Theory*, 2: 111–130.

Shenoy, P.P. (1977) 'On game theory and coalition formation', Ph.D. Thesis, Cornell University, Ithaca, N.Y. **TAG**

Shenoy, P.P. (1979) 'On coalition formation: A game-theoretical approach, *International Journal of Game* Theory, 8: 133–164. **TR**

Shenoy, P.P. (1980) 'A three-person cooperative game formulation of the world oil market', *Applied Mathematical Modelling*, 4: 301–307. **A**

Shishko, R. (1974) 'A survey of solution concepts for majority rule games', RAND Corp. paper P-5169, Santa Monica, California. **R**

Shitovitz, B. (1989) 'The bargaining set and the core in mixed markets with atoms and an atomless sector', *Journal of Mathematical Economics*, 18: 377–383. **A**

Shubik, M. (1983) *Game theory in the social sciences: Concepts and solutions.* Cambridge, Masachusetts–London: The MIT Press. **TAG**

Shubik, M. (1985) *A game-theoretic approach to political economy*, Volume 2 of *Game theory in the social sciences.* Cambridge, Massachusetts–London: The MIT Press. **TAG**

Shubik, M. and H.P. Young (1978) 'The nucleolus as a noncooperative game solution', in: P.C. Ordeshook, ed., *Game theory and political science.* New York: New York University Press, pp. 511–527. **RR**

Šimelis, Č. (1973a) 'The sum of games and *k*-kernels', *Matematicheskie Metody v Sotsial'nykh Naukakh. Trudy Respublicanskogo Nauchnogo Seminara 'Protesessy Optimal'nogo Upravleniya'. II Sektsiya. (Vilnius)*, 3: 50–86 [in Russian, English and Lithuanian summaries]. **T**

Šimelis, Č. (1973b) 'The *k*-kernel of the product of two games', *Matematicheskie Metody v Sotsial'nykh Naukakh. Trudy Respublicanskogo Nauchnogo Seminara 'Protesessy Optimal'nogo Upravleniya'. II Sektsiya. (Vilnius)*, 3: 87–97 [in Russian, English and Lithuanian summaries]. **T**

Šimelis, Č. (1973c) 'Composition of games and their *k*-kernels', *Matematicheskie Metody v Sotsial'nykh Naukakh. Trudy Respublicanskogo Nauchnogo Seminara 'Protesessy Optimal'nogo Upravleniya'. II Sektsiya. (Vilnius)*, 2: 129–174 [in Russian, English summary]. **T**

Šimelis, Č. (1975a) 'The *n*-nucleolus of a product game', *Matematicheskie Metody v Sotsial'nykh Naukakh. Trudy Respublicanskogo Nauchnogo Seminara 'Protesessy Optimal'nogo Upravleniya'. II Sektsiya. (Vilnius)*, 5: 69–90. **T**

Šimelis, Č. (1975b) 'The structure of the *k*-kernel of a certain composition of games', *Litovskiĭ Matematicheskiĭ Sbornik. (Vilnius)*, 15, (1975) 225–238, 249 [in Russian, English and Lithuanian summaries]; English translation, *Cybernetics* 15: 685–695 (1976a). **TF**

Šimelis, Č. (1976b) 'Algorithms for computing the nucleolus of a centered game', in: *Current trends in game theory*, Vilnius: Izdat. 'Mokslas', pp. 127–135 [in Russian, English summary]. **CTF**

Šimelis, Č. (1977) 'A note on the *n*-core of a monotonic game', *Litovskiĭ Matematicheskiĭ Sbornik. (Vilnius)*, 17: 105–110, 113 [in Russian, English and Lithuanian summaries]. **T**

Šimelis, Č. (1982a) 'On a modification of the algorithm for computing the nucleolus of a centered game', *Litovskiĭ Matematicheskiĭ Sbornik, 'Mokslas', Vilnius*, 22 (1983a). 193–197 [in Russian, English and Lithuanian summaries]; English translation, *Lithuanian Mathematical Journal*, 22: 193–197. **FC**

Šimelis, Č. (1983b) 'Coincidence of the kernel and the nucleolus of a centered game', *Litovskiĭ Matematicheskiĭ Sbornik, 'Mokslas', Vilnius*, 22 (1982b) 203–207 [in Russian, English and Lithuanian summaries]; English translation, *Lithuanian Mathematical Journal*, 22: 101–104. **TF**

Šimelis, Č. (1984) 'The kernel of the sum of two cooperative games', *International Journal of Game Theory*, 13: 97–126. **T**

Snijders, C. (1991) 'Axiomatization of the nucleolus', RM, part of an M.Sc. Thesis, Dept. of Mathematics, University of Utrecht, **T**

Sobolev, A.I. (1973) 'The functional equations that give the payoff of the players in an *n*-person game', in: *Advances in game theory* (Proceedings of the Second All-Union Conference on Game Theory, Vilnius, 1971), Vilnius: Izdat. 'Mintis', pp. 151–153 [in Russian, English summary]. **T**

Sobolev, A.I. (1975) 'The characterization of optimality principles in cooperative games by functional equations', in: N.N. Vorobjev, ed., *Matematischeskie metody v socialnix naukakh*, Proceedings of the Seminar, Issue 6, Vilnius: Institute of Physics and Mathematics, Academy of Sciences of the Lithuanian SSR, pp. 94–151 [in Russian, English summary]. **T**

Sobolev, A.I. (1982) 'Cooperative games', *Problemy Kibernet*, 39: 201–222 [in Russian]. **T**

Sourbis, H. (1984) 'Aspects of cooperative game theory and its applications to economics', Ph.D. Thesis, University of Florida. **R**

Sourbis, H.D. (1986) 'Security as a stability criterion for *n*-person cooperative games', RM, Dept. of Economics, University of Florida, Gainsville, Florida. **R**

Spinetto, R.D. (1971) 'Solution concepts of *n*-person cooperative games as points in game space', Ph.D. Thesis, Cornell University, Ithaca, 1971, New York; also, Technical report No. 138, Dept. of Operations Research, Cornell University, Ithaca, New York. **RR**

Spinetto, R.D. (1973/4) 'The geometry of solution concepts for *n*-person cooperative games', *Management Science*, 20: 1292–1299. **R**

Ståhl, I. (1980a) 'The application of game theory and gaming to conflict resolution in regional planning', Working paper WP–80–82, International Institute for Applied Systems Analysis, A–2361, Laxenburg, Austria. **EA**

Ståhl, I. (1980b) 'A gaming experiment on cost allocation in water resources development', Working paper WP–80–38, Int. Inst. for Applied Systems Analysis, A–2361, Laxenburg, Austria. **EA**

Stearns, R.E. (1968) 'Convergent transfer schemes for *n*-person games', *Transactions American Mathematical Society*, 134: 449–459. **D**

Stearns, R.E. (1966) 'The discontinuity of the bargaining set', Unpublished RM. **T**

Straffin, P.D. and B. Grofman (1984) 'Parliamentary coalitions: A tour of models', *Mathematics Magazine*, 57: 259–274. **AHRN**

Straffin, P.D. and J.P. Heaney (1981) 'Game theory and the Tennessee Valley Authority', *International Journal of Game Theory*, 10: 35–43. **AH**

Sussangkarn, C. (1977) 'Some stability concepts for cooperative games', Ph.D. Thesis, Cambridge University. **RN**

Sussangkarn, C. (1983) 'A unique bargaining solution based on competitive commitments', Working papers in econometric theory and econometrics, Center for Research in Management Science, Institute of Business and Economic Research, University of California, Berkeley, No. IP–2775. *Mathematics of Operations Research*, 8: 205–214. **RR**

Suzuki, M. (1973) 'The nucleolus and the responsibility in the planning', *Shūkan Tōyō Keizai*, 10: 76–85 [in Japanese]. **A**

Suzuki, M. and M. Nakayama (1976) 'The cost assignment of cooperative water resource development: A game theoretic approach', *Management Science*, 22: 1081–1086. **AH**

Suzuki, M. and M. Nakayama (1977) 'The cost assignment of cooperative water resource development: A game theoretic approach', in: R. Henn and O. Moeschlin, eds., *Mathematical economics and game theory, essays in honor of Oskar Morgenstern*. Berlin–Heidelberg–New York: Springer Verlag, pp. 616–625. **AH**

Syed, S.A. (1990) 'The bargaining set for cost allocation games', *Pure and Applied Mathematika Sciences*, 32: 1–9. **A**

Tadenuma, K. (1990a) 'Dual characterization of the core and the anti-core', Department of Economics, Hitotsubashi University, Tokyo. **RR**

Tadenuma, K. (1990b) 'Duality relations between the core and the anti-core of NTU games', Department of Economics, Hitotsubashi University, Tokyo. **RR**

Tiba, M. (1982) 'The (*m, p*)-bargaining set of the three-person games', *Analele Ştiinţifice ale Universităţii 'Al. I. Cuza' din Iaşi. Serie Nouă Secţiunea Ia Matematică. Univ. Al. I. Cuza, Iaşi*, 28: 93–100. **R**

Tijs, S.H. (1981) 'Bounds of the core and the τ-value', in: O. Moeschlin and O. Pallaschke, eds., *Game theory and mathematical economics*. Amsterdam–New York–Oxford: North-Holland, pp. 123–132. **R**

Tijs, S.H. and T.S.H. Driessen (1986) 'Game theory and cost allocation problems', *Management Science*, 32: 1015–1028. **A**

Tulkens, H. and Sh. Zamir (1979) 'Local games in dynamic exchange processes', *Review of Economics Studies*, XLVI (2) No. 143: 305–313. **AD**

Ueda, T. (1971) 'The continuity of a critical point set with applications to some decision problems', *SIAM Journal of Applied Mathematics*, 21: 145–154. **T**

Vilkas, È. Ĭ. (1976) 'Optimality concepts in game theory', in: *Contemporary directions in game theory*. Collected papers, pp. 25–43. **G**

Vilkas, È. Ĭ. (1978) 'Two game theory theorems', Mathematical Methods in the Social Sciences. Vilnius: Institute of Mathematical Cybernetics, Lithuanian Academy of Science, SSR 10: 9–17, 119 [in Russian, English summary]. **R**

Vilkas, È. Ĭ. (1979) 'An axiomatic definition of a generalized *n*-nucleolus', *Matematicheskie Metody v Sotsial'nykh Naukakh. Trudy Respublicanskogo Nauchnogo Seminara 'Protesessy Optimal'nogo Upravleniya'. II Sektsiya. (Vilnius)*, 12: 52–54, 127 [in Russian, Lithuanian and English summaries]. **R**

Vilkas, È. Ĭ. (1986) 'On the existence of a bargaining set', *Lietuvos TSR Moskłų Akademija (Lithuanian Math. J.)*, 26: 221–230 [in Russian, English and Lithuanian summaries]. **R**

Vilkov, V.B. (1974) 'The *n*-kernel in cooperative games without side payments', *Žurnal Vyčislitel'noĭ Matematiki i Matematicĕskoĭ Fiziki (Moscow)*, 14: 1327–1331, 1367 [in Russian]; English translation, *U.S.S.R. Computational Mathematics and Mathematical Physics* 14(5): 226–232 (1974). **NRT**

Vilkov, V.B. (1979) 'Some properties of the *n*-nucleolus', *Matematicheskie Metody v Sotsial'nykh Naukakh. Trudy Respublicanskogo Nauchnogo Seminara 'Protesessy Optimal'nogo Upravleniya'. II Sektsiya. (Vilnius)*, 12: 55–64, 127 [in Russian, English and Lithuanian summaries]. **NRT**

Vind, K. (1988) 'Two characterizations of bargaining sets', RM, Mathematical Sciences Research Institute and University of Copenhagen. **TR**

Vohra, R. (1987) 'On a bargaining set with restricted coalition size', Working paper, Department of Economics, Brown University, Providence, Rhode Island, *Journal of Mathematical Economics* (to appear). **RINA**

Vohra, R. (1991) 'An existence theorem for a bargaining set', *Journal of Mathematical Economics*, 20: 19–34. **RINA**

Vohra, R. (1991b) 'On the existence of a proportional bargaining set', in: *Fixed Point Theory and Applications* (Marseille 1989), Pitman Res. Notes Math., Series 252, Department of Economics, Brown University, Providence, Rhode Island: Longman Sci. Tech., Harlow, pp. 437–446. **R**

Von Neumann, J. and O. Morgenstern (1953) *Theory of games and economic behavior*. Princeton: Princeton University Press, 1944, 3rd edition.

Wallmeier, E. (1980) 'Der *f*-Nukleolus als Lösungskonzept für *n*-Personenspiele in Funktionsform', Institut für Mathematische Statistik der Universität Münster, Working paper WO1. **TR**

Wallmeier, E. (1983) 'Der *f*-Nukleolus und ein Dynamisches Verhandlungsmodell als Lösungskonzepte für kooperative *n*-Personenspiele', Ph.D. Thesis, Skripten zur Mathematischen Statistik, Nr. 5, Westfälischen Wilhelm-Universität Münster, Institut für Mathematische Statistik. **TRFC**

Wallmeier, E. (1984) 'A procedure for computing the *f*-nucleolus of a cooperative game', in: G. Hammer and D. Pallaschke, eds., *Selected topics in operations research and mathematical economics*, Lecture notes in economics and mathematical systems, 226, Proceedings of the 8th Symposium on Operations Research held at the University of Karlsruhe, Germany, pp. 288–296. **C**

Wang, J.H. (1986) 'An introduction to cooperative games', *Chinese Journal of Operations Research*, 5: 1–9. **I**

Wang, J. (1988) 'The theory of games', *Oxford Science Publications: Oxford Mathematical Monographs*. New York: Oxford University Press [translated from the Chinese]. **T**

Wang, L.S.-Y (1974) 'On dynamic theories for *n*-person games', Ph.D. Thesis, Cornell Univ., Ithaca, New York. **D**

Wang, L.S.-Y. and L.J. Billera (1974) 'Some theorems on a dynamic theory for the kernel of an *n*-person game', RC 4783 (No. 21274), IBM Research Report, IBM, Yorktown Heights, New York. **D**

Wesley, E. (1971) 'An application of non-standard analysis to game theory', part of an M.Sc. Thesis, Department of Mathematics, The Hebrew University of Jerusalem, *Journal of Symbolic Logic*, 36: 385–394. **IT**

Williams, M.A. (1988) 'An empirical test of cooperative game solution concepts', *Behavioral Science*, 3: 224–237. **E**

Wilson, R. (1971) 'Stable coalition proposals in majority-rule voting', *Journal of Economic Theory*, 3: 254–271. **TR**

Wolsey, L.A. (1976) The nucleolus and kernel for simple games or special valid inequalities for 0–1 linear integer programs', *International Journal of Game Theory*, 5: 227–238. **CF**

Xing, W.X. (1989) 'An algorithm for the nucleolus of an *n*-person cooperative game, *Beijing Daxue Xuebao*, 25: 513–526 [in Chinese, English summary]. **C**

Yanovskaya, E.B. (1978) 'Weakly $\tilde{\mathcal{M}}_1^i$-stable sets in general nonstrategic games', *Game Theory Questions of Decision Making*. Leningrad: Nauka, Otdel, pp. 19–26, 126 [in Russian]. **RR**

Yanovskaya, E.B. (1985) 'Axiomatic characterization of maximum and lexicographically maximin solutions of bargaining schemes', *Akademia Nauk SSR. Avtomatika i Telemekhanika*, 9: 128–136 [in Russian, English summary]. **NR**

Yanovskaya, E.B. (1987) 'A solution of cooperative games defined by the relation of partial lexicographic ordering on the payoff set', *Matematicheskie Setody v Sotsialnykh Naukakh*, 20: 82–97 [in Russian, English and Lithuanian summaries]. **NR**

Yanovskaya, E.B. (1988) 'The function of group choice for different measurement scales of participants' preference. The software of multicriteria optimization problems and their application', *Erevan*, 201–204 [in Russian]. **NRR**

Yanovskaya, E.B. (1989a) 'On a definition of excess in games without side payments', RM, Institute for Socio-Economic Problems, Academy of Science of USSR, Leningrad. **NR**

Yanovskaya, E.B. (1989b) 'Excess functions and nucleoli of cooperative games', RM, Institute for Socio-Economic Problems, Academy of Sciences of USSR, Leningrad. **NR**

Yarom, M (1979) 'Lexicographic kernel in characteristic function games', M.Sc. Thesis, The Hebrew University of Jerusalem [in Hebrew].

Yarom, M. (1981) 'The lexicographic kernel of a cooperative game', *Mathematics of Operations Research*, 6: 88–100. **R**

Yarom, M. (1982) 'Bargaining procedures for cooperative games', Ph.D. Thesis, The Hebrew University of Jerusalem [in Hebrew]. **TD**

Yarom, M. (1983) 'Existence of solutions to differential inclusions leading to critical points', Discussion paper No. 111, Institute für Gesellschafts- und Wirtschaftswissenschaften, Universität Bonn. **D**

Yarom, M. (1985) 'Dynamics systems of differential inclusions for the bargaining sets', *International Journal of Game Theory*, 14: 51–61. **D**

Ye, T.X. (1989) 'Structures of prekernels and pseudokernels of games with coalition structures', *Chinese Journal of Operations Research*, 2: 67–68. **T**

Younes, Y., 'Power and market relations', unknown date, unknown address.

Young, H.P. (1985a) 'Monotonic solutions of cooperative games', *International Journal of Game Theory*, 14: 65–72. **T**

Young, H.P. (1985b) 'Cost allocation', in: H.P. Young, ed., *Fair allocation*, Proceedings of Symposia in Applied Math., Vol. 33: American Mathematical Society, Providence, Rhode Island, pp. 69–94. **A**

Young, H.P. (1985c) 'The allocation of debts and taxes', in: H.P. Young, ed., *Fair allocation*, Proceedings of Symposia in Applied Mathematics, Vol. 33, American Mathematical Society, Providence, Rhode Island, pp. 95–108. **A**

Young, H.P. (1987) 'On dividing an amount according to individual claims or liabilities', *Mathematics of Operations Research*, 12: 398–414. **RRA**

Young, H.P., N. Okada and T. Hashimoto (1982) 'Cost allocation in water resources development', *Water Resources Research*, 18: 463–475. **A**

Zachow, E.W. 'Nuclear equlibrium points', RM, Westfälische Wilhelms–Universität, Institut für Mathematische Statistik, Münster. **RR**

Zhang, D. (1987) 'The kernel and the nucleolus of a tree game with an empty core', *Acta Mathematica Applicatae Sinica*, 10: 8–23 [in Chinese]. **F**

Zhang, J.G. (1990) 'The nucleolus of *n*-person cooperative games with generalized symmetric partitions' *Acta Mathematica Applicatae Sinica*, 13: 1–5 [in Chinese, English summary]. **TF**

Zhou, L. (1990) 'An equilibrium existence lemma for *n*-person games and its applications', RM, Cowles Foundation, Yale University. **TNR**

Zhou, L. (1991a) 'A refined bargaining set of an *n*-person game and endogeneous coalition formation', Cowles Foundation discussion paper no. 974, Cowles Foundation, Yale University. **R**

Zhou, L. (1991b) 'A weak monotonicity property of the nucleus', *International Journal of Game Theory*, 19: 407–411. **T**

Chapter 19

GAME AND DECISION THEORETIC MODELS IN ETHICS

JOHN C. HARSANYI*

University of California at Berkeley

Contents

*I am indebted to the National Science Foundation for supporting this research through grant SES-8700454 to the Center for Research in Management, University of California at Berkeley.

Handbook of Game Theory, Volume 1, Edited by R.J. Aumann and S. Hart

I. SOCIAL UTILITY

1. Ethics as a branch of the general theory of rational behavior

The concept of rational behavior is essentially an idealization of the common-sense notion of goal-directed behavior: it refers to behavior that is not only goal-directed but is also perfectly *consistent* in pursuing its goals, with consistent priorities or preferences among its different goals. Accordingly, rational behavior is not a *descriptive* concept but rather is a *normative* concept. It does not try to tell us what human behavior *is* in fact like, but rather tells us what it *would have to be* like in order to satisfy the consistency and other regularity requirements of perfect rationality.

We all have an intuitive idea of what rationality means. But we cannot provide a precise formal definition for it without careful study. It is the task of the various *normative disciplines*, such as utility theory, decision theory, and game theory, to supply precise formal definitions for normative rationality under different conditions. Indeed, most philosophers also regard *moral behavior* as a special form of rational behavior. If we accept this view (as I think we should) then the theory of morality, i.e., moral philosophy or ethics, becomes another normative discipline dealing with rational behavior.

To be sure, ethics is commonly regarded as a philosophical discipline, whereas utility theory, decision theory, and game theory are scientific and, more particularly, mathematical disciplines. But in actual fact, *all* these disciplines use a combination of mathematical and philosophical methods. In each of them, finding the right axioms and the right definitions for their basic concepts is a philosophical problem, whereas finding rigorous proofs for the theorems implied by these axioms and by these definitions is a logical and mathematical problem. True, in the past, students of ethics made little use of mathematics in discussing ethical problems. But in my opinion this was a serious mistake because mathematical models of moral behavior can substantially clarify many problems of ethics.

To elucidate how the various normative disciplines are related to each other, it is convenient to regard them as branches of the same *general theory of rational behavior*. This general theory can be divided into a theory of *individual* rational behavior and the theory of rational behavior in a *social setting*. The former includes the theory of rational behavior under *certainty*, under *risk*, and under *uncertainty*. We speak of *certainty* when the decision-maker can uniquely predict the outcome of any action he may take. We speak of *risk* when he knows at least the objective probabilities associated with alternative possible outcomes. We speak of *uncertainty* when even some or all of these objective probabilities are unknown to him (or are even undefined as numerical probabilities). The normative theory dealing with certainty I shall call *utility theory*, while those dealing with risk and uncertainty I shall combine under the heading of *decision theory*. (Sometimes utility theory is more broadly defined so as to include also what I am calling decision theory.)

On the other hand, the theory of rational behavior in a social setting can be divided into *game theory* and *ethics*. Game theory deals with two or more individuals often having very different interests who try to maximize *their own* (selfish or unselfish) interests in a rational manner against all the other individuals who likewise try to maximize *their own* (selfish or unselfish) interests in a rational manner. In contrast, ethics deals with two or more individuals often having very different personal interests yet trying to promote the *common interests of their society* in a rational manner.

2. The axioms of Bayesian decision theory

This chapter is about ethics, and not about decision theory. Yet, in discussing ethical problems, I shall have to refer several times to the axioms of decision theory. Therefore, in this section I shall briefly state these axioms in a way convenient for my subsequent discussion.

I shall sharply distinguish between axioms that merely refer to some facts established by general logic and mathematics and axioms proper to decision theory as such. The former I shall call *background assumptions*. The latter I shall call *rationality postulates*. Otherwise I shall follow the Anscombe–Aumann (1963) approach.

I shall use the following notations. *Strict preference*, *nonstrict preference*, and *indifference* (or *equivalence*) will be denoted by the symbols $>$, $\gtrsim$, and $\sim$, respectively. Unless otherwise indicated, these symbols will always refer to the preferences and indifferences entertained by *one* particular individual. I shall use the notation

$$L = (A_1 \mid e_1; \ldots; A_m \mid e_m) \tag{2.1}$$

to denote a *lottery* yielding the *prizes* or the *outcomes* $A_1, \ldots, A_m$ if the events $e_1, \ldots, e_m$ occur, respectively. These events are assumed to be mutually exclusive and jointly exhaustive of all possibilities. They will be called *conditioning events*. Thus, this notation is logically equivalent to m conditional statements, such as "If e_1 then A_1", etc.

In the special case of a *risky lottery*, where the decision-maker *knows* the objective probabilities $p_1, \ldots, p_m$ associated with the conditioning events

$e_1, \ldots, e_m$ and, therefore, also with the outcomes $A_1, \ldots, A_m$, I shall sometimes write

$$L = (A_1, p_1; \ldots; A_m, p_m).\tag{2.2}$$

I shall make the following two background assumptions:

Assumption 1. The conditional statements defining a lottery [as discussed in connection with (2.1)] follow the laws of the propositional calculus.

Assumption 2. The objective probabilities defining a risky lottery [as in (2.2)] follow the laws of the probability calculus.

I need Assumption 1 because I want to use Anscombe and Aumann's "Reversal of order" postulate without making it into a separate axiom. Their postulate can be restated so as to assume that the "roulette lottery" and the "horse lottery" they refer to will be conducted *simultaneously* rather than one after the other (as they assume). Once this is done, their postulate becomes a corollary to a well-known theorem of the propositional calculus. If we write $p \rightarrow q$ for the statement "If p then q", and write = for logical equivalence, then the relevant theorem can be written as

$$p \rightarrow (q \rightarrow r) = q \rightarrow (p \rightarrow r).\tag{2.3}$$

I need Assumption 2 because, in computing the final probability of any given outcome in a two-stage lottery, I want to use the addition and multiplication laws of the probability calculus without introducing them as separate axioms.
I need the following rationality postulates:

Postulate 1 (Complete preordering). The relation $\succsim$ (nonstrict preference) is a complete preordering over the set of all lotteries. (That is to say, $\succsim$ is both transitive and complete.)

Postulate 2 (Continuity). Suppose that $A > B > C$. Then there exists some probability mixture

$$L(p) = (A, p; C, 1 - p)\tag{2.4}$$

of A and C with $0 \leqslant p \leqslant 1$ such that $B \sim L(p)$.

Postulate 3 (Monotonicity in prizes). Suppose that $A_k^* \succsim A_k$ for $k = 1, \ldots, m$. Then also

$$(A_1^* | e_1; \ldots; A_m^* | e_m) \succsim (A_1 | e_1; \ldots; A_m | e_m) . \qquad (2.5)$$

(This postulate is a version of the sure-thing principle.)

Postulate 4 (Probabilistic equivalence). Let Prob denote objective probability. Define the lotteries L and L' as

$$L = (A_1 | e_1; \ldots; A_m | e_m) \quad \text{and} \quad L' = (A_1 | f_1; \ldots; A_m | f_m) . \qquad (2.6)$$

Suppose the decision-maker *knows* that

$$\text{Prob}(e_k) = \text{Prob}(f_k) \quad \text{for } k = 1, \ldots, m . \qquad (2.7)$$

Then, for this decision-maker

$$L \sim L' . \qquad (2.8)$$

In other words, a rational decision-maker must be indifferent between two lotteries yielding the same prizes with the same objective probabilities. (In particular, he must be indifferent between a one-stage and a two-stage lottery yielding the same prizes with the same final probabilities.)

We can now state:

Theorem 1. *Given Assumptions 1 and 2, an individual whose preferences satisfy Postulates 1–4 will have a utility function U that equates the utility U(L) of any lottery L to this lottery's* expected utility *so that*

$$U(L) = \sum_{k=1}^{m} p_k U(A_k) , \qquad (2.9)$$

where $p_1, \ldots, p_m$ are either *the objective probabilities of the conditioning events $e_1, \ldots, e_m$ known to him or are his own subjective probabilities for these events.*

Any utility function equating the utility of every lottery to its expected utility is said to possess the *expected-utility property* and is called a *von Neumann–Morgenstern* (vNM) *utility function.*

As Anscombe and Aumann have shown, using the above axioms one can first prove the theorem for risky lotteries. Then, one can extend the proof to all lotteries, using the theorem, restricted to risky lotteries, as one of one's axioms.

In view of the theorem, we can now extend the notation described under (2.2) also to *uncertain lotteries* if we interpret $p_1, \ldots, p_m$ as the relevant decision-maker's subjective probabilities.

3. An equi-probability model for moral value judgments

Utilitarian theory makes two basic claims. One is that all morality is based on maximizing *social utility* (also called the *social welfare funtion*). The other is that social utility is a *linear function* of all individual utilities, assigning the *same positive weight* to each individual's utility.[1]

In this section and the next I shall try to show that these two claims follow from the rationality postulates of Bayesian decision theory and from some other, rather natural, assumptions. In this section I shall propose an equi-probability model for *moral value judgments*, whereas in the next section I shall propose some axioms characterizing *rational choices among alternative social policies*.

First of all I propose to distinguish between an individual's *personal preferences* and his or her *moral preferences*.[2] The former are his preferences governing his everyday behavior. Most individuals' personal preferences will be by no means completely selfish. But they will be *particularistic* in the sense of giving *greater* weight to this individual's, his family members', and his friends' interests than giving to other people's interests. In contrast, his *moral preferences* will be his preferences governing his moral value judgments. Unlike his personal preferences, his moral preferences will be *universalistic*, in the sense of giving the *same* positive weight to everybody's interests, including his own because, by definition, moral value judgments are judgments based on impersonal and impartial considerations.

For example, suppose somebody tells me that he strongly prefers our capitalist system over any socialist system. When I ask him why he feels this way, he explains that in our capitalist system he is a millionaire and has a very interesting and rewarding life. But in a socialist system in all probability he would be a badly paid government official with a very uninteresting bureaucratic job. Obviously, if he is right about his prospects in a socialist system then

---

[1] Some utilitarians define social utility as the *sum* of all individual utilities whereas others define it as their *arithmetic mean*. But as long as the number n of individuals in the society is constant, these two approaches are mathematically equivalent because maximizing either of these two quantities will also maximize the other. Only in discussing population policies will this equivalence break down because n can no longer be treated as a constant in this context.

[2] In what follows, in similar phrases I shall omit the female pronoun.

he has very good reasons to prefer his present position in our capitalist system. Yet, his preference for the latter will be simply a *personal* preference based on self-interest, and will obviously not qualify as a *moral* preference based on impartial moral considerations.

The situation would be very different if he expressed a preference for the capitalist system *without knowing* what his personal position would be under either system, and in particular if he expected to have the *same chance* of occupying any possible social position under either system.

More formally, suppose that our society consists of n individuals, to be called individuals $1, \ldots, i, \ldots, n$. Suppose that one particular individual, to be called individual j, wants to compare various possible social situations s from an impartial moral point of view. Let $U_i(s)$ denote the utility level of individual i ($i = 1, \ldots, n$) in situation s. I shall assume that each utility function U_i is a vNM utility function, and that individual j can make interpersonal utility comparison between the utility levels $U_i(s)$ that various individuals i would enjoy in different social situations s (see Section 5).

Finally, to ensure j's impartiality in assessing different social situations s, I shall assume that j must assume that he has the *same probability* $1/n$ of ending up in the social position of any individual i with i's utility function U_i as his own utility function. (This last assumption is needed to ensure that he will make a realistic assessment of i's interests in the relevant social position and in the relevant social situation. Thus, if i were a fish merchant in a given social situation, then j must assess this fact in terms of the utility that i would derive from this occupation, and not in terms of his own (j's) tolerance or intolerance for fishy smells.)

Under these assumptions, j would have to assign to any possible social situation s the expected utility

$$W_j(s) = \frac{1}{n} \sum_{i=1}^{n} U_i(s) \tag{3.1}$$

and, by Theorem 1, this would be the quantity in terms of which he would evaluate any social situation s from an impartial moral point of view. In other words, $W_j(s)$ would be the *social utility function* that j would use as a basis for his *moral preferences* among alternative social situations s, i.e., as a basis for his *moral value judgments*.

Note that if two different individuals $j = j'$ and $j = j''$ assess each utility function U_i in the same way – which, of course, would be the case if both of them could make *correct* estimates of these utility funtions – then they will arrive at the *same* social utility function W_j. We can now summarize our conclusions by stating the following theorem.

Theorem 2. *Under the equi-probability model for moral value judgments, a rational individual j will always base his moral assessment of alternative social situations on a social utility function W_j defined as the* arithmetic mean *of all individual utility functions U_i (as estimated by him). Moreover, if different individuals j all form correct estimates of these utility functions then they will arrive at the* same *social utility function W_j.*

Note that this model for moral value judgments is simply an updated version of Adam Smith's (1976) theory of morality, which equated the moral point of view to that of an impartial but sympathetic observer (or "spectator" as he actually described him).

4. Axioms for rational choice among alternative social policies

In this section, for convenience I shall describe social situations as *pure alternatives*, and lotteries whose outcomes are social situations as *mixed alternatives*, I shall assume four axioms, later to be supplemented by a nondegeneracy (linear independence) assumption.

Axiom 1 (Rationality of individual preferences). The personal preferences of each individual $i(i = 1, \ldots, n)$ satisfy the rationality postulates of Bayesian decision theory (as stated in Section 2). Therefore his personal preferences can be represented by a vNM utility function U_i.

Axiom 2 (Rationality of the social-policy-maker's moral preferences). The moral preferences of individual j, the social-policy-maker, that guide him in choosing among alternative social policies likewise satisfy the rationality postulates of Bayesian decision theory. Therefore, j's moral preferences can be represented by a social utility function W_j that has the nature of a vNM utility function.

Axiom 3 (Use of the policy-maker's own subjective probabilities). Let π be a policy whose possible results are the pure alternatives $s_1, \ldots, s_m$. Then individual j will assess the desirability of this policy both from a *moral* point of view and from each individual's *personal* point of view in terms of the subjective probabilities $p_1, \ldots, p_m$ that *he himself* assigns to these possible outcomes $s_1, \ldots, s_m$. Thus, j will define the social utility of policy π as

$$W_j(\pi) = \sum_{k=1}^{m} p_k W_j(s_k) , \tag{4.1}$$

and will define the utility of this policy to a particular individual i as

$$U_i(\pi) = \sum_{k=1}^{m} p_k U_i(s_k) , \tag{4.2}$$

even though i himself will define the utility of this policy to him as

$$U_i(\pi) = \sum_{k=1}^{m} q_k^i U_i(s_k) , \tag{4.3}$$

where $q_1^i, \ldots, q_m^i$ are the probabilities that i himself assigns to the outcomes $s_1, \ldots, s_m$, respectively.

That is to say, a rational policy-maker will choose his subjective probabilities on the basis of the best information available to him. Therefore, once he has chosen these subjective probabilities, he will always select his policies, and will always form his expectations about the likely effects of these policies on all individuals i, on the basis of these probabilities rather than on the basis of the subjective probabilities that these individuals i may themselves entertain.

Axiom 4 (Positive relationship between the various individuals' interests as seen by the policy-maker and his moral preferences between alternative policies). Suppose that, in the judgment of individual j, a given policy π would serve the interests of every individual i *at least as well* as another policy π' would. Then, individual j will have at least a *nonstrict* moral preference for π over π'. If, in addition, he thinks that π would serve the interests of at least one individual i definitely *better* than π' would, then he will have a *strict* moral preference for π over π'. (This implies that the social utility function W_j representing j's moral preferences will be a single-valued strictly increasing function of the individual utilities $U_1, \ldots, U_n$.)

In addition to these four axioms, I shall assume:

Linear independence. The n utility functions $U_1, \ldots, U_n$ are linearly independent.
(This seems to be a natural assumption to make because any linear dependence could arise only by a very likely coincidence.) One can show that our four axioms and this linear-independence assumption imply the following theorem.

Theorem 3. *A rational policy-maker j will evaluate all social policies π in terms of a social utility function W_j having the mathematical form*

$$W_j(\pi) = \sum_{i=1}^{n} a_i U_i(\pi) , \tag{4.4}$$

with

$$a_1, \ldots, a_n > 0,$$ (4.5)

where the (expected) utilities $U_i(\pi)$ are defined in accordance with (4.2).

Of course, Theorem 3 as it stands is weaker than Theorem 2 because it does not tell us that the coefficients $a_1, \ldots, a_n$ must be *equal* to one another. But we can strengthen Theorem 3 so that it will include this requirement by adding a symmetry axiom to our four preceding axioms. Yet we can do this only if we assume *interpersonal comparability* of the various individuals' utilities (see Section 5). If we are willing to make this assumption, then we can introduce:

Axiom 5 (Symmetry). If the various individuals' utilities are expressed in the *same utility unit*, then W_j will be a *symmetric function* of the individual utilities $U_1, \ldots, U_n$.

The axiom requires interpersonal comparability (at least for utility differences) because otherwise the requirement of an identical utility unit becomes meaningless.

Yet, given Axiom 5, we can infer that

$$a_1 = \cdots = a_n = \alpha.$$ (4.6)

We are free to choose any positive constant as our α. If we choose $\alpha = 1/n$, then equation (4.4) becomes the same as equation (3.1). Another natural choice is $\alpha = 1$, which would make W_j the *sum*, rather than the *arithmetic mean*, of individual utilities.

5. Interpersonal utility comparisons

As is well known, in order to obtain a well-defined vNM utility function U_i for a given individual i, we have to choose a *zero-utility* level and a *utility unit* for him. Accordingly, full interpersonal comparability between two or more vNM utility functions $U_i, U_k, \ldots$ will obtain only if *both* the zero-utility points and the utility units of all these utility functions are comparable. Actually, utilitarian theory needs only *utility-unit* comparability. But since the same arguments can be used to establish full comparability as to establish mere utility-unit comparability, I shall argue in favor of *full* comparability.

In ordinary economic analysis the arguments of vNM utility functions are commodity vectors and probability mixtures (lotteries) of such vectors. But for ethical purposes we need more broadly defined arguments, with benefit vectors taking the place of commodity vectors. By the *benefit vector* of a given individual I shall mean a vector listing all economic and noneconomic *benefits* (or advantages) available to him, including his commodity endowments, and listing also all economic and noneconomic *discommodities* (or disadvantages) he has to face.

The vNM utility function U_i of an individual i is usually interpreted as a mathematical representation of his *preferences*. For our purposes we shall add a second interpretation. We shall say that i's vNM utility function U_i is also an indicator of the *amounts of satisfaction* that i derives (or would derive) from alternative benefit vectors (and from their probability mixtures). Indeed, any *preference* by i for one thing over another can be itself interpreted as an indication that i expects to derive *more satisfaction* from the former than from the latter.

It is a well-known fact that the vNM utility functions of different individuals tend to be very *different* in that they have very different preferences between different benefit vectors, and in that they tend to derive very different amounts of satisfaction from the same benefit vectors. Given our still very rudimentary understanding of human psychology, we cannot really explain these differences in any specific detail. But common sense does suggest that these differences between people's preferences and between their levels of satisfaction under comparable conditions – i.e., the differences between their vNM utility functions – are due to such factors as differences in their innate psychological and physiological characteristics, in their upbringing and education, in their health, in their life experiences, and other similar variables.

The variables explaining the differences between different people's vNM utility functions I shall call *causal variables*. Let $r_i, r_k, \ldots$ be the vectors of these causal variables explaining why the individuals $i, k, \ldots$ have the vNM utility functions $U_i, U_k, \ldots$, and why these utility functions tend to differ from individual to individual. I shall call these vectors $r_i, r_k, \ldots$, the *causal-variable vectors* of individuals $i, k, \ldots$.

Suppose that i's and k's benefit vectors are x and y, respectively, so that their vNM utility levels – and therefore also their levels of satisfaction – are $U_i(x)$ and $U_k(y)$. On the other hand, if their present benefit vectors were interchanged, then their vNM utility levels – and therefore also their levels of satisfaction – would be $U_i(y)$ and $U_k(x)$.

Under our assumptions, each individual's vNM utility level will depend both on his *benefit vector* and on his *causal-variable vector*. Therefore, there exists some mathematical function V such that

$$U_i(x) = V(x, r_i) , \qquad U_k(x) = V(x, r_k) ,$$

$$U_i(y) = V(y, r_i) , \qquad U_k(y) = V(y, r_k) . \tag{5.1}$$

Moreover, this function V will be the *same mathematical function* in all four equations (and in all similar equations). This is so because the differences between the utility functions U_i and U_k can be *fully explained* by the differences between the two individuals' causal-variable vectors r_i and r_k, whereas the function V itself is determined by the basic psychological laws governing human preferences and human satisfactions, *equally applying to all human beings*. This function V I shall call the inter-individual utility function.[3]

To be sure, *we do not know* the mathematical form of this function V. Nor do we know the nature of the causal-variable vectors $r_i, r_k, \ldots$ belonging to the various individuals. But my point is that *if we did know* the basic psychological laws governing human preferences and human satisfactions then we *could* work out the mathematical form of V and could find out the nature of these causal-variable vectors $r_i, r_k, \ldots$. This means that even if we do not know the function V, and do not know the vectors $r_i, r_k, \ldots$, these are *well-defined mathematical entities*, so that interpersonal utility comparisons based on these mathematical entities are a meaningful operation.

Moreover, in many specific cases we *do* have enough insight into human psychology in general, and into the personalities of the relevant individuals in particular, to make some interpersonal utility comparisons. For instance, suppose that both i and k are people with considerable musical talent. But i has in fact chosen a musical career and is now a badly paid but very highly respected member of a famous orchestra. In contrast, k has opted for a more lucrative profession. He has obtained an accounting degree and is now the highly paid and very popular chief accountant of a large company. Both individuals seem to be quite happy in their chosen professions. But I would have to know them really well before I could venture an opinion as to which one actually derives *more satisfaction* from his own way of life.

Yet, suppose I do know these two people very well. Then I *may* be willing to make the tentative judgment that i's *level of satisfaction*, as measured by the quantity $U_i(x) = V(x, r_i)$, is in fact *higher* or *lower* than is k's *level of satisfaction*, as measured by the quantity $U_k(y) = V(y, r_k)$. Obviously, even if I made such a judgment, I should know that such judgments are very hard to make, and must be subject to wide margins of error. But such possibilities of error do not make them into *meaningless* judgments.

I have suggested that, in discussing interpersonal comparisons of utilities, these utilities should be primarily interpreted as *amounts of satisfaction*, rather than as indicators of *preference* as such. My reason has been this.

Suppose we want to compare the vNM utility $U_i(x) = V(x, r_i)$ that i assigns to the benefit vector x, and the vNM utility $U_k(y) = V(y, r_k)$ that k assigns to the benefit vector y. If we adopted the *preference* interpretation, then we would have to ask whether i's situation characterized by the vector pair (x, r_i)

---

[3]In earlier publications I called V an *extended utility function*.

or k's situation characterized by the vector pair (y, r_k) was *preferred* (or whether these two situations were equally preferred). But this would be an *incomplete question* to ask. For before this question could be answered we would have to decide whether "preferred" meant "preferred by individual i" or meant "preferred by individual k" – because, for all we know, one of these two individuals might prefer i's situation, whereas the other might prefer k's situation. Yet, if this were the case then we would have no way of telling whether i's or k's situation were *intrinsically* preferable.

In contrast, if we adopt the *amount-of-satisfaction* interpretation, then no similar problem will arise. For in this case $U_i(x) = V(x, r_i)$ would become simply the *amount of satisfaction* that i derives from his present situation, whereas $U_k(y) = V(y, r_k)$ would become the *amount of satisfaction* that k derives from his present situation. To be sure, we have no way of directly measuring these two amounts of satisfaction, but can only estimate them on the basis of our – very fallible – intuitive understanding of human psychology and of i's and k's personalities. Yet, assuming that *there are* definite psychological laws governing human satisfactions and human preferences (even if our knowledge of these laws is very imperfect as yet), V will be a well-defined mathematical function, and the quantities $V(x, r_i)$ and $V(y, r_k)$ will be well-defined, real-valued mathematical quantities, in principle always comparable to each other.

6. Use of von Neumann–Morgenstern utilities in ethics

6.1. Outcome utilities and process utilities

Both Theorems 2 and 3 make essential use of vNM utility functions. Yet, the latter's use in ethics met with strong objections by Arrow (1951, p. 10) and by Rawls (1971, pp. 172 and 323) on the ground that vNM utility functions merely express people's attitudes toward *gambling*, and these attitudes have no moral significance.

Yet this view, it seems to me, is based on a failure to distinguish between the *process utilities* and the *outcome utilities* people derive from gambling and, more generally, from risk-taking. By *process utilities* I mean the (positive and negative) utilities a person derives from the *act of gambling* itself. These are basically utilities he derives from the various psychological experiences associated with gambling, such as the nervous tension felt by him, the joy of winning, the pain of losing, the regret for having made the wrong choice, etc. In contrast, by *outcome utilities* I mean the (positive and negative) utilities he assigns to various possible physical outcomes.

With respect to people's *process utilities* I agree with Arrow and Rawls: these utilities do merely express people's attitudes toward gambling and, therefore, have no moral significance. But I shall try to show that people's vNM utility functions express solely people's *outcome utilities* and completely disregard their process utilities. Indeed, what people's vNM utility functions measure are their *cardinal utilities* for various possible outcomes. Being cardinal utilities, they indicate not only people's *preferences* between alternative outcomes, as their ordinal utilities do, but also the *relative importance* they assign to various outcomes. Yet, this is morally very valuable information.

6.2. Gambling-oriented vs. outcome-oriented attitudes

I shall now define two concepts that I need in my subsequent discussion. When people gamble for entertainment, they are usually just as much interested in the process utilities they derive from their subjective experiences in gambling as they are in the outcome utilities they expect to derive from the final outcomes. In fact, they may gamble primarily for the sake of these subjective experiences. This attitude, characterized by a strong interest in these process utilities, I shall call a *gambling-oriented* attitude.

The situation is different when people engage in risky activities primarily for the sake of the expected *outcomes*. In such cases, in particular if the stakes are very high or if these people are business executives or political leaders gambling with other people's money and sometimes even with other people's lives, then they will be certainly well advised, both for moral reasons and for reasons of self-interest, to focus their attention on the *outcome utilities* and the *probabilities* of the various possible outcomes in order to achieve the best possible outcomes for their constituents and for themselves – without being diverted from this objective by their own positive or negative psychological experiences and by the process utilities derived from these experiences. This attitude of being guided by one's expected *outcome utilities* rather than by one's process utilities in risky situations I shall call an *outcome-oriented* attitude.[4]

6.3. Von Neumann–Morgenstern utility functions and outcome utilities

Now I propose to argue that vNM utility functions are based *solely* on people's outcome utilities, and make no use of their process utilities. Firstly, this can be

---

[4]Needless to say, everybody, whether he takes a gambling-oriented or a strictly outcome-oriented attitude, *does* have process utilities in risky situations. My point is only that some people intentionally take an outcome-oriented attitude and *disregard* these process utilities in order not to be diverted from their main objective of maximizing their expected *outcome utility*. (In philosophical terminology, our *first-order* preferences for enjoyable subjective experiences give rise to *process utilities*. On the other hand, an outcome-oriented attitude is a *second-order* preference for *overriding* those of our first-order preferences that give rise to such process utilities.)

verified by mere inspection of equation (2.9) defining the vNM utility of a lottery. This utility depends only on the outcome utilities $U(A_k)$ and on the probabilities p_k of the various possible outcomes A_k $(k = 1, \ldots, m)$, but does not in any way depend on the process utilities connected with gambling.

Secondly, we shall reach the same conclusion by studying von Neumann and Morgenstern's (1953) axioms defining these vNM utility functions, or by studying the rationality postulates listed in Section 2, which are simplified versions of their axioms.

For instance, consider Postulate 4 (p. 674). This postulate implies that a rational person will be indifferent between a one-stage and a two-stage lottery if both yield the same prizes with the same final probabilities. This is obviously a very compelling rationality requirement for a strictly *outcome-oriented* person, interested only in the utilities and the probabilities of the various possible outcomes. Yet, it is *not* a valid rationality requirement for a *gambling-oriented* person, taking a strong interest also in the process utilities he will obtain by participating in one of these two lotteries. For participation in a one-stage lottery will give rise to *one* period of nervous tension, whereas participation in a two-stage lottery may give rise to *two* such periods. Therefore, the two lotteries will tend to produce quite *different* process utilities so that we cannot expect a *gambling-oriented* person to be indifferent between them.

It is easy to verify that the same is true for Postulate 3: it is a very compelling rationality postulate for strictly *outcome-oriented* people but is not one for *gambling-oriented* people. [For further discussion, see Harsanyi (1987).]

Thus, only a strictly outcome-oriented person can be expected to conform to all four rationality postulates of Section 2. Yet, this means that only the behavior of such a person can be represented by a vNM utility function. But all that such a person's vNM utility function can express are his outcome utilities rather than his process utilities because his behavior is guided solely by the former.

Note that von Neumann and Morgenstern (1953, p. 28) themselves were fully aware of the fact that their axioms excluded what they called the *utility of gambling*, and what I am calling *process utilities*. It is rather surprising that this important insight of theirs later came to be completely overlooked in the discussions about the appropriateness of using vNM utility functions in ethics.

6.4. Von Neumann–Morgenstern utilities as cardinal utilities

Let me now come back to my other contention that vNM utility functions are very useful in ethics because they express the *cardinal utilities* people assign to various possible outcomes, indicating the *relative importance* they attach to these outcomes.

Suppose that individual i pays \$10 for a lottery ticket giving him $1/1000$ chance of winning \$1000. This fact implies that

$$\frac{1}{1000} U_i(\$1000) \geq U_i(\$10) . \tag{6.1}$$

In other words, even though \$1000 is only a 100 times larger amount of *money* than \$10 is, i assigns an at least 1000 times higher *utility* to the former than he assigns to the latter. Thus, his vNM utility function not only indicates that he *prefers* \$1000 to \$10 (which is all that an ordinal utility function could do), but also indicates that he attaches *unusually high importance* to winning \$1000 as compared with the importance he attaches to not losing \$10 (as he would do if he did not win anything with his lottery ticket – which would be, of course, the far more likely outcome).

To be sure, i's vNM utility function does not tell us *why* he assigns such a high importance to winning \$1000. We would have to know his personal circumstances to understand this. (For instance, if we asked him we might find out that winning \$1000 was so important for him because he hoped to use the money as cash deposit on a *very badly needed* second-hand car, or we might obtain some other similar explanation.)

Thus, in ethics, vNM utility functions are important because they provide information, not only about people's *preferences*, but also about the *relative importance* they attach to their various preferences. This must be very valuable information for any humanitarian ethics that tries to encourage us to satisfy other people's wants and, other things being equal, to give priority to those wants they themselves regard as being most important. Admittedly, a vNM utility function measures the relative importance a person assigns to his various wants by the *risks* he is willing to take to satisfy these wants. But, as we have seen, this fact must not be confused with the untenable claim that a person's vNM utility function expresses merely his like or his dislike for *gambling* as such.

II. RULE UTILITARIANISM, ACT UTILITARIANISM, RAWLS' AND BROCK'S NONUTILITARIAN THEORIES OF JUSTICE

7. The two versions of utilitarian theory

Act utilitarianism is the view that a morally right action is simply one that would maximize expected social utility in the existing situation. In contrast, *rule utilitarianism* is the view that a morally right action must be defined in two steps. First, we must define the right moral rule as the moral rule whose acceptance would maximize expected social utility in similar situations. Then, we must define a morally right action as one in compliance with this moral rule.

Actually, reflection will show that in general we cannot judge the social utility of any proposed moral rule without knowing the *other* moral rules accepted by the relevant society. Thus, we cannot decide what the moral obligations of a father should be toward his children without knowing how the moral obligations of other relatives are defined toward these children. (We must ensure that *somebody* should be clearly responsible for the well-being of every child. On the other hand, we must not give *conflicting* responsibilities to different people with respect to the same child.)

Accordingly, it seems to be preferable to make society's *moral code*, i.e., the set of all moral rules accepted by the society, rather than individual moral rules, the basic concept of rule utilitarian theory. Thus, we may define the *optimal moral code* as the moral code whose acceptance would maximize expected social utility,[5] and may define a morally right action as one in compliance with this moral code.

How should we interpret the term "social acceptance" used in these definitions? Realistically, we cannot interpret it as full compliance by all members of the society with the accepted moral code (or moral rule). All we can expect is partial compliance, with a lesser degree of compliance in the case of a very demanding moral code. (Moreover, we may expect much more compliance in the moral judgments people make about each other's behavior than in their own actual behavior.)

How will a rational utilitarian choose between the two versions of utilitarian theory? It seems to me that he must make his choice in terms of the basic utilitarian choice criterion itself: he must ask whether a rule utilitarian or an act utilitarian society would enjoy a *higher level of social utility*.

Actually, the problem of choosing between the rule utilitarian and the act utilitarian approaches can be formally regarded as a special case of the rule utilitarian problem of choosing among alternative moral codes according to their expected social utility. For, when a rule utilitarian society chooses among alternative moral codes, the act utilitarian moral code (asking each individual to choose the social-utility maximizing action in every situation) is one of the moral codes available for choice.

Note that this fact already shows that the social-utility level of a rule utilitarian society, one using the *optimal* rule utilitarian moral code, *must be at least as high* as that of an act utilitarian society, using the act utilitarian moral code.

---

[5]For the sake of simplicity, I am assuming that the social-utility maximizing moral code is unique.

8. The effects of a socially accepted moral code

When a society accepts a given moral code, the most obvious effects of this will be the *benefits* people will obtain by other people's (and by their own) compliance with this moral code. These effects I shall call *positive implementation effects*.

Yet, compliance with any moral code – together with the fact that this compliance will always be somewhat incomplete – will give rise also to some *social costs*. These include the efforts needed to comply with specific injunctions of the moral code, and in particular to do so in some difficult situations; the guilt feelings and the social stigma that may follow noncompliance; loss of respect for the moral code if people see widespread noncompliance; and the efforts needed to inculcate habits consistent with the moral code in the next generation [cf. Brandt (1979, pp. 287–289)]. These effects I shall call *negative implementation effects*. They will be particularly burdensome in the case of very demanding moral codes and may make adoption of such a moral code unattractive even if it would have very attractive positive implementation effects.

Another important group of social effects that a moral code will produce are its *expectation effects*. They result from the fact that people will not only themselves comply with the accepted moral code to some degree but will expect *other* people likewise to comply with it. This expectation may give them *incentives* to socially beneficial activities, and may give them some *assurance* that their interests will be protected. Accordingly, I shall divide the expectation effects of a moral code into *incentive effects* and *assurance effects*. As we shall see, these two classes of expectation effects are extremely important in determining the social utility of any moral code. It is all the more surprising that so far they have received hardly any attention in the literature of ethics.

9. The negative implementation effects of act utilitarian morality

In Section 3 I argued that people's personal preferences are *particularistic* in that they tend to give much greater weight to their own, their family members', and their closest friends' interests than they tend to give to other people's interests; but that they often make moral value judgments based on *universalistic* criteria, giving the same weight to everybody's interests. As a result, the utility function U_i of any individual i will be quite different from his social utility function W_i since the former will be particularistic while the latter will be universalistic. A world like ours where people's personal preferences and their utility functions are particularistic I shall call a *particularistic world*.

In contrast, imagine a world where even people's personal preferences and their utility functions would be *universalistic*. In such a world, each individual's utility function U_i would be *identical* to his social utility function W_i. Indeed, assuming that different individuals would define their society's interests in the same way, different individuals' social utility functions would be likewise *identical*. Such an imaginary world I shall call a *universalistic world*.

In a *universalistic* world, people would have no difficulty in following the act utilitarian moral code. To be sure, the latter would require them in every situation to choose the action maximizing social utility. But since for them maximizing social utility would be the same thing as maximizing their own individual utility, they could easily comply with this requirement without going against their own natural inclinations.

Yet, things are very different in our own *particularistic* world. Even those of us who try to comply with some moral code will tend in each situation to choose, among the actions permitted by our moral code, the one maximizing our individual utility. But act utilitarian morality would require a *radical shift* in our basic attitudes and in our everyday behavior. It would require complete replacement of maximizing our *individual utility* by maximizing *social utility* as our choice criterion for all our decisions.

Clearly, this would amount to suppressing our particularistic personal preferences, our personal interests, and our personal commitments to our family and our friends, for the rigidly universalistic principles of act utilitarian morality. Such a complete suppression of our natural inclinations could be done, if it could be done at all, only by extreme efforts and at extremely high psychological costs. In other words, act utilitarian morality would have *intolerably burdensome* negative implementation effects.

In contrast, compliance with a rule utilitarian moral code would not pose any such problems. The latter would be basically a greatly improved and much more rational version of conventional morality, and compliance with it would require much the same effort as compliance with conventional morality does. No doubt it would require us in many cases to give precedence to other people's interests and to society's common interests over our personal preferences, concerns, and interests. But within these limits it would let us follow our own preferences, concerns, and interests.

10. The value of free individual choice

One aspect of the negative implementation effects of act utilitarian morality would be its social-utility maximization requirement in every situation, i.e., its insistence on the *highest possible moral performance* at every instant of our life. We would not be permitted to relax, or to do what we would like to do, even for one moment. If I were tempted to read a book or to go for a leisurely walk after a tiring day, I would always have to ask myself whether I could not do something more useful to society, such as doing some voluntary work for charity (or perhaps trying to convert some of my friends to utilitarian theory) instead. This of course means that, except in those rare cases where two or more actions would equally produce the highest possible social utility, I would never be permitted a *free choice* between alternative actions.

This also means that act utilitarian theory could not accommodate the traditional, and intuitively very appealing, distinction between merely doing one's *duty* and performing a *supererogatory* action going beyond the call of duty. For we would do only our duty by choosing an action maximizing social utility, and by doing anything else we would clearly fail to do our duty.

In contrast, a rule utilitarian moral code could easily recognize the intrinsic value of *free individual choice*. Thus, suppose I have a choice between action A, yielding the social utility α, and action B, yielding the social utility β, with $\alpha > \beta$. In this case, act utilitarian theory would make it my duty to choose action A. But a rule utilitarian moral code could assign a *procedural utility* γ to free moral choice. This would make me morally free to choose between A and B as long as $\beta + \gamma \geqslant \alpha$. Nevertheless, because $\alpha > \beta$, A would remain the morally preferable choice. Thus, I would do my duty both by choosing A and by choosing B. Yet, by choosing the morally preferable action A, I would go *beyond* merely doing my duty and would perform a supererogatory action.

11. Morally protected rights and obligations, and their expectation effects

The moral codes of civilized societies recognize some individual rights and some special obligations[6] that *cannot be overriden* merely because by overriding them one could here and now increase social utility – except possibly in some very special cases where fundamentally important interests of society are at stake. I shall describe these as *morally protected* rights and obligations.

As an example of individual rights, consider a person's property rights over a boat he owns. According to our accepted moral code (and also according to our legal rules), nobody else can use this boat without the owner's permission in other than some exceptional cases (say, to save a human life). The mere fact that use of the boat by another person may *increase social utility* (because he would derive a greater utility by using the boat than the owner would) is *not* a morally acceptable reason for him to use the boat without the owner's consent.

Even though, as we have seen, the *direct* effects of such property rights will

---

[6]By special obligations I mean moral obligations based on one's social role (e.g., one's obligations as a parent, or as a teacher, or as a doctor, etc.) or on some past event (e.g., on having made a promise, or on having incurred an obligation of gratitude to a benefactor, and so on).

often be to prevent some actions that might increase social utility, their *indirect* effects, namely their *expectation effects*, make them into a socially very useful institution. They provide socially desirable *incentives* to hard work, saving, investment, and entrepreneurial activities. They also give property owners *assurance* of some financial security and of some independence of other people's good will. (Indeed, as a kind of assurance effect benefiting society as a whole, widespread property ownership contributes to social stability, and is an important guarantee of personal and political freedom.)

As an example of special obligations, consider a borrower's obligation to repay the borrowed money to the lender, except if this would cause him extreme hardship. In some cases, in particular when the borrower is very poor whereas the lender is very rich, by repaying the money the borrower will substantially *decrease social utility* because his own utility loss will greatly exceed the lender's utility gain (since the marginal utility of money to very poor people tends to be much higher than the marginal utility of money to very rich people). Nevertheless, it is a socially very beneficial moral (and legal) rule that normally loans must be repaid (even in the case of very poor borrowers and very rich lenders) because otherwise people would have a strong *incentive* not to lend money. Poor people would particularly suffer by being unable to borrow money if it were known that they would not have to repay it. The rule that loans must be repaid will also give lenders some *assurance* that society's moral code will protect their interests if they lend money.

Since a rule utilitarian society would choose its moral code by the criterion of social utility, it would no doubt choose a moral code recognizing many morally protected rights and obligations, in view of their very beneficial expectation effects. But an act utilitarian society *could not do this*. This is so because act utilitarian morality is based not on choosing between alternative *moral codes*, but rather on choosing between alternative *individual actions* in each situation. Therefore, the only expectation effects it could pay attention to would be those of individual actions and not those of entire moral codes. Yet, normally an individual action will have *negligibly small* expectation effects. If people know that their society's *moral code* does protect, or does not protect, property rights, this will have a substantial effect on the extent to which they will expect property rights to be actually respected. But if all they know is that *one* individual on *one* particular occasion did or did not respect another individual's property rights, this will hardly have any noticeable effect on the extent to which they will expect property rights to be respected in the future. Hence, an act utilitarian society would have no reason to recognize morally protected rights and obligations.

Yet, in fairness to act utilitarian theory, a consistent act utilitarian would not really regret the absence of morally protected rights and obligations from his society. For he would not really mind if what we would consider to be his individual rights were violated, or if what we would consider to be special obligations owed to him were infringed, if this were done to maximize social utility – because maximization of social utility would be the only thing he would care about.

12. The advantages of the rule utilitarian approach

To conclude, most of us *would very much prefer to live* in a rule utilitarian society rather than in an act utilitarian society. For one thing, we would very much prefer to live in a society whose moral code permitted us within reasonable limits to make our own choices, and to follow our own personal preferences and interests as well as our personal commitments to the people we most cared about. In other words, we would prefer to live under a moral code with much less burdensome *negative implementation effects* than the act utilitarian moral code would have.

For another thing, we would much prefer to live in a society whose moral code recognized individual rights and special obligations that *must not be overridden* for social-expediency considerations, except possibly in some rare and special cases. We would feel that in such a society our interests would be much better protected, and that society as a whole would benefit from the desirable expectation effects of such a moral code.

The fact that most of us would definitely prefer to live in a rule utilitarian society is a clear indication that most of us would expect to enjoy a much higher level of *individual utility* under a rule utilitarian moral code than under an act utilitarian moral code. Yet, social utility can be defined as the arithmetic mean (or as the sum) of individual utilities. Therefore, we have very good reasons to expect that the level of *social utility* would be likewise much higher in a rule utilitarian society than it would be in an act utilitarian society.

Both act utilitarianism and rule utilitarianism are *consequentialist* theories because both of them define morally right behavior ultimately in terms of its consequences with respect to social utility. This gives both versions of utilitarianism an important advantage over *nonconsequentialist* theories be-cause it gives them a clear and readily understandable *rational criterion* for the solution of moral problems – something that nonconsequentialist theories of morality altogether lack, and that they have to replace by vague references to our "moral intuitions" or to our "sense of justice" [e.g., Rawls (1971, pp. 48–51)].

As I have tried to show, even though both versions of utilitarianism are based on the same consequentialist choice criterion, the rule utilitarian ap-proach has important advantages over the act utilitarian approach. At a fundamental level, all these advantages result from the much greater *flexibility* of the rule utilitarian approach. As we have seen, whereas the rule utilitarian approach is free to choose its moral code from a *very large set* of possible moral codes, the act utilitarian approach is restricted to *one* particular moral code within this set.

Yet, this means that act utilitarianism is committed to evaluate each individual action *directly* in terms of the consequentialist utilitarian criterion of social-utility maximization. In contrast, rule utilitarianism is free to choose a moral code that judges the moral value of individual actions partly in terms of *nonconsequentialist* criteria if use of such criteria increases social utility. Thus, it may choose a moral code that judges the moral value of a person's action not only by its direct social-utility yield but also by the social relationship between this person and the people directly benefiting or directly damaged by his action, by this person's prior promises or other commitments, by procedural criteria, and so on. Even if consequentialist criteria are used, they may be based not only on the social consequences of individual actions but also on the consequences of a morally approved *social practice* of similar behavior in all similar cases. This greater flexibility in defining the moral value of individual actions will permit adoption of moral codes with significantly higher social utility.

13. A game-theoretic model for a rule utilitarian society

I shall use the following notations. A strategy of player i, whether pure or mixed, will be denoted as s_i. We can assume without loss of generality that every player has the *same* strategy set $S = S_1 = \cdots = S_n$. A strategy combination will be written as $\bar{s} = (s_1, \ldots, s_n)$. The strategy $(n-1)$-tuple obtained when the *i*th component s_i of $\bar{s}$ is omitted will be written as $\bar{s}_{-i} = (s_1, \ldots, s_{i-1}, s_{i+1}, \ldots, s_n)$.

I propose to model a rule utilitarian society as a *two-stage game*. At first I shall assume that all n players are consistent rule utilitarians fully complying with the rule utilitarian moral code. (Later this assumption will be relaxed.) On this assumption, stage 1 of the game will be a *cooperative* game in which the n players together choose a moral code M so as to maximize the social utility function W, subject to the requirement that

$$M \in \mathcal{M} , \tag{13.1}$$

where $\mathcal{M}$ is the set of all possible moral codes. On the other hand, stage 2 of the game will be a *noncooperative* game in which each player i will choose a strategy s_i for himself so as to maximize his own individual utility U_i, subject to the requirement that

$$s_i \in P(M) , \qquad (13.2)$$

where $P(M)$ is the set of all strategies permitted by the moral code M chosen at stage 1. I shall call $P(M)$ the *permissible set* for moral code M, and shall assume that, for all $M \in \mathcal{M}, P(M)$ is a nonempty compact subset of the strategy set S.

The noncooperative game played at stage 2, where the players' strategy choices are restricted to $P(M)$, will be called $\Gamma(M)$. At stage 1, in order to choose a moral code M maximizing social utility, the players must try to predict the equilibrium point $\bar{s} = (s_1, \ldots, s_n)$ that will be the actual outcome of this game $\Gamma(M)$. I shall assume that they will do this by choosing a *predictor function* π selecting, for every possible game $\Gamma(M)$, an equilibrium point $\bar{s} = \pi(\Gamma(M))$ as the likely outcome $\Gamma(M)$. [For instance, they may choose this predictor function on the basis of our solution concept for noncooperative games. See Harsanyi and Selten (1988).] For convenience, I shall often use the shorter notation $\pi^*(M) = \pi(\Gamma(M))$.

Finally, I shall assume that each player's individual utility will have the mathematical form

$$U_i = U_i(\bar{s}, M) . \qquad (13.3)$$

I am including the chosen moral code M as an argument of U_i because the players may derive some direct utility by living in a society whose moral code permits a considerable amount of free individual choice (see Section 10).

Since the social utility function W is defined in terms of the individual utilities $U_1, \ldots, U_n$, it must depend on the same arguments as the latter do. Hence it has to be written as

$$W = W(\bar{s}, M) = W(\pi^*(M), M) . \qquad (13.4)$$

How does this model represent the implementation effects and expectation effects of a given moral code M? Clearly, its *implementation effects*, both the positive and the negative ones, will be represented by the fact that the players' strategies will be restricted to the permissible set $P(M)$ defined by this moral code M. This fact will produce both utilities and disutilities for the players and, therefore, will give rise both to positive and negative implementation effects.

On the other hand, the *expectation effects* of M will be represented by the fact that some players will choose different strategies than they would choose if their society had a different moral code – not because M directly requires them to do so but rather because these strategies are their *best replies* to the other players' expected strategies, on the assumption that these *other players* will use only strategies permitted by the moral code *M*.

This model illustrates the fact, well known to game theorists, that an ability to make *binding commitments* is often an important advantage for the players in many games. In the rule utilitarian game we are discussing, it is an important advantage for the players that at stage 1 they can *commit* themselves to comply with a jointly adopted moral code. Yet, in most other games, this advantage lies in the fact that such commitments will prevent the players from disrupting some agreed joint strategy in order to increase *their own* payoffs. In contrast, in this game, part of the advantage lies in the fact that the players' commitment to the jointly adopted moral code will prevent them from violating the other players' rights or their own obligations in order to increase *social utility*.

Our model can be made more realistic by dropping the assumption that *all* players are fully consistent rule utilitarians. Those who are I shall call the *committed* players. Those who are not I shall call the *uncommitted* players. The main difference will be that requirement (13.2) will now be observed only by the committed players. For the uncommitted players, it will have to be replaced by the trivial requirement $s_i \in S$. On the other hand, some of the uncommitted players *i* might still choose a strategy s_i at least in *partial* compliance with their society's moral code *M*, presumably because they might derive some utility by at least partial compliance. [This assumption, however, requires no formal change in our model because (13.3) has already made *M* an argument of the utility functions U_i.]

The realism of our model can be further increased by making the utilitarian game into one with *incomplete information* [see Harsanyi (1967–68)].

14. Rawls' theory of justice

Undoubtedly, the most important contemporary nonutilitarian moral theory is John Rawls' theory of justice [Rawls (1971)]. Following the contractarian tradition of Locke, Rousseau, and Kant, Rawls postulates that the principles of justice go back to a fictitious *social contract* agreed upon by the "heads of families" at the beginning of history, both on their own and on all their future descendants' behalf. To ensure that they will agree on *fair* principles not biased in their own favor, Rawls assumes that they have to agree on this social contract under what he calls the *veil of ignorance*, that is, without knowing what their personal interests are and, indeed, without knowing their personal identities. He calls this hypothetical situation characterized by the veil of ignorance the *original position*.

As is easy to see, the intuitive idea underlying Rawls' original position is very similar to that underlying my own equi-probability model for moral value judgments, discussed in Section 3.[7]

Nevertheless, there are important differences between Rawls' model and mine. In my model, a person making a moral value judgment would choose between alternative social situations in a *rational manner*, more particularly in a way required by the rationality postulates of Bayesian decision theory. Thus, he would always choose the social situation with the highest *expected utility* to him. Moreover, in order to ensure that he will base his choice on impartial *universalistic* considerations, he must make his choice on the assumption that, whichever social situation he chose, he would always have the *same probability* of ending up in any one of the n available social positions.

In contrast, Rawls' assumption is that each participant in the original position will choose among alternative conceptions of justice on the basis of the highly irrational *maximin principle*, which requires everybody to act in such a way as if he were absolutely sure that, whatever he did, the *worst possible outcome* of his action would obtain. This is a very surprising assumption because Rawls is supposedly looking for that conception of justice that *rational individuals* would choose in the original position.

It is easy to verify that the maximin principle is a highly irrational choice criterion. The basic reason is that it makes the value of any possible action wholly dependent on its worst possible outcome, *regardless of how small its probability*. If we tried to follow this principle, then we could not cross even the quietest country road because there was always some very small probability that we would be overrun by a car. We could never eat any food because there is always some very small probability that it contains some harmful bacteria. Needless to say, we could never get married because a marriage may certainly come to a bad end. Anybody who tried to live this way would soon find himself in a mental institution.

Yet, the maximin principle would be not only a very poor guide in our everyday life, it would be an equally poor guide in our moral decisions. As Rawls rightly argues, if the participants of the original position followed the maximin principle, then they would end up with what he calls the *difference principle* as their basic principle of justice. The latter would ask us always to give *absolute priority* to the interests of the most disadvantaged and the poorest social group over the interests of all other people *no matter what* – even if this group consisted of a mere handful of people whose interests were only minimally affected, whereas the rest of society consisted of many millions with

---

[7]Rawls first proposed his concept of the original position in 1957 [Rawls (1957)]. I proposed my own model in 1953 and 1955 [Harsanyi (1953, 1955)]. But both of us were anticipated by Vickrey, who suggested a similar approach already in 1945 [Vickrey (1945)]. Yet all three of us arrived quite independently at our own models.

very important interests at stake. This principle is so extreme and so implaus-
ible that I find it hard to take seriously the suggestion to make it our basic
principle of justice.

In other areas, too, Rawls seems to be surprisingly fond of such rigid and
unconditional principles of *absolute priority*. Common sense tells us that social
life is full of situations where we have to *weigh* different social values against
each other and must find morally and politically acceptable *trade-offs* between
them: we must decide how much individual freedom or how much economic
efficiency to give up for some possible increase in economic equality; how to
balance society's interest in deterring crime against protecting the legitimate
interests of defendants in criminal cases; how to balance the interests of gifted
children against the interests of slow learners in schools; etc.

Utilitarian theory suggests a natural criterion for resolving such trade-off
problems by asking the question of what particular compromise between such
conflicting social values would *maximize social utility*. (Even if we often cannot
really calculate the social-utility yields of alternative social policies with any
reasonable degree of confidence, if we at least know what question to ask, this
will focus our attention in the right direction.)

In contrast, Rawls seems to think that such problems can be resolved by the
simple-minded expedient of establishing rigid *absolute priorities* between different social values, for instance by declaring that *liberty* (or, more exactly, the
greatest possible basic liberty for everybody as far as this is compatible with
equal liberty for everybody else) shall have absolute priority over solving the
problems of *social and economic inequality* [Rawls (1971, p. 60)]. In my own
view, the hope that such rigid principles of absolute priority can work is a
dangerous illusion. Surely, there will be cases where common sense will tell us
to accept a *very small* reduction in our liberties if this is a price for a *substantial*
reduction in social and economic inequalities.

15. Brock's theory of social justice based on the Nash solution and on the Shapley value

15.1. Nature of the theory

Another interesting theory of social justice has been proposed by Brock
(1978). It is based on two game-theoretic solution concepts: one is the
n-person Nash solution [see Nash (1950) for the two-person case; and see Luce
and Raiffa (1957, pp. 349–350) for the n-person case); the other is the NTU
(nontransferable utility) Shapley value [see Harsanyi (1963) and Shapley
(1969)]. The Nash solution is used by Brock to represent "need justice",
characterized by the principle "To Each According to His Relative Need";

whereas the Shapley value is used to represent "merit justice", characterized by the principle "To Each According to His Relative Contribution".

For convenience, I shall first discuss a *simplified* version of Brock's theory, based solely on the *n*-person Nash solution. Then I shall consider Brock's actual theory, which makes use of the NTU Shapley value as well.

His simplified theory would give rise to an *n*-person *pure* bargaining game. The disagreement payoffs d_i^* would be the utility payoffs the *n* players (i.e., the *n* individual members of society) would obtain in a Hobbesian "state of nature", where people's behavior would not be subject to any Constitutional or other moral or legal constraints. The *n*-vector listing these payoffs will be denoted as $d^* = (d_1^*, \ldots, d_n^*)$. The outcome of this bargaining game would be the utility vector $u^* = (u_1^*, \ldots, u_i^*, \ldots, u_n^*)$ maximizing the *n*-person Nash product

$$\pi^* = \prod_{i=1}^{n} (u_i - d_i^*) , \tag{15.1}$$

subject to the two constraints $u^* \in F$ and $u_i^* > d_i^*$ for all *i*. Here *F* denotes the convex and compact feasible set. (It is customary to write the second constraint as a *weak* inequality. But Brock writes it as a *strong* inequality because he wants to make it clear that every player will positively benefit by moving from d^* to u^*. In any case, mathematically it makes no difference which way this constraint is written.)

Let me now go over to Brock's *full* theory. This involves a *two-stage* game. At stage 1, the players choose a Constitution *C* restricting the strategies of each player *i* at stage 2 to some *subset* S_i^* of his original strategy set S_i. As a result, this Constitution *C* will define an NTU game $G(C)$ in characteristic-function form to be played by the *N* players at stage 2. The outcome of $G(C)$ will be an NTU Shapley-value vector $u^{**} = (u_1^{**}, \ldots, u_n^{**})$ associated with this game $G(C)$. The players can choose only Constitutions *C* yielding a Shapley-value vector u^{**} with $u_i^{**} > d_i^*$ for every player *i*. (If a given game $G(C)$ has more than one Shapley-value vector u^{**} satisfying this requirement, then the players can presumably choose any one of the latter as the outcome of the game.) Let F^* be the set of all possible Shapley-value vectors u^{**} that the players can obtain by adopting any such admissible Constitution *C*.

Actually, Brock assumes that the players can choose not only a *specific* Constitution *C* but can choose also some *probability mixture* of two or more Constitutions $C, C', \ldots$. If this is what they do, then the outcome will be the corresponding weighted average of the Shapley-value vectors $u^{**}, (u^{**})', \ldots$ generated by these Constitutions. This of course means that the set of possible outcomes will not be simply the set F^* defined in the previous paragraph, but rather will be the *convex hull* F^{**} of this set F^*.

Finally, Brock assumes that at stage 1 the players will always choose that particular Constitution, or that particular probability mixture of Constitutions, that yields the payoff vector $u^0 = (u_1^0, \ldots, u_n^0)$ maximizing the Nash product

$$\pi^0 = \prod_{i=1}^{n} (u_i - d_i^*),$$ (15.2)

subject to the two constraints $u \in F^{**}$ and $u_i > d_i^*$ for all i.

15.2. Brock's theory of "need justice"

Brock admits that, instead of representing "need justice" by maximization of an n-person *Nash product*, he could have represented it by maximization of a *social utility function*, defined as the sum (or as the arithmetic mean) of individual utilities in accordance with utilitarian theory [Brock (1978, p. 603, footnote)]. But he clearly prefers the former approach.

He does so for two reasons. One is that the utility vector maximizing the social utility function may have *undesirable mathematical properties* in some cases. His other reason is that the utilitarian approach makes essential use of *interpersonal utility comparisons*, whose validity has often been called into question.

I shall illustrate the first difficulty by three examples. *Example 1* will be about a society consisting of two individuals. The feasible set of utility vectors will be defined by the two inequalities u_1 and $u_2 \geqslant 0$ and by the third inequality

$$u_1 + u_2 \leqslant 10.$$ (15.3)

The social utility function to be maximized will be $W = u_1 + u_2$. In this case, maximization of W will yield an *indeterminate* result in that *any* utility vector $u = (u_1, u_2)$ with u_1 and $u_2 \geqslant 0$ and with $u_1 + u_2 = 10$ will maximize W.

Example 2 will be similar to Example 1, except that (15.3) will be replaced by

$$u_1 + (1 + \varepsilon)u_2 \leqslant 10,$$ (15.4)

where ε is a very small positive number. Now, in order to maximize $W = u_1 + u_2$, we must set $u_1 = 10 - u_2 - \varepsilon u_2$, which means that $W = 10 - \varepsilon u_2$. Hence, maximization of W will require us to choose $u_2 = 0$ and $u_1 = 1$. In other words, we must choose the *highly inequalitarian* utility vector $u = (10, 0)$.

Finally, *Example 3* will be like Example 2, except that (15.4) will be replaced by

$$(1 + \varepsilon)u_1 + u_2 \leqslant 10 .$$ (15.5)

By symmetry, maximization of W will now require choice of the utility vector $u = (0, 10)$, which will be once more a *highly inequalitarian* outcome. Moreover, arbitrarily small changes in the feasible set – such as a shift from (15.4) to (15.3) and then to (15.5) – will make the utilitarian outcome *discontinuously* jump from $u = (10, 0)$ first to an indeterminate outcome and then to $u = (0, 10)$.

I agree with Brock that, at least at an *abstract* mathematical level, these mathematical anomalies are a serious objection to utilitarian theory. But I should like to argue that they are much less of a problem for utilitarian theory as an ethical theory for *real-life* human beings, because these anomalies will hardly ever actually arise in real-life situations.

This is so because in real life we can never transfer abstract "utility" as such from one person to another. All we can do is to transfer assets *possessing* utility, such as money, commodities, securities, political power, etc. Yet, most people's utility functions are such that such assets sooner or later will become subject to the *Law of Diminishing Marginal Utility*. As a result, in real-life situations the upper boundary of the feasible set in the utility space will tend to have enough concave curvature to prevent such anomalies from arising to any significant extent.

As already mentioned, Brock also feels uneasy about use of interpersonal utility comparisons in utilitarian theory. No doubt, such comparisons are rejected by many philosophers and social scientists. But in Section 5 I already stated my reasons for considering such comparisons to be perfectly legitimate intellectual operations.

Let me now add that interpersonal utility comparisons not only are *possible*, but are also strictly *necessary* for making moral decisions in many cases. If I take a few children on a hiking trip and we run out of food on our way home, then common sense will tell me to give the last bite of food to the child likely to derive the *greatest utility* from it (e.g., because she looks like the hungriest of the children). By the same token, if I have a concert ticket to give away, I should presumably give it to that friend of mine likely to enjoy the concert *most*, etc. It seems to me that we simply could not make sensible moral choices in many cases without making, or at least trying to make, interpersonal utility comparisons.

This is also my basic reason why I feel that our moral decisions should be based on the utilitarian criterion of maximizing *social utility* rather than Brock's criterion of maximizing a particular *Nash product*.

Suppose I can give some valuable object A *either* to individual 1 *or* to individual 2. If I give it to 1 then I shall increase his utility level from u_1 to $(u_1 + \Delta u_1)$, whereas if I give it to 2 then I shall increase the latter's utility level from u_2 to $(u_2 + \Delta u_2)$. Let me assume that in a Hobbesian "state of nature" the two individuals' utility levels would be d_1^* and d_2^*, respectively.

If my purpose is to maximize *social utility* in accordance with utilitarian theory, then I have to give A to 1 if

$$(u_1 + \Delta u_1) + u_2 > u_1 + (u_2 + \Delta u_2) \,, \tag{15.6}$$

that is, if

$$\Delta u_1 > \Delta u_2 \,, \tag{15.7}$$

and have to give it to 2 if these two inequalities are reversed.

In contrast, if my purpose is to maximize the relevant *Nash product* in accordance with Brock's theory, then I have to give A to 1 if

$$(u_1 + \Delta u_1 - d_1^*)(u_2 - d_2^*) > (u_1 - d_1^*)(u_2 + \Delta u_2 - d_2^*) \,, \tag{15.8}$$

which also can be written as

$$\frac{\Delta u_1}{u_1 - d_1^*} > \frac{\Delta u_2}{u_2 - d_2^*} \,. \tag{15.9}$$

On the other hand, I have to give A to 2 if the last two inequalities are reversed. For convenience, the quantities $(u_i - d_i^*)$ for $i = 1, 2$ I shall describe as the two individuals' *net utility levels*.

The utilitarian criterion, as stated in (15.7), assesses the *moral importance* of any individual need by the importance that the relevant individual *himself* assigns to it, as measured by the *utility increment* Δu_i he would obtain by satisfying this need. In contrast, Brock's criterion, as stated in (15.9), would assess the moral importance of this need, *not* by the utility increment Δu_i as such, but rather by the *ratio* of Δu_i to the relevant individual's net utility level $(u_i - d_i^*)$.

Both (15.7) and (15.9) will tend to give priority to *poor* people's needs over *rich* people's needs. For, owing to the Law of Diminishing Marginal Utility, from any given benefit, poor people will tend to derive a *larger* utility increment Δu_i than rich people will. Yet, (15.9) will give poor people's needs an even *greater priority* than (15.7) would give. This is so because in (15.9) the two relevant individuals' net utility levels $(u_i - d_i^*)$ occur as divisors; and of course these net utility levels will tend to be *smaller* for poor people than for rich people. Obviously, the question is whether it is *morally justified* to use (15.9) as our decision rule when (15.7) would point in the opposite direction.

We all agree that in most cases we must give priority to poor people's needs over rich people's because the former's needs tend to be *more urgent*, and because poor people tend to derive *much greater utility gain* from our help. But the question is what to do in those – rather exceptional – cases where some rich people are in greater need of our help than any poor person is.

For instance, what should a doctor do when he has to decide whether to give a life-saving drug in short supply to a *rich* patient likely to derive the greatest medical benefit from it, or to give it to a *poor* patient who would derive a lesser (but still substantial) benefit from this drug. According to utilitarian theory, the doctor must give the drug to the patient who would obtain the *greatest* benefit from it, regardless of this patient's wealth (or poverty). To do otherwise would be morally intolerable *discrimination* against the rich patient because of his wealth. In contrast, under Brock's theory, the greater-benefit criterion, as expressed by (15.7), can sometimes be overridden by the higher-ratio criterion, as expressed by (15.9). I find this view morally unacceptable.[8]

To conclude: Brock's theory of "need justice" represents a very interesting alternative to utilitarian theory. There are arguments in favor of either theory. But, as I have already indicated, I regard utilitarian theory as a morally much preferable approach.[9]

15.3. Brock's theory of "merit justice"

Brock's aim is to provide proper representation both for "need justice" and for "merit justice" within his two-stage game model. Yet, it seems to me that his model is so much dominated by "need justice" considerations that it fails to provide proper representation for the requirements of "merit justice".

Take the special case where all n players have the *same* needs and, therefore, have the *same* utility functions, and also have the *same* disagreement payoffs d_i^*; but where they have very *different* productive abilities and skills. I now propose to show that in this case Brock's model would give all players the *very same payoffs* – which would in this case satisfy the requirements of "need justice" if considered in isolation, but would mean complete disregard of "merit justice".

To verify this, first consider what I have called Brock's "simplified" theory, involving maximization of the Nash product π^*, defined by (15.1). Since all players are assumed to have the *same* utility functions and the *same* disagree-

---

[8]As is easy to verify, condition (15.9) is really one version of *Zeuthen's Principle* [see Harsanyi (1977, pp. 149–166)]. As I argued in my 1977 book and in other publications, this Principle is a very good decision rule in *bargaining situations*. But, for reasons already stated, I do not think that it is the right decision rule in making *moral decisions*.

[9]A somewhat similar theory of justice, based like Brock's on the n-person Nash solution, but apparently independent of Brock's (1978) paper, has been published by Yaari (1981).

ment payoffs d_i^*, maximization of π^* would give all players *equal utility payoffs* with $u_i^* = \cdots = u_n^*$.

Let us now consider Brock's full theory. Under this latter theory, the players' payoffs would be determined by maximization of another Nash product π^0, defined by (15.2). Yet, this would again yield *equal utility payoffs* with $u_1^0 = \cdots = u_n^0$, for the same reasons as in the previous case.

Why would these payoffs u_i^0 completely fail to reflect the postulated differences among the players in productive abilities and skills? The reason is, it seems to me, that the requirement of maximizing the Nash product π^0 would force the players to choose a Constitution *preventing* those with potentially greater productivity from making *actual use* of this greater productivity within sectional coalitions. As a result, these players' Shapley values could not give them credit for their greater productive abilities and skills.

To avoid this presumably undesired results, it would have to be stipulated that no Constitution adopted by the players could do more than prevent the players from engaging in *immoral* and *illegal* activities such as theft, fraud, murder, and so on. But it could not prevent any player from making full use of his productive potential in socially desirable economic and cultural activities. Of course, in order to do this a clear criterion would have to be provided for distinguishing socially *desirable* activities that *cannot* be constrained by any Constitution, and socially *undesirable* activities that *can* and *must* be so constrained.

III. REASSESSING INDIVIDUAL UTILITIES

16. Mistaken preferences vs. informed preferences

I now propose to argue that a person's observable *actual* preferences – as expressed by his choice behavior and by his verbal statements – do not always correspond to his real interests and even to his own real preferences at a deeper level, because they may be based on incorrect, or at least very incomplete, information. For instance, suppose somebody chooses a glass of orange juice over a glass of water without knowing that the former contains some deadly poison. From this fact we obviously cannot infer that he really prefers to drink the poison, or that drinking the poison is in his real interest.

When somebody chooses one alternative A over another alternative B then he will do this on some *factual assumptions*. Typically, these will be assumptions suggesting that A has a greater *instrumental value* or a greater *intrinsic value* (or both) than B has. Thus, he may choose A because he thinks that A is a more effective means than B is for achieving a desired goal G; or because he thinks that A has some intrinsically desirable characteristic C that B lacks. His preference for *A* will be an *informed preference*[10] if these factual assumptions are *true*; and will be a *mistaken preference* if these assumptions are *false*.

More generally, I shall define a person's *informed preferences* as the *hypothetical* preferences he *would* have if he had all the relevant information and had made full use of this information. On the other hand, I shall call any preference of his *mistaken* if it conflicts with these hypothetical informed preferences of his.

Note that, under this definition, a person may entertain mistaken preferences not only because he does not know some of the relevant facts but also because he chooses to *disregard* some of the relevant facts well known to him. For instance, suppose a person is a very heavy drinker even though he knows that his drinking habit will ruin his health, his career, and his personal relationships. Suppose also that, when he thinks about it, he has a clear preference for breaking his drinking habit. Yet, his urge to drink is so strong that he is quite unable to do so. (Following Aristotle, philosophers call this predicament "weakness of the will".) Under our definitions, this person's preference for heavy drinking will be contrary to his "informed preferences" and, therefore, will be a *mistaken* preference.

Let me now describe the utility function we use to represent a given individual's interests in our social utility function as this individual's *representative* utility function. Our discussion in this section suggests that each individual's representative utility function should not be based on his possibly mistaken actual preferences but rather on his hypothetical *informed preferences*.

17. Exclusion of malevolent preferences

I now propose to suggest that a person's representative utility function must be further restricted: it must be based only on those preferences of his that can be *rationally supported* by other members of society. For by including any given preference of a person in our social utility function we in effect recommend that other members of society should assist him in satisfying this preference. But this would be an unreasonable recommendation if the other members of society could not rationally do this.

More specifically, in this section I shall argue that a person's *malevolent* preferences – those based on sadism, envy, resentment, or malice – should be excluded from his representative utility function. [Most contemporary utilitarian authors would be opposed to this suggestion; see, for example, Smart (1961, pp. 16–18) and Hare (1981, pp. 169–196).] If these preferences are not excluded, then we obtain many paradoxical implications.

[10]My term "informed preference" was suggested by Griffin's (1986) term "informed desire".

For instance, suppose that a number of sadists derive sadistic enjoyment by watching the torture of one victim. Even if the victim's disutility by being tortured is much greater than each sadist's utility by watching it, if the number of sadists in attendance is large enough, then social utility will be maximized by encouraging the sadists to go on with their sadistic enjoyment. Yet, this paradoxical conclusion will be avoided if we exclude utilities based on sadistic preferences and sadistic pleasures from our social utility function.

It seems to me that exclusion of malevolent preferences is fully consistent with the basic principles of utilitarian theory. The basis of utilitarianism is *benevolence* toward all human beings. If X is a utilitarian, then it will be inconsistent with his benevolent attitude to help one person Y to hurt another person Z just for the sake of hurting him. If Y does ask X to help him in this project, then X can always legitimately refuse his help by claiming "conscientious objection" to any involvement in such a malevolent activity.

18. Exclusion of other-oriented preferences

In actual fact, excluding malevolent preferences is merely a special case of a more general principle I am proposing, that of excluding *all* other-oriented preferences form a person's representative utility function.

Apart from terminology (I find my own terminology more suggestive), my distinction between *self-oriented* and *other-oriented* preferences is the same as Dworkin's (1977, p. 234) well-known distinction between *personal* and *external* preferences. Following Dworkin, I define a person's *self-oriented* preferences as his preferences "for [his own] enjoyment of goods and opportunities", and define his *other-oriented* preferences as his preferences "for assignment of goods and opportunities to others".

My suggestion is to exclude, from each person's representative utility function, not only his malevolent other-oriented preferences, but rather *all* his other-oriented preferences, even *benevolent* ones. My reason is that inclusion of any kind of other-oriented preferences would tend to undermine the basic utilitarian principle of assigning the *same* positive weight to every individual's interests in our social utility function. For instance, if we do not exclude benevolent other-oriented preferences, then in effect we assign *much greater* weight to the interests of individuals with *many well-wishers* (such as loving relatives and friends) than we assign to the interests of individuals without such friendly support.

Again, it seems to me that my suggestion is fully consistent with the basic principles of utilitarian theory. Benevolence toward another person does require us if possible to treat *him* as he wants to be treated. But it does not require us by any means to treat *other people* as he wants them to be treated.

(In fact, benevolence toward these people requires us to treat them as *they* want to be treated, not as *he* wants them to be treated.)

Yet, if we want to exclude other-oriented preferences from each individual's representative utility function, then we must find a way of defining a *self-oriented utility function* V_i for each individual i, based solely on i's self-oriented preferences. There seem to be two possible approaches to this problem. One is based on the notion of *hypothetical* preferences, which we already used in defining a person's informed preferences (see Section 16). Under this approach, a person's self-oriented utility function V_i must be defined as his utility function based on his preferences he *would* display if he knew that all his other-oriented preferences – his preferences about how other people should be treated – *would be completely disregarded*.

Another possible approach is to define a person's self-oriented utility function, V_i by means of mathematical operations performed on his *complete* utility function U_i, based on *both* his self-oriented and his other-oriented preferences (assuming that U_i itself is already defined in terms of i's *informed* preferences).

Let x_i be a vector of all variables characterizing i's economic conditions, his health, his job, his social position, and all other conditions over which i has self-oriented preferences. I shall call x_i i's *personal position*. Let y_i be the composite vector $y_i = (x_1, \ldots, x_{i-1}, x_{i+1}, \ldots, x_n)$, characterizing the personal positions of all $(n-1)$ individuals other than i. Then, i's *complete utility function* U_i will have the mathematical form

$$U_i = U_i(x_i, y_i) .\tag{18.1}$$

I shall assume that U_i is a von Neumann–Morgenstern utility function.

It can happen that U_i is a *separable* utility function of the form

$$U_i(x_i, y_i) = V_i(x_i) + Z_i(y_i) ,\tag{18.2}$$

consisting of two terms, one depending only on x_i the other depending only on y_i. In this case we can define i's self-oriented utility function as $V_i = V_i(x_i)$.

Yet, in general, U_i will not be a separable function. In this case we can define V_i as

$$V_i(x_i) = \sup_{y_i} U_i(x_i, y_i) .\tag{18.3}$$

This definition will make V_i always well-defined if U_i has a finite upper bound. (But even if this is not the case we can make V_i well-defined by restricting the sup operator to *feasible* y_i values.)

Equation (18.3) defines *i*'s self-oriented utility $V_i(x_i)$ as that utility level that *i* would enjoy in a given personal position x_i if his other-oriented preferences were *maximally satisfied*. In other words, my definition is based on disregarding any disutility that *i* may suffer because his other-oriented preferences may *not* be maximally satisfied. This is one way of satisfying the requirement that *i*'s other-oriented preferences should be *disregarded*.

From a purely mathematical point of view, an equally acceptable approach would be to replace the sup operator in (18.3) by the inf operator. But from a substantive point of view, this would be, it seems to me, a very infelicitous approach. If we used the inf operator, then we would define V_i essentially as the utility level that *i* would experience in the personal position x_i if he knew that all his relatives and friends, as well as all other people he might care about, would suffer the worst possible conditions.

Obviously, if this were really the case then *i* could not derive much utility from *any* personal position x_i, however desirable a position the latter may be. Yet, the purpose of the utility function $V_i(x_i)$ is to measure the desirability of any personal position from *i*'s own point of view. Clearly, a utility function $V_i(x_i)$ defined by use of the inf operator would be a very poor choice for this purpose.

19. Conclusion

I have tried to show that, under reasonable assumptions, people satisfying the rationality postulates of Bayesian decision theory must define their moral standards in terms of utilitarian theory. More specifically, they must define their social utility function as the *arithmetic mean* (or possibly as the *sum*) of all individual utilities. I also defended the use of von Neumann–Morgenstern utility functions in ethics.

I have argued that a society basing its moral standards on the *rule utilitarian* approach will achieve much higher levels of social utility than one basing them on the *act utilitarian* approach. I have also stated some of my objections to Rawls' and to Brock's nonutilitarian theories of justice.

Finally, I argued that, in our social utility function, each individual's interests should be represented by a utility function based on his *informed* preferences, and excluding his *mistaken* preferences as well as his *malevolent* preferences and, more generally, all his *other-oriented* preferences.

References

Anscombe, F.J. and R.J. Aumann (1963) 'A definition of subjective probability', *Annals of Mathematical Statistics*, 34: 199–205.

Arrow, K.J. (1951) *Social choice and individual values*. New York: Wiley.

Brandt, R.B. (1979) *A theory of the good and the right*. Oxford: Clarendon Press.

Brock, H.W. (1978) 'A new theory of social justice based on the mathematical theory of games', in: P.C. Ordeshook, ed., *Game theory and political science*. New York: New York University Press.

Dworkin, R.M. (1977) *Taking rights seriously*. Cambridge, Mass.: Harvard University Press.

Griffin, J. (1986) *Well-being*. Oxford: Clarendon Press.

Hare, R.M. (1981) *Moral thinking*. Oxford: Clarendon Press.

Harsanyi, J.C. (1953) 'Cardinal utility in welfare economics and in the theory of risk taking', *Journal of Political Economy*, 61: 434–435.

Harsanyi, J.C. (1955) 'Cardinal welfare, individualistic ethics, and interpersonal comparisons of utility', *Journal of Political Economy*, 63: 309–321.

Harsanyi, J.C. (1963) 'A simplified bargaining model for the n-person cooperative game', *International Economic Review*, 4: 194–220.

Harsanyi, J.C. (1967–68) 'Games with incomplete information played by Bayesian players', Parts I–III, *Management Science*, 14: 159–182, 320–334, and 486–502.

Harsanyi, J.C. (1977) *Rational behavior and bargaining equilibrium in games and social situations*. Cambridge: Cambridge University Press.

Harsanyi, J.C. (1987) 'Von Neumann–Morgenstern utilities, risk taking, and welfare', in: G.R. Feiwel, ed., *Arrow and the ascent of modern economic theory*. New York: New York University Press.

Harsanyi, J.C. and R. Selten (1988) *A general theory of equilibrium selection in games*. Cambridge, Mass.: MIT Press.

Luce, R.D. and H. Raiffa (1957) *Games and decisions*. New York: Wiley.

Nash, J.F. (1950) 'The bargaining problem', *Econometrica*, 18: 155–162.

Rawls, J. (1957) 'Justice as fairness', *Journal of Philosophy*, 54: 653–662.

Rawls, J. (1971) *A theory of justice*. Cambridge, Mass.: Harvard University Press.

Shapley, L.S. (1969) 'Utility comparisons and the theory of games', in G.T. Guilbaud, ed., *La decision: Aggregation et dynamique des ordres de preference*. Paris: Centre National de la Recherche Scientifique.

Smart, J.J.C. (1961) *An outline of a system of utilitarian ethics*. Melbourne: Melbourne University Press.

Smith, Adam (1976) *Theory of moral sentiments*. Clifton: Kelley. First published in 1759.

Vickrey, W.S. (1945) 'Measuring marginal utility by reactions to risk', *Econometrica*, 13: 319–333.

Von Neumann, J. and O. Morgenstern (1953) *Theory of games and economic behavior*. Princeton: Princeton University Press.

Yaari, M.E. (1981) 'Rawls, Edgeworth, Shapley, Nash: Theories of distributive justice re-examined', *Journal of Economic Theory*, 24: 1–39.

INDEX[*]